Lecture Notes in Computer Science 4873

Commenced Publication in 1973
Founding and Former Series Editors:
Gerhard Goos, Juris Hartmanis, and Jan van Leeuwen

Srinivas Aluru Manish Parashar
Ramamurthy Badrinath Viktor K. Prasanna (Eds.)

High Performance Computing – HiPC 2007

14th International Conference
Goa, India, December 18-21, 2007
Proceedings

Springer

Volume Editors

Srinivas Aluru
Iowa State University
Department of Electrical and Computer Engineering
3227 Coover Hall, Ames, IA 50011, USA
E-mail: aluru@iastate.edu

Manish Parashar
Rutgers, The State University of New Jersey
Dept. of Electrical and Computer Engineering
94 Brett Rd., Piscataway, NJ 08854, USA
E-mail: parashar@caip.rutgers.edu

Ramamurthy Badrinath
HP - India Software Operations
29 Cunningham Road, Bangalore 560 052, India
E-mail: badrinath@hp.com

Viktor K. Prasanna
University of Southern California
Dept. of Electrical Engineering, Los Angeles, CA 90089-2562, USA
E-mail: prasanna@usc.edu

Library of Congress Control Number: 2007940903

CR Subject Classification (1998): D.1-4, C.1-4, F.1-2, G.1-2

LNCS Sublibrary: SL 1 – Theoretical Computer Science and General Issues

ISSN 0302-9743
ISBN-10 3-540-77219-7 Springer Berlin Heidelberg New York
ISBN-13 978-3-540-77219-4 Springer Berlin Heidelberg New York

Springer is a part of Springer Science+Business Media

springer.com

Printed in Germany

Typesetting: Camera-ready by author, data conversion by Scientific Publishing Services, Chennai, India
Printed on acid-free paper SPIN: 12201358 06/3180 5 4 3 2 1 0

Message from the Program Chair

Welcome to the 14th International Conference on High-Performance Computing (HiPC 2007), which took place amidst the rich culture and pristine beaches of exotic Goa! Continuing in the tradition of HiPC, this year's conference featured a high-quality peer-reviewed technical program, five keynote talks by leading experts, four workshops, two tutorials, industrial and research exhibits, mini symposium on high-performance computing, and a poster session.

We received 253 submissions from 31 countries, reflecting the international character of this conference. A majority of the submissions came from India (52%) and the USA (20.1%), but contributions came from all corners of the world including the Asia/Pacific region, Europe, Middle East, Africa, North America and Latin America. A preliminary review process was carried out by the Program Vice Chairs and myself to identify manuscripts that lacked original content. The remaining 221 submissions were put through a rigorous peer-review process. Each paper was reviewed by three Program Committee members. In some cases, external reviews were solicited. Each paper was then assessed in the context of reviews in an on-line Program Committee meeting that generated a week of lively discussion. Particular attention was paid to understanding the contribution of manuscripts that received divergent initial reviews. A decision on each manuscript was taken based on the results of this discussion, taking into account constraints imposed by the conference schedule. Only 53 out of the total 253 submissions (20.95%) were eventually accepted for presentation and publication in the proceedings. These submissions reflect the efforts of authors spanning 13 countries. Fifty-two of these papers were presented in ten technical sessions spanning the three days of the conference.

The conference featured a separate plenary session for presenting best paper and best poster awards. One outstanding paper was selected for the best paper award based on the following procedure: A pool of six papers was initially identified based on high ratings and review scores. These papers were carefully studied and discussed by the six Program Vice Chairs and myself. The final selection was based on originality and novelty of the contribution, importance of the results, likely impact on the field, interest to a diverse audience, and clarity of presentation. The paper "Distributed Ranked Search," authored by Vijay Gopalakrishnan from AT&T Research, Ruggero Morselli from Google, and Samrat Bhattacharjee, Peter Keheler and Aravind Srinivasan from University of Maryland was selected for the award. The authors presented an efficient algorithm using random sampling to rank documents in a distributed search where the documents are spread over a peer-to-peer network. The best paper award was sponsored by Infosys Corporation and the best poster awards were sponsored by the IEEE Technical Committee on Parallel Processing (TCPP).

Inspiring keynote speeches by leading experts has been a tradition at HiPC. This year, we were fortunate to have five distinguished talks: Michael Flynn on the dawn of the "parallel" future, David Keyes on petaflop computing, Vipin Kumar on high-performance data mining, Yale Patt on the era of multi-cores, and Prabhakar Raghavan on Web search.

The high-quality technical program was the highlight of the conference, and was the result of the hard work of many individuals. I would like to thank authors of all submitted papers, for choosing this conference for disseminating their technical work. The program and proceedings in your hands (or on your computer drive!) would not be possible without their hard work, and I look forward to their continued patronage in the coming years. To evaluate these contributions, we assembled a team of 98 Program Committee members, a diverse team of experts that cover the many areas of interest to the conference. It is their hard work and timely review submissions that allowed us to assemble the program. The selection of PC members and individual review assignments were handled by an able and dedicated team of six Program Vice Chairs overseeing the six tracks of the conference: Peter Sanders (Algorithms), Sivan Toledo (Applications), Peter J. Varman (Architecture), Dhabaleswar K. (DK) Panda (Communication Networks), Ahmed Helmy (Mobile and Sensor Computing), and Manish Gupta (Systems Software). They made tremendous contributions in helping me craft the technical program by leading discussions on contributed papers beyond the initial reviews, in a week-long Program Committee deliberation.

Several organizers and volunteers worked tirelessly to make the conference a successful and productive meeting and I would like to take this opportunity to express my gratitude: Manimaran Govindarasu for organizing the workshops, Rajeev Sivaram for organizing the tutorials, Rajeev Thakur for organizing the poster session, Sushil K. Prasad for putting together the proceedings and providing us with an excellent archival record, and Viraj Bhat for maintaining the conference Web site on a continual basis.

Finally, I would like to thank the primary leadership group of this year's conference: General Co-chairs Manish Parashar and Badrinath Ramamurthy, Vice General Co-chairs Rajendra V. Boppana and Rajeev Muralidhar, and Steering Committee Chair Viktor K. Prasanna. I relied on their advice and guidance throughout the process. I want to thank Viktor Prasanna for giving me the opportunity to serve you as the Program Chair. Not only did he direct the overall effort and always found time when I needed his counsel and advice, he also obliged my Vice Chair Ahmed Helmy and served as a Program Committee member in the Mobile and Sensor Computing track! It is through his dedication and continued leadership from the inception of the conference that we have this highly regarded international conference to publish our ideas and learn from each other. Please join me in learning from the keynote speakers and authors of contributed papers. I hope you enjoyed all that Goa had to offer during the breaks and conference organized events, and that you had a productive and enjoyable meeting.

December 2007 Srinivas Aluru

Message from the General Co-Chairs and the Vice General Co-Chairs

On behalf of the organizers of the 14th International Conference on High-Performance Computing (HiPC), it was our pleasure to welcome you to Goa. I do hope you found the conference exciting and rewarding.

The HiPC call for papers, once again, received an overwhelming response, attracting 253 submissions from 31 countries. Srinivas Aluru, the Program Chair, and the Program Committee worked with remarkable dedication to put together an outstanding technical program consisting of the 53 papers that appear in these proceedings.

Several events, complementing this strong technical program, made HiPC 2007 another special and exciting meeting. The HiPC 2007 keynotes were presented by internationally renowned researchers. The conference featured the mini symposium "High-Performance Computing Technologies, Applications and Experience," which aimed at bringing together the users and providers of HPC. The poster session presented hot off-the-press research results. Finally, there was a dedicated industry session and the industry and research exhibits. The meeting was preceded by a set of tutorials and workshops highlighting new and emerging aspects of the field.

Arranging an exciting meeting with a high-quality technical program is easy when one is working with an excellent and dedicated team and can build on the practices and levels of excellence established by a quality research community. HiPC 2007 would not have been possible without the tremendous efforts of the many volunteers. We would like to acknowledge the critical contributions of each one.

We would like to thank Srinivas Aluru, Program Chair, and the Program Committee for their efforts in assembling such an excellent program, and the authors who submitted the high-quality material from which that program was selected. We would also like to thank the presenters of the keynotes, posters and tutorials, the organizers of the workshops, and all the participants, who complete the program.

We would specially like to thank Viktor Prasanna, Chair of the HiPC Steering Committee, for his leadership, sage guidance, and untiring dedication, which have been key to the continued success of the conference. We would also like to welcome our new volunteers to the team – your efforts are critical to the continued success of this conference. Finally, we would like to gratefully acknowledge our academic and industry sponsors including IEEE Computer Society, ACM SIGARCH, Infosys, DELL, Google, IBM, Intel, NetApp, Satyam, Yahoo, HP, AMD and Cray.

December 2007

General Co-Chairs
Manish Parashar
Ramamurthy Badrinath

Vice General Co-Chairs
Rajendra V. Boppana
Rajeev Muralidhar

Message from the Steering Chair

It was my pleasure to welcome you to the 14th International Conference on High-Performance Computing.

This conference would not be possible without the dedicated effort of many volunteers over the past year. First, I would like to single out the contributions of Srinivas Aluru, Program Chair, for his outstanding contributions in putting together an excellent technical program. I am indebted to him for his thorough evaluation of the submitted manuscripts and his relentless efforts to further improve the quality of the technical program. Manish Parashar and Ramamurthy Badrinath as General Co-chairs provided the leadership in resolving numerous meeting-related issues and putting together the overall meeting program including the workshops and tutorials. They were ably assisted by Rajeev Muralidhar and Rajendra Boppana, Vice General Co-chairs. The industry track was coordinated by Rama Govindaraju and Raghuram Tupuri. The Poster/presentation session was organized by Rajeev Thakur. The meeting offers scholarships for India-based students. These scholarships were administered by Anu Bourgeois and Madhusudhan Govindaraju. We have several continuing as well as new workshops. These workshops were coordinated by Manimaran Govindarasu. The Web site was maintained by Viraj Bhat. Raghuram Tupuri and Rama Govindaraju coordinated the industry exhibits. Rajeev Sivaram assisted us with the tutorials. The local arrangments were handled by Ch. Kalyana Krishna and Venkatesh Kamat. Sushil Prasad liaised with the authors and Springer to bring out the proceedings. Sumir Chandra, Manisha Gajbe and Rajeev Raje handled the publicity for us. Sally Jelinek acted as the Registration Chair. Ajay Gupta and Thondiyil Venugopalan handled the meeting finances.

Tirumale Ramesh and Raghuram Tupuri with assistance from Santosh Sreenivasan put together the Mini Symposium on High-Performance Computing Technologies, Applications and Experience. They were ably assisted by Haresh Bhatt and Venkat Ramana. The intent of the mini symposium is to provide a forum for vendors as well as HPC users in India to present the technologies and user experiences.

I would like to thank all our volunteers for their tireless efforts. The meeting would not be possible without the enthusiastic commitment of these individuals.

Major financial support for the meeting was provided by several leading IT companies and multinationals operating in India. I would like to acknowledge the following individuals and their organizations for their support:

- N.R. Narayana Murthy, Infosys
- Kris Gopalakrishnan, Infosys
- Harish Grama, IBM India
- Manish Gupta, IBM Watson
- David Ford, NetApp

- Siddhartha Nandi, NetApp
- B. Rudramuni, Dell India
- Ramesh Rajagopalan, Dell India
- Reza Rooholamini, Dell
- V. Sridhar, Satyam
- Prasad Ram, Google R&D, India
- Prabhakar Raghavan, Yahoo! Inc.
- Arun Ramanujapuram, Yahoo! India R&D
- Vittal Kini, Intel Research, India
- Akshay Kadam, Intel Research, India
- Dinkar Sitaram, HP India
- Faisal Paul, HP India
- Raghuram Tupuri, AMD
- Venkat Ramana, Hinditron Infosystems

December 2007 Viktor K. Prasanna

Conference Organization

General Co-chairs

Manish Parashar, Rutgers University, USA
Ramamurthy Badrinath, HP, India

Vice General Co-chairs

Rajendra V. Boppana, University of Texas at San Antonio, USA
Rajeev Muralidhar, Intel, India

Program Chair

Srinivas Aluru, Iowa State University, USA

Program Vice-chairs

Algorithms
  Peter Sanders, Universität Karlsruhe, Germany
Applications
  Sivan Toledo, Tel Aviv University, Israel
Architecture
  Peter J. Varman, Rice University, USA
Communication Networks
  Dhabaleswar K. Panda, Ohio State University, USA
Mobile and Sensor Computing
  Ahmed Helmy, University of Florida, USA
Systems Software
  Manish Gupta, IBM Corporation, India

Steering Chair

Viktor K. Prasanna, University of Southern California, USA

Workshops Chair

Manimaran Govindarasu, Iowa State University, USA

Poster/Presentation Chair

Rajeev Thakur, Argonne National Laboratory, USA

Tutorials Chair

Rajeev Sivaram, Google, USA

Industry Liaison Co-chairs

Rama K. Govindaraju, Google, USA
Raghuram Tupuri, Advanced Micro Devices (AMD), India

HiPC User Symposium Co-chairs

T.K. Ramesh, The Boeing Company, USA
Raghuram Tupuri, Advance Micro Devices (AMD), India
Santosh Sreenivasan, Talentain, India

Cyber Chair

Viraj Bhat, Rutgers University, USA

Finance Co-chairs

Ajay Gupta, Western Michigan University, USA
Thondiyil Venugopalan, India

Registration Co-chairs

Sally Jelinek, Electronics Design Associates, USA
Mamatha Raghavendra, Dell, India

Local Arrangements Chair

Venkatesh V. Kamat, Goa University, India
Kalyana Krishna, Talentain, India
Raghavendra Buddi, NetApp, India

Awards Chair

George Westrom, Odetics Inc. and FSEA, USA

Publications Chair

Sushil K. Prasad, Georgia State University, USA

Publicity Co-chairs

Sumir Chandra, Rutgers University, USA
Manisha Gajbe, Georgia Institute of Technology, USA
Rajeev R. Raje, Indiana University Purdue University, Indianapolis, USA

Scholarships Co-chairs

Anu G. Bourgeois, Georgia State University, USA
Madhusudhan Govindaraju, SUNY Binghamton, USA

Steering Committee

(Steering Committee 2007 membership also includes the General Co-chairs, Program Chairs and Vice General Co-chairs from 2005, 2006 and 2007.)

P. Anandan, Microsoft Research, India
David A. Bader, Georgia Institute of Technology, USA
Ramamurthy Badrinath, HP, India
Rudramuni B., Dell India , Bangalore, India
Frank Baetke, HP, USA
R. Govindarajan, Indian Institute of Science, India
Harish Grama, IBM, India
Manish Gupta, India Systems and Technology Lab, IBM, India
Vittal Kini, Intel, India
N. S. Nagaraj, Infosys, India
Viktor K. Prasanna, University of Southern California, USA (Chair)
Venkat Ramana, Cray-Hinditron, India
Sartaj Sahni, University of Florida, USA
V. Sridhar, Satyam Computer Services Ltd., India
Harrick M. Vin, Tata Research, Development & Design Center (TRDDC), Pune, India

Program Committee

Algorithms

Thomas Cormen, Dartmouth College, USA
Devdatt Dubhashi, Chalmers University of Tech. and Göteborg University, Sweden
Matteo Frigo, University of Paderborn, Germany
Klaus Jansen, University of Kiel, Germany
Christos Kaklamanis, University of Patras, Greece
Samir Khuller, University of Maryland, USA
Dariuzs Kowalski, University of Liverpool, UK
Jesper Larsson Traeff, C&C Research Labs, NEC Europe Ltd., Germany

Applications

Architecture

Parthasarathy Ranganathan, HP Labs, USA
Amir Roth, University of Pennsylvania, USA
Anand Sivasubramaniam, Penn State University, USA
Daniel Sorin, Duke University, USA
Steven Swanson, University of California San Diego, USA
Jun Wang, University of Central Florida, USA
Jun Yang, University of Pittsburgh, USA

Communication Networks

Pavan Balaji, Argonne National Laboratory, USA
Mohammad Banikazemi, IBM T.J. Watson Research Center, USA
Ron Brightwell, Sandia National Laboratory, USA
Darius Buntinas, Argonne National Laboratory, USA
Wu-Chun Feng, Virginia Tech University, USA
Jose Flich, Universidad Politécnica de Valencia, Spain
Ada Gavrilovska, Georgia Tech University, USA
Rama K. Govindaraju, Google, USA
Darren Kerbyson, Los Alamos National Laboratory, USA
Nectarios Koziris, National Technical University of Athens, Greece
Olav Lysne, Simula Research Lab, Norway
Fabrizio Petrini, Pacific Northwest National Laboratory, USA
Tim Pinkston, NSF and University of Southern California, USA
Martin Schulz, Lawrence Livermore National Laboratory, USA
Yuanyuan Yang, University of Stony Brook, USA

Mobile and Sensor Computing

Kevin Almeroth, UC Santa Barbara, USA
Fan Bai, General Motors Research, USA
Debojyoti Dutta, University of Southern California, USA
Eylem Ekici, Ohio State University, USA
Polly Huang, National Taiwan University, Taiwan
Vana Kalogeraki, University of California at Riverside, USA
Iyad Kanj, DePaul University, USA
Arunesh Mishra, University of Wisconsin, USA
Viktor Prasanna, University of Southern California, USA
Tirumale K. Ramesh, Boeing, USA
Karim Seada, Nokia Research, USA
Raghupathy Sivakumar, Georgia Institute of Technology, USA
My Tra Thai, University of Florida, USA
Ye Xia, University of Florida, USA
Yanchao Zhang, New Jersey Inst. of Tech., USA

Systems Software

Gianfranco Bilardi, University of Padova, Italy
Rajkumar Buyya, University of Melbourne, Australia
Bronis De Supinski, Lawrence Livermore National Laboratory, USA
Beniamino DiMartino, Seconda Università di Napoli, Italy
R. Govindarajan, IISc Bangalore, India
Rinku Gupta, Dell, USA
Lawrence Rauchwerger, Texas A&M University, USA
Yogish Sabharwal, IBM India Research Laboratory, India
P. Sadayppan, Ohio State University, USA
Vijay Saraswat, IBM T.J. Watson Research Center, USA
Henk Sips, Delft University of Technology, Netherlands
Jordi Torres, Universidad Politecnica de Catalunya, Spain
Nalini Venkatasubramanian, University of California at Irvine, USA
Rajeev Wankar, University of Hyderabad, India
Yanyong Zhang, Rutgers University, USA

Workshop Organizers

Workshop on New Horizons in Compilers

Co-chairs
R. Govindarajan, IISc, Bangalore, India
Uday Khedker, IIT Mumbai, India
Rahul Simha, The George Washington University, USA
Bhagi Narahari, GWU, Washington, USA

Workshop on Service-Oriented Engineering and Optimization

Co-chairs
Amol Bakshi, University of Southern California, USA
Alok Choudhary, Northwestern University, USA

Workshop on Grid and Utility Computing

Co-chairs
Anirban Chakrabarti, Infosys Technologies, India
Shubhashis Sengupta, Infosys Technologies, India

Workshop on Storage Technologies in Computing Clusters and Datacenter Environments

Co-chairs
Vittal Kini, Intel India Research Center, Bangalore, India
Rajat Moona, IIT Mumbai, India

Tutorials

Sensor Networks

Dharma P. Agrawal, University of Cincinnati, USA

Programming Models and Compiler Optimizations for GPUs and Multi-core Processors

J. Ramanujam, Louisiana State University, USA
P. Sadayappan, Ohio State University, USA

List of Reviewers

In addition to the PC members, the following colleagues provided reviews for HiPC 2007 papers. Their help is gratefully acknowledged.

Francisco Alfaro
Nawab Ali
K.V.S. Arya
Rocco Aversa
Alex Babansky
Bhargav Bellur
Johannes Blömer
Olaf Bonorden
Humberto Calderon
Andrey Chernikov
Francisco Alfaro Cortés
Maurizio D'Arienzo
Stefan Donath
Suryanarayana Durbhakula
Andriy Fedorov
Mruggesh R. Gajjar
Carlo Galuzzi
Naveen Garg
Georgios Goumas
Tobias Gradl
Jordi Guitart
Pavol Hell
Martin Hirt
Gavin Holland
Wei-jen Hsu
Mauro Iacono
Klaus Iglberger
Alexandru Iosup
Shyam Iyer
Michael Klemm
Andriy Kot
Sriram Krishnamoorthy
Volker Krumel
Daniele Ludovici
Amith Mamidala
R. Manikantan
Rajit Manohar
Stefano Marrone
Faisal Ghias Mir
Pabitra Mitra
Naveen Muralimanohar
Tamer Nadeem
Bharath Narasimha Swamy
Rupesh Nasre
Leandro Navarro-Moldes
Shivaraj Nidoni
Ramon Nou
Sreepathi Pai
Sagar Pandit
Ali Pinar
Antoniu Pop
Massimiliano Rak
Sebastian Isaza Ramirez
Toby Sebastian
Daniele Scarpazza

Ahsan Shabbir	Nils Thuerey	Rajesh Vivekanandham
A.P. Shanthi	Ravi Tiwari	Nawaporn Wisitpongphan
Tim Smith	Jan Treibig	George Zagaris
Thomas Sodring	Ana Lucia Varbanescu	Yan Zhang
Dimitris Theodoropoulos	Anitha Varghese	Youtao Zhang
Nathan Thomas	Salvatore Venticinque	

Table of Contents

Keynote Addresses (Abstracts)

Plenary Session - Best Paper

Session I - Applications on I/O and FPGAs

Session II - Microarchitecture and Multiprocessor Architecture

Session III - Applications of Novel Architectures

Session VI - Energy-Aware Computing

Session VII - P2P and Internet Applications

Session VIII - Communication and Routing

Session IX - Cluster and Grid Applications

Session X - Mobile Computing

Petaflop/s, Seriously

David Keyes

Applied Physics and Applied Mathematics, Columbia University, USA
Acting Director, Institute for Scientific Computing Research, LLNL, USA

Abstract. Sustained floating-point rates on real applications, as tracked by the Gordon Bell Prize, have increased by over five orders of magnitude from 1988, when 1 Gigaflop/s was reported on a structural simulation, to 2006, when 200 Teraflop/s were reported on a molecular dynamics simulation. Various versions of Moore's Law over the same interval provide only two to three orders of magnitude of improvement for an individual processor; the remaining factor comes from concurrency, which is of order 100,000 for the BlueGene/L computer, the platform of choice for the majority of recent Bell Prize finalists. As the semiconductor industry begins to slip relative to its own roadmap for silicon-based logic and memory, concurrency will play an increasing role in attaining the next order of magnitude, to arrive at the long-awaited milepost of 1 Petaflop/s sustained on a practical application, which should occur around 2009. Simulations based on Eulerian formulations of partial differential equations can be among the first applications to take advantage of petascale capabilities, but not the way most are presently being pursued. Only weak scaling can get around the fundamental limitation expressed in Amdahl's Law and only optimal implicit formulations can get around another limitation on scaling that is an immediate consequence of Courant-Friedrichs-Lewy stability theory under weak scaling of a PDE. Many PDE-based applications and other lattice-based applications with petascale roadmaps, such as quantum chromodynamics, will likely be forced to adopt optimal implicit solvers. However, even this narrow path to petascale simulation is made treacherous by the imperative of dynamic adaptivity, which drives us to consider algorithms and queueing policies that are less synchronous than those in common use today. Drawing on the SCaLeS report (2003-04), the latest ITRS roadmap, some back-of-the-envelope estimates, and numerical experiences with PDE-based codes on recently available platforms, we will attempt to project the pathway to Petaflop/s for representative applications.

Biography: David E. Keyes is the Fu Foundation Professor of Applied Mathematics in the Department of Applied Physics and Applied Mathematics at Columbia University, an affiliate of the Computational Science Center (CSC) at Brookhaven National Laboratory, and Acting Director of Institute for Scientific Computing Research (ISCR) at Lawrence Livermore National Laboratory. Keyes graduated summa cum laude with a B.S.E. in Aerospace and Mechanical Sciences and a Certificate in Engineering Physics from Princeton University

S. Aluru et al. (Eds.): HiPC 2007, LNCS 4873, pp. 2–3, 2007.

in 1978. He received his Ph.D. in Applied Mathematics from Harvard University in 1984. He then post-doc'ed in the Computer Science Department at Yale University and taught there for eight years, as Assistant and Associate Professor of Mechanical Engineering, prior to joining Old Dominion University and the Institute for Computer Applications in Science and Engineering (ICASE) at the NASA Langley Research Center in 1993. At Old Dominion, Keyes was the Richard F. Barry Professor of Mathematics and Statistics and founding Director of the Center for Computational Science. Author or co-author of over 100 publications in computational science and engineering, numerical analysis, and computer science, Keyes has co-edited 10 conference proceedings concerned with parallel algorithms and has delivered over 200 invited presentations at universities, laboratories, and industrial research centers in over 20 countries and 35 states of the U.S. With backgrounds in engineering, applied mathematics, and computer science, and consulting experience with industry and national laboratories, Keyes works at the algorithmic interface between parallel computing and the numerical analysis of partial differential equations, across a spectrum of aerodynamic, geophysical, and chemically reacting flows. Newton-Krylov-Schwarz parallel implicit methods, introduced in a 1993 paper he co-authored at ICASE, are now widely used throughout engineering and computational physics, and have been scaled to thousands of processors.

The Future Is Parallel But It May Not Be Easy

Michael J. Flynn

Maxeler Corporation
Stanford University, USA

Abstract. Processor performance scaling by improving clock frequency has now hit power limits. The new emphasis on multi core architectures comes about from the failure of frequency scaling not because of breakthroughs in parallel programming or architecture. Progress in automatic compilation of serial programs into multi tasked ones has been slow. A look at parallel projects of the past illustrates problems in performance and programmability. Solving these problems requires both an understanding of underlying issues such as parallelizing control structures and dealing with the memory bottleneck. For many applications performance comes at the price of programmability and reliability comes at the price of performance.

Biography: Michael Flynn is Senior Advisor to the Maxeler Corporation, an acceleration solutions company based in London. He received his Ph.D. from Purdue University and joined IBM working there for ten years in the areas of computer organization and design. He was design manager System 360 Model 91 Central Processing Unit. Between 1966 and 1974 Prof. Flynn was a faculty member of Northwestern University and the Johns Hopkins University. From 1975 until 2000, he was a Professor of Electrical Engineering at Stanford University and served as the Director of the Computer Systems Laboratory from 1977 to 1983. He was founding chairman of both the ACM Special Interest Group on Computer Architecture and the IEEE Computer Society's Technical Committee on Computer Architecture. Prof. Flynn was the 1992 recipient of the ACM/IEEE Eckert-Mauchley Award for his technical contributions to computer and digital systems architecture. He was the 1995 recipient of the IEEE-CS Harry Goode Memorial Award in recognition of his outstanding contribution to the design and classification of computer architecture. In 1998 he received the Tesla Medal from the International Tesla Society (Belgrade), and an honorary Doctor of Science from Trinity College (University of Dublin), Ireland. He is the author of three books and over 250 technical papers, and he is also a fellow of the IEEE and the ACM.

S. Aluru et al. (Eds.): HiPC 2007, LNCS 4873, p. 1, 2007.

High Performance Data Mining - Application for Discovery of Patterns in the Global Climate System

Vipin Kumar

William Norris Professor, Head of the Computer Science and Engineering Department University of Minnesota, USA

Abstract. Advances in technology and high-throughput experiment techniques have resulted in the availability of large data sets in commercial enterprises and in a wide variety of scientific and engineering disciplines. Data in terabytes range are not uncommon today and are expected to reach petabytes in the near future for many application domains in science, engineering, business, bioinformatics, and medicine. This has created an unprecedented opportunity to develop automated data-driven techniques of extracting useful knowledge. Data mining, an important step in this process of knowledge discovery, consists of methods that discover interesting, non-trivial, and useful patterns hidden in the data. This talk will provide an overview of a number of data mining research in our group for understanding patterns in global climate system and computational challenges in addressing them.

Biography: Vipin Kumar is currently William Norris Professor and Head of Computer Science and Engineering at the University of Minnesota. His research interests include High Performance computing and data mining. He has authored over 200 research articles, and co-edited or coauthored 9 books including the widely used text book "Introduction to Parallel Computing", and "Introduction to Data Mining" both published by Addison-Wesley. Kumar has served as chair/co-chair for over a dozen conferences/workshops in the area of data mining and parallel computing. Currently, he serves as the chair of the steering committee of the SIAM International Conference on Data Mining, and is a member of the steering committee of the IEEE International Conference on Data Mining. Kumar is founding co-editor-in-chief of Journal of Statistical Analysis and Data Mining, editor-in-chief of IEEE Intelligent Informatics Bulletin, and series editor of Data Mining and Knowledge Discovery Book Series published by CRC Press/Chapman Hall. Kumar is a Fellow of the AAAS, ACM and IEEE. He received the 2005 IEEE Computer Society's Technical Achievement Award for contributions to the design and analysis of parallel algorithms, graph-partitioning, and data mining.

S. Aluru et al. (Eds.): HiPC 2007, LNCS 4873, p. 4, 2007.

The Transformation Hierarchy in the Era of Multi-core

Yale Patt

Professor of Electrical and Computer Engineering,
Ernest Cockrell, Jr. Centennial Chair in Engineering,
University of Texas at Austin, USA

Abstract. The transformation hierarchy is the name I have given to the mechanism that converts problems stated in natural language (English, Spanish, Hindi, Japanese, etc.) to the electronic circuits of the computer that actually does the work of producing a solution. The problem is first transformed from a natural language description into an algorithm, and then to a program in some mechanical language, then compiled to the ISA of the particular processor, which is implemented in a microarchitecture, built out of circuits. At each step of the transformation hierarchy, there are choices. These choices enable one to optimize the process to accomodate some optimization criterion. Usually, that criterion is microprocessor performance. Up to now, optimizations have been done mostly within each of the layers, with artifical barriers in place between the layers. It has not been the case (with a few exceptions) that knowledge at one layer has been leveraged to impact optimization of other layers. I submit, that with the current growth rate of semiconductor technology, this luxury of operating within a transformation layer will no longer be the common case. This growth rate (now more than a billion trnasistors on a chip is possible) has ushered in the era of the chip multiprocessor. That is, we are entering Phase II of Microprocessor Performance Improvement, where improvements will come from breaking the barriers that separate the transformation layers. In this talk, I will suggest some of the ways in which this will be done.

Biography: Yale Patt is a teacher at The University of Texas at Austin, where he also directs the research of nine PhD students, while enjoying an active consulting practice with several microprocessor manufacturers. He teaches the required freshman intro to computing course to 400 first year students every other fall, and the advanced graduate course to PhD students in microrchitecture every other spring. His research ideas (HPS, branch prediction, etc.) have been adopted by almost every microprocessor manufacturer on practically every high end chip design of the past ten years. Yale Patt has earned the appropriate degrees from reputable universities and has received more than his share of prestigious awards for his research and teaching. More detail on his interests and accomplishments can be obtained from his web site: www.ece.utexas.edu/ patt.

S. Aluru et al. (Eds.): HiPC 2007, LNCS 4873, p. 5, 2007.

Web Search: Bridging Information Retrieval and Microeconomic Modeling

Prabhakar Raghavan

Head, Yahoo! Research
Consulting Professor, Computer Science Department, Stanford University, USA

Abstract. Web search has come to dominate our consciousness as a convenience we take for granted, as a medium for connecting advertisers and buyers, and as a fast-growing revenue source for the companies that provide this service. Following a brief overview of the state of the art and how we got there, this talk covers a spectrum of technical challenges arising in web search- ranging from spam detection to auction mechanisms.

Biography: Prabhakar Raghavan has been Head of Yahoo! Research since 2005. His research interests include text and web mining, and algorithm design. He is a Consulting Professor of Computer Science at Stanford University and Editor-in-Chief of the Journal of the ACM. Raghavan received his PhD from Berkeley and is a Fellow of the ACM and of the IEEE. Prior to joining Yahoo, he was Chief Technology Officer at Verity; before that he held a number of technical and managerial positions at IBM Research.

S. Aluru et al. (Eds.): HiPC 2007, LNCS 4873, p. 6, 2007.

Distributed Ranked Search

Vijay Gopalakrishnan[1], Ruggero Morselli[2], Bobby Bhattacharjee[3], Pete Keleher[3], and Aravind Srinivasan[3]

[1] AT&T Labs – Research
[2] Google Inc
[3] University of Maryland

Abstract. P2P deployments are a natural infrastructure for building distributed search networks. Proposed systems support locating and retrieving all results, but lack the information necessary to rank them. Users, however, are primarily interested in the most relevant results, not necessarily all possible results.

Using random sampling, we extend a class of well-known information retrieval ranking algorithms such that they can be applied in this decentralized setting. We analyze the overhead of our approach, and quantify how our system scales with increasing number of documents, system size, document to node mapping (uniform versus non-uniform), and types of queries (rare versus popular terms). Our analysis and simulations show that a) these extensions are efficient, and scale with little overhead to large systems, and b) the accuracy of the results obtained using distributed ranking is comparable to that of a centralized implementation.

1 Introduction

Search infrastructures often order the results of a query by application-specific notions of *rank*. Users generally prefer to be presented with small sets of ranked results rather than unordered sets of all results. For example, a recent Google search for "HiPC 2007" matched over 475,000 web pages. The complete set of all results would be nearly useless, while a very small set of the top-ranked results would likely contain the desired web site. Moreover, collecting fewer results reduces the network bandwidth consumed, helping the system scale up—to many users, hosts, and data items— and down—to include low-bandwidth links and low-power devices.

Ranking results in a decentralized manner is difficult because decisions about which results to return are made locally, but the basis of the decisions, rank, is a global property. Technically, we could designate one node as being responsible for ranking all the search results. Such an approach, however, would result in this peer receiving an unfair amount of load. Further, there are the issues of scalability and fault-tolerance with using just one node.

The main contribution of this paper is the design and evaluation of a decentralized algorithm that efficiently and consistently ranks search results over arbitrary documents. Our approach is based on approximation techniques using uniform random sampling, and the classic centralized Vector Space Model (VSM) [1]. Our results apply to both structured and unstructured networks.

S. Aluru et al. (Eds.): HiPC 2007, LNCS 4873, pp. 7–20, 2007.

Our analysis shows that the cost of our sampling-based algorithm is usually small and *remains constant as the size of the system increases*. We present a set of simulation results, based on real document sets from the TREC collection [2], that confirm our analysis. Further, the results show that the constants in the protocol are low, e.g., the protocol performs very well with samples from 20 nodes per query on a 5000 node network, and that the approach is robust to sampling errors, initial document distribution, and query location.

The rest of the paper is organized as follows. We first present some background on ranking in classical information retrieval in Section 2. We then discuss our design for ranking results in Section 3 and analyze its properties. In Section 4, we present experimental results where we compare the performance of the distributed ranking scheme with a centralized scheme. We discuss other related work in Section 5 before concluding in Section 6.

2 The Vector Space Model (VSM)

The Vector Space Model (VSM) is a classic information retrieval model for ranking results. VSM maps documents and queries to vectors in a T-dimensional term space, where T is the number of *unique* terms in the document collection. Each term i in the document d is assigned a weight $w_{i,d}$. The vector for a document d is defined as $\boldsymbol{d} = (w_{1,d}, w_{2,d}, \ldots, w_{T,d})$. A query is also represented as a vector $\boldsymbol{q} = (w_{1,q}, w_{2,q}, \cdots, w_{T,q})$, where q is treated as a document.

Vectors that are similar have a small angle between them. VSM uses this intuition to compute the set of relevant documents for a given query; relevant documents will differ from the query vector by a small angle while irrelevant documents will differ by a large angle. Given two vectors X and Y, the angle θ between them can be computed using $\cos\theta = \frac{\sum_{i=1}^{n} x_i y_i}{\sqrt{\sum_{i=1}^{n} x_i^2}\sqrt{\sum_{i=1}^{n} y_i^2}}$. This equation is also known as the cosine similarity, and has been used in traditional information retrieval to identify and rank relevant results.

2.1 Generating Vector Representation

The vector representation of a document is generated by computing the *weight* of each term in the document. The key is to assign weights such that terms that capture the semantics in the document and therefore help in discriminating between the documents are given a higher weight.

Effective term weighting formulae have been an area of much research (e.g., [3,4]), unfortunately with little consensus. While any of the commonly used formulae can be used with our scheme, we use the weighting formula used in the SMART [5] system as it has shown to have good retrieval quality in practice:

$$w_{t,d} = (\ln f_{t,d} + 1) \cdot \ln\left(\frac{D}{D_t}\right) \tag{1}$$

where $w_{t,d}$ is the weight of term t in document d, $f_{t,d}$ is the raw frequency of term t in document d, D is the total number of documents in the collection, and D_t is the number of documents in the collection that contain term t.

3 Distributed VSM Ranking

In this section, we present our distributed VSM ranking system for keyword-based queries. There are three main components needed for ranking: generating a vector representation for exported documents, storing the document vectors appropriately, and computing and ranking the query results. We first describe our assumed system model and then discuss each of these components in detail.

3.1 System Model

Our ranking algorithm is designed for both structured and unstructured P2P systems. Our algorithm constructs an inverted index for each keyword and these indexes are distributed over participating nodes (which are assumed to be cooperative). An *inverted index* of a keyword stores the list of all the documents having the keyword. We assume that the underlying P2P system provides a *lookup* mechanism necessary to map indexes to nodes storing them. While APIs for lookup are available in all structured systems, we rely on approaches such as LMS [6] and Yappers [7] for lookup over unstructured systems. The underlying P2P system dictates how the indexes are mapped to nodes; structured P2P systems store indexes at a single location, while an index may be partitioned over many locations in unstructured systems. Each node *exports* a set of documents when it joins the system. A set of keywords (by default, all words in the document) is associated with the document. The process of exporting a document consists of adding an entry for the document in the index associated with each keyword. When querying, users submit queries containing keywords and may specify that only the highest ranked K results be returned. The system then computes these K results in a distributed manner and returns the results to the user.

3.2 Generating Document Vectors

Recall that to generate a document vector, we need to assign weights to each term of the document. Also recall Equation (1), which is used to compute the weight of each term t in a document. The equation has two components: a local component, $\ln f_{t,d}+1$, which captures the relative importance of the term in the given document, and a global component, $\ln(D/D_t)$, which accounts for how infrequently the term is used across all documents. The local component can be easily obtained by counting the frequency of the word in the document. The global component is stated in terms of the number of documents D in the system, and the number of documents D_t that have the term t. We use random sampling to estimate these measures.

Let N be the number of nodes in the system, and D and D_t be as above. We choose k nodes uniformly at random. This can be done either with random walks, in unstructured systems [6], or routing to a random point in the namespace in structured systems [8]. We then compute the total number $\tilde{D}$ of documents and $\tilde{D}_t$ of documents with term t at the sampled nodes. For simplicity, we accept that the same node may be sampled more than once. It is easy to see that $\mathrm{E}[\tilde{D}] = k\frac{D}{N}$ and $\mathrm{E}[\tilde{D}_t] = k\frac{D_t}{N}$ where E indicates expectation of a random variable. The intuition is that if we take enough samples, then $\tilde{D}$ and $\tilde{D}_t$ are reasonably close to their expected value. If that is the case, then we can estimate D/D_t as $\frac{D}{D_t} \approx \frac{\tilde{D}}{\tilde{D}_t}$

To derive a sufficient condition for this approximation, we introduce two new quantities. Let M and M_t be the maximum number of documents and maximum number of documents with the term t, respectively, on a node. We call the estimate $\tilde{D}$ (resp. $\tilde{D}_t$) "good", if it is within a factor of $(1 \pm \delta)$ of its expected value. The estimate can be "bad" with a small probability (ϵ). As usual, we let e denote the base of the natural logarithm.

Theorem 1. *Let D, N, k, M be as above. For any* $0 < \delta \leq 1$ *and* $\epsilon > 0$*, if*

$$k \geq \frac{3}{\delta^2} \frac{M}{D/N} \ln(2/\epsilon) \tag{2}$$

then the random variable $\tilde{D}$ *(as defined above) is very close to its mean, except with probability at most* ϵ*. Specifically:*

$$\Pr[(1-\delta)kD/N \leq \tilde{D} \leq (1+\delta)kD/N] \geq 1-\epsilon. \tag{3}$$

Proof. The proof is an application of the Chernoff bound [9]. For $i = 1, \ldots, k$, let Y_i be the random variable representing the number of documents found during the i-th sample. Note that $\tilde{D} = \sum_{i=1}^{k} Y_i$. In order to apply the Chernoff bound, we need random variables in the interval [0, 1]. Let $X_i = Y_i/M$ and let $X = \sum_{i=1}^{k} X_i = \tilde{D}/M$. Define:

$$\mu = \mathrm{E}[X] = \frac{kD}{MN}.$$

Since the X_i are in $[0, 1]$ and are independent, we can use the Chernoff bound, which tells us that for any $0 < \delta \leq 1$.

$$\Pr[|X - \mathrm{E}[X]| > \delta\,\mathrm{E}[X]] \leq 2e^{-\frac{\mu\delta^2}{3}},$$

which can be rewritten as:

$$\Pr[(1-\delta)kD/N \leq \tilde{D} \leq (1+\delta)kD/N] \geq 1 - 2e^{-\frac{\mu\delta^2}{3}}.$$

We now impose the constraint that the probability above is at least $1-\epsilon$:

$$2e^{-\frac{\mu\delta^2}{3}} \leq \epsilon,$$

from which we derive the bound on k:

$$k \geq \frac{3}{\delta^2} \frac{M}{D/N} \ln(2/\epsilon). \qquad \square$$

If we replace $D, M, \tilde{D}$ with D_t, M_t, $\tilde{D}_t$, the theorem also implies that if

$$k \geq \frac{3}{\delta^2} \frac{M_t}{D_t/N} \ln(2/\epsilon) \tag{4}$$

then the random variable $\tilde{D}_t$ is also a good estimate.

The following observations follow from Theorem 1:

- Theorem 1 tells us that for a good estimate, the number of samples needed does not depend on N directly, but on the quantities D/N and D_t/N and, less importantly, on M and M_t. This means that as the system size grows, we do *not* need more samples as long as the number of exported documents (with term t) also increases.
- If the number of documents D is much larger than the system size N and queries consist of popular terms ($D_t = \Omega(N)$), then our algorithm provides performance with ideal scaling behavior: Sampling a constant number of nodes gives us provably accurate results, *regardless of the system size.*
- In practice, documents and queries will contain rare (i.e., not popular) terms, for which $\ln(D/D_t)$ may be estimated incorrectly. However, we argue that such estimation error is both unimportant and inevitable. The estimation is relatively unimportant because if the query contains rare terms, then the entire set of results is relatively small, and ranking a small set is not as important. In general, sampling is a poor approach for estimating rare properties and alternative approaches are required.
- The number of samples is proportional to the ratios between the maximum and the average number of documents stored at a node (i.e., $\frac{M}{D/N}$ and $\frac{M_t}{D_t/N}$). This means that, as the distribution of documents in the system becomes more imbalanced, more samples are needed to obtain accurate results.

Special case: uniform distribution. We next restrict our attention to the special case in which the underlying storage system randomly distributes documents to the nodes, uniformly at random and independently. Such a distribution approximately models the behavior of a DHT. In this special case, a stronger version of Theorem 1 holds, provided we do sampling without replacement (alternatively, we do *rejection sampling* where, if we sample a node that has already been sampled, we reject this sample and sample again):

Theorem 2. *Let D, N, k be as above and assume each document is stored independently and uniformly at random at one node. For any $\epsilon > 0$, if we choose k samples* ***without replacement*** *with*

$$k \geq \frac{3}{\delta^2} \frac{1}{D/N} \ln(2/\epsilon), \tag{5}$$

then the random variable $\tilde{D}$ (as defined above) is very close to its mean kD/N, except with probability at most ϵ. Specifically:

$$\Pr[(1-\delta)kD/N \leq \tilde{D} \leq (1+\delta)kD/N] \geq 1-\epsilon. \tag{6}$$

This probability is taken over the random distribution of the documents, as well as the randomness in our sampling.

Proof. As opposed to the proof of Theorem 1, we now crucially use the fact that the documents have been mapped *independently*. For every document d, define the random variable X_d as the indicator of the event that the document d is stored at one of the k sampled nodes (i.e., X_d is 1 if this event happens, and is 0 otherwise). So, $\mathrm{E}[X_d] = k/N$. Defining $X = \sum_d X_d$, we see that $\mathrm{E}[X] = kD/N$.

We note that the different random variables X_d are *independent*, which is a consequence of the fact that we are conducting sampling without replacement. To see this, suppose we have distinct documents $d_1, d_2, \ldots, d_i$. We want to show that $\Pr[\bigwedge_{j=1}^{i} X_{d_j} = 1] = \prod_{j=1}^{i} \Pr[X_{d_j} = 1]$. Let S be the random variable representing the set of k distinct samples and let U be the family of all sets of k nodes. Exposing the value of S, we can write:

$$\Pr[\bigwedge_{j=1}^{i} X_{d_j} = 1] = \sum_{S^* \subset U} \Pr[S = S^*] \Pr[\bigwedge_{j=1}^{i} X_{d_j} = 1 | S = S^*] =$$

$$= \sum_{S^* \subset U} \Pr[S = S^*](k/N)^i = (k/N)^i = \prod_{j=1}^{i} \Pr[X_{d_j} = 1],$$

where we exploited the facts that the documents are stored independently and that the probability that d_i has been stored in S^* is exactly k/N. Since the X_d are all binary-valued, it is known that the above implies that the random variables X_d are all mutually independent. This is because the probability of any possible assignment for the X_d can be written as a linear combination of probabilities such as above. (This follows, for instance, from basic Fourier analysis.) As an example, note that

$$\Pr[X_{d_1} = 1,\ X_{d_2} = 0] = \Pr[X_{d_1} = 1] - \Pr[X_{d_1} = X_{d_2} = 1].$$

We now apply the Chernoff bound to X to yield our result; the calculation-details are the same as in the proof of Theorem 1. □

Hence, for the uniform case, the number of samples does not need to be proportional to the maximum number M of documents at any node. Therefore, the cost of our sampling algorithm is significantly decreased.

3.3 Storing Document Vectors

Document vectors need to be stored such that a query relevant to the document can quickly locate them. We store document vectors in distributed inverted indexes. As mentioned previously, an *inverted index* for a keyword t is a list of all the documents containing t. For each keyword t, our system stores the corresponding inverted index like any other object in the underlying P2P lookup system. This choice allows us to efficiently retrieve vectors of all documents that share at least one term with the query.

Figure 1 shows the process of exporting a document. We first generate the corresponding document vector by computing the term weights, which uses the procedure described in section 3.2. Next, using the underlying storage system API, we identify the node storing the index associated with each term in the document and add an entry to the index. Such entry includes a pointer to the document and the document vector.

The details of storing document vectors in inverted indexes depend on the underlying lookup protocol. In structured systems, given a keyword t, the index for t is stored at the node responsible for the key corresponding to t. The underlying protocol can be used to efficiently locate this node. A similar approach using inverted indexes has previously been used by [10,11,12,13] for searching in structured systems. In unstructured networks, indexes would need to be partitioned or replicated [7,6].

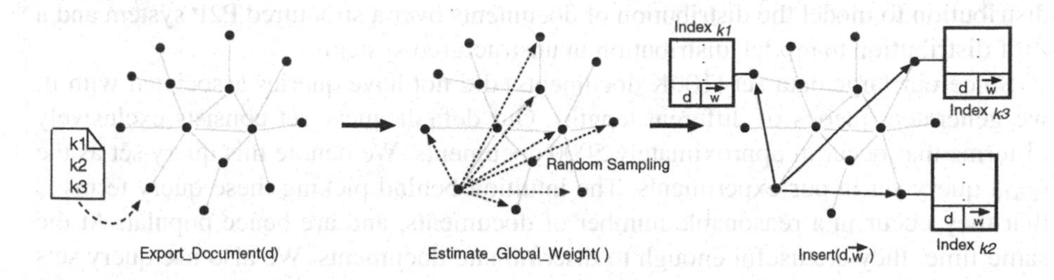

Fig. 1. Various steps in exporting documents and their vector representation

Reducing storage cost. So far, we have assumed that each word in the document is a keyword. Hence an entry is added for the document in the indexes of all the words in the document. A document, however, will not appear among the top few results when its weights for the query terms are low. Hence, not having these low-weight entries in the index does not reduce the retrieval quality of the top few results. We use this intuition to reduce the cost of propagating and storing vectors in indexes. We assume that there is a constant threshold w_{min}, that determines if the document entry is added to an index. The vector is not added to the index corresponding to the term t if the weight of t is below the threshold w_{min}. Note that the terms with weights below this threshold are still part of the vector. This heuristic has also been successfully used in eSearch [10].

3.4 Evaluating Query Results

A query is evaluated by converting it into a vector representation, and then computing the cosine similarity with respect to each "relevant" document vector. We compute query vectors using the same techniques used to generate the document vector. The next step is to locate the set of relevant documents. For each keyword in the query, we use the lookup functionality provided by the underlying system to identify the node storing the index of that keyword. We then compute the cosine similarity between the query and each of the document vectors stored in the index. This gives us a ranking of the documents available in this index. Finally, we fetch the top-K results computed at each of the indexes and compute the *union* of these result sets. The top-K documents in this union, sorted in the decreasing order of cosine similarities, give us our final result set.

4 Evaluation

In this section, we validate our distributed ranking system via simulation. We measure performance by comparing the quality of the query results returned by our algorithm with those of a centralized implementation of VSM.

Experimental setup. We use the TREC [2] Web-10G data-set for our documents. We used the first 100,000 documents in this dataset for our experiments. These 100K documents contain approximately 418K unique terms. Our default system size consists of 1000 nodes. We use two different distributions of documents over nodes: a uniform distribution to model the distribution of documents over a structured P2P system and a Zipf distribution to model distribution in unstructured systems.

Since our large data set (100K documents) did not have queries associated with it, we generated queries of different lengths. Our default query set consists exclusively of terms that occur in approximately 5000 documents. We denote this query set as the Q_{5K} query set in our experiments. The intuition behind picking these query terms is that they occur in a reasonable number of documents, and are hence popular. At the same time, they are useful enough to discriminate documents. We also use query sets that exclusively contain keywords that are either very popular (occur in more than 10K documents) or those that are very rare (occur in less than 200 documents). We denote these query sets as Q_{pop} and Q_{rare} respectively. Each result presented (except for details from individual runs) is an average of 50 runs. We use three metrics to evaluate the quality of distributed ranking:

1. **Coverage:** We define coverage as the number of top-K query results returned by the distributed scheme that are also present in the top-K results returned by a centralized VSM implementation for the same query. For example, if we're interested in the top 3 results, and the distributed scheme returns documents (A, C, D) while the centralized scheme returns (A, B, C), then the coverage for this query is 2.
2. **Fetch:** We define fetch as the minimum number R' such that, when the user obtains the set of $\mathcal{R}'$ results as ranked by the distributed scheme, $\mathcal{R}'$ contains all the top-K results that a centralized implementation would return for the same query. In the previous example, if the fourth result returned by the distributed case had been B, then the fetch for $K = 3$ would be 4.
3. **Consistency:** We define consistency as the similarity in the rank of results, for the same query, for different runs using different samples.

4.1 Coverage

In the first experiment, we measure the coverage of the distributed retrieval scheme. We show that by sampling only a few nodes even on a reasonably large system, our scheme produces results very close to a centralized implementation.

In our base result, we use a 1000 node network. The documents are mapped uniformly to nodes. To compute the global weight of term t, we sample 10, 20 and 50 nodes in different runs of the experiment. The queries consist of keywords from the Q_{5K} query set, i.e. the keywords occur in approximately 5000 documents.

The results are presented in Table 1. It is clear from Table 1 that the distributed ranking scheme performs very similar to the centralized implementation. On a 1000 node network with documents distributed uniformly, the mean accuracy for the top-K results is close to 93% with 50 random samples. Even with 10 random samples, the results are only slightly worse at 85% accuracy.

With 5000 nodes, the retrieval quality is not as high as a network with 1000 nodes. With 20 random samples, the mean accuracy is 77% for top-K results. There is a 8% increase in mean accuracy when we increase the sampling level and visit 1% (50) of the nodes. This result is a direct consequence of Theorem 2. Here, the number of documents has remained the same, but the number of nodes has increased. Hence, higher number of nodes sampled leads to better estimates.

Table 1. Mean and Std. Deviation of coverage with the distributed ranking scheme

Network Setup	Number of Samples	Top-K results 10	20	30	40	50
1000 uniform	10	8.49 (1.08)	16.99 (1.20)	25.30 (1.55)	33.68 (2.07)	42.28 (2.01)
	20	8.90 (0.99)	17.81 (1.04)	26.44 (1.26)	35.23 (1.87)	44.30 (1.82)
	50	9.28 (0.82)	18.63 (0.82)	27.66 (1.04)	36.08 (1.45)	46.30 (1.46)
5000 uniform	10	6.78 (1.39)	13.58 (1.74)	20.43 (2.39)	27.35 (2.99)	34.59 (3.40)
	20	7.74 (1.29)	15.41 (1.46)	22.92 (1.96)	30.50 (2.47)	38.49 (2.58)
	50	8.52 (1.09)	16.96 (1.18)	25.20 (1.56)	33.59 (2.11)	42.34 (1.98)
1000 Zipf	10	8.27 (1.15)	16.52 (1.26)	24.66 (1.71)	32.82 (2.21)	41.20 (2.27)
	20	8.82 (0.99)	17.63 (1.06)	26.22 (1.35)	34.83 (1.93)	43.70 (1.88)
	50	9.26 (0.80)	18.54 (0.88)	27.52 (1.12)	36.71 (1.49)	46.12 (1.56)
5000 Zipf	10	6.09 (1.54)	12.29 (1.97)	18.58 (2.68)	25.01 (3.39)	31.67 (3.97)
	20	7.34 (1.31)	14.71 (1.62)	21.89 (2.10)	29.34 (2.64)	36.93 (2.90)
	50	8.41 (1.13)	16.73 (1.22)	24.92 (1.61)	33.22 (2.08)	41.71 (2.03)

Table 1 also shows the retrieval quality for documents mapped to nodes using a Zipf distribution with parameter 0.80. With 1000 nodes and 50 samples, the retrieval quality is similar to that of the uniformly distributed case. With 10 samples, however, the mean accuracy drops a few percentage points to between 82–83%. With 5000 nodes and 50 samples, we see similar trends. While the quality is not as good as it is with the uniformly distributed data, it does not differ by more than 2%. With 10 samples, the results worsen by about much as 7%. Hence, we believe our scheme can be applied over lookup protocols on unstructured networks without appreciable loss in quality.

4.2 Fetch

In this experiment, we measure how many results need to be fetched before all the top-K results from the centralized implementation are available (we called this measure Fetch). We experiments with both 1000 and 5000 nodes with the documents uniformly distributed. We used the Q_{5K} query set for our evaluation. We plot the result in Figures 2 and 3. The x-axis is the top-K of results from the centralized implementation, while the y-axis represents the corresponding average fetch.

With a 1000 node network, we see that fetch is quite small even if only ten nodes are sampled. For instance, sampling 10 nodes, we need 13 results to match the top-10 results of the centralized case. With samples from 50 nodes, fetch is minimal even for less relevant documents: we need 11 entries to match the desired top-10 results and 63 to match the top-50 results from the centralized implementation.

As expected, with increasing network size, but same document set, the fetch increases. When we sample 1% of the 5000 nodes, we need 13 results to cover the top-10 and 88 to cover the top-50. With lesser sampling, however, we need to fetch a lot more results to cover the top-K. This behavior, again, is predicted by Theorem 2: when the number of nodes increases without a corresponding increase in the number of documents, the samples needed to guarantee a bound on sampling error also increases.

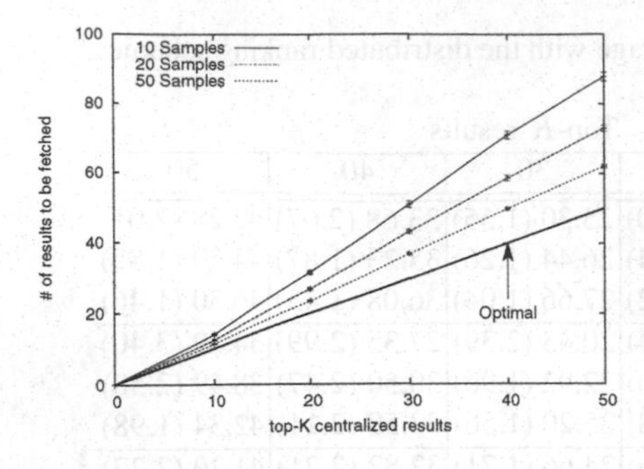

Fig. 2. Average fetch of the distributed ranking scheme with 1000 nodes. The error bars correspond to 95% confidence interval.

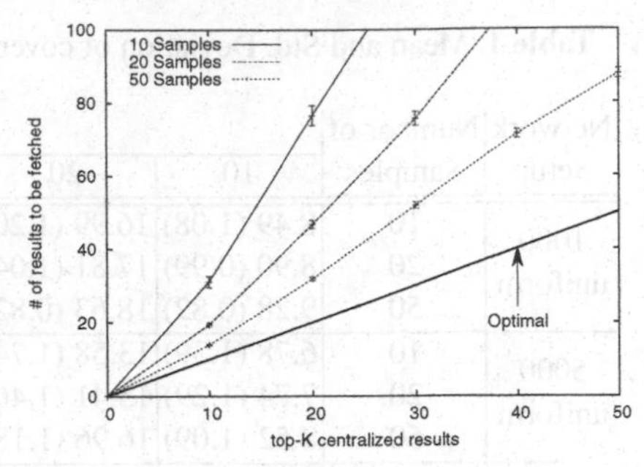

Fig. 3. Average fetch of the distributed ranking scheme with 5000 nodes. The error bars correspond to 95% confidence interval.

Other experiments indicate similar results when the document distribution is skewed. We merely summarize those results here. With a 1000 node network and 10 random samples, the fetch increases by 10% compared to the network where documents are mapped uniformly to nodes. In a 5000 node network, this increases by 35% compared to the uniform case. The results in both the network sizes with 50 random samples, however, are comparable to the uniform case.

4.3 Consistency

In our system, a new query vector is generated each time a query is evaluated. This leads to different weights being assigned to the terms during different evaluations of the same query. This can increase the variance in ranking, and potentially lead to different results for different evaluations of the query. In this experiment, we show that is not the case, and that the results are minimally affected by the different samples.

We use a network of 1000 nodes with documents mapped uniformly to these nodes. We sample 20 random nodes while computing the query vector. We use Q_{5K} and record the top-50 results for different runs and compare the results against each other and against the centralized implementation.

Figure 4 shows the results obtained during five representative runs for three representative queries each. For each run, the figure includes a small box corresponding to a document ranked in the top-50 by centralized VSM if and only if this document was retrieved during this run. For example, in Figure 4, query 1, run 2 retrieved documents ranked $1 \ldots 25$, but did not return the document ranked 26 in its top 50 results. Also, note that the first 25 centrally ranked documents need not necessarily be ranked exactly in that order, but each of them were retrieved within the top-50.

There are two main observations to be drawn: first, the sampling does not adversely affect the consistency of the results, and different runs return essentially the same results. Further, note that these results show that the coverage of the top results is uniformly good, and the documents that are not retrieved are generally ranked towards the bottom of the top-50 by the centralized ranking. In fact, a detailed analysis of our data shows that this trend holds in our other experiments as well.

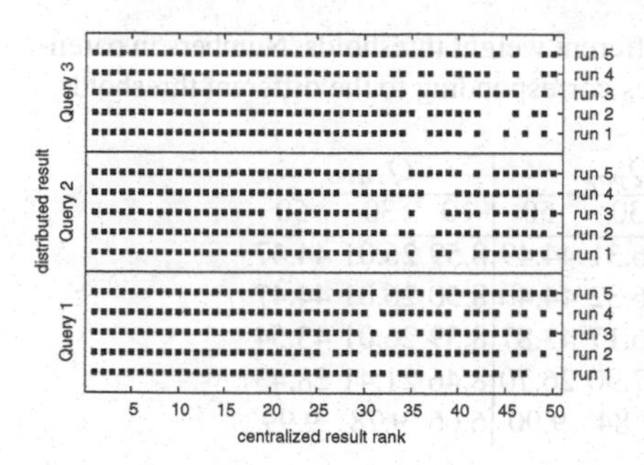

Fig. 4. Consistency of top-50 results in distributed ranking for three different queries from Q_{5K} set

Table 2. Mean coverage when the number of nodes and documents scale proportionally

Network	Top-K results				
Setup	10	20	30	40	50
500 nodes	7.75	16.11	25.08	33.62	42.24
1000 nodes	7.99	16.33	24.58	32.98	41.59
2000 nodes	7.67	15.85	23.96	32.00	40.11
5000 nodes	6.95	15.21	22.99	30.66	38.85

4.4 Scalability

In this experiment, we evaluate the scalability of our scheme with increasing system size. Theorems 1 and 2 states that the number of samples required is independent of the system size, under the condition that the size of the document set grows proportionally to the number of nodes. We demonstrate this fact by showing that coverage remains approximately constant as we increase the system size ten-fold (from 500 to 5000), while sampling the same number of nodes (20).

The number of documents in each experiment is 20 times the number of nodes in the system. For all the configurations, the terms used in queries occur in more than 10% of the total documents. For the 5000 node network, this corresponds to the Q_{pop} query set. In each case, we sample 20 random nodes to estimate the global weights.

Table 2 shows the mean coverage of our distributed scheme. As the table shows, the coverage of the distributed retrieval is very similar in most cases. This result confirms that our scheme depends almost entirely on the density of the number documents per node, and that it scales well as long as the density remains similar.

4.5 Reducing Storage Cost

Recall our optimization to store document vectors only in the indexes of keywords whose weights are greater than a threshold w_{min}. In this experiment, we quantify the effect of this optimization. For this experiment, we used a network of 1000 nodes with documents distributed uniformly at random over the nodes. We use all the three query sets and sample 20 nodes to estimate the weights. Note that we normalize the vectors; so the term weights range between 0.00 and 1.00. We present results for thresholds ranging from 0.00 to 0.30. We compare the results retrieved from the centralized implementation with $w_{min} = 0.00$.

The results of this experiment are tabulated in Table 3. Coverage of distributed ranking is not adversely affected when the threshold is set to 0.05 or 0.10. However, larger thresholds (say 0.20 and above) discard relevant entries, and consequently decrease rank quality appreciably. In order to understand the reduction obtained by using the threshold, we recorded the total number of index entries in the system for each threshold.

Table 3. Mean coverage of distributed ranking for different weight thresholds. Numbers in parenthesis indicate percentage reduction in size of indexes corresponding to the different thresholds.

Weight	Q_{5K}			Q_{pop}			Q_{rare}		
Threshold	10	30	50	10	30	50	10	30	50
0.0 (0.0)	8.90	26.44	44.30	8.32	26.31	44.49	8.59	26.01	44.47
0.05 (55.5)	8.90	26.44	44.34	8.33	26.32	44.40	8.50	26.01	44.47
0.10 (85.0)	8.90	26.40	44.22	8.32	26.17	43.87	8.59	26.01	43.54
0.20 (97.2)	7.64	20.43	30.97	6.39	17.90	26.70	8.46	21.41	28.43
0.30 (99.3)	4.53	7.98	8.88	2.79	6.84	9.90	6.66	9.78	9.99

The total number of index entries in our system is 15.9M when the threshold is 0.0. Our experiments show a reduction of 55.5% entries when we use a threshold of 0.05. Increasing the threshold to 0.1 leads to an additional 30% reduction in index size. A threshold value of 0.1 seems to be a reasonable trade-off between search quality and decreased index size.

5 Related Work

Our paper builds on prior work on efficient lookup and storage schemes. We assume the existence of a lookup protocol provided by the underlying system. Such lookup protocols have been studied in detail both in a structured setting (e.g., Chord [14] and Pastry [15]) and in an unstructured setting (e.g., Yappers [7] and LMS [6]). Providing a useful search facility has been an important area of research. Prior work in searching can broadly be classified into two categories: traditional centralized approaches, and search strategies over structured P2P networks.

Classic Information Retrieval. A lot of effort has gone into the area of information retrieval and ranking. We discussed the Vector Space Method [1] in Section 2. Latent Semantic Indexing (LSI) [16] is an extension to VSM that attempts to eliminate the issues of synonyms and polysemy. LSI employs singular value decomposition (SVD) to reduce the matrix generated by VSM. It is still an open question as to how LSI could be implemented in a completely distributed manner. There has also been work on implementing PageRank [17] in a distributed setting (e.g., [18]), but it cannot be applied on an arbitrary document set because of the lack of hyper-links.

Fagin et al.'s Threshold Algorithm (TA) [19] can also be used to compute the top-K results. Cao et al. [20] and [21] provide optimizations to TA in a distributed setting. In this paper, we store the entire vector in each index entry and use the union operation to group the results. Instead, we could just store the term weights in the index and use TA to compute the final result set. We plan to explore this option in the future.

Distributed Search over P2P systems. The idea of using Vector Space Methods has been applied previously in the context of P2P search. PlanetP [22] is a content-based search scheme that uses VSM. Nodes store vectors locally, but gossip digests of their local content. Queries are evaluated by ranking the *nodes* first and then evaluating the query using VSM at the top-ranked nodes. pSearch [23] uses VSM and LSI to generate document and query vectors, and maps these vectors to a high-dimension P2P system. Bhattacharya et al. [24] use similarity-preserving hashes (SPH) and the cosine similarity to compute similar documents over any DHT. Odissea [21], a P2P distributed search infrastructure, proposes to make use of TA to rank search results. None of these schemes, however, discuss how to compute the vectors. The work presented in this paper can be applied in all these settings to generate the document and query vectors.

There have also been proposals that use distributed inverted indexes to support the boolean query model. Reynolds et al. [12] and eSearch [10] use inverted indexes for searching in structured P2P systems. Loo et al. [13] design a hybrid solution that uses a DHT with inverted indexes to locate rare documents and Gnutella-style flooding for popular documents. None of these directly support ranking search results.

6 Conclusions

In this paper, we have presented a distributed algorithm for ranking search results. Our solution demonstrates that distributed ranking is feasible with little network overhead. Unlike previous work, we do not assume that the document vectors are provided to the system. Instead, our algorithm computes such vectors by using random sampling to estimate term weights. Through simulations and formal analysis, we show that the retrieval quality of our approach is comparable to that of a centralized implementation of VSM. We also show that our approach scales well under the reasonable condition that the size of the document set grows with the number of nodes.

Acknowledgments

We thank Divesh Srivastava and the referees for their comments and suggestions. This work was partially supported by NSF awards CCR-0208005, CNS-0626636, and NSF ITR Award CNS-0426683. Bobby Bhattacharjee was also supported in part by a fellowship from the Sloan Foundation. Part of Aravind Srinivasan's work was done while on sabbatical at the Network Dynamics and Simulation Science Laboratory of the Virginia Bioinformatics Institute, Virginia Tech.

References

1. Salton, G., Wong, A., Yang, C.: A vector space model for information retrieval. Journal of the American Society for Information Retrieval 18(11), 613–620 (1975)
2. TREC: Text REtrieval Conference. `http://trec.nist.gov/()`
3. Dumais, S.T.: Improving the retrieval of information from external sources. Behavior Research Methods, Instruments, and Computers 23(2), 229–236 (1991)
4. Salton, G., Buckley, C.: Term-weighting approaches in automatic text retrieval. Information Processing and Management 24(5), 513–523 (1988)
5. Buckley, C.: Implementation of the SMART information retrieval system. Technical report, Dept. of Computer Science, Cornell University, Ithaca, NY, USA (1985)
6. Morselli, R., Bhattacharjee, B., Srinivasan, A., Marsh, M.A.: Efficient lookup on unstructured topologies. In: PODC 2005. Proceedings of the 24th symposium on Principles of distributed computing, New York, NY, USA, pp. 77–86 (2005)

7. Ganesan, P., Sun, Q., Garcia-Molina, H.: Yappers: A peer-to-peer lookup service over arbitrary topology. In: INFOCOM. 22nd Annual Joint Conf. of the IEEE Computer and Communications Societies, San Francisco, USA (2003)
8. King, V., Saia, J.: Choosing a random peer. In: PODC 2004. Proceedings of the 23rd symposium on Principles of distributed computing, New York, NY, USA, pp. 125–130 (2004)
9. Chernoff, H.: A measure of asymptotic efficiency for tests of a hypothesis based on the sum of observations. Annals of Mathematical Statistics 23, 493–509 (1952)
10. Tang, C., Dwarakadas, S.: Hybrid global-local indexing for efficient peer-to-peer information retrieval. In: Proceedings of USENIX NSDI 2004 Conference, San Fransisco, CA (2004)
11. Gopalakrishnan, V., Bhattacharjee, B., Chawathe, S., Keleher, P.: Efficient peer-to-peer namespace searches. Technical Report CS-TR-4568, University of Maryland, College Park, MD (2004)
12. Reynolds, P., Vahdat, A.: Efficient peer-to-peer keyword searching. In: Endler, M., Schmidt, D.C. (eds.) Middleware 2003. LNCS, vol. 2672, pp. 21–40. Springer, Heidelberg (2003)
13. Loo, B.T., Hellerstein, J.M., Huebsch, R., Shenker, S., Stoica, I.: Enhancing P2P file-sharing with an internet-scale query processor. In: VLDB 2004. Thirtieth International Conference on Very Large Data Bases, Toronto, Canada, pp. 432–443 (2004)
14. Stoica, I., Morris, R., Karger, D., Kaashoek, M.F., Balakrishnan, H.: Chord: A scalable peer-to-peer lookup service for internet applications. In: Proceedings of the ACM SIGCOMM 2001, San Diego, California (2001)
15. Rowstron, A., Druschel, P.: Pastry: Scalable, distributed object location and routing for large-scale peer-to-peer systems. In: Guerraoui, R. (ed.) Middleware 2001. LNCS, vol. 2218, pp. 329–350. Springer, Heidelberg (2001)
16. Deerwester, S., Dumais, S., Landauer, T., Furnas, G., Harshman, R.: Indexing by latent semantic analysis. Journal of the American Society for Information Science 41(6), 391–407 (1990)
17. Page, L., Brin, S., Motwani, R., Winograd, T.: The pagerank citation algorithm: bringing order to the web. Technical report, Dept. of Computer Science, Stanford University (1999)
18. Wang, Y., DeWitt, D.J.: Computing PageRank in a distributed internet search engine system. In: VLDB 2004. Thirtieth International Conference on Very Large Data Bases, Toronto, Canada, pp. 420–431 (2004)
19. Fagin, R., Lotem, A., Naor, M.: Optimal aggregation algorithms for middleware. Journal of Computer and System Sciences (JCSS) 66(4), 614–656 (2003)
20. Cao, P., Wang, Z.: Efficient top-k query calculation in distributed networks. In: PODC 2004. Proceedings of the twenty-third annual ACM symposium on Principles of distributed computing, pp. 206–215. ACM Press, New York (2004)
21. Michel, S., Triantafillou, P., Weikum, G.: KLEE: A framework for distributed top-k query algorithms. In: VLDB 2005. Proceedings of the 31st International Conference on Very Large Data Bases, Trondheim, Norway, pp. 637–648 (2005)
22. Cuenca-Acuna, F.M., Peery, C., Martin, R.P., Nguyen, T.D.: PlanetP: Using Gossiping to Build Content Addressable Peer-to-Peer Information Sharing Communities. In: HPDC-12. Proceedings of the 12th Symposium on High Performance Distributed Computing, IEEE Press, Los Alamitos (2003)
23. Tang, C., Xu, Z., Dwarkadas, S.: Peer-to-peer information retrieval using self-organizing semantic overlay networks. In: Proceedings of ACM SIGCOMM 2003, pp. 175–186. ACM Press, New York (2003)
24. Bhattacharya, I., Kashyap, S.R., Parthasarathy, S.: Similarity searching in peer-to-peer databases. In: ICDCS 2005. Proceedings of the 25th International Conference on Distributed Computing Systems, pp. 329–338 (2005)

ROW-FS: A User-Level Virtualized Redirect-on-Write Distributed File System for Wide Area Applications

Vineet Chadha and Renato J. Figueiredo

Advanced Computing and Information Laboratory
Department of Electrical Engineering, University of Florida
{chadha,renato}@acis.ufl.edu

Abstract. We propose a virtualization approach to implement redirect-on-write capabilities that overlay a traditional distributed file system. The redirect-on-write distributed file system (ROW-FS) is implemented via a user-level proxy that is able to selectively steer Network File System (NFS) RPC calls to one of two servers: a "main" read-only server, and a "shadow" read-write server. By employing virtualization by means of a user-level proxy and using the de-facto standard NFS protocol, ROW-FS can be mounted as an NFS file system by existing, unmodified clients from a variety of platforms, and requires no changes to existing kernels. Its primary application is in supporting wide-area computing environments, where ROW-FS can provide improved performance and fault-tolerance (file system modifications can be check-pointed along with application state). Results show that benchmark applications including Linux kernel compilation and instantiation of virtual machines across wide-area networks achieve substantially better performance with ROW-FS as compared to NFS.

Keywords: File System, Virtual Machine, Distributed Computing, Redirect-on-write, Grid Computing, Virtualization.

1 Introduction

A key challenge arising in wide-area, Grid computing infrastructures is that of data management – how to provide data to applications, seamlessly, in environments spanning multiple domains. In these environments, data movement and sharing is often mediated by middleware that schedules applications and workflows [25], and data management is achieved by means of explicit file transfers [26][29][28]. This paper presents a novel approach that enables wide-area applications to leverage on-demand block-based data transfers and a de-facto distributed file system (NFS) to *access* data stored remotely and *modify* it in the local area.

The approach is based on user-level redirect-on-write virtualization techniques that address two important needs. First, the ability of accessing and caching file system data and meta-data from remote servers, on-demand, on a per-block basis, while buffering file system modifications locally. This is key in supporting applications that rely on the availability of a file system or operate on sparse data – a representative example is the instantiation of customized execution environment containers such as system virtual

S. Aluru et al. (Eds.): HiPC 2007, LNCS 4873, pp. 21–34, 2007.

machines (VMs) [19][14][23][14][3] or physical machines provisioned on demand [17]. Second, the ability to *checkpoint* filesystem modifications to facilitate application recovery and restart in the event of a failure. Checkpointing/migration in wide-area computing systems is often achieved at the level of operating system processes by means of library interposition or system call interception [11], which limits the applicability of checkpointing to a restricted set of applications. In contrast, ROW-FS enables checkpoint/restart of modifications made by a client to a mounted distributed file system.

The approach is unique in supporting this functionality on top of existing, widely available kernel distributed file system clients and servers that implement the NFS protocol. The paper describes the organization of the ROW proxy and the techniques used to virtualize NFS remote procedure calls, and evaluate the performance of a user-level implementation of these techniques in a variety of micro-benchmarks and applications. Results show that ROW-FS mounted file systems can achieve better performance than non-virtualized NFS in wide-area setups by steering data and metadata calls to a local-area shadow server, and that it enables an unmodified application running on a VM container and operating on data within a ROW-FS file system to be successfully restarted from a checkpoint following a failure.

This paper is organized as follows. Section 2 discusses motivations, background and applications of ROW-FS. Section 3 describes its architecture and approaches to deploying ROW-FS in conjunction with file system caching proxies. Section 4 details the virtualization of a representative subset of NFS remote procedure calls. Section 5 presents experimental results and analysis of the performance of ROW-FS. Section 6 discusses consistency considerations, and Section 7 gives a brief survey of related work. Section 8 concludes with a summary and discussion of future directions.

2 Motivation and Background

There are three goals which motivate the approach of this paper. First, with ROW-FS a primary server can be made read only, thus preventing the integrity of data mounted from the primary server from unintentional user modification. Second, since heterogeneity and dynamism in distributed computing makes failure recovery a difficult task, we provide a consistent point-in-time view of a recently modified file system. Third, to facilitate deployment, ROW-FS leverages capabilities provided by the underlying file system (e.g. NFS) without requiring kernel-level modifications.

An important class of Grid applications consists of long-running simulations, where execution times in the order of days are not uncommon, and mid-session faults are highly undesirable. Systems such as Condor 11 have dealt with this problem via application check-pointing and restart. A limitation of this approach lies in that it only supports a restricted set of applications - they must be re-linked to specific libraries and cannot use many system calls (e.g. fork, exec, mmap). ROW-FS, in contrast, supports unmodified applications; it uses client-side virtualization that allows for transparent buffering of all modifications produced by DFS clients on local storage.

ROW-FS is well-suited for systems where execution environments are created, allocated for an application to host their execution, and then terminated after the application finishes. This approach is taken by several projects in Grid/utility computing (e.g. In-VIGO[14], COD[17] and Virtual Workspaces[23]). In this context, ROW-FS complements capabilities provided by "classic" virtual machine (VMs [18][19]) to support flexible, fault-tolerant execution environments in distributed computing systems. Namely, ROW-FS enables mounted distributed file system data to be periodically check-pointed along with a VM's state during the execution of a long-running application. ROW-FS also enables the creation of non-persistent execution environments for non-virtualized machines. For instance, it allows multiple clients to access in read/write mode an NFS file system containing an O/S distribution exported in read-only mode by a single server. Local modifications are kept in per-client "shadow" file systems that are created and managed on-demand.

2.1 NFS-Mounted Virtual Machine Images and O/S File Systems

One important application of ROW-FS is supporting read-only access of shared VM disks or O/S distribution file systems to support rapid instantiation and configuration of nodes in a network. The ROW capabilities, in combination with aggressive client-side caching, allow many clients to efficiently mount a system disk or file system from a single image – even if mounted across wide-area networks.

One particular use case is the on-demand provisioning of non-persistent VM environments. In this scenario, the goal is to have thin, generic boot-strapping VMs that can be pushed to computational servers without requiring the full transfer or storage of large VM images. Upon instantiation, a diskless VM boots through a pre-boot execution environment (PXE) using one out of several available shared non-persistent root file system images, stored potentially across a wide area network.

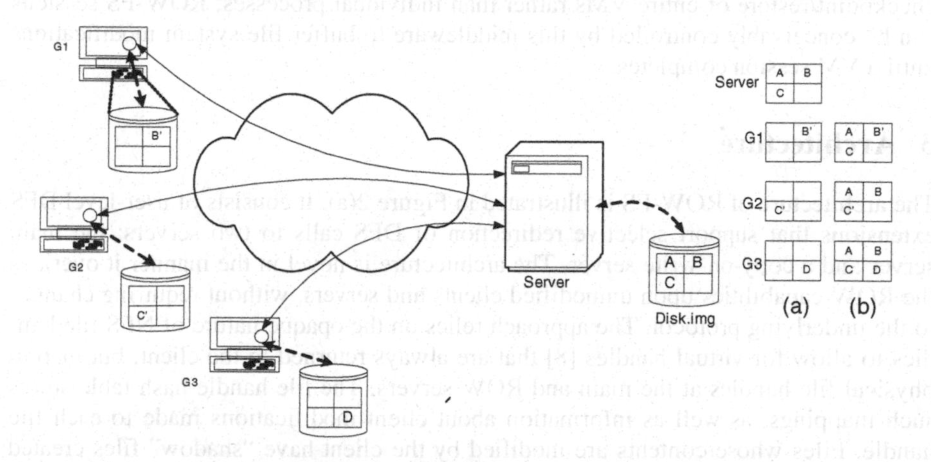

Fig. 1. Middleware Data Management: - Grid Users G1, G2 and G3 accesses file disk.img from server and customize for personal use through ROW proxy. G1 modifies second block B to B', G2 modifies block C to C' and G3 extends the file with additional block D (a) Modifications are stored locally at each shadow server (b) virtualized view.

This approach delivers capabilities that are not presently provided by VM monitors themselves. Without an NFS-mounted file system *on the host*, on-demand transfer of VM image files is not possible, thus the entire VM image would need to be brought to the client before a non-persistent VM could start. In shared Grid computing environments it is difficult to acquire privileges on the host to perform such file system mounts; in contrast, with ROW-FS, the NFS-mounted file system can be kept inside a guest, and no host configuration or privileges are required to deploy the bootstrapping VM and the diskless VM. Section 6.3 discusses experiments that evaluate the performance of ROW-FS in this environment.

2.2 Fault Tolerant Distributed Computing with Virtual Machines

Existing VM monitors have support for check-pointing/resuming of VM state – a key capability upon which many fault-tolerant techniques can be built. However, check-pointing of the VM state alone is not sufficient to cover the scenarios envisioned for a VM-based distributed computing environment. Consider a virtual machine based client/server session using traditional NFS and ROW-FS. A long-running application may take hours to complete; if it operates on data mounted over a distributed file system, a failure in the client may require restarting the entire session again – even if the VM had been check-pointed. In contrast, a ROW-FS session with regular check-points provides fault tolerance additional to VM checkpointing by allowing file system-mounted data used by the application to be checkpointed with the VM.

In this use case, the role of ROW-FS in supporting checkpoint/restart is to buffer file system modifications within a VM container. The actual process of checkpointing is external and complementary to ROW-FS. It can be achieved with support from VMM APIs (e.g. vmware-cmd and Xen's "xm") and distributed computing middleware. For instance, the Condor 11[30] middleware is being extended to support checkpoint/restore of entire VMs rather than individual processes; ROW-FS sessions can be conceivably controlled by this middleware to buffer file system modifications until a VM session completes.

3 Architecture

The architecture of ROW-FS is illustrated in Figure 2(a). It consists of user-level DFS extensions that support selective redirection of DFS calls to two servers: the main server and a copy-on-write server. The architecture is novel in the manner it overlays the ROW capabilities upon unmodified clients and servers, without requiring changes to the underlying protocol. The approach relies on the opaque nature of NFS file handles to allow for virtual handles [8] that are always returned to the client, but map to physical file handles at the main and ROW servers. The file handle hash table stores such mappings, as well as information about client modifications made to each file handle. Files whose contents are modified by the client have "shadow" files created by the ROW server in a sparse file, and block-based modifications are inserted in-place in the shadow file. A presence bitmap marks which blocks have been modified, at the granularity of NFS blocks (8-32KB).

Figure 2(b) shows possible deployments of proxies enabled with user-level disk caching and ROW capabilities. For example, a cache proxy configured to cache read-only data may precede the ROW proxy; thus effectively forming a read/write cache

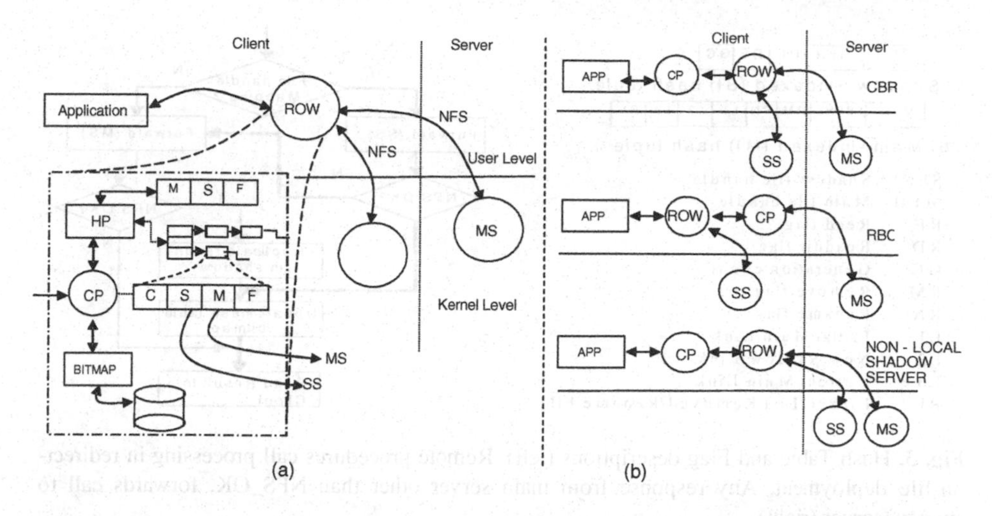

Fig. 2. (a) Architecture – The Redirect-on-write file system is implemented by means of a user-level proxy which virtualizes NFS by selectively steering calls to a main server (MS) and shadow server (SS). (b) Proxy Deployment - Cache-before-redirect (CBR), Redirect-before-cache (RBC), Non-local shadow server. MS: Main server, SS: shadow server, CP: cache proxy.

hierarchy. Such cache-before-redirect (Figure 2(b), top) proxy setup allows disk caching of both read-only contents of the main server as well as of client modifications. Write-intensive applications can be supported with better performance using a redirect-before-cache (Figure 2(b), middle) proxy setup. Furthermore, redirection mechanisms based on the ROW proxy can be configured with both shadow and main servers being remote (Figure 2(b), bottom). Such setup could, for example, be used to support a ROW-mounted O/S image for a diskless workstation.

3.1 Hash Table

The hash table processor is responsible for maintaining in-memory data structures on a per-session basis to keep mapping of file handles between the client and the two servers. Two hash tables are employed. The *shadow-indexed* (SI) hash table is used to keep mappings between the shadow and main servers. This hash table is indexed by shadow file handle because the number of file system objects in the shadow server is a superset of the file system objects in main server. The *main-indexed* (MI) table is needed to maintain state information about files in the main server. Figure 3 (left) shows the structure of the hash table and flag information. The *readdir* flag (RD) is used to indicate the occurrence of the invocation of an NFS *readdir* procedure call for a directory in the main server. Generation count (GC) is a number inserted into hash tuple for each file system object to create a unique disk based bitmap. Remove (RM) and Rename (RN) flags are used to indicate deletion/rename of a file.

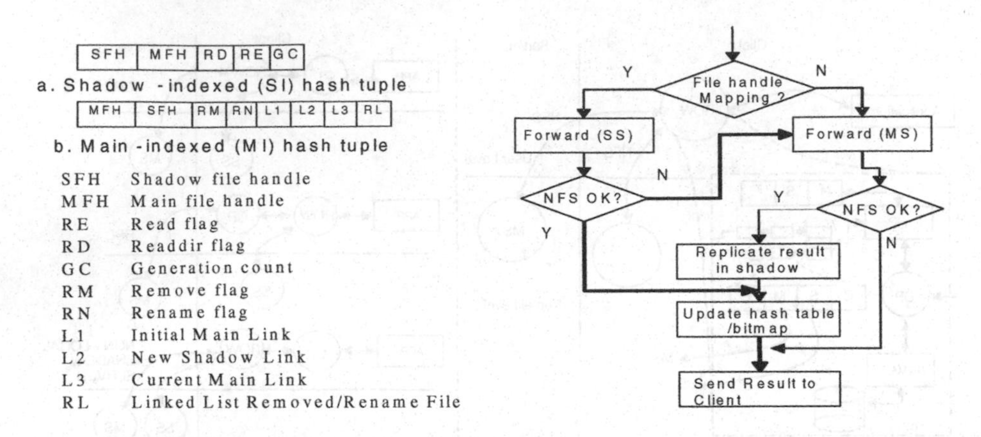

Fig. 3. Hash Table and Flag descriptions (left). Remote procedures call processing in redirect-on-file deployment. Any response from main server other than NFS_OK, forwards call to shadow server (right).

3.2 Bitmap Processor

The bitmap processor processes file handle and offset information and checks the presence bitmap data structure to determine whether read and write calls should be directed to the main or shadow server. The bitmap is a disk-based hierarchical data structure to keep information about individual blocks within a file. The parent directory in the bitmap data structure is a concatenation of the hashed value of a shadow file handle and the generation count, which results in a unique bitmap directory for each file system object. As in NFS, reads and writes are performed in a per-block basis in ROW file system. To keep track of current location of updated block, each file is represented by a two-level hierarchical data structure in disk. The first level indicates the name of file which contains information about the block. The second level indicates the location of a presence bit within the bitmap file.

4 Implementation

This section describes how a ROW-FS proxy handles NFS protocol calls. Each procedure call is handled in three phases: predicate, process and update. In the following discussion, for the sake of brevity, we explain supporting examples and scenarios for a representative subset of NFS procedure calls. The complete implementation and semantics of RPC calls for ROW-FS is presented in detail in a technical report 24. Figure 4 describes the various hash table entries stored in the proxy which are referenced throughout this section.

MOUNT: The *mount* procedure is modified to obtain initial mount file handle of shadow server. Specifically, the mount proxy forwards a *mount* call to both shadow and main server. When the mount utility is issued by a client, the shadow server is contacted first to save the file handle of the directory to be mounted. This file handle is later used by NFS procedure calls to direct RPC calls to the shadow server. The initial mapping of file handles of a mounted directory is inserted in the SI hash table during invocation of the *getattr* procedure.

LOOKUP: The *lookup* procedure returns a file handle reference (FH) to the file system object sought by the client. The indirection of *lookup* calls between shadow and main server works as follows. In the predicate phase, the proxy obtains SI hash table mapping of parent file handle. We choose to first forward the *lookup* call to shadow server since more often client session involves repeatedly accessing same file and data. Further *lookup* call semantics depend on status of state flags as explained in section 3.

READ /WRITE: Read call reads data from a file referred by a given file handle and at a given offset. Again, *read/write* call is always preceded by a *lookup* call. Hence, file handle is always valid once the RPC call invokes *read/write* procedure. Note that mounted file system block size depends on parameters specified during invocation of mount utility (after proxy initialization) and bitmap block size is specified at initialization of ROW proxy. Hence, it may be the case that proxy forwards calls to both shadow and main server if requested data is present partly in main and shadow server.

5 Consistency Considerations

It is important to consider consistency in distributed file systems because data can be potentially shared by multiple clients. In the current ROW-FS architecture, we make the assumptions that 1) ROW-FS file systems are ephemeral; they are dynamically created and terminated by middleware that oversees the scheduling of application workflows, and 2) data stored in the main server of a file system mounted as ROW-FS remains unmodified for the duration of such an ephemeral file system session. Previous work has described techniques for establishing such dynamic file system sessions and enforcing exclusive access to shared data with a service-oriented architecture [27]; it is also conceivable to integrate the logic to configure, create and teardown ROW-FS sessions with application workflow schedulers such as [25].

During the time a ROW-FS file system session is mounted, all modifications are redirected to the shadow server. For consistency, two different scenarios need to be considered. First, there are applications in which it is neither needed nor desirable for data in the shadow server to be reconciled with the main server; an example is the provisioning of system images for diskless clients or virtual machines.

For applications in which it is desirable to reconcile data with the server, the ROW-FS proxy holds state in its primary data structures that can be used to commit modifications back to the server. The approach is to remount the file system at the end of a ROW-FS session in read/write mode, and signal the ROW-FS proxy to traverse its file handle hash tables and bitmaps to commit changes (moves, removes, renames, etc) to directories and files, and commit individually each data block modified at the shadow server back to the main server by crafting appropriate NFS calls to the server.

6 Experiments

Experiments have been conducted to measure the performance of a ROW-FS user-level implementation for varied application such as VM instantiation. We simulated wide area network links using the NISTnet network emulation package 16. The NISTnet emulator is deployed as a virtual Router in a VMware VM with 256MB memory and running Linux Redhat 7.3. Redirection is performed to a shadow server running in a virtual machine in the client's local domain.

6.1 Microbenchmarks

The goal of micro-benchmarks is to measure the performance of basic file system operations. For ROW-FS, we stress important NFS procedure calls. Specifically, we conducted benchmarks for lookup, remove/rmdir and readdir to evaluate overheads in these operations. For LAN, measured TCP bandwidth (using iperf) is 40Mbits/s. For the WAN setup, the bandwidth is 5Mbits/s with round-trip latencies of 70ms. We tested the benchmarks on file system hierarchy of nearly 15700 file systems objects (the total disk space consumed is approximately 190MB). In all micro-benchmark experiments, the main server is a Linux VM with 256MB memory hosted on an Intel Pentium 4 1.7GHz workstation with 512MB memory. The WAN router is hosted in the same machine as the main server. The client machine is an Intel Pentium 1.7.Ghz workstation with 512 MB memory running cache and ROW proxies.

Lookup/Stat benchmark: Lookup is often the most frequent operation in the NFS protocol. Since the initial request for a file handle invokes a lookup request, we decided to measure individual lookup latency and a recursive stat of file system hierarchy. For a random set of individual files (in the LAN setup), the average lookup time for initial run of ROW-FS is 18ms and the second run is executed in approximately 8ms. In comparison, NFSv3 executes a lookup call in approximately 11ms. The results summarized in Table 1 show that ROW-FS performance is superior to NFSv3 in a WAN scenario, while comparable in a LAN. In the WAN experiment, recursive stat shows nearly five times improvement over second run of ROW-FS, because all the file objects are present in the shadow server during the second run of recursive stat.

Readdir: For newly created files and directories, the Readdir micro benchmark scans completely a directory to display the file system objects to the client. Results for Readdir along with lookup and recursive stats are shown in Table 1. Clearly, performance for WAN for ROW-FS during the second run is comparable with LAN performance and much improved over NFSv3. This is because once a directory is replicated at the shadow server, subsequent calls are directed to the shadow server by means of readdir status flag. The initial readdir overhead for ROW-FS (especially in LAN setup) is due to the fact that dummy file objects are being created in the shadow server during the execution.

Table 1. LAN and WAN experiments for lookup, readdir and recursive stat micro-benchmarks

Benchmark	LAN (sec)			WAN (sec)		
	ROW-FS		NFSv3	ROW-FS		NFSv3
	1st RUN	2nd RUN		1st RUN	2nd RUN	
Lookup	18msec	8msec	11msec	89msec	18msec	108msec
Readdir	67	17	41	1127	17	1170
Recursive Stat	425	404	367	1434	367	1965

Remove: To measure the latency of remove operations, we deleted a large number of files (greater than 15000 and total data size 190MB). We observed that in ROW-FS, since only remove state is being maintained rather than complete removal of file, performance is nearly 80% better than that of conventional NFSv3. It takes nearly 37 minutes in ROW-FS in comparison to 63 minutes in NFS3 to delete 190MB of data over wide area network. Note that each experiment is performed with cold caches, setup by re-mounting file systems in every new session. If the file system is already replicated in shadow server, it takes only 18 minutes (WAN) to delete the complete hierarchy.

Table 2. Remove Statistics for WAN and LAN

REMOVE	ROW-FS (sec)	NFSv3 (sec)
LAN	160	230
WAN	2250	3785

6.2 Application Benchmarks

The primary goal of the application benchmark experiments is to evaluate performance of redirect-on-write file system in comparison to traditional kernel network file system (NFSv3). Experiments are conducted for both local area and wide area networks. The client machine is a 1.7 GHz Pentium IV workstation with 512MB RAM and RedHat 7.3 Linux installed. The main and shadow servers are VMware-based virtual machines. Each VM is based on VMware GSX 3.0 and are configured with one CPU and 256 MB RAM. They are hosted by a dual-processor Intel Xeon CPU 2.40GHz server with 4GB memory.

Andrew Benchmark: We tested Andrew benchmark to gauge performance of ROW file system in local and wide area networks. In addition, we collected statistics of RPC calls going to shadow and main server to evaluate our performance. Table 3 summarizes the performance of Andrew benchmark and Figure 4 provides statistics for number of RPC calls. The important conclusion taken from the data in Figure 5 is that ROW-FS, while increasing the total number of calls routed to the local shadow server, reduces the number of RPC calls that cross domains to less than half. Note increase in number of *getattr* calls is due to invocation of *getattr* procedure to virtualize read calls to main server. We need to virtualize *Read* calls with shadow attributes (the case when blocks are read from Main server) because the client is unaware of the shadow server; file system attributes like file system statistics and file inode number has to be consistent between read and post-read *getattr* call. Nonetheless, since, all *getattr* calls go to the local-area shadow server, the overhead of extra *getattr* calls is small compared to *getattr* calls over WAN.

Table 3. Andrew Benchmark and AM-Utils execution times in local- and wide-area networks

Benchmark	ROW-FS (sec)	NFSv3 (sec)
Andrew (LAN)	13	10
Andrew(WAN)	78	308
AM Utils LAN	833	703
AM Utils WAN	986	2744

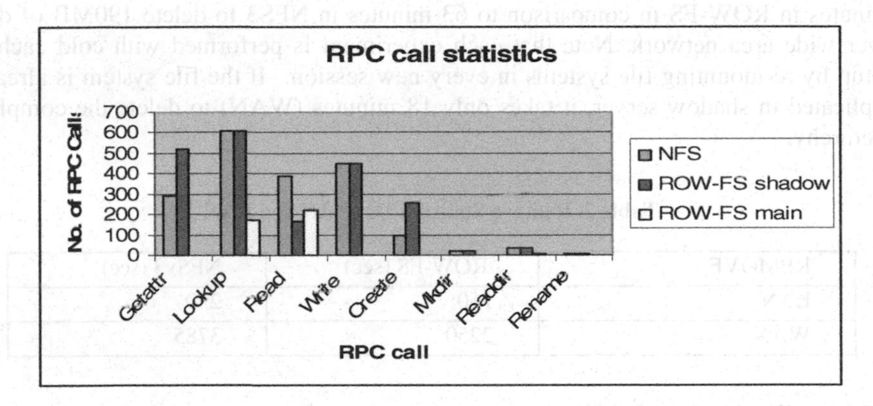

Fig. 4. Number of RPC calls received by NFS server in non-virtualized environment, and by ROW-FS shadow and main servers during Andrew benchmark execution

AM-Utils: We also used Berkeley automounter 20 build as an additional benchmark to evaluate performance. The automounter build consists of configuration tests to determine required features for build; thus generating large number of lookups, read and write calls. The second step involves compiling of am-utils software package. Table 3 provides readings for LAN and WAN. The resulting average ping time for the NIST-emulated WAN is 48.9ms in the ROW-FS experiment and 29.1ms in the NFSv3 experiment. Wide-area performance of ROW-FS for this benchmark is again better than NFSv3, even under larger average ping latencies.

Linux-Kernel Compilation: Compilation of the Linux kernel is used to benchmark the application-perceived performance of ROW for a typical software development environment application. This is a representative application with a mix of compute- and I/O-intensive features which runs for minutes and generates thousands of NFS calls of various kinds – block reads/writes, file and directory creation, and metadata lookups and modifications. The kernel used is debian 2.4.27 with compilation steps including: make "oldconfig", make "dep" and make "bzImage". Table 4 shows performance readings for both LAN and WAN environments.

We achieve comparable LAN performance and substantial improvement in the performance over the emulated WAN. Note that, for WAN, kernel compilation performance is nearly five times better with the ROW proxy in comparison with NFSv3 (current ROW-proxy is based on NFSv2). The results shown in Table 4 do not account for the overhead in synchronizing the main server. Nonetheless, as shown in Figure 4, a majority of RPC calls do not require server updates (read, lookup, getattr); furthermore, many RPC calls (write, create, mkdir, rename) are also aggregate in statistics – often the same data is written again, and many temporary files are deleted and need not be committed.

Table 4. Linux kernel Compilation execution times on a LAN and WAN

Setup	FS	Oldconfig time (s)	Dep time (s)	BzImage time (s)
LAN	NFSv3	49	120	710
	ROW-FS	55	315	652
WAN	NFSv3	472	2648	4200
	ROW-FS	77	1590	780

Fault tolerance: Finally, we tested the check-pointing and recovery of a computational chemistry scientific application (Gaussian 21). A VMware virtual machine running Gaussian is checkpointed (along with ROW-FS state in the VM's memory and disk). It is then resumed, runs for a period of time, and a fault is injected. Some Gaussian experiments take more than hour to finish and generate lot of temporary data. For example, an execution of Gaussian run generated about 300MB of data. With ROW-FS, we observe that the application successfully resumes from a previous checkpoint. With NFSv3, inconsistencies between the client checkpoint and the server state crash the application, preventing its successful completion.

6.3 Virtual Machine Instantiation

Diskless Linux: We performed experiments to boot diskless Linux nodes over the emulated WAN. We choose to measure diskless boot setup over wide area network as VM booting is often a frequent operation in dynamic system provisioning. The NISTnet delay is fixed at 20ms with measured bandwidth of 7Mbit/s. In this experiment, VM1 is the diskless virtual machine, and VM2 is a boot proxy machine configured with two NIC cards for communication with host only and public network. Both ROW and cache proxies are deployed in VM2, which proxies NFS requests to a remote file server. In addition, VM2 is configured to run DHCP and TFTP servers to provide an IP address and initial kernel image to VM1. Table 5 summarizes the performance of diskless boot times with different proxy cache configurations. The results show what pre caching of attributes before redirection and post redirection data caching deliver the best performance, reducing wide-area boot time with "warm" caches by over 300%.

Table 5. Wide Area Experimental Results for Disk Less Linux boot/Second Boot for (1) ROW proxy only (2) ROW proxy + data cache (3) attribute + ROW + data cache

WAN	Boot (sec)	2nd Boot (sec)
Client -> ROW -> Server	435	236
Client -> ROW -> Data Cache -> Server	495	109
Client -> Attr Cache -> ROW -> Data cache -> Server	409	76

VM boot/Second boot: This experiment involves running a Xen virtual machine (domU) with the root file system mounted over ROW-FS. The primary goal is to measure the overhead for additional layer of proxy indirection. The experiments are conducted in two parts. In the first part, there is only a ROW proxy and no cache proxy. The Xen domU VM is booted with and re-booted to capture the behavior of ROW-FS with the presence of data locality.

Table 6. Remote Xen boot/reboot experiment with ROW proxy and ROW proxy + cache

NISTNet Delay	ROW Proxy		ROW Proxy + Cache Proxy	
	Boot (sec)	2nd Boot (sec)	Boot (sec)	2nd Boot (sec)
1ms	121	38	147	36
5ms	179	63	188	36
10ms	248	88	279	37
20ms	346	156	331	37
50ms	748	266	604	41

We benchmarked the time to boot a Xen VM because we believe this will be a frequent operation - a container can be started for just the duration of an application run. Results for this experiment are summarized in Table 6. In the second part, we tested the setup with aggressive client side caching (Figure 2(b)). Table 6 also presents the boot/ second boot latencies for this scenario.

For delays smaller than 10 ms, the ROW+CP setup has additional overhead for Xen boot (in comparison with ROW setup); however, for delays greater than 10 ms, the boot performance with ROW+CP setup is better than ROW setup. Reboot execution time is almost constant with ROW+CP proxy setup. Clearly, the results show much better performance of Xen second boot for the ROW+CP experimental setup.

7 Related Work

The notion of network file system call indirection is not new; interposition of a proxy for routing remote procedural calls was previously addressed to provide scalable network file system services [4]. In past, researchers have used NFS shadowing technique to log users' behavior on old files in a versioning file system [31]. Emulation of NFS mounted directory hierarchy is often used as a means of caching and performance improvement [6]. Kosha provides a peer to peer enhancement of network file system to utilize redundant storage space [7]. In past, file virtualization was addressed through NFS mounted file system within the private name spaces for group of processes with motivation to migrate the process domain 9. To leverage on RAID performance, striped network file system is implemented which increases throughput by striping file between multiple servers [10]. A copy-on-write file server is deployed to share immutable template images for operating systems kernels and file-systems in [3]. The proxy-based approach presented in current paper is unique in how it not only provides copy-on-write functionality, but also provides provision for inter-proxy composition. Checkpoint mechanisms are integrated into language specific byte-code virtual machine as means of saving application's state [2]. VMware and Xen 3.0 virtual machines have provision of taking checkpoints (snapshots) and reverting back to them. These snapshots, however, do not support checkpoints of changes in a mounted distributed file system.

8 Conclusion and Future Work

The paper proposes a novel architecture that enables redirect-on-write functionality using virtualization techniques. It is designed to overlay existing NFS deployments, and can leverage virtual machine techniques to support client-side checkpointing of distributed file system modifications. For a benchmark application (Linux kernel compilation), the performance of ROW-FS across an emulated wide area network is four times better than conventional NFS. For the provisioning of non-persistent virtual machine execution environments, the performance of Xen virtual machine boot-up over wide area networks is comparable with local-area networks if the ROW-FS proxy is coupled with user-level NFS caching proxies. This is because the majority of calls are redirected to local domain machine. Initial results are encouraging; future research directions will address consistency between files shared among multiple users, and leveraging user-level redirection mechanisms to support request redirection to replicated servers for load distribution and fault tolerance.

References

1. Figueiredo, R., et al.: The PUNCH Virtual File System: Seamless Access to Decentralized Storage in a Computational Grid. In: Proceedings of the 10th HPDC (2001)
2. Agbaria, A., Friedman, R.: Virtual machine based heterogeneous checkpointing. Software-Practice & Experience 32(12), 1175–1192 (2002)
3. Kotsovinos, E., et al.: Global-scale service deployment in the XenoServer platform. In: Proc. of the USENIX WORLDS Workshop (2004)
4. Anderson Darrell, C., Chase, J.S., Vahdat, A.M.: Interposed Request for Routing for Scalable Network storage. In: Symposium on OSDI, San Diego, CA, pp. 22–25 (October 2000)
5. Figueiredo, R., Dinda, P.A., Fortes, J.A.B.: A Case for Grid Computing on Virtual Machine. In: Proceeding of the ICDCS (May 2003)
6. Minnich, R.G.: The AutoCacher: A File Cache which Operates at the NFS level. In: USENIX conference Proceedings, pp. 77–83. USENIX Association (winter 1993)
7. Butt, A.R., Johnson, T., Zheng, Y., Hu, Y.C.: Kosha: A Peer-to-Peer Enhancement for the Network File System. In: Proceedings of IEEE/ACM SC2004, Pittsburgh, PA, (59/192) (November 6-12, 2004)
8. Callaghan, B.: NFS illustrated. Addison-Wesley, London (2000)

9. Osman, S., et al.: The Design and Implementation of Zap: A System for Migrating Computing Environments. In: OSDI 2002, Boston, MA (December 9-11, 2002)
10. Hartman, J., Ousterhout, J.: The zebra striped network file system. In: Proceedings 14th symposium on operating system principles, pp. 29–43 (December 1993)
11. Litzkow, M., Livny, M., Mutka, M.W.: Condor: a Hunter of Idle Workstations. In: Proc. 8th Int. Conf. on Distributed Computing Systems, pp. 104–111 (June 1988)
12. Barham, P., Dragovic, B., Fraser, K., Hand, S., Harris, T., Ho, A., Neugebauer, R., Pratt, I., Warfield, A.: Xen and the Art of Virtualization. In: Proc of the ACM SOSP (October 2003)
13. Sugerman, J., Venkitachalan, G., Lim, B.H.: Virtualizing. I/O Devices on VMware Workstation's Hosted Virtual Machine Monitor. In: Proc of the USENIX tech Conf. (2001)
14. Adabala, S., et al.: From Virtualized Resources to Virtual Computing Grids: The In-VIGO System. Future Generation Computing Systems 21(6) (June 2005)
15. Krsul, et al.: VMPlants: Providing and Managing Virtual Machine Execution Environments for Grid Computing. In: Proc. of Supercomputing (2004)
16. http://snad.ncsl.nist.gov/itg/nistnet/ (Accessed 2006)
17. Chase, J., Irwin, D.E., Grit, L.E., Moore, J.D., Sprenkle, S.E.: Dynamic Virtual Clusters in a Grid Site Manger. In: Proc. 12th HPDC (June 2003)
18. Goldberg, R.: Survey of virtual machine research. IEEE Computer Mag. 7(6), 34–45 (1974)
19. Smith, J., Nair, R.: Virtual Machines: Versatile Platforms for Systems and Processes. Morgan Kaufmann, San Francisco (2005)
20. Spadavecchia, J., Zadok, E.: Enhancing NFS Cross-Administrative Domain Access. USENIX Annual Technical Conference, FREENIX Track 2002, 181–194 (2002)
21. http://www.gaussian.com/ (Accessed 2006)
22. Warfield, A., et al.: Parallax: Managing Storage for a Million Machines. In: Proc of the Tenth Workshop on HotOS X, Santa Fe, New Mexico (June 2005)
23. Keahey, K., Foster, I., Freeman, T., Zhang, X., Galron, D.: Virtual Workspaces in the Grid. In: Proc of the Euro-Par Conference (2005)
24. Chadha, V., Figueiredo, R.: A user level virtualization technique for redirect-on-write distributed file system. Tech, Report TR-ACIS-05-011, ACIS, Univ. of Florida (December 2005)
25. Frey, J.: Condor DAGMan: Handling Inter-Job Dependencies (Accepted 2007), http://www.cs.wisc.edu/ condor/dagman/
26. Allcock, B., et al.: Secure Efficient Data Transport and Replica Management for High-Performance Data- Intensive Computing. IEEE Mass Storage Conf. (2001)
27. Zhao, M., Chadha, V., Figueiredo, R.J.: Supporting Application-Tailored Grid File System Sessions with WSRF-Based Services. In: Proc. of the HPDC 2005 (2005)
28. Bester, J., et al.: GASS: A Data Movement and Access Service for Wide Area Computing Systems. In: Proc. of 6th IOPADS, Atlanta, GA (May 1999)
29. kosar, T., Livny, M.: Stork: Making Data Placement a First Class Citizen in the Grid. In: Proceedings of ICDCS 2004, Tokyo, Japan (March 2004)
30. Santhanam, S., Elango, P., Arpaci-Dusseau, A., Livny, M.: Deploying Virtual Machines as Sandboxes for the Grid. In: USENIX WORLDS 2004 (2004)
31. Santry, D., Feeley, M., Hutchinson, N., Veitch, A., Carton, R., Ofir, J.: Deciding when to forget in the Elephant file system. In: 17th ACM SOSP Principles (1999)

No More Energy-Performance Trade-Off: A New Data Placement Strategy for RAID-Structured Storage Systems

Tao Xie and Yao Sun

Department of Computer Science, San Diego State University,
San Diego, CA 92182, USA
{xie, sun}@cs.sdsu.edu

Abstract. Many real-world applications like Video-On-Demand (VOD) and Web servers require prompt responses to access requests. However, with an explosive increase of data volume and the emerging of faster disks with higher power requirements, energy consumption of disk based storage systems has become a salient issue. To achieve energy-conservation and prompt responses simultaneously, in this paper we propose a novel energy-saving data placement strategy, called Striping-based Energy-Aware (SEA), which can be applied to RAID-structured storage systems to noticeably save energy while providing quick responses. Further, we implement two SEA-powered RAID-based data placement algorithms, SEA0 and SEA5, by incorporating the SEA strategy into RAID-0 and RAID-5, respectively. Extensive experimental results demonstrate that compared with three well-known data placement algorithms Greedy, SP, and HP, SEA0 and SEA5 reduce mean response time on average at least 52.15% and 48.04% while saving energy on average no less than 10.12% and 9.35%, respectively.

Keywords: Data placement, energy conservation, response time, RAID.

1 Introduction

Many real-world applications intensively read data stored in large-scale parallel disk storage systems like RAID, Redundant Arrays of Inexpensive Disks. To guarantee the quality of service demanded by end-users, prompt responses to read requests are essential for these applications. For example, a Video-On-Demand (VOD) server has to quickly respond access requests from multiple users so as to provide them with continuous glitch-free video [6]. It is obvious that reducing mean response time of parallel disk storage systems is a must for these applications.

There are a wide variety of ways of reducing the mean response time or improving the system throughput for parallel I/O systems [1][6][10][12]. Data placement, or file assignment, allocating of all the data onto a disk array before they are accessed, is one of such avenues that can significantly affect the overall performance of a parallel I/O system [1][12][19]. Generally, these algorithms place data onto a parallel disk array so that a special cost function or performance metrics can be optimized. While

S. Aluru et al. (Eds.): HiPC 2007, LNCS 4873, pp. 35–46, 2007.

common cost functions include communication costs, storage costs, and queuing costs, popular performance metrics are mean response time and overall system throughput [5]. It is well-known that finding the optimal solution for a cost function or a performance metric in the context of data placement on multiple disks is an NP-complete problem [5]. Thus, heuristics algorithms became practical solutions.

Energy consumption of disk based storage systems has become a salient issue that not only raises the costs but also inversely affects our environment [18]. According to a recent industry report [17], storage devices contribute for around 27% of the total energy consumed by a data center. This problem will become much more severe with an explosive increase of data volume and the emerging of faster disks with higher power requirements. Therefore, energy-conservation and prompt response need to be achieved simultaneously through intelligent data placement. Unfortunately, traditional data placement algorithms such as Greedy [7], Sort Partition (SP) [12], and Hybrid Partition (HP) [12], for parallel disk systems only pursue minimized mean response times and normally ignore energy-conservation. Furthermore, most current energy-saving techniques adversely affect system performance [4][16]. Thus, seeking a good trade-off between energy-saving and graceful performance degradation becomes their feasible goal. Now the question is: can we develop a new data placement strategy so that energy-saving can be achieved without a trade-off of performance?

In this paper we address the problem of energy-saving yet quick-response data placement in a parallel disk storage system where data accesses exhibit Poisson arrival rates and fixed service times. Each data can be viewed as a file, which will be assigned onto an array of disks in a striping manner. We propose a novel energy-saving data placement strategy, called Striping-based Energy-Aware (SEA), which aims at minimizing mean response time and overall energy-consumption simultaneously. The basic idea of SEA is to statically place popular data onto a subset of the disks in the array and assign unpopular data onto the rest of disks. The rationale behind this idea is that the distribution of web page requests generally follows a Zipf-like distribution [12] where the relative probability of a request for the i'th most popular page is proportional to $1/i^{\alpha}$, with α typically varying between 0 and 1 [2][15]. Moreover, the request frequency and the file size are inversely correlated, i.e., the most popular files are typically small in size, while the large files are relatively unpopular [12]. Based on these workload characteristics, we divide all data into two categories: *popular* and *unpopular* according to their popularity weights [15]. Similarly, we separate disks in a disk array into two zones: *hot disk zone* and *cold disk zone*. Disks in hot disk zone are called *hot disks* with popular data, whereas disks in cold disk zone are named *cold disks* with unpopular data. As a result, the overall load balancing between two zones can be achieved, which improves the inter-request parallelism. Next, we employ multi-speed disks in the disk array to save energy. Specifically, hot disks are always running in a higher speed mode with more energy consumption, while cold disks are continuously operating in a lower speed mode with less energy dissipation. Although real multi-speed (more than 2 speeds) hard disks are not widely available in the market yet, a few simple variations of multi-speed disks, such as a two-speed Hitachi Deskstar 7K400 hard drive has recently been produced [9]. For simplicity, in this study we only utilize 2-speed hard disks. Note that once a disk was configured as a hot disk or a cold disk, its operation characteristics such as transfer (read) speed and energy consumption rate is fixed and it cannot be dynamically switched to the other mode during the process of serving requests. Further, to provide quick responses, we combine SEA with RAID structures so that each data (file) is partitioned into multiple same size stripe blocks, which are then allocated across an array of disks. This way all disks in the same zone can simultaneously serve a request. The implication is that the response time of the request can be dramatically decreased due to an enhanced intra-request parallelism.

The rest of the paper is organized as follows. In the next section we discuss the related work and motivation. A system model and an energy consumption model were built in Section 3 and Section 4, respectively. Section 5 presents the SEA strategy and introduces three existing algorithms. In Section 6 we evaluate performance of our algorithms based on synthetic benchmarks. Section 7 concludes the paper with summary and future directions.

2 Related Work and Motivation

Very recently energy-saving for parallel disk storage systems began to draw much attention from research community [3][4][8][10][16]. A multi-speed parallel disk system that can modulate disk speed dynamically was proposed by Gurumurthi et al. [8]. In [10] data replication was used to dynamically place copies of data in free blocks according to the disk access patterns.

Comparing with the energy-efficient techniques mentioned above, data placement shows its unique advantages. First, to save disk energy, it has no need to modify applications. Next, no extra hardware such as cache is necessary. Last, the overhead of data placement strategy is relatively low and it is easy to implement. Attracted by these advantages, a research group led by Son proposed an array of energy-aware disk layout algorithms very recently [18]. Based on our knowledge, their studies are the only results of energy-aware data placement for parallel disk storage systems reported in the literature so far. However, their techniques have some obvious limitations. First, they are only dedicated for array-based scientific applications. Still, there are many other types of disk I/O-intensive applications, where energy conservation and quick response need to be realized simultaneously through data placement. Therefore, a more general energy-response efficiency data placement scheme that can be applied to a wide range of disk I/O-intensive applications is needed. Further, to apply their algorithms, one has to modify compiler to make it be aware of disk layout information. This requirement prevents them from being readily used because it incurs an extra burden for system software programmers. Besides, to better exploit existing power-saving capabilities, their disk layout algorithms need to be combined with application code restructuring to increase length of idle periods. This strategy demands modifications of application's code, and thus brings users additional overhead. As a result, the need of a new energy-response efficiency data placement strategy that bridges the gap between the existing algorithms and the open problems is greatly felt.

In this paper, we are proposing a static heuristic energy-aware strategy SEA, which can be incorporated with RAID structures to generate energy-aware data placement algorithms like SEA0 and SEA5. Our schemes are orthogonal to existing disk layout strategies. First, there is no need to modifying any software using our methods.

Second, our schemes are not dedicated for some particular applications. Thus, they are more general in the sense that they can be applied in multiple application domains. Without loss of generality, we assume that (1) each data is viewed as an independent file; (2) each data is allocated in a striping manner across an array of disks; (3) communication delays between any pair of disks are identical and negligibly small [12]; (4) disk access (read) to each data is modelled as a Poisson process with a mean access rate λ_i; (5) a fixed service time s_i for each data; for example, each read on a data results in a sequential scan of the entire data. For large size data, this assumption is valid because when the basic unit of data access is entire data, seek time, rotation latency, and controller overhead are negligible in comparison with data transfer time.

3 System Model

Data placement algorithms such as Greedy, SP, HP, SEA0, and SEA5 allocate a set of data (hereafter file) onto a group of 2-speed disks so that the mean response time can be minimized. The set of files is represented as $F = \{f_1, ..., f_u, f_v, ..., f_m\}$, which is further categorized into a popular file set $F_h = \{f_1, ..., f_h, ..., f_u\}$ and au unpopular file set $F_c = \{f_v, ..., f_c, ..., f_m\}$ ($F = F_h \cup F_c$ and $F_h \cap F_c = \emptyset$). Since each file will be allocated onto a set of disks in a striping manner, let *sp* denote the size of a stripe in Mbyte and it is assumed to be a constant in the system. A file f_i ($f_i \in F$) is modeled as a set of rational parameters, e.g., $f_i = (s_i, \lambda_i)$, where s_i, λ_i are the file's size in Mbyte and its access rate. In this paper, requests to a file f_u are modeled as a Poisson process with a mean access rate λ_i. Also, we assume each access to file f_i is a sequential read of the entirc file, which is a typical scenario in most file systems or WWW servers [11]. Besides, we assume that the distribution of file access requests is a Zipf-like distribution with a skew parameter $\theta = \log \frac{X}{100} / \log \frac{Y}{100}$, where X percent of all accesses were directed to Y percent of files [12]. The value of $X{:}Y$ is called *skew degree* (SD) in this paper and $\alpha = 1 - \theta$ (see Section 1 for α). In addition, the file access frequency is inversely correlated to the file size. The number of popular files in F is defined as $|F_h| = (1-\theta) * m$. Similarly, the number of unpopular files is $|F_c| = \theta * m$. Thus, the ratio between the number of popular files and the number of unpopular files in F is defined as η

$$\eta = \frac{1 - \theta}{\theta}. \tag{1}$$

A disk array storage system consists of a linked group $D = \{d_1, ..., d_e, d_f, ..., d_n\}$ of n independent 2-speed disk drives, which can be divided into a hot disk zone $D_h = \{d_1, ..., d_h, ..., d_e\}$ and a cold disk zone $D_c = \{d_f, ..., d_c, ..., d_n\}$ ($D = D_h \cup D_c$ and $D_h \cap D_c = \emptyset$). Disks in the hot zone are all configured to their high speed modes, which always run in the high transfer rate t^h (Mbyte/second) with the high active power consumption rate p^h (Joule/Mbyte) and the high idle power consumption rate i^h (Watt). Similarly, disks in cold zone are set to their low speed modes, which continuously operate in the low transfer rate t^l (Mbyte/second) with the low active power consumption rate p^l (Joule/Mbyte) and the low idle power consumption rate i^l (Watt). In the system, a hot disk d_h ($d_h \in D_h$) is modeled as a tuple $d_h = (c, t^h, p^h, i^h)$ where c is the capacity of d_h in GByte. Similarly, a cold disk d_c ($d_c \in D_c$) is modeled as a tuple $d_c = (c, t^l, p^l, i^l)$ where c is the capacity of d_c in GByte. Since we only consider homogeneous disks, all disks have the same capacity c. We assume that disks are always large enough to accommodate files to be assigned on them. Each popular file $f_h \in F_h$ is partitioned in multiple units with the size of each unit equal to sp. All units of f_h will be allocated across the hot disks in a RAID-0 (striping without parity) or a RAID-5 fashion (striping with parity). Similarly, each unpopular file $f_c \in F_c$ is also partitioned into multiple size sp units and then allocated across the cold disks in a RAID-0 or a RAID-5 manner. Let sv_i be the expected service time of file f_i ($f_i \in F$). It can be computed by

$$sv_i = \begin{cases} s_i / t^h, \text{if } f_i \text{ is popuplar} \\ s_i / t^l, \text{if } f_i \text{ is unpopular} \end{cases} \tag{2}$$

Since the combination of λ_i and sv_i accurately gives the load of f_i, we define the load h_i of f_i as follows [12]:

$$h_i = \lambda_i \cdot sv_i. \tag{3}$$

The ratio between the number of hot disks and the number of cold disks is defined as γ, which is decided by the ratio between the total load of popular files and the total load of unpopular files as below

$$\gamma = \frac{\sum_{i=1, f_i \in F_h}^{(1-\theta)*m} h_i}{\sum_{j=1, f_j \in F_c}^{\theta*m} h_j} \tag{4}$$

We employ the First-Come-First-Serve (FCFS) scheduling heuristic to schedule arrival requests. Suppose there are totally u requests in a request set, which visits a file set that has been allocated on a disk array. The request set is designated as $R = \{r_1, ..., r_k, ..., r_x\}$, which can be separated into a hot request set $R_h=\{r_b, ..., r_h, ..., r_o\}$ and a cold request set $R_c=\{r_p, ..., r_c, ..., r_s\}$ ($R = R_h \cup R_c$, $R_h \cap R_c = \emptyset$). Each request is modeled as $r_k = (fid_k, a_k)$, where fid_k is the file identifier targeted by the request and a_k is the request's arrival time. For each arrival request, the FCFS scheduler uses the allocation scheme X generated in data placement stage to find the disks on which the target file of the request resides. In fact, the request workload is an m-class workload with each class of requests having its fixed λ_i.

To obtain the response time of a request r_k, two important parameters, the earliest start time and the latest finish time of r_k must be computed. We denote the earliest start time and the latest finish time of r_k by $est(r_k)$ and $lft(r_k)$, respectively. In what follows we present derivations leading to the final expressions for these two parameters. Since each file is distributed across multiple disks in a striping manner, we need to compute the start time and the finish time for each stripe of the file that request r_k is targeting on. Suppose r_k is visiting file f_i, which was distributed on a disk set $\{d_a, ..., d_g, ..., d_w\}$ ($a \leq g \leq w$, $1 \leq a, g, w \leq e$ or $f \leq a, g, w \leq n$). The stripe set of f_i is represented as $\{ s_i^1, ..., s_i^k, ..., s_i^z \}$, where $z = \frac{s_i}{sp}$. Also, a disk d_g has its own local queue Q_g in the set $\{Q_a, ..., Q_g, ..., Q_w\}$. There are three cases when r_k arrives on disk d_g. First, d_g is idle and Q_g is empty. Second, d_g is busy but Q_g is empty. Third, d_g is busy and Q_g is not empty. Thus, the start time for a strip s_i^k on disk d_g is

$$st_g^k(r_k) = \begin{cases} a_k, \text{ if } d_g \text{ is idle, } Q_g \text{ is empty} \\ a_k + r_g, \text{ if } d_g \text{ is busy, } Q_g \text{ is empty} \\ a_k + r_g + \sum_{r_p \in Q_g, a_p \le a_k} t_{fid_p}, \text{ otherwise} \end{cases} \quad (5)$$

where r_g represents the remaining service time of a request currently running on d_g, and $\sum_{r_p \in Q_g, a_p \le a_k} t_{fid_p}$ is the overall service time of requests in Q_g whose arrival times are earlier than that of r_k. Consequently, $ft_g^k(r_k)$ can be calculated by

$$ft_g^k(r_k) = st_g^k(r_k) + ts_i, \quad (6)$$

where ts_i is the service time of the stripe s_i^k on disk d_g and it can be computed using the following formula

$$ts_i = \begin{cases} sp / t^h, \text{ if } d_g \text{ is hot} \\ sp / t^l, \text{ if } d_g \text{ is cold} \end{cases} \quad (7)$$

As a result, the earliest start time of request r_k can be obtained by

$$est(r_k) = \min\{ st_g^1(r_k), .., st_g^k(r_k), .., st_g^z(r_k) \}. \quad (8)$$

Consequently, the latest finish time of r_k can be calculated by

$$lft(r_k) = \max\{ ft_g^1(r_k), .., ft_g^k(r_k), .., ft_g^z(r_k) \}. \quad (9)$$

Hence, the response time of r_k can be obtained

$$t(r_k) = lft(r_k) - est(r_k). \quad (10)$$

Thus, the mean response time of the request set R is expressed as below

$$mrt(R) = \sum_{k=1}^{x} rt(r_k) \Big/ x \quad (11)$$

4 Energy Consumption Model

For a request r_h in the hot request set R_h, assume it accesses a popular file f_h in the popular file set F_h, which is allocated in the hot disk zone. The energy consumed by r_h can be written as below

$$e_h^{active} = \frac{s_h}{p^h} \quad (12)$$

The service time for r_h provided by a set of hot disks, where file f_h were allocated can be computed as follows

$$at_h^{active} = \frac{s_h}{t^h} \tag{13}$$

Thus, the energy consumption of the whole hot request set can be derived by

$$e_{R_h}^{active} = \sum_{h=1}^{|R_h|} e_h^{active} \tag{14}$$

Similarly, the total service time imposed by the whole hot request set R_h in the hot disk zone is

$$at_{R_h}^{active} = \sum_{h=1}^{|R_h|} at_h^{active} \tag{15}$$

In addition, we define rft_k as the finish time of request r_k. Then, we obtain the analytical formula for the energy consumed by the hot disks when they are idle:

$$e_{hot}^{idle} = i^h * (|D_h| * \max_{k=1}^{x}(rft_k) - at_{R_h}^{active}) \tag{16}$$

Hence, the total energy consumed by the hot disk zone can be computed by

$$e_{hot} = e_{R_h}^{active} + e_{hot}^{idle} = \sum_{h=1}^{|R_h|} e_h^{active} + i^h * (|D_h| * \max_{k=1}^{x}(rft_k) - at_{R_h}^{active}) \tag{17}$$

Similarly, the total energy consumed by the cold disk zone can be obtained by

$$e_{cold} = e_{R_c}^{active} + e_{cold}^{idle} = \sum_{c=1}^{|R_c|} e_c^{active} + i^l * (|D_c| * \max_{k=1}^{x}(rft_k) - at_{Rc}^{active}) \tag{18}$$

Therefore, the total energy consumption for the whole storage system is:

$$e_{total} = e_{hot} + e_{cold} \tag{19}$$

5 The SEA Strategy

In this section, we first present a detailed description of the SEA strategy. Then we briefly introduce the three baseline algorithms Greedy, SP, and HP.

Fig. 1 outlines SEA with some detailed explanations. Note that the input F has been sorted in an ascending order in terms of popularity before it is fed into SEA. In other words, file f_1 is the most popular file with the smallest file size, whereas file f_m is the most unpopular one with the largest file size. First, SEA uses the skew parameter θ to derive the number of popular files and the number of unpopular files in F based on Eq. 1 (Step 1). Second, Step 2 calculates γ, the ratio between the number of hot disks and the number of cold disks, based on Eq. 4, which in turn results in the number of hot disks HD and the number of cold disks CD. Consequently, HD of n disks are configured to their high speed modes and CD of n disks are set to their low speed modes (Step 4). Next, SEA assigns all popular files onto the hot disk zone in a striping manner (Step 5–Step 16). Similarly, all unpopular files are allocated onto the cold disk zone in a striping fashion (Step 17 – Step 28).

Input: A disk array D with n 2-speed disks, a collection of m files in the set F, and the skew parameter θ

Output: A file allocation scheme $X(m, k)$, where $k = \max_{i=1}^{m}\left(\frac{s_i}{sp}\right)$

1. Use **Eq. 1** to compute the number of popular files and number of unpopular files in F
2. Use **Eq. 4** to compute γ
3. Hot disk number $HD = \frac{\gamma * n}{\gamma + 1}$, cold disk number $CD = n - HD$, d_h=1, d_c=1
4. Configure HD of n disks to high speed mode and set CD of n disks to low speed mode
5. **for** each popular file $f_p \in F_h$ **do**
6. $p = 1$;
7. **for** each stripe sp_p of f_p **do**
8. $X(f_p, p) = d_h$
9. $p = p + 1$
10. **if** $d_h == HD$
11. $d_h = 1$
12. **else**
13. $d_h = d_h + 1$
14. **end if**
15. **end for**
16. **end for**
17. **for** each unpopular file $f_u \in F_c$ **do**
18. $u = 1$;
19. **for** each stripe sp_u of f_u **do**
20. $X(f_u, u) = d_c$
21. $u = u + 1$
22. **if** $d_c == CD$
23. $d_c = 1$
24. **else**
25. $d_c = d_c + 1$
26. **end if**
27. **end for**
28. **end for**

Fig. 1. The SEA strategy

The average disk load ρ can be obtained by the following equation:

$$\rho = \frac{1}{n} \cdot \sum_{i=1}^{m} h_i \qquad (20)$$

Note that all the three existing algorithms assign nonparitioned files onto a disk array. In other words, each file must be allocated entirely onto one disk. In addition, since they only pursue minimized mean response times, all disks in the disk array are set to hot disks with high speed. The three algorithms are briefly described below.

(1) *Greedy*: It first calculates the mean load of all files and then assigns a consecutive set of files whose total load is equal to the mean load onto each disk.
(2) *SP (Sort Partition)*: It first computes the average disk utilization using Eq. 13. Next, it sorts all files into a list I in descending order of their service times. Finally, it allocates each disk d_j the next contiguous segment of I until its load $load_j$ reaches the maximum allowed threshold ρ. The remainder files (if any) after one round allocation will be assigned to the last disk d_n.
(3) *HP (Hybrid Partition)*: For each batch, HP assigns files to disks in distinct allocation intervals. It selects, for each allocation intervals l, a different disk d_k as the allocation target. It chooses the disk with the smallest accumulated load. A number of files are allocated to d_k until its load reaches the threshold T_k.

6 Performance Evaluation

6.1 Simulation Setup

We adopt the same strategy used in [16] to derive corresponding low speed mode disk statistics from parameters of a conventional Cheetah disk. The main characteristics of the 2-speed disk are shown in Table 1. The performance metrics by which we evaluate system performance include:

(1) *Mean response time:* Average response time of all file access request submitted to the simulated parallel disk storage system. Note that the mean response times are normalized in the scale [0, 1].
(2) *Energy consumption*: Energy (in Joules) consumed by the disk systems during the process of serving the entire request set.
(3) *Mean slowdown:* The ratio between average request turnaround time and average request service time.

Table 1 summarizes the configuration parameters of a simulated parallel disk array system used in our experiments and characteristics of the synthetic workload. All synthetic workload used were created by our trace generator.

Table 1. Characteristics of system parameters

Parameter	Value
Transfer rate in low mode	9.3 Mbytes/second
Idle power at low mode	2.17 Watts
Active energy at low mode (8-KB read)	43 mJoules
Transfer rate in high mode	31 Mbytes/second
Idle power at high mode	5.26 Watts
Active energy at high mode (8-KB read)	61 mJoules
Number of files	(5000)
Simulation duration	(1000) seconds
Aggregate access rate	(35) – (21~45) (1/second)
γ	3:13 ~ 10:6

6.2 Overall Performance Comparison

The goal of this experiment is to compare the proposed SEA0 and SEA5 algorithms against the three well-known file assignment schemes, and to understand the sensitivity of the five heuristics to the aggregate access rate in a parallel disk storage system, where an array of 2-speed disk drives serve requests simultaneously. The aggregate access rate varies from 21 (1/second) to 45 (1/second). The file sizes were generated according to a Zipf-like distribution with skew degree 70:30 and *file size base* is set to 1 Mbyte.

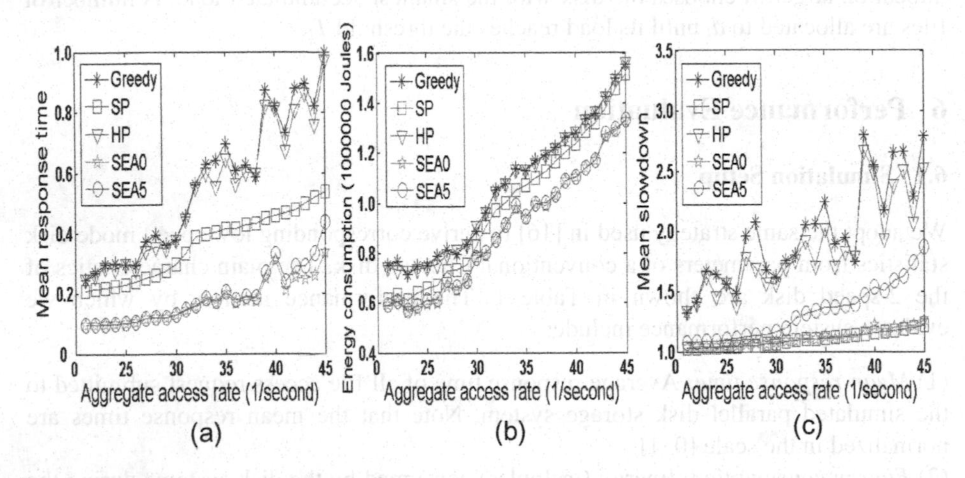

Fig. 2. Impact of aggregate access rate

Fig. 2 shows the simulation results for the five algorithms on a parallel disk array with 16 disk drives, where 5 of them are hot disks and 11 of them are cold disks. We observe from Fig. 2a that SEA0 and SEA5 consistently outperform the three exiting approaches in terms of mean response time. This is because they employ a striping-based data placement scheme, where intra-request parallelism is very high. Compared with the SP algorithm, SEA0 and SEA5 can reduce mean response time on average by 52.15% and 48.04%, while saving energy on average no less than 10.12% and 9.35%, respectively (see Fig. 2b). Although we only test a relatively light physical read workload (in the range [21, 45] 1/second), the actual system workload can be 10 times heavier (in the range [210, 450] 1/sccond) because of very low miss rates (lcss than 10%) provided by the high speed buffers on the data servers. The implication is that both SEA0 and SEA5 can be applied in applications where system workload is heavy. One example of such applications is OLTP (Online Transaction Processing).

7 Summary and Future Work

In this paper, we developed a new energy-saving strategy, called <u>s</u>triping-*b*ased <u>e</u>nergy-<u>a</u>ware (SEA), to generate optimized file allocation schemes. SEA0 and SEA5, two SEA-powered RAID-based data placement algorithms are implemented to evaluate the effectiveness and practicality of SEA. Comprehensive experimental results show that both SEA0 and SEA5 consistently improve the performance of parallel disk storage systems in terms of mean response time and save energy over three well-known data placement algorithms. Normally, there are two inherent drawbacks of current multi-speed disk based energy-saving approaches. First, disk speed mode transitions bring extra overhead in terms of transition time and transition energy [16], which is against their original goals. Second, frequent disk speed mode transitions are detrimental to the lifetime of hard disks [3]. SEA0 and SEA5 avoid these two shortcomings by statically configuring all disks to one of the multiple modes prior to serving requests. Furthermore, there is no speed mode transition during the process of serving the requests.

In summary, the SEA strategy realizes energy-saving not at the cost of performance degradation. Rather, it delivers much shorter mean response times compared with existing non-energy-aware data placement algorithms. Besides, it can provide fault-tolerance because of the RAID structures that it relies on. We will extend our scheme to a fully dynamic environment, where file access characteristics are not known in advance and may vary over time. As a result, a dynamic energy-saving data placement strategy is mandatory so that dynamically arrived files can be re-allocated by migrating files from one disk to another. File migration, however, incurs a relatively heavy overhead. How to make a good trade-off between migration cost and algorithm efficiency is a problem that needs to be solved.

Acknowledgments. This work was supported by the National Science Foundation under grant number CCF-0702781.

References

1. Akyürek, S., Salem, K.: Adaptive block rearrangement. ACM Trans. on Computer Systems 13(2), 89–121 (1995)
2. Breslau, L., Cao, P., Fan, L., Phillips, G., Shenker, S.: Web Caching and Zip-like Distributions: Evidence and Implications. In: Proc. IEEE INFOCOM, pp. 126–134 (1999)
3. Carrera, E.V., Pinheiro, E., Bianchini, R.: Conserving disk energy in network servers. In: Proc. 17th Supercomputing, pp. 86–97 (2003)
4. Colarelli, D., Grunwald, D.: Massive arrays of idle disks for storage archives. In: Proc. 16th Annual Int'l Conf. Supercomputing, pp. 1–11 (2002)
5. Dowdy, W., Foster, D.: Comparative Models of the File Assignment Problem. ACM Computing Surveys 14(2), 287–313 (1982)
6. Ghandeharizadeh, S., Kim, S.H., Shababi, C.: On disk scheduling and data placement for video servers. Sigmetrics Performance Evaluation 23(1), 37–46 (1995)
7. Graham, R.L.: Bounds on Multiprocessing Timing Anomalies. SIAM Journal Applied Math. 7(2), 416–429 (1969)
8. Gurumurthi, S., Sivasubramaniam, A., Kandemir, M., Franke, H.: DRPM: Dynamic speed control for power management in server class disks. In: Proc. Int'l Symp. Computer Architecture, pp. 169–179 (2003)
9. Hitachi Corp.: Hitachi Power & Acoustic Management – quietly cool. White paper (2004)

10. Huang, H., Hung, W., Shin, K.G.: FS2: dynamic data replication in free disk space for improving disk performance and energy consumption. In: Proc. 12th ACM SOSP, pp. 263–276 (2005)
11. Kwan, T., Mcgrath, R., Reed, D.: Ncsas World Wide Web Server Design and Performance. Computer 28(11), 67–74 (1995)
12. Lee, L.W., Scheuermann, P., Vingralek, R.: File assignment in parallel I/O systems with minimal variance of service time. IEEE Trans. Computers 49(2), 127–140 (2000)
13. Merialdo, P., Atzeni, P., Mecca, G.: Design and development of data-intensive web sites: The Araneus approach. ACM Trans. Internet Technology 3(1), 49–92 (2003)
14. Narris, M., Obal, J.: Performance Analysis of the Linux Buffer Cache While Running an Oracle OLTP Workload. Worcester Polytechnic Institute (2002)
15. Nishikawa, N., Hosokawa, T., Mori, Y., Yoshida, K., Tsuji, H.: Memory-based architecture for distributed WWW caching proxy. In: Proc. 7th Int'l Conf. World Wide Web, pp. 205–214 (1998)
16. Pinheiro, E., Bianchini, R.: Energy conservation techniques for disk array-based servers. In: Proc. 18th Supercomputing, pp. 68–78 (2004)
17. Power, heat, and sledgehammer (2002), http://www.max-t.com/downloads/whitepapers/SledgehammerPowerHeat20411.pdf
18. Son, S.W., Chen, G., Kandemir, M.: Disk layout optimization for reducing energy consumption. In: Proc. 19th Supercomputing, pp. 274–283 (2005)
19. Triantafillou, P., Christodoulakis, S., Georgiadis, C.: Optimal data placement on disks: a comprehensive solution for different technologies. IEEE Trans. Knowledge and Data Engineering 12(2), 324–330 (2000)

Reducing the I/O Volume in an Out-of-Core Sparse Multifrontal Solver*

Emmanuel Agullo[1], Abdou Guermouche[2], and Jean-Yves L'Excellent[3]

[1] LIP-ENS Lyon, France
[2] Université de Bordeaux and LaBRI, France
[3] INRIA and LIP-ENS Lyon, France

Abstract. High performance sparse direct solvers are often a method of choice in various simulation problems. However, they require a large amount of memory compared to iterative methods. In this context, out-of-core solvers must be employed, where disks are used when the storage requirements are too large with respect to the physical memory available. In this paper, we study how to minimize the *I/O* requirements in the multifrontal method, a particular direct method to solve large-scale problems efficiently. Experiments on large real-life problems also show that the volume of *I/O* obtained when minimizing the storage requirement can be significantly reduced by applying algorithms designed to reduce the *I/O* volume.

1 Introduction

We are interested in solving a sparse system of linear equations of the form $Ax = b$ by a so-called direct method. Such methods work in three phases: (i) an analysis phase, that orders the variables of the problem to limit the computations and prepares the work for the factorization; (ii) a numerical factorization phase, where A is factored under the form LU, LL^t or LDL^t; and (iii) a solve phase, where triangular factors are used to obtain the solution of the problem. Because of their large memory requirements, several authors have worked on out-of-core sparse direct solvers [1,3,8,12,15,16,17]. Left-looking and multifrontal methods are two main classes of sparse direct methods that can be extended to an out-of-core context. In that case, a left-looking approach allows to reduce significantly the minimal memory requirements, while the multifrontal method may lead to large frontal matrices that prevent processing arbitrarily large problems [16] if frontal matrices are not assembled and factored with out-of-core algorithms. On the other hand, for problems in which the largest frontal matrix fits in memory or can be treated reasonably using an out-of-core algorithm, the multifrontal method remains interesting [7,14] and motivates the design of robust software solutions [2,15].

In the multifrontal method, the factorization of a sparse matrix A is done by a succession of partial factorizations of small dense matrices called *frontal*

* Partially supported by ANR project SOLSTICE, ANR-06-CIS6-010.

S. Aluru et al. (Eds.): HiPC 2007, LNCS 4873, pp. 47–58, 2007.

matrices. Since the frontal matrices are dense, this method allows an efficient use of memory hierarchy and caches, where optimized dense kernels (BLAS) can be applied. For matrices with a symmetric structure (or in approaches like [6] when the structure of matrix A is unsymmetric), each frontal matrix is associated with a node of a so-called *assembly tree* which represents the dependencies of the tasks in the factorization algorithm. Before a partial factorization of a parent node can be performed, temporary data (so-called *contribution blocks*) extracted from the frontal matrices of children are assembled into the frontal matrix of the parent. The parent is then factored and the contribution block it produces is in turn kept in memory for later use at the upper layer of the tree. Since the factors are terminal data for the factorization phase, it appears natural to write them to disk as soon as they are produced. Focusing on memory handling issues, the multifrontal algorithm may be presented as follows:

```
foreach node k in the tree (postorder traversal) do
    Allocate memory for the frontal matrix of k
    if k is not a leaf then
        Assemble and free contributions from children
    Perform a partial factorization of the frontal matrix of k, writing factors to
    disk on the fly
    Keep the contribution block of k for later use
```

Note that, because we rely on a post-order traversal, the multifrontal algorithm can use a stack mechanism to store the contribution blocks: the contribution blocks produced last are the first ones assembled. Still, there is a lot of freedom to order the siblings at each level of the tree so that the tree traversal can have a significant impact on both the number of contribution blocks stored simultaneously and the memory usage. Liu [13] (and, more recently, [11,10]) have shown the impact of the tree traversal on the memory behaviour and proposed tree traversals that minimize the storage requirements of the multifrontal method when factors are systematically written to disk. With this assumption, Liu suggested in the conclusion of [13] that minimizing the storage requirements was well adapted to an out-of-core execution.

In this paper we focus on the volume of I/O related to the stack of contribution blocks and we aim at designing optimal tree traversals with respect to minimizing the volume of I/O. By expressing this volume in a formal way, we show that minimizing the storage requirements is different from minimizing the volume of I/O.

Note that we consider several minor variants of the multifrontal algorithm. We call *last-in-place* a variant of the assembly scheme (available, for example, in a code like MA27 [9]) where the memory of the frontal matrix at the parent node is allowed to overlap with the contribution block of the last child. In that case, we save memory by not summing the memory of the frontal matrix of the child with the memory of the frontal matrix of the parent (a maximum between these two values is enough). We also propose a new variant, where we overlap the memory for the frontal matrix of the parent with the memory of the child having the largest contribution block (even if that child is not processed last). For each variant, we present the tree traversal that minimizes memory (algorithms so called `MinMEM`); then, we show by how much the volume of *I/O* can be reduced (depending on the physical memory available) with new algorithms (called `MinIO`) that minimize the *I/O* volume.

The paper is organized as follows. In Sections 2 and 3, we explain how to model and minimize the volume of *I/O* induced by the *classical* and *last-in-place* schemes, respectively. In Section 4, we discuss the new variant of the *in-place* algorithm. Section 5 illustrates the difference between `MinMEM` and `MinIO` on matrices arising from real-life problems, and shows the interest of the new *in-place* variant proposed.

2 Limiting the Amount of I/O

Before discussing the volume of *I/O*, we introduce some general notations. In a limited memory environment, we define M_0 as the volume of core memory available. As described in the introduction, the multifrontal method is based on a tree in which a parent node is allocated in memory after all its child subtrees have been processed. When considering a generic parent node and its n children numbered $j = 1, \ldots, n$, we note:

- cb_j, the storage for the contribution block passed from child j to the parent;
- m / m_j, the storage of the frontal matrix associated to the parent node / to child j (note that $m_j > cb_j$ and $m_j - cb_j$ is the size of the factors produced by child j);
- S / S_j, the storage required to process the subtree rooted at the parent / at child j (note that if $S_j < M_0$, no *I/O* is necessary to process the whole subtree rooted at j);
- $V^{I/O}$ / $V_i^{I/O}$ the volume of *I/O* required to process the subtree rooted at node j given an available memory of size M_0.

2.1 Illustrative Example

To illustrate the memory behaviour, let us take the toy example described in Figure 1(left): we consider a root node (e) with two children (c) and (d). The frontal matrix of (e) requires a storage $m_e = 5$. The contribution blocks of (c) and (d) require a storage $cb_c = 4$ and $cb_d = 2$, while the storage requirements for their frontal matrices are $m_c = 6$ and $m_d = 8$ respectively. (c) has itself two children with characteristics $cb_a = cb_b = 3$ and $m_a = m_b = 4$. We assume that the core memory available is $M_0 = 8$.

To respect a postorder traversal, there are two possible ways to process this tree: (a-b-c-d-e) and (d-a-b-c-e). (Note that (a) and (b) are identical and can be swapped.) For each sequence we now describe the memory behaviour and *I/O* operations. We first consider sequence (a-b-c-d-e). (a) is first allocated ($m_a = 4$) and factored (we write its factors of size $m_a - cb_a = 1$ to disk), and $cb_a = 3$ remains in memory. After (b) is processed, the memory contains $cb_a + cb_b = 6$.

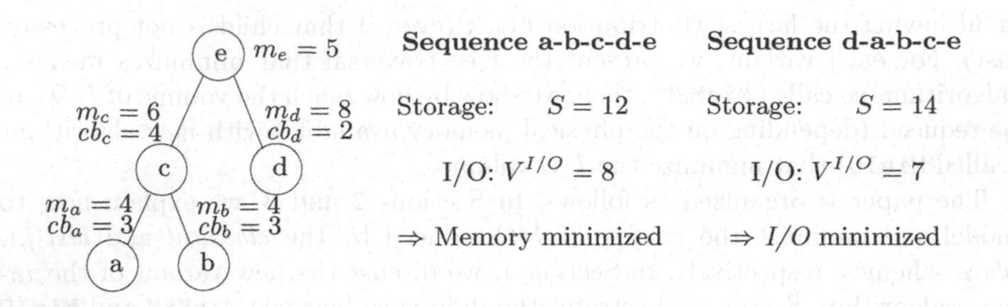

Fig. 1. Influence of the tree traversal on the storage requirement and on the volume of *I/O* (with $M_0 = 8$)

Then a peak of storage $S_c = 12$ is reached when the frontal matrix of (c) is allocated. Since only 8 (MegaBytes, say) can be kept in core memory, this leads to write to disk a volume of data equal to 4. During the assembly process we first assemble contributions that are in memory, and then read (r_4) data from disk to assemble them in turn in the frontal matrix of (c). After the factors of (c) of size $m_c - cb_c = 2$ are written to disk, its contribution block $cb_c = 4$ remains in memory. When leaf node (d) is processed, the peak of storage reaches $cb_c + m_d = 12$. This leads to a new volume of *I/O* equal to 4 (and corresponding to cb_c). After (d) is factored, the storage requirement is equal to $cb_c + cb_d = 6$ among which only $cb_d = 2$ is in core (cb_c is already on disk). Finally, the frontal matrix of the parent (of size $m_e = 5$) is allocated, leading to a storage $cb_c + cb_d + m_e = 11$: after cb_d is assembled in core (into the frontal matrix of the parent), cb_c is read back from disk and assembled in turn. Overall the volume of data written to (and read from) disk[1] is $V_e^{I/O}$(a-b-c-d-e)= 8 and the peak of storage was S_e(a-b-c-d-e)= 12.

When the tree is processed in order (d-a-b-c-e), the storage requirement successively takes the values $m_d = 8$, $cb_d = 2$, $cb_d + m_a = 6$, $cb_d + cb_a = 5$, $cb_d + cb_a + m_b = 9$, $cb_d + cb_a + cb_b = 8$, $cb_d + cb_a + cb_b + m_c = 14$, $cb_d + cb_c = 6$, $cb_d + cb_c + m_e = 11$, with a peak S_e(d-a-b-c-e)= 14. Nodes (d) and (a) can be processed without inducing *I/O*, then 1 unit of *I/O* is done when allocating (b), 5 units when allocating (c), and finally 1 unit when the frontal matrix of the root node is allocated. We obtain $V_e^{I/O}$(d-a-b-c-e)= 7.

We observe that the postorder (a-b-c-d-e) minimizes the peak of storage, while (d-a-b-c-e) minimizes the volume of *I/O*. This shows that minimizing the peak of storage is different from minimizing the volume of *I/O*.

2.2 Expressing the Volume of I/O

Since contribution blocks are stored thanks to a stack mechanism, some contribution blocks (or parts of contribution blocks) may be kept in memory and

[1] Remember that we do not count *I/O* for factors since factors are written to disk systematically in all variants considered.

consumed without being written to disk [3]. Assuming that the contribution blocks are written only when needed (possibly only partially), that factors are written to disk as soon as they are computed, and that a frontal matrix must fit in core memory, we focus on the computation of the volume of *I/O* on this stack of contribution blocks.

When processing a child j, the contribution blocks of all previously processed children have to be stored. Their memory size sums up with the storage requirements S_j of the considered child, leading to a global storage equal to $S_j + \sum_{k=1}^{j-1} cb_k$. After all the children have been processed, the frontal matrix (of size m) of the parent is allocated, requiring a storage equal to $m + \sum_{k=1}^{n} cb_k$. Therefore, the storage required to process the complete subtree rooted at the parent node is given by the maximum of all theses values, that is:

$$S = \max\left(\max_{j=1,n}(S_j + \sum_{k=1}^{j-1} cb_k), m + \sum_{k=1}^{n} cb_k\right) \quad (1)$$

Knowing that the storage requirement S for a leaf node is equal to the size of its frontal matrix m, applying this formula recursively (as done in [13]), allows to determine the storage requirement for the complete tree.

In our out-of-core context, we now assume that we are given a core memory of size M_0. If $S > M_0$, some *I/O* will be necessary. Since the contribution blocks are accessed with a stack mechanism, writing the bottom of the stack first results in an optimal volume of *I/O*. To simplify the discussion we first consider that $S_j \leq M_0$ for all children j. The volume of contribution blocks that will be written to disk corresponds to the difference between the memory requirement at the moment when the peak S is obtained and the size M_0 of the memory allowed (or available). Indeed, each time an *I/O* is done, an amount of temporary data located at the bottom of the stack is written to disk. Furthermore, data will only be reused (read from disk) when assembling the parent node. More formally, the expression of the volume of *I/O*, $V^{I/O}$, using Formula (1) for the storage requirement, is:

$$V^{I/O} = \max\left(0, \max(\max_{j=1,n}(S_j + \sum_{k=1}^{j-1} cb_k), m + \sum_{k=1}^{n} cb_k) - M_0\right) \quad (2)$$

Suppose now that $\exists j : S_j > M_0$. We know that child j will have an intrinsic volume of *I/O* $V_j^{I/O}$ (recursive definition based on a bottom-up traversal of the tree). In addition, we know that it cannot occupy more than M_0 in memory. Thus, we can consider it as a child using exactly M_0 memory ($A_j \stackrel{def}{=} \min(S_j, M_0)$), and inducing an intrinsic volume of *I/O* equal to $V_j^{I/O}$. With this definition of A_j as the *active memory*, *i.e.* the amount of core memory effectively used to process the subtree rooted at child j, we can now generalize Formula (2). We obtain:

$$V^{I/O} = \max\left(0, \max(\max_{j=1,n}(A_j + \sum_{k=1}^{j-1} cb_k), m + \sum_{k=1}^{n} cb_k) - M_0\right) + \sum_{j=1}^{n} V_j^{I/O} \quad (3)$$

To compute the volume of I/O on the whole tree, we apply recursively Formula (3) at each level of the tree (knowing that $V^{I/O} = 0$ for leaf nodes). The volume of I/O for the factorization is then given by the value of $V^{I/O}$ value at the root.

2.3 Tree Traversals

It results from Formula (3) that minimizing the volume of I/O is equivalent to minimizing the expression $\max_{j=1,n}(A_j + \sum_{k=1}^{j-1} cb_k)$, since it is the only term sensitive to the order of the children.

Theorem 1. (Liu, 86) *Given a set of values* $(x_i, y_i)_{i=1,\ldots,n}$*, the minimal value of* $\max_{i=1,\ldots,n}(x_i + \sum_{j=1}^{i-1} y_j)$ *is obtained by sorting the sequence* (x_i, y_i) *in decreasing order of* $x_i - y_i$*, that is,* $x_1 - y_1 \geq x_2 - y_2 \geq \ldots \geq x_n - y_n$*.*

Thanks to Theorem 1 (proved in [13]), we deduce that we should process the children nodes in decreasing order of $A_j - cb_j = \min(S_j, M_0) - cb_j$. (This implies that if all subtrees require a storage $S_j > M_0$ then MinIO will simply order them in increasing order of cb_j.) An optimal postorder traversal of the tree is then obtained by applying this sorting at each level of the tree, constructing Formulas (1) and (3) from bottom to top. We will name MinIO this algorithm.

Note that, in order to minimize the peak of storage (defined in Formula (1)), children had to be sorted (at each level of the tree) in decreasing order of $S_j - cb_j$ rather than $A_j - cb_j$. The corresponding algorithm (that we name MinMEM and that leads to sequence (a-b-c-d-e) on the example from Section 2.1) is thus different from MinIO (that leads to (d-a-b-c-e)).

3 In-Place Assembly of the Last Contribution Block

In this variant (used in of the MA27 [9] and its successors, for example) of the *classical* multifrontal algorithm, the memory of the frontal matrix of the parent is allowed to overlap with (or to include) that of the contribution block from the last child. The contribution block from the last child is then expanded (or assembled *in-place*) in the memory of the parent. Since the memory of a contribution block can be large, this scheme can have a strong impact on both storage and I/O requirements. In this new context, the storage requirements needed to process a given node (Formula (1)) becomes:

$$S = \max\left(\max_{j=1,n}(S_j + \sum_{k=1}^{j-1} cb_k), m + \sum_{k=1}^{\boxed{n-1}} cb_k\right) \tag{4}$$

The main difference with Formula (1) comes from the *in-place* assembly of the last child (see the boxed superscript in the sum in Formula (4)). In the rest of the paper we will use the term *last-in-place* to denote this scheme. Liu has shown[13] that Formula (4) could be minimized by ordering children in decreasing order of $\max(S_j, m) - cb_j$.

In an out-of-core context, the use of this *in-place* scheme induces a modification of the amount of data that has to be written to/read from disk. As previously for the memory requirement, the volume of I/O to process a given node with n children (Formula (3)) becomes:

$$V^{I/O} = \max\left(0, \max(\max_{j=1,n}(A_j + \sum_{k=1}^{j-1} cb_k), m + \sum_{k=1}^{\boxed{n\text{-}1}} cb_k) - M_0\right) + \sum_{j=1}^{n} V_j^{I/O}$$

Once again, the difference comes from the *in-place* assembly of the contribution block coming from the last child. Because $m + \sum_{k=1}^{n-1} cb_k = \max_{j=1,n}(m + \sum_{k=1}^{j-1} cb_k)$, this formula can be rewritten as:

$$V^{I/O} = \max\left(0, \max_{j=1,n}(\max(A_j, m) + \sum_{k=1}^{j-1} cb_k) - M_0\right) + \sum_{j=1}^{n} V_j^{I/O} \quad (5)$$

Thanks to Theorem 1, minimizing this quantity can be done by sorting the children nodes in decreasing order of $\max(A_j, m) - cb_j$, at each level of the tree.

4 In-Place Assembly of the Largest Contribution Block

In order to do better than equation (4), one is tempted to try to overlap the memory of the parent not with the contribution from the last child, but with the largest child contribution block. Compared to Equation (1) corresponding to the *classical* scheme, cb_{max} must be subtracted from the term $m + \sum_j cb_j$. Since cb_{max} is a constant that does not depend on the order of children, minimizing the storage (MinMEM) is done by using the same tree traversal as for the classical scheme (decreasing order of $S_j - cb_j$). We call this new scheme *max-in-place*. From an implementation point of view, note that the *in-place* assembly of the largest contribution block requires storing it in a particular area, rather than in the main stack. While processing the tree using a postorder, we thus need to use two stack mechanisms: one for the normal contribution blocks (for example on the left of a workarray), and one for the largest contribution blocks of each family (for example in the right part of the same workarray). The second one is used to extend the adequate contribution block into the frontal matrix of the parent.

In an out-of-core context, it is not immediate or easy to generalize MinIO. Indeed, there is no guarantee that we will be able to keep the largest contribution block of a family in core memory to enable its *in-place* assembly (suppose, for example, that a subtree ordered after that which induces the largest contribution block forces us to write this contribution to disk). Therefore, we propose the following heuristic. We first try to apply MinMEM + *max-in-place* to a given family (in a bottom-up process). If this leads to a storage smaller than M_0, we keep this approach to process this family. Otherwise, we *switch* to MinIO + *last-in-place* to process this family and any parent family. In the following we name this heuristic MinIO + *max-in-place*.

5 Experimental Results

In this section we experiment the behaviour of strategies presented in Sections 2, 3, and 4 on different matrices issued from the Parasol, Rutherford-Boeing or university of Florida collections. The matrices used, numbered from 1 to 30, are: AUDIKW_1, BCSSTK, BMWCRA_1, BRGM, CONESHL_MOD, CONV3D_64, GEO3D-20-20-20, GEO3D-50-50-50, GEO3D-80-80-80, GEO3D-20-50-80, GEO3D-25-25-100, GEO3D-120-80-30, GEO3D-200-200-200, GUPTA1, GUPTA2, GUPTA3, MHD1, MSDOOR, NASA1824, NASA2910, NASA4704, SAYLR1, SHIP_003, SPARSINE, THERMAL, TWOTONE, ULTRASOUND3, ULTRASOUND-80, WANG3 and XENON2. Matrices GEO3D*, BRGM and CONV3D_64 come from Geosciences Azur, BRGM, and CEA-CESTA (code AQUILON), respectively. We used several ordering heuristics, that, for a given matrix, define the task dependency graph (or assembly tree) and impact the computational complexity. The volumes of *I/O* were computed by instrumenting the analysis phase of MUMPS [5], which allowed us to experiment four ordering heuristics: AMD, AMF, METIS and PORD. The matrices have a size from very small up to very large (a few million equations) and can lead to huge factors (and storage requirements). For example, the factors of matrix CONV3D_64 with AMD ordering represent 53 *GB* of data.

As previously mentioned, the *I/O* volume depends on the amount of core memory available. Figure 2 illustrates this general behaviour on a sample matrix, TWOTONE ordered with PORD, for the 3 assembly schemes presented above, for both MinMEM and MinIO algorithms. For all assembly schemes and algorithms used, we first notice that exploiting all the available memory is essential to limit the *I/O* volume. Before discussing the results we remind the reader that the *I/O* volumes presented are valid under the hypothesis that the largest frontal matrix may hold in-core. With a core memory lower than this value (*i.e.* the

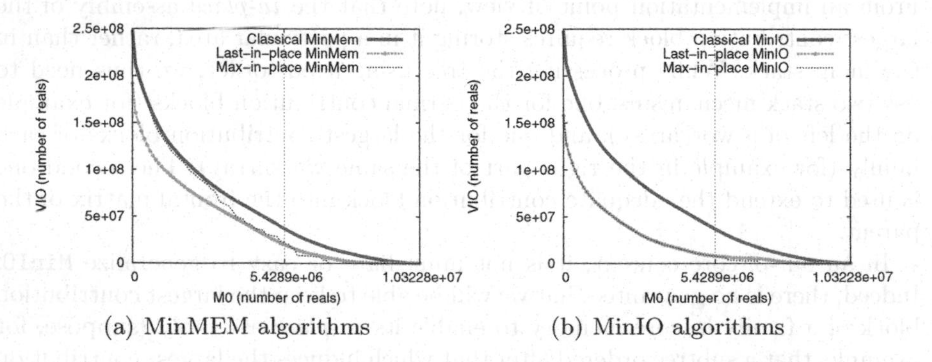

(a) MinMEM algorithms (b) MinIO algorithms

Fig. 2. I/O volume on matrix TWOTONE with PORD ordering as a function of the core memory available, for the 3 assembly schemes presented above, for both MinMEM and MinIO algorithm. The vertical bar represents the size of the largest frontal matrix.

area on the left of the vertical bar in Figure 2), the *I/O* volumes presented are actually lower bounds on the effective *I/O* volume. They are computed as if we could process the out-of-core frontal matrices with a read-once write-once scheme. They however remain meaningful because the extra-cost due to the specific treatment of frontal matrices will be independent of the assembly scheme used. We first notice that the *last-in-place* assembly schemes strongly decrease the amount of *I/O* compared to the *classical* assembly schemes. In fact, using an *in-place* assembly scheme is very useful in an out-of-core context: it divides the *I/O* volume by more than 2 on most of our test matrices. With the *classical* assembly scheme (presented in Section 2) we observe (on this particular matrix) that the `MinIO` and `MinMEM` algorithms produce the same *I/O* volume (their graphs are identical). Coming back to Formula (3), we have minimized the term $\max\left(\max_{j=1,n}(A_j + \sum_{k=1}^{j-1} cb_k), m + \sum_{k=1}^{n} cb_k\right)$ by minimizing the first member (the second one is constant); unfortunately on this particular matrix the second term is usually the largest and there is nothing to gain. From the list of matrices presented above, we have extracted four cases (one for each ordering strategy) for which the gains are significant and we report them in Figure 3(a). To better illustrate the gains resulting from the `MinIO` algorithm, we analyze

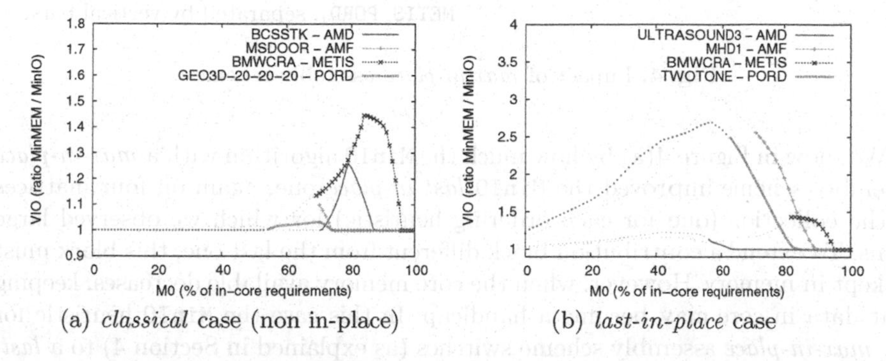

(a) *classical* case (non in-place) (b) *last-in-place* case

Fig. 3. *I/O* volume obtained with `MinMEM` divided by the one obtained with `MinIO`. For each matrix/ordering, the filled (right) part of the curve matches the area where the amount of core memory is larger than the size of the largest frontal matrix, while the dotted (left) part matches the area where this amount is lower. For each matrix, we normalized the memory (x-axis) to the in-core minimum requirement.

the *I/O* volume ratios as a function of the amount of core memory available (in percentage of the core memory requirements). For instance, the point of $(x = 80\%, y = 1.3)$ (obtained with both `BCSSTK` and `BMWCRA`) means that `MinMEM` leads to 30% more *I/O* when 80% of the in-core memory requirement is provided.

We now focus on the *in-place* assembly scheme (described in Section 3). As we cannot show the graphs obtained for our whole collection of matrices, we again decided to present in Figure 3(b) four cases (one for each ordering strategy) for which `MinIO` was significantly more efficient than `MinMEM` (*I/O* volume was divided for instance by more than 2 for a large range of core memory amounts on MHD1-AMF matrix). An extensive study has shown that the largest gains from MinIO are obtained when matrices are preprocessed with orderings that tend to build irregular assembly trees (AMF and PORD and to a lesser extent AMD - see [11] for more details). On such trees, there is a higher probability to be sensitive to the order of children.

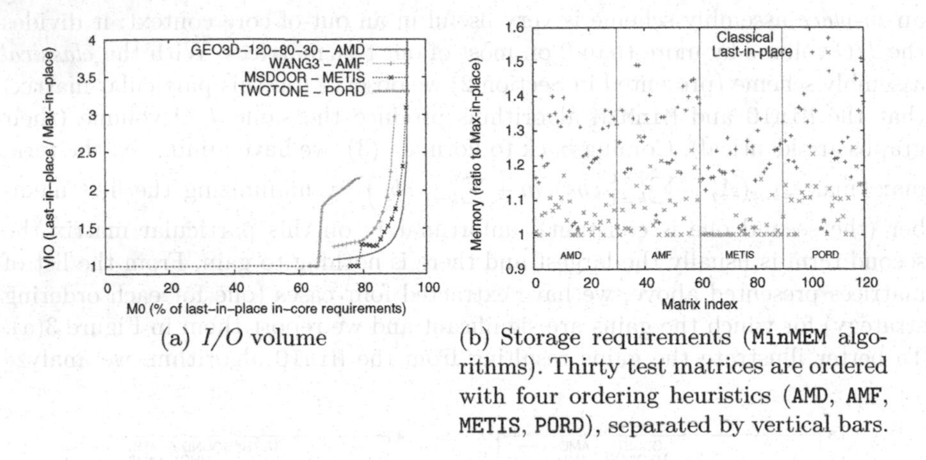

(a) *I/O* volume

(b) Storage requirements (MinMEM algorithms). Thirty test matrices are ordered with four ordering heuristics (AMD, AMF, METIS, PORD), separated by vertical bars.

Fig. 4. Impact of *max-in-place* assembly scheme

We show in Figure 4(a) by how much the MinIO algorithm with a *max-in-place* assembly scheme improved the MinIO *last-in-place* one, again on four matrices of the collection (one for each ordering heuristic) for which we observed large gains. To extend a contribution block different from the last one, this block must be kept in memory. However, when the core memory available decreases, keeping that data in-core may become a handicap. In this case the MinIO heuristic for the *max-in-place* assembly scheme switches (as explained in Section 4) to a *last-in-place* scheme. Thus, with a small amount of core memory, the *last-in-place* and *max-in-place* MinIO heuristics have a similar behaviour (the left part of their curves are identical in Figure 2(b); the ratio is equal to 1 in Figure 4(a)).

Figure 4(b) shows that the peak of storage (critical for the in-core case) can also be decreased significantly. This allows us to interpret the extreme right parts of the curves in Figure 4(a) which tend to (or are equal to) infinity: the *max-in-place* assembly scheme does *not* induce *I/O* while the *last-in-place* scheme *does*.

6 Conclusion and On-Going Work

Table 1 summarizes the contributions of this paper. We have reminded the existing memory-minimization algorithms for the *classical* and *last-in-place* assembly schemes. We have shown that these algorithms are not optimal to minimize the

Table 1. Summary. Contributions of this paper are in bold.

Assembly scheme	Algorithm	Objective function: Memory minimization	Objective function: **I/O minimization**
classical	MinMEM	• Optimum ([11], adapting[13])	• **Reasonable but not optimum**
	MinIO	• Not suited	• **Optimum**
last-in-place	MinMEM	• Optimum[13]	• **Bad especially on irregular trees**
	MinIO	• Not suited	• **Optimum**
max-in-place	**MinMEM**	• **Optimum**	• Not suited
	MinIO	• **Optimum**	• **Efficient heuristic**

I/O volume. [2] We have proposed optimal algorithms for the *I/O* volume minimization and have shown that significant gains could be obtained on real problems (especially with the *in-place* assembly scheme). We have then presented a new assembly scheme (which consists in extending the child with the largest contribution block) and a corresponding tree traversal which is optimal to minimize memory and leads to an efficient heuristic when the objective is to minimize the *I/O* volume.

This work is particularly important when applied to large-scale problems (millions of equations) in limited-memory environments (which is actually always the case, even on high-end platforms). It is applicable for shared-memory solvers relying on threaded BLAS libraries. In a parallel distributed context, it will help to limit memory requirements and to decrease the *I/O* volume in the sequential (often critical) parts of the computations.

We are currently working on adapting this work to a more flexible task allocation scheme, where the parent node is allowed to be allocated before all children have been processed [10]. Again, instead of limiting the storage requirement of the methods, the goal consists in minimizing the volume of *I/O* involved. The work presented in this paper is a basis to this new and more difficult flexible context.

Acknowledgement. We are grateful to Patrick Amestoy for his comments on a first version of this paper.

References

1. The BCSLIB Mathematical/Statistical Library. http://www.boeing.com/phantom/bcslib/
2. Agullo, E., Guermouche, A., L'Excellent, J.-Y.: A preliminary out-of-core extension of a parallel multifrontal solver. In: Nagel, W.E., Walter, W.V., Lehner, W. (eds.) Euro-Par 2006. LNCS, vol. 4128, pp. 1053–1063. Springer, Heidelberg (2006)
3. Agullo, E., Guermouche, A., L'Excellent, J.-Y.: Towards a parallel out-of-core multifrontal solver: Preliminary study. Research report 6120, INRIA, 02, Also appeared as LIP report RR2007-06 (2007)

[2] Note that we have proved in [4] that minimizing the storage requirement can be arbitrarily worse in terms of *I/O* volume than directly minimizing the *I/O* volume.

4. Agullo, E., Guermouche, A., L'Excellent, J.-Y.: Reducing the I/O volume in an out-of-core sparse multifrontal solver. Research Report 6207, INRIA, 05, Also appeared as LIP report RR2007-22 (2007)
5. Amestoy, P.R., Guermouche, A., L'Excellent, J.-Y., Pralet, S.: Hybrid scheduling for the parallel solution of linear systems. Parallel Computing 32(2), 136–156 (2006)
6. Amestoy, P.R., Puglisi, C.: An unsymmetrized multifrontal LU factorization. SIAM Journal on Matrix Analysis and Applications 24, 553–569 (2002)
7. Dobrian, F.: External Memory Algorithms for Factoring Sparse Matrices. PhD thesis, Old Dominion University (2001)
8. Dobrian, F., Pothen, A.: Oblio: a sparse direct solver library for serial and parallel computations. Technical report, Old Dominion University (2000)
9. Duff, I.S., Reid, J.K.: MA27—a set of Fortran subroutines for solving sparse symmetric sets of linear equations. Technical Report R.10533, AERE, Harwell, England (1982)
10. Guermouche, A., L'Excellent, J.-Y.: Constructing memory-minimizing schedules for multifrontal methods. ACM Transactions on Mathematical Software 32(1), 17–32 (2006)
11. Guermouche, A., L'Excellent, J.-Y., Utard, G.: Impact of reordering on the memory of a multifrontal solver. Parallel Computing 29(9), 1191–1218 (2003)
12. Hénon, P., Ramet, P., Roman, J.: PaStiX: A High-Performance Parallel Direct Solver for Sparse Symmetric Definite Systems. Parallel Computing 28(2), 301–321 (2002)
13. Liu, J.W.H.: On the storage requirement in the out-of-core multifrontal method for sparse factorization. ACM Transactions on Mathematical Software 12, 127–148 (1986)
14. Liu, J.W.H.: The multifrontal method and paging in sparse Cholesky factorization. ACM Transactions on Mathematical Software 15, 310–325 (1989)
15. Reid, J.K., Scott, J.A.: An out-of-core sparse Cholesky solver. Technical report, Rutherford Appleton Laboratory (2006)
16. Rothberg, E., Schreiber, R.: Efficient methods for out-of-core sparse Cholesky factorization. SIAM Journal on Scientific Computing 21(1), 129–144 (1999)
17. Rotkin, V., Toledo, S.: The design and implementation of a new out-of-core sparse Cholesky factorization method. ACM Trans. Math. Softw. 30(1), 19–46 (2004)

Experiments with a Parallel External Memory System*

Mohammad R. Nikseresht[1], David A. Hutchinson[2], and Anil Maheshwari[1]

[1] School of Computer Science, Carleton University
[2] Dept. of Systems and Computer Engineering, Carleton University

Abstract. The theory of bulk-synchronous parallel computing has produced a large number of attractive algorithms, which are provably optimal in some sense, but typically require that the aggregate random access memory (RAM) of the processors be sufficient to hold the entire data set of the parallel problem instance. In this work we investigate the performance of parallel algorithms for extremely large problem instances relative to the available RAM. We describe a system, Parallel External Memory System (PEMS), which allows existing parallel programs designed for a large number of processors without disks to be adapted easily to smaller, realistic numbers of processors, each with its own disk system. Our experiments with PEMS show that this approach is practical and promising and the run times scale predictable with the number of processors and with the problem size.

1 Introduction

In this work we investigate the performance of parallel algorithms for extremely large problem instances relative to the available Random Access Memory (RAM). Using theoretical results of [1,2], we transform parallel algorithms designed for a large number of processors without disks to smaller, realistic numbers of processors, each with its own disk system.

External Memory (EM) Algorithms: These algorithms are designed so that their run times scale predictably even as the size of their data increases far beyond the size of internal RAM. This huge data size requires that such algorithms optimize the transfer of data between RAM and some sort of secondary memory devices, typically moveable-head disk drives. The time to access an element of data at an arbitrary position on a moveable-head disk is several orders of magnitude greater than for RAM access. In addition, the time to set up the data transfer (disk head movement, rotational delay) is much larger than the time to actually transfer the data. Data items are grouped into *blocks* and accessed block-wise by efficient external memory algorithms in order to amortize the setup time over a large number of data items.

* This work was partially supported by the National Sciences and Engineering Research Council of Canada (NSERC) and by the High Performance Computing Virtual Laboratory (HPCVL).

S. Aluru et al. (Eds.): HiPC 2007, LNCS 4873, pp. 59–70, 2007.

A critical quality in a good EM algorithm is its locality of reference to items in a disk block (or collection of disk blocks). EM algorithms are designed to use all, or a significant fraction, of the elements in a block while the block is in memory, avoiding the cases where the same block must be brought into memory many times. The operating system's data cache may interfere with the operation of an EM algorithm, as it may make useless copies of data blocks, and occupy RAM that could be used more effectively for other purposes. EM algorithms manipulate huge data volumes relative to their RAM size, and so the time to move data between RAM and disk dominates the running time in most cases. For this reason, the Parallel Disk Model (PDM) [3] uses the number of distinct disk block input or output (I/O) operations as a measure of the goodness of an EM algorithm. If D disks are present and the block size is B items, a single I/O operation can transfer BD items in parallel in this model. An optimal EM algorithm is one that achieves the minimum number of I/O operations possible.

Coarse Grained Parallel Algorithms: These algorithms have a number of interesting and useful properties for our purposes: (a) the processors perform multiple rounds of computation separated by communication of interim results, (b) during each computation round, the processors perform a chunk of computational work. For this period of time the processors operate completely independently, and are restricted to accessing the data in their own local memories. and (c) between computation rounds the processors synchronize and exchange information via a communication network.

The simulation techniques described in [1] permit coarse grained parallel algorithms to be executed efficiently on machines with p *real* processors rather than the number of processors v required by the original algorithm, where $1 \leq p \leq v$. We refer to the original v processors as *virtual* processors. The tradeoff is that each of the real processors must have enough disk space to store the internal memory for v/p of the virtual processors plus the messages that would be sent between the virtual processors in each communication round. The real processors must have enough RAM to represent at least one virtual processor at a time, and that should be at least DB items in size so swapping between virtual processors is I/O efficient. Finally, the communication should be I/O efficient, meaning that at least DB blocks should be sent and received by each virtual processor in each communication round. Intuitively, the simulation technique works because the coarse grained property of the original parallel algorithm ensures that the requisite locality of reference is present in the resulting EM algorithm. We can "trade-off" some parallelism for blocked I/O but retain some parallelism to take advantage of multiple real machines and the scalability of multiple disk systems. The theoretical guarantees of asymptotically optimal parallelism and I/O in the resulting algorithms makes them interesting for practical use, since their parameters (v, p, B, D) can be scaled to fit the parallel hardware at hand.

Previous Work: The work of this paper extends the results of previous implementation work reported in [2] from a single processor to multiple real processors. LEDA-SM [4], TPIE [5] and STXXL [6] are I/O workbenches developed to explore external memory algorithm implementations for sequential machines. TPIE provides prototype implementations of many contemporary EM algorithms but currently it does not offer algorithms for multiple processors or multiple disks. STXXL provides a mechanism for asynchronous I/O and therefore allows overlapping of I/O and computation. It handles multiple disks but it does not itself support multiple processor algorithms. SSCRAP is a framework for implementing parallel coarse grained algorithms and has been extended [7] to support the simulation of certain parallel algorithms in external memory with reference to the theoretical framework of [1]. However, it is not clear whether SSCRAP handles communication traffic between virtual processors in an I/O-optimal way, or whether very large problem instances relative to RAM size have been tested.

Organization and Contributions of this Paper: This paper takes a step forward in determining the practicality of the simulation approach. In Sect. 3, we describe a framework that allows existing MPI-based implementations of CGM algorithms (or BSP algorithms with appropriate parameters) to be executed efficiently on a machine with fewer real processors but with disk storage. Such MPI-based programs are modified in the following ways: (a) calls to the MPI library are replaced by calls to corresponding procedures in the PEMS library and (b) calls to the C dynamic memory management routines are replaced by calls to corresponding PEMS routines. In Sect. 4, we report preliminary timing results for sorting that are comparable with TPIE and STXXL, two contemporary workbenches for high performance I/O experiments. We compare single processor instances of sorting on these workbenches with sorting using various numbers of real processors running a CGM sample sort implementation. While our timing results lagged those of TPIE and STXXL for a single processor, we are able to surpass both by adding more processors. Our experiments show that this approach is practical and promising and the run times scale predictable with the number of processors and with the problem size.

2 Preliminaries

In this section we present the main ideas behind the simulation technique proposed in [1]. It optimizes blockwise data access and disk I/O and at the same time utilizes multiple processors connected via a communication network or shared memory. The Bulk-Synchronous Parallel (BSP) model [8] consists of v processor/memory components, a *router* that delivers messages in a point to point fashion, and a facility to synchronize all processors. Each processor has a unique label in the range $0, 1, \ldots, v-1$. Computation proceeds in a succession of *supersteps* separated by synchronizations, usually divided into *communication* and *computation supersteps*. In computation supersteps processors perform local computations on data that is available locally at the beginning of the superstep and issue send operations. Between computation supersteps, a communication superstep is performed, where each processor exchanges data with its peers, via the router. This is done through an *h-relation*, where $O(h)$ data are sent and received by every processor in a superstep. In addition to the parameters v and h, BSP uses two additional parameters. The parameter g is the time required to send a single word of data between two processors, where time is measured in number of CPU operations, and the parameter L is the minimum setup time or latency of a superstep, measured in CPU operations.

The technique in [1] simulates a v (virtual) processor BSP algorithm $\mathcal{A}'$, executing a problem of size N, which communicates via h-relations of size $h = O(\frac{N}{v})$ with λ supersteps/rounds, local memory size μ, computation time $\beta + \lambda L$, communication time $g\alpha + \lambda L$ as a p (real) processor EM-BSP algorithm $\mathcal{A}$ with $\frac{v}{p}\lambda$ rounds, computation time $\frac{v}{p}(\beta + O(\lambda\mu)) + \frac{v}{p}\lambda L$, communication time $\frac{v}{p}g\alpha + \frac{v}{p}\lambda L$, and I/O time $\frac{v}{p}G \cdot O(\lambda\frac{\mu}{B}) + \frac{v}{p}\lambda L$ for $M = \Theta(\mu)$, $N = \Omega(vB)$, $B = O(\frac{N}{v^2})$, $p < v$. The parameter G is the ratio between the local computational capacity and the local I/O capacity and the parameter B is the disk block size.

Next we sketch the main steps of the simulation. We distribute v virtual processors evenly on p real processors. Each real processor i, $0 \leq i \leq p$ executes the following steps (for a single processor simulation, we set $p = 1$, $i = 0$ and omit Step 5):

For j = 0 to $\frac{v}{p} - 1$ do in parallel on each real processor i

1. Read the context of virtual processor $i\frac{v}{p} + j$ from the local disk.
2. Read any messages addressed to virtual processor $i\frac{v}{p} + j$ from the local disk.
3. Simulate the computation superstep of virtual processor $i\frac{v}{p} + j$, collecting all generated messages in the local internal memory.
4. Send all generated messages to the required (real) destination processors.
5. Receive all messages addressed to real processor i on behalf of virtual processors $i\frac{v}{p}$ to $(i+1)\frac{v}{p} - 1$ in local internal memory and write them to the local disk.
6. Write the changed context for virtual processor $i\frac{v}{p} + j$ back to the local disk.

---

3 Software Design

In this section we explain our software design and show how different components work and interact together in the Parallel External Memory System (PEMS). In PEMS a virtual processor is represented by a user space thread. The input to our system is an existing MPI program implementing a coarse grained BSP algorithm. In such a program MPI is responsible for interprocessor communications. Most of the MPI functions are for sending and receiving messages. In PEMS, each MPI call is replaced by call to a corresponding PEMS service. These PEMS calls may in turn incorporate MPI calls. The communication between virtual processors is managed using the disk, memory buffers and communication network. This is the main challenge in the design and implementation of PEMS.

To date, our focus has been on confirming the high level behavior of the simulation approach, that is to confirm that the behavior of PEMS scales predictably

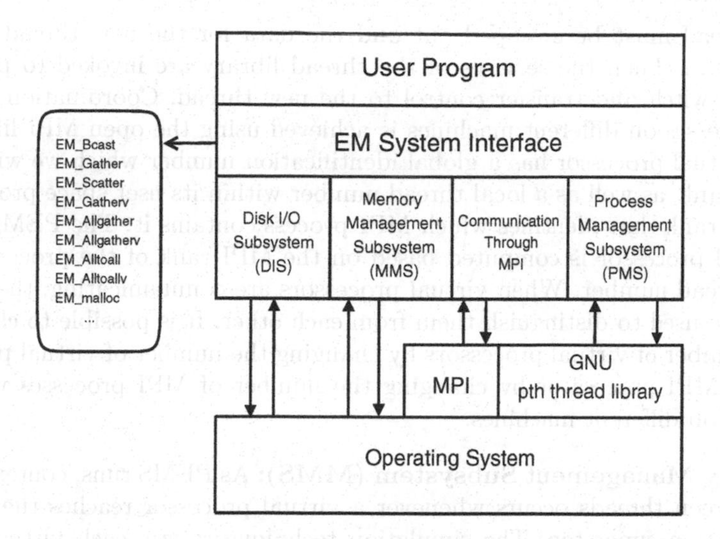

Fig. 1. PEM System Software Layers

as problem sizes and the number of real processors vary. This involves both running time and disk space usage behaviors. We have not been overly concerned about optimization in this initial version of PEMS.

Figure 1 shows the layered software design of PEMS. The user program is shown as the top layer in the diagram. The EM System Interface (Layer 2) provides the external memory communication primitives. All MPI and memory allocation calls in the original user program are replaced by a call to a similar PEMS function in this layer. Layer 3 has four major components. The Disk I/O Subsystem (DIS) is responsible for reading and writing data blocks to disk efficiently. The Memory Management Subsystem (MMS) is responsible for allocating memory for virtual processor data and for swapping this data when required. The Process Management Subsystem (PMS) is responsible for creating, scheduling and synchronizing virtual processors. The Open MPI library [9] is used for communication between real processors and also for starting the software on multiple machines. In subsequent subsections we discuss layers 2 and 3 in more detail.

Virtual Processor Simulation and Process Management Subsystem (PMS): The GNU pth thread library [10] is used for representing virtual processors as threads. Each virtual processor is a user level thread. Each thread runs a copy of the modified MPI-based user program. Modifications include replacement of MPI and memory allocation calls with the corresponding PEMS calls. The PMS consists of a set of functions to initialize, start, synchronize, manage, and schedule threads. These functions are implemented using the GNU pth thread library which is responsible for the creation and cooperative scheduling of threads. At the end of every computation superstep each virtual processor calls _EM_yield(). This first invokes the services of MMS. The data for the current thread must be swapped out and the data for the new thread must be swapped in. Then the services of the thread library are invoked to perform a context switch and transfer control to the new thread. Coordination of different processes on different machines is achieved using the open MPI library [9]. Each virtual processor has a global identification number which we will call its PEMS rank, as well as a local thread number within its user space process, and an MPI rank that identifies which MPI process contains it. The PEMS rank of a virtual processor is computed based on the MPI rank of the process and its local thread number. When virtual processors are communicating, their PEMS ranks are used to distinguish them from each other. It is possible to change the total number of virtual processors by changing the number of virtual processors in each MPI process or by changing the number of MPI processes which are running on different machines.

Memory Management Subsystem (MMS): As PEMS runs, context switching between threads occurs whenever a virtual processor reaches the end of a computation superstep. The simulation technique resizes each virtual processor (thread) to use the available physical memory. To accommodate many such threads it would normally be necessary for the operating system to allocate memory from its swap space or virtual memory. Each time a thread starts execution, the operating system would swap in the needed pages of data from the disk swap area. This is done based on a page fault mechanism which may not be efficient for our purposes. Even if the operating system was able to perform swapping efficiently, the total amount of memory needed for all virtual processors could be well beyond the virtual address space of the combined operating system and the hardware. This is especially true for 32-bit machines. Therefore, we need complete control over the PEMS physical memory and all of the disk activities related to the execution of the PEMS program.

The Memory Management Subsystem determines how much physical memory is available at the beginning of the program execution (initialization phase). Half of this memory is used as a buffer for communication purposes and for assembling and disassembling messages. The other half is reserved for use by virtual processors; they can allocate memory from this area by calling EM_malloc().

Disk I/O Subsystem (DIS): This layer allows the higher layers of the system to be independent of specific operating system I/O calls. This makes the system more flexible and portable, and potentially allows PEMS to interface to third party I/O packages. The DIS currently contains implementations of direct I/O based on the native Linux direct I/O which is supported in kernel versions 2.6 and above, asynchronous I/O and synchronous I/O. Currently, PEMS uses direct synchronous I/O for swapping and buffered synchronous I/O for messaging.

EM System Interface and Messaging: This layer is the most important layer in our implementation. While the contexts (local data) of each virtual processor can be swapped between RAM and disk using efficient streaming I/O, this is not genereally true for delivering messages between virtual processors in the simulation. All of the PEMS functions have the same number and type of parameters as their MPI counterparts. This makes it easier to transform an MPI program into a PEMS program. In fact we believe that there are simple automated ways of doing this. We have also implemented EM data types corresponding to the MPI data types (e.g. MPI_INT, MPI_CHAR). As mentioned earlier, communication is done through the disk, memory buffer, network, or a combination of them depending on the communication function. If two communicating virtual processors are on the same machine then they communicate through the disk or memory buffer, depending on the communication function invoked. If they are on two different machines they communicate through the network as well. PEMS splits the available physical memory into two partitions, each equal to the data memory size of a virtual processor. One partition is used by the virtual processors and the other partition is used by PEMS as a large buffer. Any message which fits into this buffer is kept there instead of being written to the disk. Since no virtual processor can receive a message bigger than its data memory size, it is possible to keep most of the messages in this buffer. The exceptions are messages generated by the functions EM_Alltoall and EM_Alltoallv. In this type of communication, potentially all of the virtual processors are generating messages which are destined to all other virtual processors and each virtual processor may receive different sizes and numbers of messages. Since it is not possible to keep all of these messages in the shared memory buffer, they are written to the disk. The destination virtual processor will read the relevant messages from the disk when it becomes active. This requires maintaining information pertaining to where these messages are stored on the disks, as well as efficient retrieval mechanisms of the disk blocks corresponding to these messages.

We now turn to communication between virtual processors running on different physical machines. The sending virtual processor generates a message and calls the relevant PEMS function for communication. PEMS determines the MPI rank of the destination virtual processor and sends the message to the real processor on which it is hosted. The destination virtual process (or thread) may not be running at the time that the messages arrives, however. The current running virtual processor at the destination machine receives the messages on behalf of the destination virtual processor and saves the messages in the memory buffer or on the disk. The actual receiver can read the messages later whenever it becomes active.

The most complicated communication primitive is EM_Alltoallv. This primitive allows all virtual processors to send messages to all other virtual processors. The messages can be of different sizes and may be made up of multiple packets. The only restriction is that they need to fit into the local buffers of each virtual processor. As real processor p_i simulates $\frac{v}{p}$ virtual processors, in each round of communication it will receive all messages addressed to the virtual processors which it simulates. It is possible, in our current implementation, that a user thread representing virtual processor v_{ij}, associated with the real processor p_i, is called on to handle many messages addressed to the virtual processors in p_i. The total size of these messages may be more than the total physical memory of

p_i. More precisely, let the data memory of each virtual processor be μ_d, then the maximum message size is bounded by μ_d. There are $\frac{v}{p}$ virtual processors in each real processor p_i, so p_i may receive a total of $\min(p, \frac{v}{p}) \times \mu_d$ messages which may be beyond the real processor's RAM capacity. If we use the personalized communication algorithm by Bader et.al. [11] then the message size for each virtual processor is at most $\frac{\mu_d}{v} + \frac{v-1}{2}$. Then the maximum message size for each real processor is $O(\min(p, \frac{v}{p}) \times \frac{\mu_d}{v})$ which is $O(\mu_d)$. This is within reasonable limits and can be handled by a real processor. During the personalized communication the memory buffer is used as a working area for assembling and disassembling messages and packets.

3.1 Discussion on Design Choices

In this section we highlight some of the design alternatives for different components of the PEM system and provide some reasoning behind our choices.

Kernel Space Threads versus User Space Threads: In our implementation each virtual processor is simulated as a user space thread. We could have used kernel space threads. There are advantages and disadvantages in using one over the other. We decided to use the user space threads to shorten development time; they are easier to manage and synchronize, switching between threads is faster and coding and debugging is easier. We chose the GNU pth thread library for our implementation. This library is portable, has a Posix pthread interface as well as has its own specific interface.

Choices for Communication: We categorize two types of communication. Communication between virtual processors within a real processor is called *internal communication*, whereas communication between virtual processors on different processors is called *external communication.*

Option 1: We assign the job of communication (either internal or external) to a separate process (a kernel level thread). As virtual processors generate data for communication, they send them to the communication process, which decides if those data should be communicated internally or externally. Here we also need an interprocess communication mechanism to send data from the simulating process to the communication process.
Option 2: As the internal communication is fairly simple, each virtual processor can submit its internal communication to the DIS and send its external communication to the communication process as in Option 1.
Option 3: Internal communication is done as in Option 2 but external communication is done by the main thread of the process which simulates virtual processors. This avoids the need for interprocess communication. In this approach the main thread also has a synchronizing role. Before allowing the next superstep to start, it waits until communication data from all virtual processors are written to the disk and all external communication data are received by peer processors and written to the disk or the memory buffer.

Option 4: Internal and external communication is done by the current running threads on all machines. Each running virtual processor calls the communication routines as subroutines. Communication routines classify messages, communicate with other running virtual processors and gather all the messages sent to their real processor (on behalf of all virtual processors being simulated in this real processor). We chose to use this approach in the current version of PEMS for the following reasons: First, messages to other machines are sent out as they are generated so there is no additional disk I/O or buffer activity for them. Second, there is no need for a separate process to do the communication. Third, each time virtual processors communicate through MPI, they can also go through a synchronization phase with other virtual processors and in fact all real processors can be synchronized without the need to wait for the main threads of the processors.

Direct and Asynchronous I/O: Direct I/O allows an I/O operation to be performed directly on a user space buffer without additional buffering by the operating system. Incorporating Linux direct I/O imposes constraints on the buffer size, alignment and also on the file offset especially when we want to use disk I/O functions such as pread or pwrite. The buffer size must be a multiple of the disk block size, it must also be aligned on a memory address which is a multiple of the disk block size. The same alignment constraint applies to file offset on a read or write operation. This complicates the PEMS versions of MPI-like collective primitives, as we often need to read and write at arbitrary file offsets and at arbitrary addresses in the buffer space. As a result, we decided to use direct I/O for swapping the context data and buffered I/O for communication purposes.

4 Experiments

Objectives: The main objective of our present work has been to see whether the simulation technique as proposed in [1,2] is practical. At this stage of PEMS development, our experiments focus primarily on scalability of performance when the problem size and number of processors are varied. In order to shorten development time we have not yet placed much emphasis on optimizing our use of low level services such as asynchronous I/O and and kernel threads. While our experiments show running times for sorting with TPIE and STXXL, we include these measurements only as general reference points that highlight room for improvement in our single processor results.

Experimental Setup: Single processor tests are performed on the following configuration: AMD Opteron 2.4GHz CPU with 2GB of RAM, 3 Hard Disks each with at least 30GB of free space, two partitions across two disks configured as software RAID 0, hard disk bandwith is 71MB/Sec, RAID bandwith is 112 Mb/Sec (measured by hdparm utility), file system is EXT3, GNU/Linux 64bit v. 2.6.15 operating system and we used gcc v.4.1.1. Multiple processor experiments are performed on a cluster of Linux machines with the following hardware and software configuration[1]. Intel 2 × Xeon 2.0 GHz CPUs with 1.5 GB of RAM (only one processor was used in practice), one hard disk each with at least 46GB of free space, hard disk bandwith of 45Mb/Sec (measured by hdparm utility), GNU/Linux 32bit kernel version 2.6.9-42.0.2. ELsmp, EXT2 file system, gcc v 3.4.6 compiler and Gigabit ethernet for the communication between machines.

We use parallel sample sort to test our implementation. In our experiments we have simulated parallel sample sort [12] as an external memory algorithm. For comparison, we include timings from TPIE's [5] test_sort and STXXL's [6] test_sort1. We have slightly modified test_sort and test_sort1 functions to restrict them to one round of sorting (with no extra tests). All of the test programs use 128MB of RAM for sorting. The record size is 4 bytes and timing includes the data generation, but none of the test programs create a separate output file. PEMS uses 128MB of memory but 64MB of this memory is used as buffer and shared memory for communication, and only 64MB is used for sorting. (In PEMS, we have disabled the extra memory so that the operating system cannot use it for caching.)

Test data was generated using the C or C++ standard random generator or the package specific random generator method. We do not use any special integer sorting techniques in the parallel sample sort algorithm or in other test programs. Operating system swap has been disabled in all of the tests.

Experimental Results: Figure 2 shows running times for PEMS sample sort on 1, 2, 4, 8 and 16 real processors. It also includes running time of TPIE and STXXL external memory sort test programs on a single processor. We include the TPIE and STXXL results only as general reference points that highlight room for improvement in our single processor results. The reader should not draw conclusions about the relative running times of TPIE and STXXL sort, for instance, as we have not ensured that this is a fair comparison. The TPIE and STXXL sort programs are 3 to 5 times faster than our single processor sort with this version of PEMS, but as more real processors are added, PEMS becomes faster. At 7 billion integers, the running time of PEMS with 16 processors reduces by 65%, 53%, and 48%, in comparison to TPIE, STXXL and PEMS with 8 processors, respectively. The PEMS running times increase almost linearly with problem size when the number of real processors is fixed.

Limitations: An important limitation of PEMS is its internal disk usage. It needs $\frac{v}{p} \times \mu_d$ for swapping of data on each real processor. Here μ_d represents data memory of each virtual processor. It also reserves $2\frac{v}{p} \times \mu_d$ for communication on each real processor. We are currently investigating ways to reduce this requirement on the reservation of the disk space for the messages.

Concluding Remarks: PEMS is runtime library for creating parallel external memory programs from implementations of BSP-like coarse grained parallel algorithms. The primary application area is problems that require processing of massive amounts of data. We see our work as relevant from several practical

[1] This is part of HPCVL Lab www.hpcvl.org.

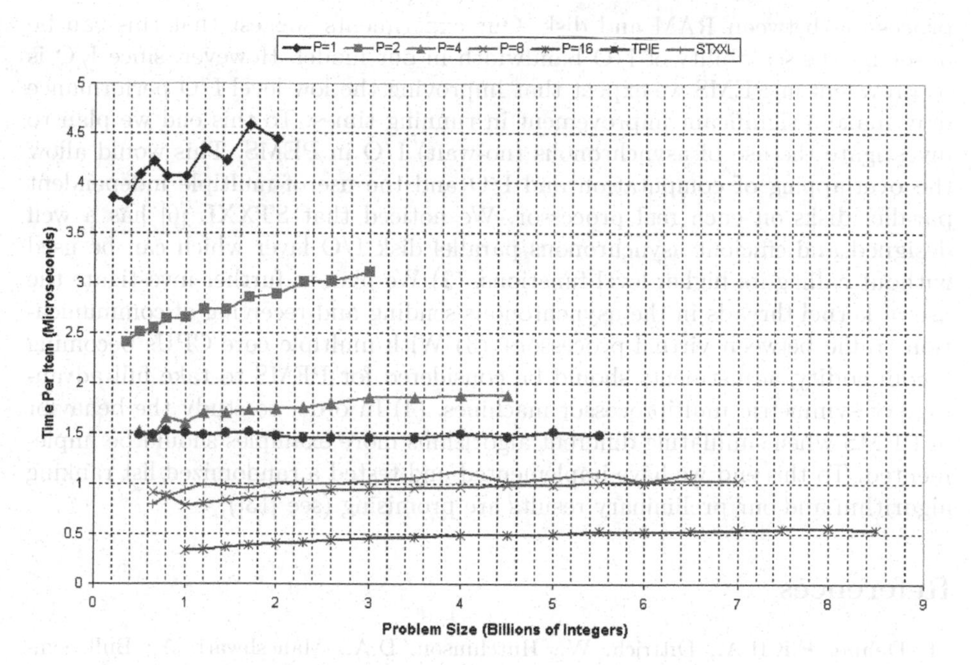

Fig. 2. Wall clock timings for sorting. The X-axis is the problem size in billions of integers. The Y-axis is wall clock time per sorted item in microseconds.

perspectives: a class of theoretically optimal parallel algorithms can be scaled to fit the hardware at hand, using both processors and disks; for a large set of problems for which suitable parallel algorithms exist, I/O optimal parallel algorithms can be used instead of single processor EM algorithms; and, both computational and I/O load are spread over multiple machines, contributing to the scalability.

Our experiments show several important properties of the PEMS approach. First of all the methodology is practical. Secondly, increasing the number of real processors decreases the running time in a predictable way. The ability to exploit parallel machine resources such as disks gives the ability to handle extremely large data sets on practical architectures such as a network of workstations. On such a system, one can inexpensively add computational power, RAM, disks, and bandwidth between RAM and disk by adding machines to a network. Using a coarse grained parallel algorithm and PEMS, our experiments in this paper suggest that one can take advantage of all of these resources, as well as adapting the theory of coarse grained parallel algorithms to the reality of a smaller number of real processors, each with a disk system. We believe that our experiments with sorting, for instance, showed good speedups in parallel running time primarily due to disk parallelism. More computationally intensive applications may be able to also make use of the additional computation power.

This brings us to several suggestions for further work: (1) PEMS algorithms have the disadvantage that they may do more I/O, than a conventional single processor EM algorithm due to the need to swap the contexts of virtual processors between RAM and disk. Our experiments suggest that this can be offset by the scalability of I/O bandwidth in our model. However, since I/O is so prevalent in PEMS we expect that improving the low level I/O performance may make a significant improvement in running times. To this end we plan to investigate the use of asynchronous (no-wait) I/O in PEMS. This would allow the overlapping of computation and I/O and the use of multiple independent parallel disks on each real processor. We noticed that STXXL [6] has a well designed and efficient asynchronous parallel disk I/O layer which can be used without calling its higher level functions. (2) We plan to further investigate the use of kernel threads in the asynchronous sending and receiving of communication traffic between virtual processors. (3) With multiple core CPUs becoming a commodity, adjustments should be considered for PEMS to take full advantage of symmetric multiprocessor machines. (4) In order to study the behavior of PEMS when simulating different algorithms more examples should be implemented. To this end we have implemented and tested a randomized list ranking algorithm and our preliminary results are promising (see [13]).

References

1. Dehne, F.K.H.A., Dittrich, W., Hutchinson, D.A., Maheshwari, A.: Bulk synchronous parallel algorithms for the external memory model. Theory Comput. Syst. 35(6), 567–597 (2002)
2. Hutchinson, D.A.: Parallel Algorithms in External Memory. PhD thesis, School of Computer Science, Carleton University (1999)
3. Vitter, J.S., Shriver, E.A.M.: Algorithms for parallel memory, I: Two-level memories. Algorithmica 12(2–3), 110–147 (1994)
4. Crauser, A., Mehlhorn, K.: LEDA-SM: Extending LEDA to secondary memory. In: Vitter, J.S., Zaroliagis, C.D. (eds.) WAE 1999. LNCS, vol. 1668, pp. 228–242. Springer, Heidelberg (1999)
5. TPIE, `http://www.cs.duke.edu/TPIE/`
6. Dementiev, R., Kettner, L., Sanders, P.: STXXL: Standard template library for XXL data sets. In: Brodal, G.S., Leonardi, S. (eds.) ESA 2005. LNCS, vol. 3669, pp. 640–651. Springer, Heidelberg (2005)
7. Gustedt, J.: Towards realistic implementations of external memory algorithms using a coarse grained paradigm. In: Kumar, V., Gavrilova, M., Tan, C.J.K., L'Ecuyer, P. (eds.) ICCSA 2003. LNCS, vol. 2668, pp. 269–278. Springer, Heidelberg (2003)
8. Valiant, L.G.: A bridging model for parallel computation. Communications of the ACM 33(8), 103–111 (1990)
9. Open MPI, `http://www.open-mpi.org/`
10. GNU Pth - The GNU Portable Threads, `http://www.gnu.org/software/pth/`
11. Bader, D.A., Helman, D.R., JáJá, J.: Practical parallel algorithms for personalized communication and integer sorting. ACM JEA 1, 3 (1996)
12. Shi, H., Schaeffer, J.: Parallel sorting by regular sampling. Journal of Parallel and Distributed Computing 14, 361–372 (1992)
13. Nikseresht, M.R.: A parallel external memory system. Master's thesis, School of Computer Science, Carleton University (2007)

An FPGA-Based Accelerator for Multiple Biological Sequence Alignment with DIALIGN

Azzedine Boukerche[1], Jan Mendonca Correa[2],
Alba Cristina Magalhaes Alves de Melo[2], Ricardo Pezzuol Jacobi[2],
and Adson Ferreira Rocha[3]

[1] SITE-School of Information Technology and Engineering, University of Ottawa, Canada
[2] Department of Computer Science, University of Brasilia, Brazil
[3] Department of Electrical Engineering, University of Brasilia, Brazil
boukerch@site.uottawa.ca, {jan,albamm,rjacobi}@cic.unb.br,
adson@ene.unb.br

Abstract. Multiple sequence alignment (MSA) is a very important problem in Computational Biology since it is often used to identify evolutionary relationships among the organisms and predict secondary/tertiary structure. Since MSA is known to be a computationally challenging problem, many proposals were made to accelerate it either by using parallel processing or hardware accelerators. In this paper, we propose an FPGA based accelerator to execute the most compute intensive part of DIALIGN, an iterative method to obtain multiple sequence alignments. The experimental results collected in our 200-element FPGA prototype show that a speedup of 383.41 was obtained when compared with the software implementation.

1 Introduction

In the last decade, genome projects have produced a very huge amount of biological data. In order to better understand newly sequenced organisms, biologists compare their sequences against other organisms contained in genomic databases, in order to infer properties. Nowadays, this comparison is done millions of times a day, all over the world.

Sequence alignment (or sequence comparison) is in fact a problem of finding an approximate pattern matching between the sequences [21]. It can involve only two sequences (pairwise alignment) or more than two sequences (multiple sequence alignment) [9]. In a multiple sequence alignment (MSA), similar residues among a set of *nseq* sequences are aligned together. Usually, sequences compared with MSA are known to be biologically related and the goal is to obtain conserved subpatterns [5].

MSAs are often scored with the sum-of-pairs (SP) objective function [4] and the exact SP MSA problem is known to be NP-complete [25]. Therefore, heuristic methods are usually used to solve this problem, even when the number of sequences is small.

In general, an MSA problem can be solved with progressive or iterative methods [15]. Progressive methods are executed in three steps. First, the NW algorithm [17] is

S. Aluru et al. (Eds.): HiPC 2007, LNCS 4873, pp. 71–82, 2007.

used to perform pairwise alignments with all sequences. After that, a phylogenetic tree is constructed with the information obtained in phase 1 and, finally, the tree is used to guide the alignment of the sequences sequentially, from the most closely related to the less related ones. CLUSTALW [24] and T-COFFEE [16] and examples of progressive MSA methods. Iterative methods also use dynamic programming but, unlike the progressive methods, iterative methods periodically evaluate the quality of the scores produced and realign subgroups of already aligned sequences. PRRP [8] and DIALIGN [14] are examples of iterative methods.

Many dedicated architectures [2,10,12,18] and parallel applications [3] have been proposed to tackle pairwise sequence alignment by accelerating the dynamic programming matrix computation. Fewer examples [11,19] do exist that accelerate MSA algorithms. In this case, the hardware is not used to execute the whole algorithm, but only the most compute intensive part of it. [19] and [11] proposed an FPGA-based accelerator to execute the first phase of CLUSTALW [24], that executes pairwise sequence comparisons among all the sequences.

In this article, we present and evaluate an FPGA-based architecture to execute the most compute intensive part of the DIALIGN algorithm for multiple sequence alignment. Our architecture is designed as a systolic array which is able to compare sequences of any size using the DIALIGN recurrence relations [14]. As far as we know, this is the first hardware-based approach to execute DIALIGN.

The results obtained on a 200-element prototype synthesized for the FPGA Altera Stratix 2 EP2S180F1508I4 show that a speedup of 383.41 is achieved when comparing real DNA sequences of size 194439 bp (*base pairs*) and 169786 bp, respectively. In this case, the software implementation took 3 hours and 4 minutes and our FPGA implementation took 28.839 seconds.

The rest of this paper is organized as follows. Section 2 describes the MSA problem and the DIALIGN algorithm to solve it. In Section 3, related work in the area of FPGA architectures for sequence alignment is discussed. Section 4 describes our FPGA-based architecture. Some results are discussed in section 5. Section 6 concludes the paper.

2 Biological Sequence Comparison with DIALIGN

2.1 The Sequence Alignment Problem

To compare two sequences, we need to find the best alignment between them, which is to place one sequence above the other making clear the correspondence between similar characters from the sequences [21]. We define *alignment* as the insertion of spaces in arbitrary locations along the sequences so that they finish with the same size.

Given a pairwise alignment between two sequences *s* and *t*, an score can be associated for it as follows. For each two bases in the same column, we associate, for instance, *+1* if the two characters are identical (*match*), *-1* if the characters are different (*mismatch*) and *–2* if one of them is a space (gap). The score is the sum of the values computed for each column. The maximal score is called the similarity between the sequences.

One of the first exact methods to globally compare two sequences was NW [17]. It is based on dynamic programming and calculates a similarity matrix of size *m* x *n*, where *m* and *n* are the sizes of the sequences. NW has time and space complexity O(*mn*). The NW algorithm was modified to deal with local alignments (SW)[22]. An algorithm based on SW that uses an affine gap function is proposed in [7].

```
 G    A    -    C    G    G    A    T    T    A    G
 G    T    -    C    G    G    -    T    T    A    -
 G    A    T    C    G    G    A    A    T    A    G
--------------------------------------------------------
+3   -1   -6   +3   +3   +3   -3   -1   +3   +3   -3   = 4
```

Fig. 1. Alignment of sequences s, v and t, with the SP score for each column

An MSA involves more than 2 sequences. In this case, the scoring function to be used is not straightforward. Often, MSAs are scored with the *Sum-of-Pairs* (SP) function, where every pair of bases is scored with the pairwise scoring function and the score is the addition of all these values [9]. Figure 1 shows an example of a MSA and its score.

2.2 The DIALIGN Algorithm

DIALIGN (DIAGonal ALIGNment) [14] is a method for sequence alignment that can be either used to pairwise alignment or multiple sequence alignment. This method searches for *fragments* (or *diagonals*) that have no gaps and aligns them. In DIALIGN, a pairwise alignment is defined to be a chain of fragments [13].

When applied to the MSA problem, DIALIGN is executed in three phases. In the first phase, all pairwise alignments are computed, i.e., there are *nseq*(*nseq*-1)/2 chains of fragments, one for each pairwise alignment, where *nseq* is the number of sequences [13]. In the second phase, the diagonals that compose the pairwise alignments are sorted by their weight and the degree of overlap with other diagonals. This sorted list is used to obtain a multiple alignment with a greedy algorithm, generating alignment *Al*. In the last phase, the alignment *Al* is completed with an iterative procedure where the parts of the sequences that are not yet aligned with *Al* are realigned by executing phase 2 again, in such a way that consistent non-aligned diagonals are included in *Al* [14]. This phase is repeated until no diagonal with a positive weight can be included in *Al*.

Now, we will explain in detail the first phase, which is the core of this algorithm.

For each pairwise alignment, it is necessary to calculate the relevance of each diagonal found before attempting to align it [13]. This is done by $E(l,sm) = -ln(P(l,sm))$, where $P(l,sm)$ is the probability of a diagonal D of size l have at least sm matches.

For each candidate diagonal D_i, a weight $w(D_i)$ is assigned as $E(l,sm)$ if $E(l,sm)$ is above a given threshold T and 0, otherwise.

When the algorithm obtains a new significant diagonal, it tries to align it consistently with other previously calculated significant diagonals [14]. In an alignment of k diagonals $D_1, D_2, ..., D_k$ the total score S is given by the addition of all weights $w(D_i)$, $i=1$ to k.

To discover the score *S*, a dynamic programming based strategy is used. Consider two sequences *A* and *B*, having sizes *m* and *n*, respectively. For each pair (i,j), it will be determined all integers k with $k \leq \min(i,j)$ where the diagonal $(a_{i-k}b_{i-k},...,a_ib_j)$ beginning at position $(i-k,j-k)$ and ending in position (i,j) has a positive weight *w*. For each position (i,j) is defined a score(i,j) for the alignment in the prefixes $(a_1a_2...a_i)$ and $(b_1b_2...b_j)$.

The last diagonal D_k aligned in position (i,j) is recovered by function prec$(i,j)= D_k$ (formula 2). For each diagonal D_k aligned in position (i,j), *prec*(i,j) chooses the chain of diagonals with the greatest score so far. The score is calculated as in formula 1, where $\sigma(D_{i,j})$ is defined as the largest score chain of diagonals that ends in point (i,j).

$$\text{score}(i,j) = \max\{\text{score}(i-1,j),\ \text{score}(i,j-1),\ \sigma(D_{i,j})\} \quad (1)$$

$$prec(i,j) = \begin{cases} \text{prec}(i,j-1), \text{ If score}(i,j)=\text{score}(i,j-1) \\ \text{prec}(i-1,j), \text{ If score}(i,j-1) < \text{score}(i,j) = \text{score}(i-1,j) \\ D_{i,j}, \text{ If score}(i,j-1),\ \text{score}(i-1,j) < \text{score}(i,j) = \sigma(D_{i,j}) \end{cases} \quad (2)$$

Two dynamic programming matrices are calculated. One for scores (formula 1) and other for the preceding diagonal (*prec* in formula 2). Once these matrices are calculated, the reverse path on the *precs* matrix gives the alignment. One example of such alignment is given in figure 2. In figure 2(a), the subsequences belonging to diagonals are shown in gray and the aligned diagonals are shown as lines. Figure 2(b) shows the final alignment.

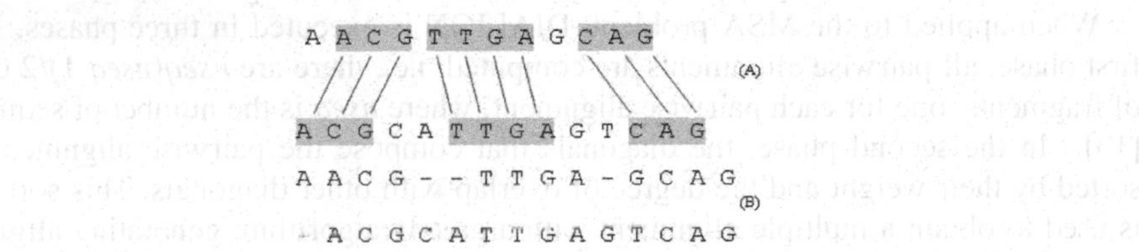

Fig. 2. Example of a pairwise DIALIGN alignment

DIALIGN-P [20] is a parallel version of DIALIGN that executes the first phase of the algorithm in parallel, with an strategy that tries to distribute evenly the pairs of sequences to be compared among the processors. An optimization called anchored alignment is introduced to reduce the execution time of each pairwise alignment. Nevertheless, this optimization potentially reduces the quality of the alignment produced [20]. Speedups of 19.32 were obtained in a 64-processor cluster, when comparing 20 sequences.

3 Related Work

There are many proposals in the literature of FPGA-based architectures to accelerate pairwise sequence alignment applications [2,10,12,18] by calculating the similarity matrix antidiagonals in hardware. In this approach, each element is capable of calculating one matrix score per turn. Thus, an N elements array can generate N scores at a time.

Figure 3 shows how each anti-diagonal of the dynamic programming matrix is calculated in parallel by a 5-element systolic array. The query sequence (ACGAT) is previously stored in the elements of the array and the database sequence (CTTAG) flows through the systolic array. Each element calculates one cell in the current anti-diagonal (shown in gray in figure 3) at the same time.

Most of the hardware solutions do not store the entire similarity matrix, obtaining only the similarity score [2]. Besides that, there is a limited number of computing elements that can be put in the systolic array. To deal with it, the smallest sequence being aligned is often stored on the computing elements as a query sequence. The other sequence can be of any size, since it "passes" through the FPGA (figure 3).

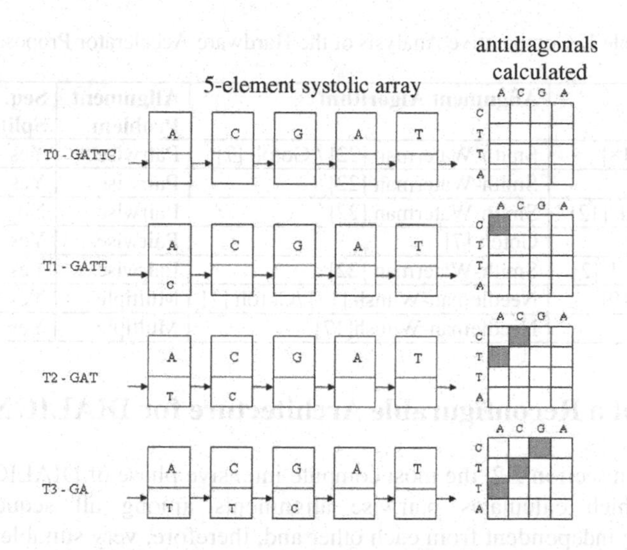

Fig. 3. Generic systolic array to calculate the similarity matrix

Frequently, it happens that even the query sequence is greater than the number of computing elements contained in the FPGA. In this case, a partitioning technique is used. To break query sequences, it is necessary to keep some scores onboard to allow new scores to be calculated. Some designs avoid this problem by putting many query bases on the same computing element. The drawback is that it requires more registers per element and thus decreases the maximum number of computing elements in the systolic array.

As an alternative, dynamic reconfiguration can be used. In this case, the first part of the query sequence is put directly in the processing elements using the dynamic reconfiguration capability. After that, the FPGA is reconfigured to contain the next part of the query sequence and the database sequence passes again through the FPGA. This procedure continues until the last part of the query sequence is processed.

A drawback of this approach is that reconfiguration time normally takes a few milliseconds.

Table 1 presents some hardware approaches to accelerate biological sequence comparison applications. Most of these accelerator proposals tackle the pairwise sequence comparison problem. The only ones that deal with multiple sequence alignment [11,19] accelerate the most compute intensive phase of Clustal-W [24], which does NW[17]-based pairwise sequence alignment among all sequences. Most of the proposals do query sequence splitting either by using reconfiguration or storing many characters at the same systolic cell. The speedups obtained range from 5.6 to 246.9 over the software implementation. Finally, all proposals implement in hardware variations of the NW or SW with constant gap functions [17,22] or affine gap functions [7]. As far as we know, there is no proposal of hardware accelerator for DIALIGN.

Table 1. Comparative Analysis of the Hardware Accelerator Proposals

Paper	Alignment Algorithm	Alignment Problem	Seq. Split	Speedup
Oliver et al. [18]	Smith-Waterman [22] / Gotoh [7]	Pairwise	Yes	170 / 125
Lavenier [10]	Smith-Waterman [22]	Pairwise	Yes	83
Marongui et al. [12]	Smith-Waterman [22]	Pairwise	No	5.6
Anish [1]	Gotoh [7]	Pairwise	Yes	170
Boukerche et al. [2]	Smith-Waterman [22]	Pairwise	Yes	246.9
Oliver et al. [19]	Needleman-Wunsh [17] /Gotoh [7]	Multiple	Yes	50.9
Lin et al. [11]	Needleman-Wunsh[17]	Multiple	Yes	34

4 Design of a Reconfigurable Architecture for DIALIGN

As discussed in section 2.2, the most compute intensive phase of DIALIGN is the first one [20], which calculates pairwise alignments among all sequences. These alignments are independent from each other and, therefore, very suitable for hardware parallelization.

As most of the previous works (section 3), we will parallelize the antidiagonal calculation of the dynamic programming matrix using a systolic array (figure 3). However, since the recurrence relations of DIALIGN (formulae 1 and 2) are different from the ones in NW and SW, an entirely distinct design must be made for each systolic element.

The goal of our architecture is to find the best DIALIGN score and its position. To do that, the following modifications were applied. First, we set *sm=l* in the probability calculation. Second, the *ln* logarithm (section 2.2) was replaced by a base 2 logarithm.

Our linear systolic array calculates the antidiagonals as shown in figure 4, using as a basis the generic systolic array (figure 3). In figure 4, the scores already calculated are shown in gray. The border between the gray and white part shows the antidiagonal being calculated. Diagonals greater than the threshold *T* are shown in black. For a diagonal that ends in position *(i,j)*, the architecture decides if it will be extended or ended and, in this case, whether it can be consistently aligned to other diagonal or not (section 2.2).

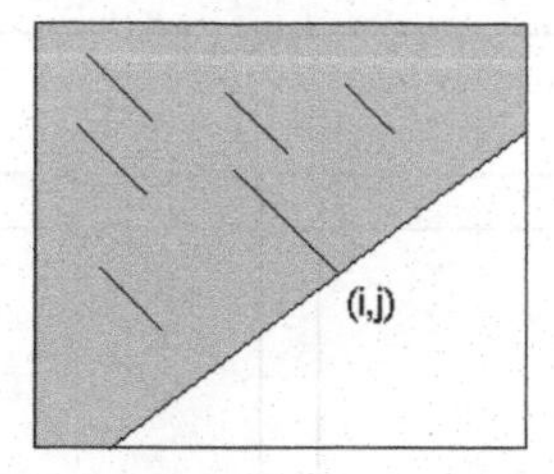

Fig. 4. Dynamic programming matrix calculation

An architecture that performs DIALIGN must contain formulae 1 and 2. To improve the performance, diagonal finding and diagonal alignment can be done simultaneously in the systolic vector. The algorithm for each systolic element is shown in figure 5.

```
calculate_recurrence_systolic (db_pair, prec(i, j-1), score(i, j-1), i)
begin
                    prec(i,j) = find_prec(prec,score);
 if (match(i,j))
  D(i,j)=extend_current_diagonal();
 else
                             if (w(D(i,j)) < T)
   discard (D(i,j));
  else
   if (consistent (D(i,j), prec(i-1,j), prec(i, j-1)))
    prec(i,j) = D(i,j);
    score(i,j) = σ(D(i,j));
   else
    if (w(D(i,j)) > prec(i-1,j) and w(D(i,j)) > prec(i,j-1))
     prec(i,j) = D(i,j);
                            score(i,j) = w(D(i,j));
    endif
   endif
  endif
 endif
 best_diagonal_systolic = prec(i,j);
 best_score_sytolic = score(i,j);
 send_to_next_systolic(prec(i,j),score(i,j),w(D(i,j)),flags);
end
```

Fig. 5. Algorithm executed in each systolic element

Figure 6 shows the systolic array diagram. The database sequence base pairs are input on the left side and the scores and their respective positions are output on the right side of the circuit. A handshake protocol is included to transfer scores and positions between the elements (blocks marked as “I” (input) and “O” (output)). Clk stands for the clock and Rst for reset signal. The DIALIGN recurrence relations are processed in the DAC (*Diagonal Alignment Circuit*) block.

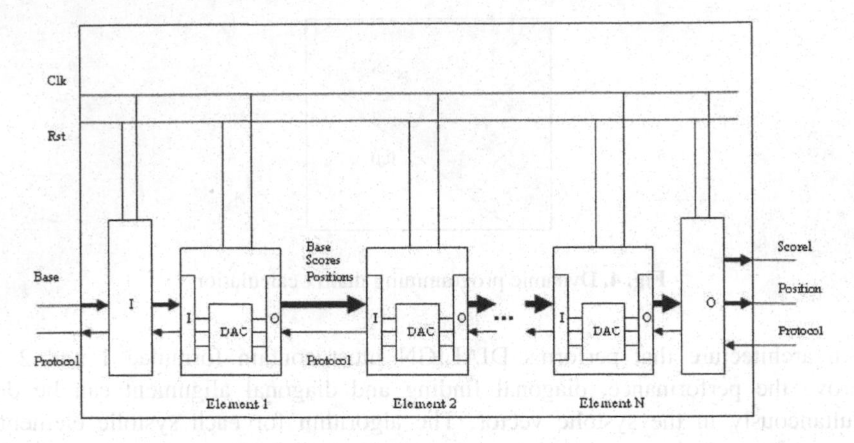

Fig. 6. Systolic Array Design

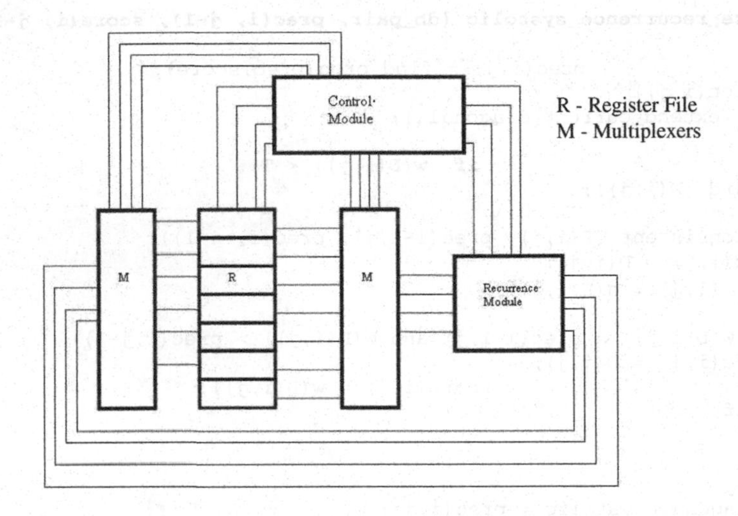

Fig. 7. Diagonal Alignment Circuit (DAC)

Figure 7 shows the DAC element. The register bank (R) contains values used in recurrence relations. They are selected by a network of multiplexers (M) to enter the "Recurrence Module". The results are stored in registers by another set of multiplexers. The control part is done by the "Control Module".

Figure 8 shows the recurrence module circuit. Inputs (from In1 to In15) and control lines (C1 to C9) for the multiplexers are on left side and outputs(Out1 to Out6) are on right side. This circuit is utilized many times to perform all relations. The adder (+), In15 and C9 are utilized to extend weights *w(D)* of current diagonal *D* by *1* when a match happens. In15, In14 and "+" are used to calculate the sum of scores $\sigma(D)$. The comparator "=" verifies if the bases are equal and whether some flag values are equal to zero or one (In5 to In8 depend on C4 and C5, giving Out3). The comparator

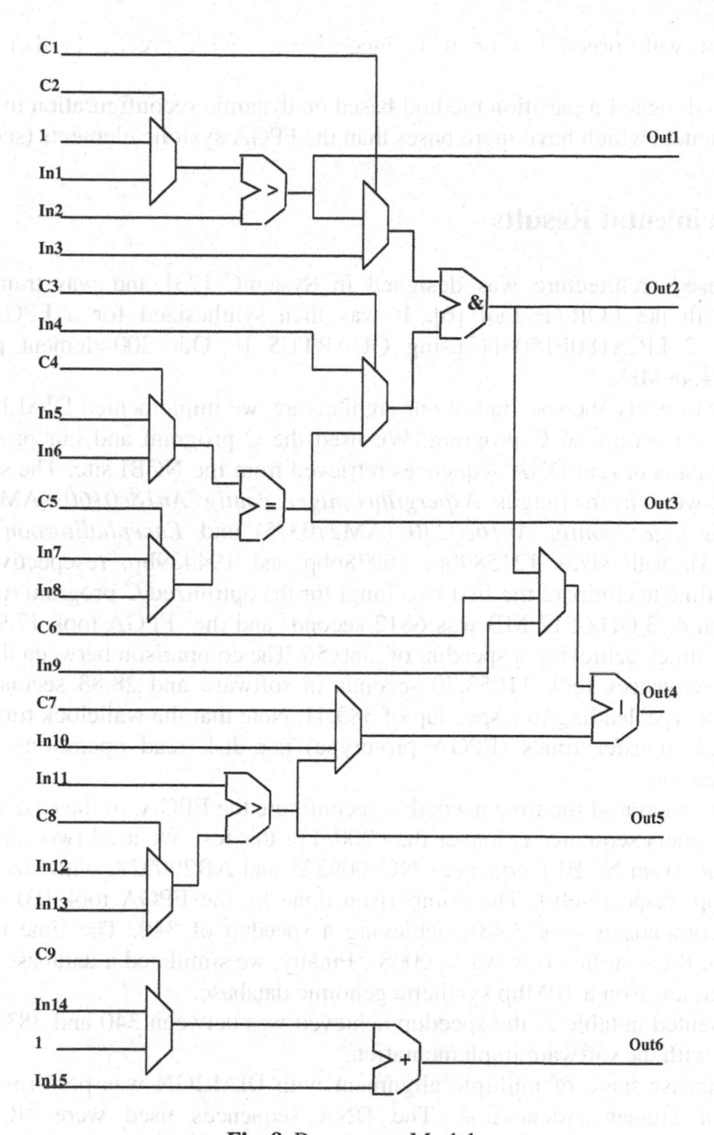

Fig. 8. Recurrence Module

">" decides if $w(D)$ is above T and is also used to find score(i,j) (formula 1). The recurrence relation in formula 2 is implemented by ">", "="and "&". The first line in formula 2 is computed by the "=" comparator. The second and third are translated to ">", "=" and "&" by the expression (score(i,j) > score(i,j-1) & (score(i,j)=score(i-1,j))).

To eliminate current diagonal $D_{i,j}$ if it is inconsistent, we must test if the ending position of the previously aligned diagonal is greater than the starting point of the current diagonal. To calculate this, an OR ("|") and two ">" are used. If $D_{i,j}$ is inconsistent with prec(i-1,j) or it is inconsistent with prec(i,j-1) then $D_{i,j}$ is inconsistent.

We also designed a partition method based on dynamic reconfiguration to compare query sequences which have more bases than the FPGA systolic elements (section 3).

5 Experimental Results

Our proposed architecture was designed in SystemC [23] and was translated to Verilog with the FORTE tool [6]. It was then synthesized for a FPGA Altera STRATIX 2 EP2S180F1508I4 using QUARTUS II. Our 200-element prototype works at 74.48 MHz.

In order to verify the speedup of our architecture, we implemented DIALIGN in C, generating an optimized C program. We used the C program and our prototype to compare 2 pairs of real DNA sequences retrieved from the NCBI site. The sequences compared were from fungus *Aspergillus niger contig An18c0160* (AM270408), *Aspergillus niger contig An16c0230* (AM270375) and *Encephalitozoon cuniculi* (AL590443), with sizes 121589bp, 169786bp and 194439bp, rescpectively. The wallclock time to compare the first two fungi for the optimized C program running on a Pentium 4 3 GHz 512 MB was 6812 seconds and the FPGA took 17.9 seconds (wallclock time), achieving a speedup of 380.56. The comparison between the second and third sequences took 11053.70 seconds in software and 28.83 seconds in our FPGA prototype, leading to a speedup of 383.41. Note that the wallclock times do not include data transfer times (FPGA prototype) nor disk read operations (software implementation).

Also, we measured the time needed to reconfigure the FPGA, in the case where the size of the query sequence is longer than 200. For this test, we used two variations of *annellovirus* from NCBI (sequences NC_009225 and AB290918, with sizes 3245bp and 3242bp, respectively). The comparison done by the FPGA took 0.01s and the software comparison took 3.48s, achieving a speedup of 348. The time needed to reconfigure the systolic array was 0.0008s. Finally, we simulated a database search of a 200bp sequence on a 10Mbp synthetic genomic database.

As presented in table 2, the speedup achieved was between 340 and 383.41, when compared with the software implementation.

The pairwise stage of multiple alignment with DIALIGN was performed with 4 variants of Human Adenovirus. The DNA sequences used were NC_004001, NC_001405, NC_002067 and NC_003266 with sizes of 34794bp, 35937bp, 35100bp and 35994bp respectively. Each cell in table 3 shows the time in seconds for a given pairwise alignment for both the FPGA architecture and the software implementation.

Table 2. Speedups Achieved by Our Architecture

Query Seq size	Database Seq size	Time FPGA (s)	Time software (s)	Speedup
169,786	194,439	28.83	11,053.70	383.41
121,589	169,786	17.9	6812.00	380.55
3245	3242	0.01	3.48	348.00
200	10,000,000	1.74	661.39	343.03

Table 3. Pairwise Aligment Times for the FPGA and the software implementation

	Time (s) FPGA / Software		
Sequences	**34794bp**	**35937bp**	**35100bp**
35937bp	1.09 / 415.31	---	---
35100bp	1.07 / 406.81	1.11 / 419.35	---
35994bp	1.10 / 416.33	1.14 / 431.42	1.11 / 420.12

In table 3, six comparisons were made. The total time for software alignment and the FPGA prototype were 2509.34 seconds and 6.62 seconds, respectively. The speedup achieved was 379.05.

The FPGA STRATIX 2 EP2S180F1508I4 has an estimated price of $10,688. Comparing against a Pentium 4 3 GHz costing $1000, the price/performance ratio is 10688/ 383.41 = 27.87 against 1000/1 for the Pentium. So the FPGA's price/performance ratio is 35.88 times lower than the Pentium.

6 Conclusions and Future Work

In this paper, we proposed and evaluated a new hardware architecture that performs multiple sequence alignment. Our architecture was designed to accelerate the pairwise step of DIALIGN that is the most compute intensive part of this algorithm for multiple sequence alignment. The proposed architecture was designed to handle large sequences by splitting the query sequence in blocks of 200. It was then successfully synthesized in an Altera FPGA STRATIX 2 EP2S180F1508I4.

As results for real DNA sequences of sizes 121 Kbp and 169 Kbp, we obtained a speedup of 383.41 against an optimized C implementation, indicating it can be very useful to accelerate the multiple sequence alignment problem. The speedups achieved with 3 very different sizes of sequences were between 343 and 383 and that indicates that the speedup achieved in not very dependent on the size of the sequences. As future work, we intend to integrate our architecture, which implements the first phase of DIALIGN, with a software algorithm that implements phases 2 and 3, leading to an integrated hardware/software approach. Also, we intend to investigate if the iterative phase of the algorithm (phase 3) can be implemented partially or fully in an FPGA.

References

1. Anish, A.: Hardware Accelerated Protein Identification, MsC Thesis, Univ. Toronto (2003)
2. Boukerche, A., et al.: Reconfigurable Architecture for Biological Sequence Comparison in Reduced Memory Space. In: IEEE IPDPS/NIDISC (2007)
3. Boukerche, A., et al.: Parallel Strategies for the Local Biological Sequence Alignment in a Cluster of Workstations. Journal of Parallel and Distributed Computing 67, 170–185 (2007)
4. Carrillo, H., Lipman, D.: The Multiple Sequence Alignment Problem. SIAM Journal of Applied Math. 48, 1073–1082 (1988)

5. Durbin, R., Eddy, S., Krogh, A., Mitchison, G.: Biological Sequence Analysis, p. 356. Cambridge Univ Press, Cambridge (1998)
6. Forte Design Systems, Cynthesizer User's Guide For Cynthesizer 2.4.0. (2005)
7. Gotoh, O.: An improved algorithm for matching biological sequences. Journal of Molecular Biology 162, 705–708 (1982)
8. Gotoh, O.: Significant Improvement in Accuracy of Multiple Protein Sequence Alignments by Iterative Refinements as Assessed by Reference to Structural Alignments. J. Mol. Biol. 264, 823–838 (1996)
9. Gusfield, D.: Algorithms on Strings, Trees and Sequences, p. 534. Cambridge Univ Press, Cambridge (1977)
10. Lavenier, D.: Speeding up genome computations with a systolic accelerator. SIAM news 31 (1998)
11. Lin, X., Peiheng, Z., Dongbo, B., Shengzhong, F., Ninghui, S.: To Accelerate Multiple Sequence Alignment using FPGAs. In: HPCASIA (2005)
12. Marongiu, A., Pallazari, P., Rosato, V.: Designing Hardware for Protein Sequence Alignment. Bioinformatics 19, 1739–1740 (2003)
13. Morgenstern, B., et al.: Multiple DNA and protein sequence alignment based on segment-to-segment comparison. In: Proc. Natl Acad. Sci., USA, pp. 12098–12103 (1996)
14. Morgenstern, B., et al.: DIALIGN: finding local similarities by multiple sequence alignment. Bioinformatics (1998)
15. Mount, D.: Bioinformatics: Sequence and Genome Analysis. C. S. Harbor Lab Press (2004)
16. Notredame, C., Higgins, D., Heringa, J.: T-Coffee: A novel method for fast and accurate multiple sequence alignment. J. Mol. Biol. 302, 205–217 (2002)
17. Needleman, S., Wunsch, C.: A general method applicable to the search for similarities in the amino acid sequence of two protein. J. Mol. Biol. 48, 443–453 (1970)
18. Oliver, T., Schmidt, B., Maskell, D.: Reconfigurable Architectures for Bio-sequence Database Scanning on FPGAs. IEEE Transactions on Circuits and Systems II 52(12), 851–855 (2005)
19. Oliver, T., Schmidt, B., Nathan, D., Clemens, R., Maskell, D.: Using Reconfigurable Hardware to Accelerate Multiple Sequence Alignment with ClustalW. Bioinformatics 21, 3431–3432 (2005)
20. Schmollinger, M., Nieselt, K., Kaufmann, M., Morgenstern, B.: DIALIGN P: Fast pair-wise and multiple sequence alignment using parallel processors BMC Bioinformatics (2004)
21. Setubal, J., Meidanis, J.: Introduction to Computational Molecular Biology. PWS Publishing Company, Boston (1997)
22. Smith, T., Waterman, M.: Identification of common molecular sub-sequences. J. Mol. Biology 147, 195–197 (1981)
23. Open SystemC Initiative Draft, Standard SystemC Language Reference Manual (2005)
24. Thompson, J., Higgins, D., Gibson, T.: Clustal W: improving the sensitivity of progressive multiple sequence alignment through sequence weighting. Nucleic Ac. Res. 22, 4673–4680 (1994)
25. Wang, T., Jiang, T.: On the Complexity of the Multiple Sequence Alignment. J. Comp. Biol. 1, 337–348 (1994)

A Speed-Area Optimization of Full Search Block Matching Hardware with Applications in High-Definition TVs (HDTV)

Avishek Saha and Santosh Ghosh

Department of Computer Science and Engineering,
IIT Kharagpur, WB, India, 721302
{avishek,santosh}@cse.iitkgp.ernet.in

Abstract. HDTV based applications require FSBM to maintain its significantly higher resolution than traditional broadcasting formats (NTSC, SECAM, PAL). This paper proposes some techniques to increase the speed and reduce the area requirements of an FSBM hardware. These techniques are based on modifications of the Sum-of-Absolute-Differences (SAD) computation and the MacroBlock (MB) searching strategy. The design of an FSBM architecture based on the proposed approaches has also been outlined. The highlight of the proposed architecture is its split pipelined design to facilitate parallel processing of macroblocks (MBs) in the initial stages. The proposed hardware has high throughput, low silicon area and compares favorably with other existing FPGA architectures.

1 Introuction

Rapid growth in digital video applications accompanied with the demand for better video quality has resulted in increasing popularity of high-definition TVs (HDTV) in the consumer market. An aspect of this trend is the increased interest in designing portable devices capable of encoding HD quality video data. However, the typical HD-compatible video encoders are based on MPEG2 MP@HL. MPEG2 MP@HL encoder uses the exhaustive Full Search Block Matching Algorithm (FSBMA) based motion estimation. In this case, the power consumption of the encoder is prohibitively high, particularly for portable implementations. Again, in a typical video encoder, the ME module occupies more than 80% of its computational complexity. Software based methods are unable to meet the real-time constraints of FSBM-ME implementations [1]. Hence, a highly efficient ME processor core is required to realize portable HD video encoding applications.

FSBM architectures can be broadly classified into FPGA [2,3,4,5,6,7] and ASIC [8,9,10,11,12,13,14,15,16,17,18] implementations. This work focusses on using FPGA technology to implement a high-performance ME hardware. A systolic array architecture for FSBM implementing realtime video encoding on a single FPGA chip was proposed in [3]. A novel OnLine Arithmetic (OLA) based design, where each bit is processed in successive clock cycles operating with the most significant bit (MSB) at first, was proposed in [4]. [5] proposed low-power core-based architectures for real-time motion estimation on FPGAs, that

S. Aluru et al. (Eds.): HiPC 2007, LNCS 4873, pp. 83–94, 2007.

are customizable for different coding parameters and hardware resources. Some FSBM hardware architectures proposed in [8] were implemented and their performance evaluated in [6]. The results show that, real-time motion estimation for CIF (352 × 288) sequences can be achieved with 2-D systolic arrays and moderate capacity (250 k gates) FPGA chip. Finally, [7] implements an adder-tree model based 16 × 1 SAD operation in FPGAs and also extends the 16 × 1 SAD implementation to perform the 16 × 16 SAD operations.

This paper proposes two approaches for speed-area optimization of the Full Search Block Matching Algorithm(FSBMA) hardware. The novelty of this work lies in the combined optimization of the mutually conflicting design parameters of high throughput and low silicon area. The first approach uses a modification of the SAD operation so as to reduce the overall computational complexity of the ME module. This modification reduces the number of operations that need to be performed for each SAD based block matching within a pre-defined search window. Subsequently, an MB scan technique has been proposed which takes advantage of the SAD modification in a manner so as to further enhance the performance of the hardware implementation. The proposed hardware design uses a pipelined architecture which reduces the processing cycle count for each MB and thus increases the overall throughput. The initial stages of the pipeline have been split to facilitate parallel processing of MBs.

The paper is organized is as follows. The next section provides a background on FSBM-based motion estimation. Section 3 describes in detail the SAD modifications and the MB search strategy. Based on the approaches proposed in Section 3, the design outline of an FSBM hardware has been sketched in Section 4. The hardware implementation results and it's comparison with existing FPGAs are presented in Section 5. Finally, Section 6 concludes this paper.

2 Full Search Block Matching

In video compression, motion-compensated prediction assumes that the pixels within the current picture can be modeled as a translation of those within a previous picture. This motion information is represented by two dimensional displacement vectors or motion vectors. Due to the block-based picture representation, many ME algorithms employ block-matching techniques. In such techniques, the motion vector is obtained by minimizing a cost function measuring the mismatch between a current MB and the reference MB. Several cost measures are available to measure the amount of distortion between the block in the current frame and candidate block in the reference frame, such as, mean-of-absolute-differences (MAD), sum-of-absolute-differences (SAD), mean-square-error (MSE) etc. SAD, the most commonly used matching criterion, between the pixels of the current MB $x(i,j)$ and the search region $y(i,j)$ can be expressed as,

$$SAD(u,v) = \sum_{i=0}^{N-1}\sum_{j=0}^{N-1} |x(i,j) - y(i+u, j+v)| \quad (1)$$

where, (u, v) is the displacement between these two blocks. Thus, each search requires N^2 absolute differences and $(N^2 - 1)$ additions.

To find the MB producing the minimum mismatch error, we need to calculate SAD at several locations within a search window. The simplest but the most computationally intensive search method, known as the FSBM, evaluates SAD at every possible pixel location in the search area. In FSBM-based motion estimation, each N×N macroblock of the current frame is compared with all candidate MBs in the (N+2p)×(N+2p) search window defined within the previously processed frame, where p is the maximum displacement of the N×N MB in all four directions around its boundary. The motion vector is determined by identifying a best matching MB. The FSBMA exhaustively evaluates all possible search locations and hence is optimal [19] in terms of reconstructed video quality and compression ratio. High computational requirements, regular processing scheme and simple control structures make the hardware implementation of FSBM a preferred choice.

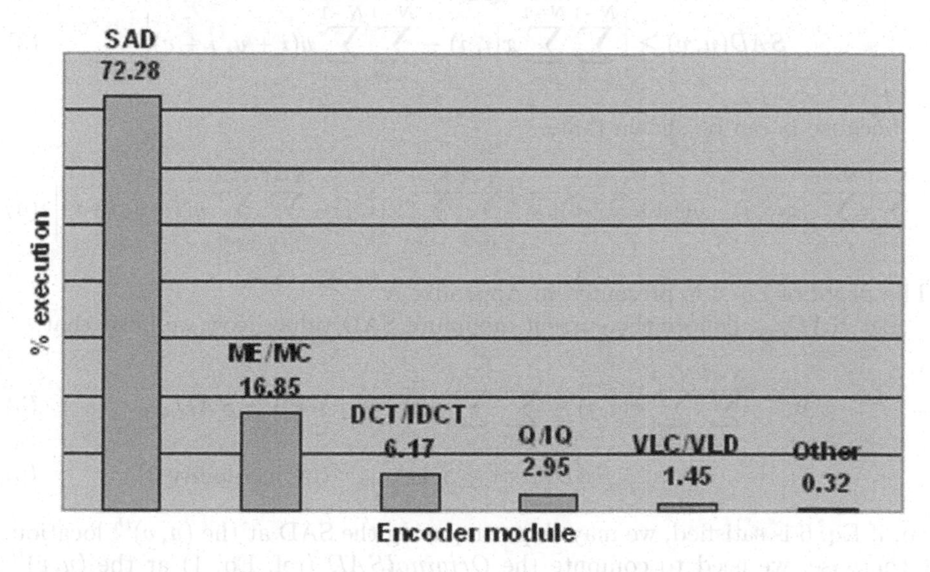

Fig. 1. Execution profile of a typical video encoder

Fig. 1 shows the execution profile of a standard video encoder, obtained using the GNU gprof tool. As can be seen, among the various afore-mentioned modules of a typical video encoder, the motion estimation is the most computationally expensive. Furthermore, it is to be noted that, the SAD computations are the most time consuming due to the complex nature of the absolute operation and the subsequent multitude of additions.

3 Proposed Approaches

This section gives a detailed description of the speed-optimized architecture. The first subsection explains the modification of the SAD equation. The MB searching technique adopted to facilitate the SAD sum derivation in Subsection 3.1 has been presented in Subsection 3.2.

3.1 Modified SAD Based Fast Block Matching

In this section, we try to modify the SAD computation so as to constrain the computational complexity of the FSBM search process, while preserving the optimal solution for the motion vector. Let us again consider the SAD Eq. 1,

$$SAD(u,v) = \sum_{i=0}^{N-1}\sum_{j=0}^{N-1} |x(i,j) - y(i+u, j+v)| \quad (2)$$

The above equation can be re-written as,

$$SAD(u,v) \geq \left| \sum_{i=0}^{N-1}\sum_{j=0}^{N-1} x(i,j) - \sum_{i=0}^{N-1}\sum_{j=0}^{N-1} y(i+u, j+v) \right| \quad (3)$$

because it can be shown that,

$$\sum_{i=0}^{N-1}\sum_{j=0}^{N-1} |x(i,j)-y(i+u,j+v)| \geq \left| \sum_{i=0}^{N-1}\sum_{j=0}^{N-1} x(i,j) - \sum_{j=0}^{N-1}\sum_{i=0}^{N-1} y(i+u,j+v) \right| \quad (4)$$

The proof of Eq. 4 is presented in Appendix A.

Let SAD_{min} denote the current minimum SAD value. Now we posit that,

$$\text{if, } \left| \sum_{i=0}^{N-1}\sum_{j=0}^{N-1} x(i,j) - \sum_{i=0}^{N-1}\sum_{j=0}^{N-1} y(i+u, j+v) \right| \geq SAD_{min} \quad (5)$$

$$\text{then, } \quad SAD(u,v) \geq SAD_{min} \quad \text{(by inequality 3)} \quad (6)$$

So, if Eq. 6 is satisfied, we may skip computing the SAD at the $(u, v)^{th}$ location. Otherwise, we need to compute the *OriginalSAD* (ref. Eq. 1) at the $(u, v)^{th}$ location and compare it with SAD_{min}. The initial SAD_{min} can be obtained by calculating the *OriginalSAD* for the first search location only. Thereafter, Eq. 6 can be used to decide on whether or not to peform the *OriginalSAD* on a particular search location. If the *OriginalSAD* needs to be calculated for some particular search location and the newly obtained *OriginalSAD* is less than the exisiting SAD_{min}, then the *OriginalSAD* is set as the new SAD_{min}. At this point, it is to be noted that, this approach is not an approximation and always finds the minimum SAD without making any compromise on compression ratio and/or video quality. This is because the algorithm tries to take an initial decision of whether to compute the *OriginalSAD*. The decision is based on comparison with SAD_{min}. Again, all SAD_{min} values are obtained after $OriginalSAD$ calculations only. Thus, no decisions are made based on approximate computations and the video quality with this SAD modification is same as that of FSBM with $OriginalSAD$ for all search locations.

Again, the right hand term of Eq. 4 can be expressed as,

$$\left|\sum_{i=0}^{N-1}\sum_{j=0}^{N-1} x(i,j) - \sum_{i=0}^{N-1}\sum_{j=0}^{N-1} y(i+u,j+v)\right| = |X - Y(u,v)| \quad (7)$$

where, $X = \sum_{i=0}^{N-1}\sum_{j=0}^{N-1} x(i,j)$ and $Y(u,v) = \sum_{i=0}^{N-1}\sum_{j=0}^{N-1} y(i+u,j+v)$, i.e., X is the sum of the intensity values of the pixels in the current MB of the current frame and $Y(u,v)$ is the sum of the pixel intensities in the $(u,v)^{th}$ MB location in the search region of the previous frame. It is to be noted that, for an entire search region the sum X for the current MB has to be calculated only once. For each search location, the sum $Y(u,v)$ needs to be calculated. Moreover, the sum $Y(u,v)$ at the $(u,v)^{th}$ location can be derived from its immediately previous value $Y(u-1,v)$ at $(u-1,v)^{th}$ by subtracting from $Y(u-1,v)$ the sum of pixel intensities of the first column at the $(u-1,v)^{th}$ MB location and adding to the result, the summation of the pixel values at the last column of the $(u,v)^{th}$ MB location.

3.2 Macro Block Searching Strategy

The FSBM algorithm primarily searches an $N \times N$ macroblock within the corresponding $(2p+1)\times(2p+1)$ search locations, where p is the search range. The traditional FSBMA requires N^2 absolute differences and (N^2-1) additions to compute every SAD value. Hence, the total operations required to find the best match of an MB within a search range is $(2p+1)^2 \times (2N^2-1)$. However, our modified SAD equation requires only (N^2-1) additions for the current MB + (N^2-1) additions and 1 absolute difference for each of the $(2p+1)^2$ search locations = $(N^2-1) + (N^2-1+1)(2p+1)^2$ = a total of $(N^2-1) + N^2(2p+1)^2$ operations.

Let, the search locations in the search region be scanned in a manner shown in Fig. 2. As mentioned in subsection 3.1, the sum of the pixel intensities at each search location can be derived from the pixel intensity sum at the previous search location. Compared to the traditional raster scan, the proposed scan technique facilitates this derivation of the SAD sums, particularly in situations where the search locations moves to a row below the current row position. As shown in Fig. 2, the sum at search location $(2,2p+1)$ can be easily derived from the sum at search location $(1,2p+1)$. However, this derivation is not possible if we compute the sum at location $(2,1)$ immediately after computing the sum at location $(1,2p+1)$.

Let, the kth search location is represented by SR^k and it's right and bottom adjacent search locations are represented by SR^{k+r} and SR^{k+b} then the SAD^{k+r} and SAD^{k+b} can be calculated by following equations.

$$SAD^{k+r} = \left|\{SR^k + \left(\sum_{i=0}^{N-1} SR_{i,j+N} - \sum_{i=0}^{N-1} SR_{i,j}\right) - MB^c\right| \quad (8)$$

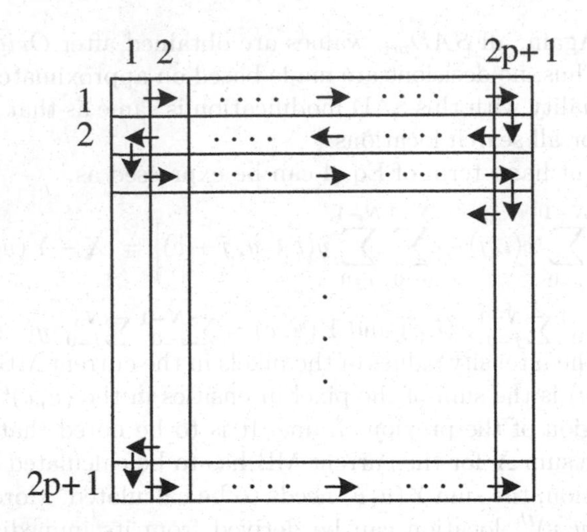

Fig. 2. Movement of search locations in the search region

$$SAD^{k+b} = \left| \{SR^k + \left(\sum_{j=0}^{N-1} SR_{i+N,j} - \sum_{j=0}^{N-1} SR_{i,j} \right) - MB^c \right| \quad (9)$$

where SR^k and MB^c represents the sum of the pixel values of kth search location and current (c^{th}) MB respectively. Eq. 8 has been used to derive the SAD sums when the scan control moves towards right or left in a column-wise manner and Eq. 9 has been used to derive the SAD sums when the scan control moves downward in a row-wise manner.

Example: *The sum of the second search location (SL_2) can be derived from the sum of the first search location (SL_1) by subtracting from SL_2, the sum of the pixel values of the 1st column and then adding to it, the sum of the pixel values of the 17th column. Again, to derive SL_{34} (assume $p = 16$) from it's previous sum at the 33rd search location (SL_{33}), we need to subtract the sum of the pixel values of the 1st row from SL_{33} and then add to it the sum of the pixel values of the 17th row.*

Each derivation of the SAD sum requires $2(N-1)$ additions [to find the sum of one old and one new column] + 2 additions/subtractions and 1 AD operation [Eq. 8 and Eq. 9] = a total of $(2(N-1)+2)$ adds/subs and 1 AD = $2N$ operations and 1 AD. Thus, an entire search region of size $(2p+1) \times (2p+1)$ requires (N^2-1) operations and 1 AD for the first search location + $(2N+1)$ operations and 1 AD for the remaining $(2p+1)^2-1$ search locations each = $[(N^2-1) + [(2p+1)^2-1]2N]$ operations and $(2p+1)^2$ ADs. For $N = 16$, $p = 16$, the proposed technique requires only 35071 addition/subtraction operations and 1089 ADs, as against the traditional raster search scan, which requires 277695 addition/subtraction operations and 278784 AD operations.

4 Hardware Design for FSBM

Fig. 3 shows the hardware architecture of the SAD calculation unit. The hardware unit consist of one pipelined datapath, two memory banks, datapath and memory controller and some registers. The modified SAD calculation for FSBM algorithm is performed by the datapath in eight independent sequential steps.

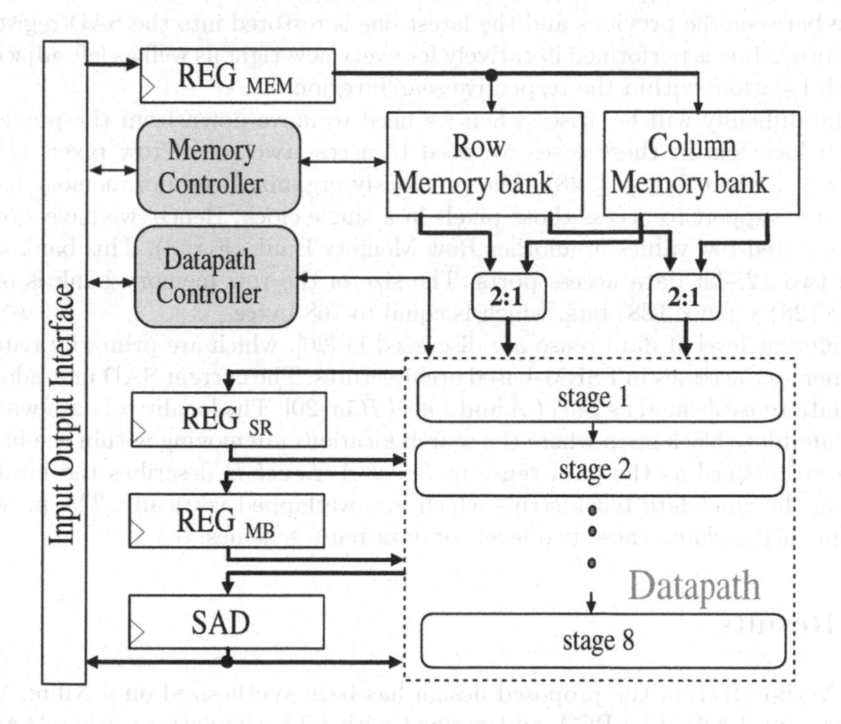

Fig. 3. Architecture of the proposed SAD unit

The proposed hardware adopts the scanning technique shown in Fig. 2. A $p = 16$ search region consist of 48×48 pixels ($P_{i,j}$, where $0 \leq i, j \leq 48$), which are formed $(2p+1)^2 = 33^2 = 1089$ different search locations. The SAD unit first loads one macro block and the respective search region into the on-chip memory. The memory controller is responsible to store the pixels into the right place by following some special organization procedure. The pixels are organized into the memory banks in such a way that the consecutive SAD calculation could be performed by only one memory access. The pixels are stored into the *Column Memory Bank* (Fig. 3) in column-major format so that one column of a search location (16 pixels) can be accessed in a single clock.

The SAD unit first computes the sum of the macro-block and the sum of the first search location and stores the resultant values into the respective REG_{MB} and REG_{SR} registers (Fig. 3). It computes the first SAD value by performing an absolute difference operation between REG_{SR} and REG_{MB} and store it into the SAD-register. It is to be explained in the previous section that the next right-adjacent search location has only one column difference from the previous location. Hence, to compute the sum of the new search location from the previous REG_{SR} value we need to access all the pixels of 1st and 17th columns ($P_{i,1}$ and $P_{i,17}$, where $0 \leq i \leq 48$). Thus our column memory bank have two $16 \times 8 = 128$-bit ports. The second SAD value is computed by the absolute difference operation between new REG_{SR} and the respective REG_{MB} values. Then the least SAD value between the previous and the latest one is restored into the SAD-register. This procedure is performed iteratively for every new right as well as left adjacent search locations within the respective search region.

The difficulty will be arises when we need to move down from the previous search location. In these cases we need to access two set of row pixels ($P_{i1,j}$ and $P_{i2,j}$, where $0 \leq j \leq 48$). The previously organized column memory bank does not support to access those pixels in a single clock. Hence, we have stored the required row values in another Row Memory Bank (Fig. 3). This bank also have two 128-bit data access ports. The size of the row memory bank is only $(32 \times 128) + (16 \times 128)$ bits, which is equal to 768 bytes.

Different level of data reuse are discussed in [20], which are primarily reduce the memory accesses in FSBM-based architectures. The current SAD unit adopts the data reuse defined as *Level A* and *Level B* in [20]. The locality of data within the candidate block strip where the search locations are moving within the block strip are defined as the data reuse in *Level A*. *Level B* describes the locality among the candidate block strips which are overlapped vertically. The present design easily adopts these two levels of data reuse schemes.

5 Results

The Verilog RTL of the proposed design has been synthesized on a Xilinx Virtex IV 4vlx100ff1513 FPGA and verified with RTL simulations using Mentor Graphics ModelSim SE. The synthesis results for a macroblock (MB) of size 16×16 and a search range of $p = 16$ show that the design can achieve a highest frequency of 221.322 MHz. In addition, the design requires 333 CLB Slices, 416 DFFs/Latches and a total of 278 input/output pins. The area required by the implementation is 380 look-up tables (LUTs). It is to be noted that, given the high operating frequency of our architecture, the area required by this design is substantially low. The modification of the SAD operation contributes to this high speed and small area and low hardware complexity. The use of memory banks has led to higher on-chip bandwidth. However, this has also led to the only drawback of our design, which is the high number of input/output pins.

The first SAD result is generated by the SAD unit after 23 clock cycles. Thereafter, every successive clock cycle generates one SAD value. For a search range of $p = 16$, which has $(2p + 1)^2 = 1089$ search locations, the number of cycles required by the proposed hardware to find the best matching block is, 23 (for the first search location) + (1089-1) (for the remaining search locations) = 1111 cycles. Thus, our proposed FPGA implementation processes a 16×16 MB in, 1111 clock cycles per MB * 4.52 ns per clock cycle = 5.022 usec. Similarly, a 720p HDTV frame of dimension 1280×720 can be processed in, 3600 MBs per frame * 5.022 usecs per MB = 18.078 msec. At this speed, 55.33 number of 720p HDTV frames can be processed by the proposed hardware every second. Thus, the number of frames processed per second by our design is much higher than other existing architectures, which is evident from Table 1. Modification of the SAD computation, the proposed MB search strategy and the split-pipeline design contributes to this high speed and throughput of our proposed hardware design.

Table 1. Comparison of hardware performance with N=16 and p=16

Design	Frequency (in MHz)	CLBs (in slices)	HDTV 720p (fps)	Throughput (MBs/sec)	Throughput/Area
Loukil et al. [2] (Altera Stratix)	103.8	1654	3.4	12237.7	7.4
Mohammad et al. [3] (Xilinx Virtex II)	191.0	300	2.09	7536.3	25.1
Olivares et al. [4] (Xilinx Spartan3)	366.8	2296	3.71	13347.4	5.8
Roma et al. [5] (Xilinx XCV3200E)	76.1	29430	7.55	27178.6	0.92
Ryszko (AB2) et al. [6] (Xilinx XC40250)	30.0	948	5.26	18939.4	11.9
Wong et al. [7] (Altera Flex20KE)	197.0	1699	1.2	4307.1	2.5
Our (Xilinx Virtex IV)	221.322	333	55.33	199209.7	598.2

Table 1 compares the performance of various existing architectures for a 16×16 MB with a search range of $p = 16$. This paper aims toward the combined speed-area optimization of FSBM hardware. Hence, a new performance criteria of throughput/area has been used to compare the speed-area optimized performance of different architectures. High speed-area optimization of an architecture is denoted by its high values of the throughput/area parameter. The architectures have been compared in terms of (a) operating frequency, (b) CLB slices, (c) number of HDTV 720p (1280x720) frames that can be processed per second, (d) throughput or MBs processed per second, (e) throughput/area, and (f) the I/O bandwidth. As can be seen, the proposed design has a very high throughput and can process the maximum number of HDTV 720p frames per second (fps). The fps value of 55.33 is close to that of 60 fps, which denotes that the proposed architecture can support both frame (25fps or 30 fps) and field (50 fps or 60 fps) processing. This is a big advantage over other existing FPGA designs. Moreover, the superior speed-area optimization in the proposed design is exhibited by its substantially high throughput/area value of 598.2.

As can be seen, in Table 1, the different implementations have been carried out on different platforms having varying perfomance levels. Xilinx Virtex IV, designed for higher performance as compared to already exisitng FPGAs, can inherently make implemented designs faster. To overcome this drawback in comparison results of Table 1, Table 2 makes a comparison of the proposed design on different FPGA implementation platforms of Xilinx, namely, Spartan3, Virtex II and Virtex IV.

Table 2. Performance comparison of our hardware on different FPGA platforms with N=16 and p=16

Platform	Frequency (in MHz)	CLBs (in slices)	HDTV 720p (fps)	Throughput (MBs/sec)	Throughput/Area
Xilinx Virtex II	147	333	36.75	132300	397.3
Xilinx Spartan 3	121	333	30.25	108900	327.03
Xilinx Virtex IV	221.322	333	55.33	199209.7	598.2

Table 2 shows that the area requirements of our design are similar in Spartan3, Virtex II and Virtex IV. However, the MBs processed per second is different for different platorms with Virtex IV resulting in the highest throughput. Hence, among the three compared platforms, Virtex IV yields the best throughput/area ratio. It is to be noted that, Mohammad et al. [3], whose design was also implemented on Virtex II, has much lesser throughput/area value of 25.1, as compared to our Virtex II implementation with a throughput/area value of 397.3. Similarly, the Spartan3 implementation by Olivares et al. [4] has a throughput/area value of only 5.8. This is substantially lesser than our Spartan3 implementation value of 327.03.

6 Conclusions

This paper has presented some approaches toward throughput-area optimization of FSBM architectures. The first approach proposed a modification of the SAD computation. This modification reduced the total number of addition/subtraction operations involved in macroblock-matching within a pre-defined search window. In addition, this approach has been utilized to derive the SAD sum at the current MB location from the already computed SAD sum at the previous MB location. Finally, an FPGA hardware design to implement the proposed approaches has been outlined. The highlight of this design is the initial splitting of its pipeline to facilitate parallel processing of MBs. In addition, our hardware has used the proposed MB scan technique so as to take further advantage of the SAD modification. Experimental results demonstrate the higher throughput and smaller area requirements of our design when compared to other existing FPGA architectures.

References

1. Ghanbari, M.: Standard Codecs: Image Compression to Advanced Video Coding. IEE (2003)
2. Loukil, H., Ghozzi, F., Samet, A., Ben Ayed, M., Masmoudi, N.: Hardware implementation of block matching algorithm with fpga technology. In: Proc. Intl. Conf. on Microelectronics, pp. 542–546 (2004)
3. Mohammadzadeh, M., Eshghi, M., Azadfar, M.: Parameterizable implementation of full search block matching algorithm using fpga for real-time applications. In: Proc. 5th IEEE Intl. Caracas Conf. on Dev., Circ. and Sys., Dominican Republic, pp. 200–203 (2004)
4. Olivares, J., Hormigo, J., Villalba, J., Benavides, I., Zapata, E.: Sad computation based on online arithmetic for motion estimation. Jrnl. Microproc. and Microsys. 30, 250–258 (2006)
5. Roma, N., Dias, T., Sousa, L.: Customisable core-based architectures for real-time motion estimation on fpgas. In: Proc. of 3rd Intl. Conf. on Field Prog. Logic and Appl., pp. 745–754 (2003)
6. Ryszko, A., Wiatr, K.: An assesment of fpga suitability for implementation of real-time motion estimation. In: Proc. IEEE Euromicro Symp. on DSD, pp. 364–367 (2001)
7. Wong, S., S., V., Cotofona, S.: A sum of absolute differences implementation in fpga hardware. In: Proc. 28th Euromicro Conf., pp. 183–188 (2002)
8. Komarek, T., Pirsch, P.: Array archtectures for block matching algorithms. IEEE Circ. and Sys. 36(10), 1301–1308 (1989)
9. Vos, L., Stegherr, M.: Parameterizable vlsi architectures for the full- search block-matching algorithm. IEEE Circ. and Sys. 36(10), 1309–1316 (1989)
10. Yang, K., Sun, M., Wu, L.: A family of vlsi designs for the motion compensation block-matching algorithm. IEEE Circ. and Sys. 36(10), 1317–1325 (1989)
11. Hsieh, C., Lin, T.: Vlsi architecture for block-matching motion estimation algorithm. IEEE Tran. Circ. and Sys. Video Tech. 2(2), 169–175 (1992)
12. Jehng, Y., Chen, L., Chiueh, T.: Efficient and simple vlsi tree architecture for motion estimation algorithms. IEEE Tran. Sig. Pro. 41(2), 889–899 (1993)
13. Yeo, H., Hu, Y.: A novel modular systolic array architecture for full-search block-matching motion estimation. In: Proc. Intl. Conf. on Acou. Speech, and Sig. Proc., vol. 5, pp. 3303–3306 (1995)
14. Lai, Y., Chen, L.: A data-interlacing architecture with two-dimensional data-reuse for full-search block-matching algorithm. IEEE Tran. Circ. and Sys. Video Tech. 8(2), 124–127 (1998)
15. Yeh, Y., Lee, C.: Cost-effective vlsi architectures and buffer. size optimization for full-search block matching algorithms. IEEE Tran. VLSI Sys. 7(3), 345–358 (1999)
16. Sousa, L., Roma, N.: Low-power array architectures for motion estimation. In: IEEE 3rd Workshop on Mult. Sig. Proc., pp. 679–684 (1999)
17. Do, V., Yun, K.: A low-power vlsi architecture for full-search block-matching. IEEE Tran. Circ. and Sys. Video Tech. 8(4), 393–398 (1998)
18. Lin, S., Tseng, P., Chen, L.: Low-power parallel tree architecture for full search block-matching motion estimation. In: Proc. of Intl. Symp. Circ. and Sys., vol. 2, pp. 313–316 (2004)
19. Salomon, D.: Data Compression: The Complete Reference, 3rd edn. Springer, New York (2004)

20. Tuan, J., Jen, C.: An architecture of full-search block matching for minimum memory bandwidth requirement. In: Proceedings of the IEEE GLSVLSI, pp. 152–156 (1998)
21. Weblink: Famous equations and inequalities. `http://www.math.utah.edu/pa/math/equations/equations.html` (2006)
22. Efimov, A., Zolotarev, Y., Terpigoreva, V.: Mathematical Analysis (Advanced Topics). Mir Publishers, Moscow (1985)

A Proof of Eq. 4

Lemma 1

$$\left|\ ||\boldsymbol{A}||_1 - ||\boldsymbol{B}||_1\ \right| \leq ||\boldsymbol{A}-\boldsymbol{B}||_1, \quad \textit{where,} \quad ||\boldsymbol{A}||_1 = \sum_k |a_k|$$

Proof. We know that, by triangle inequality [21] and reverse triangle inequality [21],

$$|\ \mathbf{a}+\mathbf{b}\ | \leq |\mathbf{a}| + |\mathbf{b}| \tag{10}$$

$$|\ \mathbf{a}-\mathbf{b}\ | \geq |\mathbf{a}| - |\mathbf{b}| \tag{11}$$

Again, by Minkowski's inequality [22],

$$|\ ||\mathbf{A}+\mathbf{B}||_1\ | \leq ||\mathbf{A}||_1 + ||\mathbf{B}||_1 \tag{12}$$

$$\text{Let,} \quad ||\mathbf{A}||_1 = \sum |a| = \sum |b+(a-b)| \leq \sum |b| + \sum |a-b| \quad \textbf{[by Eq. 12]}$$

$$\text{or,} \sum |a| \leq \sum |b| + \sum |a-b| \quad \text{or,} \sum |a| - \sum |b| \leq \sum |a-b|$$

$$\text{which implies,} \quad ||\mathbf{A}||_1 - ||\mathbf{B}||_1 \leq ||\mathbf{A}-\mathbf{B}||_1 \tag{13}$$

$$\text{Analogously, we can show that,} \quad ||\mathbf{B}||_1 - ||\mathbf{A}||_1 \leq ||\mathbf{A}-\mathbf{B}||_1 \tag{14}$$

Hence, from Eq. 13 and Eq. 14, we have,

$$||\mathbf{A}||_1 - ||\mathbf{B}||_1 \leq ||\mathbf{A}-\mathbf{B}||_1,$$

$$\text{and} \quad ||\mathbf{B}||_1 - ||\mathbf{A}||_1 \leq ||\mathbf{A}-\mathbf{B}||_1, \quad \text{i.e.,} ||\mathbf{A}||_1 - ||\mathbf{B}||_1 \geq -||\mathbf{A}-\mathbf{B}||_1$$

$$\text{which gives,} \quad \left|\ ||\mathbf{A}||_1 - ||\mathbf{B}||_1\ \right| \leq ||\mathbf{A}-\mathbf{B}||_1 \tag{15}$$

Hence, the result follows. □

Evaluating ISA Support and Hardware Support for Recursive Data Layouts

Won-Taek Lim and Mithuna Thottethodi

School of Electrical and Computer Engineering, Purdue University
{wlim,mithuna}@purdue.edu

Abstract. Recursive data layouts for matrices (two dimensional arrays) have been proposed to ameliorate the poor data locality caused by traditional layouts like row-major and column-major [3][12]. However, recursive data layouts require non-traditional address computation which involves bit-level manipulations that are not supported in current processors. As such, a number of software-based address computation techniques have been developed ranging from table-lookup based techniques to arithmetic-and-logic-operation based techniques. This effectively creates a tradeoff of extra computation for locality. In this paper, we design the appropriate instruction set architecture (ISA) support and hardware support to achieve address computation for recursive data layouts. Our technique captures the benefits of locality of the sophisticated data layouts while avoiding the cost of software-based address computation. Simulations reveal that our hardware approach improves the performance of matrix multiplication by factors ranging 30% to 59% over software-computed Morton-ordered indexing, especially at larger matrix sizes.

1 Introduction

The performance of many applications is limited by the memory bottleneck. Reorganizing the computation or data to improve locality and hence improve memory system performance can yield significant performance benefits for such applications. Linear algebra kernels which operate on large matrices, in particular, often suffer from poor locality, and hence poor overall performance. Naturally, significant research effort has been expended to improve locality by transforming the computation and the data layout for such applications[1][2][3][4][5][6][7][8][10][11][13]. This paper focuses on one particular recursive data-layout – Morton ordering – that has been known to improve locality [3].

Morton-ordered data layout for two dimensional matrices [1] is an alternative to traditional array layouts like row-major or column-major layout. Unfortunately, Morton index computation—the computation of the linear index from the multidimensional indices—requires bit-interleaving operations which are not readily supported on modern processors. In contrast, the index computation in row-major and

[1] The technique is applicable to higher dimension arrays. For ease of exposition, we limit discussion to two dimensional arrays.

S. Aluru et al. (Eds.): HiPC 2007, LNCS 4873, pp. 95–106, 2007.

column-major use addition and multiplication that are supported on all processors. There exist several software-based techniques that rely on arithmetic and logical manipulations and/or table-lookup [3][5] to reduce the cost of Morton index computation to a level comparable to that of row-major index computation. However, the row-major index computation often cannot be eliminated since it may serve other purposes (such as testing loop termination conditions). As such, Morton index computation imposes a software overhead cost to exploit the locality benefits of Morton ordered layout.

This paper addresses the challenge of hardware-based Morton address computation in order to capture the benefits of improved locality without suffering the cost of software-overheads. It makes three key contributions.

First, this paper proposes the first hardware-based Morton address computation and demonstrates that our technique can eliminate the software overheads resulting in significant speedups ranging from 30% to 59% for the matrix multiply kernel.

Second, we offer two key insights that enable low-overhead, hardware-based Morton address computation. We adopt the approach of translating full row-major addresses[2] to Morton order addresses which eliminates the need to compute the Morton index in addition to the row-major based index. This is enabled by (a) the use of aligned matrices in which the index bits can be easily extracted and (b) hardware-based Morton conversion of row-major addresses by using a single crossbar based bit permutation unit.

Finally, we evaluate two variants of the hardware-based Morton address computation. One version is purely hardware driven and transparently translates row-major addresses to Morton addresses. The second version uses instruction set support by adding a new instruction ("`morton`") that can be used to convert row-major addresses to Morton addresses.

Evaluations with two variants of tiled, loop-unrolled matrix multiply kernels reveal that the hardware based Morton indexing scheme outperforms both the row-major case and the software-based Morton indexing schemes by factors ranging from 30% to 59% at larger matrix sizes.

The rest of this paper is organized as follows. Section 0 provides the background on recursive layouts and describes related work on recursive data layouts. Section 3 describes two versions of hardware supported Morton indexing we propose. Section 4 describes our evaluation methodology. Section 5 presents the experimental results. Finally, Section 6 concludes this paper.

2 Background and Related Work

The layout of two dimensional arrays in linear memory can have a significant impact on the performance of algorithms that operate on such matrices. In this section, we offer a brief background on the Morton layout [12] for two dimensional arrays and contrast it with the traditional and well-understood row-major layout. (The column-major layout is symmetrical. As such, we omit column-major layout from our discussion.)

---

[2] Full addresses (as opposed to index computation) include the base address of the array, the index which acts as an offset into the array and the data-size which controls the size of array elements.

Figure 1 illustrates the contrast between row-major and Morton indexing. On the left, we have a traditional row-major matrix of size 8x8. Each matrix element displays the corresponding linear index corresponding to the matrix's layout in memory. To illustrate the differences with an example, we focus on the element at the Cartesian coordinates (1,3). As expected, the location of the element is computed as 11 (=8x1+3) in the row-major layout. In the special case of matrix dimensions that are powers of two, the row-major index is simply a bit-wise concatenation of the two indices as shown in the lower half in the center of Figure 1.

The array layout in the right half of the figure illustrates one flavor of the Morton order layout called the Z-Morton layout. There are other variants of the Morton layout. We restrict our discussion to the Z-Morton flavor referring to it as "the Morton order" without distinguishing the exact variant. The intuitive description of the Morton order can be described as a recursive layout order where the four quadrants are laid out in Z-order shown in the figure. Each quadrant is recursively subdivided using the same Z-layout till the recursion terminates at individual elements. As can be seen in Figure 1, the same element (1,3) maps to index 7 in the Morton layout. Unlike the row-major index which is equivalent to bit-wise concatenation (for square matrices of powers of two dimensions), the Morton index is simply a bit-wise interleaving of 1 and 3 as shown in the middle part of Figure 1. The interest in Morton layout arises because of its superior locality compared to row-major layout. Note, Morton order is only defined for square matrices whose dimensions are powers of two. Others have demonstrated that this restriction is not a problem since non-square matrices and/or non-powers of two dimensions can be accommodated via a combination of (a) padding and (b) decomposition to sub-matrix multiplications [10]. One caveat is that our technique imposes some alignment restrictions in addition to the padding required by previous techniques (as we describe later in Section 3.1).

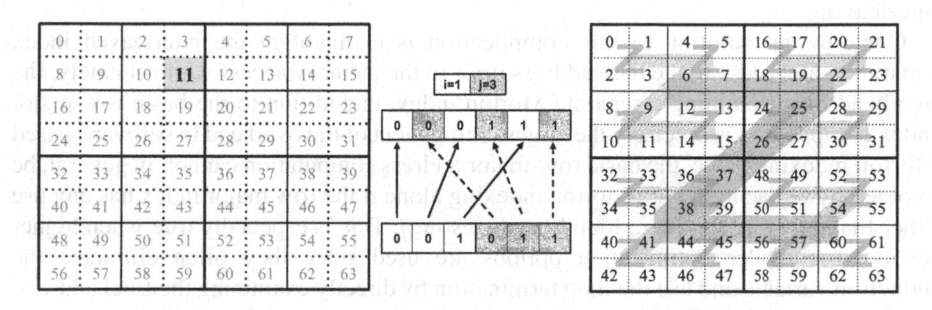

Fig. 1. Row Major vs. Morton Ordered Layout (Adapted from [10])

Given that Morton indexing uses bit-interleaving and that such bit-wise interleaving operations are not directly supported in modern processors, existing solutions typically use arithmetic/logical operations to achieve bit interleaving or use a table-lookup from a table that contains the bit-interleaved values at each corresponding entry. We demonstrate that software overheads associated with these approaches are expensive. Further, we demonstrate in Section 3 that merely providing hardware for bit-wise interleaving is not adequate to avoid the software overhead.

2.1 Related Work

Wise *et al.* proposed the application of recursive layouts [11] to achieve improved locality for matrix multiplication and also developed the compiler toolset – OPIE – to automatically generate Morton-ordered layouts [3][4][5]. While there is a large body of work examining the locality benefits of Morton-ordered matrices [2][3][4][5][6][7][8][14] in the context of matrix multiplication, our work is orthogonal to the entire portfolio since we are the first to evaluate the benefits of hardware-based Morton indexing. Cache oblivious algorithms, which use recursion to achieve cache-optimality also aim to exploit locality benefits of recursion [13]. However, cache-oblivious algorithms use control recursion and not recursive data-layouts as in this work. Pre-fetching may be seen as an alternate technique to improve the cache behavior of traditional layouts with significantly less overhead than Morton ordering. However, unlike Morton ordering, prefetching only hides the latency (by prefetching data that would otherwise miss in the cache) and does not reduce memory bandwidth requirements. In contrast, Morton ordering improves locality and reduces bandwidth requirements as well.

3 Hardware-Assisted Morton indexing

One may think, from the discussion in Section 2, that the hardware support for Morton indexing is a trivial bit-interleaving unit. Unfortunately, there are two key complications. First, the bit-interleaving shown in Figure 1 is required only for the index computation. The address computation of each matrix access involves three components: the base address of the array, the index in the array and a component that accounts for the word-size. The other two components must not be affected by the interleaving.

One obvious solution to this complication is to maintain the interleaved index separately and add it to the base address prior to the memory access. Unfortunately, the overhead of maintaining a separate Morton index, in addition to the basic row-major index computation is precisely the largest component of the overhead in software-based Morton indexing. Note, the basic row-major address computation cannot, in general, be avoided by replacing it with Morton indexing alone if the row major index has any use other than array access (e.g., loop bounds testing). This is especially true when higher levels of compiler optimization options are used since they often eliminate the induction variable and test the loop termination by directly examining the final address. As such, we skirt this issue by performing *full-address conversion* wherein we convert entire row-major addresses (including base address and word size) to corresponding Morton index-based addresses.

This design choice directly leads to the second complication. In the full-address conversion scheme, any hardware-assisted Morton indexing must carefully operate on the index component without affecting the base address and the word-offset. Isolating the word-offset is trivial since word sizes are typically powers of two. However, in general isolating the index requires a subtraction to eliminate the base-address from the overall adder. The index will then have to be bit-wise interleaved and added to the base address to get the final address. Typically, this adds three instructions for each memory access. (We ignore the alternate option of a single instruction that does the subtraction, interleaving and addition as it represents a step toward complex instruction sets.) We overcome this complication by requiring matrices to be aligned such that the index-field is *bit-separable*. This means that any 2^n x 2^n matrix of 2^w sized words each must begin at an address that has the lower (2n+w) bits set to zero. This index-separability by bit extraction eliminates the need for subtraction to isolate the index as well as the addition to get the final address.

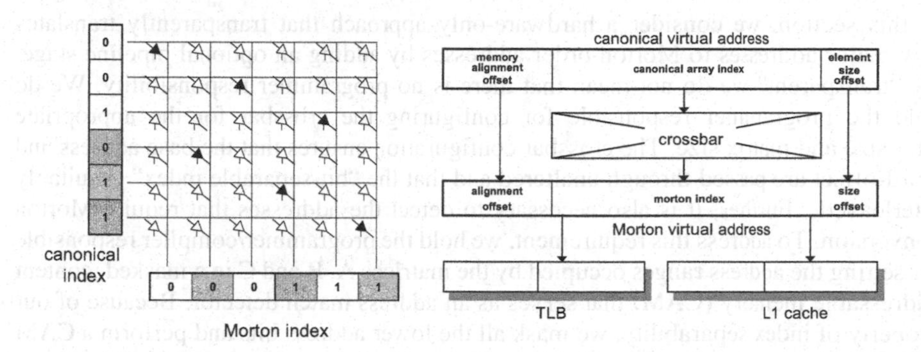

Fig. 2. Logical view of Bit-Permutation Unit (Crossbar actually extends to all bits)

Prior work has proposed aligning Morton-ordered matrices at page boundaries[14] to improve locality. Our bit-separable alignment is different on two counts. First, our alignment is not to improve locality but to facilitate reduction of Morton-conversion software overhead. Second, our alignment granularity is significantly larger than a page since it corresponds to the matrix size.

The rest of our design flows naturally from these two design choices. Section 3.1 describes the basic hardware unit to achieve full address conversion.

3.1 Hardware Support for Morton Indexing: The Bit-Permuting Unit

The basic operation of arbitrary bit-permutation can be achieved by a crossbar that accepts one address as the input and permutes them in any permutation. The left half of Figure 2 illustrates the crossbar setting for bit-interleaving the row-major index of (1,3) to achieve a Morton index conversion. Recall that we make an assumption that the arrays are address aligned to be index-separable. This assumption is central to our technique because the unchanged bits which correspond to the base address and the word offset must pass through without being permuted as shown in right half of Figure 2. However, since the size of the index can vary depending on the matrix size and the word size, we use a crossbar that is as wide as the entire address word and configure the crossbar to pass the base address and word-offset through. (The figure showing only the index part being permuted in a crossbar is purely illustrative.) The crossbar must be configured specifically for a specific matrix-size and word-size. The crosspoint buffers in the crossbar can each be controlled by a single bit. As such, the process of configuring a crossbar amounts to copying a bit for each crosspoint in the crossbar and can be achieved with tens of instructions with memory-mapped control bits and wide-word writes.

The above described hardware unit serves as the basic Morton-conversion hardware called the Bit-Permuting Unit (BPU). In the remainder of this section, we examine two designs that use the BPU to achieve hardware-based Morton-indexing.

3.2 Transparent Address Computation

In this section, we consider a hardware-only approach that transparently translates row-major addresses to Morton-order addresses by adding an optional pipeline stage. By "transparent" we do not mean that there is no programmer responsibility. We do hold the programmer responsible for configuring the crossbar for the appropriate data-size and matrix size. The crossbar configuration ensures that the base address and word-offset are passed through unaltered and that the "bit-separable index" is suitably interleaved. Further, it is also necessary to detect the addresses that require Morton conversion. To address this requirement, we hold the programmer/compiler responsible for setting the address ranges occupied by the matrices A, B and C in a masked, content addressable memory (CAM) that serves as an address match detector. Because of our property of index separability, we mask all the lower address bits and perform a CAM search for the upper bits alone. Thus any element in the matrix is guaranteed to match against the CAM entry. Further, because each entry (say the matrices A, B and C) associated with the BPU will have the same mask, we do not store a per-entry mask as done in ternary content-addressable memories (TCAMs). If a match is detected, the virtual address is Morton-converted before TLB access as also L1 access if L1 is virtually indexed. We add a one cycle penalty to TLB hits and L1 cache hits to account for the Morton-conversion delay. Note, the 1-cycle penalty applies whenever the Morton converter is active since each memory access has to be range-checked irrespective of whether the location accessed by the dynamic instruction has to be Morton-converted or not. We use the term "transparent" in the limited sense that once the address ranges and the crossbar configurations have been finalized, there is no responsibility on the software to individually demarcate individual load instructions as matrix accesses that need Morton conversion.

3.3 Instruction-Based Morton Address Translation

One weakness of the transparent version is that once configured, the range-checking imposes a 1-cycle penalty on *all* TLB and L1 cache accesses. Here we consider an alternate design where matrix accesses can be identified, possibly by the programmer or compiler. For that case, we add a single instruction with the mnemonic opcode "**`morton`**" with the following specification. It has one input register operand and one destination register operand. The instruction's input is a row-major address and its output is the corresponding Morton address that is saved to a reserved register. It performs a bit-permute operation (assuming a pre-configured bit-permuter) and places the Morton-converted address in the output register.

Figure 3 illustrates the use of the additional instruction. Assuming that a load instruction (`ld.d`) in the unmodified code is identified as an instruction that accesses a Morton-ordered matrix, the **`morton`** instruction is inserted as illustrated. Two other practical considerations with the instruction supported version are: (1) Load/store instructions targeted by the **morton** instruction may only be used with a zero immediate offset. This is because addition of the immediate offset to a Morton layout address will result in incorrect results. If a non-zero offset is used, the load instruction is further split into two instructions: one to add the non-zero immediate offset and another load instruction with a zero-offset. This results in additional instructions when unrolling loops where non-zero immediate offsets are typically used. (2) The crossbar configuration information is part of the process' state and must be swapped out and in when the process is context switched out and in.

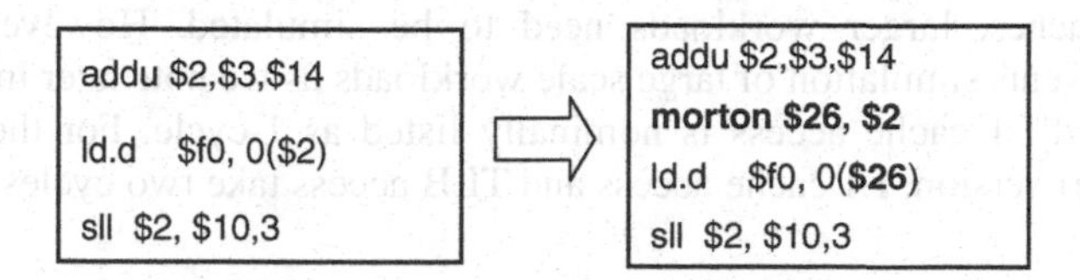

Fig. 3. Instruction-based Morton Conversion: An Example

One may think that adding another Morton-load/store instruction is a superior alternative instead of adding a special instruction for bit-permuting and then performing a normal load. This approach reduces the fetch bandwidth requirements since the **morton** instruction does not need to be fetched. However, this approach has two drawbacks. First, this extends the clock cycle since the bit-interleaving has to be interposed between the effective address computation and memory access. (Note, this may not be a concern on architectures which dynamically decompose complex instructions to RISC-like micro-ops for pipelining. We do not consider dynamic instruction processing in our experiments.) Second, with properly pipelined execution in a RISC processor, there should be no difference in instruction latencies since the bit-permuted address will be bypassed to the load instruction without unnecessary delay. We do not consider this approach a practical approach due to the clock cycle constraint. However, we do include experiments that consider an idealized case which "Mortonizes" the address for free purely for comparison purposes.

Summary: In summary, our design choice of using full-address conversion necessitates the use of aligned arrays in which the index field is bit separable. A crossbar-based bit-permuter can then achieve the necessary bit-interleaving on the index bits alone. We then describe two schemes that use this bit-permuter to achieve Morton-conversion (A) transparently on each access to specified memory address ranges OR (B) by programmer/compiler inserted Morton conversion instructions.

4 Evaluation Methodology

We use a modified version of the Simplescalar 3.0 [9] simulator for our experiments. For the ISA-supported version, we added the "**morton**" instruction to the PISA instruction set. We were able to identify an unused register to serve as the intermediate register. For the transparent version, we assumed an additional pipeline stage for the load instruction between the effective address computation and translation lookaside buffer (TLB) lookup. The pipeline stage (a) verifies that the address is within the specified range(s) and (b) if so, performs the Morton-indexing operation on the address.

We do not include the one-time cross-bar configuration costs for both the ISA-supported version and transparent version and the cost for address range(s) initialization for the transparent version. Note that these costs will not exceed tens of instructions compared to the millions of workload instructions simulated.

We simulated a common 4-way issue superscalar processor configuration with memory hierarchy as shown in Table 1. The L2 cache size is deliberately kept smaller than typical L2 of current processors because of the problems of workload scaling. With large caches, larger workloads need to be simulated. However simulation slowdown prevents simulation of large scale workloads as we note later in this section. Note, TLB and L1 cache access is nominally listed as 1-cycle. For the transparent Morton support version, L1 cache access and TLB access take two cycles.

Table 1. Processor and Memory Hierarchy Configuration

CPU	Instr. Fetch Queue	4
	Instr. Issue	4 way out-of –order
	Branch Predictor	Bimod, 2K entry BTB
	RUU/LSQ size	RUU: 32, LSQ: 8
	FU	ALU: 4 int. 4 FP , Multiplier/Divider: 1 int. 1 FP
Cache	L1 D-cache	4-way, 32KB, 32B cache block
	Unified L2 cache	8-way 512KB, 32B cache block
	Latency(cycles)	L1 hit: 1, TLB hit: 1, L2 hit: 8, Memory: 300 TLB miss: 30

```
for(i = 0; i < MSize; i++) {
  for(j = 0; j < MSize; j++) {
    temp = 0;
    for(k = 0; k < MSize; k++) {
      temp += a[i*MSize +k]*b[k*MSize +j];
    }
    c[i*MSize +j] += temp;
  }
}
```

(a) COUT version

```
for(k = 0; k < MSize; k++) {
  for(i = 0; i < MSize; i++) {
    temp = a[i* MSize +k];
    for(j = 0; j < MSize; j++) {
      c[i* MSize+j] += temp*b[k*MSize+j];
    }
  }
}
```

(b) AOUT version

Fig. 4. The Basic Matrix Multiply Versions (We use blocked, loop unrolled versions)

We use two versions of the tiled, loop-unrolled, matrix multiply kernels. The first is the traditional version that iterates over the elements of the product matrix (matrix C in C = A x B) in the outermost iterations. We refer to this version as the "Cout" version. The second version [1] iterates over elements of A in the outermost loop. Figure 4 shows the basic 3-loop version of the two kernels. Note, the three-loop versions are shown for brevity due to lack of space. Our experiments use **tiled (for both L1 and**

L2), loop unrolled versions of these two kernels which are included in a technical report [15].

We compare the hardware supported Morton indexing schemes against the version generated by Opie[5]. Opie uses purely software mechanisms to convert the Cartesian indices to a single Morton index. We also include the software morton versions of the code in the technical report [15].

We used square matrix sizes of dimensions 128, 256, 512, 1024 and 2048 for all three (i.e., A, B and C) matrices. The workloads were compiled with the "–O3" option on the gcc version 2.7.2.3, cross compiler for the SimpleScalar PISA. For the instruction-based version, the relevant loads and stores that need the additional Morton instructions were identified manually by code inspection. Our simulation runs are run to completion for all matrix sizes. We used multilevel tiling corresponding to the 32kB L1 cache and the 128kB L2 cache configuration. We used 32x32 level 1 tile and 128x128 level 2 tile size for allocating three tiles (A, B and C matrix).

5 Results

The three primary conclusions from the experimental evaluation of this paper are as follows:

[1] Canonical layouts are hard to beat at smaller matrix sizes even with free Morton addressing.
[2] Hardware-assisted Morton-indexing offers significant speedup and is clearly faster than purely software based Morton-indexing at larger matrix sizes with speedups ranging from 30% to 59%.
[3] The software overhead of Morton indexing is significant. Opie incurs nearly twice as many instructions as our transparent version. The ISA-supported version reduces the number of instructions by 17%-26%.

The remainder of this section presents detailed simulation results that quantify these three conclusions.

Figure 5 includes 2 graphs in all, one for each multiply version (aout or cout). Each graph has five bars each for the row major (xbase), Opie (xopie), instruction-based morton conversion (xmorton) transparent morton conversion (xhw) and the impractical free-morton-conversion (xideal). The X-axis has clusters of bars for each matrix size and the Y-axis plots the speedup of each configuration normalized to the Opie configuration.

Two observations are immediately apparent. First, across all the larger (>512) matrix sizes and matrix multiply versions, the hardware based Morton conversion techniques offer speedup varying between 30% and 59% over Opie with the transparent version outperforming all other techniques. Interestingly, at the smaller matrix sizes, the row-major version outperforms Opie consistently. This is not surprising because smaller matrices do not suffer from internal conflict misses and thus the locality benefits of Morton indexing do not matter while the overhead costs are still suffered. Interestingly, in the cout, configuration, the row major configuration outperforms all Morton configurations including the transparent version. In contrast, the row-major configuration is clearly outperformed by Opie and more so by our hardware-assisted Morton indexing schemes at the larger matrix sizes. Finally we observe that our transparent version captures most of the opportunity as compared to the ideal version.

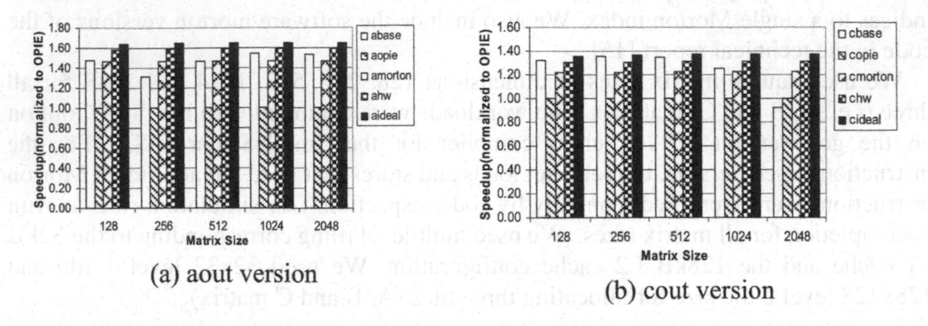

Fig. 5. Overall Speedup (Normalized to Performance of Opie)

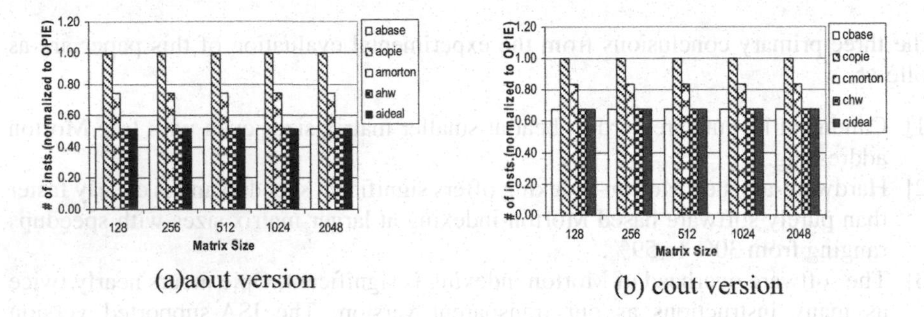

Fig. 6. Number of Instructions (Normalized to Opie)

5.1 Overhead

The graphs in Figure 6 plot the total number of instructions normalized to the Opie version for the matrix multiply versions and cache configurations. As expected, the basic row-major version has the least number of instructions (between 51% and 68% of Opie) which is exactly matched by the transparent versions. Including the small overhead of crossbar configuration, the overhead would remain negligible. The instruction-based version which includes additional (a) morton instructions and (b) instructions to eliminate non-zero immediate offsets in load instructions has between 74% and 84% of the instructions in Opie. Note, there is no difference in the matrix access patterns of the Morton-indexing schemes (Opie, instruction-based and transparent versions) since they all use the same matrix layout.

One possible reason for the performance gap is the difference in the number of instructions as the additional instructions may cause increased contention for fetch bandwidth, superscalar processor resources (e.g., ROB entries) and commit bandwidth. Note, our instruction-based version represents a lower bound on overhead since it incurs exactly one instruction overhead per access (not counting the overhead of handling loop unrolling which is also incurred in all other software versions). There are other Morton-conversion methods that aim to reduce the overhead incurred by Opie[8]. However, they also incur more than one instruction overhead per access because an additional operation is required to compute the final address after the Morton-index is looked-up in a table.

6 Conclusions and Future Work

The locality benefits of recursive layouts have been discussed widely in the literature. Unfortunately, the software overheads for computing the Morton index and the complexity of including hardware support has resulted in a failure to capture a significant fraction of the benefits of recursive layouts. This paper makes the following three contributions. First, we demonstrate that software based Morton indexing, though clearly superior to the base row-major indexing at larger matrix sizes, incurs significant performance penalties due to software overheads. Our technique demonstrates that hardware-support for Morton indexing can improve the performance of matrix multiplication by as much as 59% for the aout and as much as 31% for the cout version. Second, this paper offers the insight that "full-address conversion" and "index-bit separability" are key enablers that reduce software overhead which cannot be eliminated by simple bit-interleaving hardware. Finally, we evaluate two different versions of the hardware-based Morton indexing. The first version is an all-hardware version that transparently achieves full-address Morton conversion for all addresses in a specified address range. The second version introduces a special instruction to achieve full-address Morton-conversion. The transparent version out-performs the instruction-based version. The difference is not due to locality benefits since the memory access patterns of both versions are identical. We conjecture that the increased number of instructions in the instruction-based versions directly cause the performance loss due to contention for processor resources like fetch bandwidth and ROB entries. We conclude that the benefits of hardware-supported Morton indexing are compelling enough to merit serious consideration of its inclusion in processors (or in FPGA based application specific co-processors) targeting high-performance computing.

Future Work: Evaluation of the hardware-based Morton address mechanism for other applications/kernels (such as LU decomposition) and evaluation of support for other non-linear layouts (such as Hilbert layout) is part of our ongoing research.

Acknowledgements. We thank the anonymous reviewers for their helpful suggestions to improve this paper. This work was supported in part by Purdue University and Purdue Research Foundation.

References

[1] Lam, M., Rothberg, E., Wolf, M.: The Cache Performance and Optimizations of Blocked Algorithms. In: Proc. Fourth Int'l Conf. Architectural Support for Programming Languages and Operating Systems (April 1991)

[2] Wise, D.S.: Ahnentafel Indexing into Morton-Ordered Arrays, or Matrix Locality for Free. In: Bode, A., Ludwig, T., Karl, W.C., Wismüller, R. (eds.) Euro-Par 2000. LNCS, vol. 1900, pp. 774–783. Springer, Heidelberg (2000)
[3] Frens, J.D., Wise, D.S.: Auto-blocking matrix-multiplication or tracking BLAS3 performance from source code. In: Berman, M.A. (ed.) PPOPP 1997. Proceedings of the Sixth ACM SIGPLAN Symposium on Principles and Practice of Parallel Programming, Las Vegas, Nevada, United States, June 18 - 21, 1997, pp. 206–216. ACM Press, New York (1997)
[4] Wise, D.S., Frens, J.D., Gu, Y., Alexander, G.A.: Language support for Morton-order matrices. In: PPoPP 2001. Proceedings of the Eighth ACM SIGPLAN Symposium on Principles and Practices of Parallel Programming, Snowbird, Utah, United States, pp. 24–33. ACM Press, New York (2001)
[5] Gabriel, S.T., Wise, D.S.: The Opie compiler from row-major source to Morton-ordered matrices. In: WMPI 2004. Proceedings of the 3rd Workshop on Memory Performance Issues: in Conjunction with the 31st international Symposium on Computer Architecture, vol. 68, pp. 136–144. ACM Press, New York (2004)
[6] Chatterjee, S., Jain, V.V., Lebeck, A.R., Mundhra, S., Thottethodi, M.: Nonlinear array layouts for hierarchical memory systems. In: ICS 1999. Proceedings of the 13th international Conference on Supercomputing, Rhodes, Greece, June 20 - 25, 1999, pp. 444–453. ACM Press, New York (1999)
[7] Chatterjee, S., Lebeck, A.R., Patnala, P.K., Thottethodi, M.: Recursive array layouts and fast parallel matrix multiplication. In: SPAA 1999. Proceedings of the Eleventh Annual ACM Symposium on Parallel Algorithms and Architectures, Saint Malo, France, June 27 - 30, 1999, pp. 222–231. ACM Press, New York (1999)
[8] Thiyagalingam, J., Kelly, P.H.J.: Is Morton layout competitive for large two-dimensional arrays? In: Monien, B., Feldmann, R.L. (eds.) Euro-Par 2002. LNCS, vol. 2400, pp. 280–288. Springer, Heidelberg (2002)
[9] Austin, T., Larson, E., Ernst, D.: SimpleScalar: An infrastructure for computer system modeling. IEEE Computer 35(2), 59–67 (2002)
[10] Chatterjee, S., Lebeck, A.R., Patnala, P.K., Thottethodi, M.: Recursive Array Layouts and Fast Parallel Matrix Multiplication. IEEE Transactions on Parallel and Distributed Systems (IEEE TPDS) 13(11), 1105–1123 (2002)
[11] Thottethodi, M.S., Chatterjee, S., Lebeck, A.R.: Tuning Strassen's Matrix Multiplication For Memory Efficiency. In: Supercomputing 1998 (November 1998)
[12] Morton, G.M.: A computer oriented geodetic data base and a new technique in file sequencing. Technical report, IBM Ltd., Ottawa, Ontario, Mar. 1 (1966)
[13] Frigo, M., et al.: Cache-Oblivious Algorithms (Extended Abstract), http://citeseer.ist.psu.edu/307799.html
[14] Thiyagalingam, J., Beckmann, O., Kelly, P.H.J.: Improving the Performance of Morton Layout by Array Alignment and Loop Unrolling: Reducing the Price of Naivety. In: Proceedings of the 16th Intl. Workshop on Languages and Compilers for Parallel Computing, pp. 241–257 (October 2003)
[15] Lim, W.-T., Thottethodi, M.: Evaluating ISA support and Hardware support for Recursive Data Layouts. Technical Report TR-ECE 07-21, Purdue University (2007)

qTLB: Looking Inside the Look-Aside Buffer

Omesh Tickoo[1], Hari Kannan[2], Vineet Chadha[3], Ramesh Illikkal[1], Ravi Iyer[1], and Donald Newell[1]

[1] Intel Corporation, 2111 NE 25th Ave., Hillsboro OR, USA,
[2] Stanford University, Stanford CA, USA,
[3] University of Florida, Gainesville FL, USA

Abstract. Rapid evolution of multi-core platforms is putting additional stress on shared processor resources like TLB. TLBs have mostly been private resources for the application running on the core, due to the constant flushing of entries on context switches. Recent technologies like virtualization enable independent execution of software domains leading to performance issues because of interesting dynamics at the shared hardware resources. The advent of TLB tagging with application and VM identifiers, however, increases the lifespan of these resources. In this paper, we demonstrate that TLB tagging and refraining from flushing the hypervisor TLB entries during a VM context switch can lead to considerable performance benefits. We show that it is possible to improve the TLB performance of an important application by protecting its TLB entries from the interference of other low priority VMs/applications and providing differentiated service. We present our QoS architecture framework for TLB (qTLB) and show its benefits.

1 Introduction

CMP architectures are increasingly used for server and workload consolidation [8][9]. Industry trend is moving towards sharing the on-die and off-die platform resources across multiple heterogeneous applications or VMs running simultaneously on multiple cores of CMP systems. The success of CMP platforms depends not only on the number of cores but also heavily on the other platform resources (cache, memory, etc) available and their efficient usage. Traditionally, processor and platform architectures have been designed to perform well while running a single application. However, with the evolving software use models, CMP platforms are being geared towards running multiple applications simultaneously. The rapid deployment of virtualization [11][6] as a means to consolidate multiple applications on a platform is a prime example. When these disparate applications run simultaneously on CMP architectures, the quality of service (QoS) that the platform provides to each individual application will be non-deterministic (or chaotic) because it depends heavily on the behavior of the other simultaneously running workloads. As expected, recent studies [11][16][6] have indicated that contention for critical platform resources (e.g. cache, memory, I/O) is the primary cause for this lack of determinism. In this work we focus on the impact of virtualization on another major processor resource: translation look-aside buffer (TLB).

S. Aluru et al. (Eds.): HiPC 2007, LNCS 4873, pp. 107–118, 2007.

In order to design efficient virtualized systems on a CMP platform, the key challenge is to understand how micro-architectural features impact the performance of workloads in such environments. Recent studies [3] [2] show that significant performance overhead can be attributed to increased cache and TLB misses. TLBs are used to reduce the overhead of address translations in paging systems such as the x86. The TLB semantics mandate almost complete TLB flushes after context switches, in order to maintain consistency. While previous studies have relied on measurements to assess the performance impact of virtualization of existing workloads and systems, it is important to understand the impact of this new use model in the context of upcoming processor features like TLB tagging.

Typically, in a virtualized environment, process switches between different virtual machines (VMs) lead to complete TLB flushes. In typical consolidation environments, VM switching is often a very frequent event. Even though VMs in a virtualized environment are often scheduled based on different schedulers (such as BVT, SEDF in Xen) [11], the fun5damental problem of performance degradation due to TLB flushing on a context switch remains the same due to uncontrolled assignment and removal of TLB resources for applications running in different virtual machines. In fact, the TLB flushing behavior during frequent VM switches mitigates the advantage of faster address translations [2]. Our experimental results with SPEC CPU 2000 benchmarks support this argument. In the past, TLBs have been tagged with a global bit to prevent flushing of global pages such as shared libraries and kernel data structures. In some of the current system architectures, context switch overhead can be reduced by tagging TLB entries with address-space identifiers (ASID). A tag based on the virtual machine's ID (VMID) could be further used to improve I/O performance for virtual machines. New processor architectures, with hardware virtualization support, incorporate features such as virtual-processor identifiers (VPID) to tag entries in the TLB[4][10]. This level of tagging increases the longevity of TLB entries, and mitigates the performance penalty currently incurred on context switches.

Recent studies on shared resource management have either advocated the need for fair distribution between threads and applications, or unfair distribution with the purpose of improving overall system performance. The work presented here aims to extend these concepts to TLBs with a goal of improving the performance of an individual application at the cost of the potential detriment of others, with guidance from the operating environment. This is motivated by usage models such as server consolidation where service level agreements motivate the degree of performance differentiation [15] desired for some applications. Since the relative importance of the deployed applications is best managed by the operating software environment, we experiment with software-guided priorities (e.g. assigned by server administrators) to efficiently manage hardware resources. We compare the use of software-guided priorities (qTLB - QoS-aware TLBs) against non QoS-aware schemes. We also present the effect of scaling the TLB sizes (instruction and data), on application performance. Our full system simulation infrastructure is supplemented with detailed performance models for the caches and TLBs with QoS tuning knobs to be used by the system soft-ware. To our knowledge, this is the first study using full-system simulation to evaluate quality of service for TLBs using virtualized workloads for a CMP platform.

2 Analysis Methodology

In this section, we present an overview of our full system simulation analysis methodology. We choose to employ Xen VMM for workload characterization because it is a de-facto para-virtualized (split I/O) open source VMM. In our test framework we ported Xen VMM to run on a full system simulation environment. To identify the hardware TLB entries belonging to different VMs we needed to pass the VM information for each specific memory access to the hardware modules. We accomplished this by modifying the Xen hypervisor to provide this information on each context switch.

The Xen virtualized environment includes the Xen hypervisor, the service domain (Dom0) with its O/S kernel and applications, and a guest, "user" domain (DomU) with its O/S kernel and applications (Figure 1). This environment allows us to characterize different applications for workload characterization. The DomU guest uses a front end driver to communicate with a backend driver inside Dom0, which controls the I/O devices.

To evaluate the TLB dynamics in virtualized environments, we need an experimental framework that allows us visibility into both the hardware system architecture and the software stack. We chose SoftSDV[17] simulator for our studies. In the past, SoftSDV has been deployed to measure hardware resource usage under virtualized execution environments [19]. The simulation setup is shown in Figure 1. The execution-driven simulation environment combines functional and performance models of the platform. For this study, we chose to abstract the processor performance model and focus on detailed TLB models with QoS support to enable the coverage of multiple phases and a long execution period of the workload.

Figure 2 summarizes the profiling methodology and the tools we used. The following sections describe the individual steps in detail; these include (1) Full system simulation with virtualized workload, and (2) Performance simulation with QoS

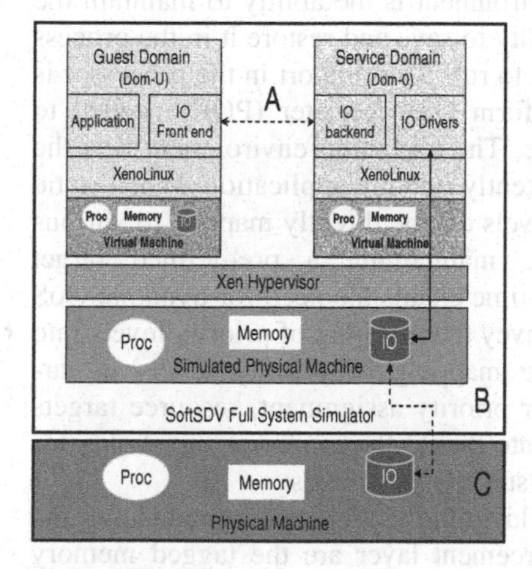

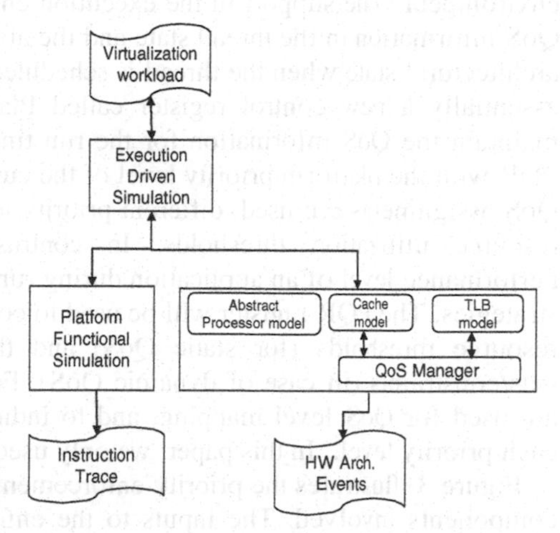

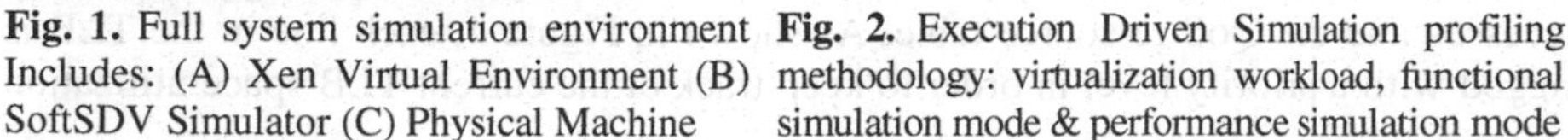

Fig. 1. Full system simulation environment Includes: (A) Xen Virtual Environment (B) SoftSDV Simulator (C) Physical Machine

Fig. 2. Execution Driven Simulation profiling methodology: virtualization workload, functional simulation mode & performance simulation mode services. The SoftSDV simulation framework was extended to support TLB QoS [19]. The simulation environment provides us with the capability of changing the underlying hardware architecture to evaluate architecture enhancements and their impact on workload performance. We tagged the TLB entries with their corresponding VMIDs, and added TLB QoS by enhancing the replacement algorithm. Our environment supported monitoring of TLB resources per VM, enforcement of QoS at the TLB level and an interface for software to communicate information about the currently running VM and individual VM priorities.

We calculated TLB utilizations while concurrently running applications in a virtualized environment. Figure 2 shows the QoS management module used to communicate with the abstract TLB model to provide QoS services. We employed a simple LRU based TLB replacement policy to evaluate the performance of various applications. QoS analysis is performed by using application level priorities to determine the percentage of TLB flushing and reservations. In addition, we considered the locality and working sets of the benchmarks for evaluation of our prototype. Section 3 discusses our proposed architecture in detail.

3 QOS-Aware Architecture

We propose a layered QoS architecture that implements static and dynamic QoS policies. Our proposed QoS-aware TLB architecture consists of three primary layers: priority enforcement, pri-ority assignment and priority classification.

The priority classification layer is responsible for identifying and providing e QoS information i.e. priority levels of each running application (e.g. 0 for high and 1 for low) and the associated targets/constraints. As shown in Figure 3, this layer requires support in the execution environment (either OS or hypervisor) as well as the processor architecture. Operationally, support (in the form of a QoS API) is required for the user or administrator to supply the required QoS in-formation to the execution environment. The support in the execution environment is the ability to maintain the QoS information in the thread state and the ability to save and restore it in the process architectural state when the thread is scheduled to run. The support in the processor is essentially a new control register called Platform QoS Register (PQR) needed to maintain the QoS information for the run time. The execution environment sets the PQR with the platform priority level of the currently running application. When static QoS assignments are used, different priority levels can be directly mapped to various resource utilization thresholds. In contrast, maintaining a pre-defined target performance level of an application during run-time entails the need for dynamic QoS strategies. The PQR register will be used to convey the mapping of priority levels into resource thresholds (for static QoS) and the mapping of priority levels to tar-gets/constraints (in case of dynamic QoS). For priority assignment, resource targets are used for QoS level mapping, and to indicate the TLB occupancy thresholds for each priority level. In this paper, we only used static QoS policies.

Figure 3 illustrates the priority enforcement layer in the architecture and shows the components involved. The inputs to the enforcement layer are the tagged memory accesses and the QoS re-source table. As shown in Figure 4, each line in the TLB is tagged with a priority level in order to keep track of the current TLB space utilization per priority level. The QoS resource table uses this information to store the TLB utilization per priority level. This is done simply by incrementing the resource usage when a new line is allocated in the TLB and decrementing the resource usage when a replacement or eviction occurs.

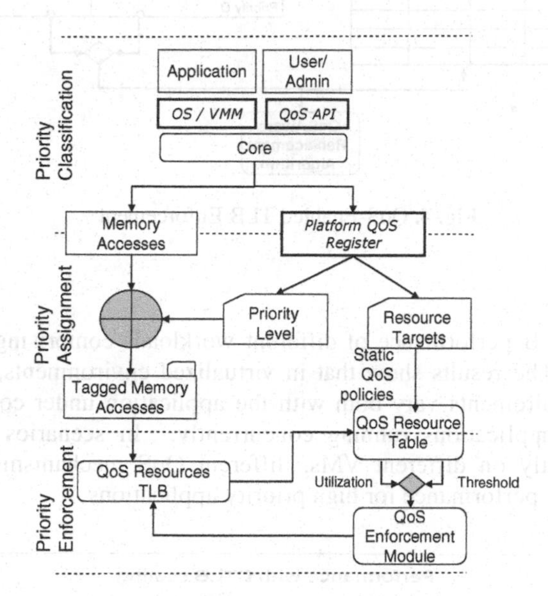

Fig. 3. QoS Architecture for TLB Resources

The static QoS policy is implemented by modifying the TLB replacement policy to be QoS aware. For each priority level, the utilization and the static QoS thresholds are available (in the QoS Resource Table - QRT) on a per priority level basis. If utilization of a priority level is lower than the specified threshold, then the replacement policy works in normal mode using the base policy (like LRU). When the utilization reaches the static QoS threshold, the QoS based replacement policy overrides the LRU policy to ensure that a victim is found within the same priority level (thus keeping the utilization for the priority level constant).

4 Experiments and Results

The goal of our experiments is to study the impact of various TLB configurations on virtualized workload performance. Keeping in mind that different applications have different working set sizes, our experiments are designed to evaluate the interaction between different applications at the TLB level. We also investigate the effect on individual applications due to different TLB QoS management policies in virtualized systems.

In following sections, we present the data TLB (DTLB) and instruction TLB (ITLB) study results. It is important to evaluate the performance of ITLB and DTLB separately because TLB access dynamics vary for data and instructions. We will first focus on the DTLB.

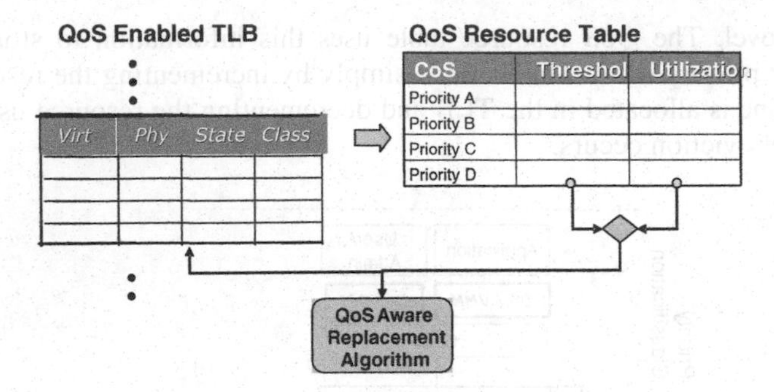

Fig. 4. QoS-Enabled TLB Enforcement

4.1 Data TLB

We evaluated DTLB performance of different workloads comprising the *CPU 2000* benchmark suite. The results show that in virtualized environments, per-application TLB resource requirements vary both with the application under consideration, and the set of other applications running concurrently. In scenarios of applications running concurrently on different VMs, different QoS mechanisms are needed to achieve the desired performance for high priority applications.

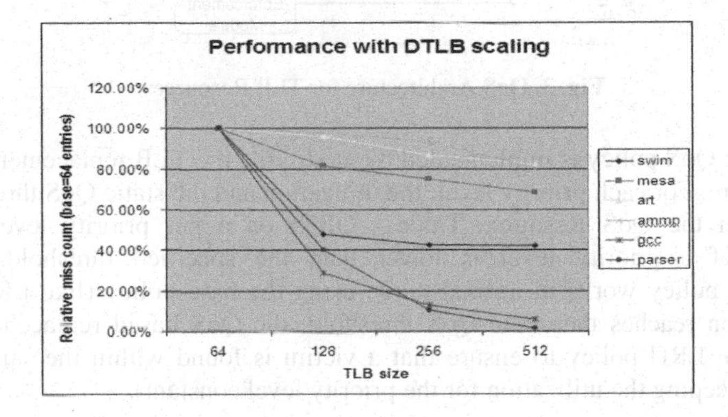

Fig. 5. Relative Change in data TLB miss rate with changing TLB size

To understand the TLB requirements for different applications, we broadly categorize the workloads on the basis of their working set sizes (Figure 5). Depending on the working set profile and data access locality, applications can fall into TLB friendly or TLB un-friendly categories. For description purposes we will categorize them into highly friendly, medium friendly and minimal friendly.

Highly TLB Friendly: These applications are characterized by a high degree of temporal and spatial locality. The applications tend to benefit highly with increased TLB resources. The more you give the better the performance is. Of the *SPEC CPU 2000* workloads we studied *parser* and *gcc* show this behavior of TLB locality.

Medium TLB friendly: The applications in this category show localization over a long range of memory addresses. Thus, while performance gains are evident at lower TLB sizes, increasing the TLB sizes leads to proportional decrease of the TLB miss count. Figure 5 illustrates this behavior for the *swim* benchmark.

Minimal TLB friendly: The working set for these applications exhibits a high degree of randomness in terms of addresses accessed. Therefore, TLB scaling has very little or no impact on the performance of such applications. From the *SPEC CPU 2000* suite, *ammp* and *art* (Figure 5) exhibit such behavior.

Next we will look into simultaneous execution of workloads and the impact of VMID tagging.

4.2 Impact of VM Tagging (VMID)

In current virtualization environments, a context switch from one VM to another leads to a complete TLB flush and subsequent repopulation of the TLBs from a clean state. Major processor manufacturers are employing TLB tagging with VMIDs in their new processor offerings. Tagging the TLB entries with global VMIDs and subsequently avoiding the flushing of these entries on a VM context switch will potentially improve the TLB performance considerably. The results from our experiments with VMID tagging are shown in Figure 6. We observe that depending on the nature of applications, significant reductions in DTLB miss count can be obtained by tagging the TLB lines with VMIDs which prevents flushing of the hypervisor mappings on context switches. Note that the graphs in Figure 6 show the percentage change in the miss count and not the absolute values of the miss count. In terms of absolute values, the miss count (or misses per instruction - MPI) of different workloads vary widely from each other depending on the nature of the individual workloads. But it can be observed that at small number of TLB entries, the impact of VMID is not that significant. As we increase the number of TLB entries, combinations with lower DTLB utilization benefit from tagging. This is due to the fact that a destructive application running after the context switch wipes the TLB out before the VM is scheduled again. One solution to this problem is to reduce the interference from the destructive VM through QoS as shown in the next section.

4.3 Impact of DTLB QoS

Our next step is to understand the TLB level interactions between multiple applications with different working set sizes and the effect of TLB QoS on performance. In our simulation setup, two different applications are run under the Xen virtualization environment with one workload running in Domain-0 and another one in a dedicated virtual machine. The TLB footprint obtained for Domain-0 is influenced by the combination of the test workload running in Domain-0 and other Xen related processes running in the administrative domain. To understand the effect of TLB QoS on an individual workload performance, we assign higher priority to the workload running in the isolated VM. The exact QoS metrics are tunable and are described in detail below. We use VMIDs to tag the TLB entries for QoS enforcement purposes. In the graphs presented below, we plot the miss ratios for high priority applications with changing occupancy limits for the lower priority application. The miss counts are plotted relative to the miss count when no QoS is enforced. Table 1 shows the different TLB configurations analyzed.

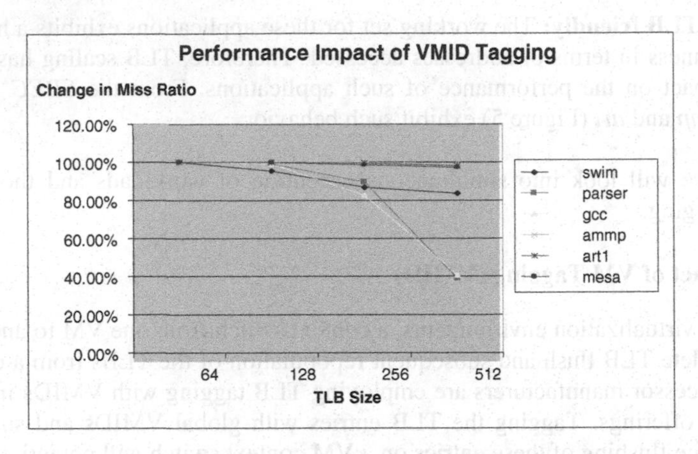

Fig. 6. DTLB performance impact of VMID tagging

Consider a scenario where the high priority application exhibits characteristics of highly TLB friendly workload. Since, the applications benefit from being allocated more entries in the TLB, restricting the background app will provide considerable performance improvement. This is more significant when the TLB size is small. Better management of the TLB can provide better results for the important application in this scenario. We will look at two sets of results to demonstrate this behavior. The first set of results (Figure 7) uses SWIM as the background process. It may be noted that the highly TLB friendly applications *gcc* and *parser* benefit highly from the increased TLB resources provided by TLB QoS. On the other hand, *art* and *ammp* which are minimal TLB friendly get minimal benefit out of TLB QoS.

Another important observation is that the VMID tagging benefits dwarfed by the excessive TLB resource utilization are now moderated by employing TLB QoS.

Table 1. TLB configurations supported

System Scenarios	TLB Semantics
Legacy System	TLBs are flushed on each context switch.
VMID tagging (No application TLB QoS)	VMID tagging and TLB entries are not flushed on VM switches. LRU is used to replace the TLB entries across VMs.
X% (preferential Resource allocation)	VMID tagging with QoS Aware replacement. Low priority application gets at most X% of the TLB capacity. X= 40, 30, 20, 10, 0 (examples)

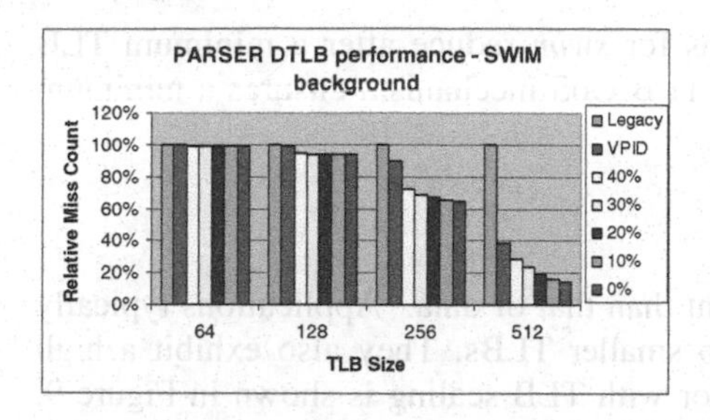

(a) Miss Rates for *parser* in *swim* vs. *parser* (*parser* has higher priority)

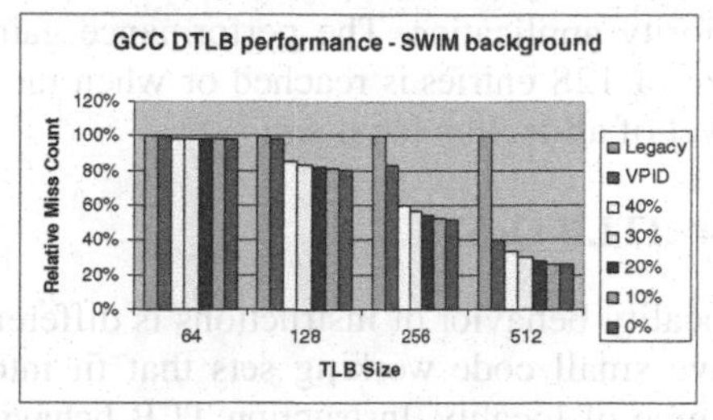

(b) Miss Rates for *gcc* in *swim* vs. *gcc* (*gcc* has higher priority)

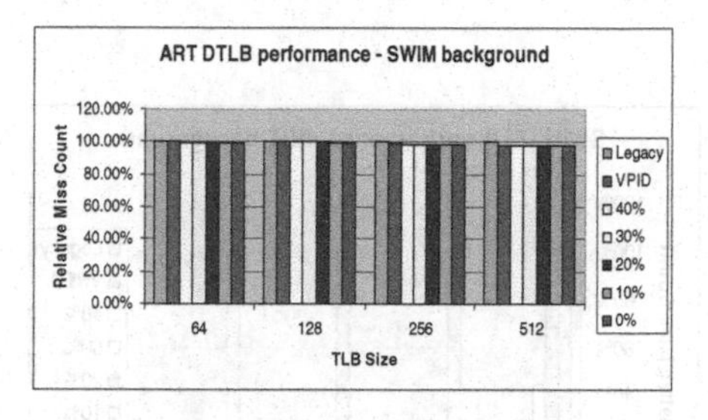

(c) Miss Rates for *art* in *swim* vs. *art* (*art* has higher priority)

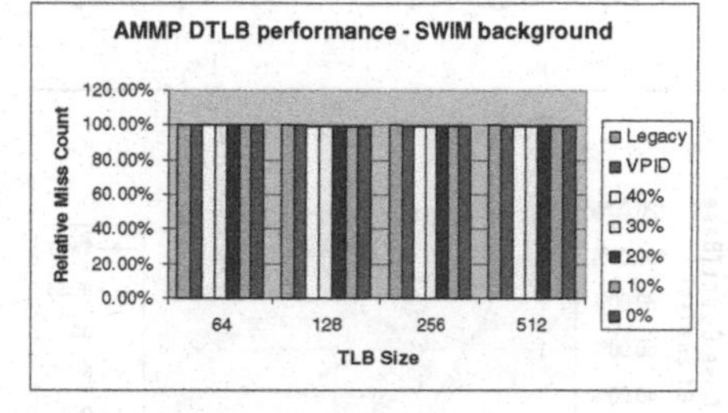

(d) Miss Rates for *ammp* in *swim* vs. *ammp* (*ammp* has higher priority)

Fig. 7. Impact of VMID and TLB QoS on various applications with SWIM in background

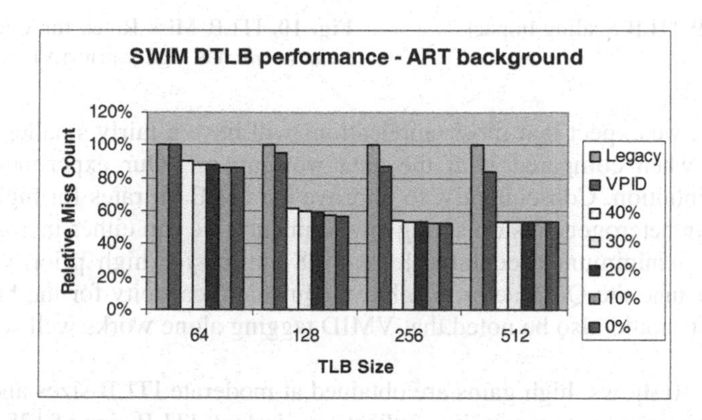

Fig. 8. Miss Rates for *swim* in *art* vs. *swim* (*swim* has higher priority)

It should be noted that the impact of TLB QoS depends both on the foreground as well as on the background application. Results with *art* as a background application are shown in Figure 8. Since *art* is less TLB intensive than *swim*, the impact of *art* on the foreground application is considerably less. This results in better QoS results even with smaller TLB sizes. The *art* vs *swim* plot shows that TLB QoS is needed to ensure that the a minimum number of entries must necessarily stay dedicated for the high priority application. The performance gains for *swim* reduce after a minimum TLB size of 128 entries is reached or when the TLB QoS mechanism ensures a minimum level of allocation for *swim*.

4.4 ITLB QoS

Locality behavior of instructions is different than that of data. Applications typically have small code working sets that fit into smaller TLBs. They also exhibit a high degree of locality. Instruction TLB behavior with TLB scaling is shown in Figure 9. We note that with increase in the size of the TLB, relative miss ratio decreases and is almost constant after size of 128. We infer that an ITLB size of 128 entries is sufficient to incorporate almost all possible address translations during the TLB stage, hence reducing the performance penalty.

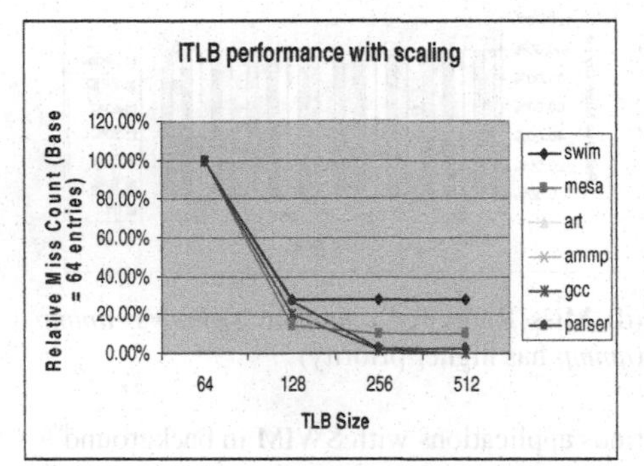

Fig. 9. ITLB Scaling Impact

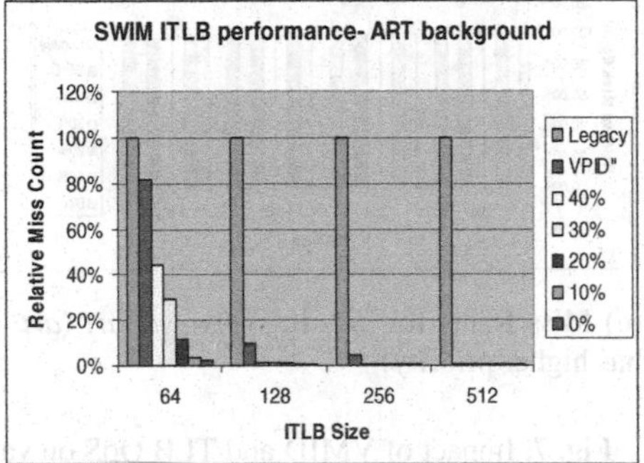

Fig. 10. ITLB Miss Rates for *swim* in *art* vs. *swim* (*swim* has higher priority)

Intuitively, we expect that most applications will have a fairly smaller instruction working set when compared with the data working set. Our experimental results support this intuition. Consequently, to improve the ITLB hit rates for higher priority applications in heterogeneous operating environments, we can either increase the size of ITLBs to a minimum acceptable level (128 entries for high priority VM from Figure 9), or tune the QoS factor to achieve equivalent capacity for the high priority application. It should also be noted that VMID tagging alone works well with all TLB sizes.

As Figure 10 shows, high gains are obtained at moderate ITLB sizes and moderate capacity restrictions for low priority applications. In fact, ITLB size of 128 entries and a QoS factor close to 0.5 ensuring fair distribution of the ITLB provides close to maximum performance boost. QoS tuning beyond this point does not produce proportionate results.

This type of behavior was observed in all the studied workloads leading us to conclude that providing ITLB QoS in virtualized systems is less application sensitive than the DTLB QoS and may amount to ensuring a fair TLB distribution in most cases.

5 Conclusion and Future Work

Virtualization and multi-core architectures are two complementary upcoming paradigms that throw open interesting workloads and applications scenarios. In this paper we analyzed the TLB level interactions of different applications operating in virtualized settings. Our execution driven simulation based results show that modifications to default TLB management policies are needed for efficient operation in such settings. We show that using VMIDs to avoid flushing the global (VMM) entries from TLBs on VM context switches leads to significant drops in TLB miss rates.

We also investigated the effect of prioritizing the applications and providing QoS in terms of TLB capacity. Our investigations show that different applications display different TLB related behaviors depending on the working set sizes and access locality. Running multiple applications within different virtual machines raises interesting TLB sharing scenarios. In such conditions, our experiments show that an administrator can potentially provide a preferential performance boost to high priority applications using TLB QoS. The knowledge of application working-set sizes and access locality can be used to determine the QoS factors needed for a targeted TLB miss count.

We are investigating how QoS services will affect TLB coherence protocols in context of performance and overhead. We are currently in the process of designing a dynamic TLB QoS policy that tunes the QoS factor during run-time to achieve a guaranteed minimum performance level for high priority applications. We are also investigating hardware and software enhancements for architecting QoS aware multi-core platforms.

References

[1] Foong, A., Fung, J., Newell, D.: An In-Depth Analysis of the Impact of Processor Affinity on Network Performance. In: Proceeding of IEEE Int'l Conf. Networks, IEEE Press, Los Alamitos (2004)

[2] Menon, A., Cox, A., Zwaenepoel, W.: Optimizing Network Virtualization in Xen, USENIX Annual Technical Conference (2006)

[3] Chandra, D., Guo, F., Kim, S., Solihin, Y.: Predicting inter-thread cache contention on a chip multiprocessor architecture. In: HPCA. Proc. 11th International Symposium on High Performance Computer Architecture (February 2005)

[4] Neiger, G., Santoni, A., Leung, F., Rodgers, D., Uhlig, R.: Intel Virtualization Technology: Hardware Support for Efficient Processor Virtualization. Intel Technology Journal (August 2006)

[5] Kannan, H., Guo, F., Zhao, L., Illikkal, R., Iyer, R., Newell, D., Solihin, Y., Kozyrakis, C.: From Chaos to QoS: Case Studies in CMP Resource Management. In: dasCMP/MICRO. From Chaos to QoS: Case Studies in CMP Resource Management, 2nd Workshop on Design, Architecture and Simulation of CMP platforms (December 2006)

[6] Intel Virtualization. Technology Specification for the IA-32 Intel Architecture (April 2005)

[7] Hsu, L., Reinhardt, S., Iyer, R., Makineni, S.: Communist, Utilitarian, and Capitalist Cache Policies on CMPs: Caches as a Shared Resource. In: 15th International Conference on Parallel Architectures and Compilation Techniques (PACT) (September 2006)
[8] Nesbit, K.J., et al.: Fair Queuing Memory Systems. MICRO (2006)
[9] Marty, M.R, Hill, M.D.: Virtual hierarchies to support server consolidation. In: proceedings of ISCA 2007 (2007)
[10] Pacifica, – Next Generation Architecture for Efficient Virtual Machines (Accessed, April 2007), http://developer.amd.com/assets/WinHEC 2005_Pacifica_Virtualization.pdf
[11] Barham, P., Dragovic, B., Fraser, K., Hand, S., Harris, T., Ho, A., Neugebauer, R., Pratt, I., Warfield, A.: Xen and the Art of Virtualization. In: Proceedings of the ACM symposium on operating systems principles (October 2003)
[12] Illikkal, R., Iyer, R., Newell, D.: Micro-Architectural Anatomy of a Commercial TCP/IP Stack. In: WWC-7. 7th IEEE Annual Workshop on Workload Characterization (October 2004)
[13] Iyer, R.: CQoS: A Framework for Enabling QoS in Shared Caches of CMP Platforms. In: ICS 2004. 18th Annual International Conference on Supercomputing (July 2004)
[14] Iyer, R.: On Modeling and Analyzing Cache Hierarchies using CASPER. In: Calzarossa, M.C., Gelenbe, E. (eds.) MASCOTS 2003. LNCS, vol. 2965, Springer, Heidelberg (2004)
[15] Iyer, R., Zhao, L., Guo, F., Illikkal, R., Makineni, S., Newell, D., Solihin, Y., Hsu, L., Reinhardt, S.: QoS Policies and Architecture for Cache/Memory in CMP Platforms. In: ACM SIGMETRICS 2007 (2007)
[16] Goldberg, R.P.: Survey of virtual machine research. IEEE Computer, 34–45 (1974)
[17] Uhlig, R., Fishtein, R., Gershon, O., Hirsh, I., Wang, H.: SoftSDV: A Presilicon Software Development Environment for the IA-64 Architecture. Intel Technology Journal. Q4 (1999), http://www.intel.com/ technology/itjf
[18] Makineni, S., Iyer, R.: Performance Characterization of TCP/IP Packet Processing in Commercial Server Workloads. In: 6th IEEE Workshop on Workload Characterization (October 2003)
[19] Chadha, V., Illikkal, R., Moses, J., Iyer, R., Newell, D., Figueiredo, R.J.: I/O Processing in a Virtualized Platform: A Simulation-Driven approach. In: Proceedings of VEE, San Diego (June 2007)

Analysis of x86 ISA Condition Codes Influence on Superscalar Execution

Virginia Escuder, Raúl Durán, and Rafael Rico

Department of Computer Engineering, Universidad de Alcalá
28871 Alcalá de Henares, Spain
{virginia.escuder, raul.duran, rafael.rico}@uah.es

Abstract. Instruction sets may have particular characteristics that produce a negative impact into the amount of available parallelism. The x86 instruction set architecture includes some of those characteristics. In particular, it is well know the negative impact of condition codes usage. In a coarse approximation, they can be considered responsible for a greater code coupling. Moreover, several in-depth works show that they introduce additional complexity in the procedures both to perform microcode binary translation and to support for precise exception mechanisms among others. To the extent of our knowledge no quantitative evaluation has been carried out that may determine the impact of condition codes usage on the x86 processors performance. In this work we will present a proposal of such quantification based on Graph Theory.

Keywords: Condition codes; Instruction set architecture; Instruction level parallelism; Graph theory.

1 Introduction

Instruction set design has always been a fundamental issue in Computer Science. Design criteria for building instruction sets have evolved in time as some theoretical studies [2, 12] show. However, evaluation of instruction set architecture has not been explored as much as it could be expected, given the importance of the subject. We propose the analysis of instruction set architectures emphasizing that they have a definitive influence in the final performance.

In-depth analysis of the impact that instruction sets on their own have on performance has been abandoned in favor of considering a single unit for study (the instruction set and the hardware that should interpret it), under the assumption that this is a sounder computational approach. Another circumstance that has also contributed to the lack of research in this type of analysis is the extensive use (sometimes abuse) of simulation as the performance evaluation method. Simulation does not differentiate between the impact on performance arising by the language itself and the impact of limited physical resources [15].

Performance in the field of superscalar execution depends on many factors: the intrinsic parallelism of algorithms, the capabilities of the used high level language, the compilation process, the target machine instruction set and, of course, the physical layer. Indeed, the ISA has a significant impact in the availability of fine-grain

S. Aluru et al. (Eds.): HiPC 2007, LNCS 4873, pp. 119–132, 2007.

parallelism before reaching the physical layer, which can reduce exploitable parallelism degree at run time.

Nowadays, one of the most important objectives in Computer Engineering is code decoupling, that is, avoid data dependency among instructions in order to obtain maximum concurrency in superscalar processing of code. However, the machine language layer can be responsible for an over-ordering of the code that has no solution in the physical layer or that may cause increased execution complexity and power consumption. It is, therefore, important to move the focus from the physical layer to the machine language layer by itself.

The x86 Instruction Set Architecture (ISA) was designed to fulfill basically two objectives: decrease the semantic gap between the high level languages and the machine level languages and obtain a compact executable code. These criteria are now obsolete but the instruction set has been maintained for binary compatibility reasons. Unfortunately, it behaves inefficiently in superscalar implementations. The x86 ISA shows many features that may compromise the intrinsic concurrency of the original computational task such as dedicated use of registers, condition code dependent branching and effective address computation where up to three registers may be involved. The sources of potential code coupling in x86 ISA have been identified from the distribution of data utilization in programs [7, 14].

Moreover, the x86 ISA performs poorly in superscalar environments compared with non-x86 sets for different architectural proposals. The IPC (Instructions Per Cycle) is 0.5 to 3.5 in different x86 execution models [13, 17]; compared to an IPC of 2.5 to 15 (and peaks of 30) of non-x86 processors [18, 19]. This seems to confirm that indeed the architecture of the instruction set is a limiting factor on its own.

In particular, it is well know the negative impact of condition codes usage. The main purpose of the condition codes is to communicate some information between process instructions and conditional branch instructions but they have additional functions in the case of x86 ISA. In a coarse approximation, they can be deemed responsible for a greater code coupling. In this sense, the condition codes have been considered accountable for the appearance of output dependence chains in the Literature [7]. Moreover, several in-depth works show that they introduce additional complexity in the procedures both to perform microcode binary translation and to support for precise exception mechanisms among others.

Regarding binary translation, it is well known that most of condition codes writings are never used. In the setting of static binary translation the proposed solution is the deferred materialization [6] but in the setting of dynamic binary translation no similar solution has been proposed so far.

Regarding precise exceptions, it can be said that patents have been filed providing solutions that include condition codes renaming (*e.g.*, USPatent 5659721: "*Processor structure and method for checkpointing instructions to maintain precise state*"). Nevertheless, the renaming capacity is pretty limited (no more than 10 registers are reserved for this purpose).

In fact, there is no available information from vendors about the degree of usage in the real world of the described techniques. Oddly enough, for example, no vendor has reported the practical use of the renaming solution for the complete instruction window, which in principle could be considered simpler. Several reasons could explain this fact: the unconditional materialization of condition codes which in practice are seldom read [7]; the extra cost due to the additional information that must be stored if precise exceptions are to be maintained [10, 16]; the impact on the ROB due to different life span of processed data and their associated condition codes; etc.

Popular wisdom states that the negative impact due to condition codes is minimized at the micro-architectural level for the case of x86 ISA. Though appealing this turns out to be not exactly true. The binary translation results in a increased number of instructions (micro-operations) to be scheduled. Since the scheduling block does not scale well, the pressure on it becomes larger.

To relax the pressure on the scheduling block several solutions have been proposed [8, 11]. Essentially they consist in fusing instructions. The pairs formed by the instruction producing the condition code and the consumer conditional branch are among the best candidates to be fused. However, it is clear that this mechanism is orthogonal with respect to the treatment of dependences produced by condition codes.

Finally, although the condition codes represent a problem in the superscalar execution setting, as far the authors know, no quantitative evaluation has been carried out that permits to determine their effect on the performance of x86 processors. In this work we present a proposal for such quantification.

2 The Analytical Model and Quantification Tools

Applying Graph Theory [1, 5] to evaluate instruction sets architectures has several advantages:

- it provides a simple description of the problem,
- it allows to predict behavior,
- it simplifies the transmission of knowledge,
- it allows an easy quantification,
- it separates the study of instruction set characteristics from the hardware that should interpret it.

We represent data dependences found in code sequences as directed graphs and then use a matrix representation for the graph to apply mathematical relations. Details about the mathematical development leading to the following summary of definitions can be found in [3, 4].

- For a sequence of n instructions the n x n dependence matrix D is defined as:

$$d_{ij} = \begin{cases} 1, & \text{if } i \text{ instruction depends on } j; \\ 0, & \text{otherwise.} \end{cases} \quad (1)$$

This matrix should comply with several properties and restrictions in the ILP setting. From this matrix we can derive a set of metrics: code coupling, critical path length, and degree of parallelism:

- We call *total coupling* C_T the total number of dependences registered by matrix D.

$$C_T = \sum_{i=1}^{n-1} \sum_{k=0}^{n-1} d_{ik} . \quad (2)$$

To obtain a coupling measurement independent of the amount of instructions in the sequence, we define a normalized coupling, C_{TN}, as the ratio C_T *vs.* the binomial coefficient n over 2.

- The *critical path length L* measures the longest data dependence path found in the n instructions window. The units are computing steps. It is an asynchronous concept, *i.e.*, defined as the process of eliminating all the nodes in the graph with no data dependences. The first power of D (in the set of integer numbers) that is identically zero indicates the length of the critical data path in computation steps:

$$L = l \text{ computation steps if and only if } D^{l-1} \bullet \mathbf{0} \text{ and } D^{l} = \mathbf{0}. \quad (3)$$

To obtain a critical path length measurement independent of the amount of instructions in the sequence, we define a normalized critical path length, L_N, as the ratio L *vs.* the number n of instructions in the instruction window.

- The *parallelism degree* G_p is inversely related to the critical path length L: the longer the length, the stricter the partial ordering of the code sequence, limiting the ability of concurrent processing. Consequently, we define the parallelism degree, G_p, as:

$$G_p = \frac{n}{L}. \quad (4)$$

- One of the most powerful properties of the method is the compositional nature of matrix D, which states that this matrix corresponds to the resultant of the contribution from different sources of dependence S_i which are, in turn, individual matrices representing isolated sources of dependences. So, it is possible to isolate and estimate the impact produced by different data types on the whole set and also perform several interesting combinations and obtain their specific contributions to the total. The expression is:

$$D = D_{s1} \text{ OR } D_{s2} \text{ OR } \ldots \text{ OR } D_{sn} \quad (5)$$

- The following equations bound C_T and L as a function of the values of their components C_{Tsi} and L_{si} and the size of the instruction window n:

$$\max_i \{C_{Tsi}\} \le C_T \le \min\left\{\sum_i C_{Tsi}, \binom{n}{2}\right\}. \quad (6)$$

$$\max_i \{L_{si}\} \le L \le \min\{C_T + 1, n\}. \quad (7)$$

While analytic Graph Theory is very rich, our model in its current version assigns a computational meaning to a limited number of its concepts. However modest the use of Graph Theory may seem up to now, the full power of the theory is available, since we have carefully established a link from superscalar setting to Graph Theory [3, 4].

3 Condition Codes in Instruction Set Architectures

Using condition codes is an alternative for implementing conditional control flow. The evaluation of the branch condition is performed using one or more condition-code bits. These bits are grouped, for practical reasons, into a status register where they get updated upon the execution of processing instructions; setting or unsetting each individual bit, the collection completely describes the result. A processing instruction typically precedes a conditional branch and therefore it creates a dependence which requires serial execution. Architectures using this schema are called status register architectures; the x86 is one of them. Considering superscalar execution, condition codes increment the ordering of instructions as they pass information from one instruction to the next.

Theoretically, there are other two alternatives for implementing conditional control flow: evaluation of the contents of a register named in the branch instruction against a criteria also contained in the branch instruction, and atomization of the comparison and branching actions into a single instruction. The first alternative, commercially adopted in the *Alpha* and *MIPS* architectures for instance, is simple and also optimal for superscalar execution while the second one, used in the *PA-RISC* and *VAX* processors, makes the pipeline design more complex as it results from the union of two operations in one.

The *PowerPC* is another status register architecture but, in contrast to the x86, its instruction set was designed to avoid the negative effect produced by output dependences due to condition codes: data processing instructions format includes a bit used to indicate whether the condition bits must be materialized or not. This effectively limits the coupling produced by condition codes to the cases where it really has computational meaning, and the compiler is in charge of driving the decision.

The x86 ISA has not solved the output dependences problem because it needs to keep backward binary compatibility.

4 Condition Codes in the x86 ISA

Condition codes are used for conditional branching and they are located into the status register. This register hosts bits with different meaning that can be classified into one of the following two groups: *control flags* and *status flags*.

Control flags include miscellaneous information related to the operation modes of the processor. These control flags do not contain computational information and therefore are not taken into account in our analysis. Status flags qualify the result of processing operations. A single or a combination of status flags correspond to what we generally refer to as a condition code. Status register has thus this dual consideration being a unique storage location while each bit has its own independent meaning and management procedure.

Condition codes are typically used for conditional branching in status register architectures. In the case of the x86 ISA, the status flags can also be used as input operands for some operations. In these particular cases, where information flows from one operation to the next, it exists a true dependence and the instructions involved cannot be executed independently. It is necessary therefore to include these cases into the analysis and evaluate their influence into the general code coupling.

We should not forget that condition codes in the x86 ISA are used by different instructions with other purposes as well: status register moving, processing instructions using condition codes as an extra input operand, and some special instructions such as repetition prefixes.

In all cases, the access to condition codes is done implicitly, it can not be avoided by the programmer and there is no mechanism to disable the access when it has no computational meaning.

5 Micro-operation Level Impact

Processors of the x86 family use a 2-level microarchitecture to improve performance. The top level acts as an interface to the CISC instruction set, translating instructions into RISC-type micro-operations which are executed in the low level machine. Obviously, computational semantics must be preserved across the transformation. Following our formalism, the RISC code corresponding to a CISC code will also have a graph representing it.

Decoding is performed in three different units: the simple, the general and the sequencer units. Instructions decoded by the sequencer unit are executed serially, while instructions decoded by the other two units are executed in a superscalar fashion.

Huang and Peng have analyzed the distribution of the number of micro-operations a single CISC instruction gets decomposed [9]. Most instructions (67%) are translated into only one micro-operation and from the rest almost 90% get translated into two micro-operations. The reported weighted average value is 1.47 micro-operations per CISC instruction.

According to these numbers, we can conclude that the transformation from CISC to RISC only increases the number of nodes in the dependence graph by a factor of 1.5.

Once the graph nodes analyzed, the question now is how many arcs the new graph will have after the transformation. To answer this question we need to analyze how a CISC instruction gets decomposed into several RISC instructions. Basically it depends on addressing modes. When a CISC operand is in memory, it gets automatically translated into two operations: a RISC load/store operation used to transfer the operand to/from memory from/to the CPU registers and a RISC processing instruction that gets the operands from the CPU registers and performs the operation. Consequently, dependence chains experiment an enlargement that basically corresponds to the increase of the number of nodes in the graph.

The result is a negligible increase of the parallelism compared with that present in the sequence of CISC code given that G_p is the relation between n (nodes) and L (chain length). Since the equivalent RISC graph is essentially the same as the CISC graph we can conclude that the binary translation does not yield by itself any new parallelism opportunity.

6 Methodology

We apply the analytical model based on Graph Theory proposed in Section 2 to obtain code coupling quantification. The input data set is the execution traces of a set of programs used as testbench for the experiment.

6.1 Defining Compositions

The selection of contributing data dependence sources depends on the objective of the ongoing analysis. Then, applying our metrics to the compositions resulting from the inclusion or exclusion of particular sources of data dependences, we obtain figures of their relevance.

In our case, we are focusing on condition codes as source of dependences, and, therefore, our space is divided into two data types: condition codes and the rest. So, we will study the following contributions:

a) contribution from all data types
b) contribution from condition codes only
c) contribution from non condition codes only

We also wish to distinguish among the different types of dependences: true dependences (RAW: Read After Write), anti-dependences (WAR: Write After Read) and output dependences (WAW: Write After Write). We build meaningful compositions by combining data types and dependence type contributions. So for each dependence type we have information about the contribution from all data, from condition codes solely and from data other than condition codes.

By interpreting values in different compositions we can reach valuable information about the effective impact of condition codes on real programs.

Whereas the statistical analysis based on instruction distributions provide a qualitative knowledge of the impact caused by condition codes to superscalar execution, the analytical method provides a quantification for this impact.

6.2 Experimental Framework

The proposed analysis follows the typical methodology of the trace driven simulation but replacing the simulator block by an application that implements our mathematical model (available at: CVSROOOT:pserver:anoncvs@atc2.aut.uah.es:2401/home/cvsmgr/repositorio). This application performs an automatic computation of the metrics proposed in Section 2. It allows the use of instruction windows of variable size. Given a profile for the dependence contributions, it builds the relevant dependence matrices and stores the desired results.

The first step is to generate a set of traces where, obviously, loops are unrolled and branch targets are exactly known since they are written in the trace file (this could be considered the omniscient predictor sometimes used in simulators).

The testbench is a set of DOS utility programs (*comp*, *find* and *debug*) compiled in real mode as well as some popular common applications such as file compressor *rar* (v.1.52) working as compress (trace *rar(c)*) and uncompress (trace *rar(d)*) modes, and the *tcc* C-language compiler (v. 1.0). Program *go* from the SPECint95 suite has also been included using two different compilation options: one optimizes for size (trace *go(t)*) and the other optimizes for speed (trace *go(v)*). For simplicity, the 16 bit subset of x86 ISA has been selected.

The programs were run in step-by-step mode and under a specific workload conditions to avoid excessively long traces. Nevertheless, more than 190 million instructions were executed.

We selected static 512 instruction sequence windows. Sliding windows, the typical mode used for the physical layer of processors and simulators, is an excessively heavy load for the computation and it adds no additional precision compared with a scenario where sufficiently large static windows are used. We tested window sizes up to 2,048 instructions and found no significant change in the results obtained, while computing time substantially increased.

7 Quantifying the Impact of Condition Codes Accesses

7.1 Condition Codes in the Basic Block

A basic block is defined as a sequence of linear instructions without any branches. This structure is frequently used in compiler theory as it is a basic unit to apply local optimizations to. The structure is also a good scenario to identify and understand different data coupling patterns produced by condition codes.

Table 1 shows the average size (in number of instructions) of the basic block found for each testbench program. It also shows the average number of processing instructions in the block. The complete work about the distribution of instructions and data usage called "*Analysis of x86 Data Usage (16 bit subset)*" is available at: http://atc2.aut.uah.es/~gap/.

Table 1. Average block size and average processing instructions per basic block for each program in the testbench

Program	Average instructions per BB	Average processing instructions per BB
comp	6.00	1.94
find	7.68	1.56
go(t)	10.31	3.14
go(v)	10.16	2.89
rar(c)	3.19	1.25
rar(d)	12.56	5.52
debug	3.92	1.35
tcc	8.98	2.28

Considering condition code data dependences only, each basic block necessarily contains a true dependence between the last instruction updating the condition codes and the branch instruction that reads them. This communication holds real computational meaning. From a statistical point of view, the number of true dependences will increase with the number of basic blocks present in the code. In other words, the smaller the basic block, the higher the number of true dependences in the trace. Table 1 shows that programs *rar(d)* and *debug* have the smallest basic blocks.

Another typical coupling in the basic block is due to output dependences: the order imposed by processing instructions performing successive writes to the same resource, in this case, the status register. The limitation on the amount of available parallelism due to this type of dependence is a direct consequence of the instruction set architecture and it has no computational meaning at all.

Statistically, the average length of output dependence chains grows with the number of processing instructions in a basic block. Large basic blocks also tend to contain a large number of processing instructions and, consequently, may increase the length of output dependence chains caused by condition codes.

7.2 Coupling

Figure 1 shows for each considered program some information about the normalized total coupling C_{TN} due to different contributions. Two types of graphs have been used: stacked area chart and line chart. First of all, the normalized total coupling due to contributions from non condition codes $C_{TNno\text{-}cc}$ (cross hatched), and then, the normalized total coupling due to contributions from condition codes C_{TNcc} (gray dotted); both of them are represented by stacked areas. Finally, the normalized total coupling C_{TN} due to contribution from all data is represented by a solid line.

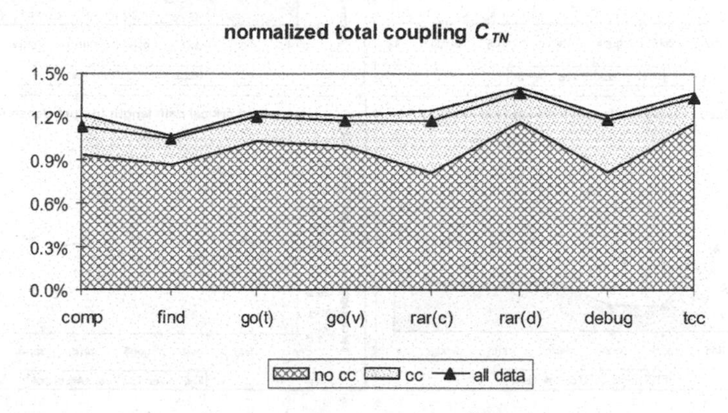

Fig. 1. Normalized total coupling: in stacked areas the contributions from non condition codes plus condition codes; in solid line, the normalized total coupling due to the contribution from all data

It is noticeable that the solid line is slightly below the total area. This means that the contributions from both condition and non condition codes are essentially non overlapping, thus proving that the contribution from condition codes is relevant.

Figure 2 shows the normalized total coupling due to the contribution from condition codes (C_{TNcc}) split into the three types of data dependences: true, anti- and output dependences.

It is apparent that the biggest contribution of conditions codes to the coupling comes from output dependences for all the programs. This quantification confirms the observations already made by some authors [7].

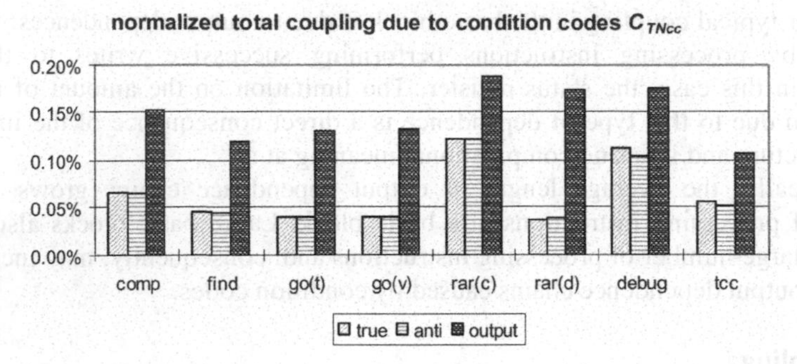

Fig. 2. Normalized total coupling due to the contribution from condition codes. Each column depicts the contribution of each different type of dependence.

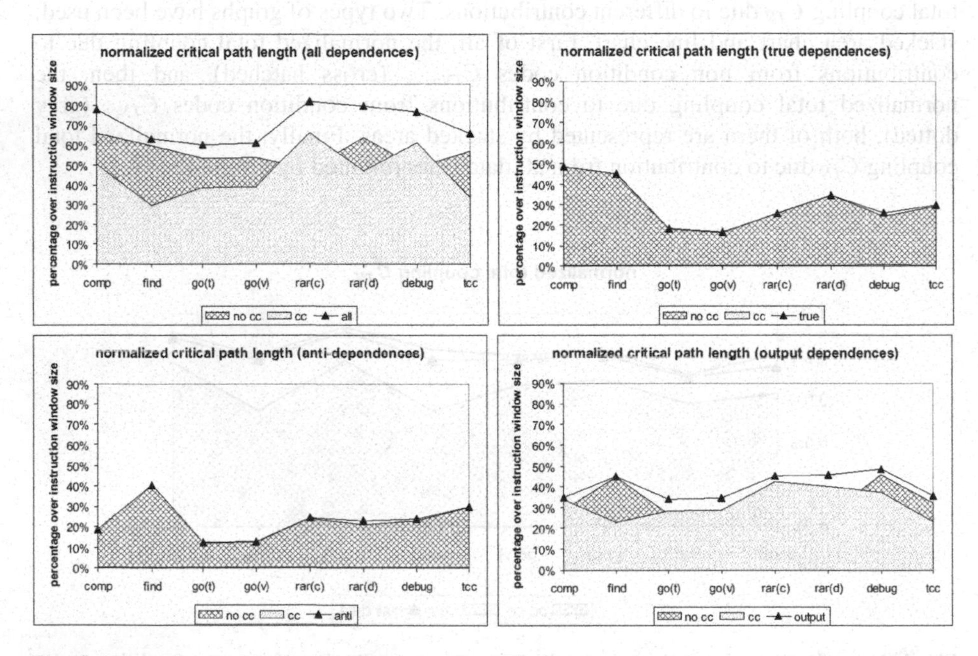

Fig. 3. Normalized critical path length for the different data dependences types: in overlapped areas the contributions from non condition codes and condition codes; in solid line, the critical path length due to the contribution from all data

7.3 Critical Path Length

Figure 3 presents the normalized critical path length L_N for the different data dependence types for all the programs in the testbench. The contributions from condition codes and non condition codes are shown in the overlapped areas. The normalized critical path length due to the contributions from all data is shown in solid line.

The first graph shows the compositions for all types of dependences. We can see that the normalized critical path length due to the contribution from condition code is in the range of the contribution from the other sources. Sometimes, it is even larger: in the case of *rar(c)* and *debug* the normalized critical path length L_{Ncc} exceeds the value obtained from the composition of the rest of sources L_{Nncc}. Indeed, for these two programs, the normalized critical path length L_{Ncc} is a low bound limit to the complete composition of dependences. The solid line seems to reflect the influence of the contributions due to condition codes especially on the mentioned programs.

The other graphs show that the only relevant contribution to the normalized critical path length from condition codes is the one due to output dependences.

Figure 4 provides a view of the contribution to L_N of the condition codes source for each dependence type on all program traces.

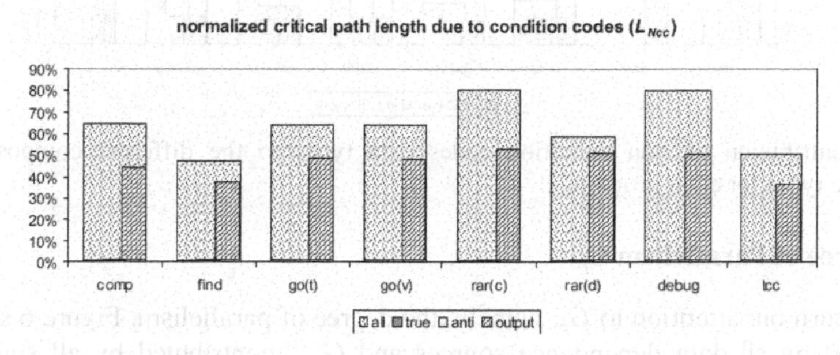

Fig. 4. Normalized critical path length due to condition codes: in stacked columns the contributions from the different types of data dependences; in the column in the back, the normalized critical path length due to the complete composition

We observe that condition codes contribution to normalized critical path length arises, mainly, from output dependences in agreement with the corresponding coupling measurements that can be observed in Fig. 2. However, the columns for true and anti-dependences are negligible; this is in contrast with the corresponding coupling measurements where the contribution due to true and anti- dependences are indeed significant (see Fig. 2).

Moreover, in general, the combination of true dependences and anti-dependences with the output dependences seems to enlarge the overall dependence chains. In fact, as we can see in Fig. 4, the normalized critical path length due to the complete composition is greater than the sum of the contributions from the different types of data dependences. This seems to suggest that the few existing true and anti-dependences would link two or more output dependence chains, thus producing a new longer chain.

Observe also that for the traces of the programs *rar(c)* and *debug* the resulting total dependences are much larger than for the rest of program traces. As both of these programs exhibit a block size quite small compared with that of the others programs from the testbench, there seems to be a correlation between the size of the basic block and this effect of "irregular enlargements."

Figure 5 shows the same information as the previous figure but for the rest of dependence sources. It is remarkable that this figure shows a more equilibrated contribution among the different dependence types when condition codes contribution is excluded. Composing all dependence types increases the normalized critical path length, although in a rather heterogeneous manner. Contrary to what happened in Fig. 4, we can see in Fig. 5 that the normalized critical path length due to the complete composition is smaller than the sum of the contributions from the different types of data dependences.

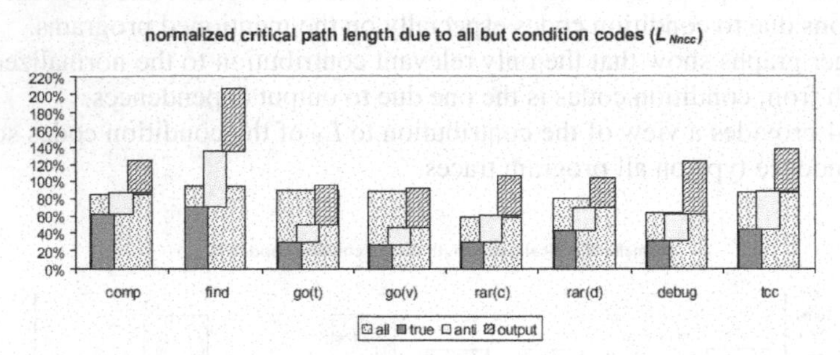

Fig. 5. Contribution of non condition codes data types to the different compositions of dependence types for each program

7.4 Degree of Parallelism G_p

Now, we turn our attention to G_p, namely, the degree of parallelism. Figure 6 shows G_p contributed by all data dependence sources and G_{pncc} contributed by all sources but condition codes. The per-program basic block size has been also presented in vertical bars.

As far as the parallelism degree by all data dependences is concerned, we observe that since G_p ranges from 1.22 to 1.67, the conclusion is that it is only possible to obtain a global parallelism of just about 2 instructions per computation step. This result is in agreement with the results obtained in other research work about the x86 ISA [9, 13, 14, 17], which is an important fact to validate our methodology.

If we avoid the impact of condition codes, we observe that G_{pncc} shows a greater degree of parallelism than G_p. In fact, according to vertical bars in Fig. 6, the programs *rar(c)* and *debug* display the smallest basic block size, and black squared

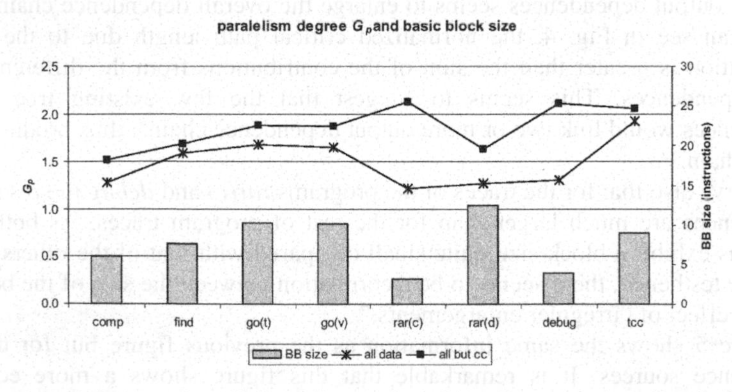

Fig. 6. Per program degree of parallelism G_p contributed by all sources (stars) and all sources but condition codes (black squares). Vertical bars show the per-program basic block size.

line displays that precisely those programs exhibit the highest potential for parallelism. This seems to suggest that some correlation between these two characteristics may be present.

8 Conclusions and Future Work

Focusing our analysis to dependences caused by condition codes, it becomes apparent that a larger basic block decrements the hazard of coupling due to true dependences but, at the same time, may also produce lengthening of output dependence chains.

The quantification based on coupling leads to the conclusion that condition codes show their coupling essentially through output dependences. In spite of this, when we measure critical path lengths, the contribution to it due to true and anti dependences increases substantially the length of the critical path due just to output dependences (see Fig. 4). Condition codes decrease the amount of available parallelism.

Figure 6 seems to suggest a correlation between the length of the dependence chains and the basic block size so that when the latter is short, the former is found to be higher. We conjecture that this may be due to the contribution of true dependences but this should be confirmed by further studies.

Register renaming techniques are not a suitable solution since they mean a waste of resources. For example, additional information should be stored if precise exceptions are to be maintained [10, 16].

Transforming the stream of CISC instructions to RISC instructions does not produce a substantial modification on the impact caused by condition codes to the instruction level parallelism, even at the micro-operation level.

Though this work has been done with the 16 bit subset of x86 ISA, we think that the conclusions are meaningful. As a future work we plan to extend the analysis to the full x86 ISA.

Since most condition codes have no computational meaning and are only originated due to the architecture of the x86 ISA, the conditional materialization could be a solution to be explored in future works. Actually, we are already evaluating a software conditional materialization that seems to be suitable in the dynamic binary translation setting performed in x86 processors. The method could be as good as the hardware conditional materialization used by *PowerPC* but maintaining binary compatibility of x86 family.

Acknowledgments. This work was partially supported by the Vicerrectorado de Investigación de la Universidad de Alcalá under Grant UAH PI2005/072.

References

1. Biggs, N.L.: Algebraic Graph Theory, 2nd edn. Cambridge University Press, Cambridge (1993)
2. Bose, P.: Instruction Set Design for Support of High-Level Languages. Ph.D. Thesis, University of Illinois at Urbana-Champaign (1983)
3. Durán, R., Rico, R.: Quantification of ISA Impact on Superscalar Processing. In: Proceeding of EUROCON 2005, pp. 701–704 (2005)

4. Durán, R., Rico, R.: On applying graph theory to ILP analysis. IEEE Latin America Transactions, 289–296 (2006)
5. Godsil, C.D., Royle, G.F.: Algebraic Graph Theory. Springer, Heidelberg (2001)
6. Gschwind, M.: Method for the deferred materialization of condition code information. Research Disclosures (1999)
7. Gschwind, M., Ebcioglu, K., Altman, E., Sathaye, S.: Binary Translation and Architecture Convergence Issues for IBM System/390. In: Proceedings of the 14th International Conference on Supercomputing, pp. 336–347 (2000)
8. Hu, S., Smith, J.E.: Using Dynamic Binary Translation to Fuse Dependent Instructions. In: Proceedings of the International Symposium on Code Generation and Optimization CGO, pp. 213–224 (2004)
9. Huang, I.J., Peng, T.C.: Analysis of x86 Instruction Set Usage for DOS/Windows Applications and Its Implication on Superscalar Design. IEICE Transactions on Information and Systems E85-D(6), 929–939 (2002)
10. Hwu, W.W., Patt, Y.N.: Checkpoint repair for out-of-order execution machines. IEEE Transactions on Computers C-36, 1496–1514 (1987)
11. Kim, I., Lipasti, M.H.: Macro-op Scheduling: Relaxing Scheduling Loop Constraints. In: Proceedings of the 36th International Symposium on Microarchitecture, pp. 1496–1514 (2003)
12. Maurer, W.D.: A theory of computer instructions. Journal of the ACM 13(2), 226–235 (1966)
13. Mutlu, O., Stark, J., Wilkerson, C., Patt, Y.N.: Runahead Execution: An Alternative to Very Large Instruction Windows for Out-of-order Processors. In: Proc. of the 9th Intl. Symp. on High-Performance Computer Architecture, pp. 129–140 (2003)
14. Rico, R., Pérez, J.I., Frutos, J.A.: The impact of x86 instruction set architecture on superscalar processing. Journal of Systems Architecture 51.1 (2005)
15. Skadron, K., Martonosi, M., August, D.I., Hill, M.D., Hill, D.J., Pai, V.S.: Challenges in Computer Architecture Evaluation. IEEE Computer 36.8 (2003)
16. Sohi, G.S.: Instruction Issue Logic for High-Performance, Interruptible, Multiple Functional Unit, Pipelined Computers. IEEE Transactions on Computers, 349–359 (1990)
17. Stark, J., Brown, M.D., Patt, Y.N.: On Pipelining Dynamic Instruction Scheduling Logic. In: Proc. of the 33rd Annual ACM/IEEE Intl. Symp. on Microarchitecture, pp. 57–66 (2000)
18. Stefanovic, D., Martonosi, M.: Limits and Graph Structure of Available Instruction-Level Parallelism. In: Proceedings of the European Conference on Parallel Computing (2000)
19. Wall, D.W.: Limits of instruction-level parallelism. In: Proc. of the 4th Intl. Conference on Architectural Support for Programming Languages and Operating Systems, pp. 176–188 (1991)

Efficient Message Management in Tiled CMP Architectures Using a Heterogeneous Interconnection Network

Antonio Flores, Juan L. Aragón, and Manuel E. Acacio

Departamento de Ingeniería y Tecnología de Computadores
University of Murcia, 30100 Murcia (Spain)
{aflores, jlaragon, meacacio}@ditec.um.es

Abstract. Previous studies have shown that the interconnection network of a Chip-Multiprocessor (CMP) has significant impact on both overall performance and energy consumption. Moreover, wires used in such interconnect can be designed with varying latency, bandwidth and power characteristics. In this work, we present a proposal for performance-and energy-efficient message management in tiled CMPs by using a heterogeneous interconnect. Our proposal consists of *Reply Partitioning*, a technique that classifies all coherence messages into critical and short, and non-critical and long messages; and the use of a heterogeneous interconnection network comprised of low-latency wires for critical messages and low-energy wires for non-critical ones. Through detailed simulations of 8- and 16-core CMPs, we show that our proposal obtains average improvements of 8% in execution time and 65% in the Energy-Delay^2 Product metric of the interconnect over previous works.

Keywords: Chip-Multiprocessor, Energy-Efficient Architectures, Heterogeneus On-Chip Interconnection Network, Parallel Scientific Applications.

1 Introduction

High performance processor designs are evolving toward architectures that implement multiple processing cores on a single die. Chip-multiprocessors (CMPs) can provide higher throughput, more scalability and greater energy-efficiency compared to wider-issue, single-core processors. Furthermore, energy-efficient architectures are currently one of the major goals pursued by designers in both high performance and embedded processor domains.

On the other hand, tiled architectures provide a scalable solution for supporting families of products with varying computational power, managing the design complexity, and effectively using the resources available in advanced VLSI technologies. Therefore, it is expected that future CMPs will be designed as arrays of replicated tiles connected over a switched direct network [1,2]. In these architectures, the design of the on-chip interconnection network has been shown to have significant impact on overall system performance and energy consumption, since

S. Aluru et al. (Eds.): HiPC 2007, LNCS 4873, pp. 133–146, 2007.

it is implemented using global wires that show long delays and high capacitance properties. Recently, Wang *et al.* [3] reported that the on-chip network of the Raw processor consumes 36% of total chip power. Magen *et al.* [4] also attribute 50% of overall chip power to the interconnect. Most of this power is dissipated in the point-to-point links of the interconnect [5,3].

By tuning wire's characteristics such as wire width, spacing or repeater size, it is possible to design wires with varying latency, bandwidth and energy properties [6]. Using links that are comprised of wires with different physical properties, a heterogeneous on-chip interconnection network is obtained. In [7], the authors propose the use of links that are comprised of two wire implementations apart from baseline wires (*B-Wires*): power optimized wires (*PW-Wires*) with fewer and smaller repeaters, and latency optimized wires (*L-Wires*) that have high widths and spacing. Then, coherence messages are mapped to the appropriate set of wires taking into account their latency and bandwidth requirements, obtaining a reduction in both execution time and energy consumption for a CMP with a two-level tree interconnect topology. Unfortunately, the authors report insignificant performance improvements for the direct network topologies employed in tiled CMPs (such as a 2D-mesh).

In this work, we present a proposal for efficient message management (from the point of view of both performance and energy) in tiled CMPs. Our proposal consists of two approaches. The first one is *Reply Partitioning*, a technique that allows all messages used to ensure coherence between the L1 caches of a CMP to be classified into two groups: a) critical and short, and b) non-critical and long messages. The second approach uses a heterogeneous interconnection network comprised of only two types of wires: low-latency wires for critical messages and low-energy wires for non-critical ones.

The main contribution of our proposal is the partitioning of reply messages that carry data into a short critical message containing the sub-block of the cache requested by the core as well as a long non-critical message with the whole cache line. This partitioning allows for a more energy-efficient use of the heterogeneous interconnect since now *all* short messages have been made critical whereas *all* long messages have been made non-critical. The former can be sent through the *L-Wires* whereas the latter can be sent through the *PW-Wires*. Differently to proposals in [7], our partitioning approach first, eliminates a complex logic for choosing the correct set of wires (we need a single bit in the message length field instead of checking the directory state or the congestion level of the network) and second, it obtains significant energy-delay improvements when direct topologies are used. Additionally, our proposal allows for a more balanced workload across the heterogeneous interconnect.

The rest of this paper is organized as follows. Our proposal for efficient message management in tiled CMPs is presented in section 2. Section 3 describes the evaluation methodology and presents the results of the proposed mechanism. Section 4 reviews some related work, and finally, section 5 summarizes the main conclusions of the work.

2 A Proposal for Efficient Message Management in Tiled CMPs

In this section we present our proposal for efficient message management in tiled CMPs. This section starts with a description of the tiled CMP architecture assumed in this paper, followed by a classification of the messages in terms of both their criticality and size and, finally, the description of the proposed *Reply Partitioning* mechanism.

2.1 Tiled CMP Architectures

A tiled CMP architecture consists of a number of replicated *tiles* connected over a switched direct network (Figure 1). Each tile contains a processing core with primary caches (both I- and D-caches), a slice of the L2 cache, and a connection to the on-chip network. The L2 cache is shared among the different processing cores, but it is physically distributed between them. Therefore, some accesses to the L2 cache will be sent to the local slice while the rest will be serviced by remote slices (L2 NUCA architecture). In addition, the L2 cache tags store the directory information needed to ensure coherence between the L1 caches. On a L1 cache miss, a request is sent down to the appropriate tile where further protocol actions are initiated based on that block's directory state. In this paper we assume a process technology of 65 *nm*, tile area of approximately 25 mm^2, and a die size in the order of 400 mm^2 [2]. Note that this area is similar to the largest die in production today (Itanium 2 processor – around 432 mm^2). Note also that, due to manufacturing costs and form factor limitations, it would be desirable to keep die size as low as possible. Further details about the evaluation methodology and the simulated CMP configuration can be found in section 3.1.

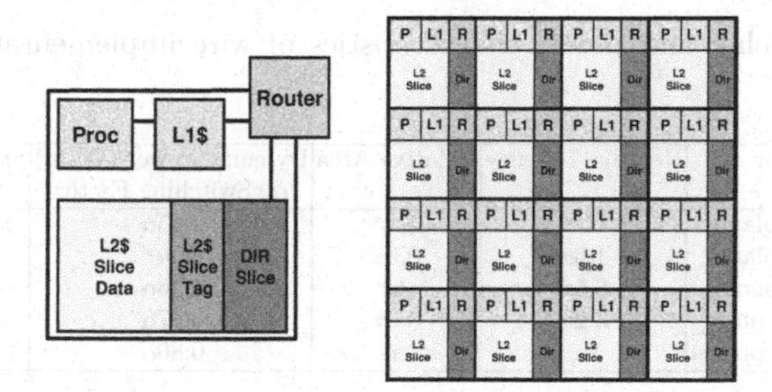

Fig. 1. Tiled CMP architecture overview

2.2 Classification of Messages in Tiled CMP Architectures

There are a variety of message types travelling on the interconnect of a CMP, each one with properties that are clearly distinct. In general, we can classify messages into the following groups: *Request messages*, that are generated by cache controllers in response to L1 cache misses and sent to the corresponding home L2 cache to demand privileges over a memory line. *Response messages* to these requests, generated by the home L2 cache controller or, alternatively, by the remote L1 cache that has the single valid copy of the data, and they can carry the memory line or not. *Coherence commands*, that are sent by the home L2 cache controller to the corresponding L1 caches to ensure coherence. *Coherence responses*, sent by the L1 caches back to the corresponding home L2 in response to coherence commands. *Replacement messages*, that the L1 caches generate in case of exclusive or modified lines being replaced.

Messages involved in the L1 cache coherence protocol can be classified according to their criticality into critical and non-critical messages. We say that a message is critical when it is in the critical path of the L1 cache miss, otherwise the message is non-critical. As expected, delaying a critical message will result in longer L1 cache miss latencies. On the other hand, slight slowdowns in the delivery of non-critical messages will not cause any performance degradation.

Using this criterion, we can observe that all message types but replacement messages and some coherence replies (such as revision messages) are critical. It is clear that performance is increased if critical messages are sent through low-latency *L-Wires*. At the same time energy is saved, without affecting performance, when non-critical messages travel on slower, power-efficient *PW-Wires*.

On the other hand, coherence messages can also be classified according to their size into short and long messages. Coherence responses do not include the address or the data block and just contain control information (source/destination, message type, MSHR id, etc). In this way, we can say that they are short messages. Other message types, in particular requests and coherence commands, also contain address block information but they are still narrow enough to be classified as short messages. Finally, replacements with data and data block transfers also carry a cache line and, therefore, they are long messages.

Table 1. Area, delay and power characteristics of wire implementations (extracted from [7])

Wire Type	Relative Latency	Relative Area	Dynamic Power (W/m) α=Switching Factor	Static Power W/m
B-Wire (8X plane)	$1x$	$1x$	2.65α	1.0246
B-Wire (4X plane)	$1.6x$	$0.5x$	2.9α	1.1578
L-Wire (8X plane)	$0.5x$	$4x$	1.46α	0.5670
PW-Wire (4X plane)	$3.2x$	$0.5x$	0.87α	0.3074
PW-Wire (8X plane)	$2x$	x	0.80α	0.2720

Table 1 shows the relative area, delay and power characteristics of *L-* and *PW-Wires* compared to baseline wires (*B-Wires*), as reported in [7]. A 65 *nm* process technology is considered, where 4X and 8X metal planes are used for global inter-core wires that are routed over memory arrays, as in [8]. It can be seen that *L-Wires* yield a two-fold latency improvement at a four-fold area cost. On the other hand, *PW-Wires* are designed to reduce power consumption with twice the delay of baseline wires (and the same area cost).

Regarding the power dissipated by each message type, Figure 2 plots their power breakdown for the baseline configuration using only *B-Wires*. As it can be seen, most of the power in the interconnect is associated to reply messages that carry L2 cache lines (55%-65%). As previously commented, most of this power is dissipated in the point-to-point links and, therefore, message size plays a major role.

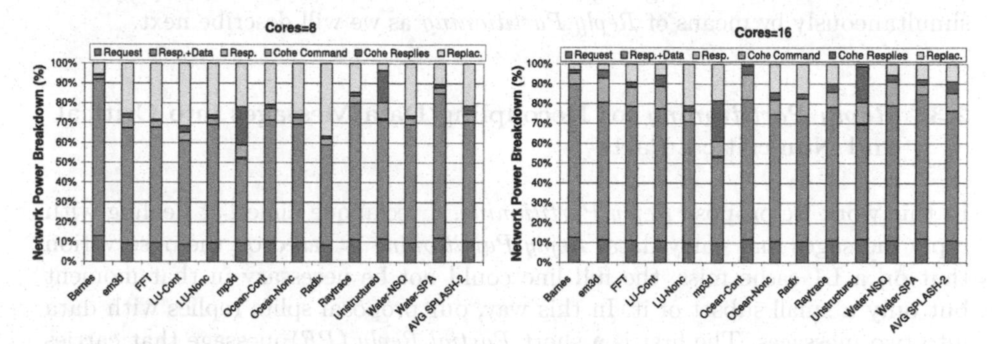

Fig. 2. Percentage of the power dissipated in the interconnection network by each type of message for an 8-core (left) and a 16-core CMP (right)

The use of a heterogeneous interconnect comprised of low-latency *L-Wires* and power-efficient *PW-Wires* allows for a more energy-efficient interconnect utilization. However, as the number of *L-Wires* is smaller because of their four-fold area cost (relative to baseline wires) only short messages can take full advantage of them. On the other hand, since message size has direct impact on the power dissipated in the interconnect, significant energy savings can be obtained when long messages are sent using *PW-Wires*.

Table 2. Classification of the messages that travel on the interconnection network according to their criticality and length

Message Type	Critical?	Length	Preferred Wires (assuming unbounded number)
Request	Yes	Short	L-Wires
Response	Yes	Short/Long	L-Wires (Performance) / PW-Wires (Energy)
Cohe Command	Yes	Short	L-Wires
Cohe Replies	Yes/No	Short	L-Wires (Critical) / PW-Wires (Non-critical)
Replacements	No	Short/Long	PW-Wires

Table 2 summarizes the characteristics of each message type and points out the links that would be preferred in every case. In general, short messages are critical and, therefore, should be sent using *L-Wires*. On the other hand, long messages can be critical (responses with data) or non-critical (replacements with data), and the choice of wires is not clear. If we pursuit performance (criticality), then *L-Wires* might be the best choice. Contrarily, if energy savings are more important (length), then *PW-Wires* should be utilized. In this way, the policy of sending critical messages on *L-Wires* and non-critical on *PW-Wires* leaves the latter links underutilized since only replacements would make use of them, and small energy savings would be obtained. On the other hand, the policy of sending long messages on *PW-Wires* and short ones on *L-Wires* causes important delays to responses with data, which would finally translate into intolerable performance degradation. Differently to the policies proposed in [7], which are mainly based on message criticality, in this work we present an efficient message management mechanism based on taking advantage of both criticality and length properties simultaneously by means of *Reply Partitioning* as we will describe next.

2.3 *Reply Partitioning* for Decoupling Data Messages into Critical and Non-critical Parts

In this work we propose *Reply Partitioning*, a technique aimed at dealing with reply messages that carry data. *Reply Partitioning* is based on the observation that on a L1 cache miss, the full line could not be necessary in that moment but only a small subset of it. In this way, our proposal splits replies with data into two messages. The first is a short *Partial Reply* (*PR*) message that carries a sub-block of the cache line that includes the word requested by the processor. And the second message, called *Ordinary Reply* (*OR*), is the original message and includes the full cache line.

This division of replies with data into *PRs* and *ORs* makes *all* critical messages short (note that *PRs* are critical since they contain the word requested by the processor) and, therefore, they can be sent using the low-latency *L-Wires*. At the same time, *all* long messages are non-critical (note that *ORs* become non-critical as the requested word also travels on a short message that hopefully will arrive sooner) and they can be sent using the power-efficient *PW-Wires* without hurting performance.

Additionally, splitting reply messages into a critical *PR* and a non-critical *OR* has slight implications on the coherence protocol. Recall that, in a non-blocking cache, MSHR registers are used to keep track of outstanding misses. In our mechanism, we have two different replies, and we need to define the actions required after the arrival of each one. Furthermore, with direct networks, the arrival order is not guaranteed and, although unlikely, the non-critical *OR* could be received before the critical *PR*.

When a *Partial Reply* arrives we are sure that all coherence actions have been done. Therefore, after its arrival all waiting requests that can be satisfied are processed (e.g., read requests that hit in the cache line sub-block and all write requests). For a read request, the corresponding value is sent to the processor and, for a write request, the value is held in the MSHR but the rest of hardware resources are released. In both cases, appropriate processor instructions are committed. Only MSHR with no processed read requests and all the write requests, if any, are kept until the arrival of the *Ordinary Reply*. At the *OR* arrival time, the rest of read requests are performed and the block is modified with the corresponding write values held in the MSHR. In case of receiving the *OR* before the *PR*, all requests waiting in the MSHR register are processed and the corresponding instructions are committed; all hardware resources are released but the MSHR, which is released when both replies arrive.

2.4 Interconnect Design for Efficient Message Management

As discussed previously, *L-Wires* have a four-fold area cost compared to baseline wires and, therefore, the number of *L-Wires* is quite limited. Considering that they will be used for sending short, critical messages, the number of wires should be fixed by considering the typical size of short messages. The remaining area will be used by *PW-Wires* for sending long, non-critical messages.

In this work, we use the same main parameters for the interconnect as in [7]. In particular, message sizes and the width of the original links of the interconnect are the same. Short messages can take up to 11 bytes. Requests, coherence commands and partial replies are 11-byte long since beside control information (3 bytes) they also carry address information (in the first two cases) or the sub-block of data of one word size (for *PRs*). On the other hand, coherence replies are just 3-byte long. Finally, *OR* reply messages are 67-byte long since they carry control information (3 bytes) and a cache line (64 bytes).

In order to match the metal area of the baseline configuration, each original 75-byte unidirectional link (600 *B-Wires*) is designed to be made up of 88 *L-Wires* (11 bytes) and 248 *PW-Wires* (31 bytes). For a discussion regarding the implementation complexity of heterogeneous interconnects refer to [7].

The resulting design is similar to that proposed in [7], but with some important differences. First of all, the election of the right set of wires for a message does not require any additional logic since it can be made based exclusively on one bit in the length field (some of the proposals developed in [7] require checking the directory state or tracking the congestion level in the network). Secondly, the routing logic and the multiplexers and de-multiplexers associated with wires are simpler since we only consider two types of wires instead of three. Finally, our proposal achieves better levels of utilization of each set of wires (as we will discuss in section 3.2).

3 Experimental Results

This section shows the results that are obtained for our proposal and compares them against those achieved by two different configurations of the interconnect. The first is the configuration that employs just *B-Wires*, which is taken as baseline. The second configuration is an implementation of the 3-subnetwork heterogeneous interconnect proposed in [7] that uses *L-*, *B-* and *PW-Wires*.

3.1 Evaluation Methodology

The results presented in this work have been obtained through detailed simulations of a full CMP. We have employed a cycle-accurate CMP power-performance simulation tool, called *Sim-PowerCMP* [9], that estimates both dynamic and leakage power and is based on RSIM [10]. RSIM is a detailed execution-driven simulator that models out-of-order superscalar processors (although in-order issue is also supported), several levels of caches, an aggressive memory and the interconnection network, including contention at all resources. In particular, our simulation tool employs as performance simulator a modified version of RSIM that models the architecture of the tiled CMP presented in section 2. We have incorporated into our simulator power models already proposed and validated for both dynamic power (from Wattch [11]) and leakage power (from HotLeakage [12]) of each processing core, as well as the interconnection network (from Orion [13]). Further details about the implementation and validation of *Sim-PowerCMP* can be found in [9].

Table 3. Configuration of the baseline CMP architecture (left) and applications evaluated (right)

CMP Configuration	
Process technology	65 *nm*
Tile area	25 mm^2
Number of tiles	8, 16
Cache line size	64 bytes
Core	4GHz, in-order 2-way model
L1 I/D-Cache	32KB, 4-way
L2 Cache (per core)	256KB, 4-way, 10+20 cycles
Coherence Protocol	MESI (with $-$ transfers)
Directory access time	10 cycles (tags access)
Memory access time	400 cycles
Network configuration	2D mesh (BW of 75 GB/s)
Router latency	1 cycle
Link width	75 bytes (*8X-B-Wires*)
Link latency/length	4 cycles / 5 *mm*

Application	Problem size
Barnes-Hut	16K bodies, 4 timesteps
EM3D	9600 nodes, 5% remote links, 4 timesteps
FFT	256K complex doubles
LU-cont	256 × 256, B=8
LU-noncont	256 × 256, B=8
MP3D	50000 nodes, 2 timesteps
Ocean-cont	258 × 258 grid
Ocean-noncont	258 × 258 grid
Radix	2M keys
Raytrace	car.env
Unstructured	mesh.2K, 5 timesteps
Water-nsq	512 molecules, 4 timesteps
Water-spa	512 molecules, 4 timesteps

Table 3 (left) shows the architecture configuration used along this paper. It describes an 8- and 16-core CMP built in 65 *nm* technology. The tile area has been fixed to 25 mm^2, including a portion of the second-level cache [2]. With this configuration, links that interconnect routers configuring the 2D mesh topology would measure around 5 *mm*. Reply messages are 67-byte long since they carry control information (3-bytes) and a cache line (64 bytes). On the contrary, request, coherence and coherence reply messages that do not contain data are, at most, 11-byte long (just 3-byte long for coherence replies).

Table 3 (right) shows the applications used in our experiments. *EM3D* and *Unstructured* are from the Berkeley suite, the rest of them are from the SPLASH/SPLASH-2 benchmark suites. Problem sizes have been chosen commensurate with the size of the L1 caches and the number of cores used in our simulations. All experimental results reported in this work are for the parallel phase of these applications. Data placement in our programs is either done explicitly by the programmer or by our simulator which uses a first-touch policy on a cache-line granularity. Thus, initial data-placement is quite effective in terms of reducing traffic in the interconnection network.

In order to match the metal area of the baseline configuration, each original 75-byte unidirectional link (600 *B-Wires*) is designed to be made up of 88

L-Wires (11 bytes) and 248 *PW-Wires* (31 bytes) using the 8X metal plane. For comparison purposes, the 3-subnetwork heterogeneous interconnect described in [7] was also implemented. In that configuration, each link is comprised of three types of wires in two metal planes. Each wire type has the area, delay, and power characteristics described in Table 1, so each original link is designed to be made up of 24 *L-Wires* (3 bytes), 512 *PW-Wires* (64 bytes), and 256 *B-Wires* (32 bytes).

3.2 Simulation Results and Analysis

In this section, we report on our simulation results. First of all, we show how messages distribute between the different types of wires of the heterogeneous networks evaluated in this work. Then, we analyze the impact of our proposal on execution time and on the energy dissipated by the inter-core links. Finally, we report the energy and energy-delay^2 product (ED^2P) metrics for the full CMP. As in [7], all results have been normalized with respect to the baseline case where only *B-Wire*, unidirectional 75-byte wide links are considered.

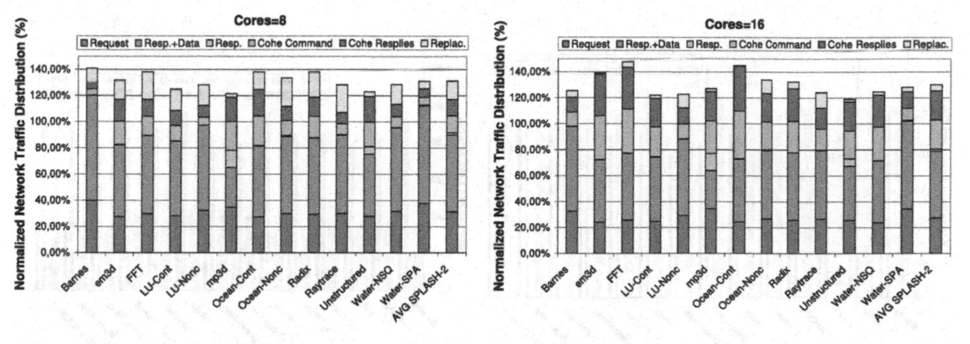

Fig. 3. Breakdown of the messages that travel on the interconnection network for an 8-core (left) and 16-core CMP (right) when an *L-Wire/PW-Wire* heterogeneous network is used and long critical messages are splitted

Figure 3 plots the percentage of each message type over the total number of messages sent in the baseline configuration for the 8- and 16-core configurations. It is important to note that *Reply Partitioning* increases the total number of messages that travel on the interconnect. The reason is that replies with data are converted into two messages, the *Partial Reply* and the *Ordinary Reply*. In our particular case, we have observed that the number of messages increases around 30% on average. This extra traffic has been considered in all our evaluations.

Figure 4 plots the workload distribution between the different types of wires for both the 3- and 2-subnetwork heterogeneous interconnects. This figure is obtained measuring the traffic observed for each type of wire, normalized to the width of the wire. As it can be seen in the left graph, there is an underutilization of *L-* and *PW-* wires that leads to an imbalanced workload distribution when

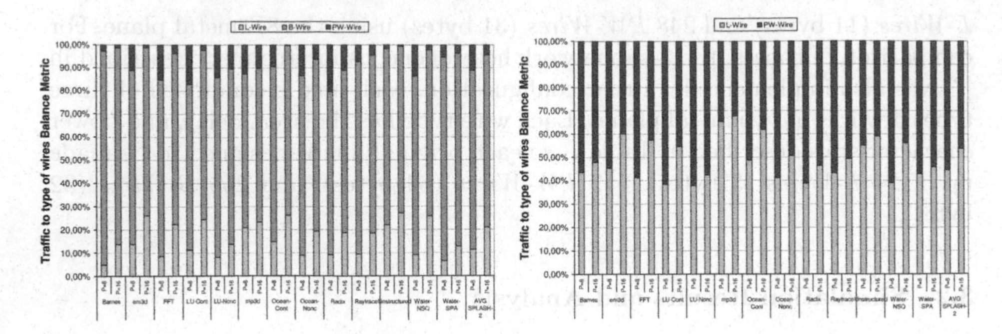

Fig. 4. Workload distribution for the 3-subnetwork approach (as in [7]) (left) and for the 2-subnetwork approach (right)

the 3-subnetwork configuration is used. However, the use of a 2-subnetwork interconnect, where *B-Wires* have been replaced by wider *L-Wires*, in conjunction with the *Reply Partitioning* technique, leads to a much more balanced workload distribution (Figure 4, right).

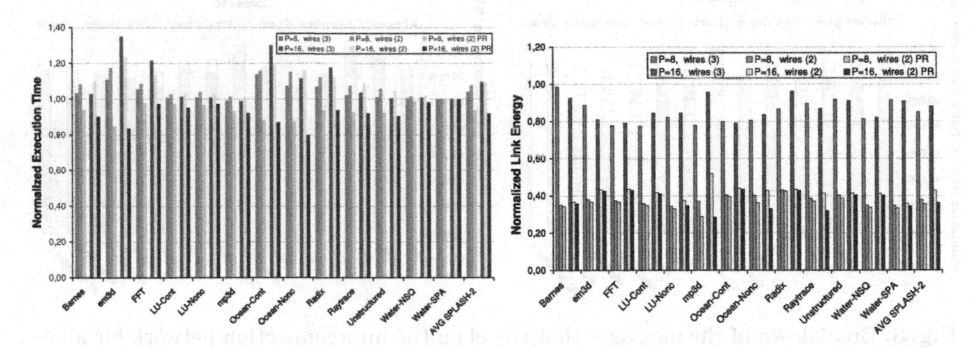

Fig. 5. Normalized execution time (left) and link energy (right) when heterogeneous links are used

Figure 5 (left) depicts the normalized execution time with respect to that obtained for the baseline configuration for an 8- and a 16-core CMP. The first barline (*P=8/16, wires (3)*) shows the normalized execution time for the 3-subnetwork interconnect (as in [7]). An average performance degradation of 4-8% is observed, which is a trend also reported in [7] when a 2D torus topology is employed. The reason of this degradation is the low use of the *L-Wires* as it was shown in Figure 4. Similar results are obtained when a 2-subnetwork interconnect (*L-Wire/PW-Wire*) is considered without using the proposed *Reply Partitioning* mechanism, as shown by the second barline (*P=8/16, wires (2)*). The reason of this performance degradation is the increased latency of the reply messages that carry data (sent through the slower *PW-Wires*) which cannot be hidden by using faster *L-Wires* for critical messages. This degradation has high variability, ranging from almost negligible for MP3D and Water-NSQ applications to almost 40% for Ocean-Cont application. This result is quite interesting because it shows that the increased latency imposed by the use of *PW-Wires* for replies with data can be hidden in some applications, whilst in others, as Barnes or Ocean-Cont, it translates into significant performance degradation. This high variability is related with the access pattern and utilization of the cache lines. A sequential access pattern to the whole cache lines leads to a performance degradation whereas with a more irregular access pattern or an underutilization of the cache lines the increased latency is hidden. Finally, the third barline (*P=8/16, wires (2) PR*) shows the case when reply messages are split into critical, short *Partial Replies* (PR) and non-critical *Ordinary Replies.* On average, we observe performance improvements of 16% over the two previous options for a 16-core CMP as a direct consequence of the better distribution of the messages between *L-Wires* and *PW-Wires* that *Reply Partitioning* allows for. Again, high variability is found, with improvements ranging from 1-2% in some applications to 50-55% in other. Compared with the baseline configuration, where no heterogeneous network is used, an average performance improvement of 8% is obtained.

Figure 5 (right) plots the normalized link energy. Our proposed *Reply Partitioning* approach results in an average reduction of 60%-65% in the energy dissipated by the inter-core links. This reduction shows little variability. The ED^2P metric shows average improvements close to 75% although, in this case, the variability between applications is higher because in the ED^2P metric the execution time gains importance. Note also that, although *Reply Partitioning* increases the total number of messages that travel on the interconnect around 30% on average, the use of *PW-Wires* for sending long reply messages leads to important energy savings that overcome this drawback.

Finally, Figure 6 presents both the normalized energy and ED^2P product metrics for the full CMP. As it can be observed, important energy savings are obtained for our proposal. The magnitude of these savings depends on the total number of cores of the CMP, ranging from 12% for the 8-core configuration to 17% for the 16-core configuration. On the other hand, when the ED^2P metric

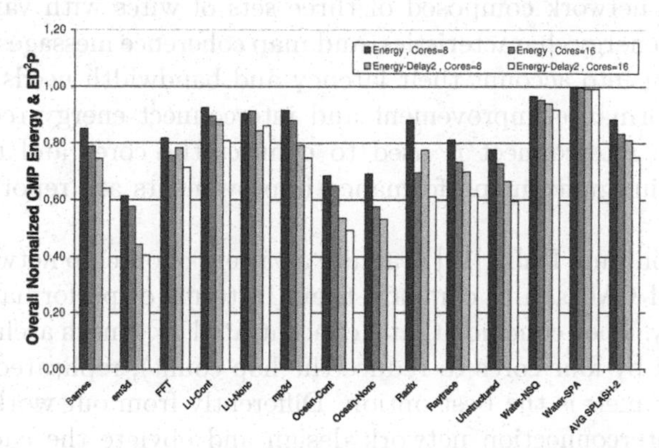

Fig. 6. Normalized energy and ED^2P for the full CMP

is considered, we find an increased improvement which ranges from 19% for the 8-core CMP to 25% for the 16-core one, due to bigger emphasis on the execution time.

4 Related Work

The on-chip interconnection network is a critical design element in a multi-core architecture and, consequently, it is the subject of several recent works. Among others, Kumar *et al.* [8] analyze several on-chip interconnection mechanisms and topologies, and quantify their area, power, and latency overheads. The study concludes that the design choices for the interconnect have a significant effect on the rest of the chip, potentially consuming a significant fraction of the real estate and power budget.

A reduced number of works have attempted to exploit the properties of a heterogeneous interconnection network at the microarchitectural level in order to reduce the interconnect energy share. Beckmann and Wood [14,15] propose the use of transmission lines to access large L2 on-chip caches in order to reduce the required cache area and the dynamic power consumption of the interconnection network. Nelson *et al.* [16] propose the use of silicon-based on-chip optical interconnects for minimizing the performance gap that the separation of the processing functions creates in a clustered architecture in an effort to alleviate power density. In [17], Balasubramonian *et al.* make the first proposal of wire management at the microarchitectural level. They introduce the concept of a heterogeneous interconnect that is comprised of wires with varying area, latency, bandwidth and energy characteristics, and they apply it to register communication within a clustered architecture. Subsequently, they extend their proposal in [18] with new techniques aimed at accelerating cache accesses in large L2/L3 splitted caches by taking advantage of a lower-bandwidth, lower-latency network.

Very recently, Cheng *et al.* [7] applied the heterogeneous network concept to the cache coherence traffic problem in CMPs. In particular, they propose an interconnection network composed of three sets of wires with varying latency, bandwidth and energy characteristics, and map coherence messages to the appropriate set taking into account their latency and bandwidth needs. They report significant performance improvement and interconnect energy reduction when a two-level tree interconnect is used to connect the cores and the L2 cache. Unfortunately, insignificant performance improvements are reported for direct topologies.

Finally, Balfour and Dally [19] evaluate a variety of on-chip networks designed for 64-core tiled CMPs, and compare them in terms of performance, area and energy efficiency. They conclude that a concentrated 4×4 mesh architecture (each router is shared by four cores to reduce the hop count), replicated subnetworks and express channels is the best option. Differently from our work, the authors focus on the interconnection network design and obviate the cache coherence protocol (they assume an abstract communication protocol).

5 Conclusions

In this work we propose a performance- and energy-efficient message management mechanism for tiled CMPs that consists of two approaches. The first one is *Reply Partitioning*, a technique that allows all coherence messages to be classified into two groups: critical and short, and non-critical and long. In particular, *Reply Partitioning* concentrates on replies that carry data and splits them into a critical and short *Partial Reply* message that carries the word requested by the processor and a non-critical *Ordinary Reply* with the full cache block. The second approach of our proposal is the use of a heterogeneous interconnection network comprised of low-latency wires for critical messages and low-energy wires for non-critical ones which also allows for a more balanced workload.

Results obtained through detailed simulations of 8- and 16-core CMPs show that the proposed on-chip message management mechanism can reduce the power dissipated by the links of the interconnection network about 65% with an additional reduction in execution time of 8% over previous works. Finally, these reductions translate into overall CMP energy savings ranging from 12% for the 8-core configuration to 17% for the 16-core one (from 19% to 25% if the ED^2P metric is considered). These results reveal that correctly organizing the interconnection network and properly managing the different types of messages through it have significant impact on the energy consumed by CMPs, especially for next-generation dense CMP architectures.

Acknowledgments. The authors would like to thank the anonymous reviewers for their helpful insights. This work has been jointly supported by the Spanish MEC and European Comission FEDER funds under grants "Consolider Ingenio-2010 CSD2006-00046" and "TIN2006-15516-C4-03".

References

1. Taylor, M.B., Kim, J., et al.: The Raw Microprocessor: A Computational Fabric for Software Circuits and General-Purpose Programs. IEEE Micro 22(2), 25–35 (2002)
2. Zhang, M., Asanovic, K.: Victim Replication: Maximizing Capacity while Hiding Wire Delay in Tiled Chip Multiprocessors. In: Proc. of the 32nd Int'l Symp. on Computer Architecture, pp. 336–345 (2005)
3. Wang, H., Peh, L.S., Malik, S.: Power-driven Design of Router Microarchitectures in On-chip Networks. In: Proc. of the 36th Int'l Symp. on Microarchitecture, pp. 105–111 (2003)
4. Magen, N., Kolodny, A.W., et al.: Interconnect-power dissipation in a microprocessor. In: Proc. of the 2004 Int'l Workshop on System Level Interconnect Prediction, pp. 7–13 (2004)
5. Shang, L., Peh, L., Jha, N.: Dynamic voltage scaling with links for power optimization of interconnection networks. In: Proc. of the 9th Int'l Symp. on High-Performance Computer Architecture, pp. 91–102 (2003)
6. Banerjee, K., Mehrotra, A.: A power-optimal repeater insertion methodology for global interconnects in nanometer designs. IEEE Trans. on Electron Devices 49(11), 2001–2007 (2002)

7. Cheng, L., Muralimanohar, N., et al.: Interconnect-Aware Coherence Protocols for Chip Multiprocessors. In: Proc. of the 33rd Int'l Symp. on Computer Architecture, pp. 339–351 (2006)
8. Kumar, R., Zyuban, V., Tullsen, D.M.: Interconnections in Multi-Core Architectures: Understanding Mechanisms, Overheads and Scaling. In: Proc. of the 32nd Int'l Symp. on Computer Architecture, pp. 408–419 (2005)
9. Flores, A., Aragón, J.L., Acacio, M.E.: Sim-PowerCMP: A Detailed Simulator for Energy Consumption Analysis in Future Embedded CMP Architectures. In: Proc. of the 4th Int'l Symp. on Embedded Computing, pp. 752–757 (2007)
10. Hughes, C.J., Pai, V.S., et al.: RSIM: Simulating Shared-Memory Multiprocessors with ILP Processors. IEEE Computer 35(2), 40–49 (2002)
11. Brooks, D., Tiwari, V., Martonosi, M.: Wattch: a framework for architectural-level power analysis and optimizations. In: Proc. of the 27th Int'l Symp. on Computer Architecture, pp. 83–94 (2000)
12. Zhang, Y., Parikh, D., et al.: HotLeakage: A Temperature-Aware Model of Subthreshold and Gate Leakage for Architects. Technical report, University of Virginia (2003)
13. Wang, H.S., Zhu, X., et al.: Orion: a power-performance simulator for interconnection networks. In: Proc. of the 35th Int'l Symp. on Microarchitecture, pp. 294–305 (2002)
14. Beckmann, B.M., Wood, D.A.: TLC: Transmission Line Caches. In: Proc. of the 36th Int'l Symp. on Microarchitecture, pp. 43–54 (2003)
15. Beckmann, B.M., Wood, D.A., et al.: Managing Wire Delay in Large Chip-Multiprocessor Caches. In: Proc. of the 37th Int'l Symp. on Microarchitecture, pp. 319–330 (2004)
16. Nelson, N., Briggs, G., et al.: Alleviating Thermal Constraints while Maintaining Performance via Silicon-Based On-Chip Optical Interconnects. In: Workshop on Unique Chips and Systems (2005)
17. Balasubramonian, R., Muralimanohar, N., et al.: Microarchitectural Wire Management for Performance and Power in Partitioned Architectures. In: Proc. of the 11th Int'l Symp. on High-Performance Computer Architecture, pp. 28–39 (2005)
18. Muralimanohar, N., Balasubramonian, R.: The Effect of Interconnect Design on the Performance of Large L2 Caches. In: 3rd IBM Watson Conf. on Interaction between Architecture, Circuits, and Compilers (P=ac2) (2006)
19. Balfour, J., Dally, W.J.: Design tradeoffs for tiled CMP on-chip networks. In: Proc. of the 20th Int'l Conf. on Supercomputing, pp. 187–198 (2006)

Direct Coherence: Bringing Together Performance and Scalability in Shared-Memory Multiprocessors

Alberto Ros, Manuel E. Acacio, and José M. García

Departamento de Ingeniería y Tecnología de Computadores
Universidad de Murcia, 30100 Murcia (Spain)
{a.ros,meacacio,jmgarcia}@ditec.um.es

Abstract. Traditional directory-based cache coherence protocols suffer from long-latency cache misses as a consequence of the indirection introduced by the home node, which must be accessed on every cache miss before any coherence action can be performed. In this work we present a new protocol that moves the role of storing up-to-date coherence information (and thus ensuring totally ordered accesses) from the home node to one of the sharing caches. Our protocol allows most cache misses to be directly solved from the corresponding remote caches, without requiring the intervention of the home node. In this way, cache miss latencies are reduced. Detailed simulations show that this protocol leads to improvements in total execution time of 8% on average over a highly optimized MOESI directory-based protocol.

1 Introduction

Shared-memory multiprocessors are quite popular since the communication between the processors that conform the machine occurs implicitly as a result of conventional memory access instructions (i.e. loads and stores), which makes them easier to program than message-passing multiprocessors. In most of these architectures, memory accesses are accelerated using one or several levels of private caches to each processor. Caches are made transparent to software through a cache coherence protocol. Supporting cache coherence in hardware, however, requires important engineering efforts.

In general, there are several approaches to solve the cache coherence problem in hardware. Snoopy protocols [5] typically rest on one or several buses to broadcast coherence operations. In this way, coherence messages go directly from the requesting caches to their proper recipients (those caches that hold a copy of the corresponding memory block), which reduces cache miss latencies. TokenB [9] removes the requirement of using buses and enable low-latency cache-to-cache transfer misses on unordered interconnection networks. Unfortunately, the fact that the latter two alternatives are based on broadcasting coherence actions restricts their scalability. Currently, scalable cache coherence is based on a distributed directory that keeps the location and state of cached blocks (directory-based protocols [5]). In these protocols, each memory block is assigned to the

S. Aluru et al. (Eds.): HiPC 2007, LNCS 4873, pp. 147–160, 2007.

home node which keeps the directory information for the memory block and acts as an *intermediary* for it. When a cache miss takes place, a request is sent over an unordered interconnection network to the corresponding home node, which performs the coherence actions necessary to satisfy the miss. In this way, apart from providing main memory storage for every memory block and keeping the associated directory information, the home node acts as an ordering point for the different requests that several caches issue over the block.

The fact that cache misses must reach the home node before any coherence action can be performed introduces indirection, which adds unnecessary hops (and thus, cycles) into the critical path of cache misses, finally resulting in long cache miss latencies. Moreover, the increasing gap between processor and memory speeds (the memory wall problem [16]) and the availability of low-latency interconnects make that cache coherence protocols that exploit cache-to-cache transfers for blocks in shared state (MOESI-like protocols) will be preferable to those that obtain them from main memory (MESI-like protocols)[1]. This results in a very significant fraction of the cache misses suffering from indirection.

In this work, we address the design of a solution to the cache coherence problem that avoids this indirection without using any brute-force method (as broadcasting requests) or requiring particular network topologies. The later two aspects compromise scalability. In particular we present *Direct Coherence*, a novel cache coherence protocol that based on MOESI decouples the role of providing main memory storage for every memory block, which is still responsibility of the home, from the role of storing up-to-date sharing information (and thus ensuring totally ordered accesses) for every memory block, which is moved from the home to one of the nodes that actually shares the block, particularly the node that provides the block on a cache miss. We call this node the *owner* node, and that copy of the block will be the *primary* copy.

In Direct Coherence, each cache keeps up-to-date sharing information for every primary copy of a block stored on it and every miss is solved by sending the request to the owner node instead of the home node. We have found that for most cache misses the owner node is the last node that invalidated the copy from the rest of caches. Hence, this information can be stored in a small structure to find the owner node when a subsequent miss takes place. Moreover, as the owner node changes on write misses, the requests sent by several caches for a particular block could be distributed among different nodes, thus helping prevent potential bottlenecks at the home node, and therefore, helping scalability.

Direct Coherence, therefore, reduces the latency of cache misses by avoiding the indirection added by the access to the home node. In this way, our proposal offers both the performance advantage of snoopy-based protocols, since coherence messages are directly sent from the requesting caches to those that must observe them, and the scalability of directory-based ones, since our proposal is

[1] Cache-to-cache transfers of clean data has also been recently used as a simple form of cooperation that reduces the number of off-chip accesses in CMPs [3]. In the context of cc-NUMAs, it has been also shown that cache-to-cache transfer for clean blocks can reduce average cache miss latency [14].

not based on any brute-force method or requires any particular network topology. Detailed simulations using a modified version of RSIM and several scientific applications demonstrate that using Direct Coherence most of the cache misses can be completed without requiring indirection, which leads to improvements in total execution time of 8% on average over a highly optimized MOESI protocol. In this work, Direct Coherence has been evaluated in the context of cc-NUMAs, although it is equally applicable to other domains, such as CMPs.

The rest of the paper is organized as follows. Direct Coherence is described in Section 2. Section 3 introduces the methodology employed in the evaluation. In Section 4 we show the performance results obtained for our proposal. In Section 5 we present a review of the related work. Finally, Section 6 concludes the paper.

2 Direct Coherence

2.1 The Owner Node and the Home Node

In directory-based protocols the home node maintains cache coherence and all the misses must go through it to obtain the directory information. Direct Coherence avoids this indirection by storing the directory information in the node that must provide the block in case of cache misses, the *owner* node, and by assigning the role of keeping cache coherence to this node. Then, when a cache miss takes place the request is sent to the owner node instead of the home node. Since the owner node is no longer fixed and can change on write misses, it is necessary to keep the identity of the current owner node in some place. In particular, the *home* node has the role of storing the identity of the owner and it is notified of every change.

The owner node of a block is either main memory when the block is not stored in any cache, an L2 cache in exclusive state, or the last L2 cache that wrote the block when there are multiple sharers. In this way, it is easy to find out the owner node because the other nodes can easily store the identity of the last node that invalidated their copy. Moreover, being the owner node the last one that wrote the block, many upgrades avoid indirection for some common sharing patterns. For the producer-consumer pattern, the node that updates the block is always the same one, and therefore the upgrades always take place in the owner node. For the migratory-sharing pattern, upgrades that follow the load misses just need two hops since the identity of the owner is known once the load misses have been completed, and the owner is the only node that must be invalidated in this case.

2.2 Changes to the Structure of the L2 Caches

Direct Coherence requires the L2 caches included in each node of the system to store extra coherence information. This information can be divided into the following three categories:

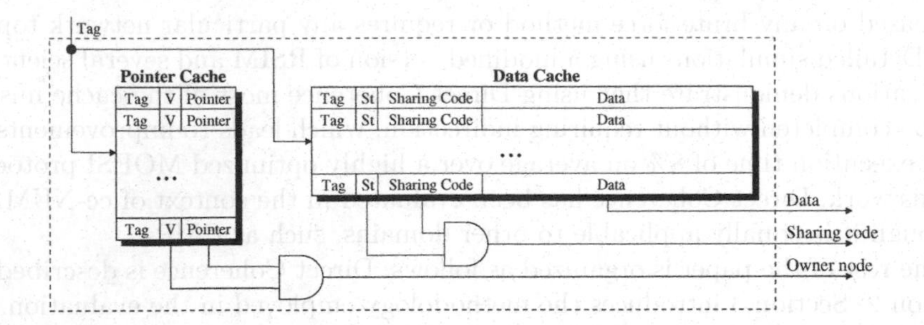

Fig. 1. Organization of the L2 caches required by Direct Coherence

- *Updated Sharing Information* (USI): This information is necessary for all the primary copies of all the blocks stored in any L2 cache, since the node that holds one of these copies in its L2 cache is responsible for keeping coherence between the accesses to this memory block. The USI must identify all the sharers of the block.
- *Current Owner Information* (COI): For each block stored in any cache, its home node must maintain a pointer which identifies the owner node. This information must be updated whenever the owner node changes and it is accessed when the requesting cache is not able to locate the current owner.
- *Extra Owner Information* (EOI): This information is stored in any node except the home and the owner. It is used for avoiding the access to the home node on a cache miss. Particularly, each node keeps a pointer in its L2 cache that identifies the last node that invalidated its previous copy of a memory block. Future misses will use the value of this pointer to send the request directly to the owner node, thus removing indirection. Our cache coherence protocol can perform correctly in absence of EOI, but performs more efficiently when this information is included.

For storing this information, we propose the L2 cache organization shown in figure 1. The *sharing code* field is used to store the USI for the primary copies of the blocks held in the *data cache*. The *pointer cache* is used for storing the identity of the owner (COI if it is the home node or EOI in other case). Note that our proposal does not need to keep directory information in main memory nor the use of additional directory caches.

2.3 Description of the Coherence Protocol

Requester node. When a cache miss takes place in a node (requester node), the identity of the owner node must be obtained. If the identity of the owner is found in the pointer cache (COI for local misses or EOI in other case) the request is sent to this node. Otherwise (first reference to a block or replacement in the pointer cache), the request is sent to the home node which subsequently redirects the miss to the current owner.

Request received by a node that is not the owner. When a request arrives to a node that is not the owner of the block, the request must be resent to another node. If the former node is the home and the COI is found in the pointer cache, the request is sent to the owner node. The COI is recent since this information is updated whenever the owner changes. Hence, in absence of race conditions the request will reach the owner node. On the other hand, if the home node does not find the COI in the pointer cache, the owner of the block is main memory because the block is not held by any cache. Then the miss is solved by providing the block from main memory, and the home node allocates a new COI entry in its pointer cache. Finally, if the request reaches any other remote node, it is resent to the home node.

Request received by the owner node. Every time a request reaches the owner node, it is necessary to check whether this node is currently processing a request from a different processor for the same block. In this case, we can say that the block is in busy state, and the request must be returned to the requester node asking it to try again.

On the other hand, if the block in the owner node is not in busy state, the miss can be solved. Read misses are completed by sending a copy of the block to the requester node and adding it to the sharing code. For write misses, the owner node must invalidate all the copies from all the caches before it can send the block to the requester. If the miss is an upgrade the owner node checks the sharing code field to know whether the requester still holds a copy of the block (note that a previous write miss from a different processor could have invalidated its copy and in this case the owner node should also provide a new copy of the block). In this case, the owner node replies to the requester with the ownership of the block once the rest of the copies have been invalidated. Note that upgrade misses that take place in the owner node just need to send invalidations and receive acknowledgements (two hops in the critical path).

Moreover, as the home node must have up-to-date information of the owner of the block (COI), every time that an owner node gives its ownership to other node, it must send a control message to the home node indicating the identity of the new owner. Note that messages reporting ownership changes for a particular block should be processed by the home node in the same order in which they were generated. Otherwise, the COI could fail to store the identity of the current owner. Although there are other alternatives to ensure this order, in our particular implementation we associate a version number to every primary copy. This version number is stored in both the home node and the current owner of the block, and is increased on every ownership change. The idea is that when a message reporting an ownership change arrives to the home node, it is only processed (the identity of the new owner is stored) if the version number in the message has the same value than the one stored in the home, along with the COI. In other case, the message could be buffered or NACKed to the processed later. In practice, we have found that this version number could be stored using a small 3-bit wrapping counter.

Replacements. In our particular implementation the replacement of a block stored in any cache only requires coherence actions when it is a primary copy. In other case, the replacement is performed transparently to the rest of the sharers. The replacement algorithm used in the data caches is LRU, but the age of the primary copy of every block is updated every time that it is accessed by any local or remote request.

When the primary copy of a block is evicted, it is looked for another node that will receive the responsibility of keeping coherence for the block (the owner property is moved to one of the sharers). Since the owner node knows the current set of the sharers, it sends the request to one of them (chosen randomly). If the new owner node had previously invalidated its copy, it resends the request to another node (note that the request includes directory information for the block as well). The node that receives this request and has a valid copy of the block will be the new owner node, and therefore, must notify the home node of the change of the owner. On the other hand, if all the nodes had replaced its copy, the request is finally sent to the home node which removes the COI from the pointer cache and stores the block in main memory.

On the other hand, replacements in the pointer cache also follow the LRU algorithm, but it is distinguished between COI and EOI. EOI entries are preferably evicted for two reasons. First, we have found that keeping EOI entries too much time is not worthy since this information gets obsolete (it would cause a significant number of misses when finding the identity of the owner node). Second, when a COI entry is replaced, the home node must ask the owner node to invalidate all the copies of the block and main memory must be updated.

2.4 Preventing Deadlock and Starvation

Direct Coherence ensures that not deadlock can occur by returning back to the issuing nodes those requests that cannot be solved instead of enqueuing these requests in a buffer.

On the other hand, in directory-based protocols starvation can be easily avoided if the requests are buffered in FIFO order at the home node. In Direct Coherence each write miss implies that the identity of the owner node changes. If a memory block is repeatedly written by several nodes, a request could take some time to find the owner node, even when it is sent by the home node. Hence, some nodes could be solving their misses while other misses are starved. Figure 2 shows an example of a scenario in which starvation appears. The nodes N_1 and N_2 are issuing write requests repeatedly, and therefore, the owner node is continually moving from N_1 to N_2 and vice versa. Each time that the owner changes, the home node is notified. However, at the same time, the home node is trying to send the request issued by the node N_3 to the owner node, but this request could always be returned to it whenever the write request issued by the other node arrives before.

Since this kind of scenario is very infrequent, we think that it is more important for the starvation avoidance mechanism to be simple rather than efficient. In particular, each time that a request must be retried, a counter is increased.

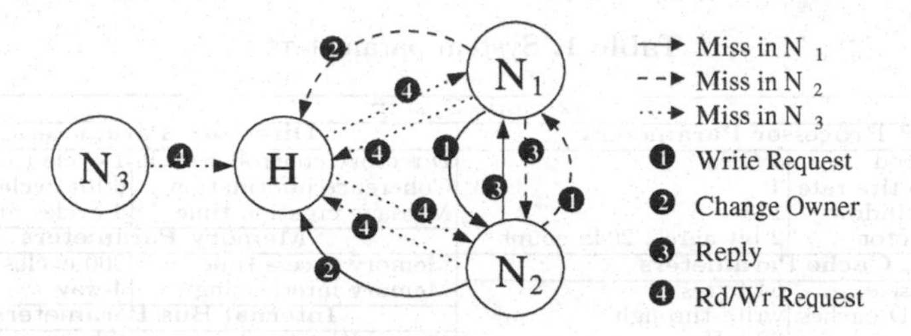

Fig. 2. Example of a starvation scenario

Starvation is detected when this counter reaches a certain maximum value. To guarantee that the identity of the owner node does not change during certain period of time, and therefore, the owner can be reached by the request, each time that a starved request arrives to a node that cannot solve it, the node is blocked and it cannot issue new requests for the same block. Once the starved request is solved by the owner node, some messages are sent to the nodes that were blocked to allow them to continue issuing requests for the block. Through experimentation we have found that a value of 100 hops works fine in most cases.

3 Simulation Environment

We have implemented and evaluated the Direct Coherence protocol through RSIM, a detailed execution-driven simulator that models cc-NUMA multiprocessors [7]. Our proposal is compared against a highly optimized MOESI directory-based protocol that employs unbounded directory caches in each node to cut down the number of accesses to main memory (base configuration from now on). Moreover, this MOESI protocol has been optimized to allow that a read miss can be directly solved by providing the requesting block from the home node whenever this node has a copy of it in its cache (either in shared or owned state). MOESI states allow that for most cache misses the corresponding memory block is provided by a remote cache instead of main memory.

In both cases, bit-vector has been used as the sharing code (4 bytes per entry in a 32-node system). Therefore, the L2 cache used in Direct Coherence protocol employs 32 KB of additional storage for the sharing code field. The size of the pointer cache (5 bits per pointer) is just 2 KB. We have found that this small size avoids replacements of COI entries. In this way, the total memory overhead of our proposal is very small (6.6% of the L2 cache storage). Remember that for the base configuration we consider unbounded directory caches.

We have simulated multiprocessors with 32 uniprocessor nodes. Table 1 shows the main parameters used for our proposal. Simulations have been performed using an optimized version of the sequential consistency model with speculative load execution [7]. The nine scientific programs used in our simulations cover a variety of computation and communication patterns. Barnes (8192 bodies, 4 time steps), Cholesky (tk16.O), FFT (256K complex doubles), Ocean (258x258 ocean), Radix (1M keys, 1024 radix) and Water-NSQ and Water-SP

Table 1. System parameters

32-Node System			
ILP Processor Parameters		**Directory Parameters**	
Processor speed	5 GHz	Directory controller cycle	1 cycle (on-chip)
Max. fetch/retire rate	4	Coherence information	6 hit cycles
Instruction window	128	Message creation time	4 cycles first, 2 next
Branch predictor	2 bit agree, 2048 count	**Memory Parameters**	
Cache Parameters		Memory access time	300 cycles
Cache block size	64 bytes	Memory interleaving	4-way
Split L1 I & D caches:	write-through	**Internal Bus Parameters**	
Size	32 KB	Bus width	8 bytes
Associativity	direct mapped	Bus cycles	1 cycle
Hit time	2 cycles	**Network Parameters**	
Unified L2 cache:	write-back	Topology	2D mesh (4x8)
Size	512 KB	Flit size	8 bytes
Associativity	4-ways	Non-data message size	2 flits
Hit time	6 + 9 cycles (tag + data)	Flit delay	4 cycles
Pointer cache	2 KB, 4-ways, 6 hit cycles	Arbitration delay	5 cycles

(512 molecules, 4 time steps) are from the SPLASH-2 benchmark suite [15]. Unstructured (Mesh.2K, 5 time steps) is a computational fluid dynamics application. Finally, EM3D (38400 nodes, 15% remotes, 25 time steps) is a shared memory implementation of the Split-C benchmark. All the programs were run to completion, but all experimental results reported in this paper are for the parallel phase of these benchmarks. The size of the L2 caches (512KB in our simulations) has been chosen taking into account both current L2/L3 cache sizes and the characteristics of the applications used for the evaluation.

4 Evaluation Results

In this section, we present and analyze the simulation results obtained for the Direct Coherence protocol (*DiCo* configuration) presented in this work. Our proposal is compared against the base system described in the previous section (*Base* configuration).

4.1 Impact on the Number of Hops Needed to Solve Cache Misses

In general, Direct Coherence can reduce the number of hops needed to solve a miss by avoiding the indirection that the access to the home node introduces. The extent of the reductions varies depending on the cache miss type (read miss, write miss or upgrade miss). Therefore, we study separately how the number of hops is reduced according to the miss type. Figure 3 shows how each type of cache miss is solved in both the base protocol and Direct Coherence protocol. These results are normalized with respect to the base case. Each cache miss can be classified in one of the following types:

- *2-hop misses*: This miss type does not suffer indirection. Read and write misses are solved using two hops when the identity of the owner node is stored at the requesting cache (and invalidation messages are not necessary). Upgrade misses fall into this category when they take place in the owner

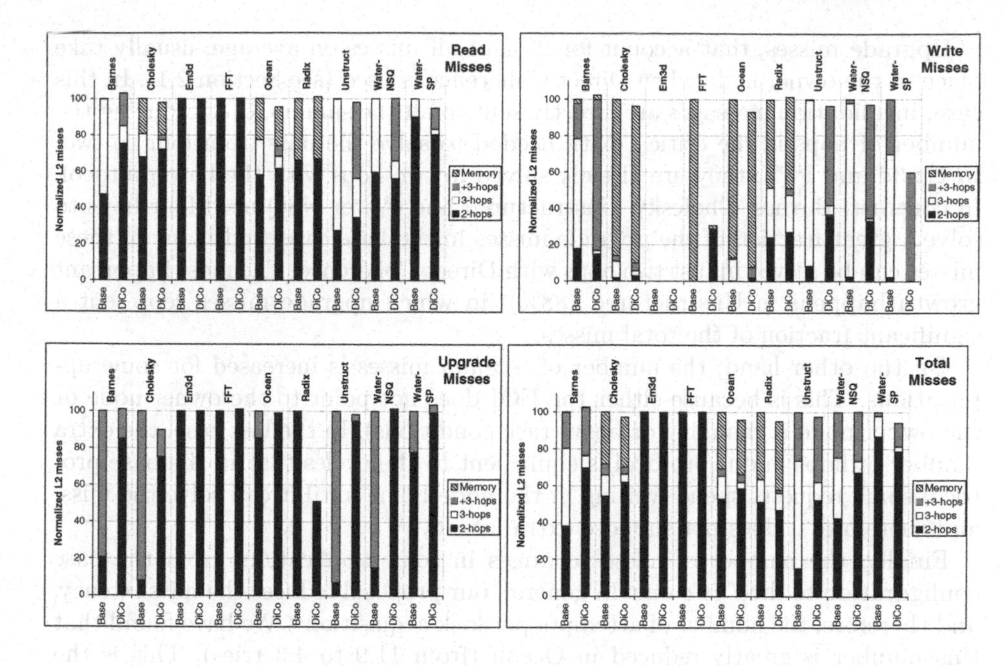

Fig. 3. How each miss type is solved

node, or alternatively, when the block is just shared by the node that issues the miss and the owner node.

- *3-hop misses*: A miss belongs to this type when the requesting cache has not EOI and the home node resends the request to the proper owner, which solves the miss without invalidation messages.
- *+3-hop misses*: We include in this category misses that need more than three hops to be solved.
- *Memory misses*: When the block is provided by main memory since it is not held by any L2 cache.

As shown in figure 3, in general, our protocol increases the number of misses solved in two hops. The number of read misses that need only two hops increases in some applications, especially in Barnes. In EM3D and FFT applications, all the read misses are already solved in two hops. Finally, in other applications the number of 2-hop read misses does not increase since the optimized MOESI protocol already increases the number of read misses solved in two hops by providing the block from the home node's cache whenever clean data is found in it (even when it is not the current owner of the block).

The percentage of two-hop write misses is smaller than the percentage of two-hop read misses, but fortunately, write misses are less frequent than read misses. This lower percentage is because some blocks are continuously written by different nodes, and therefore, the EOI becomes obsolete quite soon. Nevertheless, Direct Coherence increases the number of two-hop write misses with respect to the base protocol for all the applications except Barnes and Water-SP.

Upgrade misses, that account for 23% of all misses on average, usually take place in the owner node when Direct Coherence is used (see Section 2.1). In this case, invalidation messages are directly sent to all the sharers, thus reducing the number of hops in the critical path needed to solve the miss from four to two. In Em3d and FFT, they are already solved in two hops with the base protocol. In contrast, Barnes, Cholesky, Unstructured and Water-NSQ need four hops to solve a great fraction of the upgrade misses in the base case, and many of these misses can be solved in just two hops with Direct Coherence. The most important growth happens in Unstructured (88%), in which upgrade misses represent a significant fraction of the total misses.

On the other hand, the number of *+3-hop* misses is increased for some applications. This is because either the EOI does not point to the owner node or the owner node is changing or busy (race conditions). In the last case, the extra number of hops in our protocol is equivalent to the cycles that in the base protocol some requests spend waiting in the node home until it can solve the miss, and therefore, it does not suppose extra latency.

Finally, the number of misses changes in some applications from the base configuration to the *DiCo* one. In general, our proposal reduces the miss latency, and therefore, the number of attempts per lock acquisition[2]. We have found that this number is greatly reduced in Ocean (from 11.9 to 4.3 tries). This is the reason for the lower number of misses observed in applications like Cholesky, Ocean, Unstructured and Water-SP. On the other hand, our proposal increases the number of misses in Barnes. This is because in our protocol owner blocks cannot be evicted from cache when they have pending requests (busy state). If a cache set has several busy blocks for long time, the rest of blocks stored in the same set will be evicted quite frequently, even when they are frequently requested by the local processor. This growth can be easily avoided in L2 caches with higher associativity. Radix is also affected by this fact, but the total number of misses does not increase because many upgrade misses are removed when Direct Coherence is used. The latter is because in our protocol replacements of the primary copy of memory blocks are sent to another node, thus informing of the replacement and changing the identity of the owner of the block. In this way, when the new owner subsequently upgrades the block and finds that no other cache holds it, the miss is avoided. In the base protocol, only when the upgrade miss reaches the directory is when it is known that the requesting cache is the only sharer for the block. Finally, some applications like FFT and Water-SP convert some write misses into upgrades, since they keep the primary copy of some blocks in cache longer.

4.2 Impact on L2 Cache Miss Latencies

For each miss type, figure 4 shows the speed-up obtained for Direct Coherence with respect to the base protocol. We can observe that the latency of read misses

[2] Note that locks in RSIM are implemented using the well-known test-and-test&set method.

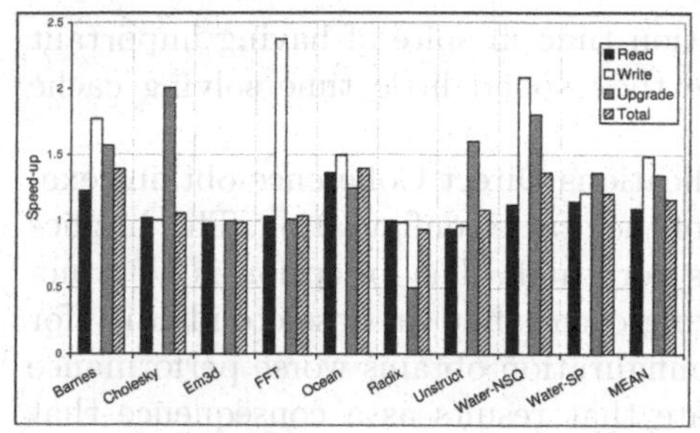

Fig. 4. Percentage improvements for cache miss latencies

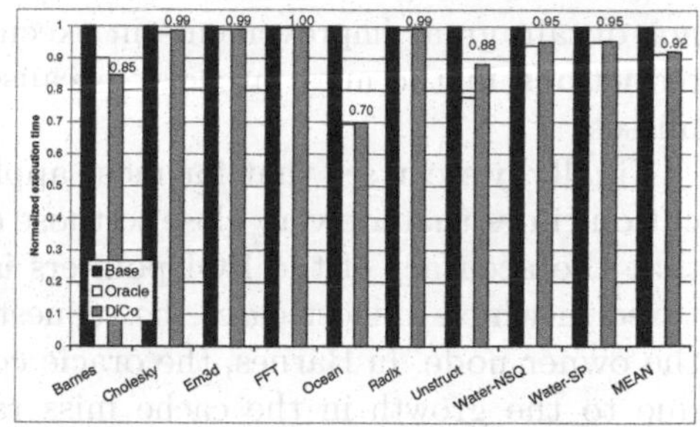

Fig. 5. Normalized execution time

is reduced in all the applications except Unstructured. In this application, we have found that Direct Coherence increases the number of *+3-hop* read misses due to that the EOI gets obsolete. On the other hand, read misses are significantly accelated in Barnes and Ocean (1.22 and 1.36 respectively). Both applications increase the number of two-hop misses, and contrary to Unstructured, the increase in the number of *+3-hop* misses is due to race conditions that do not increase the latency of the misses.

The important speed-up (2.37) for write misses found in FFT is due to almost all the write misses are solved in two hops instead of accessing memory. Barnes, Ocean and Water-NSQ also obtain important speed-ups ranging from 1.49 to 2.07.

For upgrade misses some applications like Barnes, Cholesky, Unstructured and Water-NSQ increase very significantly the total number of two-hop misses, and therefore, obtain speed-ups ranging from 1.56 in Barnes to 2 in Cholesky. Radix reaches a reduction in the number of upgrade misses. As these misses do not have to invalidate any copy in the base case, they have low miss latencies. This is why there is a growth in the average miss latency in our protocol.

4.3 Impact on Execution Time

Finally, the percentage improvements in terms of L2 miss latency translate into reductions on applications' execution time. Figure 5 plots the execution times that are obtained for the base configuration (*Base*), the oracle configuration (*Oracle*) which shows the improvements in total execution time that would be obtained by Direct Coherence if the identity of the owner were known on every miss, and Direct Coherence (*DiCo*). Results have been normalized with respect to the base case.

Important reductions are observed for Barnes (15%), Ocean (30%) and Unstructured (12%). In Barnes and Ocean important reductions have been reported for the L2 cache misses. Unstructured reduces considerably the latency of upgrade misses that are the bottleneck of this application. For the rest of the applications (except for FFT), reductions range from 1% for Cholesky, Em3d and Radix to 5% for Water-NSQ and Water-SP. Water-NSQ and Water-SP do not obtain great improvements in execution time in spite of having important reductions in the miss latencies because they spend little time solving cache misses.

Finally, we can see that for most applications Direct Coherence obtains execution times that are very close to those of the oracle configuration. This implies that the accuracy of the EOI pointers is very high. The exception is Unstructured in which in most cases the requesting caches find an obsolete identity for the owner node. In Barnes, the oracle configuration obtains worse performance due to the growth in the cache miss rate that results as a consequence that Direct Coherence do not replace owner blocks in busy state (see Section 4.1).

5 Related Work

Snoopy protocols do not introduce indirection because they are based on a totally-ordered interconnection network. Unfortunately, these interconnection networks are not scalable. Some proposals have focused on using snoopy protocols with arbitrary network topologies. Martin. *et al.* [10] present a technique that allows SMPs to utilize unordered networks (with some modifications to support snooping). Bandwidth Adaptive Snooping Hybrid (BASH) [11] is an hybrid coherence protocol that dynamically decides whether to act like snoopy protocols (broadcast) or directory protocols (unicast) depending on the available bandwidth. TokenB coherence protocol [9] avoids both the need of a totally ordered network and the indirection caused by the directory by assigning N tokens to every memory block. In this way, a node can read a block if it has at least one token and can update the block if it has all the tokens. Subsequently, TokenM [8] was proposed to reduce the demand of interconnect bandwidth by using destination-set prediction. However, TokenB and TokenM increase network traffic becoming a bottleneck for large-scale systems. In contrast, our proposal keeps network traffic low by sending only one message per cache miss.

Acacio *et al.* propose to avoid the indirection for cache-to-cache transfer misses [1] and upgrade misses [2] separately by predicting the current holders of every cache block. In contrast, our protocol avoids the indirection for cache-to-cache transfer misses by using recent information about the node that must solve the miss, and for upgrade misses by removing the directory information from the home node and by storing it in the node that issues the upgrade request. In this way, our proposal does not need extra hardware to predict neither the owner nor the sharers of the block.

Recently, Cheng *et al.* have proposed converting 3-hop read misses into 2-hop read misses for memory blocks following the producer-consumer sharing pattern [4]. They need extra hardware to detect when a block is accessed according to this pattern. In contrast, our proposal obtains 2-hops misses for read, write and upgrade misses without taking into account sharing patterns.

Finally, directory caches (originally proposed in [6] for cutting down directory memory overhead) can be also used for reducing the latency of cache misses by obtaining directory information from a much faster structure than main memory

[12]. In [13], we evaluated the impact that completely removing the directory structure from main memory and storing directory information at the last-level caches has in terms of cache miss rate and performance. In this proposal, the directory information is only stored in the home node, but in Direct Coherence this information is stored in the owner node for avoiding indirection.

6 Conclusions

In this work we have presented *Direct Coherence*, a novel cache coherence protocol that avoids the indirection introduced by the directory-based protocols. Direct Coherence moves the role of storing up-to-date sharing information (and ensuring totally ordered accesses) from the home node to the owner node. In this way, indirection is avoided by directly sending the requests to the owner node.

Direct Coherence offers both the performance advantage of snoopy-based protocols, as coherence messages are directly sent from the requesting caches to those that must observe them, and the scalability of directory-based ones, as our proposal is not based on broadcasting or any other brute-force method.

We have described the implementation of Direct Coherence and we have evaluated it using the RSIM simulator. Simulation results show that our proposal can increase the number of misses without indirection. The reduction in the number of hops translate into an average reduction in the latency of the L2 misses of 20.7%, which finally leads to improvements in applications' execution time up to 30% (8% on average) when compared with a MOESI directory-based protocol. In this way, Direct Coherence is revealed as a promising alternative to current cache coherence protocols, bringing together performance and scalability.

Acknowledgments. The authors would like to thank the anonymous reviewers for their helpful insights. This work has been jointly supported by the Spanish MEC and European Comission FEDER funds under grants "Consolider Ingenio-2010 CSD2006-00046" and "TIN2006-15516-C04-03". A. Ros is supported by a research grant from the Spanish MEC under the FPU national plan (AP2004-3735).

References

1. Acacio, M.E., González, J., García, J.M., Duato, J.: Owner prediction for accelerating cache-to-cache transfer misses in cc-NUMA multiprocessors. In: SC2002. High Performance Networking and Computing (November 2002)
2. Acacio, M.E., González, J., García, J.M., Duato, J.: The use of prediction for accelerating upgrade misses in cc-NUMA multiprocessors. In: PACT 2002. 11th Int'l Conference on Parallel Architectures and Compilation Techniques, pp. 155–164 (September 2002)
3. Chang, J., Sohi, G.S.: Cooperative caching for chip multiprocessors. In: ISCA 2006. 33th Int'l Symp. on Computer Architecture, pp. 264–276 (June 2006)
4. Cheng, L., Carter, J.B., Dai, D.: An adaptive cache coherence protocol optimized for producer-consumer sharing. In: HPCA-13. 13th Int'l Symp. on High Performance Computer Architecture, pp. 328–339 (February 2007)

5. Culler, D.E., Singh, J.P., Gupta, A.: Parallel Computer Architecture: A Hardware/Software Approach. Morgan Kaufmann Publishers, Inc, San Francisco (1999)
6. Gupta, A., Weber, W.-D., Mowry, T.C.: Reducing memory traffic requirements for scalable directory-based cache coherence schemes. In: ICPP 1990. Int'l Conference on Parallel Processing, pp. 312–321 (August 1990)
7. Hughes, C.J., Pai, V.S., Ranganathan, P., Adve, S.V.: RSIM: Simulating shared-memory multiprocessors with ILP processors. IEEE Computer 35(2), 40–49 (2002)
8. Martin, M.M.: Token Coherence. PhD thesis, University of Wisconsin-Madison (December 2003)
9. Martin, M.M., Hill, M.D., Wood, D.A.: Token coherence: Decoupling performance and correctness. In: ISCA 2003. 30th Int'l Symp. on Computer Architecture, pp. 182–193 (June 2003)
10. Martin, M.M., Sorin, D.J., Ailamaki, A., Alameldeen, A.R., Dickson, R.M., Mauer, C.J., Moore, K.E., Plakal, M., Hill, M.D., Wood, D.A.: Timestamp snooping: An approach for extending SMPs. In: ASPLOS IX. 9th Int'l Conference on Architectural Support for Programming Languages and Operating Systems, pp. 25–36 (November 2000)
11. Martin, M.M., Sorin, D.J., Hill, M.D., Wood, D.A.: Bandwidth adaptive snooping. In: HPCA-8. 8th Int'l Symp. on High-Performance Computer Architecture, pp. 251–262 (January 2002)
12. Nanda, A.K., Nguyen, A.-T., Michael, M.M., Joseph, D.J.: High-throughput coherence controllers. In: HPCA-6. 6th Int'l Symp. on High-Performance Computer Architecture, pp. 145–155 (January 2000)
13. Ros, A., Acacio, M.E., García, J.M.: A novel lightweight directory architecture for scalable shared-memory multiprocessors. In: Cunha, J.C., Medeiros, P.D. (eds.) Euro-Par 2005. LNCS, vol. 3648, pp. 582–591. Springer, Heidelberg (2005)
14. Ros, A., Acacio, M.E., García, J.M.: An efficient cache design for scalable glueless shared-memory multiprocessors. In: ACM Int'l Conference on Computing Frontiers, pp. 321–330 (2006)
15. Woo, S.C., Ohara, M., Torrie, E., Singh, J.P., Gupta, A.: The SPLASH-2 programs: Characterization and methodological considerations. In: ISCA 1995. 22nd Int'l Symp. on Computer Architecture, pp. 24–36 (June 1995)
16. Wulf, W., McKee, S.: Hitting the memory wall: Implications of the obvious. Computer Architecture News 23(1), 20–24 (1995)

Constraint-Aware Large-Scale CMP Cache Design

L. Zhao, R. Iyer, S. Makineni, R. Illikkal, J. Moses, and D. Newell

Systems Technology Group, Intel Corporation
srihari.makineni@intel.com

Abstract. Within the next decade, we expect that large-scale CMP (LCMP) platforms consisting of 10s of cores to become mainstream. The performance and scalability of these architectures is highly dependent on the design of the cache hierarchy. In this paper, our goal is to explore the cache design space for LCMP platforms, which can be vast with several constraints. We approach this exploration problem by developing a constraint-aware analysis methodology (CAAM). CAAM first considers two important constraints and limitations -- cache area constraints and on-die / off-die bandwidth limitations. We determine a viable range of cache hierarchy options. We then estimate the bandwidth requirements for these by running server workload traces on our LCMP performance model. Based on allowable bandwidth constraints, we narrow the design space further to highlight a few cache options. Finally, we compare these options based on performance, area and bandwidth trade-offs to make recommendations.

Keywords: Large Scale CMP, constraint-aware design, cache hierarchy, CAAM, LCMP.

1 Introduction

The momentum behind CMP architectures [5] is pushing architects and designers to consider integrating more and more cores on the die resulting in large-scale CMP (LCMP) architectures. However, for LCMP architectures to be scalable, it is critical that the on-die cache/memory hierarchy be designed to support many cores / threads efficiently. In this paper, our focus is on exploring the cache hierarchy design for LCMP platforms.

When investigating cache hierarchy design for LCMP platforms, there are several important factors to consider (die area, power consumption, etc.). Keeping these constraints in mind, in this paper, we attempt to answer the following key questions: 1) how do we go about pruning the cache design space for LCMP architectures? What methodology needs to be put in place? How should the cache be sized at each level and shared at each level in the hierarchy? How much memory and interconnect bandwidth is required for scalable performance?

Previous studies on cache design space exploration have largely been focused on performance [2, 3, 4, 15], with few that have considered the implications of power and/or area [1, 8, 14]. To our knowledge, this is the first study that proposes a methodology to study area, bandwidth and performance implications on cache design

S. Aluru et al. (Eds.): HiPC 2007, LNCS 4873, pp. 161–171, 2007.

space exploration. Furthermore, this is the first study that applies such a methodology in the context of LCMP platforms.

2 Cache Hierarchy for LCMPs

In this section, we introduce the LCMP architecture and discuss the cache design considerations in more detail.

2.1 Architecture Overview

Figure 1(a) illustrates the LCMP platform architecture with a single socket. Figure 1(b) describes a potential on-die architecture of the LCMP socket. We chose this architecture because it offers the opportunities mentioned in Figure 1(c). The on-die architecture consists of several nodes (each with some number of multi-threaded cores and a shared node cache), an on-die fabric that interconnects the nodes, potentially a shared last-level cache (L3), integrated memory controllers and other external interfaces.

This study focuses on massively (64 to 128 hw threads) parallel on-die architectures. The performance scalability of such massively parallel architectures depends significantly on efficient design of the cache, interconnect and memory subsystems with sufficient capacity and/or bandwidth. In the next section, we discuss in more detail the cache hierarchy considerations for LCMP architectures.

2.2 LCMP Cache Design Considerations

We focus on the cache hierarchy design of LCMP architectures with 16 or 32 quad-threaded light weight cores. The scalar performance of each core is assumed to be low, but many cores packed together can provide a throughput computing advantage.

There are several design parameters when exploring cache hierarchy design. The typical first order questions are: (a) How many levels of cache should we employ? (b) What should the cache size be at each level? and (c) How many cores should share a cache at a given level?

There are several design considerations to account for when exploring cache hierarchy design for LCMP platforms. Some of the key considerations are: (a) Area

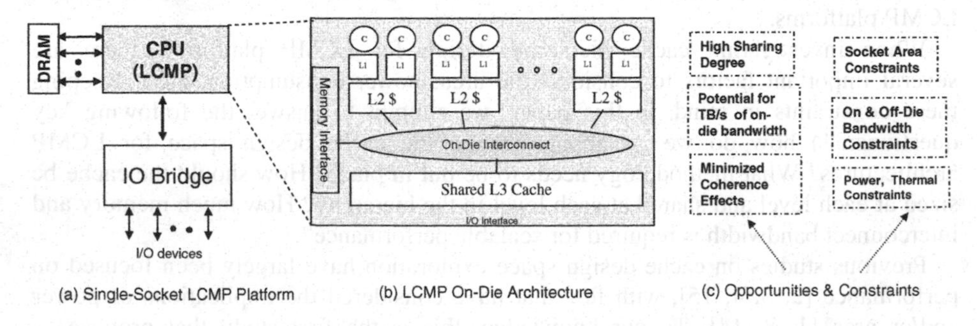

Fig. 1. LCMP Architecture Overview: Opportunities and Constraints

constraints, (b) Bandwidth Constraints, (c) Power Implications and (d) Performance of the cache hierarchy and overall platform. In this paper, we study three of the above four considerations (excluding power implications).

2.3 Constraints-Aware Analysis Methodology

In order to prune the cache design space, we propose a constraints-aware analysis methodology (CAAM). This methodology assumes that the area constraint, the on-die bandwidth constraint and the off-die bandwidth constraints are known. The CAAM methodology consists of three major steps:

(1) Area-Constrained Options: This step essentially attempts to prune the design space by the area constraints. We first estimate the area required for L2, and then apply the overall area constraints to this cache. All options that exceed the area constraints are immediately discarded. The same process is repeated for each level until the desired number of levels of cache has been covered.
(2) Bandwidth-Constrained Options: This step attempts to further prune the options of those already pruned by area constrained as above by applying the on-die and off-die bandwidth constraints.
(3) Overall Performance: Once the area and bandwidth constraints are applied, we have a pruned set of design options that are viable. The performance of these options is then compared to determine the top two or three design choices.

3 Evaluation Tools and Workloads

In this section, we describe the simulation environment, area estimation tools and the workloads used.

3.1 LSIM Simulation Environment

Analyzing the performance behavior of LCMP platforms with numerous cores and threads is challenging because simulation speed becomes a critical bottleneck. With speed and accuracy in mind, we developed the LSIM simulation environment to allow for varying degrees of fidelity.

The simulation environment starts with instruction traces (either on a real system or from a full-system simulator) that are collected for the workloads of interest. In order to simulate many cores with multiple threads, the instruction traces can then be run through a micro-architecture simulator. The micro-architecture simulator is configured with only an L1 cache and the output of this simulation is a trace that represents the execution profile -- compute delays (time spent on the CPU between memory events) and the memory events (L1 misses, L1 writebacks, synchronization events etc). For our study here, we used an abstract CPU simulator with a configurable internal CPI and a L1 cache.

The annotated CPU traces are then fed into the LSIM simulator. The LSIM core simulation mimics the execution profiles present in the traces and injects memory events into the interconnect/cache subsystem. We simulate a hierarchical interconnect

– a first level interconnect between the L1 and L2 caches as well as a 2nd level interconnect between the L2 and L3 caches (if an L3 is present). The L2 and the L3 caches are configurable to be private caches per core or shared cache per node. A detailed invalidation-based coherence protocol is implemented with appropriate ordering constraints enforced within the interconnect subsystems. The cache hierarchy is modeled to be inclusive by modeling the back-invalidation messages required to evict L2 copies of a line that is replaced in L3.

The operating frequency (of the core, interconnect, etc), queue sizes (interconnect interface and cache controller structures), bandwidths (interconnect, cache & memory) and latencies (delays between L2 and L3s, etc) are all configurable in LSIM and allows us to explore the design space sufficiently.

3.2 Workloads and Traces

Our focus in this study is largely on LCMP server platform architecture and performance. As a result, we picked a few important commercial server workloads (OLTP [16], SAP [10] and SPECjbb [13]) as well as high performance computing kernels (from SPECrate [12]).

For all of these workloads, we collected long instruction traces on Intel Xeon platforms. These instruction traces are used to drive LCMP simulation. Wherever sufficient number of instruction traces is not available, we replicate the execution profiles appropriately to feed the remaining cores/threads. When replicating traces, we make sure that the code memory accesses are shared, whereas data accesses are privatized in order to not artificially inject any incorrect data sharing. Based on detailed understanding of thc workloads as well as measurements to validate them, we already know that SAP, SPECjbb and SPECrate workloads have negligible data sharing. TPC-C is known to have significant data sharing (which we do not simulate sufficiently well due to the nature of our tracing/simulation environment), but newer databases seem to be trending towards reduced data sharing to avoid synchronization penalties.

The workload characteristics described and/or traces collected were not audited and the data presented in this paper should not be misused to represent benchmark performance of the architecture under evaluation.

3.3 Area Estimation Tools

For area estimation, we used CACTI (version 3.2), an integrated cache access time, cycle time, area, aspect ratio, and power model [11]. This model allows us to estimate the area of different cache sizes and organization. The parameters we held constant in our evaluation are the line size at 64 bytes. We vary the cache size, the number of banks and the associativity depending on L2 or L3 caches. We assume that a shared L3 cache across all of the cores/threads in the socket. However, we assume that the shared L3 cache is actually distributed in organization (somewhat like in NUCA [6]) with independent smaller caches and associated controllers. All cache area estimates are based on a 45nm process as our intention is to look at architectures around the end of the decade.

3.4 Baseline Configurations and Assumptions

In this section, we cover the baseline architecture and associated simulation configurations that we evaluate.

The simulated LCMP on-die architecture consists of 16 or 32 cores (with 4 threads each) operating at a frequency of 4GHz. As discussed in Section 3.1, we assume that the processor core has 32KB of L1 icache and dcache. The effect of the L1 cache is already captured within the execution profile trace that is fed to LSIM. The execution duration consisted of over 100 million instructions and we ensured that the caches and the platform have been sufficiently warmed up.

The on-die architecture is made up of several nodes. Each node may consist of 1, 2 or 4 cores. In LSIM, we simulate L2 cache size per node that varies from 128K to 4M. The L2 cache may be configured as either private per core or shared between all of the cores in the node. For configurations where an L3 cache appeared to be viable (based on area constraints), we simulated an L3 cache with size varied from 8M to 32M to a perfect L3 cache. All the caches have used 64-byte line size. L3 cache hit latency of 50 cycles (varied) is assumed. Interconnect bandwidth is varied between 128 GB/s to 512 GB/s. Memory access time is set to 400 cycles and bandwidth is varied from 32 GB/s to 128 GB/s. L3 MSHR is set to 16 entries, memory queue length is set to 64 or 128, coherence controller queue is set to 16 and interconnect interface is set to 8 entries.

4 Area and Bandwidth Implications

In this section, we present our evaluation of the LCMP cache hierarchy design space based on CAAM.

4.1 Implications of Area Constraints

We start applying the CAAM methodology to LCMP cache design space exploration by first considering area constraints. Figure 2(a) summarizes the L2 cache area estimates for a 32-core LCMP as a function of the L2 cache size per node (128K to 4M) and the number of cores per node (1 to 4). The CACTI data shows that as the L2 cache size increases from 128K to 512K, the space consumed by the cache does not increase linearly. However, as the cache size increases past 512K, the cache area starts showing closer to a linear increase. Also note that as the cache size per core goes to 1M and beyond, the area consumed by the cache space is about 400 mm^2 or higher.

Due to manufacturing costs as well as form factor limitations, it is important to keep the die size of the processor as low as possible. Server chips are larger than desktop processor chips and have been generally less than 400 mm^2. The largest die in production today is an Itanium 2 processor (estimated to be around 432 mm^2 [7]). In order to keep the die area under 400 mm^2, it is important to keep the cache area to a reasonable fraction of the overall area. In this paper we study the effect of constraining the cache space to 50% (200 mm^2) or 75% (300 mm^2). Figure 2(b) shows the constraint on a 32-core LCMP. The two horizontal lines in the figure represent the 200mm2 and 300mm^2 constraints.

The next step is to identify design options where there is provision for a shared L3 cache. Figure 3(b) highlights the configurations where there is a potential for a L3 cache. Since we are considering traditional inclusive cache hierarchies, it is important that the area available to an L3 cache allow for at least twice the size of the L2 cache area. By applying this criteria, we have figured L3 cache size estimates (numbers to the right of the bars) for both area constraints (200 mm^2 and 300 mm^2 in a 45nm process). For example, with the configuration of 128K per node, 1 core per node and a total of 32 cores, we cannot employ a suitable L3 cache if the area constraint is 200 mm^2. This is because the amount of area available cannot accommodate an inclusive L3 cache that is equal to or larger than twice the size of the L2 cache size (which is 4M = 128K*32 in this case). However, if the area constraint is relaxed to 300 mm2, then a L3 cache that is roughly 12 MB in size can be accommodated. A similar process is applied to 16-core LCMP configuration.

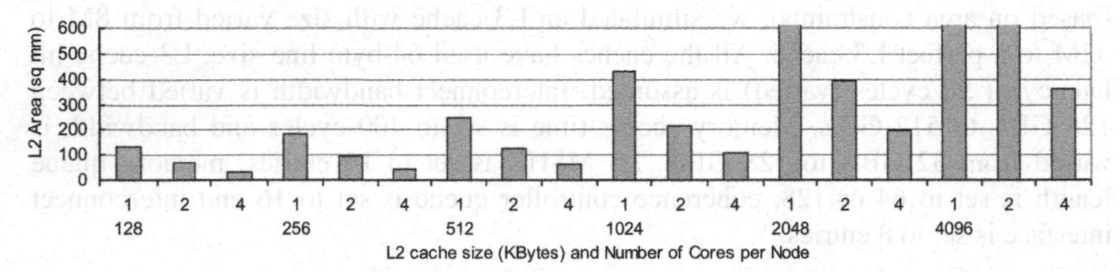

a) Area Consumed by L2 Cache on a 32-Core LCMP

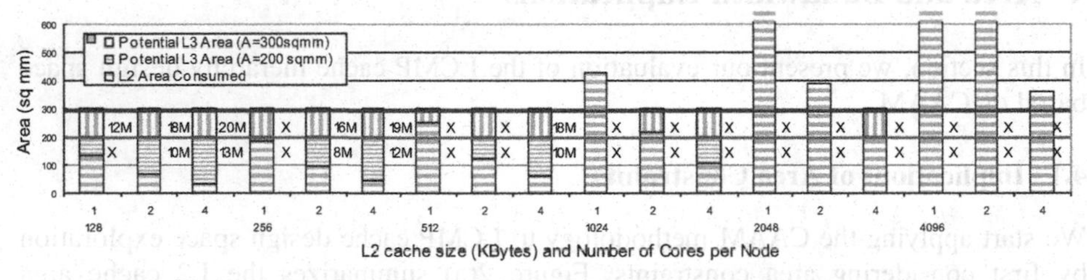

b) Applying Area Constraints to Prune LCMP Cache Design Space (L2 & L3 Cache Space)

Fig. 2. Area Considerations for a 32-Core LCMP Cache Design

4.2 Implications of Bandwidth Constraints

Another crucial step in the CAAM methodology is to apply on-die and off-die bandwidth limitations to the cache design space exploration. Note that unlike area constraints which can be applied independent of the workload running on the platform, the bandwidth constraints need to be considered along with a representative set of workloads that place bandwidth demand on the platform. For this exercise, we chose TPC-C from server workloads and Swim from SPECrate as they are more memory intensive.

To understand the bandwidth demand on the on-die interconnect (between L2 and L3), we first simulated the LCMP architecture and measured the number of L2 misses per instruction (MPI) for TPC-C and swim. Note that there is a single L2 cache per node and all of the cores share the cache.

The data shows that it is best to share the cache space across 4 cores in the node as opposed to having private caches per core. The benefits are more pronounced for TPC-C since at these small to moderate cache sizes, a large fraction of the cache is occupied by code which is shared by many threads. Replicating the code in private caches obviously wastes space; hence, shared caches provide significant performance/area benefit for CMP architectures [9]. The benefit for swim is moderate because even if there is no sharing in SPECrate workloads, it is possible for a core to perceive a larger cache when another core is not using the cache as heavily. In addition, it is worth noting that the L2 MPI reduces significantly when going from 256K to 512K. Therefore, the 512K cache size appears to be a sweet spot for this workload in such a configuration.

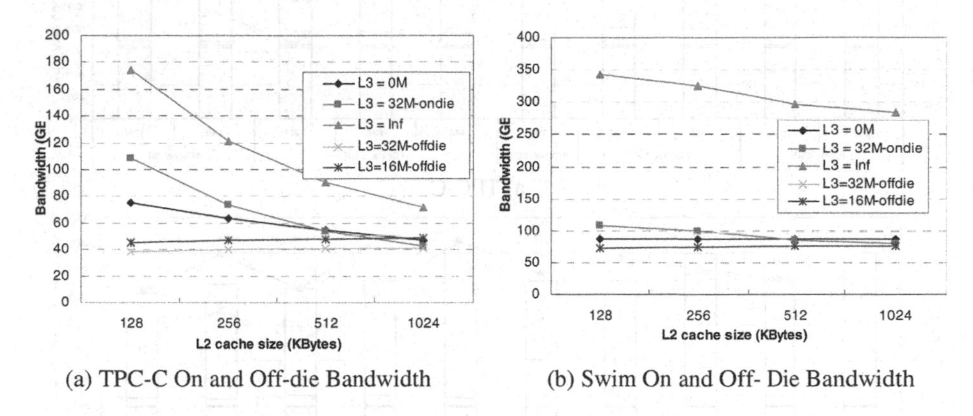

(a) TPC-C On and Off-die Bandwidth (b) Swim On and Off- Die Bandwidth

Fig. 3. MPI and Bandwidth Characteristics of LCMP Server Platforms

Figures 3 (a) and (b) show the on-die and off-die bandwidth demands of these workloads for a 32-core LCMP with 8 nodes and 4 cores per node (since this had the lowest MPI). We estimated the on-die bandwidth demand for three cases: (i) with no L3, (ii) with a 32M L3, represented by "L3=32M-ondie" line in the graph and (iii) with a perfect L3. The use of a perfect cache points to the maximum demanded on-die bandwidth. For TPC-C, the maximum bandwidth demand appears to be ~180 GB/s. Swim, on the other hand, places a maximum bandwidth demand of about 350GB/s. With a large L3 cache, the bandwidth demands reduce significantly to the range of 50 to 100 GB/s. Since it would be preferable that the interconnect utilization is low (avoiding high queuing delays or saturation), it is clear that an on-die interconnect with 200 GB/s sustainable data bandwidth or more would be sufficient for the LCMP architecture.

Off-die memory bandwidth is also a key consideration when determining the cache hierarchy. For example, if sufficient interconnect bandwidth is available, but the memory bandwidth is meager, it is desirable to allocate more space to the L3 as opposed to the L2. Figures 3(a) and 3(b) show the memory bandwidth demands of TPC-C and swim respectively for three configurations: (i) with no L3 cache (L3=0M), (ii) with a 16M L3, represented by "L3=16M-offdie" line in the graph, and (ii) with a

Table 1. LCMP Cache Options Summary

Number of Cores	Cores per Node	Number of Nodes	L2 Cache per Node	L3 cache size
32 cores	1	32	128K	~12M
	2	16	256K - 512K	8M - 16M
	4	8	512K - 1M	10M - 18M
16 cores	1	16	128K - 256K	8M - 18M
	2	8	256K - 512K	10M to 20M
	4	4	512K - 1M	10M to 20M

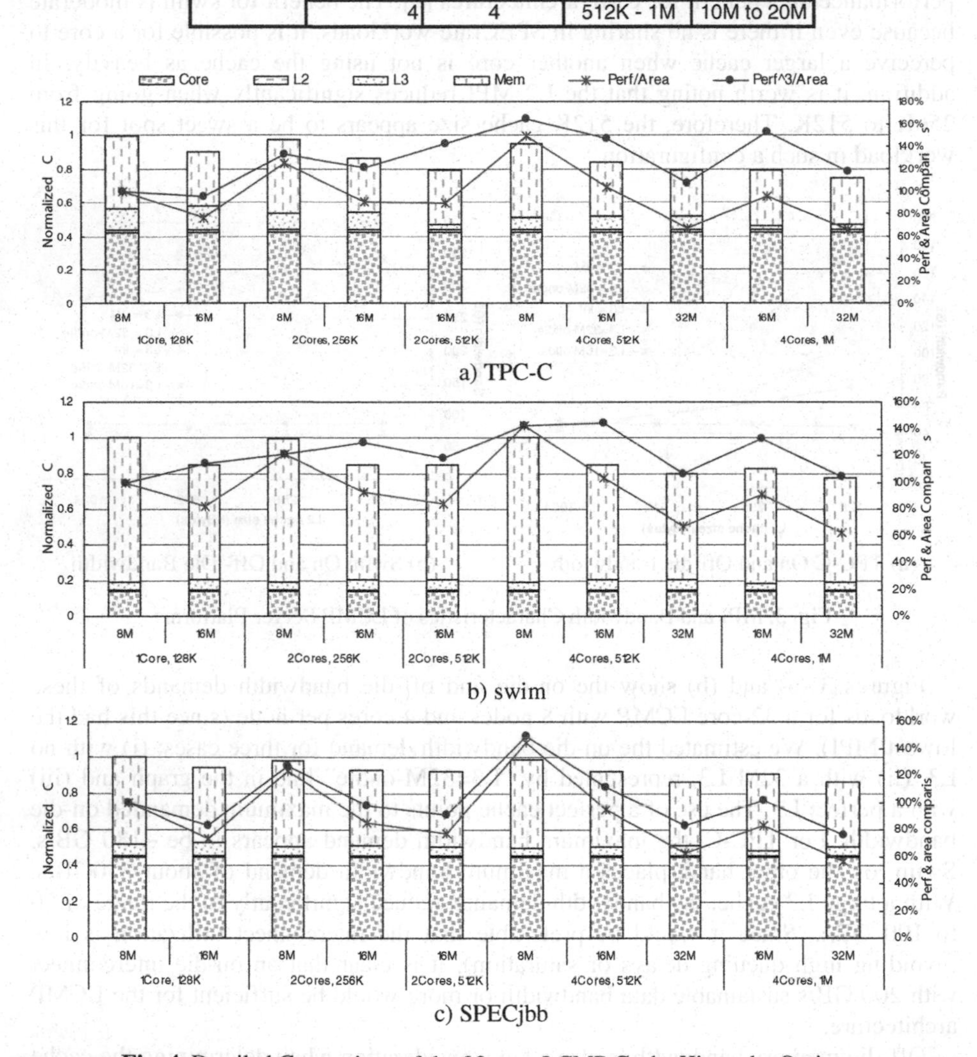

Fig. 4. Detailed Comparison of the 32-core LCMP Cache Hierarchy Options

32M L3 cache, represented by "L3=32M-offide" line in the graph. As we can see, TPC-C memory bandwidth demands range between 50GB/s and 75 GB/s. With a 32 MB L3 cache, the memory bandwidth demands reduce down to 40 GB/s. Swim shows much larger memory bandwidth demands than TPC-C (~85GB/s without L3, ~75GB/s with L3).

In order to have a low memory utilization (< 50%), it is important that a 32-core LCMP memory subsystem (with a L3 cache) provide a sustainable bandwidth of over 100 GB/s. Based on DDR/FBD memory trends [9], we expect that towards the end of the decade (when 45nm is available), the peak memory bandwidth will be around 64 to 128 GB/s. Applying a 64GB/s constraint essentially shows that the "no L3" options are not viable for TPC-C and swim workloads. Even with the options with L3 cache (as much as 32M), memory bandwidth remains a bottleneck for swim, but the TPC-C workload may scale.

As expected, the 16-core configurations showed lower sensitivity to on-die and off-die bandwidth constraints. However, in order to build a modular architecture, the observations from the 32-core analysis need to be taken into account even while building a 16-core LCMP platform.

Based on the area and bandwidth constraints, we were able to prune the design space sufficiently and summarize a smaller set of configurations as listed in Table 1. The major factors that affected the pruning process are: 1) Applying area constraints showed that around 128K to 256K per core seems viable for 16-core and 32-core LCMP, 2) Applying area constraints resulted in L3 sizes ranging from 8M to about 20M depending on the configuration being considered, and 3) Applying bandwidth constraints essentially showed that configurations without L3 cache were not viable.

5 LCMP Cache Hierarchy Performance

In this section, we study the performance of the LCMP cache hierarchy options summarized in Table 1. The metrics used in this section are both performance and performance/area, although area constraints have already been applied to prune the design space sufficiently.

5.1 Performance of LCMP Cache Options

We have collected performance data for the 32-core LCMP cache hierarchy options. Figure 4 shows this data for TPC-C, swim and SPECjbb workloads. The variation in the number of cores per node, the L2 cache size per node and the L3 cache size are shown on the x-axis. The on-die bandwidth is 512 GB/s and the maximum sustainable memory bandwidth is 64 GB/s. The vertical bars in Figure 4 show the CPI broken down into the time spent in the core, between the core and L2, between the L2 and L3 and finally in the memory subsystem. A simple observation is that the dominating factors are the performance of the core and the performance of the memory subsystem as the time spent in L2 & L3 subsystems are fairly low.

From a performance (CPI) perspective, it is not surprising that the configuration that performs the best is the one with 4 cores per node, 1M L2 per node and 32 M of L3 cache. It should be noted that in all configurations except those with 32M, the area consumed remains between 170 mm^2 and 350mm^2. If we exclude the options with 32M L3 cache, then the high performance option is the 4-core node configuration with 16M L3 cache and either 1M or even 512K of L2 cache per node.

However, this performance comes at the expense of additional area. Placing equal emphasis on performance and area shows that the configuration with the least amount of cache area (4 cores per node, 512K per node, 8M L3) turns out to be the best configuration. We then looked at performance^3/area as a potential metric. The behavior of the two metrics (performance/area and performance^3/area) is significantly different for TPC-C and Swim, but not as much for other workloads. The reason for minimal change with other workloads is their insensitivity to cache size beyond 8M and minimal performance impact as a result. However, TPC-C and Swim are very memory-intensive. As a result, the performance (CPI) is affected significantly for these workloads. Overall, since the area constraints have already been applied (except for options with 32M L3 cache), the design option that provides good performance with low area overhead consists of 4 cores per node, 512K to 1M of L2 cache and 16M L3 cache.

5.2 Bandwidth Considerations (Off/On-Die)

Memory bandwidth has a significant impact on the LCMP application performance. All applications show a significant improvement as the memory bandwidth is increased from 32 to 64GB/s. Further increase in memory bandwidth to 128 GB/s does not provide significant boost in performance for four of the six workloads (exceptions are art and swim).

We also study the impact of on-die interconnect bandwidth by varying it from 128GB/s to 512GB/s. We find that most of the workloads do not generate sufficient demand requests to stress the interconnect subsystem. The detailed data is not shown here for space limitations.

6 Conclusions

Designing large-scale CMP processors is a challenging endeavor because of the vast design space. In this paper, we performed the first study of performance, area and bandwidth implications on LCMP cache exploration. We introduced a constraints-aware analysis (CAAM) methodology for exploring the LCMP cache hierarchy options. We applied this methodology to 16-core and 32-core LCMP architectures and showed the pruning process.

References

1. Bahar, R.I., Albera, G., Manne, S.: Power & performance tradeoffs using various caching strategies. In: Proc. of Int'l Symp. on Low-Power Electronics & Design (1998)
2. Chishti, Z., et al.: Distance Associativity for High-Performance Energy-Efficient Non-Uniform Cache Architectures. In: 36th Int'l Symp. on Microarchitecture (2003)
3. Chishti, Z., Powell, M., Vijaykumar, T.N.: Optimizing Replication, Communication, and Capacity Allocation in CMPs. In: ISCA. Proceedings of the 32nd International Symposium on Computer Architecture (June 2005)
4. Hsu, L., Iyer, R., et al.: Exploring the Cache Design Space for Large-Scale CMPs. In: 1st Workshop on Design, Architecture and Simulation of CMP (dasCMP) (November 2005)

5. Intel Dual-Core: First in Multi-core Revolution, http://www.intel.com/technology/computing/dual-core/
6. Kim, C., Burger, D., Keckler, S.W.: Nonuniform Cache Architectures for Wire-Delay Dominated On-Chip Caches. IEEE Micro 23(6), 99–107 (2003)
7. Krewell, K.: Best Servers of 2004: Multicore is Norm. Microprocessor Report (January 2005), http://www.mpronline.com
8. Kumar, R., Zyuban, V., Tullsen, D.M.: Interconnections in Multi-core Architectures: Understanding Mechanisms, Overheads and Scaling. In: ISCA-32. 32nd International Symposium on Computer Architecture (June 2005)
9. Olukotun, K., Nayfeh, B.A., et al.: The case for a single-chip multiprocessor. In: 7th International Conf. on Architectural Support for Programming Languages and Operating Systems (October 1996)
10. SAP, http://www.sap.com/solutions/benchmark/index.epx
11. Shivakumar, P., Jouppi, N.: CACTI 3.0: An Integrated Cache Timing, Power & Area Model. WRL Research Report (August 2001)
12. SPECrate, http://www.spec.org/cpu2000/SPECrate
13. SPECjAppServer, http://www.spec.org/jAppServer/
14. Speight, E., et al.: Adaptive Mechanisms and Policies for Managing Cache Hierarchies in Chip Multiprocessors. In: ISCA. 32nd Int'l Symp. on Computer Architecture (June 2005)
15. Su, C., Despain, A.M.: Cache design trade-offs for power and performance optimization: a case study. In: ISLPED 1995. Int'l Symp. On Low Power Design (1995)
16. TPC-C Design Document, http://www.tpc.org/tpcc/

FFTC: Fastest Fourier Transform for the IBM Cell Broadband Engine

David A. Bader and Virat Agarwal

College of Computing
Georgia Institute of Technology
Atlanta, GA, USA 30332
{bader, virat}@cc.gatech.edu

Abstract. The Fast Fourier Transform (FFT) is of primary importance and a fundamental kernel in many computationally intensive scientific applications. In this paper we investigate its performance on the Sony-Toshiba-IBM Cell Broadband Engine, a heterogeneous multicore chip architected for intensive gaming applications and high performance computing. The Cell processor consists of a traditional microprocessor (called the PPE) that controls eight SIMD co-processing units called synergistic processor elements (SPEs). We exploit the architectural features of the Cell processor to design an efficient parallel implementation of Fast Fourier Transform (FFT). While there have been several attempts to develop a fast implementation of FFT on the Cell, none have been able to achieve high performance for input series with several thousand complex points. We use an iterative out-of-place approach to design our parallel implementation of FFT with 1K to 16K complex input samples and attain a single precision performance of 18.6 GFLOP/s on the Cell. Our implementation beats FFTW on Cell by several GFLOP/s for these input sizes and outperforms Intel Duo Core (Woodcrest) for inputs of greater than 2K samples. To our knowledge we have the fastest FFT for this range of complex inputs.

1 Introduction

The Cell Broadband Engine (or the Cell/B.E.) [15,8,9,18] is a novel high-performance architecture designed by Sony, Toshiba, and IBM (STI), primarily targeting multimedia and gaming applications. The Cell BE consists of a traditional microprocessor (called the PPE) that controls eight SIMD co-processing units called synergistic processor elements (SPEs), a high speed memory controller, and a high bandwidth bus interface (termed the element interconnect bus, or EIB), all integrated on a single chip. The Cell is used in Sony's PlayStation 3 gaming console, Mercury Computer System's dual Cell-based blade servers, and IBM's QS20 Cell Blades.

In this paper we present our design of an efficient parallel implementation of Fast Fourier Transform on the Cell Broadband Engine. FFT is of primary importance and a fundamental kernel in many computationally intensive scientific

S. Aluru et al. (Eds.): HiPC 2007, LNCS 4873, pp. 172–184, 2007.

applications such as computer tomography, data filtering and fluid dynamics. Another important application area of FFTs is in spectral analysis of speech, sonar, radar, seismic and vibration detection. FFTs are also used in digital filtering, signal decomposition, and in solution of partial differential equations. The performance of these applications rely heavily on the availability of a fast routine for Fourier transforms.

The literature contains several publications related to FFTs on the Cell/B.E. processor. Williams *et al.* [19] analyze the Cell's peak performance for FFT of various types (1D, 2D), accuracy (single, double precision) and input sizes. Cico, Cooper and Greene [7] estimate the performance of 22.1 GFLOP/s for a single FFT that is reside in the local store of one SPE, or 176.8 GFLOP/s for computing 8 independent FFTs with 8K complex input samples. (Note that all other computation rates given in this paper – except for Cico *et al.* – consider the performance of a single FFT and include the overheads when considering that the source and output of the FFT are both stored in main memory.) In another work, Chow, Fossum and Brokenshire [6] achieve 46.8 GFLOP/s for a large FFT (16 million complex samples) on the Cell that is highly-specialized for this particular input size. FFTW on the Cell [11] is a highly-portable FFT library of various types, precision and input size.

In our design of FFTC we use an iterative out-of-place approach to solve 1D FFTs with 1K to 16K complex input samples. We describe our methodology to partition the work among the SPEs to efficiently parallelize a *single FFT computation* where the source and output of the FFT are both stored in main memory. This differentiates our work from the prior literature and better represents the performance that one realistically sees in practice. The algorithm requires a synchronization among the SPEs after each stage of FFT computation. Our synchronization barrier is designed to use inter SPE communication without any intervention from the PPE. The synchronization barrier requires only $2 \log p$ stages (p: number of SPEs) of inter SPE communication by using a tree-based approach. This significantly improves the performance, as PPE intervention not only results in a high communication latency but also in sequentialization of the synchronization step. We achieve a performance improvement of over 4 as we vary the number of SPEs from 1 to 8. We attain a performance of 18.6 GFLOP/s for a single-precision FFT with 8K complex input samples and also show significant speedup in comparison with other architectures. Our implementation is generic for this range of complex inputs. The source code is freely available from our CellBuzz project in SourceForge (`http://sourceforge.net/projects/cellbuzz/`).

This paper is organized as follows. We first describe the Fast Fourier Transform and the algorithm we choose to parallelize in Section 2. The novel architectural features of the Cell processor are reviewed in Section 3. We then present our design to parallelize FFT on the Cell and optimize for the SPEs in Section 4.

2 Fast Fourier Transform

Fast Fourier Transform (FFT) is an efficient algorithm that is used for computing the Discrete Fourier Transform. Some of the important application areas of FFTs have been mentioned in the previous section. There are several algorithmic variants of the FFTs that have been well studied for parallel processors and vector architectures [1,2,3,4].

In our design we utilize the naive Cooley-Tukey radix-2 Decimate in Frequency (DIF) algorithm. The pseudo-code for an out-of-place approach of this algorithm is given in Alg. 1. The algorithm runs in $\log N$ stages and each stage requires $O(N)$ computation, where N is the input size.

Algorithm 1: Sequential FFT algorithm

```
   Input: array A[0] of size N
1  NP ← 1 ;
2  problemSize ← N;
3  dist ← 1;
4  i1 ← 0;
5  i2 ← 1;
6  while problemSize > 1 do
7      Begin Stage;
8      a ← A[i1];
9      b ← A[i2];
10     k = 0, jtwiddle = 0;
11     for j ← 0 to N − 1 step 2 * NP do
12         W ← w[jtwiddle];
13         for jfirst ← 0 to NP do
14             b[j + jfirst] ← a[k + jfirst] + a[k + jfirst + N/2];
15             b[j + jfirst + Dist] ← (a[k + jfirst] − a[k + jfirst + N/2]) * W;
16         k ← k + NP;
17         jtwiddle ← jtwiddle + NP;
18     swap(i1, i2);
19     NP ← NP * 2;
20     problemSize ← problemSize/2;
21     dist ← dist * 2;
22     End Stage;
   Output: array A[i1] of size N
```

The array w contains the *twiddle factors* required for FFT computation. At each stage the computed complex samples are stored at their respective locations thus saving a bit-reversal stage for output data. This is an iterative algorithm which runs until the parameter *problemSize* reduces to 1. Fig. 1 shows the butterfly stages of this algorithm for an input of 16 sample points (4 stages).

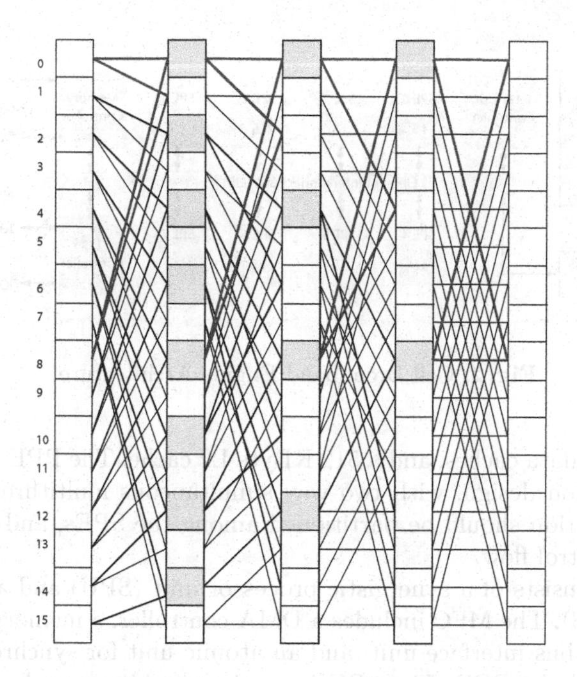

Fig. 1. Butterflies of the ordered DIF FFT algorithm

Apart from the theoretical complexity, another common performance metric used for the FFT algorithm is the floating point operation (FLOP) count. On analyzing the sequential algorithm, we see that during each iteration of the innermost *for* loop there is one complex addition for the computation of first output sample, which accounts for 2 FLOPs. The second output sample requires one complex subtraction and multiplication which accounts for 8 FLOPs. Thus, for the computation of two output samples during each innermost iteration we require 10 FLOPs, which suggests that we require 5 FLOPs for the computation of a complex sample at each stage. The total computations in all stages are $N \log N$ which makes the total FLOP count for the algorithm as $5N \log N$.

3 Cell Broadband Engine Architecture

The Cell Broadband Engine (Cell/B.E.) processor is a heterogeneous multi-core chip that is significantly different from conventional multiprocessor or multi-core architectures. It consists of a traditional microprocessor (the PPE) that controls eight SIMD co-processing units called synergistic processor elements (SPEs), a high speed memory controller, and a high bandwidth bus interface (termed the element interconnect bus, or EIB), all integrated on a single chip. Fig. 2 gives an architectural overview of the Cell/B.E. processor. We refer the reader to [17,10,16,12,5] for additional details.

The PPE runs the operating system and coordinates the SPEs. It is a 64-bit PowerPC core with a vector multimedia extension (VMX) unit, 32 KByte L1

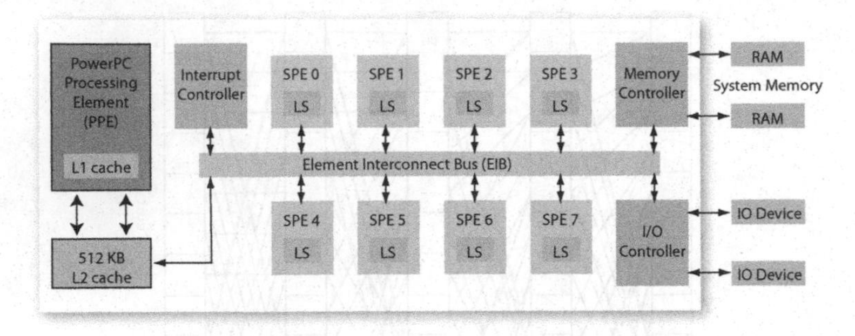

Fig. 2. Cell Broadband Engine Architecture

instruction and data caches, and a 512 KByte L2 cache. The PPE is a dual issue, in-order execution design, with two way simultaneous multithreading. Ideally, all the computation should be partitioned among the SPEs, and the PPE only handles the control flow.

Each SPE consists of a synergistic processor unit (SPU) and a memory flow controller (MFC). The MFC includes a DMA controller, a memory management unit (MMU), a bus interface unit, and an atomic unit for synchronization with other SPUs and the PPE. The SPU is a micro-architecture designed for high performance data streaming and data intensive computation. It includes a 256 KByte *local store* (LS) memory to hold SPU program's instructions and data. The SPU cannot access main memory directly, but it can issue DMA commands to the MFC to bring data into the Local Store or write computation results back to the main memory. DMA is non-blocking so that the SPU can continue program execution while DMA transactions are performed.

The SPU is an in-order dual-issue statically scheduled architecture. Two SIMD [14] instructions can be issued per cycle: one compute instruction and one memory operation. The SPU branch architecture does not include dynamic branch prediction, but instead relies on compiler-generated branch hints using *preparc-to-branch* instructions to redirect instruction prefetch to branch targets. Thus branches should be minimized on the SPE as far as possible.

The MFC supports naturally aligned transfers of 1,2,4, or 8 bytes, or a multiple of 16 bytes to a maximum of 16 KBytes. DMA list commands can request a list of up to 2,048 DMA transfers using a single MFC DMA command. Peak performance is achievable when both the effective address and the local storage address are 128 bytes aligned and the transfer is an even multiple of 128 bytes. In the Cell/B.E., each SPE can have up to 16 outstanding DMAs, for a total of 128 across the chip, allowing unprecedented levels of parallelism in on-chip communication. Kistler *et al.* [16] analyze the communication network of the Cell/B.E. and state that applications that rely heavily on random scatter and or gather accesses to main memory can take advantage of the high communication bandwidth and low latency.

With a clock speed of 3.2 GHz, the Cell processor has a theoretical peak performance of 204.8 GFLOP/s (single precision). The EIB supports a peak bandwidth of 204.8 GB/s for intrachip transfers among the PPE, the SPEs, and the memory and I/O interface controllers. The memory interface controller (MIC) provides a peak bandwidth of 25.6 GB/s to main memory. The I/O controller provides peak bandwidths of 25 GB/s inbound and 35 GB/s outbound.

4 FFTC: Our FFT Algorithm for the Cell/B.E. Processor

There are several architectural features that make it difficult to optimize and parallelize the Cooley-Tukey FFT algorithm on the Cell Broadband Engine. The algorithm is branchy due to presence of a doubly nested *for* loop within the outer *while* loop. This results in a compromise on the performance due to the absence of a branch predictor on the Cell. The algorithm requires an array that consists of the $N/2$ complex twiddle factors. Since each SPE has a limited local store of 256 KB, this array cannot be stored entirely on the SPEs for a large input size. The limit in the size of the local store memory also restricts the maximum input data that can be transferred to the SPEs. Parallelization of a single FFT computation involves synchronization between the SPEs after every stage of the algorithm, as the input data of a stage is the output data of the previous stage. To achieve high performance it is necessary to divide the work equally among the SPEs so that no SPE waits at the synchronization barrier. Also, the algorithm requires $\log N$ synchronization stages which impacts the performance. It is difficult to vectorize every stage of the FFT computation. For vectorization of the first two stages of the FFT computation it is necessary to shuffle the output data vector, which is not an efficient operation in the SPE instruction set architecture. Also, the computationally intensive loops in the algorithm need to be unrolled for best pipeline utilization. This becomes a challenge given a limited local store on the SPEs.

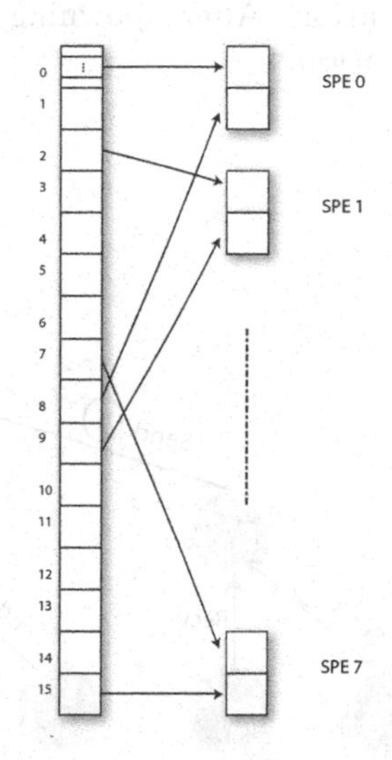

Fig. 3. Partition of the input array among the SPEs (e.g. 8 SPEs in this illustration)

4.1 Parallelizing FFTC for the Cell

As mentioned in the previous section for best performance it is important to partition work among the SPEs to achieve load balancing. We parallelize by dividing the input array held in main memory into $2p$ chunks, each of size $\frac{N}{2p}$, where p is the number of SPEs.

During every stage, SPE i is allocated chunk i and $i+p$ from the input array. The basis for choosing these chunks for an SPE lies in the fact that these chunks are placed at an offset of $N/2$ input elements. For the computation of an output complex sample we need to perform complex arithmetic operation between input elements that are separated

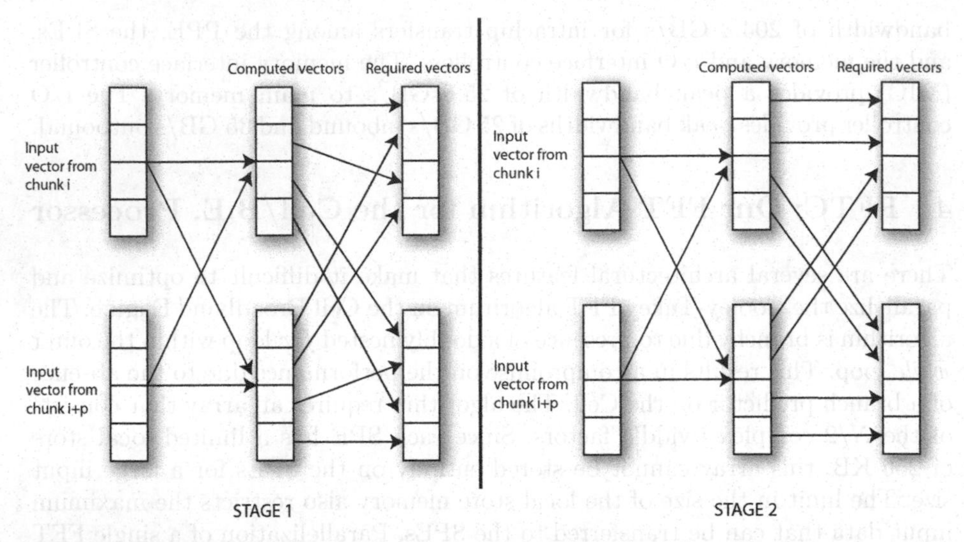

Fig. 4. Vectorization of the first two stages of the FFT algorithm. These stages require a shuffle operation over the output vector to generate the desired output.

by this offset. Fig. 3 gives an illustration of this approach for work partitioning among 8 SPEs.

The PPE does not intervene in the FFT computation after this initial work allocation. After spawning the SPE threads it waits for the SPEs to finish execution.

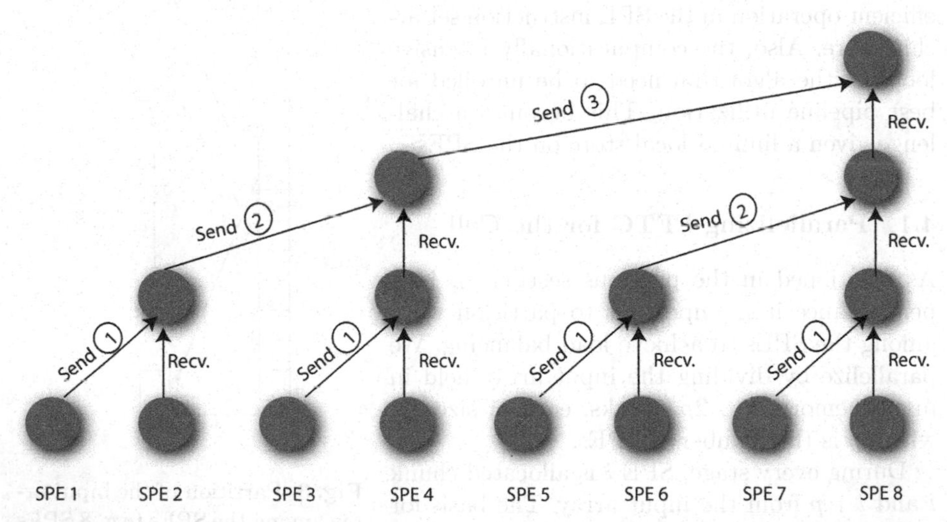

Fig. 5. Stages of the synchronization barrier using inter SPE communication. The synchronization involves sending inter SPE mailbox messages up to the root of the tree and then sending back acknowledgment messages down to the leaves in the same topology.

Algorithm 2: Parallel FFTC algorithm: View within SPE

```
   Input: array in PPE of size N
   Output: array in PPE of size N
 1 NP ⟵ 1 ;
 2 problemSize ⟵ N;
 3 dist ⟵ 1;
 4 fetchAddr ⟵ PPE input array;
 5 putAddr ⟵ PPE output array;
 6 chunkSize ⟵ N/(2*p);
 7 Stage 0 (SIMDization achieved with shuffling of output vector);
 8 Stage 1 ;
 9 while NP < buffersize && problemSize > 1 do
10     Begin Stage;
11     Initiate all DMA transfers to get data;
12     Initialize variables;
13     for j ← 0 to 2 * chunkSize do
14         Stall for DMA buffer;
15         for i ← 0 to buffersize/NP do
16             for jfirst ← 0 to NP do
17                 SIMDize computation as NP > 4;
18             Update j, k, jtwiddle;
19         Initiate DMA put for the computed results
20     swap(fetchAddr, putAddr);
21     NP ← NP * 2;
22     problemSize ← problemSize/2;
23     dist ← dist * 2;
24     End Stage;
25     Synchronize using Inter-SPE communication;
26 while problemSize > 1 do
27     Begin Stage;
28     Initiate all DMA transfers to get data;
29     Initialize variables;
30     for k ← 0 to chunkSize do
31         for jfirst ← 0 to min(NP, chunkSize − k) step buffersize do
32             Stall for DMA buffer;
33             for i ← 0 to buffersize do
34                 SIMDize computation as buffersize > 4;
35             Initiate DMA put for the computed results;
36         Update j, k, jtwiddle;
37     swap(fetchAddr, putAddr);
38     NP ← NP * 2;
39     problemSize ← problemSize/2;
40     dist ← dist * 2;
41     End Stage;
42     Synchronize using Inter SPE communication;
```

4.2 Optimizing FFTC for the SPEs

After dividing the input array among the SPEs, each SPE is allocated 2 chunks each of size $\frac{N}{2p}$. Each SPE, fetches this chunk from main memory using DMA transfers and uses double-buffering to overlap memory transfers with computation. Within each SPE, after computation of each buffer, the computed buffer is written back into main memory at the correct offset using DMA transfers.

The detailed pseudo-code is given in Alg. 2. The first two stages of the FFT algorithm are duplicated, that correspond to the first two iterations of the outer *while* loop in sequential algorithm. This is necessary as the vectorization of these stages requires a shuffle operation (*spu_shuffle()*) over the output to re-arrange the output elements to their correct locations. Please refer to Fig. 4 for an illustration of this technique for stages 1 and 2 of the FFT computation.

The innermost *for* loop (in the sequential algorithm) can be easily vectorized for $NP > 4$, that correspond to the stages 3 through $\log N$. However, it is important to duplicate the outer *while* loop to handle stages where $NP <$ *buffersize*, and otherwise. The global parameter *buffersize* is the size of a single DMA get buffer. This duplication is required as we need to stall for a DMA transfer to complete, at different places within the loop for these two cases. We also unroll the loops to achieve better pipeline utilization. This significantly increases the size of the code thus limiting the unrolling factor.

SPEs are synchronized after each stage, using *inter-SPE communication*. This is achieved by constructing a binary synchronization tree, so that synchronization is achieved in $2 \log p$ stages. The synchronization involves the use of inter-SPE mailbox communication without any intervention from the PPE. Please refer to Fig. 5 for an illustration of the technique.

This technique performs significantly better than other synchronization techniques that either use chain-like inter-SPE communication or require the PPE to synchronize between the SPEs. The chain-like technique requires $2p$ stages of inter-SPE communication whereas with the intervention of the PPE latency of communication reduces the performance of this barrier.

5 Performance Analysis of FFTC

We use the Cell SDK 2.1 for instruction level profiling and performance analysis of the code. The code was compiled using the `xlc` compiler, that is included in the SDK, with level 3 optimization.

For parallelizing a single 1D FFT on the Cell, it is important to divide the work among the SPEs. Fig. 6 shows the performance of our algorithm with varying the number of SPEs for 1K and 4K complex input samples. The performance scales well with the number of SPEs which suggests that load is balanced among the SPEs.

Our design requires a barrier synchronization among the SPEs after each stage of the FFT computation. We focus on FFTs that have from 1K to 16K complex input samples. For relatively small inputs and as the number of SPEs increases, the synchronization cost becomes a significant issue since the time

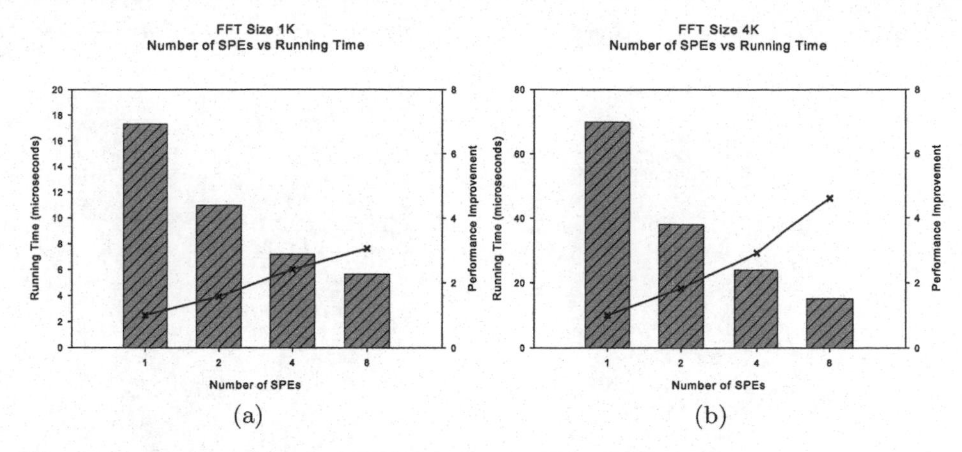

Fig. 6. Running Time of our FFTC code on 1K and 4K inputs as we increase the number of SPEs

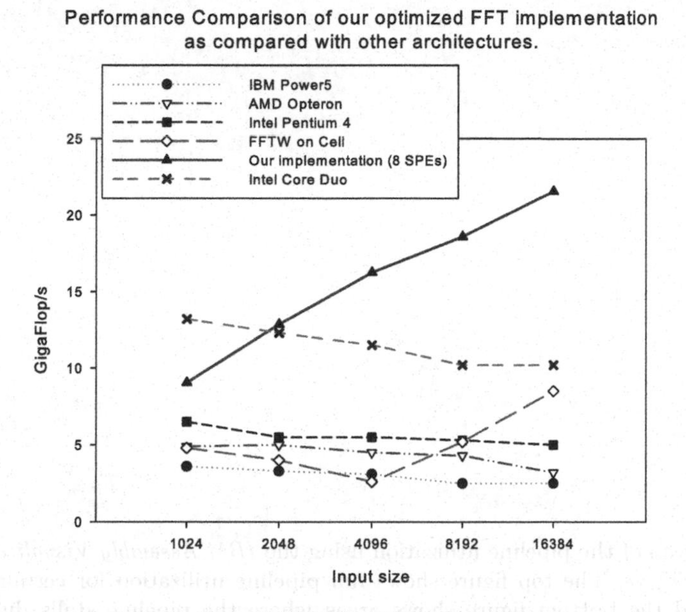

Fig. 7. Performance comparison of FFTC with other architectures for various input sizes of FFT. The performance numbers are from benchFFT from the FFTW website.

per stage decreases but the cost per synchronization increases. With instruction level profiling we determine that the time required per synchronization stage using our tree-synchronization barrier is about 1 microsecond (3200 clock cycles). We achieve a high performance barrier using inter-SPE mailbox communication which significantly reduces the time to send a message, and by using the tree-based technique we reduced the number of communication stages required for the barrier ($2 \log p$ steps).

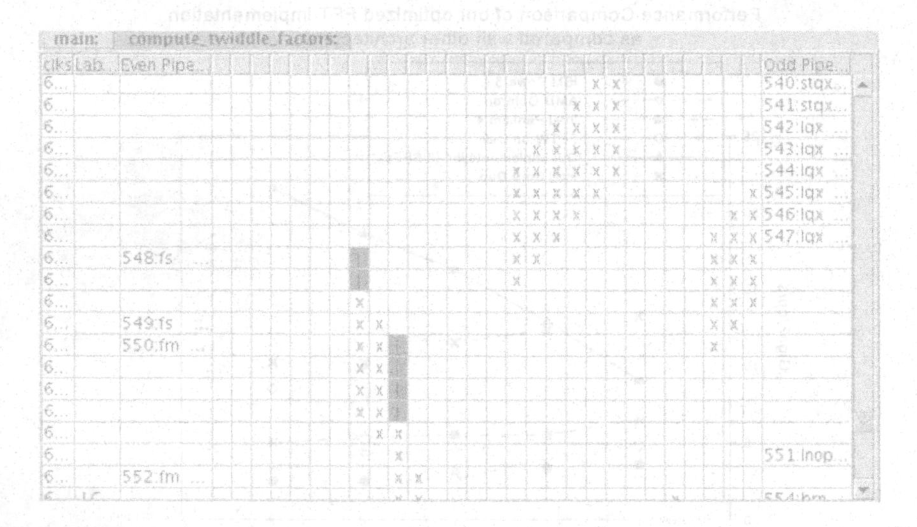

Fig. 8. Analysis of the pipeline utilization using the *IBM Assembly Visualizer for Cell Broadband Engine.* The top figure shows full pipeline utilization for certain parts of the code and the bottom figure shows areas where the pipeline stalls due to data dependency.

Fig. 7 shows the single precision performance for complex inputs of FFTC, our optimized FFT, as compared with the following architectures:

- **IBM Power 5:** IBM OpenPower 720, Two dual-core 1.65 GHz POWER5 processors.
- **AMD Opteron:** 2.2 GHz Dual Core AMD Opteron Processor 275.
- **Intel Duo Core:** 3.0 GHz Intel Xeon Core Duo (Woodcrest), 4MB L2 cache.
- **Intel Pentium 4:** Four-processor 3.06 GHz Intel Pentium 4, 512 KB L2.

We use the performance numbers from benchFFT [11] for the comparison with the above architectures. We consider the FFT implementation that gives best performance on these architectures for comparison.

The Cell/B.E. has a two instruction pipelines, and for achieving high performance it is important to optimize the code so that the processor can issue two instructions per clock cycle. This level of optimization requires inspecting the assembly dump of the SPE code. For achieving pipeline utilization it is required that the gap between dependent instructions needs to be increased. We use the *IBM Assembly Visualizer for Cell/B.E.* tool to analyze this optimization. The tool highlights the stalls in the instruction pipelines and helps the user to reorganize the code execution while maintaining correctness. Fig. 8 shows the analysis of pipeline utilization. Some portions utilize these pipelines effectively (top figure) whereas there are a few stalls in other parts of the code which still need to be optimized (bottom figure).

6 Conclusions

In summary, we present FFTC, our high-performance design to parallelize the 1D FFT on the Cell Broadband Engine processor. FFTC uses an iterative out-of-place approach and we focus on FFTs with 1K to 16K complex input samples. We describe our methodology to partition the work among the SPEs to efficiently parallelize a single FFT computation. The computation on the SPEs is fully vectorized with other optimization techniques such as loop unrolling and double buffering. The algorithm requires a synchronization among the SPEs after each stage of FFT computation. Our synchronization barrier is designed to use inter SPE communication only without any intervention from the PPE. The synchronization barrier requires only $2 \log p$ stages (p: number of SPEs) of inter SPE communication by using a tree-based approach. This significantly improves the performance, as PPE intervention not only results in a high communication latency but also results in sequentializing the synchronization step. We achieve a performance improvement of over 4 as we vary the number of SPEs from 1 to 8. We expect that the performance of FFTC will scale on the next generation of the IBM Cell Broadband Engine processor that may offer 32 SPEs [13]. We also demonstrate FFTC's performance of 18.6 GFLOP/s for an FFT with 8K complex input samples and show significant speedup in comparison with other architectures. Our implementation outperforms Intel Duo Core (Woodcrest) for input sizes greater than 2K and to our knowledge we have the fastest FFT for these range of complex input samples.

Acknowledgments

This work was supported in part by an IBM Shared University Research (SUR) award and NSF Grants CNS-0614915, CAREER CCF-0611589, and DBI-0420513. We would also like to thank Sidney Manning (IBM Corporation) for providing valuable inputs during the course of our research. We acknowledge our

Sony-Toshiba-IBM Center of Competence for the use of Cell Broadband Engine resources that have contributed to this research.

References

1. Agarwal, R.C., Cooley, J.W.: Vectorized mixed radix discrete Fourier transform algorithms. Proc. of the IEEE 75(9), 1283–1292 (1987)
2. Ashworth, M., Lyne, A.G.: A segmented FFT algorithm for vector computers. Parallel Computing 6(2), 217–224 (1988)
3. Averbuch, A., Gabber, E., Gordissky, B., Medan, Y.: A parallel FFT on an MIMD machine. Parallel Computing 15, 61–74 (1990)
4. Bailey, D.H.: A high-performance FFT algorithm for vector supercomputers. Intl. Journal of Supercomputer Applications 2(1), 82–87 (1988)
5. Chen, T., Raghavan, R., Dale, J., Iwata, E.: Cell Broadband Engine Architecture and its first implementation. Technical Report (November 2005)
6. Chow, A.C., Fossum, G.C., Brokenshire, D.A.: A Programming Example: Large FFT on the Cell Broadband Engine. In: GSPx. Tech. Conf. Proc. of the Global Signal Processing Expo. (2005)
7. Cico, L., Cooper, R., Greene, J.: Performance and Programmability of the IBM/Sony/Toshiba Cell Broadband Engine Processor. White paper (2006)
8. IBM Corporation. Cell Broadband Engine technology. `http://www.alphaworks.ibm.com/topics/cell`
9. IBM Corporation. The Cell project at IBM Research. `http://www.research.ibm.com/cell/home.html`
10. Flachs, B., et al.: A streaming processor unit for a Cell processor. In: International Solid State Circuits Conference, San Fransisco, CA, USA, vol. 1, pp. 134–135 (February 2005)
11. Frigo, M., Johnson, S.G.: FFTW on the Cell Processor (2007), `http://www.fftw.org/cell/index.html`
12. Hofstee, H.P.: Cell Broadband Engine Architecture from 20,000 feet. Technical Report (August 2005)
13. Hofstee, H.P.: Real-time supercomputing and technology for games and entertainment (keynote talk). In: Proc. SC, Tampa, FL (November 2006)
14. Jacobi, C., Oh, H.-J., Tran, K.D., Cottier, S.R., Michael, B.W., Nishikawa, H., Totsuka, Y., Namatame, T., Yano, N.: The vector floating-point unit in a synergistic processor element of a Cell processor. In: Proc. 17th IEEE Symposium on Computer Arithmetic (ARITH 2005), Washington, DC, USA, pp. 59–67. IEEE Computer Society Press, Los Alamitos (2005)
15. Kahle, J.A., Day, M.N., Hofstee, H.P., Johns, C.R., Maeurer, T.R., Shippy, D.: Introduction to the Cell multiprocessor. IBM J. Res. Dev. 49(4/5), 589–604 (2005)
16. Kistler, M., Perrone, M., Petrini, F.: Cell multiprocessor communication network: Built for speed. IEEE Micro 26(3), 10–23 (2006)
17. Pham, D., et al.: The design and implementation of a first-generation Cell processor. In: International Solid State Circuits Conference, San Fransisco, CA, USA, vol. 1, pp. 184–185 (February 2005)
18. Sony Corporation. Sony release: Cell architecture. `http://www.scei.co.jp/`
19. Williams, S., Shalf, J., Oliker, L., Kamil, S., Husbands, P., Yelick, K.: The potential of the Cell processor for scientific computing. In: Proc. 3rd Conference on Computing Frontiers (CF 2006), pp. 9–20. ACM Press, New York (2006)

Molecular Dynamics Simulations on Commodity GPUs with CUDA

Weiguo Liu[1], Bertil Schmidt[2], Gerrit Voss[1], and Wolfgang Müller-Wittig[1]

[1] School of Computer Engineering, Nanyang Technological University
Centre for Advanced Media Technology
{liuweiguo, asgerrit, askwmwittig}@ntu.edu.sg
[2] Computer Science and Engineering, University of New South Wales
bertil.schmidt@unsw.edu.au

Abstract. Molecular dynamics simulations are a common and often repeated task in molecular biology. The need for speeding up this treatment comes from the requirement for large system simulations with many atoms and numerous time steps. In this paper we present a new approach to high performance molecular dynamics simulations on graphics processing units. Using modern graphics processing units for high performance computing is facilitated by their enhanced programmability and motivated by their attractive price/performance ratio and incredible growth in speed. To derive an efficient mapping onto this type of architecture, we have used the Compute Unified Device Architecture (CUDA) to design and implement a new parallel algorithm. This results in an implementation with significant runtime savings on an off-the-shelf computer graphics card.

1 Introduction

The fast increasing power of the Graphics Processing Unit (GPU) and its streaming architecture opens up a range of new possibilities for a variety of applications. With the enhanced programmability of commodity GPUs, these chips are now capable of performing more than the specific graphics computations they were originally designed for. Recent work shows the design and implementation of algorithms for non-graphics applications. Examples include scientific computing [1], image processing [2], computational biology [3,4], and fast Fourier transform [5], just to name a few. The evolution of GPUs is driven by the computer game market. This leads to a relatively small price per unit and to very rapid developments of next generations.

Currently, the peak performance of state-of-the-art GPUs is approximately ten times faster than that of comparable CPUs. Furthermore, the growth rate of the number of transistors used on GPUs is greater than for microprocessors [6]. Consequently, GPUs will become an even more attractive alternative for high performance computing in the near future.

However, there are still a number of challenges to be solved in order to enable scientists other than computer graphics specialists to facilitate efficient usage of

S. Aluru et al. (Eds.): HiPC 2007, LNCS 4873, pp. 185–196, 2007.

these resources within their research area. The biggest challenge in order to solve a specific problem using GPUs is reformulating the proposed algorithms and data structures using computer graphics primitives (e.g. triangles, textures, vertices, fragments). Furthermore, restrictions of the underlying streaming architecture have to be taken into account, e.g. random access writes to memory is not supported and no cross fragment data or persistent state is possible (e.g. all the internal registers are flushed before a new fragment is processed).

The Compute Unified Device Architecture (CUDA) [7] is a new hardware and software architecture for issuing and managing computations on GPUs. It treats the GPU as a data-parallel computing device without the need of mapping computations to the graphics pipeline. CUDA technology gives computationally intensive applications access to the tremendous processing power of GPUs through a revolutionary new programming interface. Providing orders of magnitude more performance and simplifying software development by using the standard C language, CUDA enables developers to create innovative solutions for data-intensive problems.

Molecular dynamics (MD) is a computationally intensive method of studying the natural time-evolution of a system of atoms using Newton's classical equations of motion. In practice, MD has always been limited more by the current available computing power than by investigators' ingenuity. Researchers in this field have typically focused their efforts on simplifying models and identifying what may be neglected while still obtaining acceptable results. This has led to much skepticism on the ability of MD to be used as a predictive tool for experimental work. High-performance computing holds the key to making biologically relevant calculations tractable without compromise. In this paper we show how MD simulations can benefit from the computing power of GPUs. In order to exploit the GPU's capabilities for high performance MD simulation we present new algorithms based on the CUDA programming model. These algorithms have been implemented using C++ and CUDA and tested on a physical system of 16,384 atoms. We show that our new MD algorithms lead to a significant performance improvement on an NVIDIA GeForce 8800 GTX card.

The rest of this paper is organized as follows. In Section 2, we introduce the basic MD simulation algorithms and highlight previous work on parallelization of these algorithms on different parallel architectures. Important features of the CUDA programming model are described in Section 3. Section 4 presents our new CUDA-based MD algorithms and their efficient GPU implementation. A performance evaluation is given in Section 5. Finally, Section 6 concludes the paper with an outlook to further research topics.

2 Molecular Dynamics Simulations

Computer simulations play a very important role in scientific research. They act as bridges among microscopic length, time scales and the macroscopic world of the laboratory [8]. In very broad terms, we can identify two categories of computer simulation techniques: MD and Monte Carlo (MC). In contrast with the MC method, MD is a deterministic technique. That is, given an initial set of parameters, the subsequent time evolution is in principle completely determined [9]. In an MD simulation, the time evolution of an atomic system is followed by integrating their equations of motion described by the following classical equations of motion:

$$\begin{cases} F_i = m_i a_i \\ F_i = -\nabla_{r_i} V(r_1, \ldots, r_N) \end{cases} \quad (1)$$

In Eq.(1), the atomic system contains N atoms. m_i is the atom mass, $a_i = d^2 r_i / dt^2$ is its acceleration, and F_i is the force acting upon it. $V(r_1, \ldots, r_N)$ is the function of the positions of the atoms. It represents the potential energy of the system. In practice, function V can be written as a sum of pairwise interactions:

$$V(r_1, \ldots, r_N) = \sum_i \sum_j u_2(r_i, r_j) + \sum_i \sum_j \sum_k u_3(r_i, r_j, r_k) + \cdots \quad (2)$$

In Eq.(2), the three body (and higher order) interactions are usually neglected [10], only leaving the pair potential as the concentration of the simulation. In practice, the Lennard-Jones (LJ) potential [11] is the most commonly used interaction model. It is given by the following expression:

$$u(r) = 4\epsilon \left[\left(\frac{\delta}{r} \right)^{12} - \left(\frac{\delta}{r} \right)^{6} \right] \quad (3)$$

where r is the distance between two interacting atoms, δ is the diameter and ϵ is the well depth. Both ϵ and δ are constants and they are chosen to fit the physical properties of the material.

One of the most time-consuming parts in MD simulations is the computation of interaction forces, which takes more than 90% of the total simulation time [12]. From Eq.(2.2) and (2.3) we can see this is mainly because the force computation requires to calculate the interactions between each atom in the system with every other atom, giving rise to $O(N^2)$ evaluations of the interaction in each time step. The interaction forces decrease rapidly with increasing distance between atoms. Thus, it is possible to neglect forces between atoms separated by more than a cutoff distance r_c. This means an atom has only interaction forces with atoms that are in a sphere with a radius equal to r_c [10]. The cutoff method is also called the neighbor list method. It reduces the computational complexity to $O(N)$. Forces computed using the cutoff method are also called short-range forces.

Figure 1 illustrates how to reduce computational complexity by using the cutoff method. When the neighbor list is built, all of the nearby atoms within an extended cutoff distance $r_{list} = r_c + skin$ are stored. At the first step in a MD simulation, the neighbor list is constructed for all the neighbors of each atom. From time to time the list is reconstructed.

Because of their inherent parallelism [13], MD simulations are suitable candidates for mapping onto parallel architectures. In the past twenty years, re-

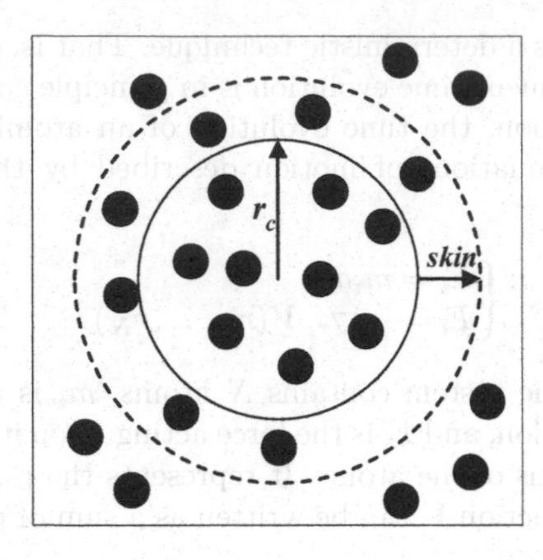

Fig. 1. Make use of r_c and *skin* to construct the neighbor list

searchers have exploited MD's parallelism on various parallel machines. In addition to architectures specifically designed for MD simulations, existing programmable sequential and parallel architectures have been used for solving them.

Special-purpose architectures can provide the fastest means of running a particular algorithm with very high processing element (PE) density. Each PE is specifically designed for the pariwise force calculation. However, such architectures are limited to one single algorithm, and thus cannot supply the flexibility necessary to run a variety of algorithms required for MD simulations. GRAPE [14] is a series of application specific processor designs, which is specially built to accelerate the MD simulations. More recent examples, better tuned to the needs of MD simulations, include ATOMS [15], FASTRUN [16], and MDGRAPE-3 [17].

Considerable effort has been spent by researchers to implement MD simulation algorithms on vector supercomputers [18]. Several other approaches are based on SIMD or MIMD parallel machines with a few dozens of processors [19,20]. SIMD and MIMD architectures are programmable and can be used for a wider range of applications. Since these architectures contain more general-purpose parallel processors, their PE density is less than the density of special-purpose architectures. Nevertheless, these solutions can still achieve significant runtime savings. However, the costs involved in designing and producing SIMD architectures are quite high. As a consequence, none of the above solutions has a successor generation, making upgrading impossible.

All these approaches can be seen as accelerators – an approach satisfying the demand for a low cost solution to compute-intensive problems. The main advantage of GPUs compared to the architectures mentioned above is that they are commodity components. In particular, most users have already access to PCs with modern graphics cards. For these users this direction provides a zero-cost solution. Even if a graphics card has to be bought, the installation of such a card is trivial (plug and play). Writing the software for such a card does still require specialist knowledge, but new high-level programming models such as CUDA [7] offer a simplified programming environment.

3 CUDA Programming Model for Computing on GPUs

Compute Unified Device Architecture (CUDA) is a novel hardware and programming model for issuing and managing computations on the GPU as a data-parallel computing device without the need of mapping them to a graphics API [21]. For now, it is available for NVIDIA 8800 series, NVIDIA Quadro FX 5600/4600, and beyond.

From the hardware point of view, CUDA treats the GPU as a set of SIMD multiprocessors. Each multiprocessor is composed of eight processors. The multiprocessor specifications of NVIDIA 8800 series and Quadro FX 5600/4600 are shown in Table 1.

Table 1. General specifications for NVIDIA CUDA-ready GPUs [21]

	Number of Multiprocessors	**Clock frequency (GHz)**	**Amount of device memory (MB)**
GeForce 8800 GTX	16	1.35	768
GeForce 8600 GTX	12	1.2	640
Quadro FX 5600	16	1.35	1500
Quadro FX 4600	12	1.2	768

A multiprocessor has on-chip memory of four types:

(1) one set of registers per processor,
(2) a parallel data cache or shared memory,
(3) a read-only constant cache,
(4) a read-only texture cache.

These on-chip memories are used to implement fast I/O operations, especially, to speed up read and write access to the non-cached device memory. Thus, applications can take advantage of them by minimizing over-fetch and round-trips to the low bandwidth device memory. Although the device memory has a low bandwidth, it is big in size and shared by all multiprocessors.

In the CUDA programming model, each multiprocessor is viewed as a multi-core device that is capable of executing a very high number of threads in parallel. These threads are organized as thread blocks. Threads in the same thread block can cooperate together by efficiently sharing data and synchronizing their execution to coordinate memory access with other threads. However, threads in different thread blocks cannot communicate or synchronize with each other. Theoretically, having more active threads per multiprocessor can help hide memory latency, and can also better fill the instruction pipeline so there are no idle processors. According to [21], the maximum number of threads that can run concurrently on a multiprocessor is 768. In practice, the number of threads is further limited by the shared on-chip memory and hence, the maximal number of threads is application-dependent.

4 CUDA-Based MD Simulation Algorithms

Many parallel algorithms for MD simulations have been proposed and implemented by different researchers. The details of these algorithms vary widely since there are numerous application-dependent and architecture-dependent characters to consider. Generally, from the point of view of data decomposition, they can be categorized into three types:

(1) **Atom-decomposition (AD):** Each processor is assigned a subset of N/P (N is the number atoms; P is the number of processors) atoms at the beginning of the simulation. As each processor must keep identical copies of atom information, it is also called replicated-data method [13]. The AD method has been widely used especially on shared memory architectures.
(2) **Force-decomposition (FD):** In this method, a subset of the force loops inherent in Eq.(2.2) is assigned to each processor. It reduces the expensive communication and memory costs by a factor $\sqrt{P}$ compared with the AD method. However, FD cannot maintain load-balance as easily as AD.
(3) **Spatial-decomposition (SD):** This method corresponds to a geometric decomposition of the physical simulation domain. Each processor computes only the forces on atoms in its sub-domain. As the simulation progresses, processors exchange atoms when they move from one sub-domain to another. SD is very well suited to large-scale MD simulations. It achieves optimal $O(N/P)$ scaling and achieves better performance on Coarse-grained architectures, such as Clusters, than AD and FD [13].

In this section we describe how MD simulations can be efficiently mapped onto a GPU using CUDA. We take advantage of the inherent parallelism of MD simulations and design parallel algorithms using the AD method. The main reasons we choose the AD method to design our algorithms is: (1) good load balancing and scalability can be easily achieved, (2) according to the CUDA model described in Section 3, the GPU hardware is viewed as a shared memory multiprocessor system, the AD method can give good performance in such a system.

The outline in Figure 2 illustrates how a sequential MD simulation works. In Figure 2, the computational complexity of each operation is listed on the end of them. In practice, the neighbor list update and force computation are the most time-consuming operations in each time step.

In the neighbor list update step (step (4) in Figure 2), a list is constructed for all neighbors of each atom. There are a large number of pairwise calculations in this step: each atom will loop over all other atoms to compute the pairwise distance between them. This corresponds to compute an $N \times N$ distance matrix D. As $r_{ij} == r_{ji}$, only the lower triangle matrix has to be computed, thus the calculation is half reduced. If the pairwise distance with the head atom of current column is within r_{list} (see Section 2), the index of current atom is added into the neighbor list array of current head atom.

There are two problems we should consider when design our CUDA-based neighbor list update algorithm. First, as mentioned in Section 2, in CUDA, thread blocks cannot communicate or synchronize with each other. This

```
1.  Initialize atoms' status and the LJ potential table;
    set parameters controlling the simulation; O(N)
2.  For all time steps do
3.        Update positions of all atoms; O(N)
4.        If there are atoms who have moved too much, do
                Update the neighbor list, including all atom pairs that are within a
                distance range (half distance matrix computation); O(N^2)
          End if
5.        Make use of the neighbor list to compute forces acted on all atoms; O(N)
6.        Update velocities of all atoms; O(N)
7.        Update the displace list, which contains the displacements of atoms; O(N)
8.        Accumulate and output target statistics of each time step; O(N)
9.  End for
```

Fig. 2. The outline of a sequential MD simulation (with the computation complexity listed, N is the number of atoms)

limitation will make the full computation of the distance matrix D necessary. For example, assume each column of the distance matrix D is assigned to a single thread and there are two threads in a thread block (see Figure 3).

In Figure 3, if we only calculate the lower triangle matrix then except for Thread 1, all other threads cannot keep the whole information of local neighbor list. For instance, as to Thread 4, the current head atom 4 will not know whether atoms 1, 2 and 3 are in the local neighbor list or not. In order get this information, Thread 4 must access the local neighbor lists of atoms 1, 2 and 3. In CUDA, this access is very expensive because it has to be done in the low bandwidth device memory. In order to solve this problem, we let each thread loop over all other atoms for current head atom. That is, in Figure 3 both the lower triangle and upper triangle matrices are calculated. Figure 4 shows our algorithm for neighbor list update using CUDA. Because the coordinates of head atoms will be reused many times in the inner loop over all other atoms in order to calculate pairwise distances, we put them into a register before the inner loop so as to speedup access for them.

After the neighbor list update step, the indices of all eligible atoms will be stored in the neighbor list array in the device memory for later usage. This is mainly because the size of the neighbor list may be very large and there is no enough on-chip memory to store it. During the compute force step, each thread will loop over the local neighbor lists to do force calculations.

Figure 5 gives our CUDA-based algorithm for the force computation. Because the coordinates of head atoms and the forces acted on them are reused many times in the inner loop over all atoms in the neighbor list, we put them into registers before the inner loop so as to increase the access efficiency for them. The results of force computation f_i will be used by other operations, such as the position and velocity update operations (step (3) and (6) in Figure 2), so we put them into a dynamically allocated shared memory to speedup access to them.

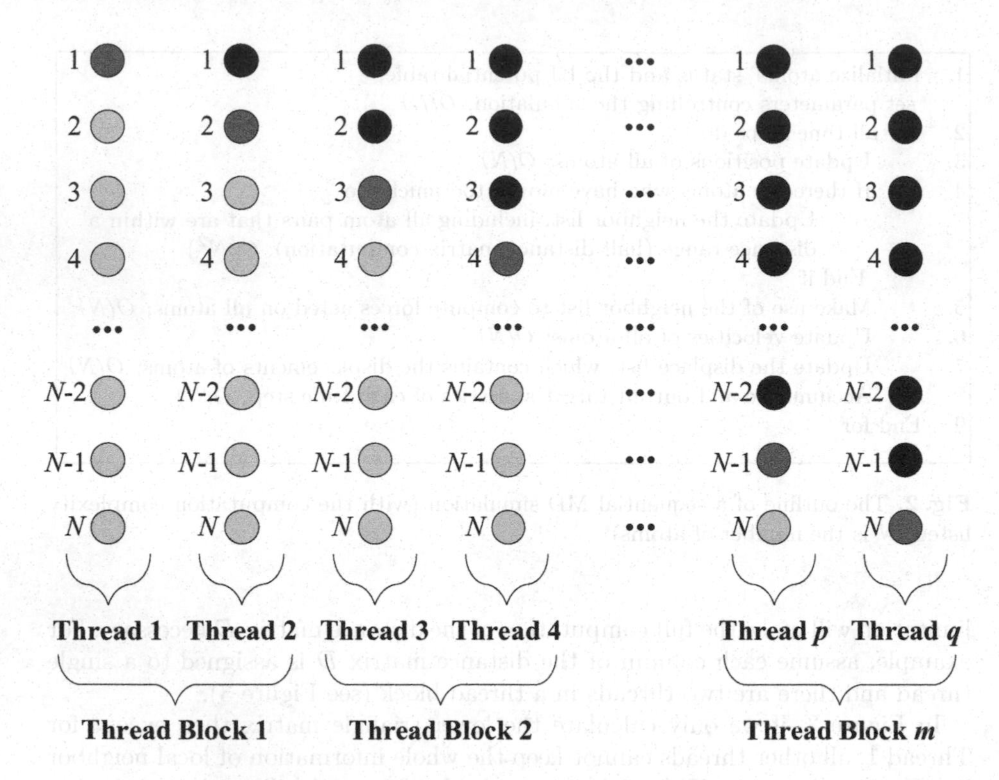

Fig. 3. Parallel neighbor list update illustration (Red circles denote head atoms). Assume each thread is allocated one column of the distance matrix and each thread block consists of two threads.

```
1.  For all allocated head atoms do
2.      Put the coordinates of current head atom into a register;
3.      For all atoms exclude the current head atom do
4.          Compute the pairwise distance between the current atom and
            head atom (full distance matrix computation);
5.          Compare the pairwise distance with r_list and put the indices
            of eligible atoms into the neighbor list in the device memory;
6.      End for
7.      Reset the displace list of current head atom with the value 0;
8.  End for
```

Fig. 4. CUDA-based neighbor list update algorithm

Figure 6 shows our CUDA-based MD simulation algorithm. In order to eliminate the overhead for launch multiple kernels, we have put all time step loops into a single kernel. As the kernel cannot output results directly, all statistics have to be read back to CPU for further processing and outputting.

```
1.  For all allocated head atoms do
2.        Put the coordinates of current head atom i into a register;
3.        Set the value of forces acted on atom i as 0
           (f_i = 0, f_i is put into a register);
4.              For atoms in the current neighbor list do
5.                    Compute the distance d_ij between the current
                        atom j and head atom i;
6.                    If d_ij < r_c do
7.                          Calculate and accumulate the force f_i acted on atom i;
8.                    End if
9.              End for
10.       Put the value of f_i into on-chip shared memory;
11. End for
```

Fig. 5. CUDA-based force computation algorithm

```
/*Host program executed on CPU*/
1.  Initialize atoms' status and the LJ potential table;
    set parameters controlling the simulation;
2.  Load data into GPU device memory and launch the kernel;
          /*Kernel program executed on GPU*/
          3.   For all time steps do
          4.         Update positions of all atoms;
          5.         If there are atoms who have moved too much, do
                           CUDA-based neighbor list update algorithm;
                      End if
          6.         CUDA-based force computation algorithm;
          7.         Update velocities of all atoms;
          8.         Update the displace list;
          9.         Put statistics of each time step into the device memory;
          10.  End for
11. Read back statistics to CPU;
12. For all time steps do
13.       Output statistics of each time step;
14. End for
```

Fig. 6. CUDA-based MD simulation algorithm

5 Performance Evaluation

We have implemented the proposed algorithm using CUDA Toolkit 0.8 [7] and evaluated it on the following graphics card:

- *nVidia GeForce 8800 GTX*: 1.35 GHz engine clock speed, 16 multiprocessors, 768 MB device memory, 16 KB shared memory/multiprocessor.

Tests have been conducted with this card installed in a PC with an AMD Opteron 2210 1.8GHz, 2GB RAM running Windows XP.

Table 2. Comparison of runtimes (in milliseconds) and speedups of MD simulation running on a single Pentium4 3GHz to our GPU-accelerated version running on an AMD Opteron 2210 1.8GHz with an NVIDIA GeForce 8800 GTX 512 for various time steps. The cutoff distance is fixed at 2.5δ.

Time steps		100	200	300	400	500
Indices in the neighbor list		1096077	1096077	1096077	1096077	1096077
MD-CPU	Overall(ms)	22468	36063	49703	63406	78078
MD-GPU (8800GTX)	Kernel(ms)	1168	1914	2656	3399	4149
	Overall(ms)	1468	2265	3078	3875	4671
Speedup	Overall	15.3	15.9	16.1	16.4	16.7

A set of performance evaluation tests have been conducted using different numbers of time steps and cutoff distances, to evaluate the processing time of the GPU implementation versus that of the original MD simulation on the PC. The MD simulation program is benchmarked on an Intel Pentium IV 3GHz processor with 1GB RAM. We have modified the MD code (md3.f90) from Ercolessi ([9], available online at http://www.fisica.uniud.it/ ercolessi/md/f90/) into a 32bit version for our evaluation. This is because for now, there is only a 32bit version of CUDA. In our experiments, there are 16,384 atoms in the simulated physical system.

Table 3. Comparison of runtimes (in milliseconds) and speedups of MD simulation running on a single Pentium4 3GHz to our GPU-accelerated version running on an AMD Opteron 2210 1.8GHz with an NVIDIA GeForce 8800 GTX 512 for various cutoff distances. The time step is fixed at 100.

r_c		2.5δ	3.0δ	3.5δ	4.0δ	4.5δ
Indices in the neighbor list		1096077	1597512	2549064	3243673	4607456
MD-CPU	Overall(ms)	22468	29984	41234	53984	69765
MD-GPU (8800GTX)	Kernel(ms)	1168	1634	2309	3111	4446
	Overall(ms)	1468	1968	2641	3437	4796
Speedup	Overall	15.3	15.2	15.6	15.7	14.5

Table 2 reports the performance of the sequential MD and our CUDA implementation for different time steps. In Table 2, we set the cutoff distance r_c 2.5δ and skin 0.5δ. Table 3 shows the performance of the sequential MD and our CUDA implementation for different cutoff distances. In Table 3, we make both programs run 100 time steps while $skin = 0.5\delta$. From Table 2 and Table 3 we can see, our GPU implementation achieves speedups of almost seventeen compared to the sequential MD simulation runtime.

6 Conclusions and Future Work

In this paper we have introduced CUDA-based MD simulation algorithms that can be efficiently implemented on modern graphics hardware. We have made use of the fast on-chip memory in CUDA to design and implement our algorithms. All key components of our algorithms have been mapped onto the GPU for execution. The evaluation of our implementation on a high-end graphics card shows a speedup of almost seventeen compared to a Pentium IV 3.0GHz. The results are especially encouraging and to our knowledge this is the first reported implementation of MD simulations on graphics hardware using CUDA.

Our implementation of the MD simulation algorithm using CUDA is quite generic. Our future work will include the extension and integration of this implementation into Gromacs [22] and Autodock [23].

Acknowledgment

This work was supported in part by the A*Star BMRC grant 04/1/22/19/375. Weiguo Liu is currently supported by the Lee Kuan Yew Postdoctoral Fellowship.

References

1. Krüger, J., Westermann, R.: Linear algebra operators for gpu implementation of numerical algorithms. ACM Trans. Graph. 22, 908–916 (2003)
2. Xu, F., Müller, K.: Ultra-fast 3d filtered backprojection on commodity graphics hardware. In: IEEE International Symposium on Biomedical Imaging 2004 (2004)
3. Liu, W., Schmidt, B., Voss, G., Müller-Wittig, W.: Streaming algorithms for biological sequence alignment on gpus. IEEE Transactions on Parallel and Distributed Systems 18(9), 1270–1281 (2007)
4. Horn, D., Houston, M., Hanrahan, P.: Clawhmmer: a streaming hmmer-search implementation. In: Proceedings of Supercomputing 2005 (2005)
5. Moreland, K., Angel, E.: The fft on a gpu. In: Proceedings SIGGRAPH/Eurographics Workshop on Graphics Hardware, pp. 112–119 (2003)
6. Owens, J., Luebke, D., Govindaraju, N., Harris, M., Kruger, J., Lefohn, A., Purcell, T.: A survey of general-purpose computation on graphics hardware. In: Eurographics 2005, pp. 21–51 (2005)
7. Nvidia: NVIDIA CUDA Homepage, `http://developer.nvidia.com/object/cuda.html`
8. Sheng, H., Guerrieri, R., Sangiovanni-Vincentelli, A.L.: Three-dimensional monte carlo device simulation for massively parallel architectures. Technical Report UCB/ERL M95/53, EECS Department, University of California, Berkeley (1995)
9. Ercolessi, F.: A molecular dynamics primer. Technical report (1997) `http://www.fisica.uniud.it/~ercolessi/md/`

10. Allen, M.: Introduction to molecular dynamics simulation. Computational Soft Matter-From Synthetic Polymers to Proteins 23, 1–28 (2004)
11. Lennard-Jones, J.: Cohesion. In: Proceedings of Physical Society, pp. 461–482 (1931)
12. Shu, J., Wang, B., Zheng, W.: Cluster-based parallel simulation for large scale molecular dynamics in microscale thermophysics. In: Cao, J., Yang, L.T., Guo, M., Lau, F. (eds.) ISPA 2004. LNCS, vol. 3358, pp. 200–211. Springer, Heidelberg (2004)
13. Plimpton, S.: Fast parallel algorithms for short-range molecular dynamics. Journal of Computational Physics 117, 1–19 (1995)
14. Hut, P., Makino, J.: Computational physics – astrophysics on the grape family of special-purpose computers. Science 283, 501–505 (1999)
15. Bakker, A., Gilmer, G., Grabow, M., Thompson, K.: A special purpose computer for molecular dynamics calculations. Journal of Computational Physics 90, 313–335 (1990)
16. Fine, R., Dimmler, G., Levinthal, C.: Fastrun: A special purpose, hardwired computer for molecular simulation. Proteins 11, 242–253 (1991)
17. Narumi, T., Ohno, Y., Okimoto, N., Suenaga, A., Yanai, R., Taiji, M.: A high-speed special-purpose computer for molecular dynamics simulations: Mdgrape-3. In: NIC Workshop 2006, vol. 34, pp. 29–36 (2006)
18. Mink, A., Bailly, C.: Parallel implementation of a molecular dynamics simulation program. In: Simulation Conference Proceedings, vol. 1, pp.13–16 (1998)
19. Nakano, P., Kalia, R.: Parallel multiple-time-step molecular dynamics with three-body interaction. Computer Physics Communications 77, 303–312 (1993)
20. Tamayo, P., Mesirov, J., Boghosian, B.: Parallel approaches to short-range molecular dynamics simulations. Supercomputing, 462–470 (1991)
21. Nvidia: NVIDIA CUDA Compute Unified Device Architecture-Programming Guide, (2007) http://developer.download.nvidia.com/compute/cuda
22. Berendsen, H., Van Der Spoel, D., Van Drunen, R.: Gromacs: A message-passing parallel molecular dynamics implementation. Computer Physics Communications 91, 43–56 (1995)
23. Morris, G., Goodsell, D., Huey, R., Olson, A.: Distributed automated docking of flexible ligands to proteins: Parallel applications of autodock 2.4. Journal of Computer-Aided Molecular Design 10(2), 293–304 (1996)

Accelerating Large Graph Algorithms on the GPU Using CUDA

Pawan Harish and P.J. Narayanan

Center for Visual Information Technology
International Institute of Information Technology Hyderabad, India
{harishpk@research., pjn@}iiit.ac.in

Abstract. Large graphs involving millions of vertices are common in many practical applications and are challenging to process. Practical-time implementations using high-end computers are reported but are accessible only to a few. Graphics Processing Units (GPUs) of today have high computation power and low price. They have a restrictive programming model and are tricky to use. The G80 line of Nvidia GPUs can be treated as a SIMD processor array using the CUDA programming model. We present a few fundamental algorithms – including breadth first search, single source shortest path, and all-pairs shortest path – using CUDA on large graphs. We can compute the single source shortest path on a 10 million vertex graph in 1.5 seconds using the Nvidia 8800GTX GPU costing $600. In some cases optimal sequential algorithm is not the fastest on the GPU architecture. GPUs have great potential as high-performance co-processors.

1 Introduction

Graph representations are common in many problem domains including scientific and engineering applications. Fundamental graph operations like breadth-first search, depth-first search, shortest path, etc., occur frequently in these domains. Some problems map to very large graphs, often involving millions of vertices. For example, problems like VLSI chip layout, phylogeny reconstruction, data mining, and network analysis can require graphs with millions of vertices. While fast implementations of sequential fundamental graph algorithms exist [4,8] they are of the order of number of vertices and edges. Such algorithms become impractical on very large graphs. Parallel algorithms can achieve practical times on basic graph operations but at a high hardware cost [10]. Bader et al. [2,3] use CRAY supercomputer to perform BFS and single pair shortest path on very large graphs. While such methods are fast, the hardware used in them is very expensive.

Commodity graphics hardware has become a cost-effective parallel platform to solve many general problems. Many problems in the fields of linear algebra [6], image processing, computer vision, signal processing [13], etc., have benefited from its speed and parallel processing capability. GPU implementations of various graph algorithms also exist [9]. They are, however, severely limited by the memory capacity and architecture of the existing GPUs. GPU clusters have also been used [5] to perform compute intensive tasks, like finite element computations [14], gas dispersion simulation, heat shimmering simulation [15], accurate nuclear explosion simulations, etc.

S. Aluru et al. (Eds.): HiPC 2007, LNCS 4873, pp. 197–208, 2007.

GPUs are optimized for graphics operations and their programming model is highly restrictive. All algorithms are disguised as graphics rendering passes with the programmable shaders interpreting the data. This was the situation until the latest model of GPUs following the Shader Model 4.0 were released late in 2006. These GPUs follow a unified architecture for all processors and can be used in more flexible ways than their predecessors. The G80 series of GPUs from Nvidia also offers an alternate programming model called Compute Unified Device Architecture (CUDA) to the underlying parallel processor. CUDA is highly suited for general purpose programming on the GPUs and provides a model close to the PRAM model. The interface uses standard C code with parallel features. A similar programming model called Close To Metal (CTM) is provided by ATI/AMD. Various products that transform the GPU technology to massive parallel processors for desktops are to be released in short time.

In this paper, we present the implementation of a few fundamental graph algorithms on the Nvidia GPUs using the CUDA model. Specifically, we show results on breadth-first search (BFS), single-source shortest path (SSSP), and all-pairs shortest path (APSP) algorithms on the GPU. Our method is capable of handling large graphs, unlike previous GPU implementations [9]. We can perform BFS on a 10 million vertex random graph with an average degree of 6 in one second and SSSP on it in 1.5 seconds. The times on a scale-free graph of same size is nearly double these. We also show that the repeated application of SSSP outscores the standard APSP algorithms on the memory restricted model of the GPUs. We are able to compute APSP on graphs with 30K vertices in about 2 minutes. Due to the restriction of memory on the CUDA device, graphs above 12 million vertices with 6 degree per vertex cannot be handled using current GPUs.

The paper is organized as follows. An overview of the CUDA programming model is given in Section 2. Section 3 presents the specific algorithms on the GPU using CUDA. Section 4 presents results of our implementation on various types of graphs. Conclusion and future work is discussed in Section 5.

2 CUDA Programming Model on the GPU

General purpose programming on graphics processing units (GPGPU) tries to solve a problem by posing it as a graphics rendering problem, restricting the range of solutions that can be ported to the GPU. A GPGPU solution is designed to follow the general flow of the graphics pipeline (consisting of vertex, geometry and pixel processors), with each iteration of the solution being one rendering pass. The GPU memory layout is also optimized for graphics rendering. This restricts the GPGPU solutions as an optimal data structure may not be available. The GPGPU model provides limited anatomy to individual processors[11]. Creating efficient data structures using the GPU memory model is a challenging problem in itself [7]. Memory size on GPU is another restricting factor. A single data structure on the GPU cannot be larger than the maximum texture size supported by it.

2.1 Compute Unified Device Architecture

On an abstract level, the Nvidia 8800 GTX graphics processor follows the shader model 4.0 design and implements the 4-stage graphics pipeline. At the hardware level, however, it is not designed as 4 different processing units. All the 128 processors of the 8800 GTX are of same type with similar memory access speeds, which makes it a massive parallel processor. CUDA is a programming interface to use this parallel architecture for general purpose computing. This interface is a set of library functions which can be coded as an extension of the C language. A compiler generates executable code for the CUDA device. The CPU sees a CUDA device as a multi-core co-processor. The CUDA design does not have memory restrictions of GPGPU. One can access all memory available on the device using CUDA with no restriction on its representation though the access times vary for different types of memory. This enhancement in the memory model allows programmers to better exploit the parallel power of the 8800 GTX processor for general purpose computing.

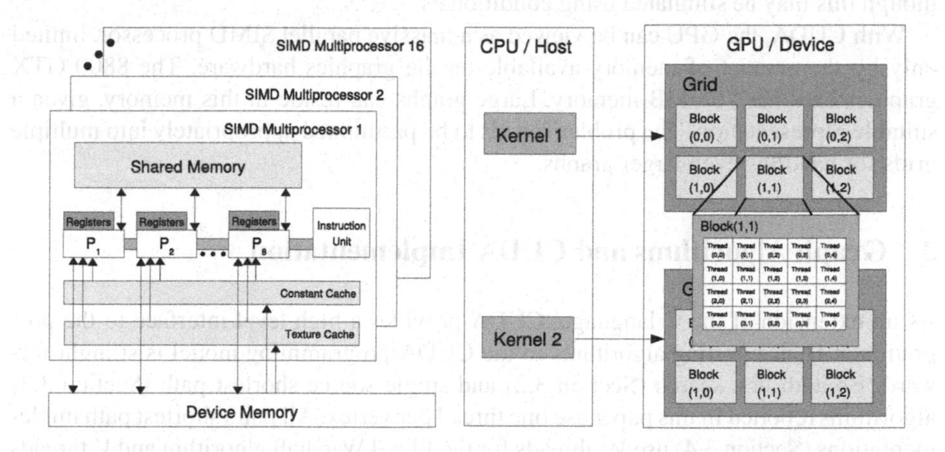

Fig. 1. CUDA Hardware interface **Fig. 2.** CUDA programming model

CUDA Hardware Model. At the hardware level, the 8800 GTX processor is a collection of 16 multiprocessors, with 8 processors each (Figure 1). Each multiprocessor has its own shared memory which is common to all the 8 processors inside it. It also has a set of 32-bit registers, texture, and constant memory caches. At any given cycle, each processor in the multiprocessor executes the same instruction on different data, which makes each a SIMD processor. Communication between multiprocessors is through the device memory, which is available to all the processors of the multiprocessors.

CUDA Programming Model. For the programmer the CUDA model is a collection of *threads* running in parallel. A *warp* is a collection of threads that can run simultaneously on a multiprocessor. The warp size is fixed for a specific GPU. The programmer decides the number of threads to be executed. If the number of threads is more than the warp size, they are time-shared internally on the multiprocessor. A collection of threads (called a *block*) runs on a multiprocessor at a given time. Multiple blocks can be assigned to a single multiprocessor and their execution is time-shared. A single execution on a device generates a number of blocks. A collection of all blocks in a single execution is called a *grid* (Figure 2). All threads of all blocks executing on a single multiprocessor divide its resources equally amongst themselves. Each thread and block is given a unique ID that can be accessed within the thread during its execution. Each thread executes a single instruction set called the *kernel*.

The kernel is the core code to be executed on each thread. Using the thread and block IDs each thread can perform the kernel task on different set of data. Since the device memory is available to all the threads, it can access any memory location. The CUDA programming interface provides an almost Parallel Random Access Machine (PRAM) architecture, if one uses the device memory alone. However, the multiprocessors follow a SIMD model, the performance improves with the use of shared memory which can be accessed faster than the device memory. The hardware architecture allows multiple instruction sets to be executed on different multiprocessors. The current CUDA programming model, however, cannot assign different kernels to different multiprocessors, though this may be simulated using conditionals.

With CUDA, the GPU can be viewed as a massive parallel SIMD processor, limited only by the amount of memory available on the graphics hardware. The 8800 GTX graphics card has 768 MB memory. Large graphs can reside in this memory, given a suitable representation. The problem needs to be partitioned appropriately into multiple grids for handling even larger graphs.

3 Graph Algorithms and CUDA Implementation

As an extension of the C language, CUDA provides a high level interface to the programmer. Hence porting algorithms to the CUDA programming model is straight forward. Breadth first search (Section 3.2) and single source shortest path (Section 3.3) algorithms reported in this paper use one thread per vertex. All pairs shortest path implementations (Section 3.4) use V^2 threads for the Floyd Warshall algorithm and V threads for other implementations. All threads in these implementations are multiplexed on 128 processors by the CUDA programming environment.

In our implementations of graph algorithms, we do not use the device shared memory, as the data required by each vertex can be present anywhere in the global edge array (explained in the following section). Finding the locality of data to be collectively read into the shared memory is as hard as the BFS problem itself.

Denser graphs with more degree per vertex will benefit more using the following algorithms. Each iteration will expand the number of vertices being processed in parallel. The worst case will be when the graph is linear which will result in one vertex being processed every iteration.

3.1 Graph Representation on CUDA

A graph $G(V,E)$ is commonly represented as an adjacency matrix. For sparse graphs such a representation wastes a lot of space. Adjacency list is a more compact representation for graphs. Because of variable size of edge lists per vertex, its GPU representation may not be efficient under the GPGPU model. CUDA allows arrays of arbitrary sizes to be created and hence can represent graph using adjacency lists.

We represent graphs in compact adjacency list form, with adjacency lists packed into a single large array. Each vertex points to the starting position of its own adjacency list in this large array of edges. Vertices of graph $G(V,E)$ are represented as an array V_a. Another array E_a of adjacency lists stores the edges with edges of vertex $i+1$ immediately following the edges of vertex i for all i in V. Each entry in the vertex array V_a corresponds to the starting index of its adjacency list in the edge array E_a. Each entry of the edge array E_a refers to a vertex in vertex array V_a (Figure 3).

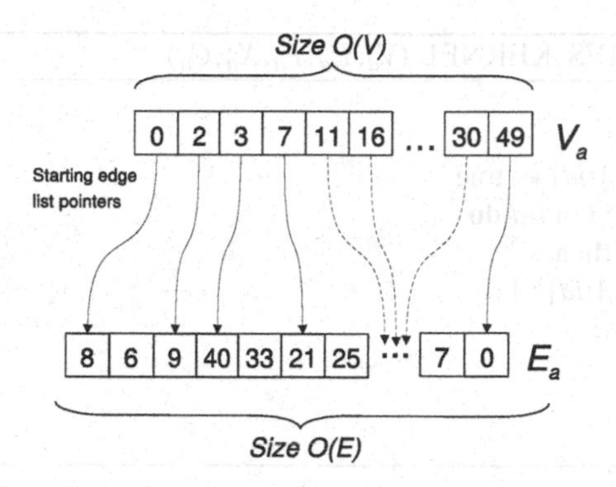

Fig. 3. Graph representation with vertex list pointing to a packed edge list

3.2 Breadth First Search

BFS finds use in state space searching, graph partitioning, automatic theorem proving, etc., and is one of the most used graph operation in practical graph algorithms. The BFS problem is, given an undirected, unweighted graph $G(V,E)$ and a source vertex S, find the minimum number of edges needed to reach every vertex V in G from source vertex S. The optimal sequential solution for this problem takes $O(V+E)$ time.

CUDA implementation of BFS. We solve the BFS problem using level synchronization. BFS traverses the graph in levels; once a level is visited it is not visited again. The BFS *frontier* corresponds to all the nodes being processed at the current level. We do not maintain a queue for each vertex during our BFS execution because it will incur additional overheads of maintaining new array indices and changing the grid configuration at every level of kernel execution. This slows down the speed of execution on the CUDA model.

For our implementation we give one thread to every vertex. Two boolean arrays, *frontier* and *visited*, F_a and X_a respectively, of size $|V|$ are created which store the BFS frontier and the visited vertices. Another integer array, *cost*, C_a, stores the minimal number of edges of each vertex from the source vertex S. In each iteration, each vertex looks at its entry in the *frontier* array F_a. If true, it fetches its cost from the *cost* array C_a and updates all the costs of its neighbors if more than its own cost plus one using the edge list E_a. The vertex removes its own entry from the *frontier* array F_a and adds

Algorithm 1. CUDA_BFS (Graph $G(V,E)$, Source Vertex S)

1: Create vertex array V_a from all vertices and edge Array E_a from all edges in $G(V,E)$,
2: Create frontier array F_a, visited array X_a and cost array C_a of size V.
3: Initialize F_a, X_a to false and C_a to ∞
4: $F_a[S] \leftarrow$ true, $C_a[S] \leftarrow 0$
5: **while** F_a not Empty **do**
6:   **for** each vertex V in parallel **do**
7:     Invoke CUDA_BFS_KERNEL(V_a, E_a, F_a, X_a, C_a) on the grid.
8:   **end for**
9: **end while**

Algorithm 2. CUDA_BFS_KERNEL (V_a, E_a, F_a, X_a, C_a)

1: $tid \leftarrow$ getThreadID
2: **if** $F_a[tid]$ **then**
3:   $F_a[tid] \leftarrow$ false, $X_a[tid] \leftarrow$ true
4:   **for all** neighbors nid of tid **do**
5:     **if** NOT $X_a[nid]$ **then**
6:       $C_a[nid] \leftarrow C_a[tid]+1$
7:       $F_a[nid] \leftarrow$ true
8:     **end if**
9:   **end for**
10: **end if**

to the *visited* array X_a. It also adds its neighbors to the *frontier* array if the neighbor is not already visited. This process is repeated until the *frontier* is empty. This algorithm needs iterations of order of the diameter of the graph $G(V,E)$ in the worst case.

Algorithm 1 runs on the CPU while algorithm 2 runs on the 8800 GTX GPU. The while loop in line 5 of Algorithm 1 terminates when all the levels of the graph are traversed and *frontier* array is empty. Results of this implementation are given in Figure 4.

3.3 Single Source Shortest Path

Single source shortest path (SSSP) problem is, given weighted graph $G(V,E,W)$ with positive weights and a source vertex S, find the smallest combined weight of edges that is required to reach every vertex V from source vertex S. Dijkstra's algorithm is an optimal sequential solution to SSSP problem with time complexity $O(V \log V + E)$. Although parallel implementations of the Dijkstra's SSSP algorithm are available [12], an efficient PRAM algorithm does not exist.

CUDA implementation of SSSP. The SSSP problem does not traverse the graph in levels. The cost of a visited vertex may change due to a low cost path being discovered later. The termination is based on the change in cost.

In our implementation, we use a *vertex* array V_a an *edge* array E_a, boolean *mask* M_a of size $|V|$, and a *weight* array W_a of size $|E|$. In each iteration each vertex checks if it is in the *mask* M_a. If yes, it fetches its current cost from the *cost* array C_a and its neighbor's weights from the *weight* array W_a. The cost of each neighbor is updated if greater

Algorithm 3. CUDA_SSSP (Graph $G(V,E,W)$, Source Vertex S)

1: Create vertex array V_a, edge array E_a and weight array W_a from $G(V,E,W)$
2: Create mask array M_a, cost array C_a and Updating cost array U_a of size V
3: Initialize mask M_a to false, cost array C_a and Updating cost array U_a to ∞
4: $M_a[S] \leftarrow$ true, $C_a[S] \leftarrow 0$, $U_a[S] \leftarrow 0$
5: **while** M_a not Empty **do**
6: **for** each vertex V in parallel **do**
7: Invoke CUDA_SSSP_KERNEL1($V_a, E_a, W_a, M_a, C_a, U_a$) on the grid
8: Invoke CUDA_SSSP_KERNEL2($V_a, E_a, W_a, M_a, C_a, U_a$) on the grid
9: **end for**
10: **end while**

Algorithm 4. CUDA_SSSP_KERNEL1 ($V_a, E_a, W_a, M_a, C_a, U_a$)

1: $tid \leftarrow$ getThreadID
2: **if** $M_a[tid]$ **then**
3: $M_a[tid] \leftarrow$ false
4: **for all** neighbors *nid* of *tid* **do**
5: **if** $U_a[nid] > C_a[tid] + W_a[nid]$ **then**
6: $U_a[nid] \leftarrow C_a[tid] + W_a[nid]$
7: **end if**
8: **end for**
9: **end if**

than the cost of current vertex plus the edge weight to that neighbor. The new cost is not reflected in the *cost* array but is updated in an alternate array U_a. At the end of the execution of the kernel, a second kernel compares *cost* C_a with *updating cost* U_a. It updates the *cost* C_a only if it is more than U_a and makes its own entry in the *mask* M_a. The *updating cost* array reflects the *cost* array after each kernel execution for consistency.

The second stage of kernel execution is required as there is no synchronization between the CUDA multiprocessors. Updating the cost at the time of modification itself can result in read after write inconsistencies. The second stage kernel also toggles a flag if any *mask* is set. If this flag is not set the execution stops. Newer version of CUDA hardware (ver 1.1) supports atomic read/write operations in the global memory which can help resolve inconsistencies. 8800 GTX is CUDA version 1.0 GPU and does not support such operations. Timings for SSSP CUDA implementations are given in Figure 4.

Algorithm 5. CUDA_SSSP_KERNEL2 ($V_a, E_a, W_a, M_a, C_a, U_a$)

1: $tid \leftarrow$ getThreadID
2: **if** $C_a[tid] > U_a[tid]$ **then**
3: $C_a[tid] \leftarrow U_a[tid]$
4: $M_a[tid] \leftarrow$ true
5: **end if**
6: $U_a[tid] \leftarrow C_a[tid]$

3.4 All Pairs Shortest Path

All pairs shortest path problem is, given weighted graph $G(V,E,W)$ with positive weights, find the least weighted path from every vertex to every other vertex in the graph $G(V,E,W)$. Floyd Warshall's all pair shortest path algorithm requires $O(V^3)$ time and $O(V^2)$ space. Since APSP requires $O(V^2)$ space, it is impossible to go beyond a few thousand vertices for a graph on the GPU, due to the limited memory size. We show results on smaller graphs for this implementation. An implementation of Floyd Warshall's algorithm on SM 3.0 GPU can be found in [9]. Another approach for all pair shortest path is running SSSP from all vertices sequentially, this approach requires $O(V)$ space as can be seen by SSSP implementation in section 3.3. For this approach, we show results on larger graphs.

CUDA implementation of APSP. Since the output is of $O(V^2)$, we use an adjacency matrix for graphs rather than the representation given in section 3.1. We use V^2 threads, each running the classic CREW PRAM parallelization of Floyd Warshall algorithm (Algorithm 6). Floyd Warshall algorithm can also be implemented using $O(V)$ threads, each running a loop of $O(V)$ inside it. We found this approach to be much slower because of the sequential access of entire vertex array by each thread. For example on a 1K graph it took around 9 seconds as compared to 1 second taken by Algorithm 6.

Algorithm 6. Parallel-Floyd-Warshall($G(V,E,W)$)

1: Create adjacency Matrix A from $G(V,E,W)$
2: **for** k from 1 to V **do**
3:     **for all** Elements in the Adjacency Matrix A, where $1 \leq i,j \leq V$ in parallel **do**
4:         $A[i,j] \leftarrow min(A[i,j], A[i,k]+A[k,j])$
5:     **end for**
6: **end for**

The CUDA kernel code implements line 4 of Algorithm 6. The rest of the code is executed on the CPU. Results on various graphs for all pair shortest path are given in Figure 6.

Another alternative to find all pair shortest paths is to run SSSP algorithm from every vertex in graph $G(V,E,W)$ (Algorithm 7). This will require only the final output size to be of $O(V^2)$, all intermediate calculations do not require this space. The final output could be stored in the CPU memory. Each iteration of SSSP will output a vector of size $O(V)$, which can be copied back to the CPU memory. This approach does not require the graph to be represented as an adjacency matrix, hence the representation given in section 3.1 can be used, which makes it suitable for large graphs. We implemented this approach and the results are given in Figure 6. This runs faster than the parallel Floyd Warshall algorithm because it is a single $O(V)$ operation looping over $O(V)$ threads. In contrast, the Floyd Warshall algorithm requires a single $O(V)$ operation looping over $O(V^2)$ threads which creates extra overhead for context switching the threads on the SIMD processors. Thus, due to the overhead for context switching of threads, the Floyd Warshall algorithm exhibits a slow down.

Algorithm 7. APSP_USING_SSSP($G(V,E,W)$)

1: Create vertex array V_a, edge array E_a, weight array W_a from $G(V,E,W)$,
2: Create mask array M_a, cost array C_a and updating cost array U_a of size V
3: **for** S from 1 to V **do**
4: $M_a[S] \leftarrow$ true
5: $C_a[S] \leftarrow 0$
6: **while** M_a not Empty **do**
7: **for** each vertex V in parallel **do**
8: Invoke CUDA_SSSP_KERNEL1($V_a, E_a, W_a, M_a, C_a, U_a$) on the grid
9: Invoke CUDA_SSSP_KERNEL2($V_a, E_a, W_a, M_a, C_a, U_a$) on the grid
10: **end for**
11: **end while**
12: **end for**

4 Experimental Results

All CUDA experiments were conducted on a PC with 2 GB RAM, Intel Core 2 Duo E6400 2.3GHz processor running Windows XP with one Nvidia GeForce 8800GTX. The graphics card has 768 MB RAM on board. For the CPU implementation, a PC with 3 GB RAM and an AMD Athlon 64 3200+ running 64 bit version of Fedora Core 4 was used. Applications were written in CUDA version 0.8.1 and C++ using Visual Studio 2005. Nvidia Graphics driver version 97.73 was used for CUDA compatibility. CPU applications were written in C++ using standard template library.

The results for CUDA BFS implementation and SSSP implementations are summarized in Figure 4 for random general graphs. As seen from the results, for graphs with millions of vertices and edges the GPU is capable of performing BFS at high speeds. Implementation of Bader et al. of BFS for a 400 million vertex, 2 billion edges graph takes less than 5 seconds on a CRAY MTA-2, the 40 processor supercomputer [2], which costs 5–6 orders more than a CUDA hardware. We also implemented BFS on CPU, using C++ and found BFS on GPU to be 20–50 times faster than its CPU counterpart.

SSSP timings are comparable to that of BFS for random graphs given in Figure 4, due to the randomness associated in these graphs. Since the degree per vertex is 6–7 and the weights vary from 1–10 in magnitude it is highly unlikely to have a less weighted edge coming back from a far away level. We compare our results with the SSSP CPU implementation, our algorithm is 70 times faster than its CPU counterpart on an average.

Many real world networks fall under the category of scale free graphs. In such graphs a few vertices are of high degree while the rest are of low degree. For these graphs we kept the maximum degree of any vertex to be 1000 and average degree per vertex to be 6. A small fraction (0.1%) of the total number of vertices were given high degrees. The results are summarized in Figure 5. As seen from the results, BFS and SSSP are slower for scale free graphs as compared to random graphs. Because of the large degree at some vertices, the loop inside the kernel (line 4 of Algorithm 2 and line 4 of Algorithm 4) increases, which results in more lookups to the device memory slowing down the kernel execution time. Loops of non-uniform lengths are inefficient on a SIMD architecture.

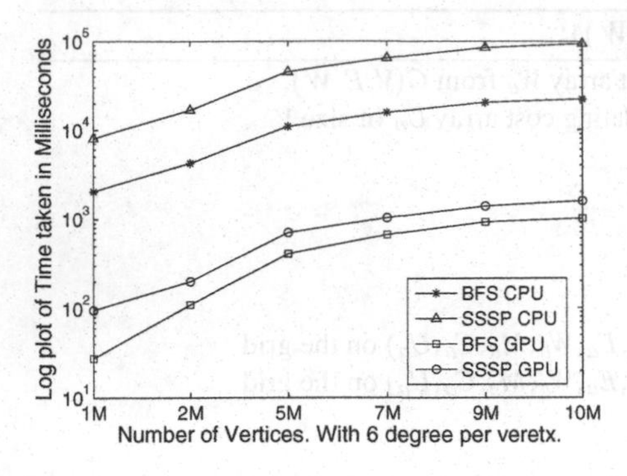

Fig. 4. BFS and SSSP times with weights ranging from 1-10

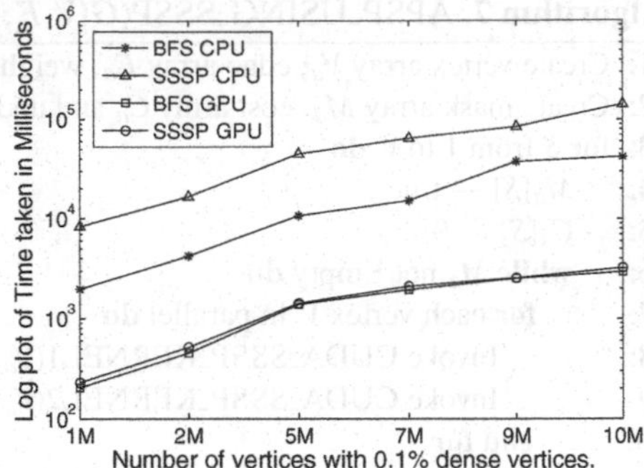

Fig. 5. BFS and SSSP times for Scale Free graphs, weights ranging from 1-10

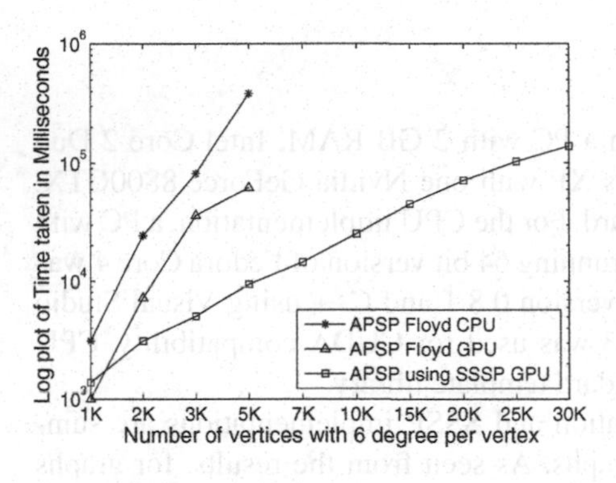

Fig. 6. APSP timings for various graphs, weights ranging from 1-10

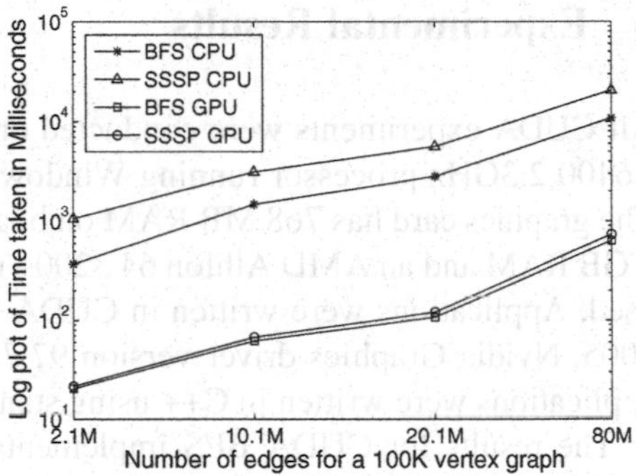

Fig. 7. Graphs with 100K vertices with varying degree per vertex, weights 1–10

Figure 6 summarizes results of all pair shortest path implementation on the CUDA architecture. The SSSP implementation of all pair shortest path requires only one vector of $O(V)$ to be copied to the CPU memory in each iteration, it does not require adjacency matrix representation of the graph and hence only $O(V)$ threads are required for its operation. For even larger graphs this approach gives acceptable results. For example on a graph with 100K vertices, 6 degree per vertex, it takes around 22 minutes to compute APSP. We also implemented CPU version of the Floyd Warshall algorithm and found an average improvement of a factor of 3 for the Floyd Warshall CUDA algorithm and a factor of 17 for the all pair shortest path using SSSP CUDA implementation. As shown by the results APSP using SSSP is faster than Floyd Warshall's APSP algorithm on the GPU, it was found to be orders of magnitude slower when implemented on the CPU.

Figure 7 summarizes the results for BFS and SSSP implementations for increase in degree per vertex. As the degree increases the time taken by both BFS and SSSP increases almost linearly, owing to the lookup cost for each vertex in the device memory.

Table 1 summarizes the results for BFS and SSSP for real world data. The graphs were downloaded from the DIMACS challenge site [1]. The results show that for both BFS and SSSP the GPU is slower than CPU on these graphs. This is due to the low

Table 1. BFS and SSSP timings for real world graphs with 2–3 degree per vertex, weights are in the range 1–300K

	Number of Vertices	Number of Edges	BFS CPU time(ms)	BFS GPU time(ms)	SSSP CPU time(ms)	SSSP GPU time(ms)
New York	250K	730K	313.117	126.04	1649.85	760.14
Florida	1M	2.7M	1055.22	1143.99	7357.83	7906.49
USA-East	3M	8M	3844.35	4005.75	27000.2	35777.52
USA-West	6 M	15M	6688.78	7853.19	48814.4	63749.54

average degree of these graphs. A degree of 2–3 makes these graphs almost linear. In the case of linear graphs parallel algorithms cannot gain much as it becomes necessary to process every vertex in each iteration and hence the performance decreases.

5 Conclusions and Future Work

In this paper, we presented fast implementations of a few fundamental graph algorithms for large graphs on the GPU hardware. These algorithms have wide practical applications. We presented fast solutions of BFS, SSSP, and APSP on large graphs at high speeds using a GPU instead of expensive supercomputers. The Nvidia 8800GTX costs $600 today and will be much cheaper before this article comes to print. The CUDA model can exploit the GPU hardware as a massively parallel co-processor.

The size of the device memory limits the size of the graphs handled on a single GPU. The CUDA programming model provides an interface to use multiple GPUs in parallel using multi-GPU bridges. Up to 2 synchronized GPUs can be combined using the SLI interface. Nvidia QuadroPlex is a CUDA enabled graphics solution with two Quadro 5600 cards each. Two such systems can be supported by a single CPU to give even better performance than the 8800GTX. Nvidia has announced its Tesla range of GPUs, with up to four 8800 cores and higher memory capacity, targeted at high performance computing. Further research is required on partitioning the problem and streaming the data from the CPU to GPU to handle even larger datasets. External memory approaches can be adapted to the GPUs for this purpose.

Another drawback of the GPUs is the lack of double or higher precision, a serious limitation for scientific applications. The regular graphics rendering applications and games – which drive the GPU market – do not require high precisions. Graphics hardware vendors have announced limited double precision support to make their hardware more appealing to high performance computing community. The use of GPUs as economical, high-performance co-processors can be a significant driving force in the future. It has the potential to bring double precision support to the GPU hardware in the future.

References

1. Nineth DIMACS implementation challange - Shortest paths
`http://www.dis.uniroma1.it/challenge9/download.shtml`
2. Bader, D.A., Madduri, K.: Designing multithreaded algorithms for breadth-first search and st-connectivity on the Cray MTA-2. In: ICPP, pp. 523–530 (2006)

3. Bader, D.A., Madduri, K.: Parallel algorithms for evaluating centrality indices in real-world networks. In: ICPP 2006. Proceedings of the 2006 International Conference on Parallel Processing, pp. 539–550. IEEE Computer Society Press, Los Alamitos (2006)
4. Cho, J.-D., Raje, S., Sarrafzadeh, M.: Fast approximation algorithms on maxcut, k-coloring, and k-color ordering for vlsi applications. IEEE Transactions on Computers 47(11), 1253–1266 (1998)
5. Fan, Z., Qiu, F., Kaufman, A., Yoakum-Stover, S.: GPU cluster for high performance computing. In: SC 2004. Proceedings of the 2004 ACM/IEEE conference on Supercomputing, p. 47. IEEE Computer Society, Los Alamitos (2004)
6. Krüger, J., Westermann, R.: Linear algebra operators for GPU implementation of numerical algorithms. ACM Transactions on Graphics (TOG) 22(3), 908–916 (2003)
7. Lefohn, A., Kniss, J.M., Strzodka, R., Sengupta, S., Owens, J.D.: Glift: Generic, efficient, random-access GPU data structures. ACM Transactions on Graphics 25(1), 60–99 (2006)
8. Lengauer, T., Tarjan, R.E.: A fast algorithm for finding dominators in a flowgraph. ACM Trans. Program. Lang. Syst. 1(1), 121–141 (1979)
9. Micikevicius, P.: General parallel computation on commodity graphics hardware: Case study with the all-pairs shortest paths problem. PDPTA, 1359–1365 (2004)
10. Narayanan, P.J.: Single Source Shortest Path Problem on Processor Arrays. In: Proceedings of the Fourth IEEE Symposium on the Frontiers of Massively Parallel Computing, pp. 553–556 (1992)
11. Narayanan, P.J.: Processor Autonomy on SIMD Architectures. In: Proceedings of the Seventh International Conference on Supercomputing, pp. 127–136 (1993)
12. Nepomniaschaya, A.S., Dvoskina, M.A.: A simple implementation of dijkstra's shortest path algorithm on associative parallel processors. Fundam. Inf. 43(1-4), 227–243 (2000)
13. Owens, J.D., Sengupta, S., Horn, D.: Assessment of Graphic Processing Units (GPUs) for Department of Defense (DoD) Digital Signal Processing (DSP) Applications. Technical Report ECE-CE-2005-3, Department of Electrical and Computer Engineering, University of California, Davis (October 2005)
14. Wu, W., Heng, P.A.: A hybrid condensed finite element model with GPU acceleration for interactive 3D soft tissue cutting: Research Articles. Comput. Animat. Virtual Worlds 15(3-4), 219–227 (2004)
15. Zhao, Y., Han, Y., Fan, Z., Qiu, F., Kuo, Y.-C., Kaufman, A.E., Mueller, K.: Visual simulation of heat shimmering and mirage. IEEE Transactions on Visualization and Computer Graphics 13(1), 179–189 (2007)

FT64: Scientific Computing with Streams

Mei Wen, Nan Wu, Chunyuan Zhang, Wei Wu, Qianming Yang, and Changqing Xun

National Laboratory for Parallel & Distributed Processing, National University of Defense Technology, Chang Sha, Hu Nan, P.R. of China, 410073
meiwen@nudt.edu.cn

Abstract. This paper describes FT64 and Multi-FT64, single- and multi-coprocessor systems designed for high performance scientific computing with streams. We give a detailed case study of porting the Mersenne Prime Search problem to FT64 and Multi-FT64 systems. We discuss several special problems associated with streamizing, such as kernel processing granularity, stream organization and workload partitioning for a multi-processor, which are generally applicable to other scientific codes on FT64. Finally, we perform experiments with eight typical scientific applications on FT64. The results show that a 500MHz FT64 achieves over 50% of its peak performance and a 4.2x peak speedup over 1.6GHz Itanium2. An eight processor Multi-FT64 system achieves 6.8x peak speedup over a single FT64.

1 Introduction

FT64 is a programmable 64 bit processor that executes scientific applications, programs structured as streams of data passing through computation kernels. Multiple boards, each consisting of a scalar host processor and eight FT64s, can be used to construct a high performance computer system for scientific computing. On several scientific computing applications, a single-chip FT64 stream processor achieves between 8% and 53% of its peak performance (between 1.3 and 8.5 64-bit GFLOPS). The multiprocessor board can get 2.2x-6.8x speedups over a single FT64. FT64 possesses four key attributes as follows:

Decoupling of memory operations and computation: A stream module running *stream level instructions* reads sequentially organized streams from main memory to on-chip RAM. An array of arithmetic clusters quickly executes *kernel level instructions* to process each stream element. Memory latency is effectively hidden by buffering and software pipelining.

A large number of arithmetic units: FT64 supports 16 fully-pipelined double-precision floating point multiply-add (FMAC) units on a chip. At an operating frequency of 500MHz, a single FT64 can support tens of billions of arithmetic operations per second.

Bandwidth hierarchy: FT64's memory hierarchy consists of DRAM, an on-chip Stream Register File (SRF) and per-cluster Local Register Files (LRF). FT64's bandwidth hierarchy increases the available bandwidth by almost an order of magnitude at each level of the hierarchy by taking advantage of the locality exposed by

S. Aluru et al. (Eds.): HiPC 2007, LNCS 4873, pp. 209–220, 2007.

the stream model and the intensive computation of scientific computing. The ratio of bandwidths provided by FT64 is 1:10:85.

High speed inter-chip Network Interface: The Network Interface on each FT64 is used to connect the SRF to other FT64 chips directly. A Multi-FT64 system can be easily constructed for scientific computing based on stream networks.

The FT64 system is based on the Imagine system from Stanford [5], it has a similar instruction set, a similar memory hierarchy and similar SIMD/VLIW arithmetic clusters. The difference is that FT64 is a 64 bit processor for scientific computing while Imagine is a 32 bit processor for media processing. FT64's scientific computing support includes symmetric FMAC units and a specialized network interface. Our prior work introduced the FT64 system and discussed compilation support [1]. This paper extends that work by describing interconnection between processors including the programming model for Multi-FT64, by detailing the mapping of an application (LUCAS) to FT64, and by analyzing additional scientific applications on both FT64 and Multi-FT64.

The rest of this paper is organized as follows. Section 2 presents the FT64 programming model. Section 3 describes the architecture of FT64. Section4 discusses how an example application, LUCAS, is mapped to FT64. Experimental results, including performances of FT64 processor and Multi-FT64 board are discussed in Section 5. Section 6 presents related work. The last section summarizes the conclusions drawn in this paper.

2 FT64 Programming Model

The stream programming model exposes the inherent parallelism of scientific applications and makes communication explicit. The FT64 model has three levels: stream-thread level, stream scheduling level and kernel execution level. Figure 1 shows an instance of this model.

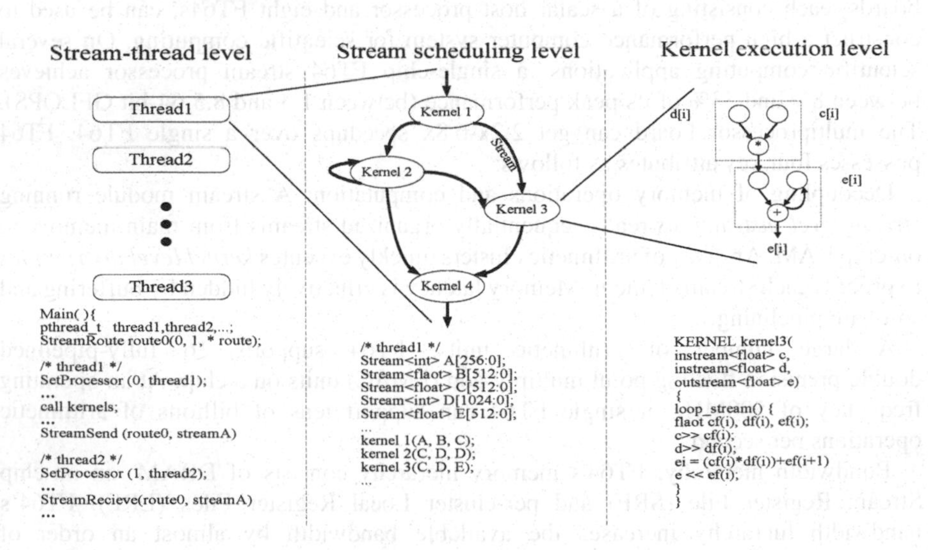

Fig. 1. FT64 Programming Model

The kernel level specifies computation. Kernel level programs are written in KernelC [7] and compiled to the FT64 kernel ISA which combines the control and communication instructions of the Imagine kernel ISA with the arithmetic instructions of IA64 [4]. The design of kernel instructions focuses on support for scientific computing, especially for double-precision floating-point calculations. All of the operands are 64 bit integer and IEEE754 double-precision floating point.

The stream scheduling level schedules kernels and manages communication between them. Stream level programs are written in streamC [7] or SF95 [1] and compiled to the FT64 stream ISA [1]. The StreamC language is a derivative of C++ that includes commands for transferring streams of data to and from the FT64 system, for defining control and data flow between kernels and for executing kernels. To map scientific applications to FT64 more effectively, we designed a new stream programming language. SF95 extends FORTRAN95 with ten compiler directives for scientific stream programming to facilitate streamization of legacy FORTRAN codes.

At the top, the stream thread level enables the construction of multithreaded streaming applications. The stream thread level is programmed using a derivative of StreamC or SF95 with MPI library support and compiled to codes that run on the Itanium2 host processor. It is responsible for kernel scheduling in Multi-FT64 and for non-stream communication between threads. Threads are explicitly declared and assigned to FT64 processors using the *SetProcessor* (index of FT64, thread) function. Within a single board, each thread execute on one of eight FT64 processors. Thread communication and synchronization is also explicit and exposed to the programmer. Threads transfer data between processors using the *StreamSend (StreamRoute, stream)* and *StreamRecieve(StreamRoute, stream)*. A *StreamRoute* variable, which is declared as a global variable in the stream program, defines a route between two FT64 processors in the 2D torus network. The pseudocode shown in Figure 1 gives an example of two threads running on two FT64s. In summary, the thread library is used for programming FT64s on one board. For programming multiple boards, MPI and StreamC/SF95 mixed programming model is used.

3 FT64 Architecture

FT64 is designed to be a stream coprocessor for a general purpose processor (Itanium 2) that acts as the host. A block diagram of the FT64 architecture is shown in Figure 2(a). FT64 consists of a 256Kbytes Stream Register File (SRF), 16 double-precision floating point multiply-add (FMAC) units in four arithmetic clusters controlled by a Microcontroller, a Network Interface (NI), a streaming memory system with two DDRRAM channels, and a Stream Controller (SC). All data stream transfers are routed through SRF. The streaming memory system transfers entire streams between the SRF and off-chip SDRAM. Kernel programs consist of a sequence of VLIW instructions and are stored in a 2K × 688-bit RAM in the Microcontroller. The Microcontroller issues kernel instructions to the four arithmetic clusters in a SIMD manner. The NI routes streams between the SRF of its node and the external network. The multiprocessor solution can take advantage of even more parallelism by using the NI.

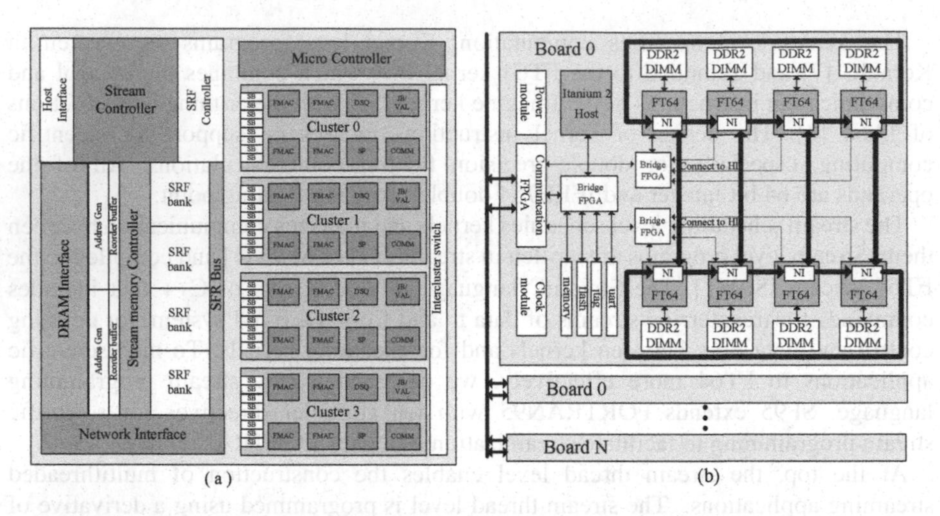

Fig. 2. (a) FT64 Architecture Block Diagram. (b) A Muti-FT64 Board Block Diagram.

The stream programming model exposes the application's bandwidth requirements to the hardware. FT64 exploits this by providing a three-level bandwidth hierarchy: off-chip memory bandwidth (51.2 Gbits per second), SRF bandwidth (512Gbits per second) and intra-cluster Local Register File (LRF) bandwidth (4352 Gbits per second). Stream programs use memory bandwidth only for application input and output and when intermediate streams cannot fit in the SRF and must spill to memory. SRF bandwidth is used when streams pass between kernels. Intra-cluster bandwidth into and out of the LRFs handles the bulk of data during kernel execution.

The Network Interface on FT64 is used to connect the SRF to other FT64 chips in multiprocessor systems or to read or write from I/O devices. NI consists of two main parts: network controller (NC) and network router (NR). NI provides 4 external bidirectional channels, each of which supports two virtual channels. Each channel is able to transfer 64 bytes each 26 internal clock cycles, for a total network bandwidth of 38.4Gbps per node. This is matched to the total bandwidth supported by the network stream buffers. Critical parameters of the FT64 network are presented in Table 1. A stream is partitioned into multiple packets. Stream *send* or *receive* instructions are used to transfer streams across the network. Basically, a message is transported from the sender to the receiver by SC giving a *send* instruction to the sender NI and a *receive* instruction to the receiver NI, which initiates the NI sides of the transaction. At the same time, SC also writes to the corresponding SCRs as well, thus opening the SRF ports at both ends. The route information is determined by the stream scheduler [7], which keeps track of link usage and tries to distribute the load in some static way. Destinations and routes are written from the host processor into an entry in the Network Routing Register File. Since source routing is used, arbitrary network topologies with up to four physical channels per node are supported. One example of a supported topology is a 2D torus network as shown in Figure 2b. A Multi-FT64 system includes multiple boards, aeach of which contains an Itanium2 host processor and eight FT64s. Within one board, the Itanium2 host processor communicates to the FT64s through the

HIs by message passing. Multiple boards are connected to a message passing network, in which the data block transfer protocol is used for the FT64s or hosts in different boards to communicate to each other.

Table 1. Critical parameters of FT64 network

Network type	Directed network	Link clock rate	Asynchronous clock
Network topology	2D torus	Switching type	Buffered wormhole switching
Number of nodes	64k	Max BW	8Gbps(one direction)
Packet size	576 bits(512 bit payload, 64 bit control)	Link width	21-bit (16bit data, 1bit parity code, 2bit control, 1bit clock, 1bit indicator)
Number of channels	4 (bi-directional)	Link flow control	Hardwired
Routing	Dimension-ordered routing	Switch	2 series-wound 5x3 crossbar
# of VCs/Channels	2	Signaling	Source synchronization

We have designed and implemented FT64 64-bit stream processor in 2006. Our goal is to validate architectural studies and provide an experimental prototype for scientific computing with stream. At a controlled voltage and temperature, the chips operate at 500 MHz, at which speed the peak performance is 16 billion 64-bit floating-point operations per second and power consumption is estimated 8.6W. The processor is implemented on a 12*12 mm^2 die in a 1.2-V (IO 3.3V/1.8V), 0.13-μm process with a full standard-cell design flow.

4 Application Study

Mersenne Prime is an important application in number theory research. LUCAS, which is one of SPEC2000 Benchmarks, distinguishes the primality of Mersenne numbers based on Lucas-Lehmer-test method [6]. Let Mersenne number tested $M(p) = 2^{p} - 1$, in which *p* is Mersenne exponent. The algorithm has two steps. First, it constructs a LUCAS sequence as follows: $L(0) = 4, \ldots, L(i + 1) = ((L(i)^2 - 2)$ (i=p-2). Then performs L(p-2) mod M(p). If the result is 0, the Mersenne number tested is a prime. Calculating the square of a large number consumes most of the time; an FFT algorithm is used to accelerate this calculation.

4.1 Basic Implementation

The program we implement on FT64 is called Stream-LUCAS. The initial inputs of the program are the execution length *n*, iteration number *iter* and Mersenne exponent *p*. The program processes a (n/8) × 8 2D array which represents large number L(i), and needs to iterate *iter* times, in which *iter* maximum value is *p-1*. Stream-LUCAS consists of 14 kernels, the simplified stream-kernel graph (n = 8192) is shown in Figure 3. The figure presents the kernel execution sequence and inter-kernel stream communication. Kernels *main_init*, *mers_init* and *fft_init* are responsible for initialization. The significant portion of calculation in Stream-LUCAS is spent calculating the square of each large number $(L(i))^2$ using an FFT algorithm (marked as fft_square by rectangle in Figure 3). Correspondingly, kernels with prefix *passes* perform the Fourier transformation or Fourier inverse transformation. The pass numbering reflects the passes performed in a standard *radix-2* algorithm. Kernel *hex_res64* is an assistance kernel that transforms the result of L(p-2) mod M(p) into hexadecimal outputs. A scalar function run on the host processor determines whether M(p) is a prime according to the result. Figure 4 shows partial pseudo-code for Stream-LUCAS. The stream program for fft_square is on the left, the kernel programs for *passes123* and *passes456* are on the right.

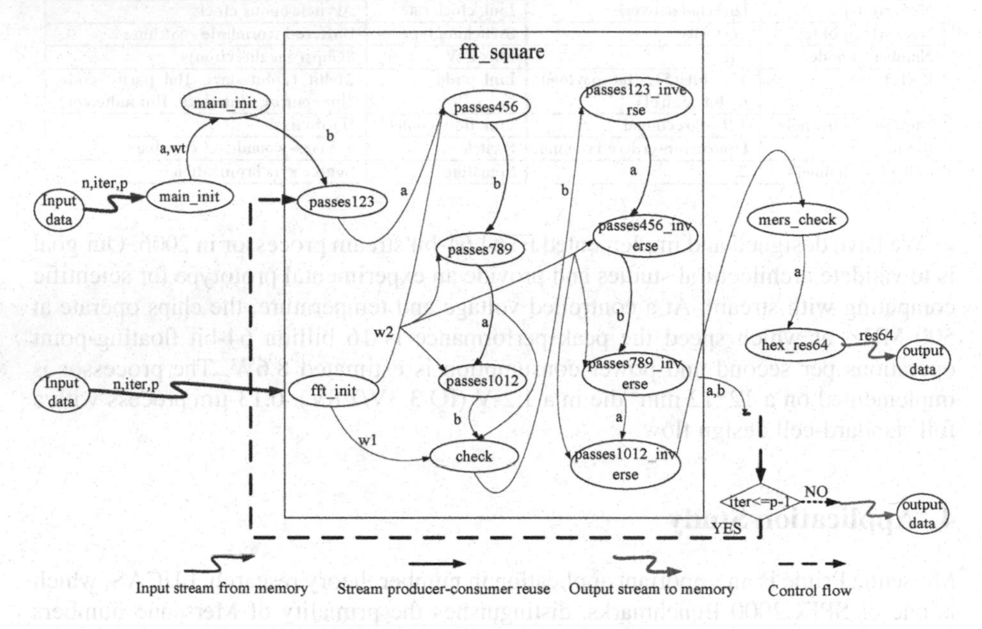

Fig. 3. Simplified Stream-Kernel Graph of Stream-LUCAS

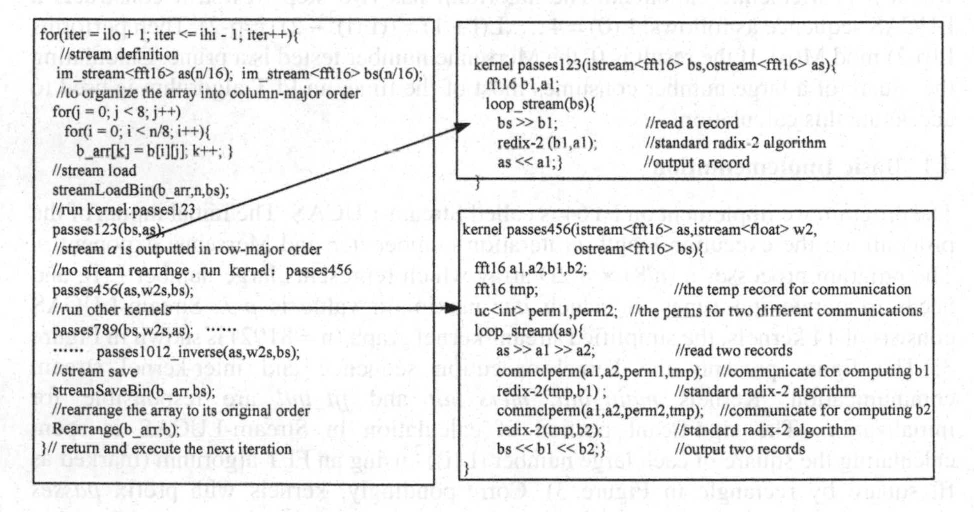

Fig. 4. Part Pseudo-code of stream-LUCAS

4.2 Kernel Processing Granularity

The core procedure of LUCAS is a 2D loop performing fft_square. The inner loop processes 16 words independently and so a natural way to organize stream-LUCAS is to divide the data streams into 16-word records and set the kernel processing granularity at 16 words. Figure 5 shows kernel *passes123* as an example. Array elements b[0][0], b[1][0] … - b[15][0] are bundled into record1 of input stream *b*. This record is the first record processed by cluster0. Records 2, 3, and 4 are organized in a similar way and processed by clusters 1, 2, and 3, respectively. Compared with 1-word at a time processing, grouping streams into 16-word records provides three advantages: (i) it enlarges the parallel granularity in each cluster, (ii) it reduces stream organization complexity, and (iii) it reduces the number of stream reads. The effect of this optimization is significant. For the LUCAS test case (p = 2203, n = 1024, $iter$ = 2202), the 16-word version of fft_square is 59% faster than the 1-word version.

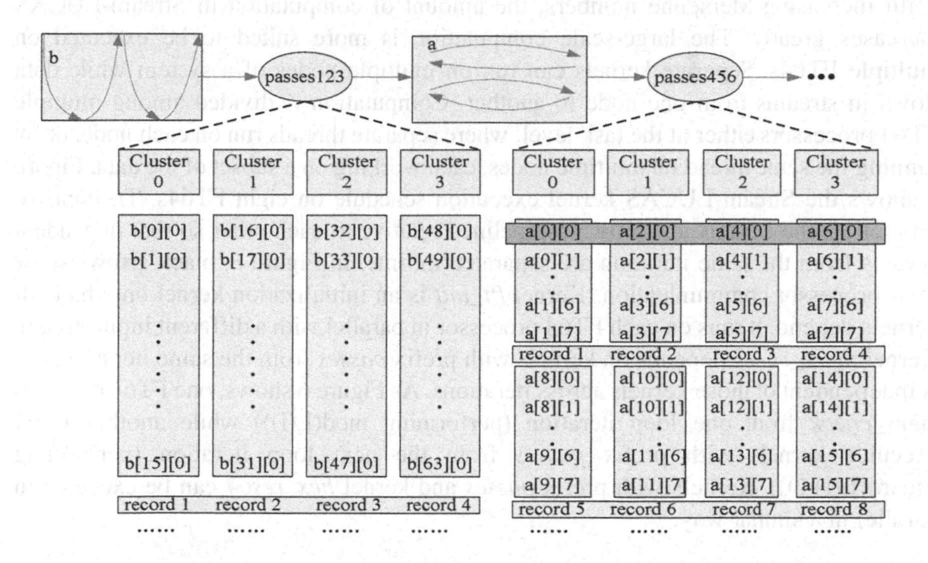

Fig. 5. Stream organization for kernel passes123 and passes456

4.3 Stream Organization and Processing

The original LUCAS processes 2D arrays. Stream-LUCAS processes 1D streams. One of the challenges of streamizing is organizing the 2D arrays into 1D streams. According to the data access sequence, kernel input streams should be organized by columns left-to-right (i.e., column major order). However, computation produces the output stream organized by rows (i.e., row major order). This means that the output stream from one kernel has to be re-organized through off-chip memory from row-major to column-major before being sent to the next kernel. Unfortunately,this makes the kernel dependent on memory access whose latency cannot be hidden.

To reduce off-chip memory access, we don't re-organize kernel output streams. Streams are read in the order in which they are produced. Because stream records are independent of each other, high degrees of data-level parallelism are still possible. Figure5 shows an example using kernels *passes123* and kernel *passes456*. Stream *a* and *b* represent 128×8 matrices. Input stream *b* to kernel *passes123* is in column-major order. Output stream *a* is in row-major order, record 1 contains the first two rows, record 2 the second two rows, etc. Kernel *passes456* reads this stream directly. Each of the four clusters reads two records so that collectively they acquire 16-word columns of array *a*. However, the result is that a each logical column is distributed among the four clusters. Because each cluster processes an entire column, clusters must communicate pieces of columns to each other. This is shown as the dark block in Figure 5. Although this organization greatly increases inter-cluster communication, it significantly reduces memory accesses.

4.4 Multiprocessors

With increasing Mersenne numbers, the amount of computation in Stream-LUCAS increases greatly. The large-scale computation is more suited to be executed on multiple FT64s. Separate kernels can run on multiple nodes of a system while data flows in streams from one node to another. Computation is divided among multiple FT64 processors either at the task level, where separate threads run on each node, or by running the same thread on multiple nodes, each working on a subset of the data. Figure 6 shows the Stream-LUCAS kernel execution schedule on eight FT64s. To improve efficiency, the kernels are software-pipelined relative to each other so that dependent kernels from the same iteration are separated in time. In Figure 6, black arrows show inter-processor communication. Kernel *fft_init* is an initialization kernel on which all kernels depend. It runs on each FT64 processor in parallel with a different input stream. Kernel *mers_check* depends on kernels with prefix *passes* from the same iteration, but is independent of those kernels across iterations. As Figure 6 shows, one FT64 executes *mers_check* from one loop iteration (performing mod(L(i))) while another FT64 executes kernels with prefix *passes* from the next loop iteration (performing square(L(i+1))). Kernels with prefix *passes* and kernel *hex_res64* can be executed in parallel in a similar way.

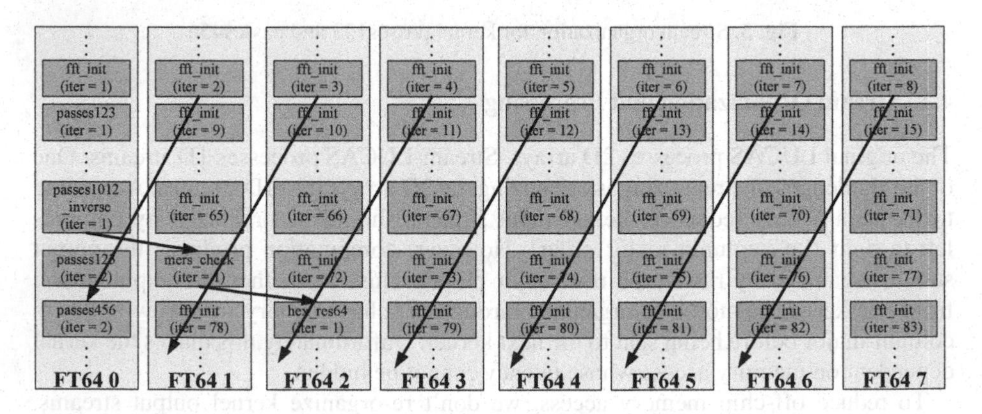

Fig. 6. Kernels on Multi-FT64

5 FT64 Applications and Performance

This section describes the performance for eight applications: Ygx2 is an IAPCM benchmark [14] which combines Lagrange and Euler method to calculate 2D detonation hydrodynamics problems with 64*603 fluid grids; LUCAS (presented in section 4) and Swim are SPEC2000 benchmarks; LU-SGS performs numerical simulations of complex steady flow in hypersonic free stream with 300 thousands input points [15]; FFT is a 512x512 2D complex FFT; QRD converts a 1536x768 complex matrix into an upper triangular and an orthogonal matrix, and is a core component of space-time adaptive processing [16]; SCPMV and SGTSV are LAPACK auxiliary routines, the result matrix of CSPMV is 128*1 while the size of coefficient matrix ins SGTSV is 128*128.

Table2. Application Performance

Applications	#Kernels	One FT64			One FT64 board (eight processors)	
		FMAC Gflops	Time (s)	Speedup over Itanium 2	FMAC Gflops	Speedup over one FT64
LUCAS	14	8.5	50.6	2.6	42.8	4.8
Ygx2	18	7.8	26.8	1.7	40.3	5.3
Swim	14	1.3	71.0	1.04	3.1	2.2
LU-SGS	8	3.4	5.6 per step	1.15	25.4	6.8
FFT	1	7.2	0.015	4.2	36.2	6.06
QRD	6	6.5	0.052 per qrd	3.2	29.5	4.4
SGTSV	1	2.3	6.6e-4 per matrix	0.7	6.8	2.7
CSPMV	1	4.9	4.7e-4 per matrix	1.5	14.9	3.1

Results for the applications above are summarized in table 2 with performance on one FT64 and an eight-processor FT64 board. The third column in the table, which lists the number of floating arithmetic operations executed per second, shows that FT64 can sustain between 1.3 and 8.5 Gflops. When compared to the conventional Itanium2 processor (1.6 GHz), four applications (FFT, QRD, LUCAS and Ygx2) perform better, three applications (Swim, CSPMV, LU-SGS) perform comparably, and one application (SGTSV) performs relatively worse. The fifth column shows the applications' real runtime on a single FT64. In fact, most achieved performance values are suitable for scientific computing systems. Furthermore, as shown in last two columns of table 2, on the eight-processor Multi-FT64 system, these applications can achieve even more operations per second and 2.2x-6.8x speedups over a single FT64. These high absolute performance numbers are the result of the stream programming model itself, which exposes the parallelism and locality in the applications, and of the FT64 architecture, which has been optimized for scientific applications.

One key to achieving high performance on FT64 applications is keeping each functional unit as busy as possible. Compared to the peak capabilities of FT64, the applications achieve between 8% and 53% of the maximum arithmetic performance. The difference between the peak performance of FT64 and the achieved performance is due to several factors, shown graphically in Figure 7. The entire execution time is composed of several parts: including cluster busy and SRF stall times (waiting for data read from or write to SRF); non-kernel overheads, including memory stalls (waiting for a stream load or store to complete), stream controller overhead incurred once per stream instruction, and stalls due to inadequate Host Interface bandwidth and scalar operations on the host. While these non-kernel overheads occupy less than 20% of the total execution time for FFT and QRD, they occupy 40%-60% of the time for LUCAS, Ygx2, LU-SGS and CSPMV and over 70% of the time for Swim and SGTSV. The two biggest culprits are memory stalls and host overhead. The large overhead caused by memory stalls occurs because a stream produced by one kernel needs to be reorganized through DRAM before being consumed by the next kernel. This memory latency cannot be hidden. The other major non-kernel overhead for some applications, host overhead, arises when control-flow decisions on the host are serialized on kernel results or when large amount of data transfer between host and FT64.

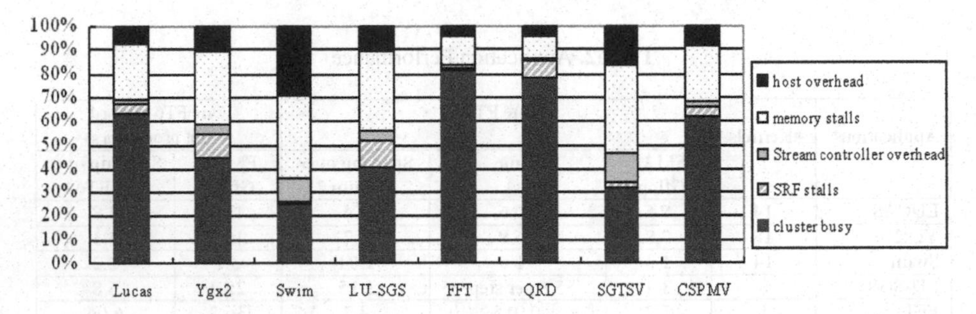

Fig. 7. Execution time breakdown of applications

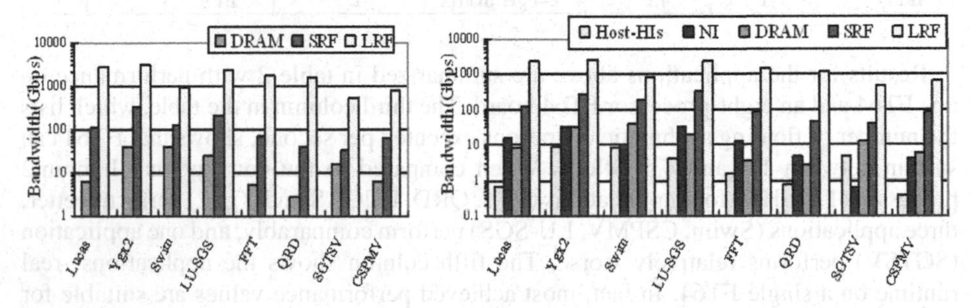

Fig. 8. Achieved bandwidths on a FT64 **Fig. 9.** Achieved bandwidths on a muti-FT64 board

A second key to FT64 performance is that applications map well to the bandwidth hierarchy. Figure 8 illustrates achieved bandwidths for selected applications on a single FT64. It can be seen that LRF bandwidth is two orders of magnitudes higher than DRAM bandwidth over all applications. This means that locality within kernels is fully captured. The difference between SRF bandwidth and DRAM bandwidth varies across applications. For LUCAS, FFT, QRD, CSPMV, SRF bandwidth is one order of magnitudes higher than DRAM bandwidth, which indicates that the bandwidth hierarchy effectively captures the producer-consumer locality exhibited by these applications. However, for other applications, SRF bandwidth is close to DRAM bandwidth. There are two main reasons: (i) inherent producer-consumer locality is limited and irregular, and (ii) a large data set causes on-chip memory spilling, resulting in pressure on memory bandwidth. Figure 9 illustrates achieved bandwidths per FT64 for several applications on a board (Host is connected to 8 FT64s, Host-HIs bandwidth is summed over all FT64s). For this configuration, application data sets are reasonably enlarged. For applications with a large amount of irregular producer-consumer locality, such as Ygx2, Swim and LU-SGS, Host-HI bandwidth is relatively high. On the multi-FT64s system, DRAM bandwidth per FT64 of is lower than that on a single FT64 system (for all applications) because spilled streams are reduced by multiple SRFs. For communication between FT64s, SRF bandwidth per FT64 is higher than that on a single FT64 system (again, for all applications). Especially for applications with large data sets or ample parallelism, such as Ygx2, LU-SGS and FFT, SRF bandwidth increases significantly. The corresponding NI bandwidth is high.

6 Related Works

Hoare's communicating sequential processor (CSP) [7] first presented stream models. Along with improvement of VLSI technology, stream models are further applied in the domains of graphics, multimedia and signal processing, where many architectures and processors supporting stream models have emerged, such as Imagine, RAW [8], VIRAM [9], TRIPS [10], SCORE [11]. Merrimac [2], MASA [12] and FT64 applied stream models to scientific computing. In addition, Cell [13] also supports stream models and is claimed to have tremendous potential for scientific computing. [1] describes the deference between FT64 and other stream processors in detail. Some papers discuss scientific applications streamization on stream processors. Erez presents a detailed case study of porting the GROMACS molecular-dynamics force calculation (kernel) to Merrimac [3]. Several techniques for dealing with the variable number of interactions of each molecule are developed. Wen discusses mapping and optimization of the fluid dynamics calculation with 2D Lagrange and Euler Method on MASA [16]. Data blocking and data reorganization are used to deal with the large 2D data set. This paper discusses selection of parallel processing granularity of a streamized LUCAS application, the exploitation of producer-consumer locality and the mapping of the application to multiple nodes.

7 Conclusions

FT64 is a 64-bit stream processor for scientific applications. The FT64 stream processor integrates 16 500MHz FMACs on a single-chip, which support peak performance of 16 Gflops. This paper describes how a multiprocessor system (Multi-FT64) is constructed using the Network Interface on each FT64 to connect the SRF to other FT64 chips directly. It also presents the mapping of scientific applications to FT64, using the program Stream-LUCAS as an example. Several problems in the streamization of Stream-LUCAS are discussed, including kernel processing granularity, stream organization and workload partitioning for a multiprocessor. Techniques for dealing with these problems are also generally applicable to other scientific codes on FT64.

We performed experiments with eight typical scientific applications on FT64. The results show that a single FT64 processor achieves over 50% of its peak performance and a 4.2x peak speedup over a 1.6GHz Itanium2. A multiprocessor board containing eight FT64 processors achieves a 6.8x peak speedup over a single FT64. The results confirm FT64's potential to deliver high performance in several scientific computing domains.

Acknowledgements. We would like to thank other members of the MASA team: Maoling Guan, Ju Ren, Yi He, Jun Cai for their efforts on this paper. We thank Zuocheng Xin, Jiang Jiang, Chiyuan Ma, Jiinwen Li, Jun Gao, Yong Li, Chunjiang Li and all other collaborators. We also thank professor Amir Roth for his earnest and helpful comments for the paper. This research was supported by NSFC No. 60673148, SRFDP No. 20069998025 and 863 project No.2007AA01Z286.

References

1. Yang, X.: A 64-bit stream processor for scientific applications. In: ISCA2007 (2007)
2. Dally, W.J., Hanrahan, P., Erez, M., Knight, T.J.: Merrimac: Supercomputing with Streams. In: SC 2003 (November 2003)
3. Erez, M., Ahn, J.H., Garg, A., Dallyet, W.J., et al.: Analysis and Performance Results of a Molecular Modeling Application on Merrimac. In: SC 2004, Pittsburg, Pennsylvania, USA (November 6-12,2004)
4. Intel Itantium Architecture Software Developer's Manual, Volume 3: Instruction Set Reference. 2001.12
5. khailany, B., Dally, W.J., Kapasi, U.J., Mattson, P., et al.: Imagine: media processing with streams. IEEE micro 3/4 (2001)
6. http://zh.wikipedia.org
7. Hoare, T.: Communicating sequential processes. Communications of the ACM 8, 21, 666–677 (1978)
8. Taylor, M.B., et al.: Evaluation of the Raw Microprocessor: An Exposed-Wire-Delay Architecture for ILP and Streams. In: ISCA 2004 (2004)
9. Kozyrakis, C.E., et al.: Scalable Processors in the Billion-Transistors Era: IRAM. IEEE Computer 30(9) (September 1997)
10. Sankaralingam, K., et al.: Exploiting ILP, TLP, and DLP with the Polymorphous TRIPS architecture. 30th Annual International Symposium on Computer Architecture (May 2003)
11. Caspiet, E., et al.: A Streaming MultiThreaded Model. The Third Workshop on Media and Stream Processors. In: Conjunction with MICRO34, Austin, Texas (2001)
12. Wen, M., Wu, N., Xun, C., Wu, W., Zhang, C.: Analysis and Performance Results of a Fluid Dynamics Application on MASA Stream Processor. In: ICIS 2006. Proceedings of International Conference on Information Systems, pp. 350–354 (2006)
13. Williams, S., Shalf, J., Oliker, L., Kamil, S., Husbands, P., Yelick, K.: The potential of the cell processor for scientific computing. In: CF 2006. Proceedings of the 3rd conference on Computing frontiers, pp. 9–20 (2006)
14. Guoxin, Y., et al.: Evaluating high performance computer for scientific computing (2003), www.ccw.com.cn
15. Tian, Z.: Numerical Simulations of Multiplex Unsteady Flow in Hypersonic Free Stream, Master Thesis, Dept. of Aerospace and Material Engineering, National University of Defense Technology (December 2003)
16. Cain, K.C., Torres, J.A., Williams, R.T.: RT STAP: Realtime space-time adaptive processing benchmark. Technical Report MTR 96B0000021, MITRE (February 1997)

Implementation and Evaluation of Jacobi Iteration on the Imagine Stream Processor

Jing Du[1], Xuejun Yang[1], Wenjing Yang[2], Tao Tang[1], and Guibin Wang[1]

[1] National Laboratory for Paralleling and Distributed Processing, School of Computer, National University of Defense Technology, Changsha, Hunan, 410073, China
[2] School of Computer, Beijing University of Aeronautics and Astronautics, Beijing, 100083, China
jingdu@nudt.edu.cn

Abstract. In this paper, we explore an efficient streaming implementation of Jacobi iteration on the Imagine platform. Especially, we develop four programming optimizations according to different stream organizations, involving using SP, dot product, row product and multi-row product methods, each highlighting different aspects of the underlying architecture. The experimental results show that the multi-row product optimization of Jacobi iteration on Imagine achieves 2.27 speedup over the corresponding serial program running on Itanium 2. It is certain that Jacobi iteration can efficiently exploit the tremendous potential of Imagine stream processor through programming optimization.

Keywords: scientific application, Imagine, Jacobi iteration, matrix-vector multiplication, computational intensiveness.

1 Introduction

The Imagine processor is designed to address the processor-memory gap through streaming technology at low cost and low power [1, 2]. It has shown tremendous effects on media applications [3]. However there is little research on using Imagine for scientific applications, which require much higher arithmetic rate and memory bandwidth. Therefore it is necessary to research the programming optimizations for scientific applications on Imagine to exploit the underlying hardware performance.

Jacobi iteration is one of the most well-studied problems in computer science since it is a fundamental problem in many scientific applications, in particular as an effective algorithm for solving linear systems. Thus it is important to research on performance optimization techniques for Jacobi iteration. In this paper, we focus on exploring efficient stream organization and kernel partition methods of Jacobi iteration on Imagine. Our specific contribution includes that we develop four programming optimizations according to different stream organizations, involving using SP, dot product, row product and multi-row product methods, each highlighting different aspects of the underlying architecture and reflecting the tradeoff among memory access, computation and communication. The experimental results on the

S. Aluru et al. (Eds.): HiPC 2007, LNCS 4873, pp. 221–232, 2007.

ISIM simulation of Imagine show that the multi-row product optimization of Jacobi iteration on Imagine achieves 2.27 speedup over the corresponding serial program running on Itanium 2. It is certain that Jacobi iteration can efficiently exploit the tremendous potential of Imagine stream processor through programming optimizations.

2 Related Work

Though media applications are becoming the main consumers of stream processors [4-6], there is an important effort to research whether scientific applications are suitable for stream processors. Examples including efficient fluid flow simulation [7, 8] and iterative solvers for sparse linear systems [9, 10] have been demonstrated to run on GPU, which is a graphic stream processor. Many linear algebra routines and scientific applications have been mapped to the Merrimac supercomputer that is also stream architecture [11, 12]. Some dense and sparse matrix applications and some mathematic algorithms such as transitive closure have been implemented on Imagine [13]. However there is little research on Jacobi iteration on Imagine, which is an important scientific kernel widely used in many applications. The core operations of Jacobi are matrix-vector multiplication and iteration process, which are hard to implement efficiently on Imagine because they are reduction operation and true dependent operation respectively. Papers [14, 15] developed some general automatic optimizations for mapping scientific programs to Imagine. Our work is a further effort to research the optimal programming approach on special scientific kernels to exploit the parallelism and high memory bandwidth within Imagine processor.

3 The Imagine Stream Processing System

Imagine developed at Stanford University is a single-chip stream processor. It consists of 48-ALUs arranged as 8 SIMD clusters and a three-level memory hierarchy including several local register files (LRFs), a 128 KB stream register file (SRF) and off-chip DRAM to keep so many ALUs saturated during stream processing [16]. Each LRF relates to a 256-word scratchpad unit (SP) used for local arrays, each SRF bank contains 8 stream buffer (SB) banks used to interface between the SRF and the 8 clusters and the memory system contains 2 address generators (AG) used to generate streams in various addressing modes. Fig. 1 diagrams the Imagine stream architecture.

The programming model of Imagine is described in two languages: the stream level and the kernel level. The stream level program executed for the host thread represents the data communication between the kernels that perform computations. However, programmers must consider the stream organization and communication using this explicit stream model, increasing the programming complexity [4]. So the programming optimization is important to achieve significant performance improvements on Imagine.

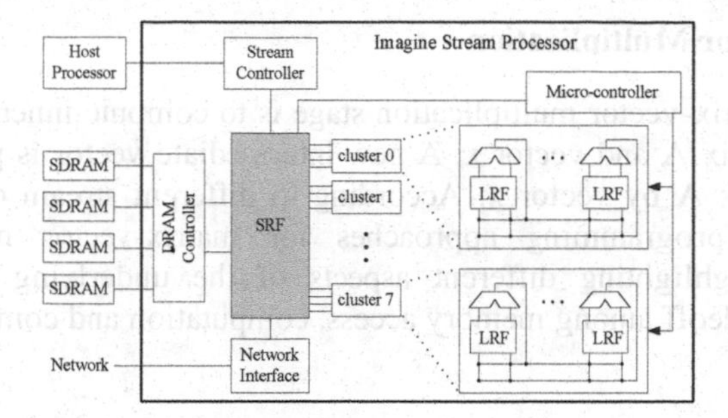

Fig. 1. The Imagine stream architecture

4 Implementation of Jacobi Iteration on Imagine

Jacobi iteration is used to solve the system of equations Ax=b, where A is a coefficient matrix, b is a determinate vector and x is the solution vector. The main operations of Jacobi iteration focus on computing inner product shown in formula (1). Fig. 2 shows the program of Jacobi iteration, which includes an imperfectly nested loop iterating on one matrix and two vectors. The computational complexity and the temporal overhead of the program are both $O(N^2)$, thus it isn't a computational intensive application. Therefore the suited stream organization and kernel partition method of Jacobi need to be studied to exploit high parallelism and fine locality.

$$x_i^{(k+1)} = \frac{1}{a_{ii}}\left(-\sum_{\substack{j=1 \\ j\neq i}}^{n} a_{ij}x_j^{(k)} + b_i\right) \tag{1}$$

```
for(k=0; k<Iter; k++){
    for(i=0; i<N; i++){
        sum = 0.0;
        for(j=0; j!= i && j<N; j++){      } Matrix-vector
            sum += A[i][j]*x[j];          } multiplication
        }
        xn[i] = (b[i] - sum) / A[i][i];   } Multi-vector
    }                                     } operation
    for(j=0; j<N; j++){                   } Vector
        x[j] = xn[j];                     } updating
    }
}
```

Fig. 2. Jacobi iteration program

According to the program structure, Jacobi iteration can be divided into three stages shown in Fig. 2: matrix-vector multiplication, multi-vector operation and vector updating. The detailed stream programming methods of the three stages are described in the following sections.

4.1 Matrix-Vector Multiplication

The kernel of matrix-vector multiplication stage is to compute inner product between every row in matrix A and vector x. A new intermediate vector is produced through multiplying matrix A by vector x. According to different stream organizations, we implement four programming approaches for matrix-vector multiplication on Imagine, each highlighting different aspects of the underlying architecture and processing the tradeoff among memory access, computation and communication.

4.1.1 Using SP

The scratchpad (SP) unit provides a 256-word indexable memory used for local arrays within the clusters [17]. In the first implementation of matrix-vector multiplication stage, we consider using the SP units to keep operators as many as possible, in order to reduce overall LRF and SRF storage overhead, and avoid SRF and DRAM bandwidth overhead of additional data transfers. First, vector x is distributed among SPs of 8 clusters to avoid memory overhead and improve computational intensiveness of the program. And then each row in matrix A is placed to a cluster, which is multiplied by vector x to produce an element. Fig. 3(a) shows the implementation of using SP version. Especially, we can adopt two different programming methods according to the matrix scale. When the matrix size is smaller than 40, all the operators can be stored in the SP units so that Jacobi iterative process would be accomplished within the kernel level. This method eliminates memory access overhead during iterative process, and thus it can achieve high computational intensiveness. Otherwise, the SP units cannot store the whole matrix A if the matrix is big, and the matrix should be partitioned at the stream level to fit the kernel computation. So, using SP to perform matrix-vector multiplication is an unscalable measure owing to the limited SP capacity despite its novelty.

4.1.2 Dot Product

The microoperation in matrix-vector multiplication is the dot product of the row in the matrix with the vector, which results in a scalar. Therefore the columns of the matrix can be executed in parallel completely during the matrix-vector multiplication stage. To exploit the powerful parallel processing ability of so many ALUs in Imagine, the vector is first loaded in the clusters and the matrix is streamed to the clusters to produce the resulting vector. This is the dot product pattern shown in Fig. 3(b). After producing partial sum in each cluster, the summation is produced through intercluster communication. The advantage of this method includes its simple stream organization and no communications during the multiplication of the corresponding elements in the matrix and the vector. But due to the restriction of SIMD parallel mode, for a $N \times N$ matrix, producing the cumulative sum of the partial sum needs $N * log_2(ClusterNum)$ communications, where *ClusterNum* denotes the number of clusters. Though the communications are few compared with all the arithmetic operations, these communications cannot be pipelined fully with the computations owing to the dependence between the cumulative sum and the partial sum, so that the communication delay cannot be overlapped. Besides computing the dot product in a tree-based fashion on Imagine, only $log_2(ClusterNum)$ of the clusters on an average do useful work, resulting in poor performance. Furthermore, the input vector is reloaded N-1 times, yielding worse LRF reuse and lower computational intensiveness, so that the stream throughput is poor. In terms of the above disadvantages, the matrix-vector multiplication of the dot product form may be inefficient on Imagine.

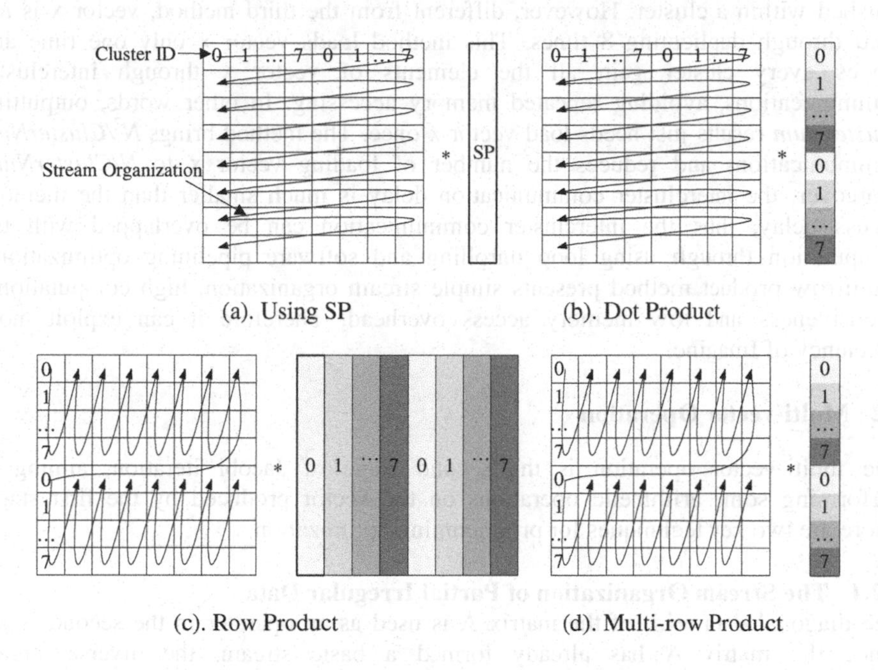

Fig. 3. Four implementations of matrix-vector multiplication

4.1.3 Row Product

To improve the operation granularity of a cluster and the efficiency of the dot product, we consider implementing the third method of matrix-vector multiplication denoted as row product shown in Fig. 3(c). From macro analysis, matrix-vector multiplication depends on every row of the matrix and the iterative vector. Thus each cluster gains the whole row of the matrix and the vector as records respectively to compute an element of the result, and the dot product is iterated within a cluster. This method can eliminate intercluster communication for the dot product computation, and exploit the instruction parallelism of so many ALUs within clusters. Especially, deriving matrix A at the stride of the length of vector x ensures that the inner loop with the true dependence is placed on the same cluster, which presents the parallel processing ability of the clusters. But the row product method needs to duplicate the vector x 8 times as a new stream, thus the stream organization is complex compared with the dot product method. And the vector is also reloaded N-1 times, so the redundant memory access overhead results in low computations per memory access.

4.1.4 Multi-row Product

The arithmetic ability of Imagine processor is more outstanding than its memory access ability. Therefore for reducing the memory access amount of the dot product method and the row product method, and enhancing the reuse ratio of the vector, we propose the fourth method of matrix-vector multiplication, namely multi-row product given in Fig. 3(d). Same as the row product mode, this method distributes each row of the matrix A to different clusters so that the dot product arithmetic of a row can be finished within a cluster. However, different from the third method, vector x is not used through duplicating 8 times. This method loads vector x only one time and makes every cluster gain all the elements of vector x through intercluster communications, avoiding repeated memory accessing. In other words, outputting *ClusterNum* results just needs load vector x once. The method brings $N^2/ClusterNum$ communications and reduces the number of loading vector x to *N/ClusterNum*. Moreover, the intercluster communication delay is much smaller than the memory access delay, thus the intercluster communication can be overlapped with the computation through using loop unrolling and software pipelining optimizations. Multi-row product method presents simple stream organization, high computational intensiveness and low memory access overhead. Therefore it can exploit more efficiency of Imagine.

4.2 Multi-vector Operation

The multi-vector operation is the second stage of Jacobi iteration, aiming at performing some arithmetic operations on the vector produced by the first stage. There are two key techniques for programming optimization.

4.2.1 The Stream Organization of Partial Irregular Data

The diagonal data a(i,i) of the matrix A is used as an operator in the second stage. Since the matrix A has already formed a basic stream, the diverse stream organizations of the diagonal data a(i,i) would influence the application performance. This problem can be amplified as the discussion on the stream organization of partial irregular data of the basic stream. Considering the tradeoff between the space overhead and the communication overhead, the stream organization for partial irregular data can be classified as three modes: forming individual basic stream, deriving from the basic stream A and using intercluster communication. The first mode can achieve fast run-time performance but needs the basic stream creating overhead in the DRAM and SRF. Compared with the first mode, deriving the diagonal stream based on the basic stream A has no spatial and temporal overhead in the DRAM, but it cannot transfer the stream data from DRAM to SRF in the burst mode, resulting in long transfer delay. The third method just loads the basic stream A, and obtains the diagonal data through intercluster communication. This method has no deriving overhead but communication spending, and its scalability is poor for the whole basic stream A needs be loaded to LRF once.

4.2.2 Improving Reuse of the Read-Only Streams

In Jacobi iteration, all the streams are read-only streams except vector x. Since the SRF capacity is limited, as the matrix A grows larger and occupies the most SRF space, the read-only streams cannot reside in the SRF permanently. Thus we need to improve reuse of the read-only streams to avoid unnecessary memory access. We adopt the following two optimizations. First, while the data size is small comparatively, the read-only vectors including vector b and a(i,i) can be stored in SPs to improve reuse on chip. But this method increases the kernel complexity, making against overlapping between computation and memory access. Furthermore, its scalability is poor. Another method partitions the matrix A to some blocks and loads each block to SRF in batches, aiming at making the block occupy relatively small SRF space to ensure reuse of read-only vectors. This method is much better than the first one for it is in favor of hiding memory access delay.

4.3 Vector Updating

The third stage of Jacobi iteration is updating vector x, that is, use the new vector just produced by the second stage as the input operator of the next iteration. Aim at improving the SRF locality, we use the same stream variant of vector x in the consecutive iterations to avoid accessing off-chip memory. If all the streams are larger than the SRF capacity, the stream operations write and read them using double-buffering, which limits the throughput of those operations and wastes the available memory bandwidth. To eliminate this bottleneck, stripmining technique partitions the input stream into segments known as strips, such that all of the intermediate state for the computation on a single strip fits in the SRF [4]. Since the consecutive iterations exhibit true dependence of vector x, a strip generated during an iteration cannot be consumed by the next iteration until all the strips are generated by the previous iteration. Therefore stripmining in Jacobi iteration cannot capture the producer-consumer locality among kernels at strip granularity but just capture the producer-consumer locality between iterations through changing the name of vector x. The iterative process cannot exploit the potential ability of stripmining technique.

4.4 Improving the Utilization of AG and SB

The powerful arithmetic ability of Imagine is limited by the relatively low bandwidth utilization. Thus the stream programming optimization focuses on reducing memory access overhead. There are 2 AGs that connect off-chip DRAM to SRF, and SRF has 8 SBs to supply streams to the clusters [2]. That is, Imagine supports 2 streams loaded from memory and 8 streams transferred to each cluster at the same time. Therefore, if the stream parameters of kernels haven't exceeded 8 streams, we should increase input streams as many as possible to improve the utilization of AG and SB.

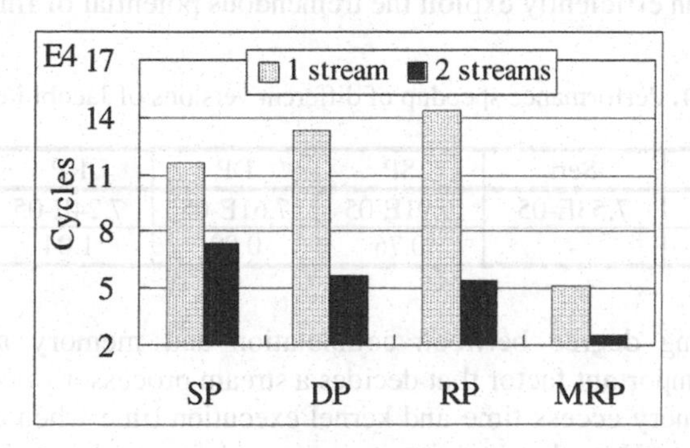

Fig. 4. Performance variety by partitioning matrix A to 2 input streams

Consider increasing the number of input streams of Jacobi iteration kernel. For the rows of matrix A can be executed in parallel, loading 2 parts of matrix A can maximize the parameters of the kernel. This method not only exploits the utilization of AG and SB effectively, but also reduces memory accesses at stream level and increases instruction-level parallelism by using so many ALUs in each cluster. Fig. 4 shows the performance improvement by partition matrix A to 2 input streams.

5 Experimental Results and Analysis

We evaluate various versions of Jacobi iteration, including serial FORTRAN version (Seri), using SP version (SP), dot product version (DP), row product version (RP) and multi-row product version (MRP). Note that the four stream versions all adopt the best programming optimizations in the multi-vector operation and vector updating stages. The matrix size used for this paper is 128×128 with 4-byte elements. The original FORTRAN programs are compiled by Intel's compiler *ifort* (version 9.0) with the optimization option *-O3*, and then executed on a single-core Itanium 2 server. Itanium 2 runs at 1.6GHz and the sizes of the caches are 16KB for the L1 cache, 256KB for the L2 cache and 6MB for the L3 cache. There is also a 4GB off-chip memory with the bandwidth of 6.4GB/s [18]. The other stream versions highly optimized run on ISIM that is a cycle-accurate simulator of Imagine [19], which works at 500MHz.

Table 1 illustrates the performance results of the different versions running on Imagine compared to the corresponding serial program running on Itanium 2. It is obvious that MRP performs better on Imagine than Itanium 2, DP and RP perform comparably, and SP performs poorer. For MRP's highest performance speedup, this is because loading vector x once can be multiplied by 16 rows in matrix A, thus the complex kernel increases ALU utilization in kernels to exploit sufficient ILP and hides memory access overhead, while Itanium 2 is highly sensitive to memory latency. DP and RP cannot overlap computation with memory access efficiently in terms of overfull redundant operations in kernel and tremendous memory access overhead respectively. For SP's lowest speedup, this is because its effective execution time is dominated by the complex kernel overhead, which is consumed in SP operation and intercluster communication. It is certain that the optimizing program of Jacobi iteration can efficiently exploit the tremendous potential of Imagine.

Table 1. Performance speedup of different versions of Jacobi iteration

Versions	Seri	SP	DP	RP	MRP
Time (s)	7.53E-05	9.91E-05	7.61E-05	7.24E-05	3.32E-05
Speedup	-	0.76	0.99	1.04	2.27

The overlapping degree between computation and memory access in stream processing is an important factor that decides a stream processor's performance. Fig. 5 demonstrates memory access time and kernel execution time when different versions running on Imagine. Note that Imagine is an access/execute decoupled processor, and thus there is no inevitable relation between memory access time and kernel execution time. It is obvious that the kernel execution time of SP and DP is much larger than memory overhead. Because the 2 programs adopt a lot of assistant operations in their kernels, such as communication and SP operation, which cannot be pipelined efficiently, thereby producing overfull stalls. Besides only RP's execution time is occupied by memory access overhead due to the duplication of vector x, and most of the kernel time are spent in waiting for data. MRP reuses vector x to process multiple rows to increase the computations in its kernel, aiming at reducing memory overhead and achieving high ILP and pipelining performance in the kernel.

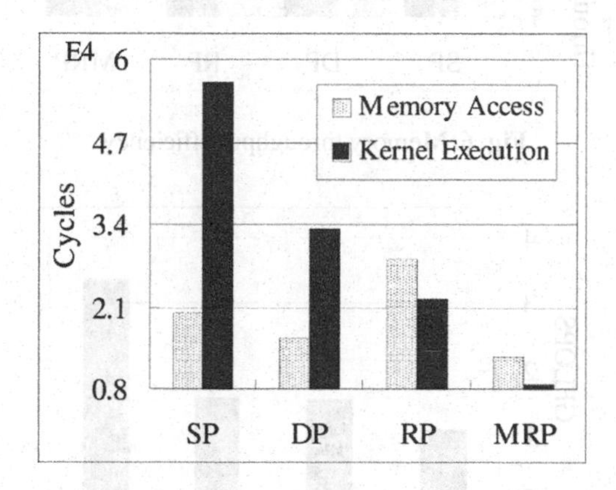

Fig. 5. Memory access time and kernel execution time

Fig. 6 shows the memory throughput efficiency on Imagine, which displays the locality exploited by different programs. It is obvious that SP and MRP can achieve high LRF throughput efficiency for dissimilar reasons respectively. SP keeps vector x in the SP units within clusters and the overfull SP operations increase the throughput efficiency of LRF. Moreover, MRP achieves high LRF throughput efficiency due to the short execution time and the effective utilization of the computational resources. Meanwhile, observing SRF's result, it can be seen that DP, RP, MRP's SRF to memory throughput ratio is larger than 4, which shows that these programs' kernel has caught the producer-consumer or producer-producer locality caused by true dependence between iterations in SRF, and as a result higher speedup is achieved. And SP's SRF throughput efficiency is as twice as that of memory, which shows SRF just transfers the data from memory to LRF resulting in low SRF reuse ratio.

Fig. 7 presents the computation rate of the various versions measured in GFLOPS. Imagine's peak performance can achieve 16GFLOPS. The results show that the sustained performance of all the stream programs except SP has reached more than 10% of the peak performance, which explains that the optimizing Jacobi iteration can efficiently exploit Imagine's potential through optimizations including reusing the iterative vector in SRF, partitioning matrix to fully utilize AGs and SBs, and so on. Though the large granularity of SP's kernel can exploit ILP and reduce memory overhead, SP still achieves low computation rate because there are tremendous assistant operations in the kernel, which is a bottleneck to performance improvement.

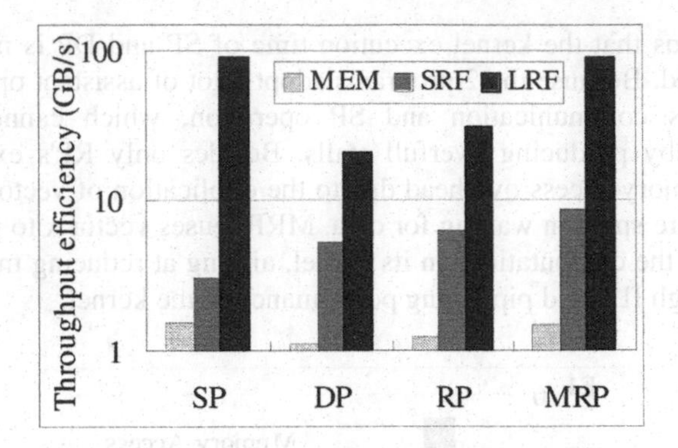

Fig. 6. Memory throughput efficiency

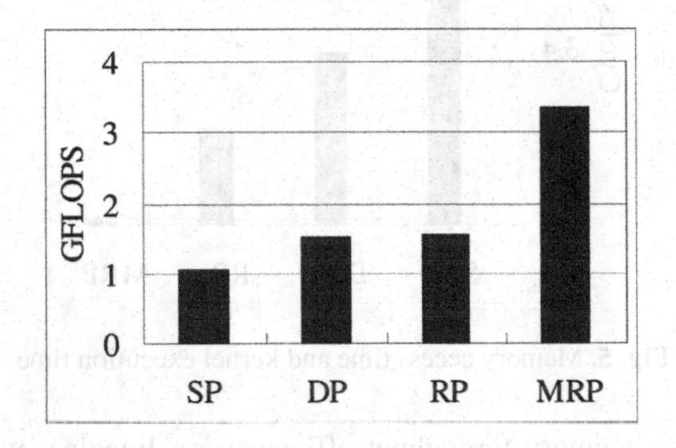

Fig. 7. Computation rate

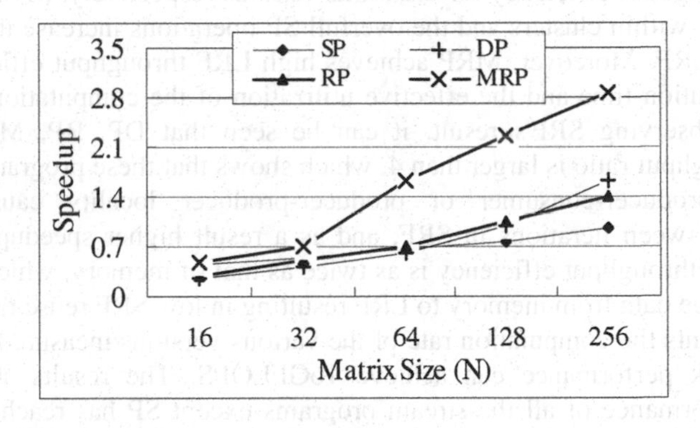

Fig. 8. Speedup variety

To evaluate the scalability of the various versions on Imagine, we adjust stream length to observe speedup's variation over Itanium 2. Fig. 8 shows that when stream is short, the performance of all the versions is poorer than that on Itanium 2. But with the increase of stream length, DP, RP and MRP's speedups are largely increased for computation and memory access can be efficiently overlapped. It can be concluded that the optimizing version of Jacobi iteration presents fine scalability on Imagine. On the contrary, SP presents unscalable feature because it is limited by the SP capacity.

6 Conclusion and Future Work

We have presented the efficient implementation of Jacobi iteration on Imagine. Especially we have developed four programming optimizations according to different stream organizations, each highlighting different aspects of the underlying architecture. The results indicate that the optimizing program of Jacobi iteration can efficiently exploit the tremendous potential of Imagine stream processor.

There are many avenues for future work. One is to research more scientific kernels mapping to stream processor to exploit more architectural features. We would like to abstract the programming optimizations as some efficient algorithms that will be implemented in the stream compiler.

Acknowledgements. We gratefully thank the Stanford Imagine team for the use of their compilers and simulators and their generous help. We also acknowledge the reviewers for their insightful comments. This work was supported by NSFC (60621003).

References

[1] Khailany, B.: The VLSI Implementation and Evaluation of Area-and Energy-efficient Streaming Media Processors. Ph.D. thesis, Stanford University (2003)
[2] Kapasi, U.J., Dally, W.J., et al.: The Imagine Stream Processor. In: Processings of the 2002 International Conference on Computer Design (2002)
[3] Khailany, B., et al.: Imagine: Media Processing with Streams. IEEE Micro 21(2), 35–46 (2001)
[4] Mattson, P.R.: A Programming System for the Imagine Media Processor. Dept. of Electrical Engineering. Ph.D. thesis, Stanford University (2002)
[5] Andrew, A.L., William, T., Saman, A.: Linear Analysis and Optimization of Stream Programs. In: Proceedings of the SIGPLAN 2003 Conference on Programming Language Design and Implementation, San Diego, CA (2003)
[6] Owens, J.D., Rixner, S., et al.: Media Processing Applications on the Imagine Stream Processor. In: Proceedings of the 2002 International Conference on Computer Design (2002)
[7] Fan, Z., Qiu, F., Kaufman, A., Yoakum-Stover, S.: Gpu Cluster for High Performance Computing. In: ACM / IEEE Supercomputing Conference 2004 (2004)
[8] Harris, M.J., Baxter, W.V., Scheuermann, T., Lastera, A.: Simulation of Cloud Dynamics on Graphics Hardware. In: Proceedings of the ACM SIGGRAPH/EUROGRAPHICS conference on Graphics hardware, Aire-la-Ville, Switzerland, pp. 92–101 (2003)
[9] Bolz, J., Farmer, I., Grinspun, E., Schröder, P.: Sparse Matrix Solvers on the Gpu: Conjugate Gradients and Multigrid. ACM Transactions on Graph 22,3, 917–924 (2003)
[10] Göddeke, D.: Gpgpu Performance Tuning. Tech. rep., University of Dortmund, Germany. http://www.mathematik.uni-dortmund.de/~goeddeke/gpgpu/ (2005)

[11] Dally, W.J., et al.: Merrimac: Supercomputing with Streams. In: ACM / IEEE Supercomputing Conference 2003 (November 2003)
[12] Erez, M., Ahn, J., Garg, A., et al.: Analysis and Performance Results of a Molecular Modeling Application on Merrimac. In: ACM / IEEE Supercomputing Conference 2004 (2004)
[13] Griem, G., Oliker, L.: Transitive Closure on the Imagine Stream Processor. In: the 5th Workshop on Media and Streaming Processors, SanDiego, CA (2003)
[14] Du, J., Yang, X., et al.: Scientific Computing Applications on the Imagine Stream Processor. In: Proceedings of the 11th Asia-Pacific Computer Systems Architecture Conference, Shanghai, China (2006)
[15] Yang, X., Du, J., et al.: Matrix-Based Programming Optimization for Improving Memory Hierarchy Performance on Imagine. In: Proceedings of the 4th International Symposium on Parallel and Distributed Processing and Applications, Sorrento, Italy (2006)
[16] Kapasi, U.J., Rixner, S., Dally, W.J., et al.: Programmable Stream Processors. IEEE Computer, 54–62 (2003)
[17] Jayasena, N.S.: Memory Hierarchy Design for Stream Computing. Ph.D. thesis, Stanford University (2005)
[18] Yang, X., Yan, X., et al.: A 64-bit Stream Processor Architecture for Scientific Applications. In: ISCA 2007 (2007)
[19] Das, A., et al.: Imagine Programming System User's Guide 2.0 (2004)

Compiler-Directed Dynamic Voltage Scaling Using Program Phases

K. Shyam[1] and R. Govindarajan[2]

[1] Sasken Communication Technologies Limited
Bangalore, India
[2] SuperComputer Education and Research Center
Indian Institute of Science, Bangalore 560 012, India
kshyam@sasken.com
govind@serc.iisc.ernet.in

Abstract. Energy consumption has become a major constraint in providing increased functionality for devices with small form factors. Dynamic voltage and frequency scaling has been identified as an effective approach for reducing the energy consumption of embedded systems. Earlier works on dynamic voltage scaling focused mainly on performing voltage scaling when the CPU is waiting for memory subsystem or concentrated chiefly on loop nests and/or subroutine calls having sufficient number of dynamic instructions. This paper concentrates on coarser program regions and for the first time uses program phase behavior for performing dynamic voltage scaling. Program phases are annotated at compile time with mode switch instructions. Further, we relate the Dynamic Voltage Scaling Problem to the Multiple Choice Knapsack Problem, and use well known heuristics to solve it efficiently. Also, we develop a simple integer linear program formulation for this problem. Experimental evaluation on a set of media applications reveal that our heuristic method obtains a 38% reduction in energy consumption on an average, with a performance degradation of 1% and upto 45% reduction in energy with a performance degradation of 5%. Further, the energy consumed by the heuristic solution is within 1% of the optimal solution obtained from the ILP approach.

1 Introduction

As the popularity of embedded systems increases, the demand for providing sophisticated applications for these devices also increases. Research has shown that power is fast becoming a first-class architecture design constraint [17]. Although significant progress has been made in the area of low-power circuit and system design [4], the research community has been focusing on a synergistic approach involving both hardware and software to achieve higher energy reduction.

It is known that dynamic power consumption can be reduced significantly by reducing the supply voltage. Dynamic Voltage Scaling (DVS) is a technique that varies CPU frequency and supply voltage during run-time to provide multiple power modes with different performance levels [9], [10], [11], [19]. Performing DVS for energy reduction

S. Aluru et al. (Eds.): HiPC 2007, LNCS 4873, pp. 233–244, 2007.

requires us to consider, in addition to the energy-performance trade-offs, the following aspects as well.

- Granularity of program regions at which DVS is applied. Changing the supply voltage (mode change) during the execution of a program is an expensive operation which could take quite a large number of clock cycles (in the range of 10,000's) [12]. Hence, identifying program regions large enough to compensate the overheads involved in switching the supply voltage is important.
- How many such regions exist in typical programs that are executed in embedded/portable systems ? How do we identify these regions?
- Given various supply voltages of a processor and various regions of a program, how do we efficiently choose the operating mode of a particular region so as to obtain the maximum energy reduction possible, while ensuring that the performance slowdown is within acceptable limits?

In this paper we address the DVS problem by combining two interesting techniques, namely (i) the use of phase behavior of programs [20] to identify program regions [21] and (ii) relating the DVS problem to a well known Multiple Choice Knapsack Problem [18]. The key contributions of this paper are:

- We show that program phases (periods of distinctive behavior) can be used for DVS. Each of these phases (identified at compile time and having coarser granularity) can be treated as a candidate region for performing DVS. We use the method described in [21] to identify and mark the phases at compile-time. This approach helps us to identify program regions at a much coarser granularity, which in turn, helps to reduce the complexity of the DVS problem.
- We formulate the problem of assigning operating mode to each of these phases as a Multiple-Choice Knapsack problem. We use a well-known heuristic described in [18], with appropriate modifications to solve the formulation. We refer to this approach as *DVS-MCKP-H*.
- The DVS problem can also be formulated as an ILP problem and solved using the commercial ILP solvers. This approach is referred to as *DVS-MCKP-ILP*. Using coarser granular program regions helps to reduce the time taken to solve the ILP problem.

We have used a power simulation tool, meant for Intel Xscale architectures, described in [8], for the purpose of our experiments. Initial experiments on five real-world media applications show that *DVS-MCKP-H* obtains, on an average, 38% reduction in energy consumption with a performance slowdown of 1%. Further, the solution obtained by using the *DVS-MCKP-H* scheme is within 1% of the optimal solution obtained from the *DVS-MCKP-ILP* scheme.

Section 2 presents the necessary background regarding dynamic voltage scaling and phase identification. Section 3 motivates our problem formulation. In Section 4 we describe our approach to solve the problem. Section 5 reports the results of our experiments. Section 6 compares our work with other related work. Finally, we conclude the paper in Section 7.

2 Background

2.1 Dynamic Voltage Scaling

The dynamic power[1] P_{dyn} dissipated is proportional to CV^2f where C, V, and f are, respectively, the capacitance, the supply voltage, and the operating frequency. The dynamic energy consumed is given by $P_{dyn} * T$ where is T is the execution time of the program. Thus, Dynamic Voltage Scaling (DVS) reduces the energy consumption by reducing the operating frequency and voltage of the CPU. DVS is accomplished by providing multiple power modes with different performance levels and support for toggling between them. In performing DVS, we must take cognizance of the fact that reducing the supply voltage and operating frequency has a negative effect on the execution time, which in turn may increase the energy consumed. There are a wide variety of processor cores like Transmeta's Crusoe [6], Intel's Xscale [12], AMD's K6-IIIE+ [1] which provide support to dynamically change the supply voltage and operating frequency.

2.2 Identifying Program Phases

Run-time behavior of programs exhibit cyclic repetitive behavior over several architecture performance metrics such as IPC, cache hits e.t.c. It has been shown that the behavior of a program depends on which region of that program is being executed [20]. Programs execute as a series of phases, each of them possibly different from the other. But within a phase, the program exhibits fairly homogeneous behavior. A phase of a program can be thought of as a sequence of dynamic instructions of the program where there exists only little variations in program characteristics like cache behavior, IPC values, etc.

We have used the method proposed in [21] to identify static program regions (a set of basic blocks) which cause the phase behavior. Their approach identifies phases of a coarser granularity extending beyond loop nests and subroutines. The approach followed in [21] uses a frequency-based filtering of basic block traces obtained using active profiling to identify program phases. Active profiling is a technique where the program is profiled with an artificial input that is designed specifically to expose the desired behavior of the program.

The process of phase identification is split into two parts. In the first part the outermost phase is detected. The program is executed with a profile input, and an instruction trace of this execution is obtained at basic-block granularity. Using this instruction trace, the number of dynamic instructions b_i between two occurrences of a basic block b is obtained. One can now calculate the average and standard deviation of number of dynamic instructions across all occurrences of this block b as r_b and σ_b. Similarly, the above metrics for all non-initial instances of each basic block p that occur in the program, are calculated and denoted by r_p and σ_p. Finally σ_q, the standard deviation of r_p across all blocks is also calculated.

[1] In addition to the dynamic power, there is a leakage power that is dissipated by the circuit. However, for the current 0.13μ technology used in embedded systems, the leakage power is not a major component and hence we do not consider it in this paper.

Now, if a basic block has to signify an outer phase marker, it must have that $r_b \approx r_p$, and $\sigma_b \approx \sigma_p$, i.e., the average and standard deviation of the number of dynamic instructions between any two occurrences of a basic block, must be approximately equal to the average and standard deviation of the number of dynamic instructions across all basic blocks that appear in the program. Those blocks for which $(r_p - r_b) > 3\sigma_p$ or $|\sigma_p - \sigma_b| > 3\sigma_q$ are filtered out and are marked as candidate blocks for outermost phase marker. Now, the original instruction trace that was obtained is searched in reverse order, and the first outermost phase marker candidate block that occurs, is marked as the outermost phase of the program. Intuitively, the outermost phase can be thought of as a marker that demarcates the end of a phase, or the end of execution.

Once the outermost phase has been identified, it is marked back into the program using binary re-writing. Now, to identify the innermost phase markers, the program is executed with the normal input. Those basic blocks that appear only once in most instances (90%) of occurrence of outermost phase marker block are the candidate blocks for innermost phase-markers. Let c_{b1} and c_{b2} be any two consecutive candidate blocks. Then the average number of dynamic instructions a_d, executed between these two blocks and the standard deviation σ_d are calculated. Similarly the average and standard deviation values between all pairs of consecutive blocks are also calculated. Now, if the number of dynamic instructions between the blocks c_{b1} and c_{b2} is greater than $a_d + 3 * \sigma_d$, then the block c_{b1} is designated as a inner phase marker. All inner phase markers are found in a similar manner.

Once these phases are marked in the basic block trace, they are mapped back to the original source code. Thus at the end of this process a set of program phases have been identified, which can be considered as candidate regions for performing Dynamic Voltage Scaling. More details on identifying and marking phases can be found in [21].

3 Motivation

In this section we motivate the Dynamic Voltage Scaling problem and our approach using an example. For the purpose of this example we choose the *tomcatv* benchmark from the SPEC'95 suite. We run this benchmark through the phase detection tool [21] and observe that it has five phases including three unique phases. We consider the Xscale core having a default operating frequency of 400MHz [12]. It provides support for DVS, with the operating frequencies of 200MHz and 300MHz apart from 400MHz[2]. The supply voltages for these frequencies are 1.0V, 1.1V and 1.3V respectively. We assume that it takes 50 microseconds[3] to perform DVS, i.e., to switch from one frequency of operation to another frequency of operation during program execution. The remaining architectural parameters of the processor are specified in Section 5.2. Table 1 gives details on the execution times (in milliseconds) and energy consumed (in microJoules) for the different phases under various frequencies of operation. These values are obtained using the XTREM [8] simulator. Note that instructions in Phase 2 and Phase 4

[2] Scaling down the frequency reduces the impact of cache misses (i.e., miss penalty in terms of CPU cycles) on program execution time. Although we do not model this in our work, the proposed DVS formulation using Multiple Choice Knapsack problem can incorporate the same.

[3] The mode switching overhead for an Xscale core as reported in [12] is 50 microseconds

Table 1. Execution Times and Energy Consumed for Various Phases for *tomcatv* benchmark

Freq.		Phase 1	Phase 2	Phase 3	Phase 4	Phase 5	Total
200MHz	Exec. Time (mSec)	151	6827	335	6828	334	14475
	Energy Cons. (μJ)	81.86	125.31	39.11	125.41	39.23	410.92
300MHz	Exec. Time (mSec)	100	4552	223	4551	223	9649
	Energy Cons. (μJ)	148.52	163.02	72.01	163.1	71.90	618.54
400MHz	Exec. Time (mSec)	76	3414	168	3413	167	7238
	Energy Cons. (μJ)	197.54	274.41	175.81	274.61	175.86	1098.23
	Frequency (MHz)	200	400	300	400	300	—
DVS	Exec. Time (mSec)	151	3414	223	3413	223	7424
	Energy Cons. (μJ)	81.86	274.41	72.01	274.61	71.90	774.79

correspond to different instances of the same static region (a set of basic blocks). Similarly Phase 3 and Phase 5 correspond to another region in the code.

The DVS problem is to choose the appropriate operating mode for each phase such that the execution time is close to what can be achieved with 400MHz frequency, but with a reduced energy consumption. Note however since we associate the different modes to different phases in the program statically (i.e., at compile time), phases which correspond to the same static region should operate in the same mode. Observe that Phase 3 and Phase 5, which consume a comparable energy as other phases, take an execution time which is an order or magnitude lower. Thus, if we can reduce the energy consumed by Phases 3 and 5 by DVS, the respective increase in the execution time of these phases, may not significantly affect the overall execution time. The last group of rows in Table 1 shows that by choosing appropriate (lower) frequency for these phases, the overall energy of the program can by reduced to 774.79 microJoules (29.5% reduction) while increasing the execution time by only 2%.

From the motivating example described, above, we observe that there are five phases, which correspond to three static regions in the code, each of which can take one of three possible modes, for a total of $3^3 = 27$ possible ways in which the regions can be assigned to modes. However, for programs with a large number of (static) program regions, the number of possible operating mode assignments is significantly higher. Hence, a brute-force enumerative approach may be prohibitively expensive. The optimal frequency assignment problem is known to be NP-Complete.

4 Our Approach

4.1 DVS as a Multiple Choice Knapsack Problem

In this section we formulate the problem of choosing the operating mode to each region so as to minimize the total energy consumed while keeping the performance degradation within certain percentage of the original execution time. Let us denote the unique program regions as $r_1, r_2, \cdots, r_m$. Each region r_i is executed n_i times in the program. We shall assume a processor which has k operating modes $o_1, o_2, \cdots, o_k$ with supply voltages $v_1, v_2, \cdots, v_k$ and corresponding operating frequencies $f_1, f_2, \cdots, f_k$. Let

e_{ij} and t_{ij} be the energy consumed and the execution time of region r_i when executed under the operating mode o_j.

We now need to choose a particular frequency assignment, and thereby a supply voltage value, for each of these regions so that the energy consumed is reduced. For each of these regions if we denote the energy consumption as a profit and the execution time as a weight, and each frequency assignment as a class, the problem can be logically mapped to the Multiple-Choice Knapsack Problem. For each region r_i, we need to choose one of the operating mods $o_1, o_2, \cdots, o_k$, hence the name Multiple-Choice. Note that in our DVS problem every region has to be selected unlike in the simple Knapsack problem. Next we use the 0-1 variable $x_{ij} = 1$ to denote that operating mode o_j is chosen for region r_i. Since one operating mode has to be assigned to each region, we have

$$\sum_{j=1}^{k} x_{ij} = 1, \text{for all } i = 1, ..., m \tag{1}$$

If o_j is the chosen operating mode for region r_i, then the energy consumed by the n_i instances of this region is $n_i * e_{ij}$. Using the above notation of x_{ij}, we can say that the energy consumed is $n_i * e_{ij} * x_{ij}$. We need to choose the appropriate operating mode such that the total energy is minimized (profit function) while keeping the performance degradation within certain percentage. Formally the objective of Multiple-Choice Knapsack Problem is to minimize the energy. That is

$$\text{minimize} \sum_{i=1}^{m} \sum_{j=1}^{k} n_i e_{ij} x_{ij} \tag{2}$$

Note that n_i and e_{ij} are constants in the objective function.

If o_1 is the default operating frequency, then the original execution time T of the program is given by

$$T = \sum_{i=1}^{m} n_i * t_{i1} \tag{3}$$

We need to keep the execution time within a certain percentage of the original execution time T. Thus if $z\%$ degradation is acceptable, then the total execution time under DVS should be within $T * (1 + z/100)$. This constraint can be stated formally as

$$\sum_{i=1}^{m} \sum_{j=1}^{k} n_i t_{ij} x_{ij} \leq T * (1 + z/100) \tag{4}$$

The value z in Equation 4 is used to control the performance impact of performing Dynamic Voltage Scaling. The constraint in Equation 4 does not capture the overheads due to mode switching. It is possible to model this as an integer constraint as in [19]. However, this will add to the complexity of the problem formulation. Fortunately, our approach of choosing coarser program regions (typically consisting of several 100,000 instructions) allows even a small degradation in performance (z = 1%) to capture the mode switching overheads of 10,000 – 20,000 cycles or instructions. Our experimental results justify such an assumption.

4.2 Solution Methods

The above formulation of the DVS problem as Multiple Choice Knapsack Problem facilitates using well-known heuristic algorithm for obtaining near-optimal solution. We use solvers described in [18] where we set the various frequency values as classes, the energy consumption as profits and execution times as weights.

Alternatively the DVS problem can also be formulated as an Integer Linear Programming Problem. Here the objective function is given by Equation 2. Also, the constraints are given by Equation 1 and Equation 4. The ILP problem can be solved using standard solvers. We have used the CPLEX [5] commercial solver.

5 Experimental Results

5.1 Implementation Details

In this section we give a brief description of our implementation details. We have built a cross-compiler tool-chain for compiling applications with Xscale as target architecture, with mode switch instructions. Our tool-chain (refer to Figure 1) consists of *gcc-2.95.2*, *binutils-2.10*, and *glibc-2.13*. We have made modifications to both *gcc-2.95.2* and *binutils-2.10* to recognize the mode switching instruction that might occur in the source code. The output of the *DVS-MCKP-H* scheme or the *DVS-MCKP-ILP* scheme is an assignment of frequencies to various phases. Appropriate frequency switching instruction is added at the boundaries of each phases if there is a mode switch.

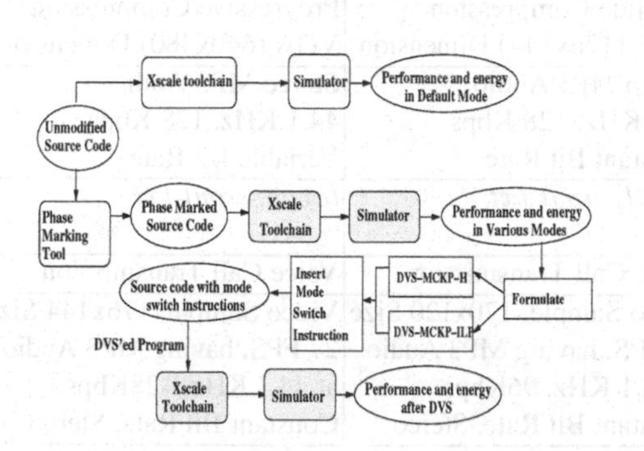

Fig. 1. Flow Diagram of our Experimental Framework

We have modified the XTREM [8] simulator which is an instruction simulator for Xscale cores. Firstly, the necessary modifications are made to recognize the mode switching instruction. Secondly modifications are made to perform DVS as follows. When a mode switching instruction is encountered, all subsequent instruction fetches are stalled until the functional units complete all previous instructions, including all pending memory requests. The mode switching instruction is executed incurring the overhead. After this, the appropriate frequency value is used for power calculations and the processor resumes instruction fetch in the new operating mode.

Figure 1 depicts an overview of our implementation. Given a benchmark's source code, we first compile it using an unmodified (no support for mode switch instructions) tool-chain and simulator to obtain the performance (execution time) and energy values in the default operating mode. We then mark the phases in the source code using the method depicted in [21]. Once phases are marked, we compile the program using the modified tool-chain and execute it using the modified simulator (shown using shaded boxes), to obtain the performance and energy values for the various operating modes supported. We provide these values to the Multiple-Choice Knapsack Problem solver, which provides us with a near optimal assignment of operating modes to various phases of the program. We add the mode switching instructions as appropriate, and compile and execute the program using the modified tool-chain and simulator to obtain the performance and energy values.

5.2 Evaluation Methodology

The details of the benchmarks, the inputs used for profiling and actual performance measurements are given in Table 2. All benchmarks are compiled with -O2 optimization using our Xscale tool-chain and are run to completion. We have used actual proprietary optimized programs which are currently in use in a large number of mobile phones, in order to get a good picture of how useful our energy reduction techniques are in real-world embedded applications. All the inputs used for phase marking and for performance measurements are part of the conformance test cases for various fora for these benchmarks. For the purposes of our experiments we have assumed a single program environment, and a system that does not provide support for virtual memory. Most of the previous studies on DVS [9,10,19] make similar assumptions about the system.

Table 2. Benchmark Inputs used in Performance Measurements

Benchmark	Source of Input	Profiling Input	Actual Input
Jpeg Decoder	JPEG Committee (ref [13])	Baseline Compression QCIF (176x144) Dimension	Progressive Compression VGA (640x480) Dimension
MP3 Decoder	MPEG Forum (ref [16])	Stereo MP3 Audio 44.1 KHz, 128 Kbps Constant Bit Rate	Stereo MP3 Audio 44.1 KHz, 128 Kbps Variable Bit Rate
Text-to-Speech	*rsynth* Program *Mibench* suite (ref [15])	*small_input.txt*	*large_input.txt*
GSM Stack	ETSI (ref [7])	Voice Call Transmission	Voice Call Transmission
MPEG-4 Decoder	MPEG Forum (ref [16])	Video Sample, 170x120 Size 30 FPS, having MP3 Audio at 44.1 KHz, 96Kbps Constant Bit Rate, Stereo	Video Sample, 176x144 Size 27 FPS, having MP3 Audio at 44.1 KHz, 128Kbps Constant Bit Rate, Stereo

The Xscale core that we have used in the simulator is a seven to eight stages single issue super-pipelined microprocessor. It has a 32KB 32-way set associative instruction cache and a 32KB 32-way set associative data cache. A 128-entry direct mapped Branch Target Buffer (BTB) with a 2-bit branch predictor is used for predicting branches. We have assumed that Level-2 caches are not present. The memory subsystem operates at a frequency of 100MHz with a supply voltage of 3.3V. The energy consumption values for various operations like cache reads and writes, are set to the defaults as specified in the simulator.

5.3 Results

Firstly, the benchmarks are executed with their inputs at each of the operating frequencies supported by the processor. Table 3 gives the Execution Times (in milliseconds) and the Energy Consumed (in microJoules) for each of the benchmarks under various operating frequencies. The number of statically identified program regions for each of

Table 3. Execution Times and Energy Consumed for Various frequencies for the benchmarks

Benchmark	Freq. = 200 MHz		Freq. = 300 MHz		Freq. = 400 MHz	
	Exec. Time (mSec)	Energy Cons.(μJ)	Exec. Time (mSec)	Energy Cons.(μJ)	Exec. Time (mSec)	Energy Cons.(μJ)
Jpeg Decoder	1227.44	150.39	818.29	194.37	613.72	278.18
MP3 Decoder	902.64	112.44	601.76	143.12	451.32	201.18
Text-to-Speech	421.04	41.44	280.69	52.61	210.52	73.87
GSM Stack	356.69	24.71	237.79	35.39	178.35	55.12
MPEG-4 Decoder	1631.05	164.75	1087.37	211.90	815.52	301.88

the benchmarks is given in Table 4. It can be observed that the number of program regions is as high as 8, which corresponds to 3^8 possible mode assignments, for the *Jpeg* benchmark. Clearly an enumerative solution is prohibitively expensive.

Table 4. Dynamic Instruction Statistics for the Benchmarks

Benchmark Name	No. Unique Phases	Dynamic Number of Instructions in a Phase			Dynamic Phase Count	Total Dynamic Instructions in Benchmark
		Min	Max	Avg		
Jpeg Decoder	8	894,541	1,192,676	1,072,088	119	137,105,818
MP3 Decoder	3	630,818	922,427	781,553	92	78,217,940
Text-to-Speech	4	367,965	523,628	389,282	73	29,362,539
GSM Stack	3	508,121	894,541	673,488	65	49,458,561
MPEG-4 Decoder	5	316,612	704,473	463,720	291	145,544,153

Figure 2 plots the energy consumed by the program normalized to the base where DVS is not performed. Figure 3 plots the execution times of the benchmarks normalized to the case when the benchmark is executed without DVS. Note that our simulation results take into account the mode switching overhead of 50 microseconds. From these figures we observe that we are able to obtain a maximum energy reduction of 38% – 40%. Also the performance degradation is within 1%, for most of the cases with the worst case being 1.6%.

We also observe that the our *DVS-MCKP-H* scheme performs as well as *DVS-MCKP-ILP* scheme. This assumes significance because the heuristic takes a few seconds (atmost two) to produce a near-optimal solution, while ILP solvers, especially for programs with large number of phases, can take hundreds of seconds to solve the same problem. However for the benchmarks that we tested, where the number of identified regions is small, thanks to the (static) program identification method which identifies

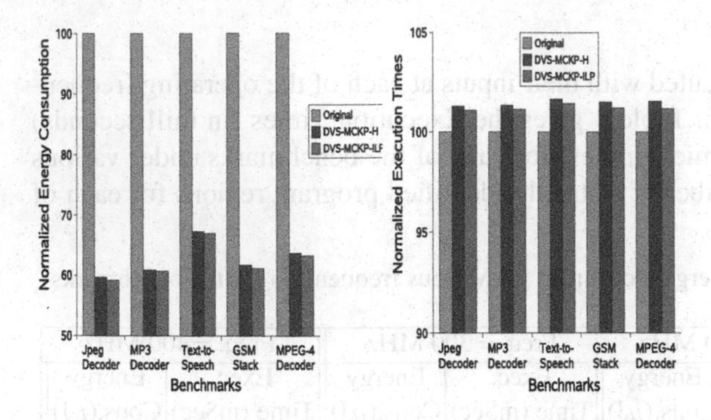

Fig. 2. Normalized Energy Consumptions of various Benchmarks

Fig. 3. Normalized Execution Times of various Benchmarks

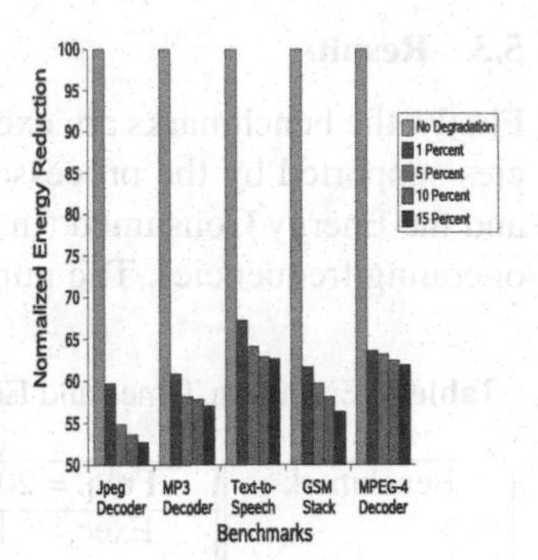

Fig. 4. Energy Reduction under Various Performance Delay Values

coarser program regions, the ILP solver also takes almost similar execution time as the heuristic. Table 4 also gives details regarding the dynamic number of instructions in a program phase. We see that our method of obtaining program phases is able to find phases that have large number of dynamic instructions.

Finally we investigate the effects of varying the z value, the percentage degradation in performance, used in Equation 4. Figure 4 plots the normalized energy reduction values for a performance delay of 1%, 5%, 10%, 15%. From the figure we observe that an energy reduction of upto 45% can achieved if we relax the performance delay to 5%. Thus for those tasks for which such a delay is acceptable, we can obtain considerable energy savings. We also note that as we relax the z value from a 5% to 10%, the achieved energy reduction is not significant.

6 Related Work

Though there are a number of approaches to performing DVS [3,14,2], in this section we compare our work with other compiler-directed approaches for DVS.

Dynamic Voltage Scaling as proposed in [10,11] is based on the idea of reducing the operating frequency of the CPU when it is waiting for the memory access to be completed. To this extent they find memory-bound regions of the program, and perform voltage scaling on one such region which gives the largest benefit. Also they do not consider performing DVS across multiple regions due to the complexity involved in assigning operating modes for these regions. In [19] Saputra et al. combine loop optimization with Dynamic Voltage Scaling. They also propose an ILP problem formulation for assigning multiple operating modes to each loop nest that occur in the source code. The phase behavior based program region approach used in this paper is more generic and and can go across loop nest boundaries to identify larger program regions, and amortize the mode switching overhead.

More recently, Xie, et al., have proposed an exact (but exponential) and a near-optimal (but linear time) algorithms for determining the upper bound of energy savings using DVS [22]. Their approach attempts to construct the state-space and efficiently prune nodes which will (may) not lead to an optimal (near-optimal) solution. Further, they model the program using a set of fine-grain scaling points, a series of events such as timer interrupts or cache misses where voltage/frequency scaling can occur.

In [9] Magklis et al. perform DVS by first isolating loop nests and subroutines having 10,000 or more dynamic instructions as their candidate regions for performing DVS. From among these regions, they chose the operating mode depending on the temporal differences involved in instruction completion in these regions which arises due to two different operating modes. A histogram of such differences is constructed and using this histogram and a predefined performance slowdown factor, an assignment of frequencies to these program regions is obtained.

7 Conclusions

Dynamic Voltage Scaling is now acknowledged as the most effective way to reduce energy consumption. In this paper we use a compiler-directed approach for DVS. We use phase behavior of programs to identify coarser grain program regions, and DVS is applied on those regions. We relate the DVS problem to Multiple-Choice Knapsack Problem and use existing heuristic approaches to obtain near optimal solution quickly. The heuristic approach that we have proposed is simpler and produces significant energy savings by assigning the appropriate operating mode to multiple regions, when compared to the ones proposed in [9,19]. We are able to obtain on an average, 38% energy reduction with a performance slowdown of just 1% and 45% reduction in energy with a performance slowdown of just 5%. Although it may not be appropriate to compare, earlier compiler directed proposals result in an energy reduction of 31.8% for significant a performance degradation of 20.8% [19] for similar workload. We have also shown that the DVS problem can be formulated as an ILP problem and the result obtained from our heuristic is within 1% of the optimal solution obtained using ILP.

References

1. AMD-K6III-E+ Processor DataSheet. http://www.amd.com/epd/processors/6.32bitproc/8.amdk6fami/29.amdk6iiie/23543/23543a.pdf
2. Azevedo, A., Issenin, I., Cornea, R., Gupta, R., Dutt, N., Veidenbaum, A., Nicolau, A.: Profile-based Dynamic Voltage Scheduling using Program Checkpoints. In: DATE 2002. Proceedings of the 2002 Design, Automation and Test in Europe Conf. and Exhibition (2002)
3. Varma, A., Ganesh, B., Sen, M., Choudhury, S.R., Srinivasan, L., Jacob, B.: A Control-Theoretic Approach to Dynamic Voltage Scheduling. In: CASES 2003. Intl. Conf. on Compilers, Architectures and Synthesis of Embedded Systems (2003)
4. Chandrakasan, A., Bowhill, W.J., Fox, F.: Design of High-Performance Microprocessor Circuits. IEEE Press, Los Alamitos (2001)
5. CPLEX®. http://www.ilog.com/products/cplex/
6. Crusoe Processor Model TM5700/TM5900 DataBook. http://www.transmeta.com/crusoe_docs/tm5900_databook_040204.pdf
7. European Telecommunication Standards Institute http://www.etsi.org

8. Contreras, G., Martonosi, M., Peng, J., Ju, R., Lueh, G.-Y.: XTREM: A Power Simulator for the Intel XScale® Core. In: LCTES 2004. Languages, Compilers and Tools for Embedded Systems (June 2004)
9. Magklis, G., Scott, M.L., Semeraro, G., Albonesi, D.H., Dropsho, S.: Profile-based dynamic voltage and frequency scaling for a multiple clock domain microprocessor. In: 30th Annual Intl. Symp. on Computer Architecture, San Diego, California (June 09-11, 2003)
10. Hsu, C.-H., Kremer, U., Hsiao, M.: Compiler-directed dynamic frequency and voltage scheduling. In: Falsafi, B., VijayKumar, T.N. (eds.) PACS 2000. LNCS, vol. 2008, Springer, Heidelberg (2001)
11. Hsu, C.-H., Kremer, U.: Compiler-directed dynamic voltage scaling based on program regions. Technical Report DCS-TR-461, Department of Computer Science, Rutgers University (November 2001)
12. Intel® PXA255 Processor Electrical, Mechanical, and Thermal Specification. http://www.intel.com/design/pca/applicationsprocessors/manuals/278780.htm
13. JPEG (Joint Photographic Experts Group) committee http://www.jpeg.org/
14. Choi, K., Soma, R., Pedram, M.: Off-Chip Latency-Driven Dynamic Voltage and Frequency Scaling for an MPEG Decoding. In: DAC 2004. Proceedings of the 41st annual conference on Design automation (June 2004)
15. Guthaus, M.R., Ringenberg, J.S., Ernst, D., Austin, T.M., Mudge, T., Brown, R.B.: MiBench: A free, commercially representative embedded benchmark suite. In: 4th Annual IEEE Workshop on Workload Characterization, Austin, TX (December 2001)
16. MPEG (Motion Picture Expert Group) Industry Forum http://www.m4if.org/
17. Mudge, T.: Power: A first class design constraint for future architectures. In: Proceedings of Intl. Conf. on High Performance Computing (December 2000)
18. Pisinger, D.: A minimal algorithm for the multiple-choice knapsack problem. European Journal of Operational Research 83, 394–410 (1995)
19. Saputra, H., Kandemir, M., Vijaykrishnan, N., Irwin, M.J., Hu, J., Hsu, C-H., Kremer, U.: Energy-conscious compilation based on voltage scaling. In: LCTES 2002. ACM SIGPLAN Conf. on Languages, Compilers, and Tools for Embedded Systems (June 2002)
20. Sherwood, T., Calder, B.: Time varying behavior of programs, UC San Diego Technical Report UCSD- CS99-630 (August 1999)
21. Shen, X., Ding, C., Dwarkadas, S., Scott, M.: Characterizing Phases in Service-Oriented Applications. Technical Report 848. Dept. of Computer Science, University of Rochester (November 2004)
22. Xie, F., Martonosi, M., Malik, S.: Bounds on power savings using runtime dynamic voltage scaling: An exact algorithm and a linear-time heuristic approximation. In: ISLPED 2005. Proc. of the Intl. Symp. on Low-Power Electronics and Design, San Diego, CA (August 2005)

Partial Flow Sensitivity

Subhajit Roy* and Y.N. Srikant

Department of Computer Science and Automation
Indian Institute of Science
{subhajit,srikant}@csa.iisc.ernet.in

Abstract. Compiler optimizations need precise and scalable analyses to discover program properties. We propose a partially flow-sensitive framework that tries to draw on the scalability of flow-insensitive algorithms while providing more precision at some specific program points. Provided with a set of critical nodes — basic blocks at which more precise information is desired — our partially flow-sensitive algorithm computes a reduced control-flow graph by collapsing some sets of non-critical nodes. The algorithm is more scalable than a fully flow-sensitive one as, assuming that the number of critical nodes is small, the reduced flow-graph is much smaller than the original flow-graph. At the same time, a much more precise information is obtained at certain program points than would had been obtained from a flow-insensitive algorithm.

Keywords: compilers, dataflow analysis, compiler optimizations, points-to analysis.

1 Introduction

Compiler optimizations largely depend on the program properties that the compiler could discover. Precision and scalability are two conflicting goals that such analyses have to meet. Control flow abstraction is an useful technique to attain high scalability, though at the cost of precision. A flow-sensitive algorithm takes the program control-flow into account to come up with a highly precise solution at each program point. On the other hand, a flow-insensitive algorithm neglects all control-flow leading to a high degree of scalability, while coming up with a summary solution for the whole flow-graph.

It is the need for scalability that forces many of the analyses to be implemented as flow-insensitive algorithms. However, it is possible that many opportunities for optimization could be exploited if precise solutions are known even at a very few program points. For instance, if a profile run for a program selects a few "hot" methods, aggressive optimizations of these few methods would give high runtime gains. However, one needs to analyze well, not only the specific method, but also the caller method to discover all such opportunities.

Fig. 1 presents an example. The function `process_node()` allocates a new element and passes it to the `populate()` method. The `populate()` method checks if the argument passed is not `NULL` and then proceeds to populate the same.

* Supported in part by doctoral fellowship provided by Philips Research, India.

S. Aluru et al. (Eds.): HiPC 2007, LNCS 4873, pp. 245–256, 2007.

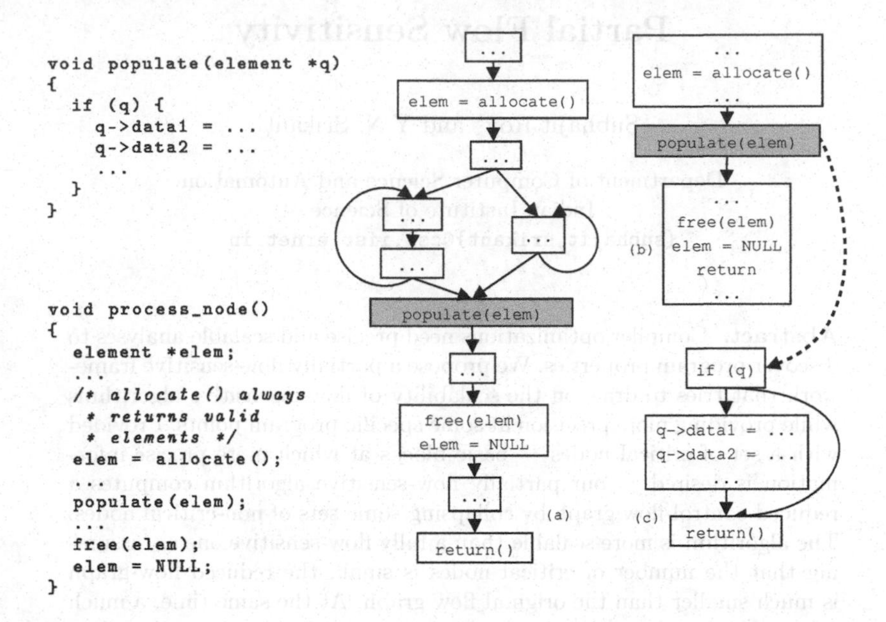

Fig. 1. A motivating example: Fig. (a) shows the original CFG for `process_node()`. Fig. (b) shows the reduced CFG for `process_node()` obtained via the PFS algorithm. Fig. (c) shows the CFG for the function `populate()`. The gray node is the critical node selected. The dotted line shows the parameter binding of 'elem' to 'q'.

The important thing to note is that it is never possible that the argument `q` passed to `populate()` is `NULL` (as the `allocate()` method is supposed to always return a valid element[1]). Hence the check "`if (q)`" could be eliminated from `populate()` if the above fact can be discovered[2]. A flow-insensitive analysis would simply miss the fact. A flow-sensitive algorithm would surely indicate that fact; however, it would be hugely expensive. This leads us to an important observation : even if `process_node()` was analyzed well enough to provide a better estimate of program properties at only the program points where `populate()` is called, our purpose would be served.

We propose a middle path between full flow-sensitivity and flow-insensitivity — that of *partial flow-sensitivity.* If we preserve some partial flow-sensitivity at a few points in a program, without losing much on scalability, we might be able to discover opportunities to optimize the program that a flow-insensitive algorithm would miss. Our partially flow-sensitive algorithm expects from the user a set of program points (currently we accept the specification in the granularity of basic-blocks) where we are interested to have a better estimate of the program properties. We term such basic-blocks as *critical nodes* in the program's flow-

[1] The code for `allocate()` is not shown here. However, such a method can be implemented simply by spinning in a loop till the `malloc()` call succeeds.

[2] We assume that the analyzer is made aware of the fact that `allocate()` will always return valid elements using suitable annotations.

graph. Our algorithm then goes about merging some sets of non-critical nodes to compute a *reduced flow-graph (r-CFG)*, on which we perform a flow-sensitive analysis. Assuming such critical nodes to be a very small fraction of all the nodes in the flow-graph, the algorithm achieves the desired scalability by running over a very small flow-graph. In fact, the scalability is then a function of the fraction of the nodes selected as critical — the algorithm reduces to a completely flow-insensitive algorithm when none of the nodes are critical, to a completely flow-sensitive one, when all the nodes are selected critical.

The reduced flow graph for fig. 1(a) is shown in fig. 1(b) with the node containing the call to `populate()` marked critical. One can see that a flow-sensitive points-to analysis on this much smaller flow-graph of the method `process_node()` can deduce the fact that the parameter `q` in `populate()` can only point to valid memory locations and hence the null-check in `populate()` is redundant.

2 Previous Work

[1] proposes a flow-sensitive interprocedural alias analysis that they claimed was, at that time, the most precise and efficient interprocedural method known. Andersen[2] described a flow-insensitive subset based algorithm based on constraint solving that computes a single solution for the whole program. [3] propose an algorithm to improve the precision of flow-insensitive interprocedural alias analysis using precomputed kill information.

[4,5,6] propose techniques towards solving a demand dataflow analysis algorithm, that answers a query about a single given dataflow fact holding at a single given program point. However, this technique differs from our work as our partial flow sensitive framework computes a solution for all program points, though of varying precision.

[7] proposed to use the SSA form to improve the precision of flow-insensitive pointer analysis. The algorithm uses repeated iterations to improve the precision of the analysis, and the final result could even be as good as that computed using a flow-sensitive analysis. However, the worst case time requirement for translating a code in SSA form is cubic. Also, the SSA translation could result in a program that is quadratic in the size of the original program. Moreover, as the algorithm has to be primed with points-to relations, it requires a points-to analysis in its initial phase. Recently, [8] reported a new approach to solving subset-based points-to analysis for Java using Binary Decision Diagrams (BDDs).

[9] proposes a client-driven pointer analysis, where the analysis adapts to the need of the client analyses. However, our work differs from this work in the way flow-sensitivity is provided; while the mentioned work looks at using the SSA from to provide flow-sensitivity to some variables at all programs points, our work looks at providing better precision to the dataflow solutions of all variables at some program points. Also, we do not need to use the SSA form, construction of which itself needs a prior pointer analysis phase [7].

3 The Reduced Control-Flow Graph (r-CFG)

The Partially Flow-Sensitive Algorithm (PFS algorithm) allows the user to specify a set of *critical nodes* from the program's flow-graph. We define *Critical Nodes* as basic blocks at which the user is interested in having a more precise information about some program properties. The selected functions could be call-sites of "hot" methods (a better estimate of the points-to sets of the passed arguments at their call-sites could enable us to drive better optimizations within these "hot" methods) or basic blocks having a high execution count identified through a profile run.

Depending on the critical nodes selected, the PFS algorithm computes a reduced control-flow graph by collapsing some sets of non-critical nodes. Finally a flow-sensitive analysis algorithm is run on the reduced flow-graph.

The Partial Flow Sensitive algorithm is safe as all the control-flow edges in the CFG are also preserved in the r-CFG; any path in the CFG being traceable in the latter[3]. Thus, the PFS algorithm does not "miss" any flow-path along which the program properties could propagate.

3.1 Yardsticks for a Reduced Control-Flow Graph (r-CFG)

Both the scalability and the precision of the Partial Flow Sensitive algorithm depends on the r-CFG. We define the notions of Precision and Size optimality to understand how good is a given r-CFG for a given flow-graph.

Precision Optimality: The reduced CFG is said to be precision optimal if the following holds for each critical node c : if there does not exist a path from any node n to c in the original flow graph, such a path would not exist even in the reduced flow-graph. [4]

Size Optimality: The reduced CFG is said to be size optimal, if there do not exist nodes n_i and n_j s.t. merging them still maintains precision optimality.

3.2 Algorithm for Computing the r-CFG

Our algorithm is shown Fig. 3. Equation 1 and 2 compute the set $\rho(n)$ for all the nodes $n \in G_{orig}$; $\rho(n)$ represents the set of all critical nodes that are reachable from the node n. This computation can be done efficiently by setting it up as a simple bit-vector dataflow problem where each bit in the bit-vector stands for a particular critical-node and an extra bit for distinguishing critical nodes from non-critical nodes.

Equation 3 represents the condition when two nodes are not mergeable — when one of them reaches a certain critical node c and the other does not. Equation 4 represents the symmetric nature of this relation. Equation 5 finally

[3] Using flow-insensitive analysis over a set of statements can be seen as a flow-sensitive analysis over a complete flow-graph formed with these set of statements.

[4] This also implies that if there exists a path, n to c, in the reduced graph, such a path surely exists in the original graph.

specifies when two nodes can actually be merged — when none of them is critical and the predicate *not_mergeable*() does not forbid their merge.

The reduced CFG is created by representing all the nodes merged together by a single aggregate node. Actually, this reduced CFG is just a conceptual graph — it need not be created. All the nodes belonging to the same aggregate node are simply assigned a common *Aggregate Node ID (anid).* The *anid* identifies each of the aggregate node in the original CFG. All the later algorithms actually work on the original CFG, identifying the aggregate nodes by the *anids*.

3.3 Analyzing the Algorithm

Let us define some notations: N and C denote the set of all nodes and the set of critical nodes respectively in the original CFG (G_{orig}). The r-CFG is denoted by $G_{reduced}$. The relation $path(n_i, n_j) \in G$ represents that there exists a path from n_i to n_j in the graph G. The relation ρ is computed over G_{orig}.

Claim: *The above algorithm produces a precision optimal reduced CFG.*

Proof. Assume $\exists n_i \in N$ s.t. for some $c \in C$, $path(n_i, c) \in G_{reduced}$ and $path(n_i, c) \notin G_{orig}$. Such a path is only possible due to a merge of n_i with some $n_j \in N$ where $path(n_j, c) \in G_{orig}$ (see Fig. 2). This implies that $\rho(n_i) = \rho(n_j)$ as otherwise the algorithm would not have merged the nodes. This causes a contradiction as the critical node $c \in \rho(n_j)$ but $c \notin \rho(n_i)$.

Claim: *The above algorithm produces a size optimal reduced CFG.*

Proof. Assume $\exists n_i, n_j \in N, n_i \neq n_j$, not merged by the above algorithm, s.t. merging them still maintains precision optimality for $G_{reduced}$. Obviously, $\exists c \in C$ s.t. $c \in \rho(n_i)$ and $c \notin \rho(n_j)$ — otherwise the nodes n_i and n_j would had been merged by the algorithm. This causes a contradiction as then precision optimality w.r.t the node $c \in C$ is compromised.

3.4 Size of the Reduced CFG

Lemma: *The above algorithm partitions the non-critical nodes into 2^n equivalence classes, where n is the number of critical nodes.*

Proof. The above algorithm allows two non-critical nodes to be collapsed iff they reach exactly the same set of critical nodes. It is obvious that the relation[5] is an equivalence relation. As each equivalence class corresponds to a possible subset of the set of critical nodes, there can only be 2^n such equivalence classes.

Claim: *The number of nodes in the reduced CFG is bounded by $2^n + n$ where n is the number of critical nodes. Also, the same is a tight bound.*

[5] Nodes a and b are related iff they reach exactly the same set of critical nodes.

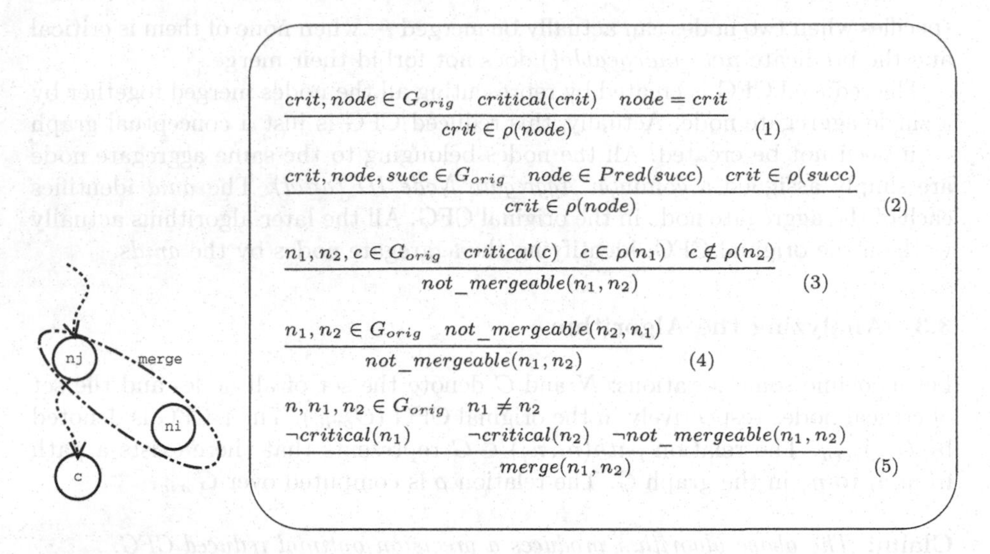

Fig. 2. Proving that the r-CFG is precision optimal

Fig. 3. The optimal algorithm for computing the r-CFG. The solution can be found by performing a fixpoint computation over the above rules.

Proof. According to the above lemma, as each of the non-critical node must belong to one of the equivalence classes and there are at most 2^n such classes — the reduced CFG can at most have 2^n aggregate nodes formed by merging of the non-critical nodes. None of the n critical nodes is merged with any other node. Hence, the reduced CFG can at most have $2^n + n$ nodes.

Also, the above bound is tight. Fig. 4 shows a case where the above bound is actually reached for **n=3**. The critical nodes are marked gray. The table shows the value of $\rho(x)$ for each non-critical node (x). Note that none of the nodes can be merged. A graph of similar structure can be constructed for any value of n.

Surely, the total number of nodes possible in the r-CFG is also bounded by the total number of nodes in the graph. If all the nodes are selected critical, then the r-CFG is same as the original CFG and the PFS algorithm reduces to a flow-sensitive algorithm.

4 The Analysis Phase in the PFS-Algorithm

We explain the analysis phase of the PFS algorithm using a classic compiler analysis — points-to analysis. We choose points-to analysis for the purpose as most compilers implement a flow-insensitive version of this analysis to attain scalability. Our experiments show that the PFS algorithm not just manages to get much better solutions at the critical nodes (arbitrarily chosen), but also improves the solution at most of the other nodes as a side-effect.

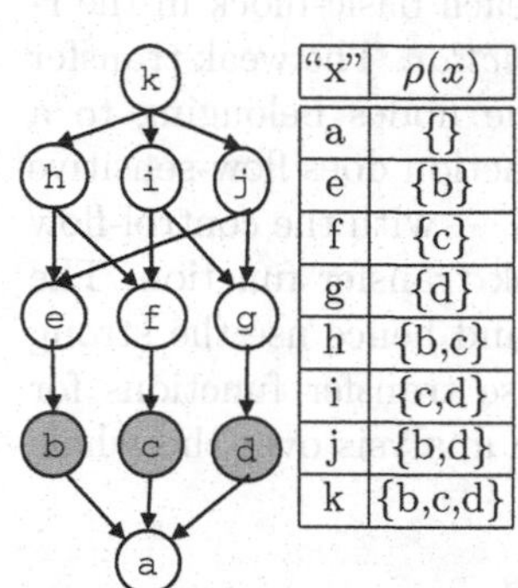

"x"	$\rho(x)$
a	{}
e	{b}
f	{c}
g	{d}
h	{b,c}
i	{c,d}
j	{b,d}
k	{b,c,d}

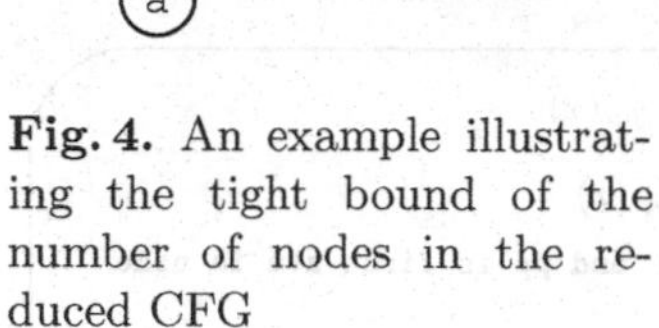

Fig. 4. An example illustrating the tight bound of the number of nodes in the reduced CFG

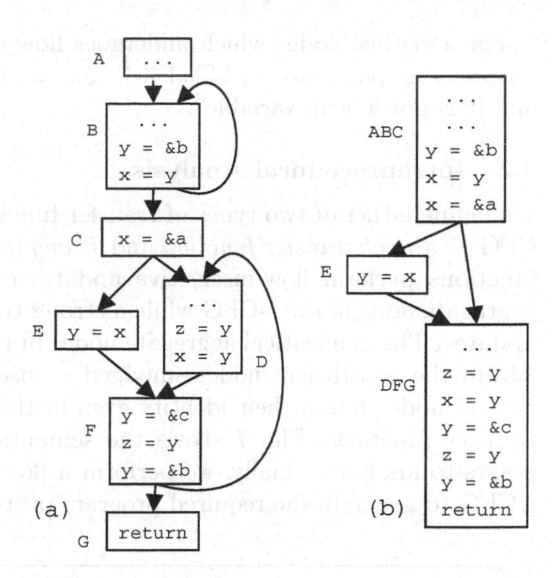

Fig. 5. An Example: The (a) original CFG and the (b) reduced CFG, taking the node 'E' as the critical node

4.1 Updating Points-To Sets for Each Statement Type

The operational semantics for updating of points-to sets is very similar to that proposed earlier in literature [2,10]. Some of the rules in the update semantics for the various statements is shown in fig. 6. The complete set of rules for all the important statement templates is given in [11]. $\pi_{cs,pp}(x, y)$ denotes the points-to relation that x points-to y at the program point pp under the call-string cs.

$$\frac{stm_{pp}(\text{"}x = y\text{"}) \quad \pi_{cs,pp}(y, z)}{\pi_{cs,pp}(x, z)} \quad (1)$$

$$\frac{stm_{pp}(\text{"}x = \&y\text{"})}{\pi_{cs,pp}(x, y)} \quad (4)$$

$$\frac{stm_{pp}(\text{"}x = *y\text{"}) \quad \pi_{cs,pp}(y, y1) \quad \pi_{cs,pp}(y1, z)}{\pi_{cs,pp}(x, z)} \quad (2)$$

$$\frac{stm_{pp}(\text{"}*x = y\text{"}) \quad \pi_{cs,pp}(x, z) \quad \pi_{cs,pp}(y, y1) \quad \neg garbage\_value(z) \quad \neg null\_value(z)}{\pi_{cs,pp}(z, y1)} \quad (5)$$

$$\frac{stm_{pp}(\text{"}x = *y\text{"}) \quad \pi_{cs,pp}(y, g) \quad garbage\_value(g)}{\pi_{cs,pp}(x, g)} \quad (3)$$

Fig. 6. Points-to set update semantics. $\pi_{cs,pp}(x, y)$ denotes the points-to relation that x points-to y at the program point *pp* under the call-string *cs*. The $stm_{pp}(S)$ indicates that the S is statement encountered at program point *pp*.

For a critical node (which undergoes flow-sensitive analysis), a variable gets its previous points-to set killed if it gets assigned a new value unambiguously and it is not a heap variable[6].

4.2 Intraprocedural Analysis

We define either of two types of transfer functions for each basic-block in the r-CFG — a *weak transfer function* and *strong transfer function*. The weak transfer functions perform flow-insensitive updates over all the nodes belonging to a aggregate node in the r-CFG while a strong transfer function does flow-sensitive updates. The non-critical aggregate nodes in the r-CFG — with the control-flow within the constituent nodes smudged — use the weak transfer function. The critical nodes retain their identity even in the r-CFG and hence use the strong transfer functions. Fig. 7 shows the semantics of these transfer functions for points-to analysis. Finally, we perform a flow-sensitive analysis over the whole r-CFG to generate the required program properties.

$$\pi^{strong}_{cs,pp}(x,y) = \pi_{cs,pp}(x,y) \vee \begin{cases} \pi^{strong}_{cs,pp-1}(x,y) \text{ if } \neg killed_{cs,pp}(x) \\ \phi \quad \text{if } killed_{cs,pp}(x) \end{cases} \vee \begin{cases} \pi^{IN}_{cs,node}(x,y) \text{ if } pp \in node \text{ and pp is first stm in node} \\ \phi \quad \text{otherwise} \end{cases}$$

$$\pi^{weak}_{cs,node}(x,y) = \sum_{\forall nodes_f(n), pp\in n, anid_f(n)=anid_f(node)} \left\{\pi_{cs,pp}(x,y) \vee \pi^{IN}_{cs,n}(x,y)\right\}$$

$$\pi^{OUT}_{cs,node}(x,y) = \begin{cases} \pi^{strong}_{cs,pp}(x,y) \text{ if } critical_f(node) \wedge \text{ pp is last stm in node} \\ \pi^{weak}_{cs,node}(x,y) \text{ if } \neg critical_f(node) \end{cases}$$

$$\pi^{IN}_{cs,node}(x,y) = \sum_{p\in pred_f(node)} \pi^{OUT}_{cs,p}(x,y)$$

Fig. 7. The semantics for the strong and weak transfer functions for points-to analysis. $\pi(x, y)$ denotes that the variable x points-to y. *cs* refers to the call-string, *pp* and *node* refer to the program point or the basic-block where the relation is being computed and f indicates that the relation is being defined for the procedure `f`. The relation *id* returns a unique identifier for all nodes. A relation $\pi(x, y)$ is killed if there exists an unambiguous definition to x. The solution can be computed by a fixpoint computation over the rules.

4.3 Interprocedural Analysis

Interprocedural analysis is performed using the k-limit call-string approach [12]. For critical nodes, the actual parameters carry the program properties existing at the call-site into the callee. For non-critical nodes, if the call-site $x \in$

[6] We summarize heap locations by summarized heap variables per allocation-site.

$$\pi^{strong}_{cs,pp}(x,y) = critical_{func\_caller}(caller\_node) \wedge (\pi^{RET}_{cs,pp}(x,z) \vee \pi^{MOD}_{cs,pp}(x,z) \vee \pi^{strong}_{cs,pp}(x,y))$$
$$\pi^{weak}_{cs,node}(x,y) = \neg critical_{func\_caller}(caller\_node)$$
$$\wedge (\pi^{RET}_{cs,pp \in node}(x,z) \vee \pi^{MOD}_{cs,pp \in node}(x,z) \vee \pi^{weak}_{cs,node}(x,y))$$

Fig. 8. Interprocedural PFS Points-to Analysis Semantics. The solution can be computed by a fixpoint computation over the rules.

$nodes(G_{orig})$ and $x \in x', x' \in nodes(G_{red})$, then the analysis information from $OUT_{x'}$ is made to pass into the callee. For procedure return at a basic block `n` in the callee, it is always the $OUT_{\{x' | n \in x', x' \in nodes(G_{red})\}}$ value that is passed back to the caller.

Fig. 8 shows the interprocedural semantics of the weak and strong transfer functions for our PFS points-to analysis. For simplicity, we assume that a function call is the first statement in a basic block, global variables are absent and parameters are passed by value. The $\pi^{RET}_{cs,pp}$ relation defines the semantics of a return statement — if the callee returns with a statement "`return(y)`" to the caller who had initiated the call with "`x=func_callee(...)`", the equation updates the points-to set of the variable `x` with that of the return parameter `y`. The π^{MOD} relation updates all points-to relations generated due to indirect references in the called procedure. The detailed description of the interprocedural semantics is given in [11].

We give an example to illustrate the effect of the PFS analysis. Fig. 5(a) shows the original flow-graph; fig. 5(b) shows the reduced flow-graph (taking the node 'E' as the critical node). The results for flow-sensitive, flow-insensitive and partially flow-sensitive analyses are shown in table 1. Note that for the PFS case, flow-insensitive analysis is done on the aggregate nodes "ABC" and on "DFG" while we perform flow-sensitive analysis on the critical node 'E' to compute the local properties. A flow-sensitive analysis is then performed on the r-CFG.

The reduced CFG is much simplified as the number of nodes drop to three and the two loops in the original CFG get dissolved. In fact, the r-CFG in this case gets reduced to a DAG; note that the number of iterations required for a flow-sensitive analysis to reach a fixpoint depends on the loop-depth. Hence, performing a flow-sensitive analysis over the r-CFG is a lot cheaper than doing the same on the original CFG.

Let us look at the results of the analysis in table 1. For the critical node 'E', the PFS solution is much better than that obtained by the flow-sensitive analysis but is not as good as that obtained by the flow-insensitive analysis. The reason for the loss in precision is the inability of the PFS algorithm to use "kill" information within the aggregate nodes reaching the critical node. In the given example, the dataflow fact `x may-point-to b`, generated at the node 'B', is killed at the node 'C'. However, the PFS algorithm is unable to use this information as a flow-insensitive analysis is carried on the aggregate node "ABC". However, the result is better than the flow-insensitive case, as the dataflow facts from the nodes 'D', 'F', and 'G' were not allowed to pollute the information at the critical

Table 1. The dataflow facts discovered for the CFG and r-CFG in fig. 5. The program points corresponding to a node 'X' is the point just after 'X'. The notation $V_1 \rightarrow V_2$ implies that any variable $v_1 \in V_1$ may point to any of the variables $v_2 \in V_2$.

BB	Flow-Sensitive	Flow-Insensitive	Partial Flow-Sensitive
A	ϕ	$\{x, y, z\} \rightarrow \{a, b, c\}$	$\{x\} \rightarrow \{a, b\}$, $\{y\} \rightarrow \{b\}$
B	$\{x, y\} \rightarrow \{b\}$	$\{x, y, z\} \rightarrow \{a, b, c\}$	$\{x\} \rightarrow \{a, b\}$, $\{y\} \rightarrow \{b\}$
C	$x \rightarrow \{a\}$, $y \rightarrow \{b\}$	$\{x, y, z\} \rightarrow \{a, b, c\}$	$\{x\} \rightarrow \{a, b\}$, $\{y\} \rightarrow \{b\}$
D	$\{x, y, z\} \rightarrow \{b\}$	$\{x, y, z\} \rightarrow \{a, b, c\}$	$\{x, y, z\} \rightarrow \{a, b, c\}$
E	$\{x, y\} \rightarrow \{a\}$	$\{x, y, z\} \rightarrow \{a, b, c\}$	$\{x, y\} \rightarrow \{a, b\}$
F	$\{x\} \rightarrow \{a, b\}$, $\{y\} \rightarrow \{b\}$, $\{z\} \rightarrow \{c\}$	$\{x, y, z\} \rightarrow \{a, b, c\}$	$\{x, y, z\} \rightarrow \{a, b, c\}$
G	$\{x\} \rightarrow \{a, b\}$, $\{y\} \rightarrow \{b\}$, $\{z\} \rightarrow \{c\}$	$\{x, y, z\} \rightarrow \{a, b, c\}$	$\{x, y, z\} \rightarrow \{a, b, c\}$

node '`E`'. As a bonus, the solutions at the nodes '`A`', '`B`' and '`C`' are also improved over the flow-insensitive one.

5 Experimental Results and Conclusion

We have implemented a framework for partial flow-sensitive points-to analysis using the Lance compiler framework [13] and the bddbddb [14] tool. The details of the implementation can be found in [11]. The results are shown in Fig. 9, 11 and 10. We selected the critical nodes arbitrarily for computing the solutions using the PFS algorithm. We used Andersen's algorithm [2] for the flow-insensitive analysis within the aggregate nodes. The results were obtained by performing an intraprocedural analysis on the respective functions by setting all the pointer arguments and the global pointer variables used in the procedure to be pointing to *undefined* (implying that they could potentially point to any location). The analysis was performed on the intermediate code generated by the Lance Compiler Framework [13].

Fig. 9 shows the effect of partial flow-sensitivity on the function "`SwapNode()`" from the "`ks`" benchmark (from [15]) with nodes `5` and `8` arbitrarily selected as the critical nodes. The results for a fully flow-sensitive and a flow-insensitive analysis are also shown. The partial flow-sensitive analysis yields a very precise solution for the critical nodes. In fact, they are as good as the flow-sensitive solution in this case. Also, as a side-effect, the solution at many of the other nodes are much better than that obtained using a flow-insensitive algorithm. However, though close, the FS and PFS solutions may not coincide in all cases as the PFS algorithm is unable to use kill information within the aggregate nodes.

Fig. 10 compares the number of nodes in the original and the reduced graphs. For instance, note that for the function `PrintChannel()`, the PFS solution is almost the same as the flow-sensitive one, even though the analysis was done on a reduced CFG with 5 nodes while the original CFG had 88 nodes. However, how close is the PFS solution to the flow-sensitive solution is hugely dictated by the choice of the critical nodes.

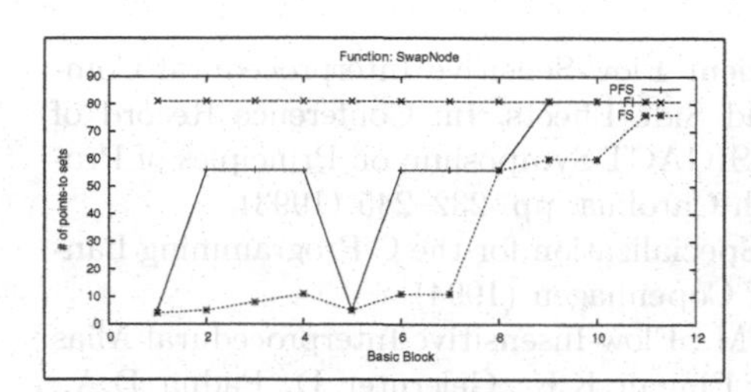

Fig. 9. The effect of Partial Flow Sensitivity (on function "SwapNode" from the "ks" benchmark suite) : The plot shows the number of `may-point-to` relations that hold at each basic-block in the program for the flow-sensitive (FS), flow-insensitive (FI) and the partially flow-sensitive (PFS) algorithms.

Function	Benchmark	N_{crit}	$\#N_o$	$\#N_r$
SwapNode	ks	{5,8}	11	5
DensityChannel	yacr2	{7,28}	32	5
PrintChannel	yacr2	{6,22,89}	88	5
HasVCV	yacr2	{6,11}	13	5
gen_bitlen	gzip	{4,15}	32	5
build_tree	gzip	{4,8}	16	5
init_block	gzip	{4,7}	10	5

Fig. 10. Details on the benchmarks used. The column 'N_{crit}' denotes the nodes selected as critical in the original flow-graph. $\# N_o$ and $\# N_r$ denotes the number of nodes in the original and the reduced flow-graphs.

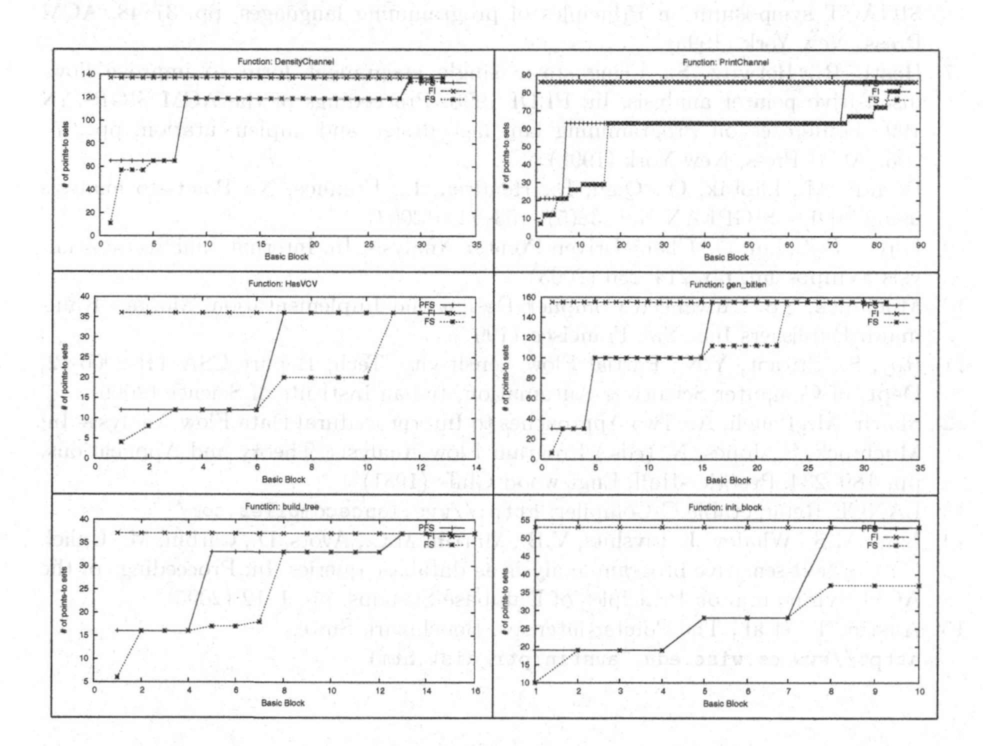

Fig. 11. Precision of the flow-sensitive (FS), flow-insensitive (FI) and the partially flow-sensitive (PFS) algorithms for various benchmarks. The plot shows the number of `may-point-to` relations that hold at each basic-block in the program.

References

1. Choi, J.D., Burke, M., Carini, P.: Efficient Flow-Sensitive Interprocedural Computation of Pointer-Induced Aliases and Side Effects. In: Conference Record of the Twentieth Annual ACM SIGPLAN-SIGACT Symposium on Principles of Programming Languages, Charleston, South Carolina, pp. 232–245 (1993)
2. Andersen, L.O.: Program Analysis and Specialization for the C Programming Language. PhD thesis, DIKU, University of Copenhagen (1994)
3. Burke, M., Carini, P., Choi, J.D., Hind, M.: Flow-Insensitive Interprocedural Alias Analysis in the Pressence of Pointers. In: Pingali, K.K., Gelernter, D., Padua, D.A., Banerjee, U., Nicolau, A. (eds.) Languages and Compilers for Parallel Computing. LNCS, vol. 892, Springer, Heidelberg (1995)
4. Babich, W.A., Jazayeri, M.: The Method of Attributes for Data Flow Analysis: Part II. Demand Analysis. Acta Inf. 10, 265–272 (1978)
5. Horwitz, S., Reps, T., Sagiv, M.: Demand interprocedural dataflow analysis. In: SIGSOFT 1995. Proceedings of the 3rd ACM SIGSOFT symposium on Foundations of software engineering, pp. 104–115. ACM Press, New York (1995)
6. Duesterwald, E., Gupta, R., Soffa, M.L.: Demand-driven computation of interprocedural data flow. In: POPL 1995. Proceedings of the 22nd ACM SIGPLAN-SIGACT symposium on Principles of programming languages, pp. 37–48. ACM Press, New York (1995)
7. Hasti, R., Horwitz, S.: Using static single assignment form to improve flow-insensitive pointer analysis. In: PLDI 1998. Proceedings of the ACM SIGPLAN 1998 conference on Programming language design and implementation, pp. 97–105. ACM Press, New York (1998)
8. Berndl, M., Lhoták, O., Qian, F., Hendren, L., Umanee, N.: Points-to analysis using BDDs. SIGPLAN Not. 38(5), 103–114 (2003)
9. Guyer, S.Z., Lin, C.: Client-Driven Pointer Analysis. In: International Static Analysis Symposium, pp. 214–236 (2003)
10. Muchnick, S.S.: Advanced Compiler Design and Implementation. Morgan Kaufmann Publishers Inc., San Francisco (1997)
11. Roy, S., Srikant, Y.N.: Partial Flow Sensitivity. Tech. Report CSA-TR-2006-12, Dept. of Computer Science & Automation, Indian Institute of Science (2006)
12. Sharir, M., Pnueli, A.: Two Approaches to Interprocedural Data Flow Analysis. In: Muchnick, S., Jones, N. (eds.) Program Flow Analysis: Theory and Applications, pp. 189–234. Prentice-Hall, Englewood Cliffs (1981)
13. LANCE Retargetable C Compiler: `http://www.lancecompiler.com/`
14. Lam, M.S., Whaley, J., Livshits, V.B., Martin, M.C., Avots, D., Carbin, M., Unkel, C.: Context-sensitive program analysis as database queries. In: Proceedings of the ACM Symposium on Principles of Database Systems, pp. 1–12 (2005)
15. Austin, T., et al.: The Pointer-intensive Benchmark Suite, `http://www.cs.wisc.edu/~austin/ptr-dist.html`

A Scalable Asynchronous Replication-Based Strategy for Fault Tolerant MPI Applications*

John Paul Walters and Vipin Chaudhary

Department of Computer Science and Engineering
University at Buffalo, The State University of New York
Buffalo, NY 14260, USA
{waltersj,vipin}@buffalo.edu

Abstract. As computational clusters increase in size, their mean-time-to-failure reduces. Typically checkpointing is used to minimize the loss of computation. Most checkpointing techniques, however, require a central storage for storing checkpoints. This severely limits the scalability of checkpointing. We propose a scalable replication-based MPI checkpointing facility that is based on LAM/MPI. We extend the existing state of fault-tolerant MPI with asynchronous replication, eliminating the need for central or network storage. We evaluate centralized storage, SAN-based solutions, and a commercial parallel file system, and show that they are not scalable, particularly beyond 64 CPUs. We demonstrate the low overhead of our replication scheme with the NAS Parallel Benchmarks and the High Performance LINPACK benchmark with tests up to 256 nodes while demonstrating that checkpointing and replication can be achieved with much lower overhead than that provided by current techniques.

1 Introduction

Computational clusters with hundreds and thousands of processors are fast-becoming ubiquitous in large-scale scientific computing. This is leading to lower mean-time-to-failure and forces the system software to deal with the possibility of arbitrary and unexpected node failure. Since MPI provides no mechanism to recover from a failure, a single node failure will halt the execution of the entire MPI world. Thus, there exists great interest in the research community for a truly fault-tolerant and transparent MPI implementation.

Several groups have included checkpointing within various MPI implementations. MVAPICH2 now includes support for kernel-level checkpointing of Infiniband MPI processes [1]. Sankaran et al. also describe a kernel-level checkpointing strategy within LAM/MPI [2,3]. However, these implementations suffer from a major drawback: a reliance on a common network file system or dedicated checkpoint servers.

We consider the reliance on network file systems, parallel file systems, and/or checkpoint servers to be a fundamental limitation of existing fault-tolerant systems. Storing checkpoints directly to network storage incurs too great an overhead. Using dedicated

* This research was supported in part by NSF IGERT grant 9987598, the Institute for Scientific Computing at Wayne State University, MEDC/Michigan Life Science Corridor, and NYSTAR.

S. Aluru et al. (Eds.): HiPC 2007, LNCS 4873, pp. 257–268, 2007.

checkpoint servers saturates the network links of a few machines, resulting in degraded performance. Even parallel file systems are quickly saturated. As such, we make the following contributions in this paper:

1. We propose and implement a checkpoint replication system that distributes the overhead of checkpointing evenly over all nodes participating in the computation. This significantly reduces the impact of heavy I/O on network storage.
2. We show that common existing strategies, including the use of dedicated checkpoint servers, storage area networks (SANs), and parallel file systems, are inadequate for even moderately-sized computations.

The remainder of this paper is outlined as follows: in Section 2 we provide a brief introduction to LAM/MPI and checkpointing. In Section 3 we discuss existing LAM/MPI checkpointing strategies. In Section 4 we compare existing checkpoint storage strategies and evaluate our proposed replication technique. In Section 5 we provide a brief overview of the work related to this project. Finally, in Section 6 we present our conclusions.

2 Background

2.1 LAM/MPI

LAM/MPI [4] is a research implementation of the MPI-1.2 standard with portions of the MPI-2 standard. LAM uses a layered software approach in its construction [5]. In doing so, various modules are available to the programmer that tune LAM/MPI's runtime functionality including TCP, Infiniband, Myrinet, and shared memory communication. The most commonly used module, however, is the TCP module which provides basic TCP communication between LAM processes. A modification of this module, CRTCP, provides a bookmark mechanism for checkpointing libraries to ensure that a message channel is clear. LAM uses the CRTCP module for its built-in checkpointing capabilities.

2.2 Checkpointing Distributed Systems

Checkpointing itself can be performed at several levels. In kernel-level checkpointing, the checkpointer is implemented as a kernel module, making checkpointing fairly straightforward. However, the checkpoint itself is heavily reliant on the operating system (kernel version, process IDs, etc.). User-level checkpointing performs checkpointing using a checkpointing library, enabling a more portable checkpointing implementation at the cost of limited access to kernel-specific attributes (e.g. user-level checkpointers cannot restore process IDs). At the highest level is application-level checkpointing where code is instrumented with checkpointing primitives. The advantage to this approach is that checkpoints can often be restored to arbitrary architectures. However, application-level checkpointers require access to a user's source code and do not support arbitrary checkpointing.

There are two major checkpointing/rollback recovery techniques: coordinated checkpointing and message logging. Coordinated checkpointing requires that all processes come to an agreement on a consistent state before a checkpoint is taken. Upon failure, all processes are rolled back to the most recent checkpoint/consistent state. Message logging requires distributed systems to keep track of interprocess messages in order to bring a checkpoint up-to-date. Checkpoints can be taken in a non-coordinated manner, but the overhead of logging the interprocess messages can limit its utility. Elnozahy et al. provide a detailed survey of the various rollback recovery protocols that are in use today [6].

3 LAM/MPI Checkpointing

We are not the first group to implement checkpointing within the LAM/MPI system. Three others [7,3,8] have added basic checkpoint/restart support. Because of the previous work in LAM/MPI checkpointing, the basic checkpointing/restart building blocks were already present within LAM's source code. This provided an ideal environment for testing our replication strategy. We begin with a brief overview of checkpointing with LAM/MPI.

Sankaran et al. first added checkpointing support within the LAM system [3] by implementing a lightweight coordinated blocking module to replace LAM's existing TCP module. The protocol begins when *mpirun* instructs each LAM daemon (*lamd*) to checkpoint its MPI processes. When a checkpoint signal is delivered to an MPI process, each process exchanges bookmark information with all other MPI processes. These bookmarks contain the number of bytes sent to/received from every other MPI process. With this information, any in-flight messages can be waited on and received before the checkpoint occurs. After acquiescing the network channels, the MPI library is locked and a checkpointing thread assumes control. The Berkeley Linux Checkpoint/Restart library (BLCR) [9] is used as a kernel-level checkpointing engine. Each process checkpoints itself using BLCR (including *mpirun*) and the computation resumes.

A problem with the above solution is that it requires identical restart topologies. If, for example, a compute node fails, the system cannot restart by remapping checkpoints to existing nodes. Instead, a new node would have to be inserted into the cluster to force the restart topology into consistency with the original checkpoint topology. This requires the existence of spare nodes that can be inserted into the MPI world to replace failed nodes. If no spare nodes are available, the computation cannot be restarted.

Two previous groups have attempted to solve the problem of migrating LAM checkpoint images. Cao et al. propose a migration scheme that parses the binary checkpoint images, finds the MPI process location information, and updates the node IDs [10]. Wang, et al. propose a pause/migrate solution where spare nodes are used for migration purposes when a LAM daemon discovers an unresponsive node [8]. Upon detecting a failure, their system migrates the failed processes via a network file system to the replacement nodes before continuing the computation.

We use the same coordinated blocking approach as Sankaran's technique described above. To perform the checkpointing, we use Victor Zandy's *Ckpt* checkpointer [11]. Unlike previous solutions, we allow for arbitrary restart topologies without relying on any shared storage or checkpoint parsing. Instead, we reinitialize the MPI library and update node and process-specific attributes in order to restore a computation on varying topologies. Due to space limitations and our focus on the replication portion of our implementation, we omit the details of the basic checkpoint/migrate/restart solution. A more detailed description is available in our extended work [12].

4 Checkpoint Storage, Resilience, and Performance

In order to enhance the resiliency of checkpointing while simultaneously reducing its overhead, we include data replication. While not typically stated explicitly, nearly all checkpoint/restart methods rely on the existence of network storage that is accessible to the entire cluster. Such strategies suffer from two major drawbacks in that they create a single point of failure and also incur massive overhead when compared to checkpointing to local disks.

A cluster that utilizes a network file system-based checkpoint/restart mechanism may sit idle should the file system experience an outage. This leads not only to wasteful downtime, but also may lead to lost data should the computation fail without the ability to checkpoint. However, even with fault-tolerant network storage, simply writing large amounts of data to such storage represents an unnecessary overhead to the application. In the sections to follow, we examine two replication strategies: a dedicated server technique, and a distributed implementation.

We acknowledge that arguments can be made in support of the use of SANs or parallel file systems for the storage of checkpoints. The most powerful supercomputers, such as the IBM Bluegene/L, have no per-node local storage. Instead, parallel file systems are used for persistent data storage in order to reduce the number of disk related node failures. We do not position our implementation for use on such massive supercomputers. Instead, we target clusters consisting of hundreds or thousands of commodity nodes, each with its own local storage.

For our implementation testing we used a Wayne State University owned cluster consisting of 16 dual 2.66 GHz Pentium IV Xeon processors with 2.5 GB RAM, a 10,000 RPM Ultra SCSI hard disk and gigabit ethernet. A 1 TB IBM DS4400 SAN was also used for the network storage tests. We evaluate both centralized-server and SAN-based storage techniques and compare them against our proposed replication strategy using the SP, LU, and BT benchmarks from the NAS Parallel Benchmarks (NPB) suite [13] and the High Performance LINPACK (HPL) [14] benchmark.

To gauge the performance of our checkpointing library using the NPB tests, we used exclusively "Class C" benchmarks. Our HPL benchmark tests used a problem size of 28,000. These configurations resulted in checkpoints that were 106MB, 194MB, 500MB, and 797MB for the LU, SP, BT, and HPL benchmarks, respectively. The LU and HPL benchmarks consisted of 8 CPUs each, while the BT and SP benchmarks required 9 CPUs. We describe the scalability tests and configuration in Section 4.4.

In order to test the overhead of our implementation we chose to checkpoint the benchmarks with much greater frequency than would otherwise be used. By checkpointing at frequencies as short as 1 minute, we are better able to demonstrate the individual components of the overhead. In a real application, users would likely checkpoint an application at intervals of several hours (or more).

As a baseline, we compare the SAN storage, dedicated server storage, and replication storage techniques against the local disk checkpoint data shown in Figure 1. Here we show the result of periodically checkpointing the NAS Parallel Benchmarks as well as the HPL benchmark along with the time taken to perform a single checkpoint. Our implementation shows very little overhead even when checkpointed at 1 minute intervals. The major source of the overhead of our checkpointing scheme lies in the time taken in writing the checkpoint images to the local file system.

In Figure 1(a) we break the checkpointing overhead down by coordination time, checkpointing time, and continue time. The percentages listed above each column indicate the overhead of a checkpoint when compared to the baseline running time of Figure 1(b). The coordination phase includes the time needed to acquiesce the network channels/exchange bookmarks (see Section 3). The checkpoint time consists of the time needed to checkpoint the entire memory footprint of a single process and write it to stable storage. Finally, the continue phase includes the time needed to synchronize the resumption of computation. The coordination and continue phases require barriers to ensure application synchronization, while each process performs the checkpoint phase independently.

As confirmed in Figure 1(a), the time required to checkpoint the entire system is largely dependent on the time needed to checkpoint the individual nodes. Writing the checkpoint file to disk represents the single largest time in the entire checkpoint process and dwarfs the coordination phase. Thus, as the memory footprint of an application grows, so too does the time needed to checkpoint. This can also impact the time needed to perform the *continue* barrier as faster nodes are forced to wait for slower nodes to write their checkpoints to disk.

4.1 Dedicated Checkpoint Servers Versus Checkpointing to Network Storage

The two most common checkpoint storage techniques presented in the literature are the dedicated server(s) [15] and storing checkpoints directly to network storage [2,1].

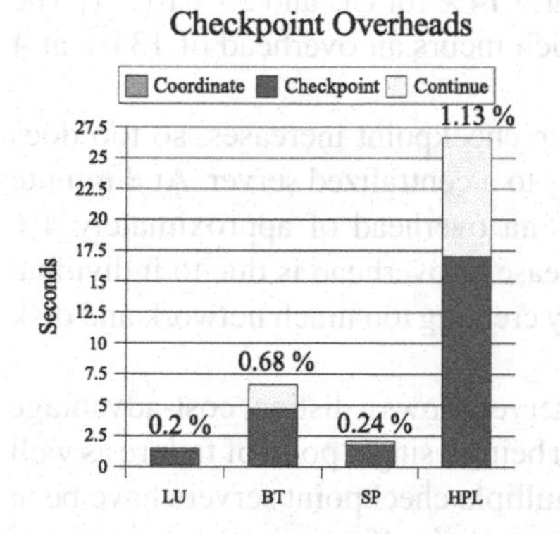

(a) % indicates the contribution of checkpointing (in terms of overhead) at 8 minute intervals over the base timings without checkpointing (from Figure 1(b)).

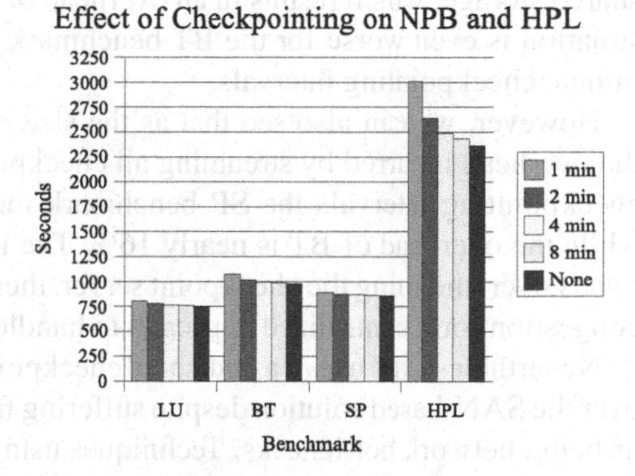

(b) Multiple checkpointing intervals.

Fig. 1. A breakdown of overheads when checkpointing to local disks

We begin our evaluation with a comparison of these two common strategies. To do so, we implemented both a dedicated checkpoint server solution as well as a SAN-based checkpoint storage solution by extending the LAM daemons and *mpirun* to collect and propagate checkpoints.

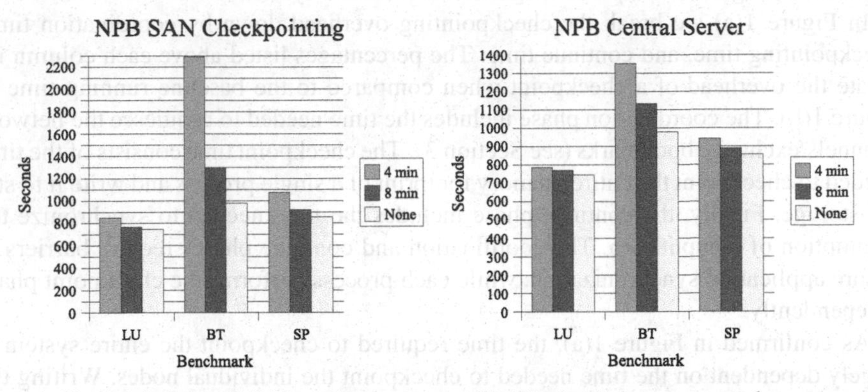

(a) Checkpointing to SAN-based storage. (b) Checkpointing to a central server.

Fig. 2. Runtime of NPB with checkpoints streamed to central checkpoint server vs. saving to SAN

In Figure 2 we show the results of checkpointing the NAS Parallel Benchmarks with the added cost of streaming the checkpoints to a centralized server or storing the checkpoints to a SAN. In the case of the LU benchmark, we notice a marked reduction in overhead when comparing the SAN data in Figure 2(a) to the checkpoint server data presented in Figure 2(b). Indeed, the overhead incurred by streaming an LU checkpoint every 4 minutes is less than 6% – a dramatic improvement over saving checkpoints to shared storage, which results in an overhead of nearly 14% for LU and 25% for SP. The situation is even worse for the BT benchmark which incurs an overhead of 134% at 4 minute checkpointing intervals.

However, we can also see that as the size of the checkpoint increases, so too does the overhead incurred by streaming all checkpoints to a centralized server. At 8 minute checkpointing intervals the SP benchmark incurs an overhead of approximately 4% while the overhead of BT is nearly 16%. The increase in overhead is due to individual *lamds* overwhelming the checkpoint server, thereby creating too much network and disk congestion for a centralized approach to handle.

Nevertheless, the use of a dedicated checkpoint server shows a distinct cost-advantage over the SAN-based solution despite suffering from being a single point of failure as well as being network bottlenecks. Techniques using multiple checkpoint servers have been proposed to mitigate such bottlenecks [15]. However, their efficacy in the presence of large checkpoint files has not been demonstrated in the literature (NPB class B results are shown).

Wang et al. propose a technique to alleviate the impact of checkpointing directly to SANs [8]. Their technique combines local checkpointing with asynchronous checkpoint propagation to network storage. However, their solution requires multiple levels of scheduling in order to prevent the SAN from being overwhelmed by the network traffic. The overhead of their scheduling has not yet been demonstrated in the literature, nor has the scalability of their approach, where their tests are limited to 16 nodes.

4.2 Checkpoint Replication

To address the scalability issues shown in Section 4.1, we implemented an asynchronous replication strategy that amortizes the cost of checkpoint storage over all nodes within the MPI world. Again we extended LAM's *lamds*, this time using a peer-to-peer strategy to replicate checkpoints to multiple nodes. This addresses both the resiliency of checkpoints to node failure as well as the bottlenecks incurred by transferring data to dedicated servers.

A variety of replication strategies have been used in peer-to-peer systems. Typically, such strategies must take into account the relative popularity of individual files within the network in order to ascertain the optimal replication strategy. Common techniques include the square-root, proportional, and uniform distributions [16]. While the uniform distribution is not used within peer-to-peer networks because it does not account for a file's query probability, our checkpoint/restart system relies on the availability of each checkpoint within the network. Thus, each checkpoint object has an equal query probability/popularity and we feel that a uniform distribution is justified for this specific case.

We opted to distribute the checkpoints randomly in order to provide a higher resilience to network failures. For example, a solution that replicates to a node's nearest neighbors would likely fail in the presence of a switch failure. Also, nodes may not fail independently and instead cause the failure of additional nodes within their vicinity. Thus, we feel that randomly replicating the checkpoints throughout the network provides the greatest possible survivability.

Figure 3(a) shows the results of distributing a single replica throughout the cluster with NPB. As can be seen, the overhead in Figure 3(a) is substantially lower than that of the centralized server shown in Figure 2(b). In each of the three NAS benchmarks, we

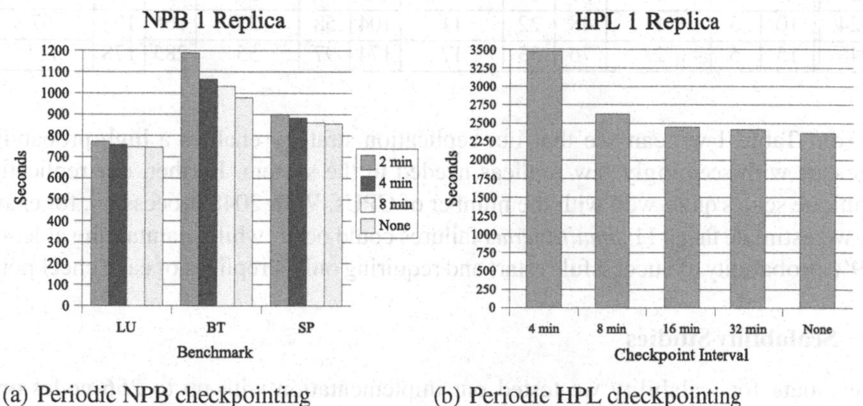

(a) Periodic NPB checkpointing with a single replica.

(b) Periodic HPL checkpointing with a single replica.

Fig. 3. Benchmark timings with one replica

are able to reduce the overhead of distributing a checkpoint by at least 50% when compared to streaming all checkpoints to a single server. For the most expensive checkpoint (BT), we are able to reduce the overhead of checkpointing to 9% at 4 minute intervals and 5.5% at 8 minute intervals (compared to 38% and 16% at 4 minute and 8 minute intervals, respectively).

In Figure 3(b) we show the results of distributing a single replica every 4, 8, 16, and 32 minutes for the HPL benchmark. We found that our network was unable to handle checkpoint distribution of HPL at intervals shorter than 4 minutes, due to the size of the checkpoint files. We notice a steady decrease in overhead as the checkpoint interval increases to typical values, with a single checkpoint resulting in an overhead of only 2.2%.

4.3 The Degree of Replication

While the replication strategy that we have described has clear advantages in terms of reducing the overhead on a running application, an important question that remains is the number of replicas necessary to achieve a high probability of restart. To help answer this question, we developed a simulator capable of replicating node failures, given inputs of the network size and the number of replicas.

Table 1. Maximum number of allowed failures with 90, 99, and 99.9% restart probability

	1 Replica			2 Replicas			3 Replicas			4 Replicas		
	Allowed Failures for			Allowed Failures for			Allowed Failures for			Allowed Failures for		
Nodes	90%	99%	99.9%	90%	99%	99.9%	90%	99%	99.9%	90%	99%	99.9%
8	1	1	1	2	2	2	3	3	3	4	4	4
16	1	1	1	2	2	2	5	4	3	7	5	4
32	2	1	1	5	3	2	8	5	4	11	8	6
64	3	1	1	8	4	2	14	8	4	19	12	8
128	4	1	1	12	6	3	22	13	8	32	21	14
256	5	2	1	19	9	5	37	21	13	55	35	23
512	7	2	1	31	14	7	62	35	20	95	60	38
1024	10	3	1	48	22	11	104	58	33	165	103	67
2048	15	5	2	76	35	17	174	97	55	285	178	111

From Table 1 we can see that our replication strategy enables a high probability of restart with seemingly few replicas needed in the system. Further, our replication technique scales quite well with the number of CPUs. With 2048 processors, for example, we estimate that 111 *simultaneous* failures could occur while maintaining at least a 99.9% probability of successful restart and requiring only 4 replicas of each checkpoint.

4.4 Scalability Studies

To evaluate for scalability we tested our implementation with up to 256 nodes on a University at Buffalo Center for Computation Research owned cluster consisting of 1600 3.0/3.2 GHz Intel Xeon processors, with 2 processors per node (800 total nodes), a 30 TB EMC SAN as well as a high performance Ibrix parallel file system. The network is connected by both gigabit ethernet and Myrinet. Gigabit ethernet was used for our tests. 21 active Ibrix segment servers are in use and connect to the existing EMC SAN.

Because our checkpointing engine, *Ckpt* [11], is only 32 bit while the University at Buffalo's Xeon processors are each 64 bit, we simulated the mechanics of checkpointing with an artificial 1 GB file that is created and written to local disk at each checkpoint interval. Aside from this slight modification, the remaining portions of our checkpointing system remain intact (coordination, continue, file writing, and replication).

In Figure 4 we demonstrate the impact of our checkpointing scheme. Each number of nodes (64, 128, and 256) operates on a unique data set to maintain a run time of approximately 1000 seconds. For comparison, we also present the overhead of checkpointing to the EMC SAN and Ibrix parallel file system in Figure 4(d). We chose to evaluate our system for up to 4 checkpoints as the results of our failure simulation (see Table 1) suggest that 4 replicas achieves an excellent restart probability with high node failures.

The individual figures in Figure 4 all represent the total run time of the HPL benchmark at each cluster size. Thus, comparing the run times at each replication level against

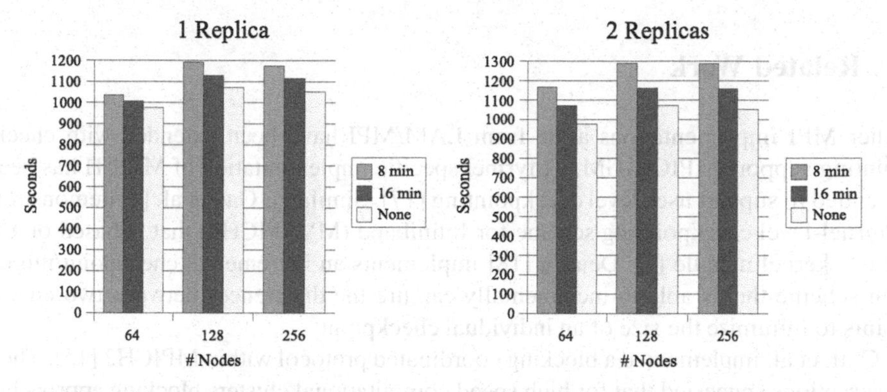

(a) HPL with one replica per checkpoint. (b) HPL with two replicas per checkpoint.

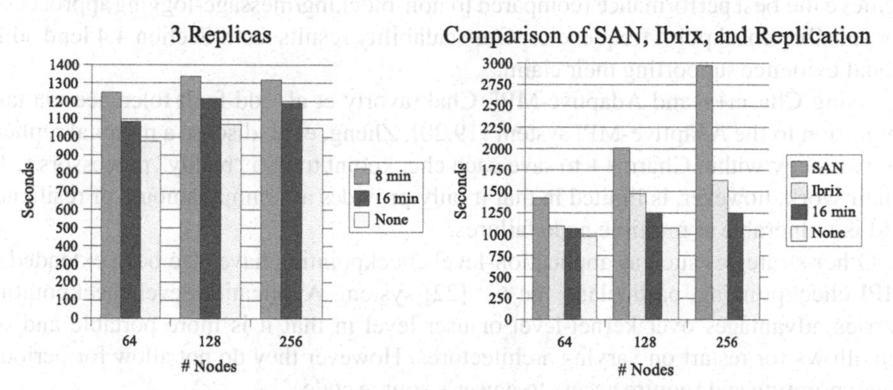

(c) HPL with three replicas per checkpoint. (d) HPL with four replicas per checkpoint compared with EMC SAN and Ibrix PFS.

Fig. 4. Scalability tests using the HPL benchmark

the base run time without checkpointing provides a measure of the overhead involved at each replication level. From Figure 4(a) we can see that the replication overhead is quite low - only approximately 6% for 256 nodes or 3% for 64 nodes (at 16 minute checkpoint intervals). Similar results can be seen at 2, and 3 replicas with only a minimal increase in overhead for each replication increase.

The most important results, however, are those shown in Figure 4(d). Here we include the overhead data with 4 replicas (labeled "16 min" in Figure 4(d)) as well as with checkpointing directly to the SAN (a common strategy in nearly all MPI checkpointing literature) and the Ibrix parallel file system. In every case, checkpoints are taken at 16 minute intervals. As can be seen, the overhead of checkpointing directly to a SAN not only dwarfs that of our distributed replication strategy but also nullifies the efficacy of additional processors for large clusters. The Ibrix file system, while scaling much better than the EMC SAN, is quickly overwhelmed as the ratio of compute nodes to segment servers increases. Indeed, the overhead of saving checkpoints to the Ibrix parallel file systems for cluster sizes of 128 and 256 nodes is 37.5% and 55% respectively, while our replication strategy results in overheads of only 15.4% and 18.7%.

5 Related Work

Other MPI implementations aside from LAM/MPI have been extended with checkpointing support. MPICH-GM, a Myrinet specific implementation of MPICH has been extended to support user-level checkpointing [17]. Similarly, Gao et al. [1] demonstrate a kernel-level checkpointing scheme for Infiniband (MVAPICH2) that is based on the BLCR kernel module [9]. DejaVu [18] implements an incremental checkpoint/migration scheme that is able to incrementally capture the differences between two checkpoints to minimize the size of an individual checkpoint.

Coti, et al. implemented a blocking coordinated protocol within MPICH2 [15]. Their observations suggested that for high speed computational clusters blocking approaches achieve the best performance (compared to non-blocking/message-logging approaches) for sensible checkpoint frequencies. Our scalability results from Section 4.4 lend additional evidence supporting their claim.

Using Charm++ and Adaptive-MPI, Chakravorty et al. add fault tolerance via task migration to the Adaptive-MPI system [19,20]. Zheng, et al. discuss a minimal replication strategy within Charm++ to save each checkpoint to two "buddy" processors [21]. Their work, however, is limited in that it only provides a minimal amount of resiliency and is vulnerable to multiple node failures.

Other strategies such as application-level checkpointing have also been extended to MPI checkpointing, particularly the C^3 [22] system. Application-level checkpointing carries advantages over kernel-level or user-level in that it is more portable and often allows for restart on varying architectures. However they do not allow for periodic checkpointing and require access to a user's source code.

Our work differs from the above in that we handle checkpoint redundancy for added resiliency in the presence of node failures. Our checkpointing solution does not rely on the existence of network storage for checkpointing. The absence of network storage allows for improved scalability and also reduced checkpoint intervals (where desired).

6 Conclusions

We have shown that it is possible to effectively checkpoint MPI applications using the LAM/MPI implementation with low overhead. Previous checkpointing implementations have typically neglected the issue of checkpoint replication. We comprehensively addressed this issue with a comparison against all major storage techniques, including commercial SAN strategies and a commercial parallel file system. Our replication implementation has proven to be highly effective and resilient to node failures.

Further, we showed that our replication strategy is highly scalable. Where previous work discussed within the literature typically tests scalability up to 16 nodes, we have demonstrated low overhead up to 256 nodes with more realistic checkpoint image sizes of 1 GB per node. Our work enables more effective use of resources without any reliance on network storage. We hope to continue this work with a greater interest in applying our replication strategies toward fault-tolerant HPC. By combining our checkpoint/restart and migration system with a fully fault-tolerant MPI, even greater resource utilization would be possible while still maintaining user-transparency.

References

1. Gao, Q., Yu, W., Huang, W., Panda, D.K.: Application-Transparent Checkpoint/Restart for MPI Programs over InfiniBand. In: ICPP 2006. Proceedings of the 35th International Conference on Parallel Processing, Columbus, OH (2006)
2. Sankaran, S., Squyres, J.M., Barrett, B., Lumsdaine, A., Duell, J., Hargrove, P., Roman, E.: The LAM/MPI Checkpoint/Restart Framework: System-Initiated Checkpointing. In: Proceedings, LACSI Symposium, Sante Fe, New Mexico, USA (2003)
3. Sankaran, S., Squyres, J.M., Barrett, B., Lumsdaine, A., Duell, J., Hargrove, P., Roman, E.: The LAM/MPI Checkpoint/Restart Framework: System-Initiated Checkpointing. International Journal of High Performance Computing Applications 19(4), 479–493 (2005)
4. Burns, G., Daoud, R., Vaigl, J.: LAM: An Open Cluster Environment for MPI. In: Proceedings of Supercomputing Symposium, pp. 379–386 (1994)
5. Squyres, J.M., Lumsdaine, A.: A Component Architecture for LAM/MPI. In: Dongarra, J.J., Laforenza, D., Orlando, S. (eds.) Recent Advances in Parallel Virtual Machine and Message Passing Interface. LNCS, vol. 2840, pp. 379–387. Springer, Heidelberg (2003)
6. Elnozahy, E.N.M., Alvisi, L., Wang, Y.M., Johnson, D.B.: A Survey of Rollback-Recovery Protocols in Message-Passing Systems. ACM Comput. Surv. 34(3), 375–408 (2002)
7. Zhang, Y., Wong, D., Zheng, W.: User-Level Checkpoint and Recovery for LAM/MPI. SIGOPS Oper. Syst. Rev. 39(3), 72–81 (2005)
8. Wang, C., Mueller, F., Engelmann, C., Scott, S.L.: A Job Pause Service under LAM/MPI+BLCR for Transparent Fault Tolerance. In: IPDPS 2007. Proceedings of 21st IEEE International Parallel and Distributed Processing Symposium, Long Beach, CA, USA (2007), Long Beach, CA, USA (2007)
9. Duell, J.: The Design and Implementation of Berkeley Lab's Linux Checkpoint/Restart (2003), http://old-www.nersc.gov/research/FTG/checkpoint/reports.html
10. Cao, J., Li, Y., Guo, M.: Process Migration for MPI Applications based on Coordinated Checkpoint. In: ICPADS 2005. Proceedings of the 11th International Conference on Parallel and Distributed Systems, pp. 306–312. IEEE Computer Society Press, Los Alamitos (2005)

11. Zandy, V.: Ckpt: User-Level Checkpointing (2005), http://www.cs.wisc.edu/~zandy/ckpt/
12. Walters, J., Chaudhary, V.: A Comprehensive User-level Checkpointing Strategy for MPI Applications. Technical Report 2007-1, University at Buffalo, The State University of New York, Buffalo, NY (2007)
13. Bailey, D., Barszcz, E., Barton, J., Browning, D., Carter, R., Dagum, L., Fatoohi, R., Frederickson, P., Lasinski, T., Schreiber, R., Simon, H., Venkatakrishnan, V., Weeratunga, S.: The NAS Parallel Benchmarks. International Journal of High Performance Computing Applications 5(3), 63–73 (1991)
14. Dongarra, J.J., Luszczek, P., Petitet, A.: The LINPACK Benchmark: Past, Present, and Future. Concurrency and Computation: Practice and Experience 15, 1–18 (2003)
15. Coti, C., Herault, T., Lemarinier, P., Pilard, L., Rezmerita, A., Rodriguez, E., Cappello, F.: MPI Tools and Performance Studies—Blocking vs. Non-Blocking Coordinated Checkpointing for Large-Scale Fault Tolerant MPI. In: Löwe, W., Südholt, M. (eds.) SC 2006. LNCS, vol. 4089, Springer, Heidelberg (2006)
16. Lv, Q., Cao, P., Cohen, E., Li, K., Shenker, S.: Search and Replication in Unstructured Peer-to-Peer Networks. In: ICS 2002. Proceedings of the 16th international conference on Supercomputing, pp. 84–95. ACM Press, New York (2002)
17. Jung, H., Shin, D., Han, H., Kim, J.W., Yeom, H.Y., Lee, J.: Design and Implementation of Multiple Fault-Tolerant MPI over Myrinet (M^3). In: Gschwind, T., Aßmann, U., Nierstrasz, O. (eds.) SC 2005. LNCS, vol. 3628, p. 32. Springer, Heidelberg (2005)
18. Ruscio, J., Heffner, M., Varadarajan, S.: DejaVu: Transparent User-Level Checkpointing, Migration, and Recovery for Distributed Systems. In: Proceedings of the 21st IEEE International Parallel and Distributed Processing Symposium (IPDPS) 2007, Long Beach, CA, USA (2007)
19. Chakravorty, S., Mendes, C., Kalé, L.V.: Proactive Fault Tolerance in MPI Applications via Task Migration. In: Robert, Y., Parashar, M., Badrinath, R., Prasanna, V.K. (eds.) HiPC 2006. LNCS, vol. 4297, Springer, Heidelberg (2006)
20. Chakravorty, S., Kalé, L.: A Fault Tolerance Protocol with Fast Fault Recovery. In: Proceedings of 21st IEEE International Parallel and Distributed Processing Symposium (IPDPS) 2007, Long Beach, CA (2007)
21. Zheng, G., Shi, L., Kalé, L.V.: FTC-Charm++: An In-Memory Checkpoint-Based Fault Tolerant Runtime for Charm++ and MPI. In: CLUSTER, pp. 93–103 (2004)
22. Bronevetsky, G., Marques, D., Pingali, K., Stodghill, P.: Collective Operations in Application-Level Fault-Tolerant MPI. In: ICS 2003. Proceedings of the 17th annual international conference on Supercomputing, pp. 234–243. ACM Press, New York (2003)

Towards a Transparent Data Access Model for the GridRPC Paradigm

Gabriel Antoniu[1], Eddy Caron[2], Frédéric Desprez[2], Aurélia Fèvre[2], and Mathieu Jan[3,*]

[1] INRIA/IRISA, Campus de Beaulieu, 35042 Rennes, France
Gabriel.Antoniu@irisa.fr
[2] LIP/ENS-Lyon/INRIA, 46 Allée d'Italie, 69364 Lyon, France
{Eddy.Caron,Frederic.Desprez,Aurelia.Fevre}@ens-lyon.fr
[3] LRI/INRIA, University of Paris South, 91405 Orsay, France
Mathieu.Jan@lri.fr

Abstract. As grids become more and more attractive for solving complex problems with high computational and storage requirements, the need for adequate grid programming models is considerable. To this purpose, the GridRPC model has been proposed as a grid version of the classical RPC paradigm, with the goal to build NES (Network-Enabled Server) environments. In this model, data management has not been defined and is now explicitly left at the user's charge. The contribution of this paper is to enhance data management in NES by introducing a *transparent data access model*, available through the concept of grid data-sharing service. Data management is totally delegated to the service, whereas the applications simply access shared data via global identifiers. We illustrate our approach using the DIET GridRPC middleware and the JUXMEM data-sharing service. Notably, our experiments performed on the Grid'5000 using a real-life application show the efficiency of using JUXMEM for managing persitent data in the GridRPC model: application execution times in a grid environment are of the same order as in a cluster environment.

1 Introduction

Computational grids have recently become increasingly attractive, as they adequately address the growing demand for resources of today's scientific applications. Thanks to the fast growth of high-bandwidth wide-area networks, grids efficiently aggregate various heterogeneous resources (processors, storage devices, network links, etc.) belonging to distinct organizations. This increasing computing power, available from multiple geographically distributed sites, increases the grid's usefulness in efficiently solving complex problems. Multi-parametric applications, for instance, which consist in applying the same algorithm to different input data, can benefit from an efficient use of grid computing infrastructures.

Running such applications on large-scale grid infrastructures requires the use of adequate programming paradigms. The *Grid Remote Procedure Call* (GridRPC) [1] approach provides such a paradigm, which extends the classical RPC model by enabling

* This author's work has mainly been done at INRIA/IRISA.

S. Aluru et al. (Eds.): HiPC 2007, LNCS 4873, pp. 269–284, 2007.

asynchronous, coarse-grained parallel tasking. GridRPC seems to be a good approach to build NES computing environments (for Network-Enabled Servers). In such systems, clients can submit problems to one (possibly distributed) agent, which selects the best server to use among a large set of candidates.

A team of researchers of the Global Grid Forum (GGF [1]) has defined a standard API for the GridRPC paradigm [2]. However, in this specification, data management has been left as an open (although fundamental) issue. For instance, data transfer in the distributed environment is left to the user, who must explicitly move them back and forth between clients and servers. This clearly increases the program complexity, especially as the number of servers used to solve a problem increases.

In this paper, we define a model for transparent access to shared data in GridRPC environments. In this model, the data-sharing infrastructure automatically manages data localization, transfer, as well as consistent data replication. We illustrate our approach with an implementation using the DIET [3] GridRPC middleware and the JUXMEM [4] grid data-sharing service. We evaluate our approach through experiments realized on the Grid'5000 [5] testbed.

The remainder of the paper is organized as follows. Section 2 introduces the GridRPC model, presents the requirements of a sample application with respect to data management, then briefly describes previous attempts to solve data management issues in NES systems. Section 3 describes our transparent data access approach provided by our concept of grid data-sharing service. Section 4 presents the implementation of our proposal, using JUXMEM and DIET. Section 5 presents and discusses our experimental results using a real-life application. Finally, Section 6 concludes the paper and suggests possible directions for additional research.

2 Data Management in the GridRPC Model

Various programming models have been proposed in order to reduce the programming complexity of grid applications. The GridRPC model is such an ongoing work carried out by the Open Grid Forum (OGF), with the goal of standardizing and implementing the Remote Procedure Call (RPC) programming model for grid computing.

2.1 The GridRPC Model

The GridRPC model enhances the classical RPC programming model with the ability to invoke asynchronous, coarse-grained parallel tasks. Requests for remote computations may indeed generate parallel processing, however this server-level parallelism remains hidden to the client.

The GridRPC approach has been defined in the GRIDRPC-WG [6] working group of the GGF. The goal of this group is to specify the syntax and the programming interface at the client level [1]. This is meant to enhance the portability of GridRPC applications to various GridRPC middleware.

[1] GGF, recently merged with EGA (Enterprise Grid Alliance) to create the OGF (Open Grid Forum).

The GridRPC model aims at serving as a basis for software infrastructures called Network-Enabled Servers (NES). Such infrastructures allow multiple applications to concurrently run on a shared set of grid resources. Examples of middleware that implement the GridRPC specification are Ninf-G [7], NetSolve [8], GridSolve [9], DIET [3], and OmniRPC [10].

Note that the GridRPC model knows the target server. Nevertheless some GridRPC middleware proposes to discover the best server automatically. In this case, *servers* register *services* to a *directory*. To invoke a service, instead of using the server given by GridRPC call function, *clients* bypass this parameter and look for a suitable, possibly "the best" server according to some performance metric. This selection is made out of a set of candidates proposed by the directory. GridRPC does not define any standard for the underlying resource discovery mechanism. The server selection is performed by one or several *agents* pr *schedulers*. The decision is usually made based on performance information provided by an information service. Informations can be static, such as processor speed or size of the memory, but also dynamic: available services, server load, input data location, etc. Based on this information, the agents make their decisions so as to optimize the overall throughput of the platform.

Two fundamental concepts in the GridRPC model are the *function handle* and the *session ID*. The function handle represents a binding between a service name and an instance of that service available on a given server. Function handles are returned by agents to clients. Once a particular function-to-server mapping has been established, all GridRPC calls of a client will be executed on the server specified by that function handle. A session ID is associated to each asynchronous GridRPC call and allows to retrieve the status of the request, wait for the call to complete, etc. Based on these two concepts, the interface of the GridRPC model mainly consists of the following two functions: `grpc_call` and `grpc_async`, which allow to make synchronous and asynchronous GridRPC calls respectively.

As regards data, most GridRPC middleware systems specify three *access modes* (also known as *access specifiers*) for parameters of a GridRPC call: 1) `in` data for input parameters that are not allowed to be modified by servers; 2)`inout` data for input parameters that can be modified by the server; 3) `out` data for output parameters produced by the server.

2.2 Requirements for Data Management in the GridRPC Model

To illustrate the requirements related to data management in the GridRPC model, we have selected the Grid-TLSE project [11]. This application aims at designing a Web portal exposing expertise about sparse matrix manipulation. Through this portal, the user may gather statistics from runs of various sophisticated sparse matrix algorithms on specific data. The input data are either submitted by the user, or picked up from a matrix collection available on the site. In general, matrix sizes can vary from a few megabytes to hundreds of megabytes. The Grid-TLSE application uses the DIET GridRPC middleware to distribute tasks over the underlying grid infrastructure. Each such task consists in executing a parallel solver, such as MUMPS [12], over a matrix, with fixed parameters. We focus on the MUMPS solver for our experiments (see Section 5.2).

When using Grid-TLSE, a typical scenario consists in determining the *ordering sensitivity* of a class of solvers, that is, how performance is impacted by the matrix traversal order. It consists of three phases. Phase 1 exercises all possible internal orderings in turn. Phase 2 computes a suitable metric reflecting the performance parameters under study for each run: effective FLOPS, effective memory usage, overall computation time, etc. Phase 3 collects the evaluation of this metric for all combinations of solvers/orderings and reports the final ranking to the user. If phase 1 requires exercising n different kinds of orders with m different kinds of solvers, then $m \times n$ executions are to be performed, using the same input data. If the server does not provide *persistent storage*, the matrix has to be sent $m \times n$ times to the server! If the server provided persistent storage, the data would be sent only once. Second, if the various pairs solvers/orderings are handled by different servers in phase 2 and 3, then *transparent* and *consistent* data transfer or replication across servers should be provided by the data management service. Finally, as the number of solvers/orderings is potentially large, many nodes are used. This increases the probability for faults to occur, which makes the use of *fault tolerant* algorithms to manage data mandatory.

Based on this application example, we can draw the requirements for a data management service for the GridRPC model.

Persistent storage. Clients should be able to invoke services on input data that is already present on the grid infrastructure, to avoid repeated data transfers to servers.

Passing arguments by reference for shared data. This is a consequence of the above requirement, as clients need a means to reference data which is shared by multiple GridRPC calls. Consequently, data consistency must be guaranteed in case of concurrent accesses.

Transparent data localization and transfer. Such a transparency would simplify the use of the GridRPC paradigm at a large scale, as developpers would no longer need to explicitly move data.

Efficient communication. An efficient use of the available bandwidth for data transfers requires to adequatly manage data granularity: only the data needed to perform computations should be copied or moved.

GridRPC interoperability. Any solution addressing the previous issues needs to be compatible with the existing core API of the GridRPC model. Thus current applications can take advantage of any improvement in data management without modifications.

2.3 Current Proposals for Data Management in the GridRPC Model

In the current GridRPC model, as defined by OGF, data persistence is not yet provided and has been left as an open issue. Therefore, output data of a computation (`inout` and `out`) are systematically sent to the client, whereas input data (`in`) are destroyed on the server. Hence, data needs to be transfered again if needed for another computation. Moreover, if data are required on multiple servers at the same time, multiple transfers from the client are needed.

The issue of data management in the GridRPC model has however been recognised as a topic of major interest. The very first proposal related to data management relies on the concept of *request sequencing* [13]. This feature consists in scheduling a sequence of GridRPC calls made by a client on a given server. In the client program, a sequence is identified by keywords `begin_sequence` and `end_sequence`. Data movements due to dependencies in calls between such keywords are then optimized. Request sequencing has been implemented in NetSolve and Ninf. To enable the calls of a sequence to be solved in parallel on two different servers, NetSolve has been enhanced [14] with data redistribution between servers (which however requires explicit calls in the NetSolve client application).

Another approach for data management relies on distributed storage infrastructure, such as Internet Backplane Protocol (IBP [15]). In this approach, clients send data to storage servers, which retrieve data as needed. NetSolve has been modified in such a way. However, data is still explicitly transfered to/from the storage servers at the application level. Besides, no support for data replication and consistency management, nor for fault tolerance is provided.

Finally, other GridRPC systems have developped ad-hoc, specific mechanisms for data management. The OmniRPC GridRPC middleware supports a static persistence model for input data of a set of GridRPC calls [16]. The user has to manually define a initialization procedure to indicate which input data should be sent and stored prior to computations. Then, these data can be reused for subsequent calls. In an earlier version, the DIET GridRPC middleware relies on an internal data management system, called Data Tree Manager (DTM), which allows to store persistent data [17] on the computing servers. However, as in both cases ad-hoc solutions are used to handle data persistence, GridRPC interoperability cannot be guaranteed, as data cannot be shared among multiple GridRPC middleware frameworks. Besides, none of these solutions addresses fault tolerance and consistent replication.

Based on such preliminary efforts, an attempt to standardize data management in NES is currently being pursued within the framework of the GridRPC working group of the OGF [2]. It relies on the concept of *data handle*, which abstracts a given data as well as its location. In addition to the possibility of referencing data stored inside external storage systems, transparent access to data is also envisioned. However, replication, consistency guarantees and fault tolerance issues have not been addressed yet.

3 Our Approach: A Transparent Data Access Model

3.1 The Concept of Data-Sharing Service

Let us recall that one of the major goals of the grid concept is to provide an easy access to the underlying resources, in a *transparent* way. The user should not need to be aware of the localization of the resources allocated to applications. When applied to the management of the data used and produced by applications, this principle means that the grid infrastructure should automatically handle data storage and data replication and/or transfer among clients, computing servers and storage servers as needed. It should also transparently provide fault tolerance and data consistency guarantees in such dynamic, large-scale, distributed environments.

In order to achieve a real virtualization of the management of large-scale distributed data, a step forward has been made by the proposal of a *transparent data access model*, as a part of the concept of *grid data-sharing service* [18]. In this *transparent data access* approach, the user accesses data via global identifiers, which allow to do argument passing by reference for shared data. The service which implements this model handles data localization and transfer without any help from the programmer. The data sharing service concept is based on a hybrid approach inspired by DSM systems (for transparent access to data and consistency management) and peer-to-peer (P2P) systems (for their scalability and volatility tolerance). An illustration of this concept has been realized through the JUXMEM software experimental platform [4]. The service specification includes three main properties.

Persistence. The data sharing service provides persistent data storage and allows the applications to reuse previously produced data, by avoiding repeated data transfers between clients and servers.

Data consistency. Data can be read, but also *updated* by the different codes. When data is replicated on multiple sites, the service has to ensure the consistency of the different replicas, based on previously defined *consistency models and protocols*.

Fault tolerance. The service has to keep data available despite disconnections and failures, e.g. through the transparent use of failure detection mechanisms and replication techniques.

Let us note that these properties match well the requirements for data management in the GridRPC model, as discussed in Section 2.2. We therefore propose to jointly use the two approaches. In this paper, we show how *persistence* can be provided in a transparent way. *Data consistency* is ensured by providing a multi-protocol framework allowing various consistency models and protocols to be implemented. JUXMEM currently supports the *entry consistency* model through a hierarchical, *fault-tolerant* protocol. A description of the concepts and technical details related to data consistency and fault tolerance is beyond the focus of this paper. The corresponding mechanisms have been detailed in [19].

3.2 Overview of the JUXMEM Data-Sharing Service

The JUXMEM [4] software experimental platform illustrates the concept of data-sharing service. The architecture of the service has been designed so as to address the properties mentioned in Section 3.1. JUXMEM's architecture mirrors a grid consisting of a federation of distributed clusters and it is therefore expressed in terms of *hierarchical* groups. The goal is to accurately take into account the latency hierarchy of the physical network topology, to take advantage of the low-latency links within the clusters and reduce higher-latency, inter-cluster communications. All nodes participating to the data-sharing service network overlay are members of the JUXMEM group. All members of the JUXMEM group that belong to the same physical cluster form a *cluster group*.

Any cluster group consists of *provider* nodes which supply memory for data storage. Each cluster group is managed by a special peer, called a *manager*. Managers make up the backbone of a given JUXMEM overlay and handle the propagation of memory allocation requests. Any node (including providers) may use the service to allocate, read or write data as *clients*, in a peer-to-peer approach. Any data stored in JUXMEM is transparently accessed through a global, location-independent identifier, which designates a

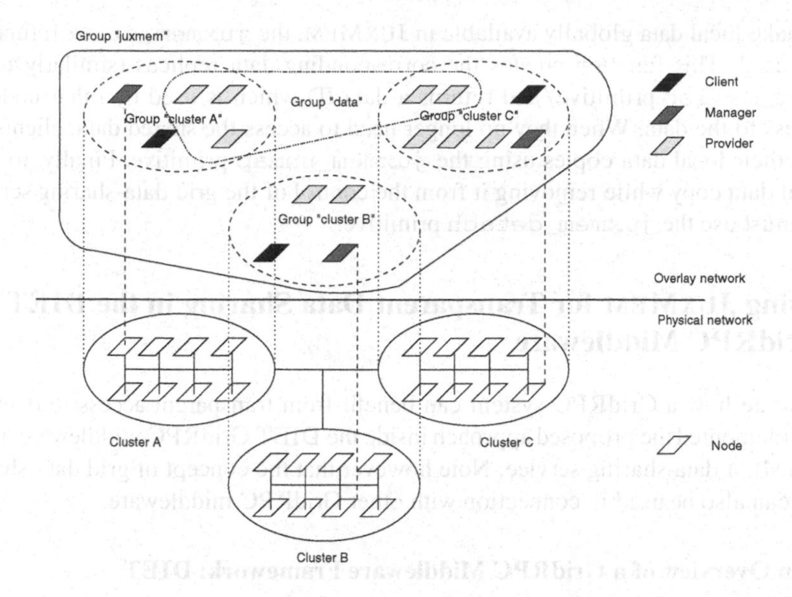

Fig. 1. Hierarchy of the entities in the network overlay defined by JUXMEM

specific *data* group that includes all replicas of that data. These replicas are kept consistent despite possible failures and disconnections [19]. This software architecture has been implemented using the JXTA [20] generic P2P platform.

3.3 JUXMEM from the User's Perspective

The programming interface proposed by the JUXMEM grid data-sharing service provides users with classical functions to allocate and map/unmap memory blocks, such as `juxmem_malloc`, `juxmem_calloc`, etc. When allocating a memory block, the client has to specify: 1) on how many clusters the data should be replicated; 2) on how many providers in each cluster the data should be replicated; 3) the consistency protocol that should be used to manage this data. The allocation operation returns a global data ID. This ID can be used by other nodes in order to access existing data through the use of the `juxmem_mmap` function. It is the responsibility of the implementation of the grid data-sharing service to localize the data and perform the necessary data transfers based on this ID. This is how a grid data-sharing service provides a transparent access to data.

According to the entry consistency model implemented by JUXMEM, processes that need to access data need to properly synchronize by acquiring a lock associated to that data. This is done by calling `juxmem_acquire_read` (prior to a read access) or `juxmem_acquire` (prior to a write access). Note that `juxmem_acquire_read` allows multiple readers to simultaneously access the same data. The `juxmem_release` primitive must be called after the access, to release the lock. These synchronization primitives allow the implementation to provide consistency guarantees according to the consistency protocol specified by the user at the allocation time of the data.

To make local data globally available in JUXMEM, the `juxmem_attach` function can be used. This function creates the corresponding data replicas (similarly to the `juxmem_malloc` primitive) and returns a data ID which is used by other nodes to get access to the data. When they no longer need to access the shared data, clients can remove their local data copies using the `juxmem_unmap` primitive. Finally, to keep the local data copy while removing it from the control of the grid data-sharing service, clients must use the `juxmem_detach` primitive.

4 Using JUXMEM for Transparent Data Sharing in the DIET GridRPC Middleware

To illustrate how a GridRPC system can benefit from transparent access to data, we have implemented the proposed approach inside the DIET GridRPC middleware, using the JUXMEM data-sharing service. Note however that the concept of grid data-sharing service can also be used in connection with other GridRPC middleware.

4.1 An Overview of a GridRPC Middleware Framework: DIET

The Distributed Interactive Engineering Toolbox (DIET) platform [3] is a GridRPC middleware, whose architecture is described on Figure 2. It relies on the following en-

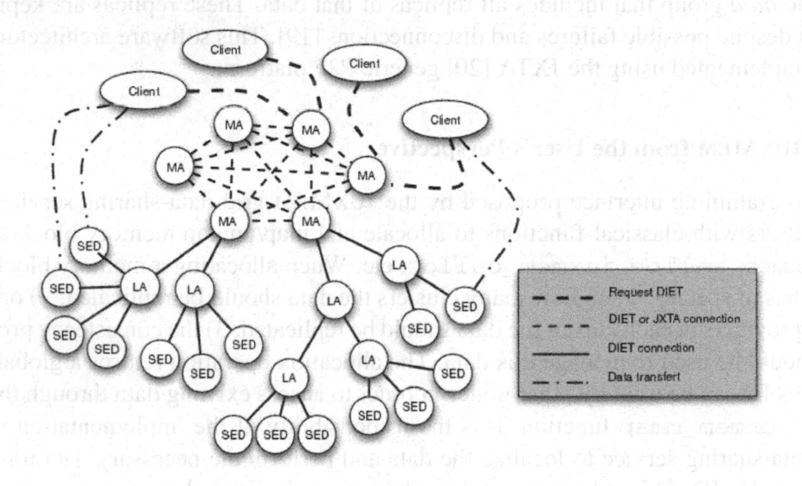

Fig. 2. The hierarchical organization of DIET

tities. A *Client* is an application which uses DIET to solve problems. *Agents* receive computation requests from clients. A request is a generic description of the problem to be solved with data information (type, size, etc.). Agents collect the computational capabilities of the available servers, and selects the *best* server according to the given request. Eventually, the reference of the selected server is returned to the client, which can then directly submit its request to this server. As opposed to other GridRPC middleware, for scalability purpose, agents can be organized in a set of trees forming a forest of local agents (LA) rooted at a master agent (MA). *The Server Daemon (SeD)* encapsulates a computational server and makes it available to its parent LA. It also provides the potential clients with an interface for submitting their requests.

Like other GridRPC middleware, DIET specifies three access modes for each data involved in a computation (see section 2.1).

4.2 How DIET Uses JUXMEM to Manage Data

In our work, DIET *internally* uses JUXMEM whenever a data is marked as persistent. However, we distinguish two cases for persistent data. If the DIET client needs to access persistent data at the end of the computation, the persistence mode is set to `PERSISTENT_RETURN`. Otherwise, it is set to `PERSISTENT`.

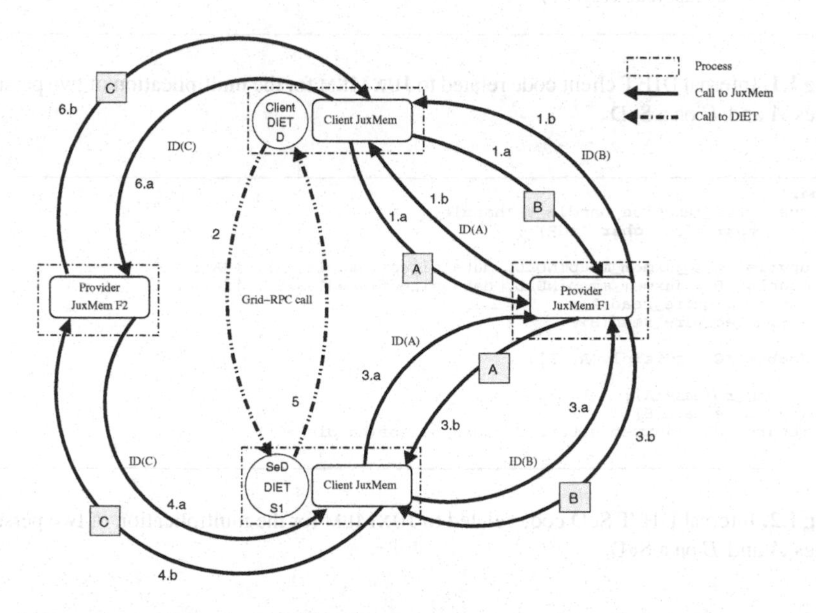

Fig. 3. Multiplication of two matrices by a DIET client configured to use JUXMEM for persistent data management

Listings 1.1 and 1.2 show an example of how DIET internally uses JUXMEM to manage data for the multiplication of two matrices A and B. The output of the computation produces the matrix C. Figure 3 presents the entities involved: one DIET client D, one DIET SeD $S1$ and two JUXMEM providers $F1$ and $F2$[2]. Let us assume that all matrices are persistent. First, input matrices are stored into JUXMEM by the client (step 1 of Figure 3, lines 5 and 6 of Listing 1.1), and their IDs $ID(A)$ and $ID(B)$ are sent in the computational request to $S1$ (step 2, line 8). On the server side, these IDs are used to locally map and acquire the input matrices in read mode (step 3, lines 5 to 8 of Listing 1.2). Then, the computation produces matrix C (line 10). Therefore, the read

[2] For the sake of clarity on the figure, however in practice $F2$ can be equal to $F1$.

lock on matrices A and B is released (lines 12 and 13), and matrix C is attached inside JUXMEM (step 4 and line 14). Its identifier ($ID(C)$) is sent back to the client D (step 5), so that it can be locally mapped and acquired in read mode by the client (steps 6, lines 10 to 12 of Listing 1.1).

```
1  grpc_error_t
2  grpc_call (grpc_function_handle_t *handle) {
3    grpc_serveur_t *SeD = request_submission(handle);
4    ...
5    char *idA = juxmem_attach(handle->A, data_sizeof(handle->A));
6    char *idB = juxmem_attach(handle->B, data_sizeof(handle->B));
7    ...
8    char *idC = SeD->remote_solve(multiply, idA, idB);
9    ...
10   juxmem_mmap(handle->C, data_sizeof(handle->C), idC);
11   juxmem_acquire_read(handle->C);
12   juxmem_release(handle->C);
13 }
```

Listing 1.1. Internal DIET client code related to JUXMEM for the multiplication of two persistent matrices A and B on a SeD

```
1  char*
2  solve (grpc_function_handle_t *handle,
3         char *idA, char *idB) {
4    ...
5    double *A = juxmem_mmap(NULL, data_sizeof(handle->A), idA);
6    double *B = juxmem_mmap(NULL, data_sizeof(handle->B), idB);
7    juxmem_acquire_read(A);
8    juxmem_acquire_read(B);
9    ...
10   double *C = multiply(A, B);
11   ...
12   juxmem_release(A);
13   juxmem_release(B);
14   return idC = juxmem_attach(C, data_sizeof(handle->C));
15 }
```

Listing 1.2. Internal DIET SeD code related to JUXMEM for the multiplication of two persistent matrices A and B on a SeD

Table 1 summarizes the interaction between DIET and JUXMEM in each case, depending on the data access mode (e.g. `in`, `inout`, `out`) on both client/server side. In the previous example, matrices A and B are `in` data, and matrix C is an `out` data. Note that for `inout` and `out` data, calls to JUXMEM are executed after the computation on the client side only if the persistent mode is `PERSISTENT_RETURN`.

Modifications performed inside the DIET GridRPC middleware to use JUXMEM for the management of persistent data are small. They consist of 200 lines of C++ code, activated whenever DIET is configured to use JUXMEM. Consequently, DIET is linked with the C/C++ binding of JUXMEM. In our setting, DIET clients or SeDs use JUXMEM's API to store/retrieve data, thereby acting as JUXMEM clients. Note also that our solution supports GridRPC interoperability, DIET simply uses JUXMEM's API, with no extra code for data management.

Table 1. Use of JUXMEM inside DIET for `in`, `inout` and `out` persistent data on client/server side, before and after a computation. The `juxmem` prefix has been omitted.

	Client side		SeD side	
Computation	Before	After	Before	After
in	attach; msync; detach;		mmap; acquire_read;	release; unmap;
inout	attach; msync;	acquire_read; release;	mmap; acquire;	
out		mmap; acquire_read; release;		attach; msync; unmap;

5 Experimental Evaluations

In this section, we present the experimental evaluation of our JUXMEM-based data management solution inside DIET.

5.1 Experimental Conditions

We performed tests using 4 clusters (Rennes, Orsay, Toulouse and Lyon) of the French Grid'5000 testbed [5], using a total number of 3 sites simultaneously for a total number of 129 nodes. Grid'5000 is an experimental grid platform consisting of 9 sites (clusters) geographically distributed in France, whose aim is to gather a total of 5,000 CPUs in the near future. The nodes used for our experiments consist of machines using dual (2.2, 2.4, 2.6 GHz) AMD Opteron, outfitted with 2 GB of RAM each, and running a 2.6 version Linux kernel; the network layer used is a Giga Ethernet (1 Gb/s) network inside each cluster of Grid'5000. Between clusters, links of 10 Gb/s are used and the latency ranges from 4,5 ms to 10 ms.

Tests were executed using JUXMEM 0.3 and DIET 2.1. All benchmarks are compiled using `gcc` 4.0 with the `-O2` level of optimization. As regards deployment, we used the ADAGE [21] generic deployment tool for JUXMEM and GoDIET [22] for DIET.

5.2 Experiments Using MUMPS: A Sparse Parallel Solver

Our goal is to demonstrate and measure the benefits of the management of persistent data by JUXMEM, in terms of impact on the overall execution time of a real-life application. We focus on MUMPS ("MUltifrontal Massively Parallel Solver"), a package for solving systems of linear equations of the form $\mathbf{Ax} = \mathbf{b}$, where $\mathbf{A}$ is a square sparse matrix that can be either asymmetric, symmetric positive definite, or general symmetric. MUMPS uses a multifrontal technique which is a direct method based on either the LU or the LDL^T factorization of the matrix. We refer the reader to the paper [23] for full details of these techniques.

We performed 3 sets of experiments using MUMPS (noted $E1$, $E2$ and $E3$). $E1$ has been performed in a one-cluster environment, whereas $E2$ and $E3$ are performed in a multi-cluster environment, using the Grid'5000 testbed, with $E3$ at a larger scale compared to $E2$. In all our experiments, a client loops 32 times on a synchronous GridRPC call to MUMPS, with the aforementioned linear equation to solve. A and b are input data (data access mode set to `in`) whereas x is an output data. Between each GridRPC call, b is changed according to the resolution needs while A is unchanged and its persistence mode is set to `PERSISTENT`. Therefore note that when JUXMEM is used by DIET for the management of matrices A, DIET calls JUXMEM primitives according to the first row of Table 1. For all experiments we used 2 different sizes for the **A** matrix: a medium size of 22 MB ($A1$) and a larger size of 52 MB ($A2$). Let us stress the difficulty of setting up such kind of experiments for a real-life, complex application, using an environment which relies on 2 different runtime software (JUXMEM and DIET), based on different technologies (JXTA and CORBA respectively) and using different deployment tools that need to interact with each other.

As a first experiment performed inside a single Grid'5000 cluster, we simply deployed a DIET hierarchy made of 1 MA, 1 LA and 1 SeD, as well as a JUXMEM network made of 1 provider and 1 manager. The goal of this experiment $E1$ is to measure the overhead of using JUXMEM for data management. Results show that if DIET is configured to use JUXMEM to store persistent data, the total execution time of all calls slightly increases, compared to DIET configured without JUXMEM: from 36.6 to 41.3 seconds with $A1$ and from 957 to 961 seconds with $A2$. We can argue that the overhead of using JUXMEM for data management of large matrices inside DIET is therefore low: it is less than 1 % with $A2$. Note however that this overhead increases when using smaller matrices, e.g. it reaches 13 % for matrix $A1$.

In a second experiment, our goal is to measure the (expected!) benefits of using JUXMEM for transparently managing persistent data in grid environment (however at a small scale). To do this, we deployed a 3-cluster configuration. In each cluster, we deploy a DIET hierarchy made of 1 LA and 1 SeD and a JUXMEM network made of 1 provider and 1 manager. We use 3 clusters of the Grid'5000 testbed, namely Lyon, Toulouse and Rennes. Results show a clear advantage to use JUXMEM, as the total execution time of the application is reduced by 42 % with $A1$ and by 38 % with $A2$, compared to results obtained for DIET configured without JUXMEM (see Table 2). Compared to the $E1$ experiment, the smaller execution time with $A2$, when JUXMEM is used, is explained by the difference of processor performance on the three sites that are used for the computations.

Finally, we performed a third experiment ($E3$) similar to $E2$, where we increased the configuration sizes, by using 32 SeDs and 8 JUXMEM providers in each cluster. The goal of this experiment is to measure the impact of an increasing number of SeDs on the performance of JUXMEM (as this leads to an increasing number of JUXMEM clients accessing the data). With $A1$, the total execution time is the same for both configurations (DIET configured with or without JUXMEM): 103 seconds. With the larger $A2$ matrix, this time is reduced by 38 % when JUXMEM is used (843 seconds), compared to DIET configured without JUXMEM (1358 seconds). Note that these results are averaged based on 4 runs, since the DIET agent may take different scheduling decisions by choosing different nodes (and clusters) from one computation to another. This also explains the

Table 2. Total execution time in seconds of a MUMPS application when DIET is configured to use JUXMEM or not, for 2 different matrices $A1$ and $A2$

Matrix	$A1$		$A2$	
DIET configured	Without JUXMEM	With JUXMEM	Without JUXMEM	With JUXMEM
Experiment $E1$	36.6	41.3	957	961
Experiment $E2$	92.6	53.7	1420	880
Experiment $E3$	103	103	1358	843

difference between results obtained for $A2$ in this experiment and experiment $E2$: the number of GridRPC calls performed on 1 site may change between runs.

Finally, Table 2 summarizes obtained results for experiments $E1$, $E2$ and $E3$ based on MUMPS. Notably, these results demonstrate the advantage to use JUXMEM for managing persistent data in the GridRPC model in a grid environment: the execution time of a MUMPS application is kept to its value as in a cluster execution, despite the high-latency WAN connections. As explained previously, the reduced times (rows 2 and 3 of last columns of Table 2) come from the difference in processor speeds of nodes used for the various computations across the sites.

6 Conclusion

Programming grid infrastructures remains a significant challenge. The GridRPC model is the grid form of the classical RPC approach. It offers the ability to perform asynchronous coarse-grained parallel tasking, and hides the complexity of server-level parallelism to clients. In its current state, the GridRPC model has not specified adequate mechanisms for efficient data management. One important issue regards data persistence, as multiple GridRPC calls with data dependencies are executed.

In this paper, we propose to couple the GridRPC model with a *transparent data access model*. Such a model is provided by the concept of *grid data-sharing service*. Data management (persistent storage, transfer, consistent replication) is totally delegated to the service, whereas the applications simply access data via global identifiers. The service automatically localizes and transfers or replicates the data as needed.

We have illustrated our approach by showing how the DIET GridRPC middleware can benefit from the above properties by using the JUXMEM grid data-sharing service. Experimental measurements on the Grid'5000 testbed show that introducing persistent storage has a clear impact on the execution time. Using a real-life application based on a sparse matrix parallel solver, experiments performed on the Grid'5000 testbed show that the use of JUXMEM allows to keep an execution time as if the application was executed in a cluster, despite the high-latency WAN connections.

The main contribution of our approach compared to related work having dealt with data persistence in GridRPC environments consists in showing that efficiency can be obtained through the use of a *generic* data-sharing service, providing *location-transparent* data access. Moreover, our approach also allows to transparently benefit from replica consistency and fault-tolerance mechanisms. We did not develop these aspects in this paper (they have been illustrated in [19] in a more general way). A

GridRPC-specific study of these aspects could be addressed in future work. For instance, it would be interesting to evaluate the impact of using the fault tolerance mechanisms provided by JUXMEM on application execution time, in presence of data storage failures.

To extend our contributions to data management in NES through the features offered by JUXMEM, several directions can be pursued. First, we plan to provide data placement information to the request scheduling algorithms. This would make it possible to balance more precisely the load among available servers. Then, we would like to implement a cache mechanism inside JUXMEM clients to avoid fetching data from providers at each GridRPC call. In addition, JUXMEM consistency and fault-tolerance mechanisms have been tested using synthetic benchmarks [19], outside the GridRPC model. We would like to further evaluate them by using other real-life DIET applications which exhibit such requirements, such as climate modeling and cosmology simulations. Besides, we also plan to compare JUXMEM with non location-transparent data access solutions, such DIET DTM for instance. Finally, the implementation of a classical file-system API over JUXMEM would allow applications based on this API to transparently leverage JUXMEM's functionalities. We have already started such a work, called JUXMEMFS, by relying on the FUSE library [24] available on Linux systems.

Acknowledgements

Experiments presented in this paper were carried out using the Grid'5000 experimental testbed, an initiative from the French Ministry of Research through the ACI GRID incentive action, INRIA, CNRS and RENATER and other contributing partners (see `https://www.grid5000.fr/`. The work has been partially supported by the GDS (ACI MD) grant and by the LEGO (ANR-CICG05-11) grant.

References

1. Seymour, K., Nakada, H., Matsuoka, S., Dongarra, J., Lee, C., Casanova, H.: Overview of GridRPC: A Remote Procedure Call API for Grid Computing. In: Parashar, M. (ed.) GRID 2002. LNCS, vol. 2536, pp. 274–278. Springer, Heidelberg (2002)
2. Nakada, H., Tanaka, Y., Seymour, K., Desprez, F., Lee, C.: The End-User and Middleware APIs for GridRPC. In: GAPI 2004. Proc. of the Workshop on Grid Application Programming Interfaces, Brussels, Belgium. In conjunction with Global Grid Forum 12 (GGF) (2004)
3. Caron, E., Desprez, F.: DIET: A Scalable Toolbox to Build Network Enabled Servers on the Grid. Intl. Journal of High Performance Computing Applications 20(3), 335–352 (2006)
4. Antoniu, G., Bougé, L., Jan, M.: JuxMem: An Adaptive Supportive Platform for Data Sharing on the Grid. Scalable Computing: Practice and Experience 6(3), 45–55 (2005)
5. Bolze, R., Cappello, F., Caron, E., Daydé, M., Desprez, F., Jeannot, E., Jégou, Y., Lanteri, S., Leduc, J., Melab, N., Mornet, G., Namyst, R., Primet, P., Quetier, B., Richard, O., Talbi, E.G., Irena, T.: Grid'5000: a large scale and highly reconfigurable experimental grid testbed. International Journal of High Performance Computing Applications 20(4), 481–494 (2006)
6. Lee, C., Nakada, H., Tanimura, Y.: Grid Remote Procedure Call WG (GRIDRPC-WG) (2003), `https://forge.gridforum.org/projects/gridrpc-wg/`

7. Tanaka, Y., Takemiya, H., Nakada, H., Sekiguchi, S.: Design, implementation and performance evaluation of gridRPC programming middleware for a large-scale computational grid. In: Proc. of the 5th Intl. Workshop on Grid Computing (GRID 2004), Pittsburgh, PA, USA, pp. 298–305. IEEE Computer Society, Los Alamitos (2004)
8. Arnold, D., Agrawal, S., Blackford, S., Dongarra, J., Miller, M., Sagi, K., Shi, Z., Vadhiyar, S.: Users' Guide to NetSolve V1.4. Computer Science Dept. Technical Report CS-01-467, University of Tennessee, Knoxville, TN (2001)
9. YarKhan, A., Seymour, K., Sagi, K., Shi, Z., Dongarra, J.: Recent Developments in GridSolve. Intl. Journal of High Performance Computing Applications 20(1), 131–141 (2006)
10. Sato, M., Boku, T., Takahasi, D.: OmniRPC: a Grid RPC System for Parallel Programming in Cluster and Grid Environment. In: Proc. of the 3rd IEEE/ACM Intl. Symp. on Cluster Computing and the Grid (CCGrid 2003), Tokyo, Japan, pp. 206–213. IEEE Computer Society, Los Alamitos (2003)
11. Daydé, M., Giraud, L., Hernandez, M., L'Excellent, J.Y., Puglisi, C., Pantel, M.: An Overview of the GRID-TLSE Project. In: Daydé, M., Dongarra, J.J., Hernández, V., Palma, J.M.L.M. (eds.) VECPAR 2004. LNCS, vol. 3402, pp. 851–856. Springer, Heidelberg (2005)
12. Amestoy, P.R., Duff, I.S., L'Excellent, J.Y., Koster, J.: A fully asynchronous multifrontal solver using distributed dynamic scheduling. SIAM Journal on Matrix Analysis and Applications 23(1), 15–41 (2001)
13. Arnold, D.C., Bachmann, D., Dongarra, J.: Request Sequencing: Optimizing Communication for the Grid. In: Bode, A., Ludwig, T., Karl, W.C., Wismüller, R. (eds.) Euro-Par 2000. LNCS, vol. 1900, pp. 1213–1222. Springer, Heidelberg (2000)
14. Desprez, F., Jeannot, E.: Improving the GridRPC Model with Data Persistence and Redistribution. In: Proc. of the 3rd Intl. Symp. on Parallel and Distributed Computing (ISPDC 2004), Cork, Ireland, pp. 193–200. IEEE Computer Society, Los Alamitos (2004)
15. Bassi, A., Beck, M., Fagg, G., Moore, T., Plank, J., Swany, M., Wolski, R.: The Internet Backplane Protocol: A Study in Resource Sharing. Future Generation Computer Systems 19(4), 551–562 (2003)
16. Nakajima, Y., Sato, M., Boku, T., Takahasi, D., Gotoh, H.: Performance Evaluation of OmniRPC in a Grid Environment. In: Proc. of the 2004 Intl. Symp. on Application and the Internet Workshops (SAINTW 2004), Tokyo, Japan, pp. 658–665. IEEE Computer Society, Los Alamitos (2004)
17. Fabbro, B.D., Laiymani, D., Nicod, J.M., Philippe, L.: Data management in grid applications providers. In: Proc. of the 1st Intl. Conference on Distributed Frameworks for Multimedia Applications (DFMA 2005), Besançon, France, pp. 315–322 (2005)
18. Antoniu, G., Bertier, M., Caron, E., Desprez, F., Bougé, L., Jan, M., Monnet, S., Sens, P.: GDS: An Architecture Proposal for a grid Data-Sharing Service. In: Getov, V., Laforenza, D., Reinefeld, A. (eds.) Future Generation Grids. CoreGRID series, pp. 133–152. Springer, Heidelberg (2005)
19. Antoniu, G., Deverge, J.F., Monnet, S.: How to bring together fault tolerance and data consistency to enable grid data sharing. Concurrency and Computation: Practice and Experience 18(13), 1705–1723 (2006)
20. Traversat, B., Arora, A., Abdelaziz, M., Duigou, M., Haywood, C., Hugly, J.C., Pouyoul, E., Yeager, B.: Project JXTA 2.0 Super-Peer Virtual Network (2003), `http://www.jxta.org/project/www/docs/JXTA2.0protocols1.pdf`
21. Lacour, S., Pérez, C., Priol, T.: Generic application description model: Toward automatic deployment of applications on computational grids. In: Proc. of the 6th IEEE/ACM Intl. Workshop on Grid Computing (Grid 2005), Seattle, WA, USA, pp. 4–8. Springer, Heidelberg (2005)

22. Caron, E., Chouhan, P.K., Dail, H.: GoDIET: A Deployment Tool for Distributed Middleware on Grid'5000. In: Workshop on Experimental Grid Testbeds for the Assessment of Large-Scale Distributed Applications and Tools (EXPGRID). conjunction with 15th IEEE International Symposium on High Performance Distributed Computing (HPDC 15), Paris, France, pp. 1–8. IEEE Compuer Society, Los Alamitos (2006)
23. Amestoy, P.R., Guermouche, A., L'Excellent, J.Y., Pralet, S.: Hybrid scheduling for the parallel solution of linear systems. Parallel Computing 32(2), 136–156 (2006)
24. Szeredi, M.: Filesystem in Userspace (FUSE) (2004), http://fuse.sourceforge.net/

A Proxy-Based Self-tuned Overload Control for Multi-tiered Server Systems

Rukma P. Verlekar and Varsha Apte

Department of Computer Science and Engineering
Indian Institute of Technology Bombay
Mumbai - 400076, India
rukma.verlekar@iitb.ac.in, varsha@cse.iitb.ac.in

Abstract. Web-sites, especially E-commerce ones, are often faced with incoming load of requests that exceeds their capacity, i.e, they are subjected to *overload*. Most existing servers show severe throughput degradation at high overload. Overload control mechanisms are required to prevent such occurrences. In this paper, we present a proxy-based overload control mechanism, which uses the drop in throughput relative to arrival rate as an indicator of overload. On overload detection, a self-clocked admission control is activated, which admits a new request only when a successful reply is observed to be leaving the server system. Thus, the mechanism is *self-tuned*, and requires no knowledge of the system. We validate our approach on an experimental testbed consisting of a two-tier Web application, and find that even at very high overload, the server operates at its maximum capacity while keeping response times within acceptable bounds.

1 Introduction

Online services of today, such as banking, shopping, stock market trading are supported by Web-based multi-tiered server systems. Such services are exposed to variable load, due to peak hour usage phenomenon, or events such as sales, holiday shopping, or headline events. Peak load during such events can sometimes be orders of magnitude higher than average load, and can exceed the capacity of the system. When the incoming request rate on a system exceeds its processing capacity, it is said to be *overloaded*.

Most server systems display unstable behaviour when overloaded. Although ideally a system should operate at its maximum capacity when overloaded, many systems experience a drop in throughput (successful request completion rate), which is often drastic. Overload results in an increase in the request response times which results in many requests timing out, or *abandoning* the server system. after being serviced for some amount of time. Abandonments (either manual or protocol-triggered) lead to retries which further elevate the effective load on the servers. The overloaded server ends up being busy serving a large number of requests which timeout, resulting in a severe drop in throughput. This "feedback phenomenon" further deteriorates the performance of the Web-site.

S. Aluru et al. (Eds.): HiPC 2007, LNCS 4873, pp. 285–296, 2007.

While the problem of overload could be partially eliminated by proper server capacity planning through server duplication, data redundancy and request redirection, it cannot fully done away with it. Unexpected peak hour usage of a Web-site can always happen; e.g., due to a major breaking news event for a popular news Web-site, or server-side failures that reduce total capacity. Since it is not prudent to size server systems for such occasional overload situations, a mechanism is required which specifically aims to keep the server system operating in a stable manner, even in the presence of overload.

Overload can be controlled using two broad approaches. The first approach is a *pro-active* approach, where the control mechanism prevents the system from getting overloaded by exercising admission control. A fair amount of knowledge of the system's capacity, a request's resource needs and monitoring of system resources is required to be able to make an accurate decision about admission. Such complex mechanisms are best employed when user QoS requirements are precisely expressed, and when the server system is required to be a QoS-aware system, that provides specific and differentiated performance guarantees..

However, most Web-sites aim for a simple "best-effort" service, where the users do not express any explicit QoS targets - thus the system goals are those of ensuring stability on overload, maintaining the throughput near capacity, and response times that do not result in a large number of abandonments. Such systems can activate an appropriate overload control mechanism only upon overload detection - a *reactive* overload control. For a reactive approach, two components are required: an overload *detection* mechanism, and an overload *control* mechanism.

A number of existing approaches [3, 2, 4] use overload detection mechanisms based on resource utilization. These mechanisms assume that the potential bottleneck resource is known and can be monitored - thus, high utilization of this resource can indicate an overload. However, system bottlenecks may not be known a-priori; they may vary based on the type of workload (CPU intensive, network I/O intensive, etc.), machine hardware configuration (CPU speed, network bandwidth, system cache and memory sizes etc.), software configuration (thread pool size, buffer size, object pool size, etc); hence determining the bottleneck resource is nearly impossible in the case of multi-tiered heterogeneous systems which support varying workload mixes. This motivates the need for an overload detection mechanism that does not require the knowledge of the bottleneck resource, and therefore does not need to monitor it.

We claim that an "absolute" indicator of a system in overload is when its *throughput* (rate of successful completion of requests) is lower than its *request arrival rate.* As long as requests arrive at a rate that the system can process them, the completion rate has to be close to the arrival rate. If the completion rate (smoothed and averaged, to ignore transient effects) drops below arrival rate, it is a clear indicator, that the server cannot process the requests at the rate they are arriving, and is hence, *overloaded.*

In this paper, we propose a proxy-based, reactive overload control mechanism which uses the ratio of the throughput to the arrival rate as an indicator of overload. Overload is flagged by the proxy when this ratio is lower than 1 by a given amount (determined by a threshold value). On overload detection, the proxy uses a "self-clocked" admission control on incoming requests that are queued at the proxy. The request at the head of the queue is admitted into the server system, only when a request is seen successfully exiting from the server, indicating that there is "room" for a new request. The mechanism is similar in concept to window-based flow control mechanisms used in networking.

Both these mechanisms, are clearly self-tuned - they require no information about the capacity of the system, the amount and type of resources required by requests, or the potential system bottlenecks. The throughput-based overload indicator and the self-clocked admission control can be expected to work without any configuration.

We validate our claim by implementing the proxy on a testbed consisting of benchmark Web applications, and provide results from our experimentation on these benchmarks. We validated our proposed mechanism against different Web servers, workloads, and hardware platforms. Experimental validation reveals that our approach is able to sustain system throughput to its peak capacity even at loads 40 times the system's maximum capacity with only 15% increase in response times.

The rest of the paper is as follows: in Section 2 we provide background and motivation for overload control. We present the details of our overload control approach in Section 3. In Section 4, we present the experimental setup and results which validate our approach. We conclude the paper in Section 5.

2 Background and Motivation

Figure 2 shows results from an experiment, in which a multi-tiered Web application consisting of a Web server and a database server (details in Section 4) was subjected to increasing load. Observe the throughput behaviour when there is no overload control ((Figures 2(a) and 2(e), plots corresponding to "Without PB-SCOC"). At low loads, the throughput of the system is equal to the offered load. As the load offered to the system exceeds the capacity, the system becomes unstable, as indicated by the drop in throughput and the increase in the rate of request timeouts and internal server errors (Figures 2(g) and 2(h)). The average response time of the completed requests also shows a sharp increase (Figures 2(b) and 2(f)).

An overload control mechanism would prevent this degradation of performance of the server. With overload control, our aim would be that the system should operate near its capacity, even after the overload sets in. A large amount of research has already been done in the area of overload prevention and control. Here we discuss a few approaches and conclude with the motivation behind the approach proposed by us.

Cherkasova and Phaal [3] propose a session based admission control policy. In this technique every new request which belongs to an already admitted session is given priority over any other newly arrived request. The number of new session requests to be admitted is based on the prediction of CPU utilization in the next control interval.

Chen and Mohapatra [2] use a session based classification of requests, where the requests are sorted into different queues which are associated with different resource consumption demands. A weighted fair sharing method is used for differentiated overload control, which selects a request to be scheduled on the basis of criteria such as estimated resource requirement of the request vs. the capacity of the system, the delay bounds specified by the request and the actual response delays that can be provided.

The approach suggested by Elnikety et al. [4] addresses the problem of overload control in multi-tiered Web sites, where the bottleneck resource is known a-priori. An external entity such as a proxy is placed in front of the server of the bottleneck resource. The proxy observes the request resource demands externally and compares them against the current remaining capacity of the system. A new request is admitted only if the system has the capacity to serve it, otherwise it is placed in a queue which is processed on implicit demand from the server, using the shortest job first (SJF) with an aging policy.

Voigt et al [8] propose a scheme based on admission control which estimates the expected resource consumption demands of a request. Each request is classified on the basis of its HTTP header as a CPU intensive, bandwidth intensive or other. At the end of each monitoring interval, based on the capacity analysis of each of these resources a threshold is set on admission rate of each class of request. A new request belonging to a particular class is admitted only if the corresponding resource has the capacity to serve it.

In addition to the above, new approaches to server architecture have also been proposed, which provide robust performance at overload (e.g. Welsh et al [9]). In our work, we concentrate on mechanisms that can be implemented using available servers and operating systems.

Our review indicates that many of the existing approaches depend on the knowledge of the utilization of a resource or estimation of the resource demands of the resource which has been identified as the bottleneck. This seems reasonable if in a system the bottleneck remains unchanged with the varying workloads. However, in a complex multi-tiered server system, determining which resource is the bottleneck resource can be very difficult. Furthermore, the bottleneck resource itself may vary with changing workloads or software and hardware configurations.

Figures 2(c) and 2(d) show values of CPU utilizations of the two servers in the testbed and the number of busy Web server threads during the experiment. We see that although the system overloads at 80 requests per second, the CPU utilization for the database and the Web server is only 21% and 49% respectively. Similarly, the number of Web server threads in use at overload is 214, but the maximum number of threads that this server is allowed to spawn is 250. Thus, the bottleneck based overload detection approaches as discussed above would have failed to detect overload in the system. These observations motivate the need for an approach that can control system overloads and ensure stability without any knowledge of the bottleneck resources.

Once overload is detected, we need an approach that would regulate entry of requests into the server system without a-priori knowledge of the capacity of the system, or the resource requirement of the request. Thus, for e.g., we would like to avoid complex mechanisms required to estimate resource requirements of a request.

In the following sections we present a platform and configuration-independent method for overload detection, and an entirely "self-clocked" admission control mechanism that is activated on overload.

3 Proposed Overload Control Approach

We propose an approach which includes a throughput based overload detection mechanism and a self-clocked admission control that activates after overload is detected.

The throughput of an overloaded system drops due to high response times at overload which result in frequent request abandonments. We use a threshold value which flags overload in the system, when the ratio of measured throughput to arrival rate (also termed as *offered load* or simply *load*) drops below this threshold. The overload detection mechanism then triggers the overload control strategy. Our overload control strategy represents a "serve on demand" scenario, where every successful reply from the server sends an implicit notification for a new admission into the system. The rationale behind this approach is as follows: at overload, the server resources can be assumed to be full, and only a successful exit from the system can indicate that there is a room in the server system for a new request. This method is self-clocked, in the sense that no explicit "admission rate" needs to be specified to the mechanism. Further, since the control is active only on overload, we avoid the situation where this control may degenerate into a "stop-and-wait" type control, where only one request may be active in the system at a time. The admission control is turned off as soon as the system comes out of overload.

Based on the above mentioned requirements, we need three interacting modules for overload detection, namely, for load calculation, throughput calculation, and for overload detection itself. We need a fourth module for overload control, which when triggered by the overload detection module, is able to control admission of every new request into the system.

Since we would like our mechanism to work without the knowledge of which server is the bottleneck server, the requests made to the system as well as the replies should be measured by an external entity. Thus, a *proxy* which sits at the "front end" of the server system can be used to implement the modules and mechanisms mentioned above.

For our implementation, we used a Java based proxy called *Muffin* as an external entity which could act as an observer for the incoming requests and outgoing replies. All the four modules discussed above were implemented in this proxy. We will now look at the proposed system architecture in detail.

3.1 Implementation Architecture

We used a Java based lightweight proxy called Muffin [5] for this implementation. It is freely available under the GNU General Public License and provides support for HTTP/0.9, HTTP/1.0, HTTP/1.1, as well as SSL (https). Source code for Muffin was obtained and modified to implement our overload detection and control mechanism. We refer to this modified proxy as *Muffin-M*.

Figure 1 shows the block diagram of the system with Muffin-M and its corresponding components, implemented by us, namely the load calculation, throughput calculation, overload detection and the overload control modules. In the figure, the functionality of each of the modules is displayed using pseudo-code. Now we discuss each of the modules in Figure 1 in detail.

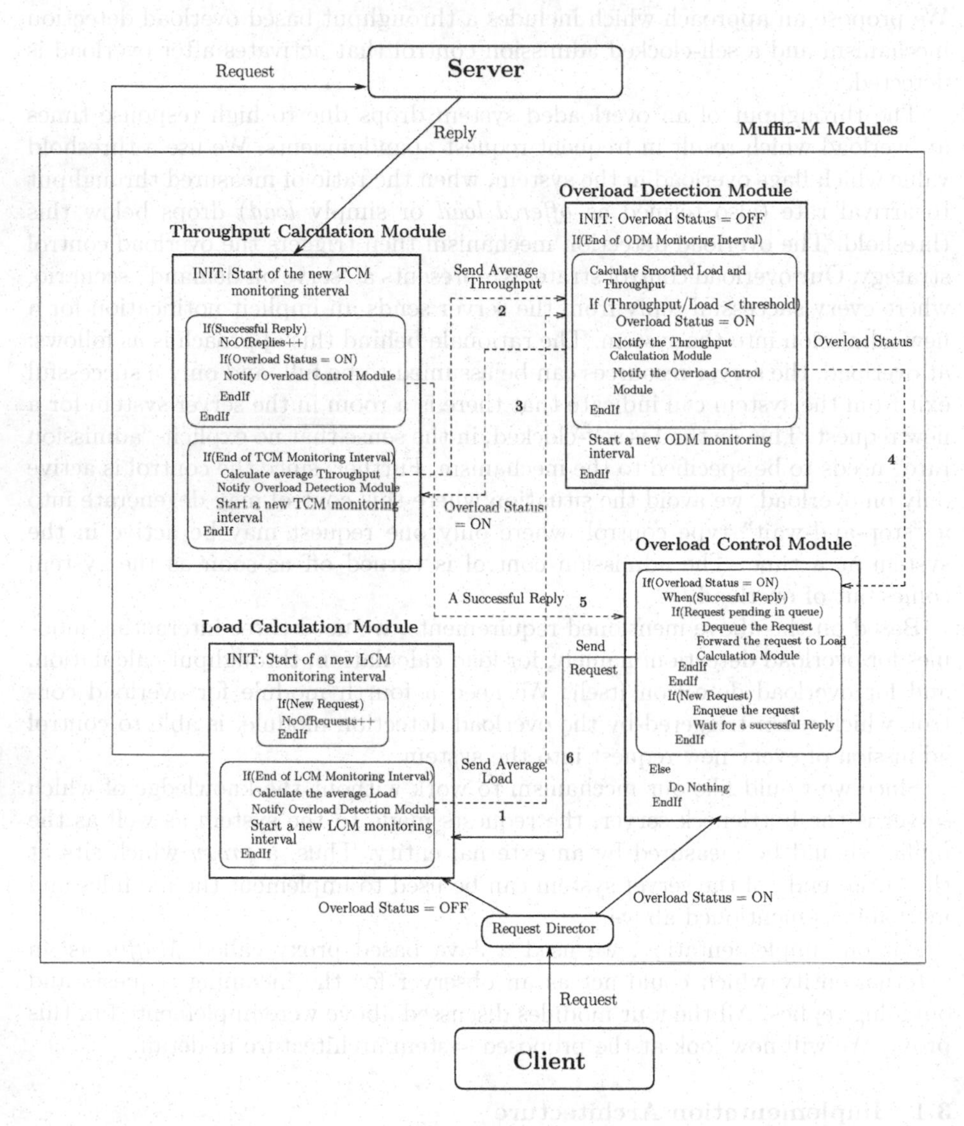

Fig. 1. Block structure of Muffin-M with its corresponding modules

Load Calculation Module (LCM): Each arriving request is routed through this module. It calculates the load at the end of each monitoring interval, and sends an exponential average to the Overload Detection Module (arrow number 1).

Throughput Calculation Module (TCM): Each successfully exiting request is observed by this module, which calculates the throughput at the end of the monitoring interval. It then sends an exponential average of the throughput to the Overload Detection Module (arrow number 2). After an overload notification is received from the overload detection module (arrow number 3), for every successful reply received from the server, it sends a notification to the overload control module, throughout the overloaded interval.

Overload Detection Module (ODM): At the end of its monitoring interval, this module calculates the average load and throughput, using the values received from the LCM and TCM at the end of their respective monitoring intervals. We use a *Threshold* value in order to flag overload in the system. The ratio of throughput to load as computed by this module is compared with this threshold. If the ratio is below the set value of Threshold, the system flags 'overload'. (Note that Threshold is the only parameter in the overload control mechanism that needs to be manually configured). On detecting overload, a notification is sent by this module to the TCM (arrow number 3) as well as the overload control module (arrow number 4). If the ratio is above the threshold, the overload status is reset to "OFF".

Overload Control Module (OCM): This module is notified by the ODM when overload is detected. On overload, all the requests which otherwise go through the LC module directly, are re-directed to this module. On arrival of a new request, the request is simply queued into its buffer, waiting for a notification from the TC module. Upon receiving a successful reply notification (arrow number 5), the module checks for the waiting but not timed out request at the head of its buffer queue; it dequeues such a request and forwards it to the LC module. If overload flag has not been set, this module does nothing and every new request passes directly to the LC module.

4 Results

In this section we present results for validation our overload detection and control approach. We begin the section with the discussion of our system setup, the type of workloads and their compositions. This is followed by the discussion of results of the experiments, which validate our approach.

4.1 The Testbed Setup

The testbed setup consisted of the following components:

Application: All experiments were carried out on the "Duke's Bank" application [1] which is an online banking application that can be used for testing purposes. Duke's Bank is a J2EE application whose components include a J2EE server with a Web container, and a database server at the second tier. Duke's Bank supports typical banking transactions such as opening a new account, closing an existing account, transferring funds from one account to another, etc.

Workload: The workload of the Duke's Bank application was used for representing *dynamic* requests. The workload mix used was as follows: open a new account (15%), close an existing account (2%), transferring funds from one account to another (13%), account enquiry (15%) and update existing account (55%).

The WebSPEC99 benchmark [7], in which the file sizes follow a Zipf's distribution, was used to represent static file workload. The static workload consisted of individual requests which retrieved files from the WebSPEC99 distribution consisting of the following four classes. *Class* 0: 100 bytes - 900 bytes (35%), *Class* 1: 1 Kbytes - 9 Kbytes (50%), *Class* 2: 10 Kbytes - 90 Kbytes (14%) and *Class* 3: 100 Kbytes - 900 Kbytes (1%).

Out of the three types of available workloads, namely: static (WebSPEC99), dynamic (Duke's Bank) and mixed (both), we used dynamic and mixed workloads in our experiments. Httperf [6], which can emulate "open" arrivals was used for load generation.

Software: Two different servers were used for the Web component of the Duke's Bank application: a JBoss application server (EJB container) with an embedded Tomcat Web server (servlet container), and the Apache Web Server. In case of the Apache Web server, PHP scripts were written to emulate the same functionality as the J2EE-based Duke's Bank application. In all experiments, MySQL is used as the database server.

Hardware: In all the experiments, the machine hosting the database server is a P4 machine with 256MB RAM and running the Linux 2.6.15 kernel. The Web servers are deployed on either of the following two machines: an AMD 64-bit machine with 1GB RAM, running Linux 2.6.15 kernel and an Intel based 32-bit machine with 256MB RAM, running Linux 2.6.16 kernel. Muffin-M is deployed on the same machine as that of the Web server (JBoss with Tomcat and Apache).

Table 1 lists the combinations of the hardware, software and workload configurations used in the eight experiments that were carried out to validate our approach (denoted by "PBSCOC": Proxy Based Self Clocked Overload Control), and summarizes the results. The main aim of the experiments was to demonstrate that the mechanism works effectively with little configuration, under varying scenarios which change the capacity of the system. The only parameter that requires setting is the *threshold* that determines whether throughput has fallen below arrival rate. We present detailed results only for Experiment 1, 2 and 3 from the above.

Experiment 1: This corresponds to the setup with Tomcat on the 64 bit AMD machine and with dynamic workload. Figure 2(a) shows that the throughput

Table 1. Hardware, server and workload configuration for various experiments

Exp. No.	Web server	Machine specs.	Workload Type	Maximum Throughput	Throughput at overload no PBSCOC	Throughput at overload with PBSCOC
1	Tomcat	64 bit AMD	dynamic	63	35	61
2	Tomcat	64 bit AMD	mixed	120	30	120
3	Apache	32 bit Intel	mixed	42	0	42
4	Apache	64 bit AMD	mixed	47	0	45
5	Tomcat	32 bit Intel	dynamic	32	28	32
6	Tomcat	32 bit Intel	mixed	40	28	40
7	Apache	32 bit Intel	dynamic	37	0	33
8	Apache	64 bit AMD	dynamic	40	0	40

drops from its maximum capacity of 63 requests per second to 30 reqs/sec and finally to a value of 9 reqs/sec on very heavy loads, in the absence of overload control. On application of the PBSCOC, the throughput saturates near its capacity at around 61 reqs/sec. Furthermore, the stability in throughput is also accompanied by only 15% increase in the net response time of every successful request, which increases exponentially otherwise as seen in Figure 2(b).

Experiment 2: This experiment uses the same hardware and software setup as in Experiment 1, but uses a mixed workload. Figure 2(e), shows the throughput drop from the maximum capacity of 120 reqs/sec to 31 reqs/sec in the system without any overload control. With overload control the throughput is sustained at 118 reqs/sec, even at loads 15 times the system capacity. Figure 2(f) compares the corresponding values of response times for the system with and without overload control. Also, as is seen from Figures 2(c) and 2(d), PBSCOC works even when none of the CPUs, nor the Web server threads, are the bottleneck.

For Experiment 1 and 2, we also compared the number of timed out requests with and without overload control (Figures 2(g) and 2(h)). We can see that the improvement in throughput is achieved mainly by preventing timeouts.

Experiment 3: This corresponds to the setup with Apache on the Intel machine with mixed workload. We observed that unlike the Tomcat server which is quite stable and reacts slowly to the increasing load, Apache server was unstable when subjected to heavy loads. The server crashes completely, with its throughput dropping to zero, when presented with heavy loads even for a small amount of time. In Figure 3(a), we can see that the server throughput drastically drops to zero on overload. The server is crashed at this point and needs reload for any further activity. However our technique prevents its crashing, by detecting overload and controlling it. The server supports a maximum throughput of 40 reqs/sec, even at loads 50 times its capacity. Figure 3(b) shows corresponding values of response time. The value of CPU utilization observed at the overload point (40 reqs/sec) was 99%, indicating in this case that the CPU is the bottleneck.

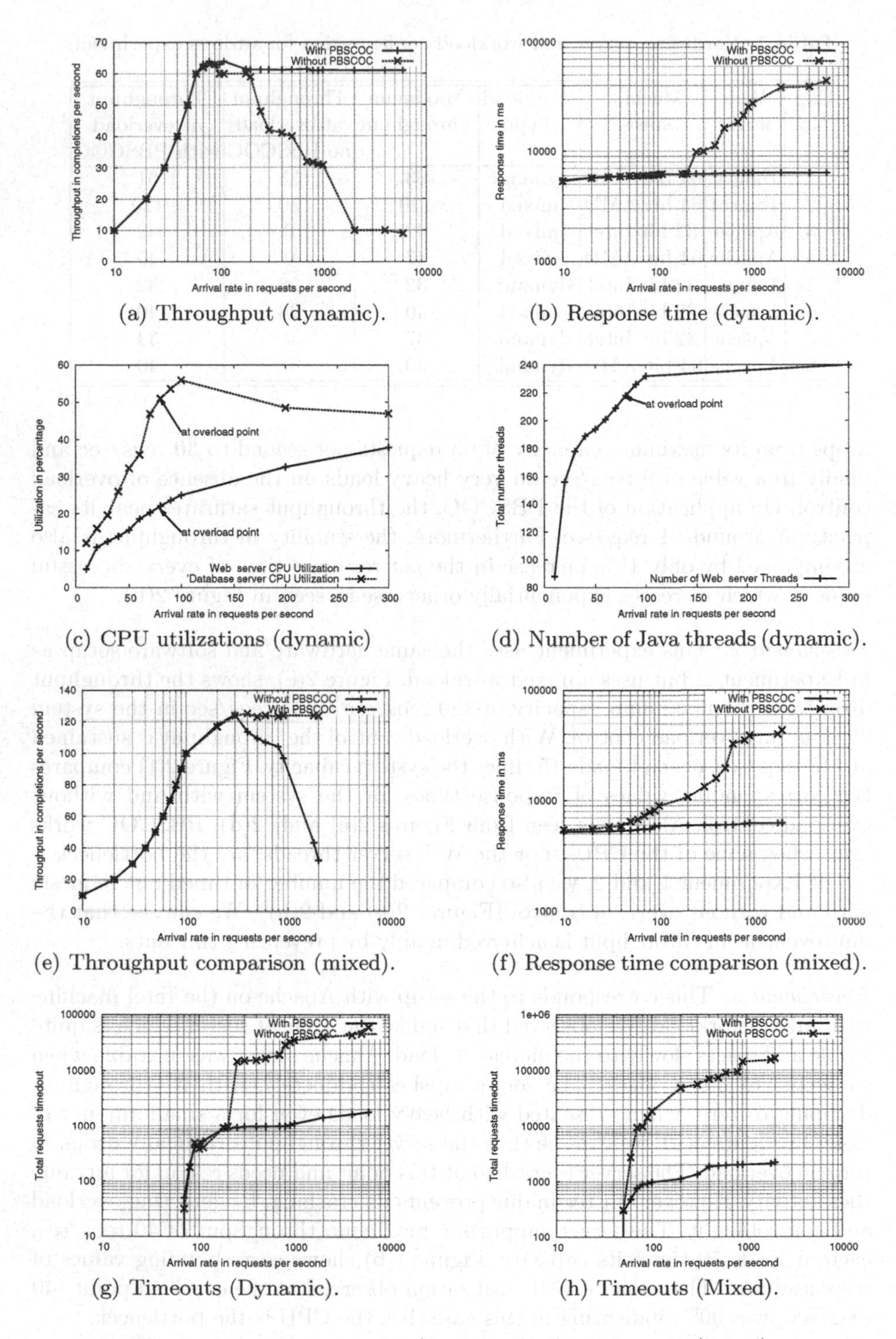

(a) Throughput (dynamic). (b) Response time (dynamic).

(c) CPU utilizations (dynamic) (d) Number of Java threads (dynamic).

(e) Throughput comparison (mixed). (f) Response time comparison (mixed).

(g) Timeouts (Dynamic). (h) Timeouts (Mixed).

Fig. 2. Experiments 1 and 2 (Tomcat+AMD64+dynamic/mixed)

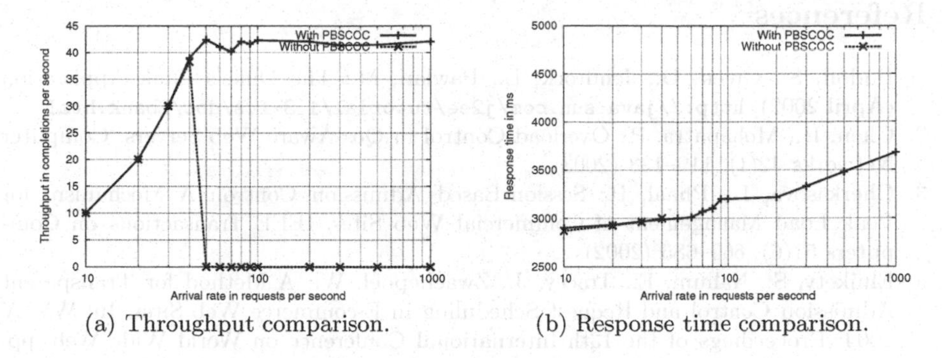

(a) Throughput comparison. (b) Response time comparison.

Fig. 3. Experiment 3 (Apache+Intel32+mixed): Performance Comparison

Recall that the value of *threshold* used by the overload detection module is not self-tuned and needs to be configured by the system administrator. In case of the servers which are highly sensitive to overloads, for e.g. the Apache server, the value of *threshold* needs to be set cautiously. We set this value to 0.7 for experiments with Apache server and to 0.8 for the Tomcat server. Apache server was observed to crash if the threshold was set to a higher value. However the value of *threshold* did not affect the performance of Tomcat server significantly. Experimental results (with the setup used in Experiment 1) show the value of saturation throughput for Tomcat *threshold* 0.6 was 60.2, with 0.7 was 60.72 and with 0.9 was 61.2. Also, in all other cases shown in Table 1, the server shows dramatic improvement in throughput with the proxy-based overload control, thus demonstrating its robustness under varying scenarios.

5 Conclusions

In this paper, we presented a simple yet effective method of overload detection and control in multi-tiered systems. The mechanism was a *reactive* overload control, which required detection of overload. We used a drop in throughput with the increasing load as an "absolute" indicator of overload. This was done by using a proxy, to observe the load and the throughput of the system, and draw a conclusion on whether the system is overloaded. Once it is detected, overload is controlled by using a "self-clocked" admission control, which admits a request only when a successful reply exits the server system. We validated the approach using a variety of hardware and software deployments, and different workloads to illustrate its generality and robustness. In all our experiments, the overload control mechanism detected the overload, and was able to sustain server throughput at nearly the maximum capacity of the server. This approach can be coupled with a more comprehensive vision of self-tuned systems, which optimize their performance by re-configuring some parameters to scale the system towards supporting higher loads.

References

1. Bodoff, S., Green, D., Jendrock, E., Pawlan, M.: The Duke's Bank Application (April 2002), http://java.sun.com/j2ee/tutorial/1_3-fcs/doc/Ebank.html
2. Chen, H., Mohapatra, P.: Overload Control in Qos-Aware Web Servers. Computer Networks 42(1), 119–133 (2003)
3. Cherkasova, L., Phaal, P.: Session-Based Admission Control: A Mechanism for Peak Load Management of Commercial Web Sites. IEEE Transactions on Computers 51(6), 669–685 (2002)
4. Elnikety, S., Nahum, E., Tracey, J., Zwaenepoel, W.: A Method for Transparent Admission Control and Request Scheduling in E-commerce Web Sites. In: WWW 2004: Proceedings of the 13th International Conference on World Wide Web, pp. 276–286. ACM Press, New York (2004)
5. Lier, T.: Muffin: World Wide Web Filtering System (April 2000), http://muffin.doit.org/
6. Mosberger, D., Jin, T.: httperf: A Tool for Measuring Web Server Performance. SIGMETRICS Performance Evaluation Review 26(3), 31–37 (1998)
7. Standard Performance Evaluation Corporation. Specweb99 benchmark (August 1999), http://www.spec.org/osg/web99/
8. Voigt, T., Gunningberg, P.: Adaptive Resource-Based Web Server Admission Control. In: Proceedings of the Seventh International Symposium on Computers and Communications (ISCC 2002), Washington, DC, USA, p. 219. IEEE Computer Society, Los Alamitos (2002)
9. Welsh, M., Culler, D., Brewer, E.: SEDA: An Architecture For Well-Conditioned, Scalable Internet Services. In: SOSP 2001: Proceedings of the Eighteenth ACM Symposium on Operating Systems Principles, pp. 230–243. ACM Press, New York (2001)

Approximation Algorithms for Scheduling with Reservations

Florian Diedrich[1,*,**], Klaus Jansen[1,**], Fanny Pascual[2], and Denis Trystram[2,***]

[1] Institut für Informatik, Christian-Albrechts-Universität zu Kiel, Olshausenstr. 40, 24098 Kiel, Germany
{fdi,kj}@informatik.uni-kiel.de

[2] LIG – Grenoble University, 51 avenue Jean Kuntzmann, 38330 Montbonnot Saint-Martin, France
{fanny.pascual,denis.trystram}@imag.fr

Abstract. We study the problem of scheduling n independent jobs on a system of m identical parallel machines in the presence of reservations. This constraint is practically important; for various reasons, some machines are not available during specified time intervals. The objective is to minimize the makespan. This problem is inapproximable in the general case unless $\mathsf{P} = \mathsf{NP}$ which motivates the study of suitable restrictions. We use an approach based on algorithms for multiple subset sum problems; our technique yields a polynomial time approximation scheme (PTAS) which is best possible in the sense that the problem does not admit an FPTAS unless $\mathsf{P} = \mathsf{NP}$. The PTAS presented here is the first one for the problem under consideration; so far, not even for special cases approximation schemes have been proposed. We also derive a low cost algorithm with a constant approximation ratio and discuss additional FPTASes for special cases and complexity results.

1 Introduction

In parallel machine scheduling, an important issue is a scenario where time intervals of machine unavailability must be taken into account. This phenomenon occurs due to periods of regular maintenance or because high-priority jobs are present. Here we obtain deterministic models capturing realistic industrial settings and scheduling problems in parallel computing. We study non-preemptive scheduling of sequential jobs on a system of identical parallel machines; however, these may be unavailable for certain periods of time which are known beforehand.

* Research supported in part by a grant "DAAD Doktorandenstipendium" of the German Academic Exchange Service. Part of this work was done while visiting the LIG, Grenoble University.

** Supported in part by EU research project AEOLUS, "Algorithmic Principles for Building Efficient Overlay Computers", EU contract number 015964, and in part by DFG priority program 1126, "Algorithmics of Large and Complex Networks".

*** Part of this work was supported by the "CoreGRID" Network of Excellence.

S. Aluru et al. (Eds.): HiPC 2007, LNCS 4873, pp. 297–307, 2007.

This setting is also called the *non-resumable* case [17,19,20]. The objective is to minimize the makespan $C_{\max}$, which is the maximum of the completion times of all jobs. As discussed below, quite restricted special cases of the model considered here have been studied. On the algorithmic side, only list scheduling algorithms (or similar approaches) and exact exponential algorithms have been analyzed and experimentally evaluated, respectively.

Contributions. We take a novel approach by using algorithms for multiple subset sum problems to govern the scheduling of jobs on identical parallel machines with reservations. We obtain a dual approximation algorithm [8], more precisely a PTAS, for the case of an arbitrary number m of machines. We show that our problem does not admit an FPTAS unless $\mathsf{P} = \mathsf{NP}$ and present additional complexity results. For $m \in \{1, 2\}$ with one reservation we obtain FPTASes; we also discuss how fast greedy algorithms can be obtained from our approach.

This article is organized as follows. Sect. 2 defines the problem and discusses the inapproximability of the general case. In Sect. 3 we present a PTAS for a suitably restricted problem as well as FPTASes for $m \in \{1, 2\}$ with one reservation. Finally we sketch how to obtain a fast approximation algorithm for our general problem. In Subsect. 3.3 our approximation algorithms are complemented by suitable hardness results. Finally we summarize the results and conclude in Sect. 4.

Related problems and previous results. Lee [16] and Lee et al. [18] studied identical parallel machines which may have different starting times; here, the LPT policy (where tasks are greedily scheduled from the largest task to the smallest task) was analyzed. Lee [17] studied the case where at most one reservation per machine is permitted while one machine is continuously available and obtained suitable approximation ratios for low-complexity list scheduling algorithms. Liao et al. [20] presented an experimental study of an exact algorithm for $m = 2$ within the same scenario. Hwang et al. [9] studied the LPT policy for the case where at most one interval of unavailability per machine is permitted. They proved a tight bound of $1 + \lceil m/(m - \lambda) \rceil /2$ where at most $\lambda \in [m - 1]$ machines are permitted to be unavailable simultaneously. The reader can find in [19], Chapt. 22, problem definitions and a survey about previous results. In [23], Scharbrodt et al. present approximation schemes and inapproximability results for a setting where the reservations also contribute to the makespan. So far, the model under consideration has not been approached with approximation schemes, not even for the special cases which have already been studied [9,17,20].

The approach taken in our work is based on multiple subset sum problems. These are special cases of knapsack problems, which belong to the oldest problems in combinatorial optimization and theoretical computer science. Hence, we benefit from the fact that they are relatively well understood. For the classical problem (KP) with one knapsack, besides the result by Ibarra & Kim [10], Lawler presented a sophisticated FPTAS [15] which was later improved by Kellerer & Pferschy [13]; see also the textbooks by Martello & Toth [21] and Kellerer et al. [14] for surveys. The case where the item profits equal their weights is called the subset sum problem and denoted as SSP. The problem with *multiple* knapsacks (MKP) is a natural generalization of KP; the case with multiple knapsacks where the item profits equal their weights is called the *multiple* subset sum problem (MSSP). Various special cases and extensions of these problems have been studied [1,2,3,4,5,11,12], finally yielding PTASes and FPTASes for the cases upon which our approach is based [2,4,12].

2 Problem Definition and Preliminaries

Now we formally define our problem. Let $m \in \mathbb{N}^*$ denote the number of machines. An instance I consists of n jobs characterized by processing times $p_1, \ldots, p_n$, and r reservations $R_1, \ldots, R_r$. For each $k \in [r]$, $R_k = (i_k, s_k, t_k)$ indicates unavailability of machine i_k in the time interval $[s_k, t_k)$, where $s_k, t_k \in \mathbb{N}, i_k \in [m]$ and $s_k < t_k$. We suppose that for reservations on the same machine there is no overlap; for two reservations $R_k, R_{k'}$ such that $i_k = i_{k'}$ holds, we have $[s_k, t_k) \cap [s_{k'}, t_{k'}) = \emptyset$. For each $i \in [m]$ let $R'_i := \{R_k \in I | i_k = i\}$ denote the set of reservations for machine i. Finally, for each $i \in [m]$ suppose that R'_i is sorted increasingly with respect to the starting times of the reservations; more precisely, $R'_i = \{(i, s_{i1}, t_{i1}), \ldots, (i, s_{ir_i}, t_{ir_i})\}$ such that $s_{i1} < \cdots < s_{ir_i}$ where we set $r_i := |R'_i|$. These assumptions are established algorithmically in $O(r \log r)$ time by sorting $\{R_1, \ldots, R_r\}$ lexicographically with respect to the first two components of its elements and partitioning it into $R'_1, \ldots, R'_m$ and finally merging adjacent reservations in R'_i for each $i \in [m] \setminus \{1\}$. In the sequel we use $P(I) := \sum_{j=1}^{n} p_j$ to denote the total processing time of an instance I and for each $S \subseteq [n]$ we write $P(S) := \sum_{j \in S} p_j$ for the total processing time of S. A schedule is a function $\sigma : [n] \to [m] \times [0, \infty)$ that maps each job to its executing machine and starting time; if σ is clear from the context it may be dropped from notation. Our goal is to compute a non-preemptive schedule of the tasks such that no task is scheduled on a machine that is unavailable, and on each machine at most one task runs at a given time; the objective is to minimize the makespan $C_{\max}$. Using the 3-field notation, we denote our problem by $\mathrm{P}m|nr\text{-}a|C_{\max}$ and show its inapproximability for $m \geq 2$.

Lemma 1. *No polynomial time algorithm for* $\mathrm{P}m|nr\text{-}a|C_{\max}$ *with* $m \geq 2$ *has a constant approximation ratio unless* $\mathsf{P} = \mathsf{NP}$.

Proof. Let $c \in \mathbb{N}^*$; for an instance I of Partition, which is NP-complete [7], given by $I = \{a_1, \ldots, a_n\}$ such that $\sum_{i \in I} a_i = 2A$, we define an instance I' of $\mathrm{P}m|nr\text{-}a|C_{\max}$ by setting $p_i := a_i$ for each $i \in [n]$, $R_1 := (1, A, A + c)$, $R_2 := (2, A, A + c)$ and $R_k := (k, 0, A + c)$ for $k \in [m] \setminus \{1, 2\}$. Then I is a yes-instance of Partition if and only if I' has an optimal makespan of $C^*_{\max} = A$. However, any suboptimal schedule of I' for a yes-instance I of Partition has a makespan $C_{\max} > A + c$; by choosing c large, any suboptimal solution can be arbitrarily bad. □

The inapproximability of the general case is due to the permission of intervals in which no machine is available. Hence it is reasonable to suppose that at each time step there is an available machine. This is not sufficient since we can prove in this case the same inapproximability result by considering, for example, the following instance: there is, for a given period p, a set of reservations which alternate on two machines in a such a way that there are no two reservations at the same time and the period between two consecutive reservations is smaller than the length of any task of the instance. In this case, no task can be put during time period p and we get the same inapproximability result as if we had on each of these machines a big reservation of length p. Hence we suppose that at least one machine is always available. If we consider that reservations are jobs with high priority which are already scheduled, then, since the machines are identical, the reservations can be put on the machines in such a way that w.l.o.g. the *first* machine is always available, hence $i_k \neq 1$ for each reservation R_k. This can be done by distributing the reservations one by one and always putting a reservation on the machine with maximum index $i \in [m]$ among the available machines.

We use $Pm, 1up|nr\text{-}a|C_{\max}$ to denote this restricted problem; $1up$ means that at least one machine is always available. This problem is NP-hard even for $m = 2$, which can be seen by following the lines of the proof of Lemma 1 using one reservation $R_1 := (2, A, A + 1)$ and arguing that I' has an optimal makespan $C^*_{\max} = A$ if and only if I is a yes-instance of Partition.

3 Approximation Algorithms and Complexity Results

We present approximation algorithms and complexity results. In Subsect. 3.1 we obtain approximation schemes; in Subsect. 3.2 we discuss fast greedy algorithms that are based on the same idea. We close the section with complexity results in Subsect. 3.3.

3.1 Polynomial Time Approximation Schemes

We explain the MSSP approach for $m \geq 2$ to obtain a PTAS for our problem. Later we discuss the cases $m \in \{1, 2\}$, which admit FPTASes for the case where one reservation is permitted. Our idea is based on obtaining a complementary representation for the periods of availability in order to reduce the problem to MSSP which admits a PTAS [2,4]; we derive a dual approximation algorithm [8] by using binary search on the makespan where a PTAS for MSSP serves as a relaxed decision procedure, as illustrated with a Gantt chart in Fig. 1. In Sect. 2 we argued how to obtain sorted sets R'_i of reservations for each $i \in [m] \setminus \{1\}$. We use the algorithm in Fig. 2 to obtain sets of inclusionwise maximal availability intervals A_i for each $i \in [m]$, each containing elements (i, s, t) indicating that machine i is available in $[s, t)$ where $s \in \mathbb{N}$ and $t \in \mathbb{N} \cup \{\infty\}$. Due to space restrictions a detailed discussion of Fig. 2 is omitted.

The running time of the algorithm in Fig. 2 is linear in m, r and independent from n; at most $2r$ intervals of availability are generated. For a fixed $i \in [m]$, we use the initial sorting of R'_i to obtain that the intervals of availability for machine

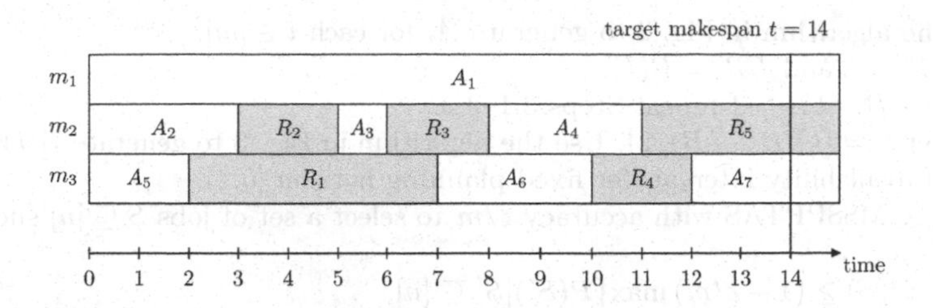

Fig. 1. Sketch illustrating the approach of the algorithm in Fig. 4. The grey zones $R_1, \ldots, R_5$ are the reservations. If the target makespan is 14, we try to fill all the jobs in knapsacks of sizes corresponding to $A_1, \ldots, A_7$; zones A_1 and A_7 end at time 14.

1. Set $A_1 := \{(1, 0, \infty)\}$ and for each $i \in [m] \setminus \{1\}$ set $A_i := \emptyset$.
2. For each $i \in [m] \setminus \{1\}$ execute Steps 2.1–2.3.
   - 2.1. If $r_i = 0$, set $A_i := A_i \cup \{(i, 0, \infty)\}$ and proceed with the next iteration of the loop in started in Step 2.
   - 2.2. Set $t := 0$.
   - 2.3. For each $r \in [r_i]$ execute Steps 2.3.1–2.3.2.
     - 2.3.1 If $s_{ir} = 0$ then proceed with the next iteration of the loop started in Step 2.1, otherwise set $A_i := A_i \cup \{(i, t, s_{ir})\}$ and $t := t_{ir}$.
     - 2.3.2 If $r = r_i$ then set $A_i := A_i \cup \{(i, t, \infty)\}$.

Fig. 2. Algorithm GenAvail

i are sorted with respect to their starting times. More important is the following subroutine that uses $A_1, \ldots, A_m$ to generate the finite intervals of *availability* for a *fixed finite* planning horizon $[0, t)$ where $t \in \mathbb{N}$.

Step 1.1 in Fig. 3 removes all intervals of availability that begin outside of $[0, t)$ while Step 1.2, if necessary, truncates the last interval on a machine to fit exactly into the planning horizon. The running time of the algorithm in Fig. 3 is independent from n and linear in m, r. We denote $A(t) := \cup_{i=1}^{m} A'_i(t)$ and will use the at most $2r$ intervals stored in $A(t)$ as knapsacks in which we like to pack the jobs in $[n]$. To this end, we use a PTAS for MSSP and for each job $j \in [n]$ define an item j with weight p_j to obtain an instance of MSSP. The algorithm is described in Fig. 4, where MSSPPTAS is a PTAS for MSSP where the capacities of the knapsacks may be different [2,4]. We suppose that MSSPPTAS does not only select a desired $S \subseteq [n]$ but also stores the feasible assignment to the knapsacks as a byproduct.

1. For each $i \in [m]$ execute Steps 1.1–1.2.
   - 1.1. Set $A'_i(t) = \{(i, s', t') \in A_i | s' < t)\}$ and $a_i := |A'_i(t)|$.
   - 1.2. If $a_i > 0$ set $t_{ia_i} := \min\{t_{ia_i}, t\}$.

Fig. 3. Algorithm GenAvailFinite

1. Use the algorithm in Fig. 2 to generate A_i for each $i \in [m]$.
2. Set $LB := 0$ and $UB := P(I)$.
3. While $UB - LB > 1$ repeat Steps 3.1–3.3.
   3.1 Set $t := \lfloor (UB - LB)/2 \rfloor$. Use the algorithm in Fig. 3 to generate $A(t)$, the set of availability intervals for fixed planning horizon $[0, t)$.
   3.2 Use MSSPPTAS with accuracy ϵ/m to select a set of jobs $S \subseteq [n]$ such that
   $$P(S) \geq (1 - \epsilon/m) \max\{P(S')|S' \subseteq [n], \\ S' \text{ permits a feasible packing into the intervals in } A(t)\}.$$
   3.3 If $P(S) < (1 - \epsilon/m)P(I)$ then set $LB := t$ else store S and set $UB := t$.
4. Schedule the jobs in the last stored set S into the interval $[0, UB)$ as indicated by the solution generated by MSSPPTAS when S was returned; schedule the jobs in $[n] \setminus S$ in the interval $[UB, \infty)$ on the first machine without unnecessary idle time.

Fig. 4. Algorithm MultiSubsetSumScheduler

Theorem 1. *The algorithm in Fig. 4 is a PTAS for* $Pm, 1up|nr\text{-}a|C_{\max}$.

Proof. Since the first machine is available at each time step $t \in [0, \infty)$, the sum of processing times $P(I)$ is an upper bound for the optimal makespan $C^*_{\max}$; hence in Step 2, the lower bound LB and the upper bound UB are initialized to have the following properties.

1. $LB < C^*_{\max}$.
2. There is a set $S \subseteq [n]$ such that the jobs in S permit a feasible schedule into the time horizon $[0, UB)$ and $P(S) \geq (1 - \epsilon/m)P(I)$.

The second property follows from the fact that, since $C^*_{\max} \leq UB$, all jobs can be scheduled in $[0, UB)$ and thus it is impossible that the algorithm MSSPPTAS returns a set $S \subseteq [n]$ such that $P(S) < (1 - \epsilon/m)P(I)$ holds; both properties are invariant under the update of LB and UB in Step 3.3. The number of iterations of the binary search in Step 3 is bounded by $\log P(I) \leq \log(np_{\max}) = \log n + \log p_{\max}$ which is polynomially bounded in the encoding length of I. On termination of the binary search in Step 3, $LB + 1 = UB$ holds, hence $UB \leq C^*_{\max}$ since $LB < C^*_{\max}$ is satisfied. This means that the set S selected in Step 4 can be scheduled in $[0, UB)$ and satisfies $P(S) \geq (1 - \epsilon/m)P(I)$; hence $P([n] \setminus S) \leq \epsilon P(I)/m$ holds. Furthermore the jobs in $[n] \setminus S$ can be scheduled on the first machine in $[UB, \infty)$ since the first machine is available. We have $P(I)/m \leq C^*_{\max}$; in total, the makespan of the schedule generated by the algorithm in Fig. 4 is bounded by $UB + \epsilon P(I)/m \leq C^*_{\max} + \epsilon C^*_{\max} = (1 + \epsilon)C^*_{\max}$ and we obtain the desired approximation ratio. □

However, since the running time of MSSPPTAS may grow exponentially in $1/\epsilon$, the running time of the algorithm in Fig. 4 may also grow exponentially in m. Furthermore, it is known that MSSP does not admit an FPTAS even for the special case of two knapsacks of equal capacity, unless $\mathsf{P} = \mathsf{NP}$ holds, as discussed in [14], Subsect. 10.4. Hence it is impossible for the approach used above to yield an FPTAS for our scheduling problem by replacing MSSPPTAS with a better algorithm, which is not surprising in the light of Corollary 1 in Subsect. 3.3.

For $m = 1$ the situation is different. Lee [17] remarked that $1|nr\text{-}a|C_{\max}$ is strongly NP-hard via reduction from 3-Partition, and the inapproximability can be seen by generalizing a suitable construction. If there is only one reservation, an FPTAS can be obtained since in [12,14] an FPTAS for SSP is available. This case corresponds to a simple knapsack problem; if all the tasks can be scheduled before the reservation then we get an optimal solution; otherwise we use the FPTAS for SSP to schedule as much as possible load before the reservation.

Now we sketch $m = 2$ with one reservation $R_1 = (2, s, t)$; for this case, an FPTAS can be obtained. Due to space constraints we omit the precise algorithmic details. The FPTAS is based on dynamic programming which yields an optimal algorithm with a pseudopolynomial runtime bound. This dynamic programming algorithm can be used to build an FPTAS by a suitable discretization of the state space; we obtain the following result.

Theorem 2. *The problem* $P2, 1up|nr\text{-}a|C_{\max}$ *with one reservation admits an FPTAS.*

3.2 Greedy Algorithms

In [5], a greedy 2-approximation algorithm for MSSP with running time $O(n^2)$ is briefly mentioned; the subject is also discussed in [14], Subsect. 10.4.1, with a slightly different approach yielding the same runtime bound. By using this algorithm instead of MSSPPTAS and changing the bound $1 - \epsilon/m$ to $1/2$ in Step 3 of the algorithm in Fig. 4 we obtain an approximation algorithm with ratio $1+m/2$ for $Pm, 1up|nr\text{-}a|C_{\max}$ by following the lines of the proof of Theorem 1. On the other hand, scheduling all jobs on the first machine here yields an m-approximation algorithm; hence the algorithm sketched above yields a better bound than this approach only if $m > 2$ holds.

In [17], Lee studied the case where at most one reservation per machine is permitted and one machine is always available; a tight approximation ratio of $(m+1)/2$ for LPT is proved. For our generalization $Pm, 1up|nr\text{-}a|C_{\max}$ we obtain the same asymptotic behaviour in m with our greedy approach. Comparing our result here with the tight bound $1+\lceil m/(m-\lambda)\rceil/2$ for LPT [9] where $\lambda \in [m-1]$ is the maximum number of machines which are permitted to be unavailable at the same time, we basically get the same ratio for our case $\lambda = m - 1$. In total, we obtain similar approximation ratios for more general problems, which comes at the cost of increased computational effort, however.

3.3 Complexity Results

We present an inapproximability result which shows that the PTAS for $Pm, 1up|nr\text{-}a|C_{\max}$ is close to best possible; hence $Pm, 1up|nr\text{-}a|C_{\max}$ is substantially harder than $Pm||C_{\max}$ which permits an FPTAS [22].

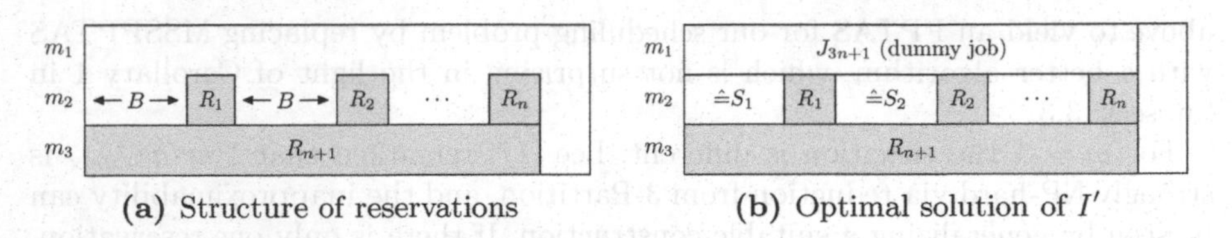

(a) Structure of reservations **(b)** Optimal solution of I'

Fig. 5. Sketch illustrating the proof of Theorem 3

Theorem 3. *The problem* $Pm, 1up|nr\text{-}a|C_{\max}$ *is strongly* NP*-hard for* $m \geq 2$.

Proof. We reduce from the strongly NP-complete problem 3-Partition [7]; see Fig. 5 for a sketch of the construction.

- *Given:* Index set $S = [3n]$, $a_i \in \mathbb{N}^*$ for each $i \in S$ and $B \in \mathbb{N}^*$ such that $B/4 < a_i < B/2$ for each $i \in S$ and $\sum_{i=1}^{3n} a_i = nB$ holds.
- *Question:* Is there a partition of the set S into $S_1, \ldots, S_n$ such that $\sum_{i \in S_j} a_i = B$ holds for each $j \in [n]$?

Given an instance I of 3-Partition we define an instance I' of the problem $Pm, 1up|nr\text{-}a|C_{\max}$ for $m \geq 2$. We set $p_i := a_i$ for each $i \in [3n]$ (*small* jobs), $p_{3n+1} := n(B+1)$ (*dummy* job) and define suitable reservations $R_i := (2, i(B+1)-1, i(B+1))$, $i \in [n]$, $R_{n+i} := (2+i, 0, n(B+1))$ for each $i \in [m-2]$. I' can be generated from I in time polynomial in the length of I and yields $C^*_{\max} = n(B+1)$ if and only if I is a yes-instance of 3-Partition by putting the small jobs according to the partition $S_1, \ldots, S_n$ in the intervals $[0, B), \ldots, [(n-1)(B+1), n(B+1)-1)$ on machine 2 and putting the dummy job on machine 1; conversely in a schedule with makespan $n(B+1)$ the dummy job must be put on machine 1 and hence the small jobs run on machine 2 which indicates the partition of S into $S_1, \ldots, S_n$ since no more than 3 small jobs can fit into an interval of length B. In total, $Pm, 1up|nr\text{-}a|C_{\max}$ is strongly NP-hard. □

Since the objective values of feasible schedules for $Pm, 1up|nr\text{-}a|C_{\max}$ are integral and $C^*_{\max} \leq P(I)$, the next result immediately follows from [6].

Corollary 1. $Pm, 1up|nr\text{-}a|C_{\max}$ *does not admit an FPTAS for* $m \geq 2$ *unless* P = NP.

It is a natural question whether the problem becomes easier if the number of reservations per machine is restricted to one. Surprisingly, this is not the case, which can be shown by adaptation of a construction from [1]. The following result implies that $Pm, 1up|nr\text{-}a|C_{\max}$ with at most one reservation per machine for $m \geq 3$ is strongly NP-hard.

Theorem 4. $Pm, 1up|nr\text{-}a|C_{\max}$ *does not admit an FPTAS, even if there is at most one reservation per machine, for* $m \geq 3$ *unless* P = NP.

Proof. We use a reduction from the following problem, Equal Cardinality Partition or ECP for short, which is NP-complete [7]; see Fig. 6 for a sketch of the construction.

- *Given:* Finite list $I = (a_1, \dots, a_n)$ of even cardinality with $a_i \in \mathbb{N}^*$ for each $i \in [n]$, $A \in \mathbb{N}^*$ such that $\sum_{i=1}^{n} a_i = 2A$ holds.
- *Question:* Is there a partition of the list I into lists I_1 and I_2 such that $|I_1| = n/2 = |I_2|$ and $\sum_{i \in I_1} a_i = A = \sum_{i \in I_2} a_i$ holds?

Given an instance I of ECP we define an instance I' of $Pm, 1up|nr\text{-}a|C_{\max}$ for $m \geq 3$ as follows. We set $p_i := 2A + a_i$ for each $i \in [n]$ (*small* jobs), $p_{n+1} := 2A(n+1)$ (*dummy* job) and $R_k := (k, A(n+1), 2A(n+1))$ for $k \in \{2, 3\}$ and $R_k := (k, 0, 2A(n+1))$ for each $k \in [m] \setminus \{1, 2, 3\}$. I' is generated from I in running time polynomial in the length of I. Furthermore I' has an optimal makespan of $C^*_{\max} = 2A(n+1)$ if and only if I is a yes-instance by executing the small jobs according to the partition I_1 and I_2 on machines 2 and 3 and putting the dummy job on machine 1; conversely in a schedule with makespan $2A(n+1)$ the dummy job is put on machine 1 and hence the small jobs run on machines 2 and 3 which indicates the partition of I into I_1 and I_2 since no more than $n/2$ jobs fit into an availability interval of length $A(n+1)$. Let I be a yes-instance of ECP and consider a suboptimal schedule of I'; the makespan of a suboptimal schedule of I' must be at least $2A(n+1) + A$ since every job in I' has a processing time larger than A and is scheduled either on machine $i \in [m] \setminus \{1\}$ or on machine 1 together with the dummy job, unless the dummy job is scheduled on a machine other than the first one. Given an FPTAS for $Pm, 1up|nr\text{-}a|C_{\max}$, choose $\epsilon \in (0, 1)$ such that

$$1 + \epsilon < \frac{2A(n+1) + A}{2A(n+1)} = \frac{2n+3}{2n+2}$$

holds, which is equivalent to $\epsilon < 1/(2n+2)$; consequently ϵ can be chosen in such a way that $1/\epsilon$ is polynomially bounded in n and hence polynomially bounded in the encoding length of I. Then, the FPTAS generates a schedule with makespan $C_{\max}$ such that

$$C_{\max} \leq (1+\epsilon)C^*_{\max} < \frac{2A(n+1) + A}{2A(n+1)} 2A(n+1) = 2A(n+1) + A$$

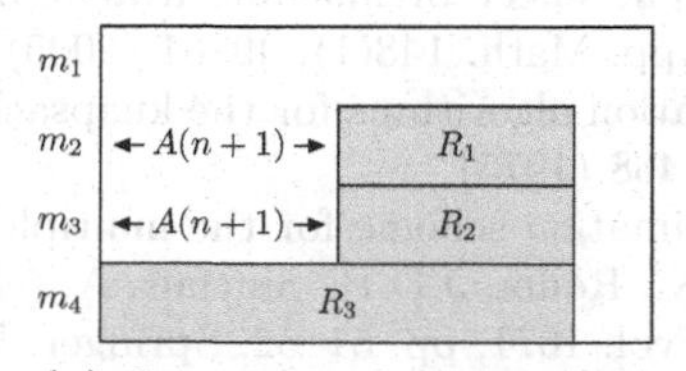

(a) Structure of reservations

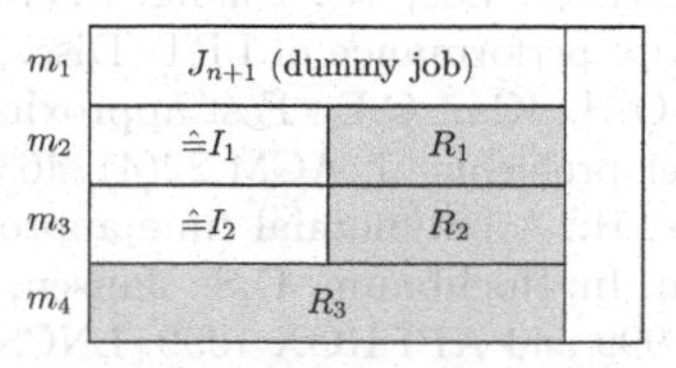

(b) Optimal solution of I'

Fig. 6. Sketch illustrating the proof of Theorem 4

holds. Hence I' is solved to optimality in polynomial time and I is identified as a yes-instance of ECP, which is impossible unless $\mathsf{P} = \mathsf{NP}$. □

4 Conclusion

We studied scheduling on a constant number of identical parallel machines with reservations and have shown that a sensible restriction to $Pm, 1up|nr\text{-}a|C_{\max}$ is necessary to obtain a bounded approximation ratio. On the algorithmic side we have taken an approach that is based on using approximation algorithms for SSP and MSSP. We obtained FPTASes for $1|nr\text{-}a|C_{\max}$ and $P2, 1up|nr\text{-}a|C_{\max}$ with one reservation, respectively. For the case of arbitrary constant m our approach yields a PTAS and we have shown that no FPTAS exists unless $\mathsf{P} = \mathsf{NP}$ holds, even if the number of reservations per machine is restricted to one.

Acknowledgements. The authors thank Érik Saule and Ulrich M. Schwarz for many fruitful discussions.

References

1. Caprara, A., Kellerer, H., Pferschy, U.: The multiple subset sum problem. Technical report, Technische Universität Graz (1998)
2. Caprara, A., Kellerer, H., Pferschy, U.: A PTAS for the multiple subset sum problem with different knapsack capacities. Inf. Process. Lett. 73(3-4), 111–118 (2000)
3. Caprara, A., Kellerer, H., Pferschy, U.: A 3/4-approximation algorithm for multiple subset sum. J. Heuristics 9(2), 99–111 (2003)
4. Chekuri, C., Khanna, S.: A polynomial time approximation scheme for the multiple knapsack problem. SIAM J. Comput. 35(3), 713–728 (2005)
5. Dawande, M., Kalagnanam, J., Keskinocak, P., Salman, F.S., Ravi, R.: Approximation algorithms for the multiple knapsack problem with assignment restrictions. J. Comb. Optim. 4(2), 171–186 (2000)
6. Garey, M.R., Johnson, D.S.: "strong" NP-completeness results: Motivation, examples, and implications. J. ACM 25(3), 499–508 (1978)
7. Garey, M.R., Johnson, D.S.: Computers and Intractability: A Guide to the Theory of NP-Completeness. W. H. Freeman (1979)
8. Hochbaum, D.S., Shmoys, D.B.: Using dual approximation algorithms for scheduling problems: theoretical and practical results. J. ACM 34(1), 144–162 (1987)
9. Hwang, H.-C., Lee, K., Chang, S.Y.: The effect of machine availability on the worst-case performance of LPT. Disc. App. Math. 148(1), 49–61 (2005)
10. Ibarra, O.H., Kim, C.E.: Fast approximation algorithms for the knapsack and sum of subset problems. J. ACM 22(4), 463–468 (1975)
11. Kellerer, H.: A polynomial time approximation scheme for the multiple knapsack problem. In: Hochbaum, D.S., Jansen, K., Rolim, J.D.P., Sinclair, A. (eds.) RANDOM 1999 and APPROX 1999. LNCS, vol. 1671, pp. 51–62. Springer, Heidelberg (1999)
12. Kellerer, H., Mansini, R., Pferschy, U., Speranza, M.G.: An efficient fully polynomial approximation scheme for the subset-sum problem. J. Comput. Syst. Sci. 66(2), 349–370 (2003)

13. Kellerer, H., Pferschy, U.: A new fully polynomial time approximation scheme for the knapsack problem. J. Comb. Optim. 3(1), 59–71 (1999)
14. Kellerer, H., Pferschy, U., Pisinger, D.: Knapsack Problems. Springer, Heidelberg (2004)
15. Lawler, E.L.: Fast approximation algorithms for knapsack problems. Math. Oper. Res. 4(4), 339–356 (1979)
16. Lee, C.-Y.: Parallel machines scheduling with non-simultaneous machine available time. Disc. App. Math. 30, 53–61 (1991)
17. Lee, C.-Y.: Machine scheduling with an availability constraint. J. Global Optimization, Special Issue on Optimization of Scheduling Applications 9, 363–384 (1996)
18. Lee, C.-Y., He, Y., Tang, G.: A note on parallel machine scheduling with non-simultaneous machine available time. Disc. App. Math. 100(1-2), 133–135 (2000)
19. Leung, J.Y.-T. (ed.): Handbook of Scheduling. Chapman & Hall (2004)
20. Liao, C.-J., Shyur, D.-L., Lin, C.-H.: Makespan minimization for two parallel machines with an availability constraint. European J. of Operational Research 160, 445–456 (2003)
21. Martello, S., Toth, P.: Knapsack Problems: Algorithms and Computer Implementations. Wiley, Chichester (1990)
22. Sahni, S.: Algorithms for scheduling independent tasks. J. ACM 23(1), 116–127 (1976)
23. Scharbrodt, M., Steger, A., Weisser, H.: Approximability of scheduling with fixed jobs. J. Scheduling 2, 267–284 (1999)

Enhanced Real-Time Divisible Load Scheduling with Different Processor Available Times

Xuan Lin, Ying Lu, Jitender Deogun, and Steve Goddard

Department of Computer Science and Engineering
University of Nebraska - Lincoln, Lincoln, NE 68588
{lxuan, ylu, deogun, goddard}@cse.unl.edu

Abstract. Providing QoS and performance guarantees for arbitrarily divisible loads in a cluster has become a significant problem. While progress is being made in scheduling arbitrarily divisible loads, some of the proposed approaches may cause Inserted Idle Times (IITs) that are detrimental to system performance. Two contributions are made in addressing this problem. First, we propose two constraints that, when satisfied, lead to an optimal partitioning in utilizing IITs. Second, we integrate the new partitioning method with a previous approach and develop an enhanced algorithm that better utilizes IITs. Simulation results demonstrate the advantages of our new approach.

Keywords: Real-Time Scheduling, Inserted Idle Time, Cluster Computing, Divisible Load.

1 Introduction

Arbitrarily divisible applications consist of an amount of data that can be divided arbitrarily into a desirable number of independent load fractions, and each sub-task (fraction) itself is arbitrarily divisible. This perfectly parallel model is a good representation of many scientific applications that consist of huge numbers of identical, low-granularity loads. For example, the CMS (Compact Muon Solenoid) [1] and ATLAS (AToroidal LHC Apparatus) [2] projects, associated with the LHC (Large Hadron Collider) at CERN (European Laboratory for Particle Physics), execute cluster-based applications with arbitrarily divisible loads. Usually, such applications require a large amount of resources and can only be deployed in commodity clusters or computational grids.

To efficiently utilize large-scale clusters, an on-line resource management system (RMS) is needed to provide real-time guarantees or QoS. This is becoming a significant issue for research computing facilities, such as the U.S. CMS Tier-2 sites [3], that execute large numbers of arbitrarily divisible loads. Thus, researchers, e.g., [4,5], have begun to investigate real-time divisible load scheduling, with significant initial progress in important theories and applications.

However, the challenge, to efficiently schedule a job when there are not enough resources, has not yet been adequately addressed for real-time divisible loads. When scheduling a parallel job, if a sufficient number of processors are available,

S. Aluru et al. (Eds.): HiPC 2007, LNCS 4873, pp. 308–319, 2007.

the processors are allocated and the job is started. But if the required number of processors is not available, the job waits for additional processors. This essentially leads to a waste of processing power as some processors sit idle waiting to start the job. This is a system inefficiency that we refer to as the Inserted Idle Times (IITs) problem [6]. To alleviate this limitation, backfilling algorithms [7] have been proposed, where small jobs could be moved ahead and run on processors that would otherwise remain idle.

Leveraging characteristics of arbitrarily divisible loads, we have previously developed a real-time scheduling algorithm that utilizes IITs [6]. Although the approach has significantly improved the system performance, it cannot fully utilize IITs. In this paper, we propose a new strategy to further make use of IITs. Not only can our enhanced algorithm schedule real-time divisible loads with different processor available times, when certain conditions hold, it can also optimally partition and schedule jobs to fully utilize IITs. Two contributions are made in this paper. First, we propose a new partitioning approach to fully utilize IITs and investigate its applicability constraints. Second, we integrate this with our previous work [6] and propose a new real-time scheduling algorithm.

The remainder of this paper is organized as follows. Related work is presented in Section 2. We describe both task and system models in Section 3. Section 4 discusses real-time scheduling algorithms investigated in this paper. We evaluate the algorithms performance in Section 5 and conclude the paper in Section 6.

2 Related Work

The scheduling models investigated for real-time distributed systems most often (e.g., [8]) assume periodic or aperiodic sequential jobs where each job is allocated to a single resource and must be executed by its deadline. With the evolution of cluster computing, researchers have begun to investigate real-time scheduling of parallel applications. However, each of these studies assume the existence of some form of task graph to describe communication and precedence relations between computational units called subtasks (i.e., nodes in the task graph). Despite the increasing importance of arbitrarily divisible applications [4], to the best of our knowledge, only a few researchers [5,9] have investigated the real-time scheduling of arbitrarily divisible loads.

Utility-driven cluster computing has been well researched [10] to improve the utility delivered to users. Proposed cluster RMSs [11] have addressed the scheduling of both sequential and parallel loads. The goal of those schemes is similar to ours: *to harness the power of resources based on user objectives.*

The most closely related work to ours is the scheduling of “scalable tasks” [9] or “moldable jobs” [12], where only a few papers [9] have considered QoS support. In [5] we investigated real-time cluster-based divisible load scheduling and proposed several algorithms for homogenous clusters. In [6], we developed a real-time scheduling approach that utilizes Inserted Idle Times (IITs). A mechanism to utilize processor idle-times, also called fragments, was investigated in [9], wherein a task is assigned a larger number of nodes to utilize more processing power. Complementary to that approach, our algorithm in [6] enables a task to utilize a processor as soon as it becomes available. While results in [9] show that the performance improvement of their approach is negligible, our previous approach [6] has led to much better performance, even though it cannot fully utilize IITs. In this paper, we propose a new partitioning approach. We prove that if certain constraints hold, any task can be partitioned optimally to fully make use of IITs. By integrating such an optimal partitioning method with algorithms we have proposed in [6], system resources are better utilized and the performance is further improved.

Divisible load theory (DLT) provides an in-depth study of distribution strategies for arbitrarily divisible loads [13]. In our previous work [5], we demonstrated that the application of DLT leads to significantly better approaches for real-time divisible load scheduling. Encouraged by its performance benefits, we again apply DLT to develop our new partitioning algorithm.

3 Task and System Models

In this paper, we adopt the same task and system models as our previous work [5,6]. For completeness, we briefly present these below.

Task Model. Similar to the classic real-time aperiodic task model, we assume each aperiodic task T_i consists of a single invocation specified by the tuple (A_i, σ_i, D_i), where A_i is the task arrival time, σ_i is the total data size of the task, and D_i is its relative deadline. The task absolute deadline is given by $A_i + D_i$. We present in Section 4 how task execution time is dynamically computed based on total data size σ_i, resources allocated (i.e., processing nodes and bandwidth), and the partitioning method applied to parallelize the computation.

System Model. We consider a common cluster model, which consists of a head node, denoted by P_0, connected via a switch to N processing nodes, denoted by $P_1, P_2, \ldots, P_N$. We assume a homogeneous model in which all processing nodes have the same computational power and all links from the switch to the processing nodes have the same bandwidth. Like a typical cluster environment, the system model assumes the head node does not participate in computation. The role of the head node is to accept or reject incoming tasks, execute the scheduling algorithm, divide the workload and distribute data chunks to processing nodes. Since different nodes process different data chunks, the head node sequentially sends every data chunk to its corresponding processing node via the switch. We assume that data transmission does not occur in parallel, although it is straightforward to generalize our model and include the case where some pipelining of communication may occur.

As with divisible load theory, we use linear models to represent processing and transmission times [14]. In the simplest scenario, the computation time of a load σ is calculated by a cost function $Cp(\sigma) = \sigma\chi$, where χ represents the time to compute a unit of workload on a single processing node. The communication time of a load σ is calculated by a cost function $Cm(\sigma) = \sigma\tau$, where τ is the time to transmit a unit of workload from the head node to a processing node.

The following notations, partially adopted from [14], are used in this paper.

- $T = (A, \sigma, D)$: A divisible task, where A is the arrival time, σ is the data size, and D is the relative deadline;
- $\alpha = (\alpha_1, \alpha_2, ..., \alpha_n)$: Data distribution vector, where n is the number of processing nodes allocated to the task, α_j is the data fraction allocated to the jth node, i.e., $\alpha_j\sigma$, is the amount of data that is to be transmitted to the jth node for processing, $0 < \alpha_j \leq 1$ and $\Sigma_{j=1}^{n}\alpha_j = 1$;
- τ: Cost of transmitting a unit workload;
- χ: Cost of processing a unit workload.

4 Algorithms

This section presents real-time divisible load scheduling algorithms that utilize Inserted Idle Times (IITs) in a cluster. Due to space limitations, we omit the proofs of the theorems in this Section.

4.1 Real-Time Divisible Load Scheduling

An admission controller, which is a part of our scheduler, executes on the head node. As is typical for dynamic real-time scheduling algorithms , when a task arrives, the scheduler dynamically determines if it is feasible to schedule the new task without compromising the guarantees for previously admitted tasks. The task's schedulability test will be described in Section 4.5.

In [5], we encapsulated the logic of a real-time divisible load scheduling algorithm in three modules. The first module determines the task execution order, which could be based on policies, such as FIFO (first in first out) or EDF (earliest deadline first). The second task partitioning module chooses a strategy to divide loads, while the third module decides the node assignment for each task. In [5] we have shown that assigning only the minimum number of nodes to a task to meet its deadline is an efficient approach. However, this strategy, as with many others, leads to the IITs problem. To utilize IITs, we focused on the second module in [6]. That is, we designed a new task partitioning module for real-time divisible load scheduling. However, that approach has limitations. First, it does not guarantee optimal partitioning [14] and cannot ensure all assigned nodes complete their computations at the same time. Second, not all IITs are fully utilized. In the following sections, we first find the scenario when the IITs can be fully utilized and propose an optimal partitioning approach to fully make use of IITs. We then integrate this partitioning method with algorithms proposed in [6] to improve the system's real-time performance.

4.2 Inserted Idle Times Problem

The IITs problem occurs when the number of processors available is less than that required by the next job. Many parallel job scheduling algorithms [7,5] lead to this problem. If no strategy is implemented to make use of IITs, the job has to wait until enough processors become available, which leads to a waste of processing power as some processors are idle waiting. Backfilling [7] is an approach proposed in the literature to alleviate this problem. It is a general approach applicable to all types of parallel jobs — whether task graph based and modularly divisible or arbitrarily divisible.

An arbitrarily divisible load, however, has a unique property, that is, it can be arbitrarily partitioned into a large number of independent subtasks of arbitrarily small size. Thus, the subtasks can be scheduled flexibly and independently. Exploiting this property of arbitrarily divisible loads, we have proposed algorithms [6] that schedule divisible loads with different processor available times and utilize IITs in a cluster.

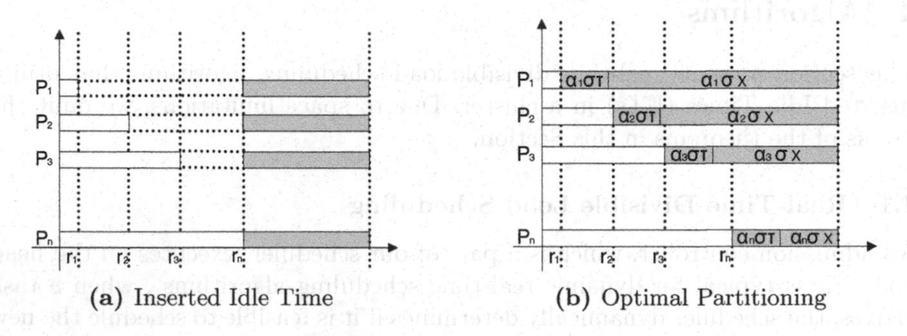

(a) Inserted Idle Time (b) Optimal Partitioning

Fig. 1. Timing Diagrams

As shown in Figure 1a, when a task arrives at time r_1, the scheduler decides to assign the task n processors. However, it is not until time r_n that all of the n processors become available. We denote the available time of the ith processor as r_i, $i = 1, \ldots, n$. Without loss of generality, we assume the processors are ordered by their available times, that is

$$r_1 \leq r_2 \leq ...r_i \leq ... \leq r_n$$

When scheduling the task, if we do not make use of the processors before time r_n, as shown in Figure 1a, the task will not start until time r_n. Let $p_i = r_n - r_i$, $i = 1, 2, \ldots, n$. Since the ith processor is idle during the time period $[r_i, r_n]$, the total wasted cycles for these n processors is $\sum_{i=1}^{n} p_i$, which leads to sub-optimal cluster performance.

4.3 Optimal Partitioning

The unique property of a divisible load provides flexibility in scheduling. We can start part of the load on a processor as soon as the processor becomes available. However, the challenge is to optimally partition the load among the n assigned processors. According to DLT, the optimal execution time is obtained when all nodes allocated to a divisible task finish their computations at the same time

[14]. Let s_i represent the start time of the ith node and $\mathcal{E}(\sigma, n)$ denote the task completion time. To ensure all nodes finish at the same time, we have

$$\begin{aligned}\mathcal{E}(\sigma, n) &= s_1 + \alpha_1\sigma(\tau + \chi) = s_2 + \alpha_2\sigma(\tau + \chi) = ... \\ &= s_n + \alpha_n\sigma(\tau + \chi).\end{aligned} \tag{1}$$

If any processor i can start as soon as it becomes available at time r_i (i.e., $s_i = r_i$, i=1,...,n), we can fully utilize IITs. Following DLT, with no IITs, an optimal partitioning will result in the task completion time of Equation (2).

$$\begin{aligned}\mathcal{E}(\sigma, n) &= r_1 + \alpha_1\sigma(\tau + \chi) = r_2 + \alpha_2\sigma(\tau + \chi) = ... \\ &= r_n + \alpha_n\sigma(\tau + \chi)\end{aligned} \tag{2}$$

Theorem 1. *If* $\forall i : 1 \leq i \leq n$, $s_i = r_i$, *Equations* (3) *and* (4) *result in an optimal partitioning of the load with no unused IITs.*

$$\alpha_1 = \frac{1}{n} + \frac{(n-1)p_1 - \sum_{i=2}^{n} p_i}{n\sigma(\tau + \chi)} \tag{3}$$

$$\alpha_i = \alpha_1 - \frac{p_1 - p_i}{\sigma(\tau + \chi)} \quad \textit{where } p_i = r_n - r_i. \tag{4}$$

Figure 1b shows the task execution time diagram following this optimal partitioning scheme. We can see in this scenario every processor is busy for either its data transmission or its subtask computation.

4.4 Constraints for Optimal Partitioning

The start time may not always be equal to the release time, as in our model data transmission is not parallel, the start time s_i of the ith node may be delayed by data transmissions to the $1^{st}, 2^{nd}, \ldots, (i-1)^{th}$ nodes. Thus, we have

$$s_1 = r_1, s_i = \max(r_i, s_{i-1} + \alpha_{i-1}\sigma\tau), i = 2, \ldots, n$$

And in the worst case when $r_1 = r_2 = \ldots = r_n$, we have

$$s_1 = r_1, s_2 = r_2 + \alpha_1\sigma\tau, s_3 = r_3 + \alpha_1\sigma\tau + \alpha_2\sigma\tau$$

$$\ldots\ldots$$

$$s_n = r_n + \sum_{i=1}^{n-1} \alpha_i\sigma\tau$$

In the dynamic scheduling process, the n processor available times could be arbitrary. Thus, the analysis to get a closed-form solution for the completion time becomes very difficult. For this reason, in [6] we cast a homogenous cluster with different processor available times to a heterogeneous cluster model and then applied a DLT heterogeneous model to guide the task partitioning and to derive a task execution time function. Although the intensive simulation results show that our previous approach does improve system performance, it cannot utilize all of the IITs ($\sum_{i=1}^{n} p_i$).

In the remainder of this section, we first prove that IITs cannot always be completely eliminated. Then we focus on identifying the scenarios where we can fully utilize IITs.

Theorem 2. *It is not always possible to eliminate all IITs.*

Constraint 1. $r_i - r_{i-1} \geq \sigma\tau,\ i = 2, 3, \ldots, n.$

Constraint 1 requires that the difference between available times of two successive processors to be no less than the total task data transmission time. This is a strong constraint. Later, we will propose another weaker constraint.

Theorem 3. *If Constraint 1 holds, Equations* (3) *and* (4) *result in an optimal partitioning of the load with no IITs.*

As we have mentioned, Constraint 1 is very strong. It is a sufficient but not a necessary condition for applying the optimal partitioning scheme. In some cases, Constraint 1 does not hold but the optimal partitioning approach is still applicable. Next we propose a weaker constraint to replace Constraint 1 and thus overcome its pessimism.

Note that for $i = 2, 3, \ldots, n$, the start time of the ith node will not be delayed if the data transmission time for the $(i-1)^{th}$ node is equal to or smaller than the difference between the two nodes' available times. Based on this observation, Constraint 2 is derived and Corollary 1 follows immediately from Theorem 1, since $\forall i : 1 \leq i \leq n,\ s_i = r_i$ in this case.

Constraint 2. $\alpha_{i-1}\sigma\tau \leq r_i - r_{i-1},\ i = 1, \ldots, n.$

Corollary 1. *If Constraint 2 holds, Equations* (3) *and* (4) *result in an optimal partitioning of the load with no IITs.*

4.5 Constraint-Based Algorithms

We first design algorithms based on Constraint 1. Figure 2 shows the general process of the schedulability test. After we detect IITs (i.e., $\exists\ r_i, r_j : r_i \neq r_j$), we first test whether Constraint 1 holds. If so, we repartition the task using our optimal partitioning approach presented in Section 4.3. Otherwise, we use the homgeneous-to-heterogeneous remodeling approach of [6] to repartition the task.

Various scheduling order policies can be integrated into our framework. In [6], we consider two such policies. One is FIFO, which is a common practice adopted by cluster administrators. The other is EDF, the Earliest-Deadline-First algorithm, which is a common real-time scheduling algorithm. The two algorithms we proposed in [6] are FIFO-DLT and EDF-DLT. They adopt different scheduling policies but apply the same partitioning and node assignment strategies. In

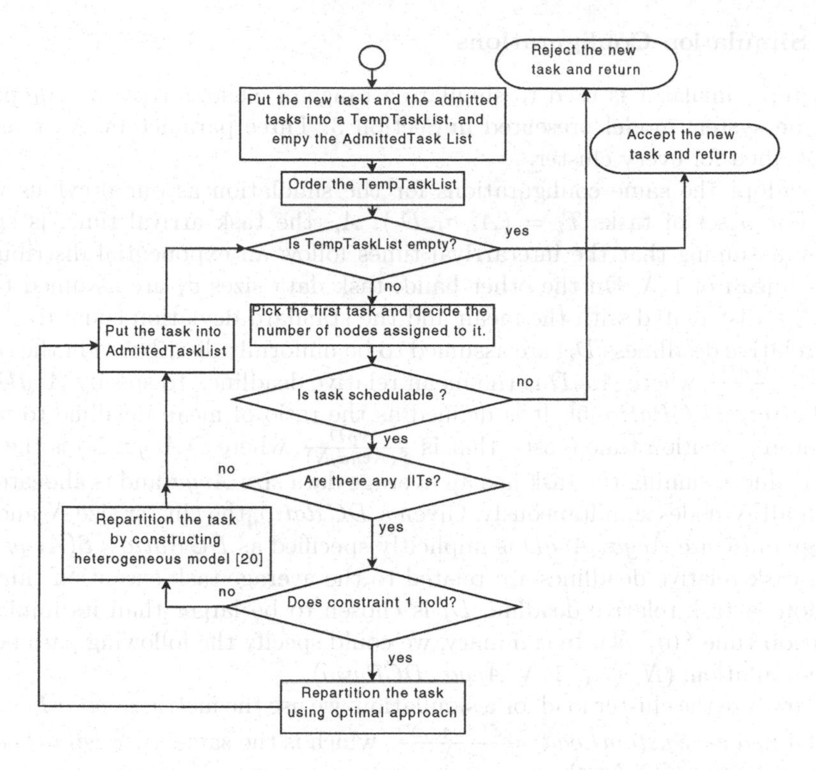

Fig. 2. Schedulability Test

this paper, by integrating our optimal partitioning approach with the two algorithms, we obtain two new algorithms: FIDO-I (FIfo Divisible load with Optimal partitioning) and EDDO-I (EDf Divisible load with Optimal partitioning).

The other two corresponding algorithms based on Constraint 2 are FIDO-II and EDDO-II. Unlike Constraint 1, Constraint 2 can only be verified after the task partition is determined because the value of α_i cannot be known until we have the partitioning scheme. Thus, we first assume that Constraint 2 will be satisfied and use the optimal approach to partition the task. We check afterwards whether or not Constraint 2 really holds. If so, we are assured that the task partition is optimal. Otherwise, we fall back to applying the non-optimal partitioning strategy defined in [6].

5 Performance Evaluation

In this section, we first evaluate the proposed two sets of real-time scheduling algorithms: Set 1 = {FIDO-I, EDDO-I} and Set 2 = {FIDO-II and EDDO-II}. We compare the two algorithms in each set with the corresponding algorithms, FIFO-DLT and EDF-DLT, that we proposed in [6], which do not apply the optimal partitioning approach presented here. Second, we evaluate how communication cost will affect the performance of our approach.

5.1 Simulation Configurations

A discrete simulator is used to simulate a range of clusters that are compliant with the system model presented in Section 3. Three parameters, N, τ and χ are specified for every cluster.

We adopt the same configurations for the simulation as our previous work [5,6]. For a set of tasks $T_i = (A_i, \sigma_i, D_i)$, A_i, the task arrival time, is specified by assuming that the interarrival times follow an exponential distribution with a mean of $1/\lambda$. On the other hand, task data sizes σ_i are assumed to be normally distributed with the mean and the standard deviation equal to $Avg\sigma$. Task relative deadlines (D_i) are assumed to be uniformly distributed in the range $[\frac{AvgD}{2}, \frac{3AvgD}{2}]$, where $AvgD$ is the mean relative deadline. To specify $AvgD$, we use the term $DCRatio$ [5]. It is defined as the ratio of mean deadline to mean minimum execution time (cost), that is $\frac{AvgD}{\mathcal{E}(Avg\sigma,N)}$, where $\mathcal{E}(Avg\sigma, N)$ is the execution time assuming the task has an average data size $Avg\sigma$ and is allocated to run on all N nodes simultaneously. Given a $DCRatio$, the cluster size N and the average data size $Avg\sigma$, $AvgD$ is implicitly specified as $DCRatio \times \mathcal{E}(Avg\sigma, N)$. Thus, task relative deadlines are related to the average task execution time. In addition, a task relative deadline D_i is chosen to be larger than its minimum execution time $\mathcal{E}(\sigma_i, N)$. In summary, we could specify the following parameters for a simulation: (N, τ, χ, $1/\lambda$, $Avg\sigma$, $DCRatio$).

To analyze the cluster load for a simulation, we use the metric $SystemLoad$ [5]. It is defined as, $SystemLoad = \frac{\mathcal{E}(Avg\sigma,N)}{\lambda}$, which is the same as, $SystemLoad = \frac{TotalTaskNumber \times \mathcal{E}(N,Avg\sigma)}{TotalSimulationTime}$. For a simulation, we could specify $SystemLoad$ instead of average interarrival time $1/\lambda$. Configuring (N, τ, χ, $SystemLoad$, $Avg\sigma$, $DCRatio$) is equivalent to specifying (N, τ, χ, $1/\lambda$, $Avg\sigma$, $DCRatio$), because, $1/\lambda = \frac{SystemLoad}{\mathcal{E}(Avg\sigma,N)}$. To evaluate the performance of the real-time scheduling algorithms, we use the metric, *Task Reject Ratio*, defined as the ratio of the number of task rejections to the number of task arrivals. The smaller the *Task Reject Ratio*, the better the real-time scheduling algorithm.

For all figures, a point on a curve corresponds to the average performance of ten simulations. For all ten runs, the same parameters (N, τ, χ, $SystemLoad$, $Avg\sigma$, $DCRatio$) are specified but different random numbers are generated for task arrival times A_i, data sizes σ_i, and deadlines D_i. For each simulation, the $TotalSimulationTime$ is 10,000,000 time units, which is sufficiently long.

5.2 Advantages of Optimal Partitioning Strategy

As discussed in Section 4.4, the optimal partitioning strategy is applicable whenever Constraint 1 or Constraint 2 is satisfied. Therefore, we first analyze the frequency of cases in which Constraint 1 or Constraint 2 holds. If most of the time Constraint 1 or Constraint 2 does not hold, then the corresponding constraint-based algorithms are not going to be very useful. For convenience, we refer to the scenario when the scheduler detects IITs as S-IITs. In addition, if Constraint 1 holds, we refer to it as Opt-IITs, and if Constraint 2 holds, we refer to it as Opt-IITs-II.

To estimate the occurrence ratio of Opt-IITs and Opt-IITs-II to S-IITs, we conducted an experiment with the baseline system configuration $N = 16$, $\tau = 1$, $\chi = 100$, $Avg\sigma = 200$, $DCRatio = 2$ [6]. We executed the FIFO-DLT algorithm [6] and measured occurrences of S-IITs, Opt-IITs and Opt-IITs-II. The results are summarized in Table 1, where the last two columns present the occurrence ratio of Opt-IITs to S-IITs and the occurrence ratio of Opt-IITs-II to S-IITs. From the data we observe that in all cases 12%-16% of S-IITs are in fact Opt-IITs, and 14% -20% of S-IITs are Opt-IITs-II. This indicates there are a large number of opportunities to apply our new approach to utilize IITs more efficiently and thus to improve system performance.

Figures 3a and 3b respectively show the comparison of algorithms FIDO-I and EDDO-I, FIDO-II and EDDO-II with their corresponding algorithms FIFO-DLT and EDF-DLT. From Figure 3a we observe that FIDO-I has a lower *Task*

Table 1. Occurrences of S-IITs and Opt-IITs. The S-IITs column indicates the number of times the scheduler detects IITs. The Opt-IITs column indicates the number of times Constraint 1 holds after the scheduler has detected IITs. The Opt-IITs-II column indicates the number of times Constraint 2 holds after the scheduler has detected IITs. The last two columns are the ratio of Opt-IITs to S-IITs and the ratio of Opt-IITs-II to S-IITs.

System Load	Number of S-IITs	Number of Opt-IITs	Number of Opt-IITs-II	Ratio of Opt-IITs to S-IITs	Ratio of Opt-IITs-II to S-IITs
0.1	131	21	22	16%	17%
0.2	262	31	36	12%	14%
0.3	665	105	113	16%	17%
0.4	1092	142	158	13%	14%
0.5	1927	229	289	12%	15%
0.6	2304	297	321	13%	14%
0.7	3560	493	601	14%	17%
0.8	4551	553	735	13%	16%
0.9	5661	757	980	14%	17%
1.0	6108	890	1203	15%	20%

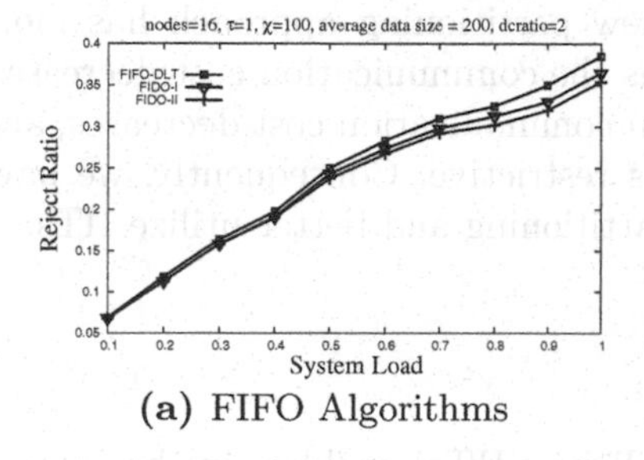

(a) FIFO Algorithms

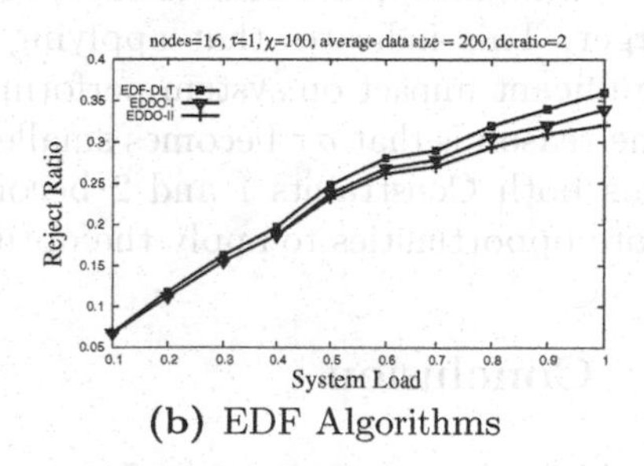

(b) EDF Algorithms

Fig. 3. Advantages of Optimal Partitioning

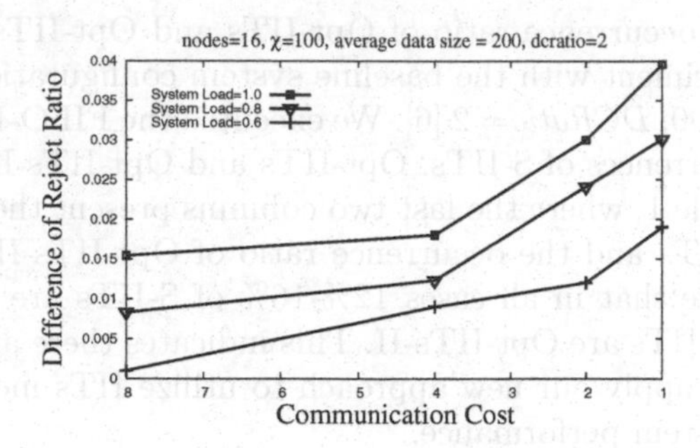

Fig. 4. Effects of Communication Cost

Reject Ratio than FIFO-DLT, which shows that applying the new partitioning approach leads to better performance. We also observe that FIDO-II has even better performance. These results confirm that because Constraint 2 is not as tight as Constraint 1, there are more chances to apply the optimal partitioning. Figure 3b shows similar results. We conclude that it is beneficial to integrate our new partitioning approaches with real-time divisible load scheduling.

5.3 Effects of Communication Cost

In this section, we run simulations with decreasing values of τ to study the effects of communication cost. With recent advancements in technology, networks are becoming faster and faster, and gigabits per second bandwidth has become commonplace. Moreover, the availability of thousands of wavelengths per fiber, and the development and deployment of all-optical switches and routers will result in further reductions in communication cost. For this reason, we investigate how decreases in communication costs will affect the performance of our approach.

For the simulation, we varied the values of τ from 8 to 4, 2 and 1, while keeping the other parameters constant as the baseline configuration presented in Section 5.2. Each point in Figure 4 represents the difference in *Task Reject Ratios* for EDF-DLT and EDDO-II algorithms. The three curves represent the cases when *SystemLoad* is equal to 1.0, 0.8 and 0.6 respectively. We can see that, as the communication cost decreases, the difference in *Task Reject Ratios* becomes larger. This indicates that applying our new partitioning approach has more significant impact on system performance as the communication cost decreases. The reason is that $\sigma\tau$ becomes smaller when communication cost decreases, and thus both Constraints 1 and 2 become less restrictive. Consequently, we have more opportunities to apply the optimal partitioning and better utilize IITs.

6 Conclusion

In this paper, we address the Inserted Idle Times (IITs) problem in the context of real-time divisible load scheduling [5]. Two contributions are made. First, we propose two constraints for the existence of the optimal partitioning that can fully utilize IITs. Second, we integrate this approach with our previous work [6] to develop new real-time scheduling algorithms that utilize IITs. Simulation results show that our approach makes use of IITs to a larger extent and significantly improves the system performance. Currently, we are working on expanding our approach to show that by adopting multi-round scheduling [13], we can further improve the IITs utilization and the system performance.

References

1. Compact Muon Solenoid (CMS) Experiment for the Large Hadron Collider at CERN (European Lab for Particle Physics): Cms web page, `http://cmsinfo.cern.ch/Welcome.html/`
2. ATLAS (AToroidal LHC Apparatus) Experiment, CERN (European Lab for Particle Physics): Atlas web page, `http://atlas.ch/`
3. Swanson, D.: Personal communication. Director, UNL Research Computing Facility (RCF) and UNL CMS Tier-2 Site (2005)
4. Robertazzi, T.G.: Ten reasons to use divisible load theory. Computer 36(5), 63–68 (2003)
5. Lin, X., Lu, Y., Deogun, J., Goddard, S.: Real-time divisible load scheduling for cluster computing. In: 13th IEEE Real-Time and Embedded Technology and Application Symposium, Bellevue, WA, pp. 303–314 (2007)
6. Lin, X., Lu, Y., Deogun, J., Goddard, S.: Real-time divisible load scheduling with different processor available times. Technical Report TR-UNL-CSE-2007-0013, University of Nebraska-Lincoln (2007)
7. Lifka, D.A.: The anl/ibm sp scheduling system. In: IPPS 1995. Proceedings of the Workshop on Job Scheduling Strategies for Parallel Processing, London, UK, Springer-Verlag, pp. 295–303. Springer, Heidelberg (1995)
8. Anderson, J.H., Srinivasan, A.: Pfair scheduling: beyond periodic task systems. In: RTCSA 2000. Proceedings of the Seventh International Conference on Real-Time Systems and Applications (RTCSA 2000), Cheju Island, South Korea, pp. 297–306 (2000)
9. Lee, W.Y., Hong, S.J., Kim, J.: On-line scheduling of scalable real-time tasks on multiprocessor systems. Journal of Parallel and Distributed Computing 63(12), 1315–1324 (2003)
10. Yeo, C.S., Buyya, R.: A taxonomy of market-based resource management systems for utility-driven cluster computing. Softw. Pract. Exper. 36(13), 1381–1419 (2006)
11. Amir, Y., Awerbuch, B., Barak, A., Borgstrom, R., Keren, A.: An opportunity cost approach for job assignment in a scalable computing cluster. IEEE Trans. on Parallel and Distributed Systems 11(7), 760–768 (2000)
12. Barsanti, L., Sodan, A.C.: Adaptive job scheduling strategies via predictive job resource allocation. In: Frachtenberg, E., Schwiegelshohn, U. (eds.) JSSPP 2006. LNCS, vol. 4376, pp. 115–140. Springer, Heidelberg (2007)
13. Bharadwaj, V., Robertazzi, T.G., Ghose, D.: Scheduling Divisible Loads in Parallel and Distributed Systems. IEEE Computer Society Press, Los Alamitos (1996)
14. Veeravalli, B., Ghose, D., Robertazzi, T.G.: Divisible load theory: A new paradigm for load scheduling in distributed systems. Cluster Computing 6(1), 7–17 (2003)

A General Distributed Scalable Peer to Peer Scheduler for Mixed Tasks in Grids

Cong Liu, Sanjeev Baskiyar*, and Shuang Li

Dept. of Computer Science and Software Engineering
Auburn University, Auburn, AL 36849
{liucong, baskisa, lishuan}@auburn.edu

Abstract. We consider non-preemptively scheduling a bag of independent mixed tasks in computational grids. We construct a novel Generalized Distributed Scheduler (*GDS*) for tasks with different priorities and deadlines. Tasks are ranked based upon priority and deadline and scheduled. Tasks are shuffled to earlier points to pack the schedule and create fault tolerance. Dispatching is based upon task-resource matching and accounts for computation as well as communication capacities. Simulation results demonstrate that with respect to the number of high-priority tasks meeting deadlines, *GDS* outperforms prior approaches by over 40% without degrading schedulability of other tasks. Indeed, with respect to the total number of schedulable tasks meeting deadlines, *GDS* outperforms them by 4%. The complexity of *GDS* is $O(n^2m)$ where n is the number of tasks and m the number of machines. *GDS* successfully schedules tasks with hard deadlines in a mix of soft and firm tasks, without a knowledge of a complete state of the grid. This way it helps open the grid and makes it amenable for commercialization.

1 Introduction

A major motivation of grid computing [5] [6] is to aggregate the power of widely distributed resources to provide services. Application scheduling plays a vital role in pro- viding such services. A number of deadline-based scheduling algorithms already exist. However, in this paper we address the problem of scheduling a bag of independent mixed tasks in computational grids. We consider three types of tasks: hard, firm and soft [8]. It is reasonable for a grid scheduler to prioritize such mission critic- al tasks while maximizing the total number of tasks meeting deadlines. *Doing so may make the grid commercially viable as it opens it up for all classes of users*.

To the best of our knowledge, *GDS* is the first attempt at prioritizing tasks according to task types as well as considering deadlines and dispatch times. It also matches tasks to appropriate computational and link bandwidth resources. Additionally, *GDS* consists of a unique shuffle phase that reschedules mission critical tasks as early as possible to provide temporal fault tolerance. Dispatching tasks to peers is based upon both

* This research was supported in part by NSF-grant OCI 048136.

S. Aluru et al. (Eds.): HiPC 2007, LNCS 4873, pp. 320–330, 2007.

computational capacity and link bandwidth. Furthermore, *GDS* is highly scalable as it does not require a full knowledge of the state of all nodes of the grid as many other algorithms do. For *GDS*'s peer to peer dispatch, knowledge of peer site capacities is sufficient. One must consider that obtaining full knowledge of the state of the grid is difficult and/or temporally intensive.

The rest of this paper is organized as follows. A review of recent related works has been given in Section 2. In Section 3, we outline the task taxonomy used in this work. Section 4 describes the grid model. Section 5 presents the detailed design of *GDS*. Section 6 presents a comprehensive set of simulations that evaluate the performance of *GDS*. Conclusions and suggestions for future work appear in Section 7.

2 Related Work

Several effective scheduling algorithms such as *EDF* [9], *Sufferage* [11], and *Min-Min* [12] have been proposed in previous works. The rationale behind *Sufferage* is to allocate a site to a task that would "suffer" most in completion time if the task is not allocated to that site. For each task, *Min-Min* tags the site that offers the earliest completion time. Among all tasks, the one that has the minimal earliest completion time is chosen and allocated to the tagged site.

Few scheduling algorithms take into account both the task types and deadlines in grids. A deadline based scheduling algorithm appears in [16] for multi-client, multi-server environment existing within a single resource site. It aims at minimizing deadline misses by using load correction and fallback mechanisms. In [2], a deadline scheduling algorithm with priority appropriate for multi-client, multi-server environment within a *single* resource site has been proposed. Since preemption is allowed, it leaves open the possibility that tasks with lower priority but early deadlines may miss their deadlines. Also, it does not evaluate the fraction of tasks meeting deadlines.

Venugopal and Buyya [17] propose a scheduling algorithm that tries to minimize the *scheduling budget* for a bag of data-intensive applications on data grid. Casanova [3] describes an adaptive scheduling algorithm for a bag of tasks in Grid environment that takes *data storage* issues into consideration. However, they make scheduling decisions centrally, assuming *full knowledge* of current loads, network conditions and topology of all sites in the grid. Liu and Baskiyar [10] propose a distributed peer to peer grid scheduler that solves the scalability issue in grid systems. Ranganathan and Foster [15] consider dynamic task scheduling along with data staging requirements. Data replication is used to suppress communication and avoid data access hotspots. Park and Kim [14] describe a scheduling model that considers both the amount of computational resources and data availability in a data grid environment. a

The aforementioned algorithms do not consider all of the following criteria: task types, dispatch times, deadlines, scalability and distributed scheduling. Furthermore, they require a full knowledge of the state of the grid which is difficult and/or expensive to maintain.

3 Task Taxonomy

We consider three types of tasks: hard, firm and soft. *GDS* uses such a task taxonomy that considers the consequence of missing deadlines, and the importance of task. Hard tasks are mission critical since the consequences of failure are catastrophic, e.g. computing the orbit of a moving satellite to make real-time defending decisions [13]. For firm tasks a few missed deadlines will not lead to total failure, but missing more may. For soft tasks, failures only result in degraded performance.

An example of mission-critical application is the Distributed Aircraft Maintenance Environment [4], a pilot project which uses a grid to the problems of aircraft engine diagnosis and maintenance. Modern aero-engines must operate in highly demanding environments with extreme reliability. As one would expect, such systems are equipped with extensive sensing and monitoring capabilities for real-time performance analysis. Catastrophic consequences may occur if any operation fails to meet its deadline.

An example of a firm task with deadline is of financial analysis and services [7]. The emergence of a competitive market force involving customer satisfaction, and reduction of risk in financial services requires accuracy and fast execution. Many corresponding solutions in the financial industry are dependent upon providing increased access to massive amounts of data, real-time modeling, and faster execution by using grid job scheduling and data access. Such applications do have deadlines; however, the consequences of missing them are not that catastrophic.

Applications which fall in the category of soft tasks include coarse-grained task-parallel computations arising from parameter sweeps, Monte Carlo simulations, and data parallelism. Such applications generally involve large-scale computation to search, optimize, and statistically characterize products, solutions, and design space but normally do not have hard real-time deadlines.

4 Grid Model

In our grid model, as shown in Fig. 1, geographically distributed sites interconnect through WAN. We define a site as a location that contains many computing resources of different processing capabilities. Heterogeneity and dynamicity cause resources in grids to be distributed hierarchically or in clusters rather than uniformly. At each site, there is a main server and several supplemental servers, which are in charge of collecting information from all machines within that site. If the main server fails, a supplemental server will take over. Intra-site communication cost is usually negligible as compared to inter-site communication.

5 Scheduling Algorithm

The following are the design goals of *GDS*:

- Maximize number of mission-critical tasks meeting their deadlines
- Maximize total number of tasks meeting their deadlines
- Provide temporal fault tolerance to the execution of mission-critical tasks
- Provide Scalability

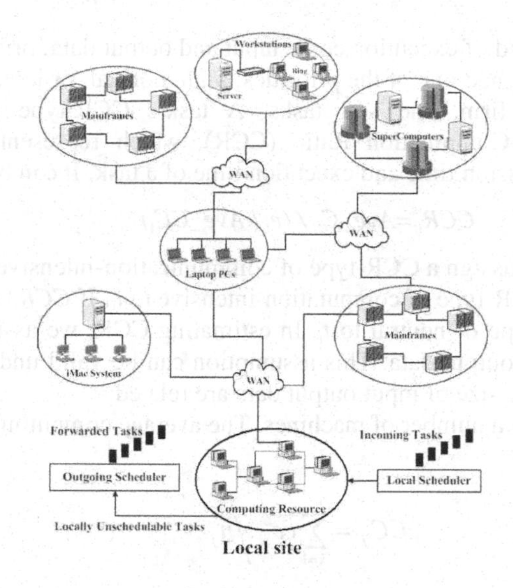

Fig. 1. Grid Model

Since neither EDF nor using priorities alone can achieve the above goals, we proposed *GDS*. *GDS* consists of three phases. First incoming tasks at each site are ranked. Second, a shuffling based scheduling algorithm is used to assign each task to a specific resource on a site, and finally those tasks that are unable to be scheduled are dispatched to remote sites where the same shuffling based algorithm is used to make scheduling decisions. The pseudo code of *GDS*'s main function is shown in Fig. 2.

5.1 Notations

The following notations have been used in this paper.

- t_i: task i
- e_{ijk}: estimated execution time of t_i on $machine_k$ at $site_j$
- c_{ij}: estimated transmission time of t_i from current site to $site_j$
- l_{ijk}: latest start time of tasks t_i on $machine_k$ at $site_j$
- e_i: instruction size of t_i
- d_i: deadline of t_i
- CCR_{ij}: communication to computation ratio of $task_i$ residing at $site_j$
- n_j: number of machines within $site_j$
- cc_{jk}: computing capacity of $machine_k$ at $site_j$
- S_{pkj}: start time of the pth slack on m_k at s_j
- E_{pkj}: end time of the pth slack on m_k at s_j
- CC_j: average computing capacity of $site_j$
- Ave_CC_i: average computing capacity of all the neighboring sites of $site_i$
- Ave_C_{ij}: estimated average transmission time of t_i from $site_j$ to all the neighbors

A task is composed of execution code, input and output data, priority, deadline, and CCR. Tasks are assigned one of the priorities: high, normal, or low, which correspond to mission-critical, firm, and soft tasks. A task's CCR-type is decided by its Communication to Computation Ratio (CCR), which represents the relationship between the transmission time and execution time of a task. It can be defined as:

$$CCR_{ij}=Ave\_C_{ij} / (e_i / Ave\_CC_i) \quad (1)$$

If CCR_{ij} >>1, we assign a CCR-type of communication-intensive to task t_i. If CCR_{ij} <<1, we assign a CCR-type of computation-intensive to t_i. If CCR_{ij} is comparable to 1, we assign a CCR-type of neutral to t_i. In estimating CCR, we assume that users can estimate the size of output data. This assumption can be valid under many situations particularly when the size of input output data are related.

Each site contains a number of machines. The average computing capacity of $site_j$ is defined as:

$$CC_j = \sum_{k=1}^{n_j} cc_{jk} \Big/ n_j \quad (2)$$

```
GDS
// Q is a task queue in site S
    Sort Q by decreasing priority then by decreasing CCR-type
    then by increasing deadline
        Schedule
    If unscheduled tasks remain in Q
        Send message to each m∈ S to execute Shuffle
        Schedule
    endif
    If unscheduled tasks remain in Q
        Dispatch
    endif
end GDS
```

Fig. 2. GDS

5.2 Multi-attribute Ranking

At each site, various users may submit a number of tasks with different priorities and deadlines. Our ranking strategy takes task priority, deadline and CCR-type into consideration. The scheduler at each site puts all incoming tasks into a task queue. First, tasks are sorted by decreasing priority, then by decreasing CCR-type and then by increasing deadline. Sorting by decreasing priority allows executing mission-critical tasks as soon as possible. Sorting by decreasing CCR-type allows executing most communication-intensive tasks locally. If we were to dispatch such tasks to a remote site, the transfer time may be negative to performance. Experimental results show that sorting by CCR-type gives us good performance.

5.3 Scheduling Tasks Within Slacks

To schedule task t_i on a site s_j, each machine m_k at s_j will check if t_i can be assigned to meet its deadline. If tasks have already been assigned to m_k, slacks of varying length will be available on m_k. If no task has been assigned, slacks do not exist, thus:

$$S_{pkj}=0 \ \&\& \ E_{pkj} = \infty \tag{3}$$

The scheduler checks whether t_i may be inserted into any slack while meeting its deadline. The slack search starts from the last to first. The criteria to find a feasible slack for t_i are:

$$e_{ijk} + max(S_{pkj}, c_{ij}) <= E_{pkj} \ \&\& \ e_{ijk} + max(S_{pkj}, c_{ij}) <= d_i \tag{4}$$

If the above conditions are satisfied, we schedule t_i to the pth slack on m_k at s_j, and set its start time to:

$$l_{ijk} = min(d_i, E_{pkj}) - e_{ijk} \tag{5}$$

Setting tasks start time to their latest start times creates large slacks, enabling other tasks to be scheduled within such slacks. Also, if s_j is the local site for t_i, the transmission time is ignorable; in other words, $c_{ijk} = 0$. The pseudo code of *Schedule* is shown in Fig. 3.

```
Schedule
  for each unscheduled task t∈ Q
    dofor each machine m∈ S //visit in random order to balance load
      Visit slacks from latest to earliest
      If t fits within slack   // while meeting deadline
         Schedule t on m at the latest possible time within the slack
         Mark t scheduled
         Update count of unscheduled tasks in Q
      endif
    until t is scheduled
  endfor
end Schedule
```

Fig. 3. Schedule

5.4 Shuffle

If after executing *Schedule*, unscheduled tasks remain, a shuffling procedure is executed on each machine of the site. *Shuffle* tries to move all mission-critical tasks as early as possible. Next, it moves other tasks as close as possible to their earliest start times. In doing so, *Shuffle* creates larger slacks for possible use by unscheduled tasks. The pseudo code of *Shuffle* is shown in Fig. 4. An example of *GDS*'s ranking, scheduling and shuffling phases are given in Fig. 5. The advantages of shuffling are two fold:

- Longer slacks may be obtained by packing tasks.
- Executing mission-critical tasks early provides temporal fault-tolerance.

```
Shuffle
    for each task t // select tasks from highest priority to lowest priority
        Re-Schedule t to the earliest available slack
    endfor
end Shuffle
```

Fig. 4. Shuffle

5.5 Peer to Peer Dispatching

Each task is assigned a ticket, which is a very small file that contains certain attributes of a task. A ticket [1] has several fields: ID, priority, deadline, CCR-type, instruction size, input data size, output data size, schedulable flag and route information. Since tickets are small they are dispatched in scheduling decisions, rather than the tasks themselves. If a task can not be scheduled locally, its ticket is dispatched to a remote site to find a suitable resource.

In dispatching, previous works have selected a remote site randomly or used a single characteristic, such as computing capacity, bandwidth, or load. *GDS* uses both the computing capacity and bandwidth in dispatching. Furthermore, *GDS* helps decrease communication overhead since each site only needs to maintain its immediate neighbors' basic information such as bandwidth and average computing capacity.

Every site always maintains three dispatching lists which are used for the three CCR-type tasks. In each list, immediate neighbors are sorted according to different attributes. The order of neighbors represents the preference of choosing a target neighboring site for dispatch. For computation-intensive tasks, the corresponding list has neighboring sites sorted by decreasing average computing capacity. For communication-intensive tasks, neighboring sites are sorted by decreasing bandwidth. For neutral-CCR tasks, neighboring sites are sorted by decreasing rank. The rank of $site_j$, a neighbor of $site_i$, is defined as:

$$Rank_{ji} = CC_j \Big/ \sum_{k=1}^{r} CC_k + BW_{ij} \Big/ \sum_{k=1}^{r} BW_{ik} \qquad (6)$$

where r is the number of neighbors of $site_i$. The three lists are available at each site and are periodically updated. A site will check whether any of its neighbors can consume a task within deadline or not. Neighbors are checked breadth-first. If none can, the most

Task	Priority	Exec. Time	Deadline
1	Mission-critical	1	3
2	Mission-critical	1.5	7
3	Mission-critical	1	11
4	Firm	2	14
5	Firm	0.5	1
6	Soft	1	4.5
7	Soft	1.5	9

Ranked Tasks at a Resource Site

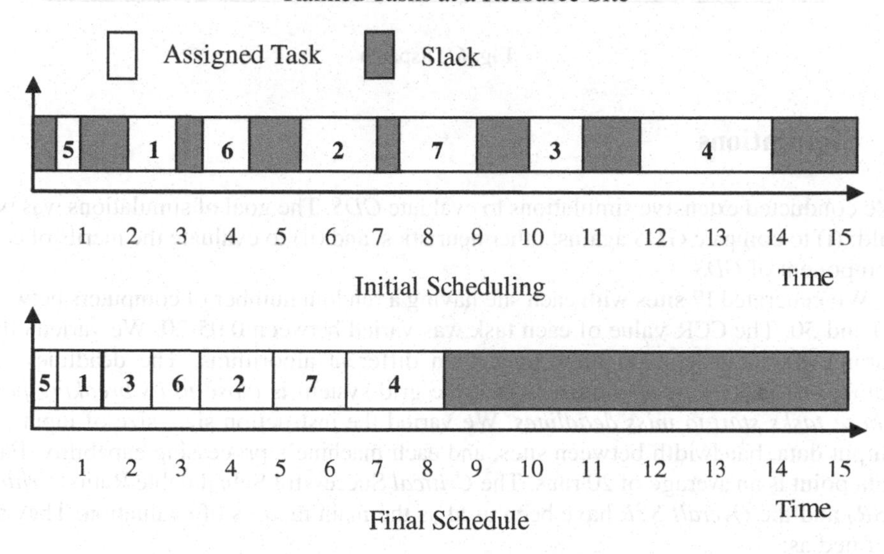

Fig. 5. An example of GDS schedule

favorite neighbor will search its neighbors. This process continues until suitable remote resource has been found, or all sites have been visited. The pseudo code of *Dispatch* is shown in Fig. 6.

5.6 Complexity

Let n be the number of incoming tasks, m the number of machines within each site, and s the number of sites. Then, the complexity of *Shuffle* is $O(n)$, of *Schedule* is $O(n^2m)$ and of *Dispatch* is $O(ns)$. The complexity of *GDS*'s ranking phase is $O(nlogn)$. Therefore, the complexity of *GDS* is $O(n^2m)$, assuming $s < nm$. If in *Schedule*, the slacks within each machine were to be evaluated in parallel by each machine in a non-blocking fashion, the complexity of *GDS* would be $O(n^2)$. We note that the complexity of *Sufferage* and *MinMin is* $O(n^2m)$.

```
Dispatch
  for each unscheduled task t∈ Q
     for each neighbor N of S
     // visit neighbors in order depending upon CCR-type of t
        Send t's ticket to N
        if N can successfully schedule t
           Send t to N
           Mark t scheduled
     endfor
  endfor
end Dispatch
```

Fig. 6. Dispatch

6 Simulations

We conducted extensive simulations to evaluate *GDS*. The goal of simulations was two fold: (i) to compare *GDS* against other heuristics, and (ii) to evaluate the merits of each component of *GDS*.

We generated 17 sites with each site having a random number of computers between 20 and 50. The CCR value of each task was varied between 0.05-20. We varied other parameters to understand their impact on different algorithms. The deadlines and number of tasks were chosen such that the grid system is *close to its breaking point where tasks start to miss deadlines*. We varied the instruction size, size of input and output data, bandwidth between sites, and each machine's processing capability. Each data point is an average of 20 runs. The *Critical* Successful Schedulable Ratio (*Critical SSR*) and the *Overall SSR* have been used as the main metrics of evaluation. They are defined as:

$$Critical\ SSR = \frac{number\ of\ mission\ critical\ tasks\ meeting\ deadlines}{total\ number\ of\ mission\ critical\ tasks}$$

$$OverallSSR = \frac{number\ of\ tasks\ meeting\ deadlines}{total\ number\ of\ tasks}$$

6.1 Performance

The first experiment set was to evaluate the performance against other algorithms. We compared *GDS* against three other heuristics: *EDF*, *Min-Min, and Sufferage*.

For *Critical SSR*, from Fig. 7, we observe that *GDS* yield 41% better performance on average than others especially when the number of tasks is high. The other three heuristics do not consider task priority, which results in a number of un-schedulable mission-critical tasks. Also, ranking tasks by CCR-type brings benefits to *GDS* through executing communication-intensive tasks locally and dispatching computation-intensive tasks to other sites.

With respect to *Overall SSR*, as shown in Fig. 8, the performance difference among the five heuristics diminishes. Although *EDF*, *Min-Min* and *Sufferage* do not consider priorities of tasks, overall they are very effective. But, the fact that *GDS* still outperforms them by 4% on average is important. Thus, *GDS* not only maximizes the number of mission-critical tasks meeting deadlines, but it does so without degrading the *Overall SSR*.

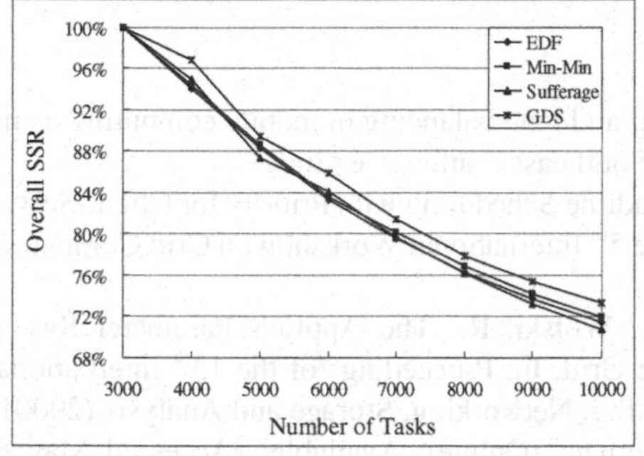

Fig. 7. Critical SSR

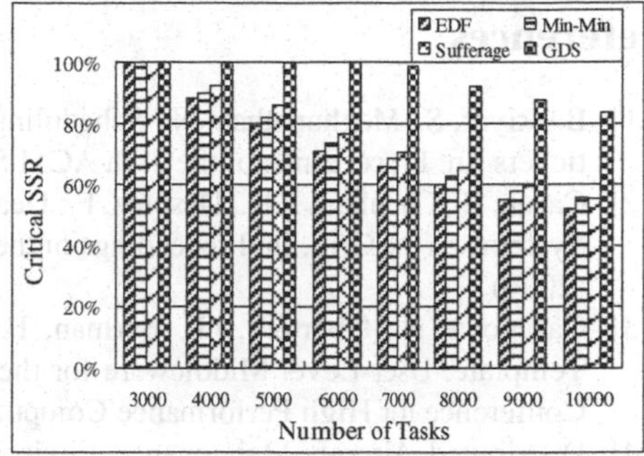

Fig. 8. Overall SSR

6.2 Impact of Shuffling

In this experiment, we investigate the use of the shuffling component of *GDS*. To do so, we use GDS_1, which is the scheduler obtained upon removing the shuffling portion from *GDS*. From Fig.9, we see that *GDS*'s *Critical SSR* is almost identical to GDS_1. However, From Fig. 10 we observe that *GDS*'s *Overall SSR* is higher than GDS_1 by 5%. In other words, *Shuffle* schedules more tasks with firm and soft deadlines while maximizing the number of mission-critical tasks that meet deadlines. It also provides temporal fault tolerance to mission-critical tasks by re-scheduling them earlier.

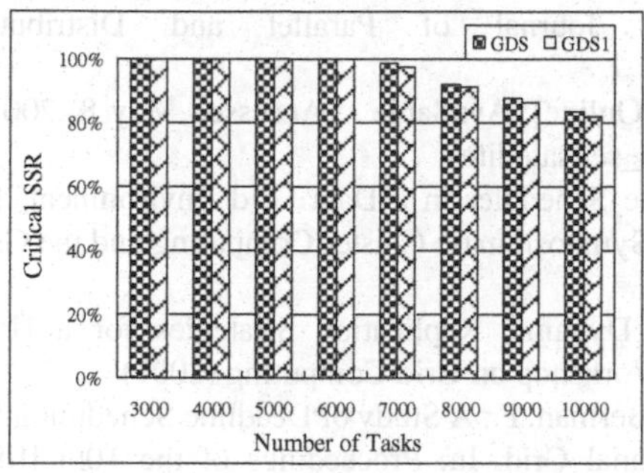

Fig. 9. Critical SSR

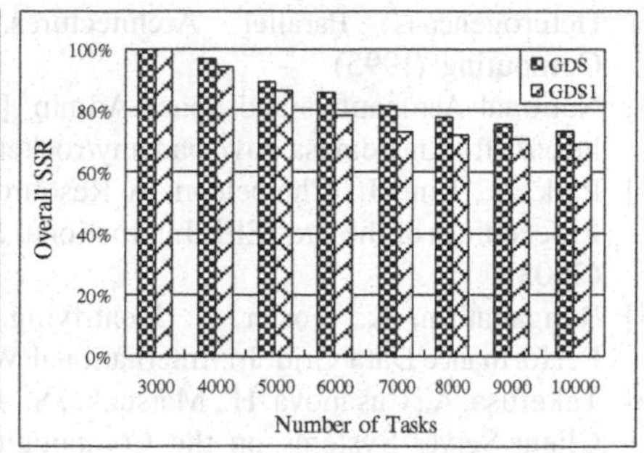

Fig. 10. Overall SSR

7 Conclusion

In this paper, we proposed a novel algorithm to schedule independent tasks with different priorities and deadlines in grid systems. Detailed simulations demonstrate that *GDS* significantly increases both the *Critical SSR* and the *Overall SSR* of all incoming tasks. In the future, we will investigate the schedulability analysis of *GDS* in order to provide deadline guarantees as well as address temporal fault tolerance.

References

[1] Baskiyar, S., Meghanathan, N.: Scheduling and load balancing in mobile computing using tickets. In: Proceedings of the 39th ACM Southeast Conference (2001)

[2] Caron, E., Chouhan, P.K., Desprez, F.: Deadline Scheduling with Priority for Client-Server Systems on the Grid. In: Proceedings of the 5th International Workshop on Grid Computing (2004)

[3] Casanova, H., Obertelli, G., Berman, F., Wolski, R.: The AppLeS Parameter Sweep Template: User-Level Middleware for the Grid. In: Proceedings of the 13th International Conference for High Performance Computing, Networking, Storage and Analysis (2000)

[4] Distributed Aircraft Maintenance Environment [Online]. Available: [Accessed May 8, 2007], http://www.cs.york.ac.uk/dame

[5] Foster, I., Kesselman, C.: The grid: blueprint for a new computing infrastructure. Morgan Kaufmann Publishers, San Francisco (1998)

[6] Foster, I., Kesselman, C.: The Grid2. Morgan Kauffmann Publishers, San Francisco (2003)

[7] Joseph, J., Fellenstein, C.: Grid Computing. Prentice Hall, Englewood Cliffs (2004)

[8] Laplante, P.A.: Real-Time Systems Design and Analysis, Wiley-IEEE Press (2004)

[9] Liu, C., Layland, J.: Scheduling Algorithms for Multiprogramming in a hard Real-Time Environment. Journal of the ACM (1973)

[10] Liu, C., Baskiyar, S., Wang, C.: A distributed peer to peer grid scheduler. In: Proceedings of the 18th International Conference on Parallel and Distributed Computing and Systems (2006)

[11] Maheswaran, M., Ali, S., Siegel, H.J., Hensgen, D., Freund, R.: Dynamic matching and scheduling of a class of independent tasks onto heterogeneous computing systems. In: Proceedings of the 8th Heterogeneous Computing Workshop (1999)

[12] Menasce, D., Saha, D., Porto, S.: Static and Dynamic Processor Scheduling disciplines in Heterogeneous Parallel Architectures. Journal of Parallel and Distributed Computing (1995)

[13] National Aeronautics and Space Admin. [Online]. Available : [Accessed May 8, 2007], http://liftoff.msfc.nasa.gov/academy/rocket_sci/satellites

[14] Park, S., Kim, J.: Chameleon: A Resource Scheduler in a Data Grid Environment. In: Proceedings of the 3rd IEEE International Symposium on Cluster Computing and the Grid (2003)

[15] Ranganathan, K., Foster, I.: Identifying Dynamic Replication Strategies for a High Performance Data Grid. In: International Workshop on Grid Computing (2001)

[16] Takefusa, A., Casanova, H., Matsuoka, S., Berman, F.: A Study of Deadline Scheduling for Client-Server Systems on the Computational Grid. In: Proceedings of the 10th IEEE Symposium on High Performance and Distributed Computing (2001)

[17] Venugopal, S., Buyya, R.: A Deadline and Budget Constrained Scheduling Algorithm for eScience Applications on Data Grids. In: Proceedings of the 6th International Conference on Algorithms and Architectures for Parallel Processing (2005)

An Energy-Aware Gradient-Based Scheduling Heuristic for Heterogeneous Multiprocessor Embedded Systems

Lee Kee Goh[1], Bharadwaj Veeravalli[2], and Sivakumar Viswanathan[1]

[1] Communication Systems Department
Institute for Infocomm Research
21 Heng Mui Keng Terrace, Singapore 119613
{lkgoh,siva}@i2r.a-star.edu.sg
[2] Computer Networks and Distributed Systems Laboratory
Department of Electrical and Computer Engineering
National University of Singapore
4 Engineering Drive 3, Singapore 117576
elebv@nus.edu.sg

Abstract. In this paper, we propose a heuristic static energy-aware scheduling algorithm for scheduling tasks with precedence constraints on a heterogeneous multiprocessor embedded system consisting of processing elements equipped with dynamic voltage scaling capabilities. While most energy-aware scheduling algorithms in the literature assume that the mapping of the tasks to the processors is known and consider only task ordering and voltage scaling, our algorithm takes into consideration all three factors using the concept of energy gradient. Higher values of energy gradient result in larger reduction in the energy consumption together with smaller increase in the makespan of the schedules. We compare our algorithm to a genetic algorithm in the literature and show that although our algorithm does not consider intra-task voltage scaling, it still provides an average energy savings of about 4% while reducing the optimization time by more than 93%. These energy savings are more significant for larger task graphs.

Keywords: Energy-aware scheduling, dynamic voltage scaling, power management, heterogeneous multiprocessor, embedded systems.

1 Introduction

Many applications today, such as multimedia streaming and medical imaging, require high-speed processing and are commonly implemented on embedded systems consisting of multiple heterogeneous processing units. These processing units may include general purpose central processing units (CPU), digital signal processors (DSP) and application specific integrated circuits (ASIC). In addition, many of these devices are portable. As a result, they require efficient energy management schemes in order to extend their battery life. Modern day devices

S. Aluru et al. (Eds.): HiPC 2007, LNCS 4873, pp. 331–341, 2007.

utilize processing units with dynamic voltage scaling (DVS) to reduce the energy consumption. DVS is a technique that lowers the supply voltage and operational frequency of the processing unit during runtime at the expense of a longer execution time. By carefully scheduling the tasks to execute at different voltage levels, an optimized schedule with minimum energy consumption can be obtained without compromising the performance. However, this problem is known to be NP-complete [21] for multiprocessor systems and there are no known algorithms that solve the problem optimally in polynomial time.

There have been several papers in the literature that focus on energy-aware scheduling using DVS. These papers can be classified into three main categories: energy-aware scheduling on uni-processors [10,12,14,18], on homogeneous multiprocessors [1,2,4,5,6,8] and on heterogeneous multiprocessors [7,9,11,15,16,3]. In this paper, we shall focus on energy-aware scheduling of dependent tasks on heterogeneous multiprocessors. In heterogeneous multiprocessors, each task requires different amount of execution time and consumes different amount of energy on different processors. [15, 16, 11, 3] are some common energy-aware scheduling algorithms that can be used to schedule dependent tasks on both homogeneous and heterogeneous multiprocessor systems. However, these algorithms assume that the assignment of tasks to processors is already known and consider only the task ordering and/or voltage scaling aspects of the schedule. In [9], Schmitz et al. proposed an energy-efficient mapping and scheduling strategy for heterogeneous multiprocessors. In their strategy, they used a genetic algorithm (GA) to determine the priorities of the tasks for list scheduling and a voltage scaling heuristic [11] to stretch the tasks in order to reclaim the available slack. This was then nested inside another GA used to determine the optimal processor mapping of the tasks. Although this approach is efficient in generating low-energy schedules, the optimization time is very high due to the nested nature of the GA and the high complexity of the voltage scaling heuristic that is being used.

In this paper, we shall look into the problem of scheduling a set of tasks with precedence constraints onto a given heterogeneous multiprocessor architecture with the objective of minimizing the total energy consumption of the system while meeting real-time constraints. As we assume that the multiprocessor architecture is already given, factors such as area constraints and monetary costs need not be taken into consideration. We propose a fast and efficient energy-aware heterogeneous embedded multiprocessor scheduling algorithm for scheduling tasks with precedence relationship as represented by a task precedence graph. A task precedence graph is a directed, acyclic graph where nodes represent sequential tasks and edges between the nodes, for example the edge between i and j requires that task i be completed before task j starts its execution. The weights on the edges represent the time required to communicate intermediate results from one task to another if they are placed on different processors. Our algorithm considers the processor mapping, the ordering of the tasks and the voltage levels at which the tasks are to be executed in order to derive an energy-efficient schedule that satisfies the deadline requirements of the tasks. The algorithm uses the concept of energy gradient to schedule a task precedence graph. Here, when a task is scheduled onto a processor at a particular voltage level, the entire task is assumed to be executed on the same processor and at the same voltage level. We compare the performance of our algorithm with the nested GA approach [9] using both hypothetical and real-life task graphs.

The remainder of the paper is organized as follows: We introduce the various models for power, system and tasks in the next section. Section 3 describes our proposed algorithm in detail. Simulation results and discussions are presented in Section 4. Lastly, we present the conclusions in Section 5.

2 Problem Formulation

Our objective is to obtain a static schedule for assigning tasks in a task precedence graph onto a heterogeneous multiprocessor system such that the total energy consumption is minimized while the task precedence constraints are observed and all the tasks meet their deadline requirements. We shall describe the power, system and task models in this section.

The dominant source of power dissipation in a digital CMOS circuit is the dynamic power dissipation [20], which is given by (1), where P denotes the dynamic power consumption, C_{ef} the effective load capacitance, V_{dd} the supply voltage and S the processor frequency. Let t denote the execution time of the task, n_c the number of execution cycles required to execute the task and E the energy consumption. Since $t = \frac{n_c}{S}$ and $E = P \cdot t$, the total energy dissipation is therefore given by (2). On the other hand, the circuit delay is given by (3), where T_D denotes the circuit delay, k a proportionality constant, V_T the threshold voltage and α the velocity saturation index. V_T and α are properties of the CMOS circuit and are constant for a particular circuit. Most papers in the literature [1,4,6,8,9,11,12,18] use the value $\alpha = 2$.

$$P = C_{ef} \cdot V_{dd}^2 \cdot S \tag{1}$$

$$E = C_{ef} \cdot V_{dd}^2 \cdot n_c \tag{2}$$

$$T_D = k \frac{V_{dd}}{(V_{dd} - V_T)^\alpha} \tag{3}$$

From the above equations, we see that when there is a reduction in the supply voltage, the energy consumption decreases quadratically while the circuit delay increases. DVS exploits this feature to reduce the energy consumption of the processor at the expense of longer execution times for the tasks.

Our system consists of a set of N_p heterogeneous processors, $\{PE_1, PE_2, \ldots, PE_{N_p}\}$, connected to a single bus. Each processor is equipped with DVS functionality. The available discrete voltage levels of PE_j are given by $V(j,k)$, $k = 1, 2, \cdots, N(j)$, where $N(j)$ denotes the total number of discrete voltage levels of PE_j. Without loss of generality, we let $N(1) = N(2) = ... = N(N_p) = N_v$ in this paper for simplicity. The power consumption and processor speed of PE_j at voltage level $V(j,k)$ are given by $P(j,k)$, and $S(j,k)$ respectively. The power consumption of the bus is denoted by P_b. We assume that negligible power is consumed by the processors and the bus when they are idle.

We consider a set of N_t dependent tasks $\{T_1, T_2, \ldots, T_{N_t}\}$ that are related by some precedence constraints as given in the task precedence graph. The amount of time required to execute a task might vary on different processors and also voltage levels. Suppose T_i is executed on PE_j at the voltage level $V(j,k)$, the worst-case execution time needed to execute T_i in this case is given by $t(i,j,k)$ while its energy consumption is given by $e(i,j,k)$. In addition, for a task T_i and its predecessor T_p, if they are executed on different processors, a communication time of $C(p,i)$ is incurred. Let d be the deadline (the latest possible time) by which all the N_t tasks in the task precedence graph must be completed. In our model, when a task is assigned to a processor, we force it to run to completion on the same processor at the same voltage level without task migration. Then, the total energy consumption of the N_t dependent tasks is given by:

$$E = P_b \cdot t_c + \sum_{i=1}^{N_t} \sum_{j=1}^{N_p} \sum_{k=1}^{N_v} (x(i,j,k) \cdot e(i,j,k)) \tag{4}$$

where t_c denotes the total duration of time for which the bus is used to transfer data and $x(i,j,k)$ is a 0-1 variable whose value is 1 if T_i is scheduled on PE_j at $V(j,k)$ and 0 otherwise.

3 Energy Gradient-Based Multiprocessor Scheduling Algorithm

In energy-aware scheduling of task precedence graphs on heterogeneous embedded multiprocessor platforms, there are three main factors that affect the quality of the solution obtained: the mapping of tasks to processors, the ordering of the tasks, and the voltage levels at which the tasks are executed. The mapping of tasks to processors plays an important role in obtaining both a feasible and energy-efficient schedule. A good mapping will ensure that tasks are assigned to low-power processors while meeting all the deadlines. The ordering of the tasks affects the makespan as well as the available slack of the schedule. The makespan of a schedule is the period of time required to completely process all the jobs while the slack is the time interval between the time when the execution of a task at a processor is completed and the deadline requirement of that task. If the tasks can be ordered in a way such that the makespan is minimized, there will be more slack available for voltage scaling. Lastly, the assignment of voltage levels affects the total energy consumption of the schedule. Based on the available slack, the tasks should be assigned the voltage levels in such a way so as to minimize the total energy consumption.

Our Energy Gradient-based Multiprocessor Scheduling algorithm (EGMS) takes into consideration all the above mentioned factors and obtains a schedule that minimizes the energy consumption while satisfying the deadlines of the tasks. We consider a system having N_p processors, N_t tasks and N_v discrete voltage levels for each processor (as described in Section 2). We use vectors M_p and M_v to denote the processor and voltage level mapping of the tasks respectively. The flow chart of our algorithm is presented in Figure 1.

In this algorithm, we first assign the tasks to the processors that can complete their execution in the shortest amount of time so that there is a higher chance of obtaining an initial schedule that is feasible. We then reorder the tasks using a generic priority-based list scheduling algorithm. Based on the schedule that is generated, the makespan m_s and energy consumption e are then calculated.

Next, we select a task to be mapped onto a new processor and/or voltage level in each iterative step so as to optimize the schedule. For each <task,processor, voltage> triplet, we define two types of priorities p_{r1} and p_{r2} as follows:

$$p_{r1}(i,j,k) = \begin{cases} \delta E \text{ if } (\delta E > 0) \text{ and } (\delta t \leq 0) \\ \quad \text{and } (m_s' \leq d) \\ -\infty \text{ otherwise} \end{cases} \tag{5}$$

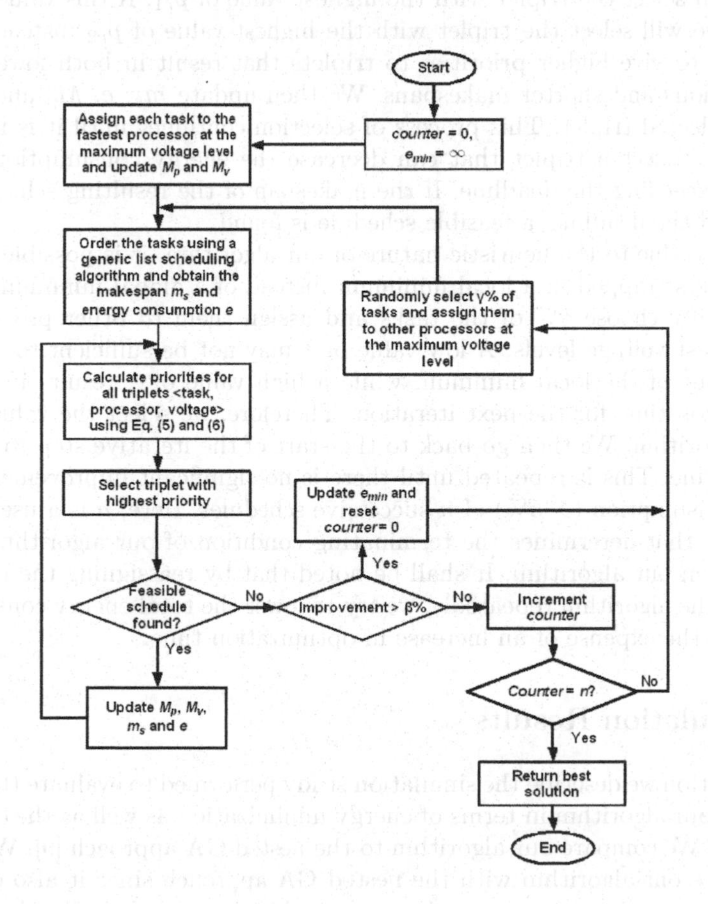

Fig. 1. Flow chart for Energy Gradient-based Multiprocessor Scheduling algorithm (EGMS)

$$p_{r2}(i,j,k) = \begin{cases} \frac{\delta E}{\delta t} \text{ if } (\delta E > 0) \text{ and } (\delta t > 0) \\ \text{and } (m'_s \leq d) \\ -\infty \text{ otherwise} \end{cases} \quad (6)$$

Here, $\delta E = e - e'$ and $\delta t = m'_s - m_s$, where e' and m'_s are the new energy consumption and makespan obtained by mapping T_i to PE_j at $V(j,k)$. From the above equations, we see that p_{r1} is used to calculate the priority when the new schedule reduces both the energy consumption and the makespan. In this case, the priority is given by the amount of decrease in energy consumption. On the other hand, p_{r2} is used to calculate the priority when the new schedule reduces the energy consumption but increases the makespan. The priority in this case is calculated using the concept of energy gradient (i.e. the ratio between the decrease in energy consumption and the increase in makespan of the schedule). In this way, the triplet that results in the largest decrease in energy consumption with the smallest increase in makespan will be assigned a higher value of p_{r2}.

We then select the triplet with the highest value of p_{r1}. If this value is equal to $-\infty$, we will select the triplet with the highest value of p_{r2} instead. We do this so as to give higher priorities to triplets that result in both lower energy consumptions and shorter makespans. We then update m_s, e, M_p and M_v using the selected triplet. This process of selection continues until it is no longer possible to select a triplet that can decrease the energy consumption further without exceeding the deadline. If the makespan of the resulting schedule does not exceed the deadline, a feasible schedule is found.

However, due to the heuristic nature of our algorithm, it is possible that the solution gets trapped in a local minimum instead of a global minimum. Hence, we randomly choose $\gamma\%$ of the tasks and assign them to other processors at their highest voltage levels. A low value of γ may not be sufficient to bring the solution out of the local minimum while a high value of γ results in a longer optimization time for the next iteration. Therefore, we select the value $\gamma = 50$ in our algorithm. We then go back to the start of the iterative step to obtain a new schedule. This is repeated until there is no significant improvement in the energy consumption ($> \beta\%$) of n successive schedules. Here, n is a user-defined parameter that determines the terminating condition of our algorithm and we set $\beta = 1$ in our algorithm. It shall be noted that by reassigning the tasks and applying the algorithm repeatedly, we try to lower the total energy consumption further at the expense of an increase in optimization time.

4 Simulation Results

In this section we describe the simulation study performed to evaluate the performance of our algorithm in terms of energy minimization as well as the optimization time. We compare our algorithm to the nested GA approach [9]. We choose to compare our algorithm with the nested GA approach since it also considers all three factors (processor mapping, task ordering and voltage level mapping) and is able to obtain good solutions as shown in their simulation results. In this approach, a GA-based task scheduling algorithm EE-GLSA was nested within another GA-based task mapping algorithm EE-GMA in order to obtain the best processor mapping, task ordering and voltage level mapping. However, it should be noted that while our algorithm considers mapping a task to a single discrete voltage level only, the nested GA approach may map a task to two voltage levels on the same processor (i.e. intra-voltage scaling is allowed). In addition, we omitted the area penalty in the calculation of the fitness function in EE-GMA, since processor area is not a constraint in our analysis.

The algorithms are implemented using C++ in a Cygwin environment on a Pentium-IV/ 3.2GHz/ 2GB RAM PC running Windows XP. We randomly generated 40 task graphs comprising a maximum of 100 tasks using TGFF [19]. The time required to execute the tasks on the processors at the highest voltage level were defined in an expected time to compute (ETC) matrix which was generated using the method described in [17]. We assumed that each processor has four voltage levels at 0.9V, 1.7V, 2.5V and 3.3V. The mean task execution time (μ_{task}) was set as 10 and the mean power consumption of each processor at maximum voltage level (μ_{power}) was set as 100. The maximum power ratings for the processors were randomly generated using a gamma distribution. The power ratings of the processors at other voltage levels were calculated using (1). As in most literature [1,4,6,8,9,11,12,18], we also set the velocity saturation index to be 2. The communication time between 2 tasks with precedence constraints was uniformly distributed between 1 and 5. Lastly, the bus power was set at 10.

For each task graph, we obtained the energy consumption and optimization time required by the nested GA approach as well as our EGMS algorithm for the case where n (the number of successive iterations in which there is no significant improvement to the solution before the algorithm terminates) is 1, 100 and 500 respectively. For the purpose of comparison, we normalized the energy consumption and the optimization time obtained using the various methods by those obtained using nested GA. However, we found that out of the 40 task graphs, the nested GA approach were unable to obtain feasible solutions for 5 task graphs. Using the results obtained from the remaining 35 task graphs, we plot the distribution of the task graphs with respect to their normalized energy consumption and optimization time for the various scheduling algorithms. The results are shown in Figure 2.

For the remaining 35 task graphs, we observe that although EGMS does not use intra-task voltage scaling, it is still able to obtain a similar or slightly better performance in terms of energy minimization compared to nested GA. On the average, EGMS consumes 1% more energy when $n = 1$, 2% less energy when $n = 100$ and 4% less energy when $n = 500$. We also observed that our algorithm is able to reduce energy consumption further when n increases.

Next, let us look at the optimization time required by the various algorithms to derive a feasible schedule. Nested GA requires about 6 - 19765 seconds to derive a feasible schedule while EGMS takes about 0.01 - 6.25 seconds when $n = 1$, 0.5 - 43 seconds when $n = 100$, and 1.7 - 97 seconds when $n = 500$ for optimization. We observe from Figure 2(b) that the optimization time required by EGMS increases when n increases from 1 to 500. On the average, EGMS

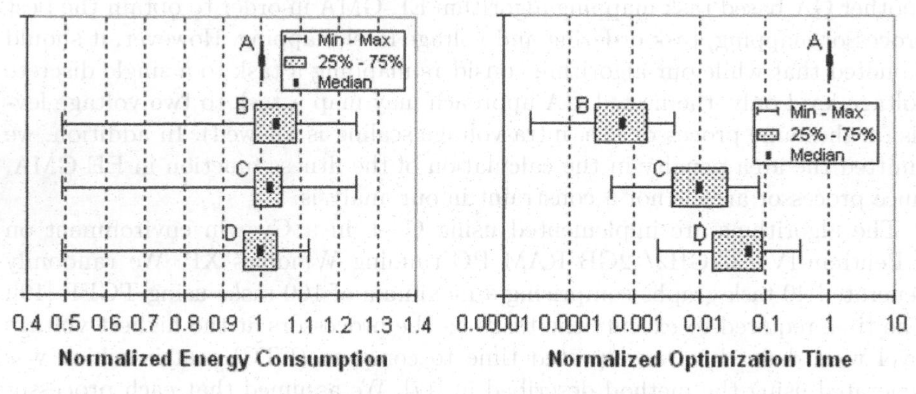

(a) Normalized Energy Consumption (b) Normalized Optimization Time

Fig. 2. Distribution of task graphs with respect to energy consumption and optimization time required by: (A) Nested GA, (B) EGMS ($n = 1$), (C) EGMS ($n = 100$) and (D) EGMS ($n = 500$) for mapping optimization. All values are normalized to the values obtained using nested GA for comparison purposes.

requires about 0.1%, 2% and 7% of the time taken by nested GA when $n = 1$, 100 and 500 respectively. From these results, we observe that when we increase the number of iterations, EGMS is able to obtain feasible schedules with lower energy consumption at the expense of a longer optimization time. However, this optimization time is still shorter than that required by nested GA.

In addition to the hypothetical task graphs generated using TGFF [19], we also applied our algorithm to some task graphs corresponding to real-life examples. We repeat the experiment using the set of task graphs used by Bambha et al. [13]. The set of task graphs consists of 2 differently implemented fast Fourier transforms (fft1, fft3), a Karplus-strong music synthesis algorithm (karp10), a quadrature mirror filter bank (qmf4), and a measurement application (meas). These applications are run on multiprocessor platforms consisting of identical processors. The normalized energy consumption and normalized optimization time of the various algorithms are shown in Table 1.

Table 1. Normalized energy consumption and optimization time required by Nested GA and EGMS for mapping optimization using real-life applications used in [13]

Task Graph	Normalized Energy Consumption/Normalized Optimization Time			
	Nested GA	EGMS(1)	EGMS(100)	EGMS(500)
fft1	1.000/1.000	0.928/0.00032	0.885/0.01349	0.885/0.06490
fft3	1.000/1.000	1.047/0.00003	0.987/0.00112	0.987/0.00497
karp10	1.000/1.000	0.981/0.00009	0.893/0.00482	0.892/0.01890
qmf4	1.000/1.000	1.049/0.00067	0.998/0.02947	0.998/0.16589
meas	1.000/1.000	1.053/0.00088	1.029/0.02236	1.029/0.09356

We achieve similar results when we compare our EGMS algorithm with the nested GA approach. For the case when $n = 500$, EGMS generates schedules that consume an average of 4% less energy when compared to the nested GA approach using less than 7% of the optimization time needed by nested GA. From the results, we observe that our EGMS algorithm can be used for homogeneous multiprocessor systems as well.

Lastly, we evaluate the performance of our algorithm with respect to the size of the task graph. We randomly generated task graphs with 10 to 50 tasks each and obtained their average normalized energy consumption. The results are shown in Figure 3. From the graph, we observe that our EGMS algorithm performs better when n is increased from 1 to 500. We also observe that for small task graphs consisting of 20 or less tasks, the nested GA performs better than EGMS in terms of energy minimization. This is because EGMS does not support intra-task voltage scaling while nested GA uses intra-task voltage scaling to reduce the energy consumption further. However, when the number of tasks increases, our algorithm performs better. This is due to the fact that when the number of tasks increases, the search space becomes exponentially larger and the genetic algorithms used in the nested GA approach are unable to converge fast enough before the terminating condition is met. In addition, we also observe that when the number of tasks is large, the average normalized energy consumption of EGMS is almost the same for all three values of n. This is due to the slow convergence of the algorithms as a result of a large search space. Hence, we conclude that the use of small values of n is sufficient for large task graphs.

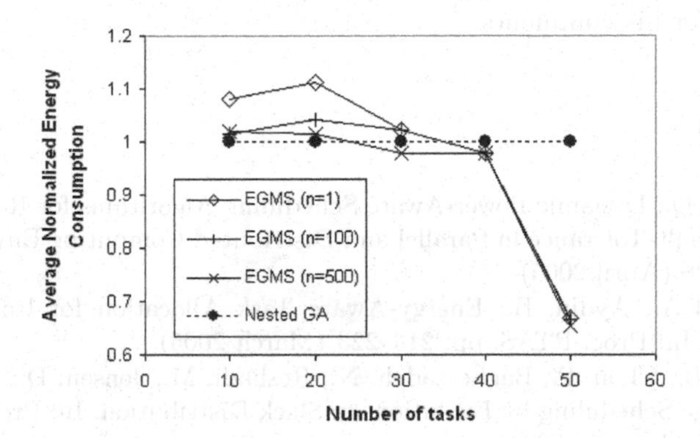

Fig. 3. Average normalized energy consumption as the number of tasks increases

5 Conclusions and Future Work

In this paper, we proposed a novel heuristic energy-aware scheduling algorithm, EGMS, for scheduling task precedence graphs on a heterogeneous multiprocessor embedded system. In EGMS, we use the concept of energy gradient to obtain an energy-efficient schedule. The algorithm is applied repeatedly until there is no improvement in the solution for n successive iterations. The larger the value of

n, the higher the probability of obtaining a feasible schedule with a lower energy consumption. However, the optimization time also increases as a result.

We compared the performance of our algorithm with the nested GA approach using both hypothetical and real-life task graphs. The simulation results showed that our algorithm is capable of obtaining energy-efficient schedules using less optimization time. In particular, we showed that for the case when $n = 500$, our algorithm is able to reduce the average energy consumption by 4% when compared to the nested GA approach even though we do not consider intra-task voltage scaling. At the same time, the average optimization time is also reduced by 93%. Our results also showed that our algorithm is able to reduce the energy consumption of larger task graphs by up to 35% when compared to the nested GA approach. This shows that our algorithm is much more effective in reducing energy consumption for larger task graphs while still meeting the deadlines.

The nested GA allows intra-task voltage scaling in order to minimize the energy consumption further. This is especially useful for processing elements with a small number of voltage levels. Therefore, we shall extend our algorithm to include intra-task voltage scaling in our future work.

Acknowledgements

The authors would like to acknowledge Marcus T. Schmitz, Bashir M. Al-Hashimi, and Petru Eles for clarifying some technical aspects and sharing the relevant data in their earlier approach [9]. The authors would also like to thank Liu Yanhong for his comments.

References

1. Han, J., Li, Q.: Dynamic Power-Aware Scheduling Algorithms for Real-Time Task Sets with Fault-Tolerance in Parallel and Distributed Computing Environment. In: Proc. IPDPS (April 2005)
2. AlEnawy, T.A., Aydin, H.: Energy-Aware Task Allocation for Rate Monotonic Scheduling. In: Proc. RTAS, pp. 213–223 (March 2005)
3. Gorji-Ara, B., Chou, P., Bagherzadeh, N., Reshadi, M., Jensen, D.: Fast and Efficient Voltage Scheduling by Evolutionary Slack Distribution. In: Proc. ASP-DAC, pp. 659–662 (January 2004)
4. Zhu, D., Melhem, R.G., Childers, B.R.: Scheduling with Dynamic Voltage/Speed Adjustment Using Slack Reclamation in Multiprocessor Real-Time Systems. IEEE Trans. Parallel and Distributed Systems 14(7), 686–700 (2003)
5. Aydin, H., Yang, Q.: Energy-Aware Partitioning for Multiprocessor Real-Time Systems. In: Proc. IPDPS (April 2003)
6. Mishra, R., Rastogi, N., Zhu, D., Mossé, D., Melhem, R.G.: Energy Aware Scheduling for Distributed Real-Time Systems. In: Proc. IPDPS (April 2003)
7. Yu, Y., Prasanna, V.K.: Power-Aware Resource Allocation for Independent Tasks in Heterogeneous Real-Time Systems. In: Proc. ICPADS, pp. 341–348 (December 2002)

8. Zhu, D., AbouGhazaleh, N., Mossé, D., Melhem, R.G.: Power Aware Scheduling for AND/OR Graphs in Multi-Processor Real-Time Systems. In: Proc. ICPP, pp. 593–601 (August 2002)
9. Schmitz, M.T., Al-Hashimi, B.M., Eles, P.: Energy-Efficient Mapping and Scheduling for DVS Enabled Distributed Embedded Systems. In: Proc. DATE, pp. 514–521 (March 2002)
10. Aydin, H., Mejía-Alvarez, P., Mossé, D., Melhem, R.G.: Dynamic and Aggressive Scheduling Techniques for Power-Aware Real-Time Systems. In: Proc. RTSS, pp. 95–105 (December 2001)
11. Schmitz, M.T., Al-Hashimi, B.M.: Considering Power Variations of DVS Processing Elements for Energy Minimisation in Distributed Systems. In: Proc. ISSS, pp. 250–255 (October 2001)
12. Gruian, F.: Hard Real-Time Scheduling for Low-Energy Using Stochastic Data and DVS Processors. In: Proc. ISLPED, pp. 46–51 (August 2001)
13. Bhamba, N.K., Bhattacharyya, S.S., Teich, J., Zitzler, E.: Hybrid Global/Local Search Strategies for Dynamic Voltage Scaling in Embedded Multiprocassors. In: Proc. CODES, pp. 243–248 (April 2001)
14. Shin, D., Kim, J., Lee, S.: Intra-Task Voltage Scheduling for Low-Energy Hard Real-Time Applications. IEEE Design and Test of Computers 18(2), 20–30 (2001)
15. Gruian, F., Kuchcinski, K.: LEneS: Task Scheduling for Low-Energy Systems Using Variable Supply Voltage Processors. In: Proc. ASP-DAC, pp. 449–455 (2001)
16. Luo, J., Jha, N.K.: Power-conscious Joint Scheduling of Periodic Task Graphs and Aperiodic Tasks in Distributed Real-time Embedded Systems. In: Proc. ICCAD, pp. 357–364 (November 2000)
17. Ali, S., Siegel, H.J., Maheswaran, M., Hensgen, D.A., Ali, S.: Task Execution Time Modeling for Heterogeneous Computing Systems. In: Proc. HCW, pp. 185–199 (May 2000)
18. Ishihara, T., Yasuura, H.: Voltage Scheduling Problem for Dynamically Variable Voltage Processors. In: Proc. ISLPED, pp. 197–202 (August 1998)
19. Dick, R.P., Rhodes, D.L., Wolf, W.: TGFF: Task Graphs for Free. In: Proc. CODES, pp. 97–101 (March 1998)
20. Chandrakasan, A.P., Sheng, S., Brodersen, R.W.: Low-Power CMOS Digital Design. IEEE Journal of Solid-State Circuits 27(4), 473–484 (1992)
21. Garey, M.R., Johnson, D.S.: Computers and Intractability: A Guide to the theory of NP-Completeness, San Francisco, CA. W. H. Freeman and Company, New York (1979)

On Temperature-Aware Scheduling for Single-Processor Systems

Deepak Rajan and Philip S. Yu

IBM T. J. Watson Research Center, Hawthorne, NY 10532, USA
{drajan,psyu}@us.ibm.com

Abstract. Power-aware operating systems/processor controllers ensure that the system temperature does not exceed a threshold by utilizing system-throttling, where the clock speed is scaled to an equilibrium load. We denote this as the Constant policy, and compare against Zig-Zag policies that alternate between phases of cooling and heating. In this paper, we characterize and calculate the best possible Zig-Zag policy, and argue that simple system-throttling rules are often optimal.

In reality, however, the system design often forces us to implement Zig-Zag policies. In particular, we consider the case where the processor can operate only at a few discrete states; thus it is required to alternate between cooling and heating phases. In such a setting, we develop an algorithm that outperforms all other Zig-Zag policies, and present computational experiments emphasizing the performance of our algorithm.

1 Introduction

Energy and temperature management of processor systems is an increasingly important problem as their power consumption rises drastically with every new generation, while the rate of technological improvements in cooling systems has not been keeping pace [1]. Naturally, this has resulted in a large body of work attempting to incorporate energy and temperature considerations into processor scheduling levels. This is now implementable at the operating system/program level since most modern day processors have interfaces that allow the user to control its speed in real-time using a mechanism called dynamic voltage scaling (DVS) [2]. For a detailed investigation into DVS and other mechanisms for implementing dynamic thermal management, see [3,4].

Processors usually ensure that the system temperature does not exceed a maximum amount by system-throttling. In [5], we showed that such simple system-throttling policies (represented by the class of Constant policies) are optimal under certain simplifying assumptions. In this paper, we are interested in developing the optimal Zig-Zag policy, for two main reasons. Firstly, such an analysis allows us to determine the exact conditions when a carefully constructed Zig-Zag policy can outperform the best Constant policy. Secondly, in most real systems, it is not possible to implement simple system-throttling using a Constant policy; one is forced to Zig-Zag because of the constraints of the system. In particular, this is true of current implementations of DVS; the processor can operate only

S. Aluru et al. (Eds.): HiPC 2007, LNCS 4873, pp. 342–355, 2007.

in a small discrete set of speeds [4]. Under such settings, an optimal Zig-Zag policy can provide significant benefits over a naive implementation. Note that all operating policies that do not maintain a constant processor speed belong to the class of Zig-Zag policies.

We begin in Section 2 by introducing the thermodynamics models. The main contributions of this paper are presented in Section 3, where we carefully deconstruct many simplifying assumptions in an attempt to analyze the effectiveness of system-throttling, and in Section 4, where we characterize the optimal Zig-Zag policy and compare it against Constant. In Section 5, we develop an efficient speed-scaling algorithm that implements our results. To illustrate our approach, we consider a practical setting where the Constant policy can not be implemented, and present the results of computational experiments in Section 6. Finally, in Section 7, we summarize the contributions of this work.

1.1 Related Work

Many other researches have looked at power management of processors. This list is by no means exhaustive, but illustrates some settings where speed-scaling has proved to be effective. In many scenarios, the tasks have different processor, memory and I/O requirements. Thus, it is possible to mix and match tasks to reduce net system power utilization [6,7]. In multi-processor systems, if it is possible to move jobs among the different processors, then a scheduling algorithm can outperform system-throttling by moving jobs between hot and cold processors [8,9]. In some settings, the net energy available is limited; studied both in a theoretical setting [10,11,12,13], and in a practical setting [14,15,16]. In other studies, the authors present a variety of scheduling and workload management strategies to develop temperature-aware computing centers [17,18,19].

2 Problem Setting

We first present the heat model for estimating the temperature of the system. We formalize system-throttling using the Constant policy, which keeps the workload constant such that the temperature threshold is not violated. We formalize all other scheduling policies using the notion of a Zig-Zag policy, since it must have alternate periods of cooling and heating. We consider a single processor system, which at speed ℓ can complete w units of work in time w/ℓ.

We assume that the system is cooled using Newton's law; $dT = -\rho T$, where dT is the instantaneous rate of change of temperature, and ρ is a positive constant [20]. To model the heat gain due to the processor, we assume that $dT = \beta P$, where P is the power dissipated and β is a positive constant [13,21]. We model the power dissipated by the processor as $P = \ell^\alpha$, where α is strictly larger than 1, and ℓ is the speed of the processor [10,22]. Combining these effects,

$$dT = \beta\ell^\alpha - \rho T. \tag{1}$$

This is an ordinary differential equation that can be easily solved [23] for constant speed. Let us define $\tau(\ell) = \beta\ell^\alpha/\rho$. We mention that the system cools/heats up exponentially until it reaches a stable temperature. This temperature is unique for speed ℓ and is given by $\tau(\ell)$. Similarly, we also define the speed that stabilizes the system at temperature T as $\ell(T)$. Let ℓ_0 be the speed at which the system operates at the maximum system design temperature T_{max}; then, $\ell_0 = \ell(T_{max})$. One way to ensure that the system never exceeds T_{max} is:

1. If the temperature hits T_{max}, the system enforces throttling (speed $= \ell_0$).
2. If the temperature is below T_{max}, the system increases the speed to 1.

In the absence of idling, the system operates at speed 1 initially, followed by speed ℓ_0. System-throttling can be represented by a policy that keeps the temperature constant (denoted as Constant); see dashed-and-dotted line in Figure 1. In general, the scheduler can decide to operate at any intermediate speed ($\ell_0 \leq \ell \leq 1$) depending on the state of the system. Furthermore, the scheduler may choose to operate at any speed $0 \leq \ell \leq l_0$ so as to increase the rate of cooling. We describe such alternate (but quite general) policies by the Zig-Zag policy, and characterize it as follows:

- The operating temperature range is $[T_m, T_{max}]$.
- The operating speeds are ℓ_b (cooling) and ℓ_a (heating), where $\ell_b < \ell_0 < \ell_a$.

We illustrate this policy in Figure 1 using solid lines. Note that a Constant policy maintaining temperature T_0 operates at a speed between the cooling and heating speeds of a Zig-Zag policy operating between temperatures T_m and T_0.

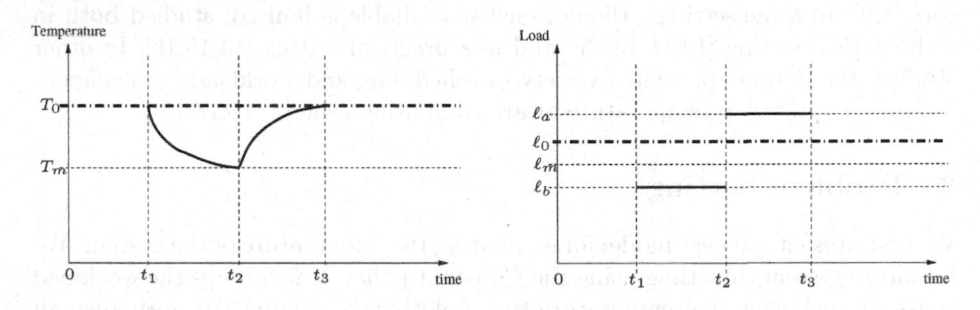

Fig. 1. Zig-Zag and Constant policies

3 Analysis of System-Throttling

In previous work [5], we showed that system-throttling is optimal for single-processor systems when the goal is to maximize the amount of work done, and this work is assumed to be a single task. We state the result here as Theorem 1. In Sections 3 and 4, we consider some of the inherent assumptions in this result. For the rest of this paper, we assume without loss of generality that the Constant policy operates at speed $\ell_0 = \ell(T_{max})$; thus maximizing the amount of work done among all possible policies that maintain a constant temperature.

Theorem 1. *The Constant policy does more work than any Zig-Zag policy.*

3.1 Multiple Tasks

All realistic scenarios involve a variety of tasks, each with its own processing requirements and importance. Let there be n tasks, each with work w_i. At speed ℓ, the time taken to complete task i is therefore w_i/ℓ. These tasks may have different levels of importance; we measure this by associating a weight γ_i with task i. Let the system begin at time 0. The jobs may arrive at different times; i.e., job i has release date r_i. Initially, we assume that all jobs arrive at time 0.

Given a schedule, we can calculate the completion time for each task in the system. We limit our attention to all objective functions that are non-decreasing functions of the completion times ("natural" objective functions). All commonly used metrics (make-span, weighted completion times, weighted flow time, maximum flow time, etc.) satisfy this property. In the absence of pre-emption, it is easy to show that any natural objective function is minimized by Constant.

We show that the Constant policy dominates any Zig-Zag policy even when pre-emption is allowed so long as we minimize a natural objective function; we state it as Theorem 2. We show that for any sequence of tasks scheduled using a Zig-Zag policy, there exists a "similar" Constant policy that dominates it. We say that two policies are similar if for all pre-empted jobs, the same fraction of its work is completed before it is pre-empted. In Figure 2, job 1 is pre-empted by job 2, and virtual job $\bar{1}$ denotes the part of job 1 completed at pre-emption.

Theorem 2. *The Constant policy minimizes any natural objective function.*

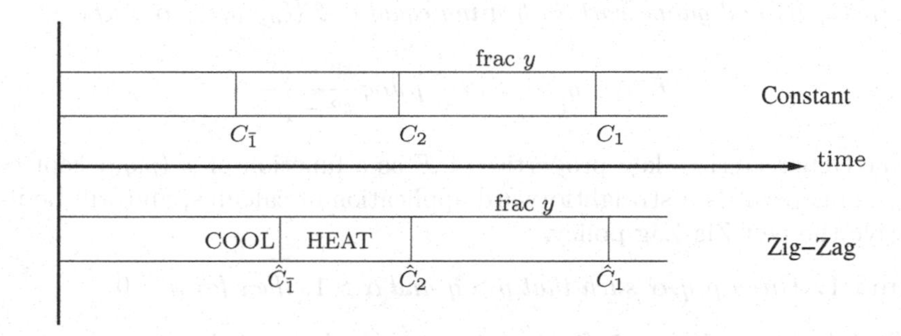

Fig. 2. Completion time: Pre-emption

3.2 Jobs with Arbitrary Release Dates

Earlier, we assumed that the Constant policy does not include any idle time; i.e., the processor would always have available work. In the absence of release dates, this is a natural assumption to make. Consider the Zig-Zag policy in Figure 2. If the release date of job 2 is the time at which it pre-empted job 1, then there exists no similar Constant policy (with fraction $1-y$ of job 1 completed before pre-emption) that has no idle time. By scaling the speed prior to a job arrival, one may be able to respond to it better, thus minimizing a variety of natural objectives. Operating at temperature T_0, if job 1 will be pre-empted by job 2 when it arrives at time t_2 (see Figure 1), there are three main decisions:

1. When to switch to a Zig-Zag policy? (Best choice of time t_1?)
2. How much to scale during cooling? (Best choice of speed ℓ_b?)
3. How much to scale during heating? (Best choice of speed ℓ_a?)

4 Generating Optimal Zig-Zag Policies

In this section, we address all the issues discussed above. We present the best Zig-Zag algorithm that minimizes any natural objective function. To develop the best Zig-Zag policy, one needs to understand the trade-offs involved in alternating between heating and cooling phases. We need to characterize (see Figure 1):

- the loss in work in the cooling phase (from time t_1 to time t_2) at speed ℓ_b, to cool from temperature T_0 to temperature T_m, and
- the gain in work in the heating phase (from time t_2 to time t_3) at speed ℓ_a, to heat from temperature T_m back to temperature T_0.

We show that this trade-off is related to the function $\mathcal{F}$; formalizing the result as Proposition 1. To illustrate, we plot the function $\mathcal{F}(x, 0.75, 0.5, 3)$ in Figure 3.

Proposition 1. *Consider a Zig-Zag policy that cools at speed ℓ_b until temperature T_m is reached and heats at speed ℓ_a until temperature T_0 is reached. Compared to the Constant policy, this Zig-Zag policy loses work in cooling equal to $\mathcal{F}(\ell_b, \ell_0, \ell_m, \alpha)$ and gains work in heating equal to $\mathcal{F}(\ell_a, \ell_0, \ell_m, \alpha)$; where*

$$\mathcal{F}(x, p, q, \alpha) = (x - p) log \frac{x^\alpha - q^\alpha}{x^\alpha - p^\alpha} \tag{2}$$

Lemma 1 characterizes key properties of $\mathcal{F}$ as a function of x (p,q,α kept constant). This result is a straightforward application of calculus, and will be used to derive the best Zig-Zag policy.

Lemma 1. *Given p, q, α such that $p > q$ and $\alpha > 1$, then for $x > 0$*

1. *$\mathcal{F}(x)$ has a maximizer $\bar{x}$ for $x > p$, and a minimizer $\hat{x}$ for $x < q$.*
2. *Furthermore, $\mathcal{F}(x_1) > \mathcal{F}(x_2)$, for all $x_1 < q$ and $x_2 > p$.*
3. *For $x > p$, $\bar{x}$ is the only local maximum of function $\mathcal{F}$, denoted by $\overline{\mathcal{F}}(p, q, \alpha)$*
4. *For $x < q$, $\hat{x}$ is the only local minimum of function $\mathcal{F}$, denoted by $\underline{\mathcal{F}}(p, q, \alpha)$*

From Proposition 1 and Lemma 1.2, we can derive an alternate proof for Theorem 1. From Lemma 1.3 and 1.4, we see that the slope of $\mathcal{F}$ changes monotonically. As a consequence of Lemma 2, it is computationally easy to implement scheduling decisions that optimize $\mathcal{F}$.

Lemma 2. *Given $\alpha > 1 > p > q$, an ϵ-approximate maximizer (minimizer) of $\mathcal{F}$ for $x > p$ ($0 < x < q$) can be calculated in* $\log(\epsilon)$ *time using bisection search.*

In [5], we showed that if the current system temperature is $T < T_{max}$, then the optimal speed (for increasing the temperature to T_{max}) can be calculated exactly.

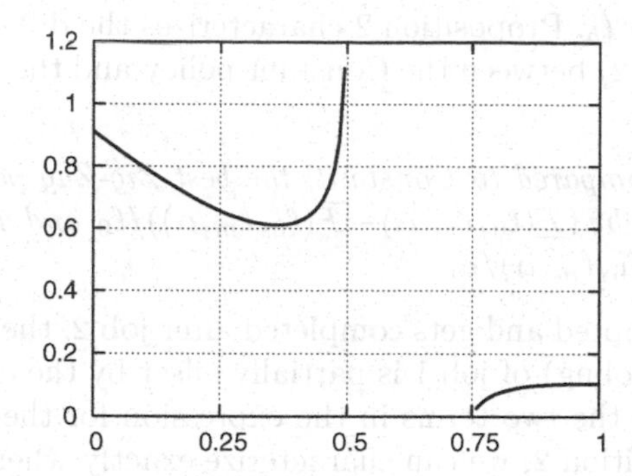

Fig. 3. Function $\mathcal{F}(x, 0.75, 0.5, 3)$

We proved that the amount of work done is maximized for a particular choice of $\ell = \bar{\ell}$, and showed how to calculate this value. Here, we derive a more general result in Theorem 3 that characterizes the optimal speed (which maximizes work done) for both cooling and heating phases of a Zig-Zag policy.

Theorem 3. *Let the current temperature be T_1. Assume that we operate the system at a speed ℓ such that the system reaches temperature T_2 in no more than t_0 time units, and then operate the system such that the temperature is maintained at T_2. Let ℓ_{exact} be the speed that reaches temperature T_2 from temperature T_1 in exactly time t_0. Let the net work done (until time infinity) be $W(\ell)$, and $\ell_i = \ell(T_i)$ for $i = 1, 2$.*

- *$T_2 > T_1$ (heating): Let $\bar{\ell}$ maximize $\mathcal{F}(\ell, \ell_2, \ell_1, \alpha)$ for $\ell > \ell_2$. Then, $W(\ell)$ has a unique maximizer ℓ^*, where $\ell^* = \min\{1, \max\{\ell_{exact}, \bar{\ell}\}\}$.*
- *$T_2 < T_1$ (cooling): Let $\hat{\ell}$ minimize $\mathcal{F}(\ell, \ell_1, \ell_2, \alpha)$ for $\ell < \ell_2$. Then, $W(\ell)$ has a unique maximizer ℓ^*, where $\ell^* = \min\{\ell_{exact}, \hat{\ell}\}$.*

4.1 Main Results

In Section 3.2, we argued that a carefully constructed Zig-Zag policy may outperform Constant if the Zig-Zag is carried out in conjunction with a pre-emption. If job 1 is to be pre-empted by job 2 at time t_2, Zig-Zag can outperform Constant if we reduce the speed (cooling) prior to time t_2, and increase the speed (heating) after time t_2. This allows us to trade-off the processing of the less important job (job 1) for the faster processing of the more important job (job 2). Analyzing the trade-offs involved, we develop and characterize the optimal Zig-Zag policy. This result is very important, for two reasons. Firstly, by comparing against the Constant policy, we describe the exact conditions for which Zig-Zag can dominate Constant. Secondly, when the Constant policy can not be implemented, we develop optimal speed-scaling algorithms.

To compare against Constant, we consider one iteration of a Zig-Zag policy that drops the temperature to T_m, and back up to T_0; see Figure 1. Let $\ell_0 = \ell(T_0)$ and $\ell_m = \ell(T_m)$. Let the speed during the cooling (heating) phase be ℓ_b (ℓ_a); we have $\ell_a > \ell_0 > \ell_m > \ell_b$. Proposition 2 characterizes the difference in completion times (of jobs 1 and 2) between the Constant policy and the best Zig-Zag policy.

Proposition 2. *Compared to Constant, the best Zig-Zag policy increases completion time of job* 1 *by* $(\underline{\mathcal{F}}(\ell_0, \ell_m, \alpha) - \overline{\mathcal{F}}(\ell_0, \ell_m, \alpha))/\ell_0$ *and decreases completion time of job* 2 *by* $\overline{\mathcal{F}}(\ell_0, \ell_m, \alpha)/\ell_0$.

Since job 1 is pre-empted and gets completed after job 2, the increase in completion time (due to cooling) of job 1 is partially offset by the earlier completion of job 2. This explains the two terms in the expression for the completion time of job 1. Using Proposition 2, we can characterize exactly when it is advantageous to Zig-Zag in conjunction with a pre-emption.

Observe that the completion time of all jobs completed after job 2 increases (not just job 1). More precisely, the completion time of all jobs processed until the next idle time increases as much as job 1. We refer to all such jobs by L. If the gains in completion time (of job 2) offset the losses in completion time (of jobs in L), then one should Zig-Zag in anticipation of the pre-emption of job 1. This analysis naturally depends on the objective function.

Make-span and Maximum Response Time: Observe that the order in which the jobs are processed does not affect the make-span. Since pre-emption is not necessary, there are no performance improvements (over Constant) that can be achieved by a Zig-Zag policy. In fact, this is true for any natural objective function where the optimal policy does not involve any pre-emption. For instance, for minimizing the Maximum Response Time, it has been shown that scheduling in increasing order of r_i (release dates) is the optimal policy. We state the general result as the following theorem.

Theorem 4. *For any objective function where the optimal schedule (sequence in which jobs are processed) does not involve pre-emption, the Constant policy dominates any Zig-Zag policy.*

Weighted Response time: Here, we consider the minimization of weighted response time of tasks. Since the response time of a job is the difference between its completion time and its release date, any algorithm that minimizes the weighted response time also minimizes the weighted completion time.

It is well known that the weighted response time is optimized by scheduling the jobs in increasing order of their remaining processing time, scaled by their weights. Thus, at any time, all available jobs are sorted in increasing order of w_i/γ_i and then processed in that sequence. Since γ_i/w_i can be interpreted as the density of job i (value of unit work), this algorithm is also referred to as HDF (Highest Density First). Implementing HDF, the optimal schedule often includes pre-emption. Suppose that job 1 is being processed (with remaining work $\hat{w}_1$) when job 2 is released. If $w_2/\gamma_2 < \hat{w}_1/\gamma_1$, then job 2 pre-empts job 1. The following theorem characterizes whether we should Zig-Zag in anticipation of a pre-emption, and is the main theoretical result in this work.

Theorem 5. *The best Zig-Zag policy should Zig-Zag in anticipation of the pre-emption of job 1 by job 2 only if there exists a speed $\ell_m < \ell_0$ for which the following condition holds: $\delta > 0$, where*

$$\delta = \overline{\mathcal{F}}(\ell_0, \ell_m, \alpha)(\gamma_2 - \sum_{i \in L} \gamma_i) - \underline{\mathcal{F}}(\ell_0, \ell_m, \alpha)(\sum_{i \in L} \gamma_i). \tag{3}$$

If the condition is satisfied, we choose ℓ_m as the speed that maximizes δ, $\tau(\ell_m)$ as the temperature to which the processor should cool down prior to pre-emption; the cooling and heating speeds are chosen as in Theorem 3.

From Lemma 1.2 and Theorem 5, we can prove that δ is no greater than 0 when all weights are 1. This proves the following result for the case of homogeneous tasks, and for the minimization of a large class of natural objective functions, including the special case of average response time.

Theorem 6. *If the tasks are of equal importance (weights $\gamma_i = 1$, $\forall i$), then the Constant policy dominates any Zig-Zag policy.*

Even though the condition in Theorem 5 may be satisfied for arbitrary weights, it is more likely that it is not. The decrease in completion time of job 2 may offset the increase in completion time of job 1, but this potential advantage is often nullified by the fact that all jobs in L complete later in the Zig-Zag policy. As a result, in practice, it does not pay to Zig-Zag unless there are idle times in the system. In the presence of idle times, L is often small, and the condition in Theorem 5 may be satisfied. Nevertheless, Theorem 5 is an exact characterization of the conditions when a Zig-Zag policy outperforms a Constant policy, and is a significant result.

5 Forced Zig-Zag

In the previous section, we showed that, in most cases, the Constant policy dominates the best Zig-Zag policy. However, our analysis of the best Zig-Zag policy is quite useful when one is forced to Zig-Zag. This often happens in practice since a processor can operate only in one of a small discrete set of operating speeds. Let this set of feasible speeds be S. If the processor can scale the speed continuously, then it could alternate between increasingly small cooling and heating phases, staying as close to ℓ_0 as possible. In practice, there is usually a cost associated with changing the speed, and can (equivalently) be modeled using a time threshold Δ between successive speed changes. We denote the largest permissible speed less than ℓ_0 as ℓ_{lb}. Formally, $\ell_{lb} = \max\{\ell \in S : \ell \leq \ell_0\}$. System-throttling can now be formalized as consisting of:

- Heating phase. Every Δ time units, set the speed to 1 (full speed).
- Cooling phase. If temperature hits T_{max}, the processor operates at speed ℓ_{lb}.

We refer to this scheduling policy as Naive. To illustrate the preceding analysis in a computational setting, we compare the Naive algorithm against the best possible Zig-Zag algorithm, which tries to minimize the work lost during the cooling phase, and maximize the work done in the heating phase. Since we are operating in a setting where a Constant policy dominates any Zig-Zag policy, there is no incentive to start a cooling phase unless forced to do so. Hence, the best Zig-Zag policy differs from Naive only in the choice of the speed during heating. To summarize, our algorithm, which we call BestZig, operates as follows.

- Heating phase. Every Δ time units, let T be the current temperature. The processor sets the speed to ℓ_{opt}, where ℓ_{opt} maximizes the work done in the heating phase; calculated as:

$$\ell_{opt} = \arg \max_{x \geq \ell_0 | x \in S} \mathcal{F}(x, \ell_0, \ell(T), \alpha) \tag{4}$$

- Cooling phase. If temperature hits T_{max}, the processor operates at speed ℓ_{lb}.

Observe that the choice of ℓ_{opt} follows directly from Theorem 3. Moreover, the computational effort involved in calculating ℓ_{opt} is trivial (can be done in $\mathcal{O}(|S|)$ time by evaluating $\mathcal{F}$ for all feasible values in S). To illustrate both algorithms, we present them pictorially in Figure 4. The algorithms BestZig and Naive differ in the choice of ℓ_{heat}, while both choose $\ell_{cool} = \ell_{lb}$. For BestZig, $\ell_{heat} = \ell_{opt}$, and for Naive, $\ell_{heat} = 1$.

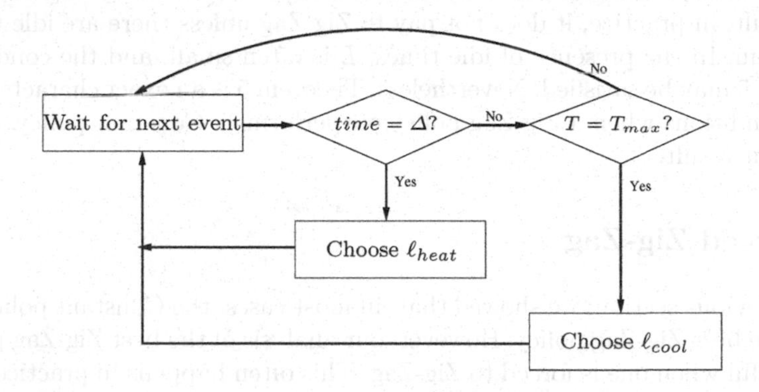

Fig. 4. Operation of system

6 Computations

Now, we present extensive computational results illustrating the performance of BestZig. The data is taken from the STG (Standard Task Graph) dataset, available at `http://www.kasahara.elec.waseda.ac.jp/schedule/index.html`. These include problems with 50, 100, 300, 500, 750, and 1000 tasks. We refer to the number of jobs in the instance as the size of the problem. There are 180 random instances for each size. We assume that the processor can operate only at speeds $\{0, G, 2G, \ldots, 1\}$, and can change the speed every Δ units of time. For the base runs of BestZig, we choose $G = 0.1$ and $\Delta = 2$. For the heat model, we choose $\rho = 0.1$ and $\beta = 12$, resulting in the maximum equilibrium speed $\ell_0 = 0.69$. Initially, we assume that all jobs are available at time 0; i.e., $r_j = 0, \forall j$.

6.1 Make-Span Minimization

For minimization of make-span on a single processor system, the sequence in which the jobs are processed is irrelevant; pre-emption is not necessary in this setting. To compare Naive and BestZig, we contrast against Constant, the theoretical best that any Zig-Zag policy can achieve. In Figure 5; we plot the percentage deviation in make-span when compared to the Constant policy in the y-axis. We present the average over all instances for a particular run; our experiments indicated that the standard deviation across instances was insignificant. We see that BestZig outperforms Naive by about 50%, and that this improvement in performance is robust to changes in data.

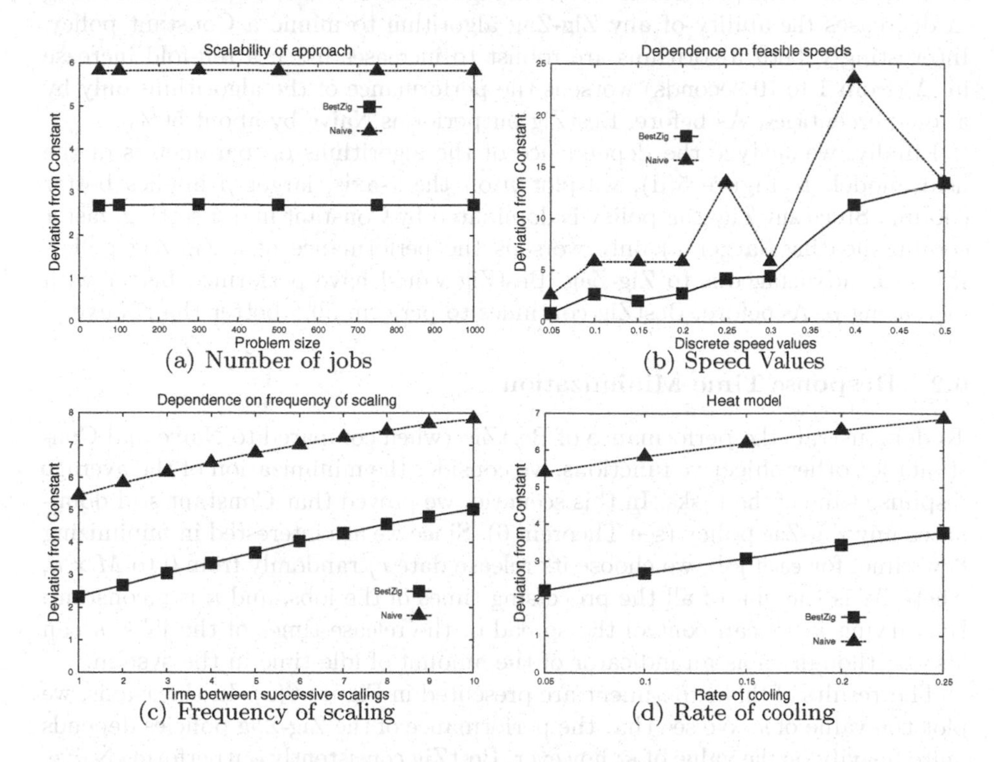

Fig. 5. Summary of computational experiments

First, we present computations which illustrate the effectiveness of our algorithm, and its robustness with respect to the number of jobs. We present a summary of the results in Figure 5(a). On the x-axis, we present the number of jobs in the instance. As we can see, both algorithms are highly insensitive to problem size. However, BestZig outperforms Naive by 50% in all the instances.

We also performed a series of experiments to understand the sensitivity of the performance of BestZig to changes in other data; we present the results of problems of size 50 (the behavior is similar for the other instances). First, we analyze the performance of the algorithm as a function of the possible operating states of the processor. This is characterized by the parameter G; the processor operates only at speeds $\{0, G, 2G, \ldots, 1\}$. In Figure 5(b), we plot G on the x-axis. We see that as G increases, the performance of both Naive and BestZig deteriorates, since larger G implies that the processor can operate at fewer number of feasible speeds. Nevertheless, the performance of BestZig is often 50% better than Naive, and is much more robust. In fact, only when the number of operating speeds is small ($G \geq 0.4$) does BestZig perform more than 5% worse than Constant. On the other hand, the performance of Naive is very sensitive to the value of G.

We also analyze the performance of the algorithm as a function of the time between successive speed changes (Δ). In Figure 5(c), we plot Δ on the x-axis. As Δ increases, the performance of both algorithms deteriorates, since increasing Δ decreases the ability of any Zig-Zag algorithm to mimic a Constant policy. Interestingly, both algorithms are robust to increases in Δ; a ten-fold increase in Δ (from 1 to 10 seconds) worsens the performance of the algorithms only by a few percentages. As before, BestZig outperforms Naive by about 50%.

Finally, we analyze the dependence of the algorithms on parameters of the heat model. In Figure 5(d), we plot ρ on the x-axis; larger ρ implies better cooling. Since any Zig-Zag policy is dominated by Constant in our setting, faster cooling/heating (larger ρ) only worsens the performance of a Zig-Zag policy. If it was advantageous to Zig-Zag, BestZig would have performed better with increasing ρ. As before, BestZig continues to perform 50% better than Naive.

6.2 Response Time Minimization

To demonstrate the performance of BestZig (when compared to Naive and Constant) for other objective functions, we consider the minimization of the average response time of the tasks. In this scenario, we proved that Constant still dominates any Zig-Zag policy (see Theorem 6). Since we are interested in minimizing flow time, for each job, we choose its release date r_j randomly from 0 to $M \times \kappa$, where M is the sum of all the processing times of the jobs, and κ is a constant. By varying κ, we can control the spread of the release times of the jobs. κ can also be thought of as an indicator of the amount of idle time in the system.

The results of this experiment are presented in Figure 6(a). In the x-axis, we plot the value of κ. We see that the performance of the Zig-Zag policies depends quite heavily on the value of κ; however, BestZig consistently outperforms Naive. Furthermore, its performance is even stronger in the cases where Naive performs poorly; BestZig performs 2% to 8% from Constant, whereas Naive is more than 20% off when $\kappa = 1$. The shape of the curve can be explained as follows. When $\kappa = 0$, all jobs are released at time 0, and there is no idle time in the system. When κ is large (≥ 2), there is a large amount of idle time in the system. In both scenarios, there is no/minimal pre-emption involved in the processing. As a result, maximizing the amount of work is a reasonable surrogate for minimizing

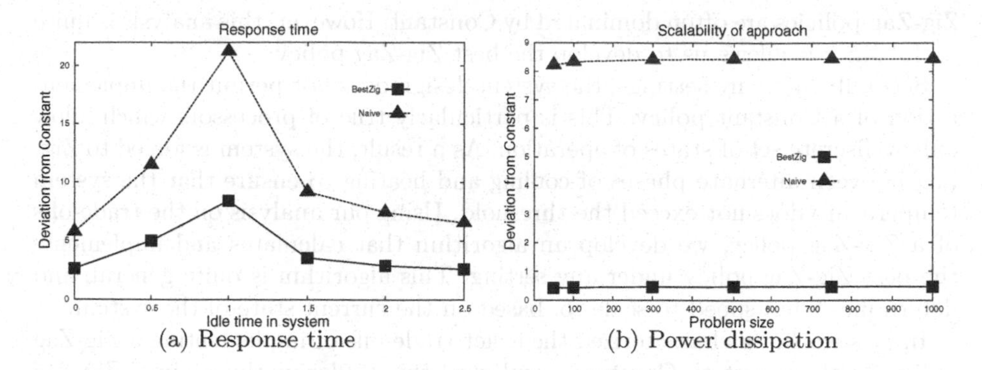

(a) Response time (b) Power dissipation

Fig. 6. Other objective functions

response time, and thus the Zig-Zag policies perform similar to the case of make-span minimization (BestZig 3% and Naive 6% from optimal). For intermediate κ, there is much more pre-emption, and thus both algorithms perform worse (compared to Constant) since they are forced to Zig-Zag.

6.3 Power Minimization

In our analysis (and in the algorithm BestZig), we do not optimize the power dissipated by the processor. However, we track the amount of heat dissipated by the processor; we present this in Figure 6(b) for both Naive and BestZig. We repeated the experiment for all problem sizes (x-axis), and plot the percentage deviation (in power dissipated) from Constant in the y-axis. We see that both Naive and BestZig dissipate more power than Constant, but the performance is very robust. Furthermore, BestZig (only 0.5% from Constant) significantly outperforms Naive (8%). These results are quite encouraging, since they indicate that the optimal Zig-Zag policy (for minimizing a variety of natural objective functions) does not increase the power dissipated (compared to Constant).

7 Conclusions

In this paper, we modeled the temperature of the system as a function of the power dissipated by the processor and the cooling system, and described how this information may be used by the scheduler to design efficient algorithms. Our analysis, and the computational illustration of such an algorithm for a real scenario are the key contributions of this paper.

We argued that simple system-throttling rules are effective scheduling policies even when few assumptions apply. These assumptions are reasonably general, but fail to accommodate two key features of real systems. Firstly, in the presence of release dates for jobs, we show that an intelligently constructed Zig-Zag policy can outperform a Constant policy (implementing system-throttling). However, by characterizing the exact necessary conditions, we show that even in this case,

Zig-Zag policies are often dominated by Constant. However, this analysis is quite useful since it allows us to develop the best Zig-Zag policy.

Secondly, in many settings, the system design does not permit the implementation of a Constant policy. This is particularly true of processors which allow only a discrete set of states of operation. As a result, the system is forced to Zig-Zag between alternate phases of cooling and heating to ensure that the system temperature does not exceed the threshold. Using our analysis on the trade-offs of a Zig-Zag policy, we develop an algorithm that calculates and implements the best Zig-Zag policy under any setting. This algorithm is quite general, and determines what speed to scale to, based on the current state of the system.

In this work, we characterized the exact trade-offs of implementing a Zig-Zag policy (with respect to Constant), and used this to derive the optimal Zig-Zag policy. In the future, we would like to extend our analysis in two directions. First, we would like to extend our analysis to real systems, by considering heterogeneous tasks (i.e., the amount of heat dissipated by the executions of a job given a certain amount of processing is different for different jobs) and multiprocessor systems. Second, we would like to incorporate power considerations into the decision making of our algorithms, thus minimizing power dissipation while ensuring that the temperature threshold is not violated.

References

1. Belady, C.: Cooling and power consideration for semiconductors into the next century. In: Macii, E., De, V., Irwin, M.J. (eds.) ISLPED, pp. 100–105. ACM, New York (2001)
2. Ma, M., Gunther, S.H., Greiner, B., Wolff, N., Deutschle, C., Arabi, T.: Enhanced thermal management for future processors. In: 2003 Symposium on VLSI Circuits, pp. 201–204 (2003)
3. Brooks, D., Martonosi, M.: Dynamic thermal management for high-performance microprocessors. In: HPCA (2001)
4. Skadron, K.: Hybrid architectural dynamic thermal management. In: DATE 2004: Proceedings of the conference on Design, automation and test in Europe, Washington, DC, USA, pp. 10–15. IEEE Computer Society, Los Alamitos (2004)
5. Rajan, D., Yu, P.S.: Temperature-aware scheduling: When is system-throttling good enough? Technical Report RC24331, IBM T. J. Watson Research Center (2007)
6. Hsu, C., Feng, W.: Reducing overheating-induced failures via performance-aware CPU power management. In: The 6th International Conference on Linux Clusters: The HPC Revolution 2005 (April 2005)
7. Weissel, A., Bellosa, F.: Process cruise control: Event-driven clock scaling for dynamic power management. In: Proceedings of the International Conference on Compilers, Architecture and Synthesis for Embedded Systems (CASES 2002), Grenoble, France (2002)
8. Merkel, A., Bellosa, F.: Balancing power consumption in multiprocessor systems. In: First Association for Computing Machinery (ACM) SIGOPS EuroSys Conference, pp. 403–414. ACM Press, New York (2006)

9. Gomaa, M., Powell, M.D., Vijaykumar, T.N.: Heat-and-run: Leveraging SMT and CMP to manage power density through the operating system. In: ASPLOS-XI: Proceedings of the 11th international conference on Architectural support for programming languages and operating systems, pp. 260–270. ACM Press, New York (2004)
10. Yao, F., Demers, A., Shenker, S.: A scheduling model for reduced cpu energy. In: FOCS 1995. Proceedings of the 36th Annual Symposium on Foundations of Computer Science, Washington, DC, USA, p. 374. IEEE Computer Society, Los Alamitos (1995)
11. Bunde, D.P.: Power-aware scheduling for makespan and flow. In: Gibbons, P.B., Vishkin, U. (eds.) SPAA, pp. 190–196. ACM, New York (2006)
12. Bansal, N., Pruhs, K., Stein, C.: Speed scaling for weighted flow time. In: SODA (2007)
13. Bansal, N., Pruhs, K.: Speed scaling to manage temperature. In: Diekert, V., Durand, B. (eds.) STACS 2005. LNCS, vol. 3404, pp. 460–471. Springer, Heidelberg (2005)
14. Lorch, J., Smith, A.J.: A new approach to dynamic voltage scaling. IEEE Transactions on Computers 53(7), 856–869 (2004)
15. Jejurikar, R., Gupta, R.K.: Dynamic voltage scaling for systemwide energy minimization in real-time embedded systems. In: Joshi, R.V., Choi, K., Tiwari, V., Roy, K. (eds.) ISLPED, pp. 78–81. ACM, New York (2004)
16. Flautner, K., Mudge, T.: Vertigo: Automatic performance-setting for linux. In: OSDI 2002: Proceedings of the 5th symposium on Operating systems design and implementation, pp. 105–116. ACM Press, New York (2002)
17. Patel, C.D., Ranganathan, P.: Enterprise power and cooling. ASPLOS Tutorial (2006)
18. Bianchini, R., Rajamony, R.: Power and energy management for server systems. Computer 37, 68–74 (2004)
19. Bradley, D.J., Harper, R.E., Hunter, S.W.: Workload-based power management for parallel computer systems. IBM J. Res. Dev. 47(5-6), 703–718 (2003)
20. Seargeant, J.E., Krum, A.: Thermal Management Handbook. McGraw-Hill, New York (1998)
21. Bellosa, F., Weissel, A., Waitz, M., Kellner, S.: Event-driven energy accounting for dynamic thermal management. In: COLP 2003. Proceedings of the Workshop on Compilers and Operating Systems for Low Power, New Orleans, LA (2003)
22. Brooks, D.M., Bose, P., Schuster, S.E., Jacobson, H., Kudva, P.N., Buyuktosunoglu, A., Wellman, J.D., Zyuban, V., Gupta, M., Cook, P.W.: Power-aware microarchitecture: Design and modeling challenges for next-generation microprocessors. IEEE Micro 20(6), 26–44 (2000)
23. Polyanin, A.D., Zaitsev, V.F.: Handbook of Exact Solutions for Ordinary Differential Equations. Chapman & Hall/CRC Press (2003)

Reuse Distance Based Cache Leakage Control

Yulai Zhao, Xianfeng Li, Dong Tong, and Xu Cheng

School of Electronics Engineering and Computer Science, Peking University
100871 Beijing, China
{zhaoyulai, lixianfeng, tongdong, chengxu}@mprc.pku.edu.cn

Abstract. As feature size shrinks, the dominant component of power consumption will be leakage. As caches represent a considerable fraction of area for many platforms, from embedded to highly paralleled systems, cache leakage control continues to become a critical issue. Drowsy cache technique is a state-preserving technique which reduces leakage by pulling down the voltages on selected lines. To exploit the temporal locality present in the data stream, existing drowsy cache policies update drowsy/active mode after an execution window of fixed clock cycles, which lack the flexibility to adapt to program behavior. We introduce a tri-mode FSM control policy, which exploits global *Reuse Distance* information and tries to keep a small set of lines in active for future references, after each *N* distinct line references. This *Reuse Distance* based policy well adapts to the temporal locality, steadily delivers better energy savings with similar performance overhead, is simple to implement, and places an upper bound on leakage power.

Keywords: Drowsy Cache Technique, Temporal Locality, Reuse Distance.

1 Introduction

Minimizing power has become a critical design issue in many platforms, from embedded to highly paralleled systems. Dynamic power is dissipated due to transistor switching activity, while leakage power continuously dissipates, even when transistors are idle. Leakage power increases exponentially as technology moves below 0.1 micron due to decreased threshold voltage, along with the improvements of transistor speed and density, and is forecasted to constitute up to 50% of overall chip power beyond 70nm processes from academic and industry data [1, 2, 3]. Leakage control at architecture and circuit level is attractive for the following reasons. First, future chips will integrate more transistors and architectural techniques can manage large groups of circuits, such as instruction windows, cache lines, banks, etc. Second, subthreshold leakage is dominant among the sources of leakage current and hard to avoid using existing technologies, although gate leakage current can be effectively reduced by using high-k dielectrics. Third, embedded and mobile systems have strict requirements in low package costs and energy efficiency. Because leakage power is dependent on the number of transistors, the majority of leakage power will come from the largest processor components. Cache sizes have grown steadily in an attempt to mask the widening gap between main memory latency and core clock frequency,

S. Aluru et al. (Eds.): HiPC 2007, LNCS 4873, pp. 356–367, 2007.

however, activities are centered on only a small set of lines are during a fixed period of time. Therefore cache leakage current is worth cutting down.

Drowsy cache technique reduces leakage power by reducing voltages on selected cache lines while preserving the contents, therefore avoiding accessing the next level of memory. The goal of the architecture level policy is to put as many lines in drowsy mode as possible while minimizing the performance loss due to hitting drowsy lines with extra wakeup cycles. Unlike instruction streams which are highly predictable [4], future references for data streams are hard to predict early, and the control policy for data caches must be very selective. The most distinguished characteristics for cache structure is its exploitation of temporal locality, and most previous drowsy cache policies have exploited this by updating the drowsy/active mode after a pre-specified number of clock cycles. However, these policies still lack the flexibility to adapt to program behavior.

In this paper, we find that the reuse distance can directly reflects the temporal locality on L1 data cache. A small distance indicates a strong likelihood of future reference, and a big distance indicates the reverse. Most cache hits are clustered to recent 5 to 10 distinct lines. We then present a tri-mode FSM control policy, which exploits the global reuse distance information and provides more fine-grained control. In this policy, mode downgrading occurs after an execution window specified by N distinct line references instead of fixed number of clock cycles. An analytical coverage model is established to measure how the window size impacts power and performance related metrics. Experimental show that the policy continuously and steadily delivers better energy savings than the best policies from the literature, suffers little performance loss, is simple to implement, and places an upper bound on leakage power consumption.

The remainder of this paper is organized as follows. Section 2 describes recent research on cache leakage power reduction. Section 3 studies the reuse distance distribution, describes and analyses the tri-mode FSM control algorithm. Section 4 evaluates our policy and state-of-the-art ones. Finally, section 5 concludes.

2 Related Work

This section reviews previous work on reducing leakage power for caches. The circuit level techniques provide an interface for architectural level control policies. Leakage power comes from transistors that are left on. Therefore the common technique to reduce leakage is to reduce the power supplied. The widely discussed techniques are classified into state-preserving and state-destroying techniques.

The gated-Vdd [6] technique uses a transistor to gate the supply of the cache SRAM cells. This technique dramatically reduces the current leakage since selected lines are powered off. The main drawback of this technique is that when a cache line is needed again after it has been put to sleep, it must be re-fetched from lower levels of memory. This re-fetch is essentially an extra miss, and this process can take many cycles. For the circuits using gated-Vdd technique, decay based policies are widely studied, including fixed- and adaptive- interval decay [7], AMC (adaptive mode control) [8], and IATAC (inter-access time per access count) tailored for L2 caches [9]. The general rationale behind these policies is that cache lines typically experience much shorter live time (the time between being brought in to the last access) than dead time (the time between the last access and next miss).

Drowsy cache is an alternative technique by making use of multiple supply voltages. When the cache line is fully on, it will dissipate too much leakage power. A lower supply voltage is used when data is not needed for a while. This will reduce the leakage power without losing the data. The tradeoff is that, while data will be preserved at this low supply voltage, it cannot be accessed while in this state. Thus there is a small wakeup time associated with changing from the lower voltage up to Vdd. The advantage of this technique is that it achieves the same L1 hit ratio as conventional caches, and therefore no additional accesses to the L2 cache are needed. Although leakage is not zero in the drowsy mode, it provides more than a 10× reduction over the regular high leakage mode. The widely studied drowsy cache policies generally execution window based. The simple policy puts all lines to sleep indistinctively after a specified number of cycles [4]. The noaccess policy only turns off lines that have not been accessed during that window [4]. The RMRO policy is tailored for set-associative caches, where if a cache way has been accessed within the time window it remains awake, otherwise it's turned off [5]. If more than two ways have been accessed, only the two MRU ways are kept awake. The simple policy has been shown to perform almost identically as noaccess policy. The RMRO is shown to improve the hit ratio for drowsy lines in some extent. The rationale behind these policies is that during a fixed interval, only limited lines are centered on.

3 Reuse Distance Based Control Policy

This section studies the reuse distance distribution for the L1 data cache. Based on this characterization, a hardware approach is proposed to capture this locality behavior.

3.1 Reuse Distance Characterization

The reuse distance of a memory access is formally defined as the number of distinct cache lines/blocks referenced since the last reference to the requested line/block. Because a new line that replaces the old one carries different data, an access that misses is assigned a reuse distance of infinity. Table 1 shows the reuse distance for an example access sequence. In this sequence, line B is continuously referenced for three times, but later A and C are still assigned small distances (1 and 2 instead of 3 and 4).

Table 1. Reuse distance for an example access sequence

address	A	B	C	A	B	B	B	A	C
distance	∞	∞	∞	2	2	0	0	1	2

To study the reuse distance distribution, we use a MRU algorithm similar to the LRU replacement algorithm in set-associative caches. This algorithm is only for the software simulator to study the cache behavior. Keeping N most recent distinct references can be implemented with a MRU structure. Each entry of the MRU contains a line ID and a counter. For set-associative caches, the line ID is composed of the set index and the way ID of the address (by matching the tags). For each cache access, the address can be translated to a unique line ID. This line ID is looked up in the MRU table. If the line exists, the corresponding counter is reset to zero. Meanwhile, all other counters are incremented. Otherwise, the LRU line which has the largest counter is evicted from the MRU table and replaced by the new line with counter reset. In this way, the MRU can keep a precise record of N most recent distinct referenced lines. Therefore, the most recent accessed line has a counter of 0, and the farmost line has the biggest counter. The reuse distance can be calculated by making a total ordering of all counters from the smallest to the biggest.

We apply this algorithm for L1 data cache by using a MRU table of 25 entries. Fig. 1 shows the percentages of cache hits with reuse distances from 0 to 24. We do not take missed access into account because drowsy policies do not incur additional misses, and they are out of our interests. Temporal locality drops significantly after most recent 5 distinct line references. After the latest 10 line references, the locality is very low for most benchmarks. We also find that if we keep a MRU table of recent accesses which may not be distinct ones, the distance distribution will not be well clustered in the lower end. This means that the reuse distance metric well matches the temporal locality behavior.

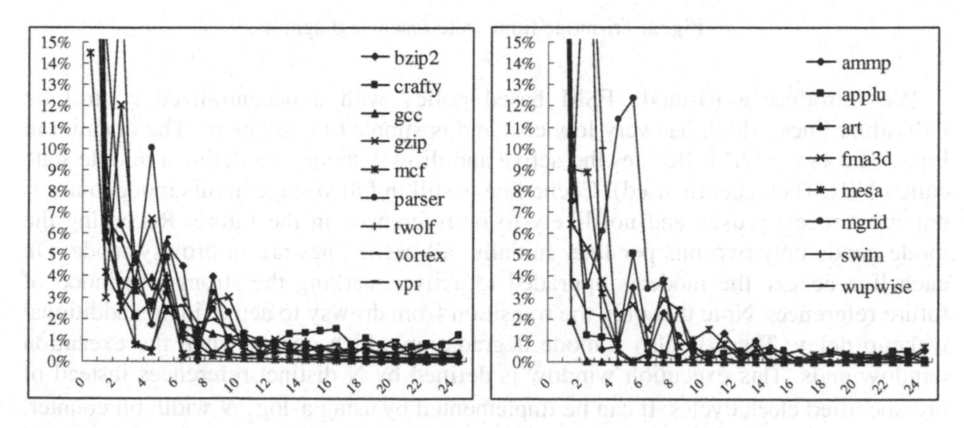

Fig. 1. Reuse distance distribution for integer (left) and floating point (right) benchmarks on L1 data cache

3.2 Tri-mode FSM Control Policy

The reuse distance distribution is clustered on recent 5 to 10 distinct line references, which indicates that we can keep 5 to 10 MRU lines active and put others into drowsy from the power perspective. Although a centralized MRU structure can keep a precise record of focused lines, it is expensive to implement in hardware. The table is indeed a small fully-associative cache, and each access must update the table with address matching and counter comparing operations. Multiple concurrent accesses may put more timing pressure on it. When the LRU line address is evicted from the MRU table, it must pass the data cache row decoder, and then select the line and put it into drowsy. This can not be performed when other cache lines are accessed.

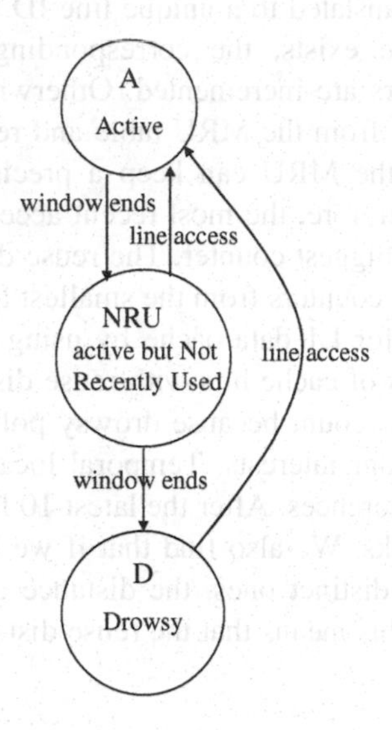

Fig. 2. Tri-mode finite state machine diagram

We introduce a tri-mode FSM based policy with a decentralized control on individual lines, which has very low cost and is simple to implement. The diagram in Fig. 2 shows the FSM. Besides the active and drowsy mode, we define a middle state called NRU (not recently used). Cache line is still in full voltage in this mode, but it is not most recently used and not likely to be referenced in the future. Recording the mode needs only two bits per line. Initially, all cache lines are in drowsy mode. On each line access, the mode is upgraded to active, marking the strong likelihood of future references. Note that only the transition from drowsy to active incurs additional wakeup delay. There is also a mode degradation which occurs when the execution window ends. This execution window is defined by N distinct references instead of pre-specified clock cycles. It can be implemented by using a $\log_2 N$ width bit counter. On each access, only if the cache line is in NRU or drowsy mode or newly brought in, the counter is incremented. Otherwise, it can be stated that the cache line is already upgraded to active mode during the execution window, and we do nothing. Thus we record only distinct references in the execution window. After the counter reaches N, the execution window ends, the counter is cleared and it signals all cache lines. This quasi-periodic event informs each cache line to downgrade according to its current mode, i.e., an active line downgrades to NRU mode, and a NRU line downgrades to drowsy mode. Because most lines are drowsy, the mode transitions are centered on a small set of lines.

As shown in Fig. 3, the tri-mode FSM policy can be implemented in a similar way as the simple policy, which only adds an extra bit in "mode" field to distinguish the

"NRU" and "Active" mode. For each cache access, the selected line is upgraded to active mode. When the global counter reaches N (execution window ends), a signal is sent to all cache lines to switch their respective mode according to current mode. For the reason of design simplicity, we assume that only data array has implemented the DVS circuit. If the tag array also implements DVS, then more leakage power can be saved but a hit on a drowsy line would incur more latencies and when the tag mismatches with the awakened line, it should be put back to drowsy again.

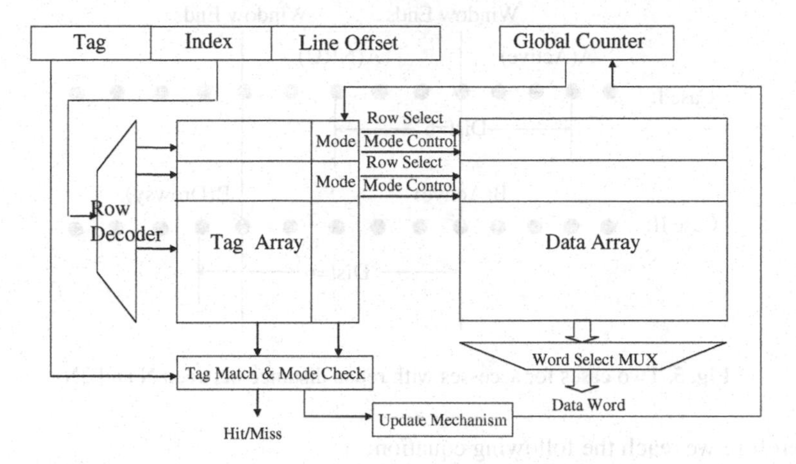

Fig. 3. Hardware implementation of tri-mode FSM policy (direct mapped cache shown)

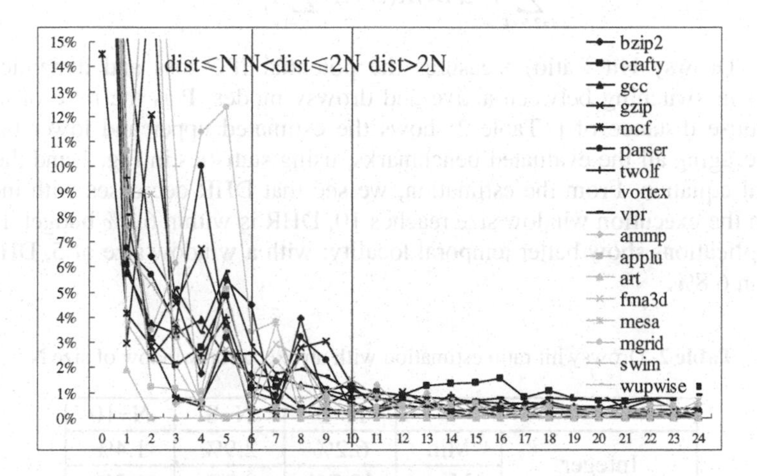

Fig. 4. Coverage model with an execution window of size N=5

Choosing the window size presents a tradeoff between power and performance. To understand how the parameter N impacts performance, we present an analytical coverage model which estimates the ratio of drowsy hits (each of which incurs additional wakeup delay) in all cache hits. As shown in Fig. 4, we choose an execution window of size 5. Cache hits with a reuse distance within 5 (including 5) will not be drowsy hits. Cache hits with a reuse distance beyond 10 are guaranteed to incur wakeup cycles because the accessed cache lines are in drowsy mode. Cache hits with a distance within 5 (not inclusive) and 10 may hit a drowsy line or a NRU line, as shown in Fig. 5. The dots in the figure represent distinct cache lines. In case I, line A experiences one mode transition from Active to NRU. In case II, line B experiences two mode transitions from Active to Drowsy.

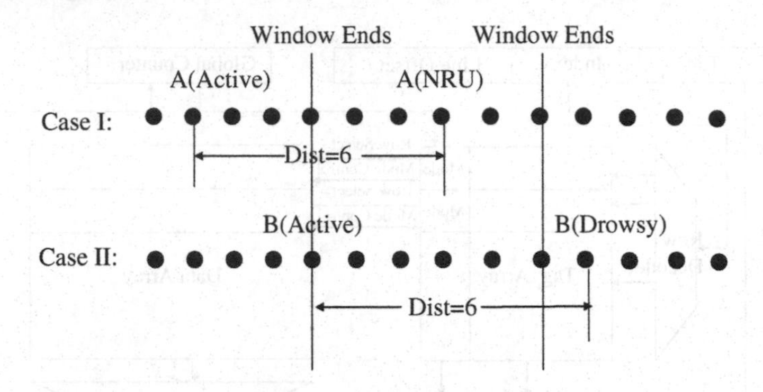

Fig. 5. Two cases for accesses with reuse distance between N and 2N

Therefore we reach the following equation:

$$\sum_{i=2N+1}^{\infty} P_i \leq DHR(N) \leq \sum_{i=N+1}^{\infty} P_i \tag{1}$$

DHR (Drowsy Hit Ratio) measures the potential IPC loss and dynamic power overhead in switching between active and drowsy modes. P_i is the ratio of accesses with a reuse distance of i. Table 2 shows the estimated upper and lower bound of DHR averaging all the evaluated benchmarks, using statistics in Fig. 1 and the above analytical equation. From the estimation, we see that DHR decreases with increased N. When the execution window size reaches 10, DHR is within 6.2% budget. Floating point applications show better temporal locality: with a window size of 5, DHR is no more than 6.8%.

Table 2. Drowsy hit ratio estimation with an execution window of size N

		N=5	N=8	N=10
Integer	Min	6.2%	2.9%	1.4%
	Max	12.7%	8.2%	6.2%
Floating Point	Min	2.6%	2.3%	1.6%
	Max	6.8%	3.4%	2.6%

DR (drowsy ratio) is the average ratio of cache lines that are in drowsy mode. This metric is to gauge the saving on cache leakage power. The tri-mode FSM based policy also permits us to estimate a lower bound of DR for a given N, as shown in equation 2.

$$DR(N) = 1 - \frac{i}{nLines} \sum_{i=0}^{2N} R_i \geq 1 - \frac{2N}{nLines} \quad (2)$$

Here nLines denotes the total number of cache lines, and R_i is the percentage of clock cycles in which exactly i cache lines remain active. For a given execution window size N, there are at most 2N active lines in any cycle, which puts a hard limit on the maximum power consumption on cache leakage power. For example, our 64K data cache has 2048 lines, and choosing a window size of 10 has a lower bound estimation of DR to be 99%. Table 3 summarizes the lower bound estimation of drowsy ratios related to window size N.

Table 3. Lower bound of drowsy ratio with an execution window of size N

N=5	N=8	N=10
99.5%	99.2%	99%

4 Evaluation

4.1 Experimental Setup

In this section, we evaluate the power and performance characteristics of our reuse distance based policy, comparing it with state-of-the-art policies using our baseline architecture configuration. We evaluate our drowsy policy and state-of-the-art ones across 17 SPEC CPU 2000 benchmarks with reference inputs on the single interval of 100 million instructions suggested by the SimPoint tool [10]. We use m-sim to model a four-issue Alpha architecture [11], and incorporated HotLeakage [12] and Wattch [13] model to evaluate leakage and dynamic power. Table 4 shows the base architecture parameters. We assume awake tag and drowsy data, and switching lines from high to low or low to high voltages incurs an extra 2-cycle transition penalty. We model the L1 cache with 2-cycle hit time, therefore the hits on drowsy lines will need 4 cycles: 2-cycle tag match and data wakeup plus 2-cycle line read/write. We model an operating temperature of 80℃, which is typical of a chip, and 70nm process technology.

Table 5 presents the 64K L1 DCache leakage power (circuit and technology related) estimated with HotLeakage using default parameters. Drowsy cache consumes only 9% leakage power of cache with full Vdd assuming awake tags. Because drowsy cache still dissipates leakage power, the total leakage power grows as cache is enlarged. The dynamic mode switch power is comparable with or larger than the drowsy leakage power of a single line, therefore frequent unwise mode switches still have considerable negative impact on leakage savings, as can be reflected by DHR metric.

We compare our policy (with a modest execution window size of 10) against simple and RMRO policy (with an optimized execution window of 2K cycles) across 17 SPEC benchmarks. From Fig. 6, it is observed that our policy continuously achieves higher drowsy ratio, which is averaged to above 99.5%. Simple and RMRO

Table 4. Architecture parameters

Technology	70nm
Frequency	1.5GHz
Temperature	80□
Voltage	1V
Issue/Commit Width	4
IQ/RF/LSQ/ROB Size	80/160/40/160
INT/FP ALU Units	4/2
L1 ICache	32KB 2-way associative 1-cycle access 32-byte line
L1 DCache	64KB 4-way associative 2-cycle access 32-byte line
L2 Unified Cache	1MB 4-way associative 8 cycles 32-byte line
Main Memory	first chunk 100 cycles inter chunk 2 cycles

Table 5. Leakage power (mW) estimated with HotLeakage

Active	Drowsy	Line Mode Switch Power (low to high/high to low)
283.2	27	0.03/0.01

achieve averaged drowsy ratio from 95% to 97.5%. Meanwhile, they achieve much lower drowsy ratio on *gcc* and *crafty* (below 90%). Higher drowsy ratio directly leads to more leakage power savings due to the fact that low-voltage line can achieve more than 10× power reductions. On the other hand, our policy continuously achieves much lower drowsy hit ratio as shown in Fig. 7, which is averaged to below 4%. Simple and RMRO achieve averaged drowsy hit ratio from 8% to15%. They even incur above 30% drowsy hit ratio for *mcf* and *art*. That means our policy has much less unwise mode switches.

From Fig. 8, it is observed that our policy cuts down leakage power to only 11% of the baseline, while simple and RMRO have normalized leakage of 20% and 17% respectively. Due to the lower drowsy ratio, simple and RMRO dissipates much more power on *gcc* and *crafty*. That means our policy improves the other two by 45% and 35% on average respectively. The results are close to the original drowsy paper which reports simple policy to achieve 26% normalized leakage on 32K cache. The improvements are mainly due to the high drowsy ratio, and partially due to the lower ratio of mode switches. We also tried different number of clock cycles for simple and RMRO, and found that not a single choice is optimal for all benchmarks. Some favor 2K cycles while others favor 4K cycles. This also agrees with the original drowsy

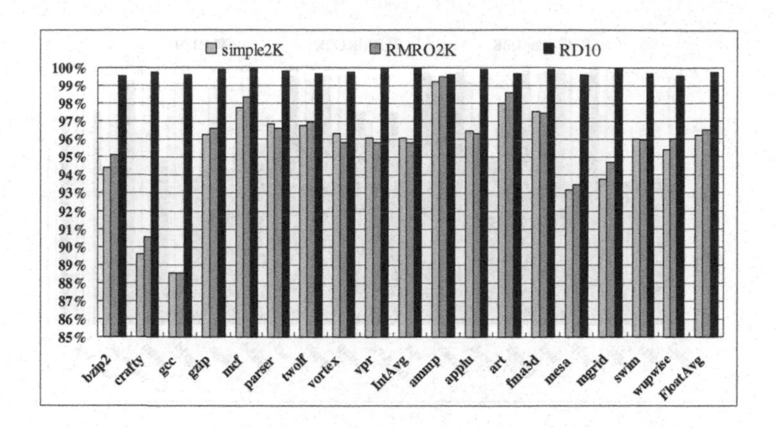

Fig. 6. Drowsy ratio

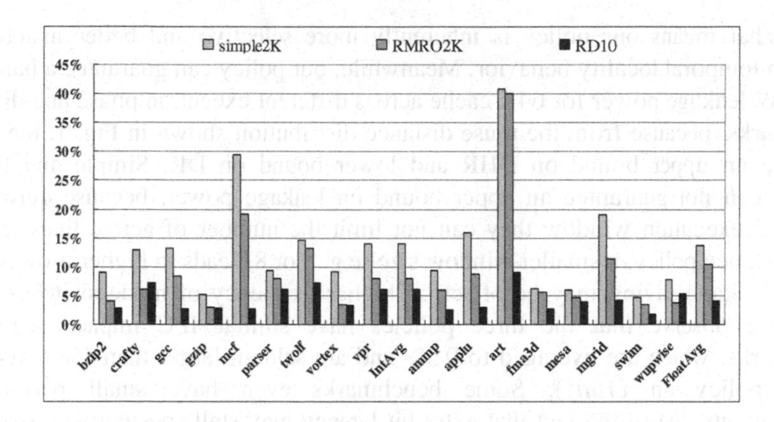

Fig. 7. Drowsy hit ratio

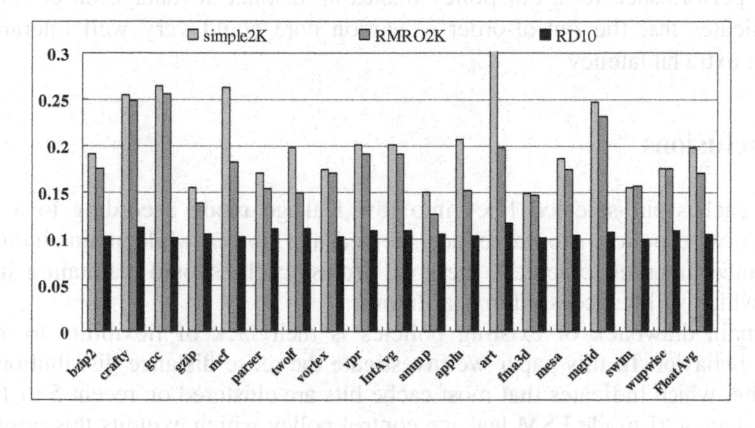

Fig. 8. Normalized leakage power

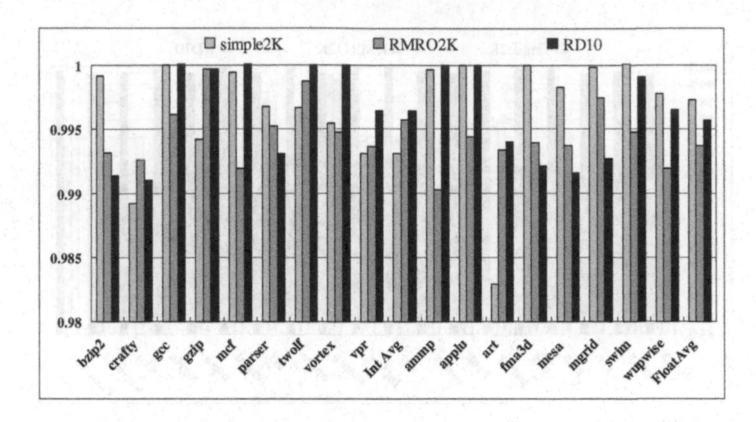

Fig. 9. Normalized IPC

paper. That means our policy is inherently more selective and better matches the program temporal locality behavior. Meanwhile, our policy can guarantee a hard limit of 30mW leakage power for 64K cache across different execution phase and different benchmarks, because from the reuse distance distribution shown in Fig. 1, the policy can give an upper bound on DHR and lower bound on DR. Simple and RMRO policies can not guarantee an upper bound on leakage power, because during any particular execution window they can not limit the number of active lines. For the reuse distance policy, a smaller window size (e.g. 5 or 8) leads to higher drowsy ratio, but the leakage saving may be offset by higher frequency of mode switches. From Fig.9, we observe that the three policies have similar IPC impacts across all benchmarks, which are averaged to 0.5% and are seldom larger than 1% (except for simple policy on *crafty*). Some benchmarks even have small performance improvements due to the fact that extra hit latency may stall speculative instructions from accessing cache which may incur additional misses. Anyway, with such a small range of performance loss, our policy makes no distinct advantage on performance. This indicates that the out-of-order execution core could very well tolerate some degree of extra hit latency.

5 Conclusions

Drowsy caches put selected lines into low leakage mode according to a drowsy policy. A wise policy should reduce the leakage power while maintaining cache performance. In order to achieve the goal, drowsy cache should maintain a small set of lines which will be accessed in near future.

The main drawback of existing policies is their lack of flexibility to adapt to program behavior. In this paper we investigate the reuse distance distribution on L1 data cache, which indicates that most cache hits are clustered on recent 5 to 10 lines. We introduce a tri-mode FSM leakage control policy which exploits this observation and is very simple to implement. We present an analytical coverage model on how the window size influences drowsy hit ratio and drowsy ratio. Experiments show that this policy improves drowsy ratio while reducing drowsy hit ratio over state-of-the-art policies. It cuts down leakage of 64K L1 data cache to only 11% of the original. The performance loss is limited to under 0.5% comparing with a standard cache. Meanwhile, it can guarantee an upper bound on power consumption.

We also find that reuse distances on unified L2 cache are not well clustered, due to the influences from the instruction streams. In future work we will cast deeper insight on better exploiting L2 cache behavior for leakage control considering existing circuit techniques.

References

[1] Flautner, K., Kim, N., Martin, S., Blaauw, D., Mudge, T.: Drowsy caches: Simple techniques for reducing leakage power. In: Proc. 29th IEEE/ACM International Symposium on Computer Architecture, pp. 529–551. New York (2002)

[2] Doyle, B., Arghavani, R., Barlage, D., Datta, S., Doczy, M., Kavalieros, J., Murthy, A., Chau, R.: Transistor elements for 30nm physical gate lengths and beyond. Intel Journal. J. 6, 42–54 (2002)

[3] Puri, R., Stok, L., Bhattacharya, S.: Keeping hot chips cool. In: Proceedings of the 42th DAC, pp. 285–288 (2005)

[4] Kim, N., Flautner, K., Blaauw, D., Mudge, T.: Drowsy instruction caches: Leakage power reduction using dynamic voltage scaling and cache sub-bank prediction. In: Proc.36th IEEE/ACM International Symposium on Microarchitecture, pp. 219–230 (2002)

[5] Petit, S., Sahuquillo, J., Such, J., Kaeli, D.: Exploiting temporal locality in drowsy cache policies. In: Proc. ACM Computing Frontiers Conference, pp. 371–377 (2005)

[6] Powell, M., Yang, S.H., Falsafi, B., Roy, K., Vijaykumar, T.: Gated-Vdd: A circuit technique to reduce leakage in deep-submicron cache memories. In: Proc. IEEE/ACM International Symposium on Low Power Electronics and Design, pp. 90–95 (2000)

[7] Kaxiras, S., Hu, Z., Martonosi, M.: Cache decay: Exploiting generational behavior to reduce cache leakage power. In: Proc. 36th IEEE/ACM International Symposium on Computer Architecture, pp. 240–251 (2001)

[8] Huiyang, Z., Toburen, M.C., Rotenberg, E., Conte, T.M.: Adaptive mode control: A static-power-efficient cache design. In: Proc. 10th IEEE International Symposium on Parallel Architectures and Compilation Techniques, pp. 61–70 (2001)

[9] Abella, J., et al.: IATAC: A smart predictor to turn-off L2 cache lines. ACM Transactions on Architecture and Code Optimization 2, 55–77 (2005)

[10] Sherwood, T., Perelman, E., Hamerly, G., Calder, B.: Automatically characterizing large scale program behavior. In: Proc. 10th ACM Symposium on Architectural Support for Programming Languages and Operating Systems, pp. 45–57 (2002)

[11] Sharkey, J.: M-Sim: A flexible, multi-threaded simulation environment. Technical Report CS-TR-05-DP1, Dept. of CS, SUNY Binghamton (2005)

[12] Zhang, Y., Parikh, D., Sankaranarayanan, K., Skadron, K., Stan, M.: Hotleakage: A temperature-aware model of subthreshold and gate leakage for architects. Technical Report CS-2003-05. Dept. of CS, University of Virginia (2003)

[13] Brooks, D., Tiwari, V., Martonosi, M.: Wattch: A framework for architectural-level power analysis and optimizations. In: Proc. 27th IEEE/ACM International Symposium on Computer Architecture, pp. 83–94 (2000)

Self-optimization of Performance-per-Watt for Interleaved Memory Systems

Bithika Khargharia[1], Salim Hariri[1], and Mazin S. Yousif[2]

[1] University of Arizona, Tucson, AZ, USA
bithika_k @ece.arizona.edu
[2] Intel Corporation, Hillsboro, OR, USA

Abstract. With the increased complexity of platforms coupled with data centers' servers sprawl, power consumption is reaching unsustainable limits. Memory is an important target for platform-level energy efficiency, where most power management techniques use multiple power state DRAM devices to transition them to low-power states when they are "sufficiently" idle. However, fully-interleaved memory in high-performance servers presents a research challenge to the memory power management problem. Due to data striping across all memory modules, memory accesses are distributed in a manner that considerably reduces the idleness of memory modules to warrant transitions to low-power states. In this paper we introduce a novel technique for dynamic memory interleaving that is adaptive to incoming workload in a manner that reduces memory energy consumption while maintaining the performance at an acceptable level. We use optimization theory to formulate and solve the power-performance management problem. We use dynamic cache line migration techniques to increase the idleness of memory modules by consolidating the application's working-set on a minimal set of ranks. Our technique yields energy saving of about 48.8 % (26.7 kJ) compared to traditional techniques measured at 4.5%. It delivers the maximum performance-per-watt during all phases of the application execution with a maximum performance-per-watt improvement of 88.48%.

1 Introduction

With the increased computing demand coupled with server sprawl in data centers, power consumption is reaching unsustainable limits. Memory is a major consumer of the overall system energy [1]. Recently, researchers [1,4,5] have explored multi-power state Rambus DRAM (RDRAM) [2] and Fully-Buffered DIMM (FB-DIMM) [3] that provide the ability to transition individual memory modules to low-power modes. Since memory is often configured to handle peak performance, it is possible to save power and simultaneously maintain performance by allocating the required memory to applications at runtime and moving the un-needed memory capacity to low power states.

However, existing techniques fall short when applied to servers with interleaved memory sub-systems. Interleaving does not offer much opportunity for energy saving because memory accesses are symmetrically distributed across all memory modules

S. Aluru et al. (Eds.): HiPC 2007, LNCS 4873, pp. 368–380, 2007.

and thus providing less opportunity for idleness. For example, we ran SPECjbb2005 [6] on our server with fully-interleaved (16-way) memory and observed that the memory was idle for less than 5% of the total runtime of SPECjbb2005. Applying existing power management techniques [1,4,5] to this memory sub-system would yield only 4.5% total saving. We also ran SPECjbb2005 with smaller degree of interleaving (12-way) and noticed little impact on performance but the idleness of few memory modules increased long enough to yield energy saving of 25% (14.7 kJ). This demonstrated an opportunity to reduce power and maintain performance by dynamically scaling the degree of interleaving to adapt to the application's memory requirements. This requires us to detect the application's memory requirements at runtime and appropriately reconfigure the degree of interleaving such that we can maximize the server's *performance-per-watt*.

In this paper, we propose a dynamic interleaving technique that intelligently interleaves data across selected memory modules and thereby increases the idle period for the remaining modules. Hence the other memory modules can transition to the really low-power states and can remain in that state for longer periods of time. This delivers more performance by expending the same amount of energy. We model the memory sub-system as a set of *states* and *transitions*. A *state* represents a specific memory configuration and is defined by a fixed base *power* consumption and a variable end-to-end memory access *delay*. Whenever, the application memory requirement changes and/or the *delay* exceeds a threshold value, a Data Migration Manager (DMM) within the Memory Controller (MC) determines a target *state* among all possible system *states* that consumes the minimum power and yet maintains the *delay*. It searches for this target *state* by solving an efficient *performance-per-watt* optimization problem. Inorder to reconfigure the interleaving to the desired degree as required by the target *state*, the DMM migrates the application's *working set* to the memory configuration in the target *state*. It works in collaboration with a local Power Manager (PM) per memory module that can implement any fine-grained power management technique to transition memory modules to low-power states based on their idleness.

The rest of the paper is organized as follows. In Section 2, we present a motivational example for our research approach. Section 3 discusses related work. In Sections 4, we discuss the power and performance model for the memory sub-system. Section 5 discusses the MC model for *performance-per-watt* management. In Section 6 we discuss some results and finally conclude in Section 7.

2 Motivational Example

Let us consider a memory sub-system with 8 memory modules as shown in Figure 1 where each module is individually power-managed. Let us consider two time instants t_i and t_{i+1} during the application execution such that the application requires n_i *pages* at t_i and n_{i+1} *pages* at t_{i+1} to achieve the maximum *hit ratio*. If we consider a sequential *page* allocation scheme [1], *pages* are allocated to one memory module completely filling it up before going to the next module. Hence, if $n_{i+1} < n_i$, we can save power by transitioning the modules, that contain the unused ($n_i - n_{i+1}$) *pages*, into a low-power state. However with full-interleaving, a single *page* would be striped across all memory modules. Consequently, the n_{i+1} *pages* would occupy all the modules. Hence there are no memory modules with unused *pages* that can be transitioned to a low-power state. We propose to create the opportunity for power saving in fully-interleaved memory by dynamically varying the degree of interleaving without hurting performance. For the example shown in Figure 1, we reduce the degree of interleaving from 8-way to 4-way by migrating the data from 8 to 4 modules. In this manner we can transition the remaining 4 modules to a low-power state. However, reducing the degree of interleaving also reduces the parallelization in memory accesses which in turn may impact *delay*. In our scheme, before we reconfigure the interleaving we ensure that this impact on *delay* is within acceptable bounds. One way of reducing the impact on *delay* is to migrate the data in a manner that exploits any unique characteristics of the underlying memory architecture. For example in Figure 1 the memory modules in 'block A' can be accessed in parallel to those in 'block B' (similar to our experimental server unit). Figure 1 shows two different migration strategies. In strategy I, data is migrated onto memory modules A1 and A2 in block A and B1 and B2 in block B. However in strategy II data is migrated onto memory modules A1, A2, A3, A4 all in the same memory block A. Naturally strategy II would have a higher impact on *delay* compared to strategy I because it did not exercise both the blocks. Since cache lines with very high spatial reference affinity would lie within the same block, they would experience sequential access pattern as compared to strategy I where they can be accessed in parallel. Since most programs demonstrate a high spatial locality of reference, this would lead to a significant reduction in the parallelism of accesses for strategy II and hence severely impact *delay*. We have experimentally verified this observation where we noticed a 5.72% drop in SPECjbb2005 performance for migration strategy II.

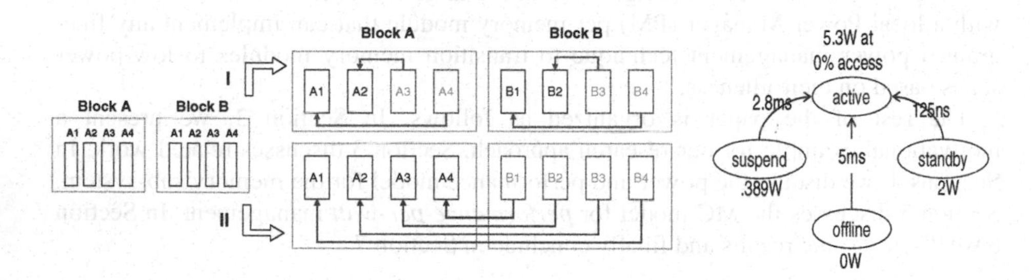

Fig. 1. Data migration strategies

Fig. 2. FB-DIMM power states

3 Related Work

Researchers have exploited hardware features such as multiple power states RDRAMs and FB-DIMMs for dynamic power management of the memory subsystem. Delaluz et al. proposed various threshold predictors to determine the maximum amount of time that a memory module must remain idle before it is transitioned into a low power state [8]. Fan et al. investigated MC policies in cache-based systems and concluded that the simple policy of immediately transitioning the DRAM chip to a lower power state as soon as it becomes idle is superior compared to more sophisticated policies that try to predict the idle time [4]. Researchers have also looked at co-operative hardware-software schemes for memory power management. Lebeck et al. studied page allocation techniques to cluster the application's *pages* onto a minimum number of memory modules thereby increasing the idleness for the other modules [1]. Zhou et al. used such page allocation schemes combined with the page Miss Ratio Curve metric to determine the optimal memory size that would give the maximum *hit ratio* for the application [5]. Delaluz et al. proposed a scheduler-based policy that used prior knowledge of memory modules used by a specific process to allocate the same memory modules the next time the process is scheduled [9]. Huang et al. built on this idea to develop Power-Aware Virtual Memory [PAVM] where the OS and the MC communicate to enhance memory energy savings through leveraging NUMA memory infrastructure to reduce energy consumption on a per-process basis [10]. Delaluz et al. [11] migrated arrays in multi-bank memory based on temporal locality to consolidate the arrays on a small set of banks. There has also been a plethora of work that addresses memory power management while maintaining performance. Li et al. [13] proposed a *Performance-directed Dynamic* (PD) algorithm that dynamically adjusts the thresholds for transitioning devices to low-power states, based on available slack and recent workload characteristics. A departure to this approach is provided by the work of Diniz et al. [14] that shows that limiting power is as effective an energy-conservation approach as techniques explicitly designed for performance-aware energy conservation.

Our scheme differs from these techniques because we address power/performance management of interleaved memory sub-systems where current techniques cannot be applied. We use migration to dynamically reduce the size of the interleaving in order to reduce power while maintaining performance. Our scheme incorporates knowledge about the underlying memory architecture in performing migrations. Our scheme is application and OS agnostic because it is closer to the hardware.

4 Memory Power and Performance Model

We consider FB-DIMM as our memory model which is popular in high-performance servers because of its reliability, speed and density features. An FB-DIMM packages multiple DDR DRAM devices and an Active Memory Buffer (AMB) in a single module. The AMB is responsible for buffering and transferring data serially between DRAM devices on the FB-DIMM and the MC.

Figure 3 shows the model of our memory sub-system. It is based on the architecture found in Intel Xeon series servers. It consists of multiple *branches* where each *branch* consists of

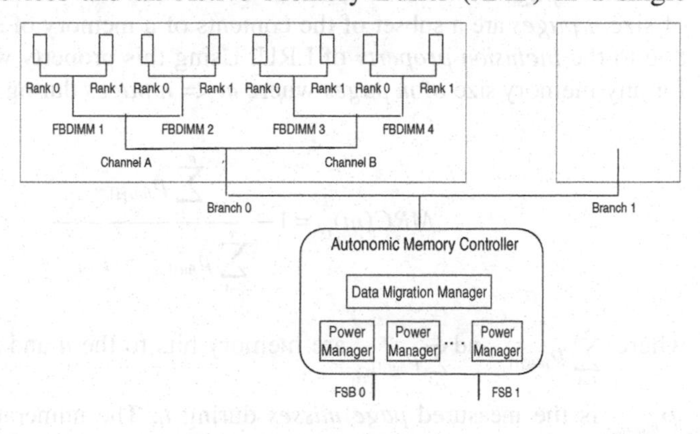

Fig. 3. Memory sub-system model

multiple *channels*, each *channel* contains multiple FB-DIMMs and each FB-DIMM contains multiple *ranks*. *Ranks* on separate *branches* can be accessed in parallel. *Ranks* on separate *channels* can be accessed in lock-step. *Ranks* within the same *channel* can be accessed sequentially. The number of DRAM devices accessed simultaneously to service a single memory request defines a *rank*. In our model we consider a *rank* as the smallest unit for power management.

An FB-DIMM includes four power states – *active*, *standby*, *suspend* and *offline* as shown in Figure 2. Memory requests are serviced only in the *active* power state. A state that consumes less power also has a high reactivation time. The AMB brings in another set of challenges for memory power management because it consumes power even when the FB-DIMM is idle in order to maintain the data transfer link between the MC and neighboring FB-DIMMs on the *channel*.

We model the performance of the memory sub-system in terms of end-to-end *delay* d. It is defined as the time from the instant the request arrives at the MC to the time when the data becomes available at the MC.

5 Memory Controller Model for *Performance-per-Watt* Management

In this Section we first discuss how we can track the dynamic memory requirements of the application. We then discuss the impact of a specific memory configuration on the *delay*. This information is used by the MC to search for a target memory configuration (*state*) that gives the maximum *performance-per-watt* among all possible memory configurations (*state*).

5.1 Dynamic Tracking of Application Memory Requirement

The DMM uses the MRC metric [5] to predict the dynamic memory requirements of the application. Let us consider that during epoch t_i, memory of size of *n* (*pages*) is in an *active* power state. Let us assume that we measured the number of hits going to each *page* and *pages* are maintained in a strict LRU [7] order where the most recently referenced *page* is moved to the head of the LRU list. Now, the contents of a memory of size *n pages* are a subset of the contents of a memory of size *n + 1 pages* or larger due to the *inclusion property* of LRU. Using this property we can calculate the MRC for any memory size of *m pages* where *m* <= *n pages* during time epoch t_i as follows

$$MRC(m)_{t_i} = 1 - \frac{\sum_{i=1}^{m} p_{hits[i]_{t_i}}}{\sum_{i=1}^{n} p_{hits[i]_{t_i}} + p_{miss_{t_i}}} \quad (1)$$

where $\sum_{i=1}^{n} p_{hits[i]_{t_i}}$ and $\sum_{i=1}^{m} p_{hits[i]_{t_i}}$ are memory hits to the *n* and *m pages* respectively and $p_{miss_{t_i}}$ is the measured *page misses* during t_i. The numerator in Equation (1) is the number of misses for a memory of size *m (pages)*, where *m*<*n*. As we increase the memory size, the corresponding *MRC* reduces. However, it would stay the same for memory of size *m (pages)* or m+1 *(pages)* if the misses stay the same for both the cases. This would mean that the $(m+1)^{th}$ *page* was over-provisioned and the ideal memory size is *m (pages)*. If however the *MRC* keeps reducing until we have covered all the *n pages* we increase the size of the memory by *bufPages* in anticipation that it would bring down the *pages misses* further. *bufPages* is determined at the end of each epoch t_i based on the measured $p_{miss_{t_i}}$, the higher the $p_{miss_{t_i}}$ the higher the *bufPages.*

We denote the total memory size *(pages)* required to attain the maximum *hit ratio* by N_{ws} *(pages)*. In fully-interleaved memory, these N_{ws} *pages* are striped across all memory *ranks*, mainly to improve performance-efficiency. In non-interleaved memory, N_{ws} *pages* are consolidated on one memory *rank* completely filling it up before going to the next, mainly to improve power-efficiency. In what follows, we discuss how at runtime we determine a degree of interleaving that stripes the N_{ws} *pages* on memory *ranks* to improve both power and performance efficiency.

5.2 Interleaving of Working Set Pages for Performance-per-Watt Management

In this Section we first discuss about data placement in a fully-interleaved memory. We then discuss how we vary the degree of interleaving to adapt to the incoming workload by using a temporal affinity prediction technique that effectively exploits the internal memory architecture.

5.2.1 Data Placement in Fully-Interleaved Memory

Let us revisit the memory architecture shown in Figure 3 that interleaves at the cache line granularity. Cache lines are allocated to *ranks* such that spatially adjacent cache lines reside on *ranks* that are furthest away from one another in order to increase the parallelization in accesses. For example, in Figure 3 cache line 0 is allocated to *rank* 0 on *branch* 0 while cache line 1 is allocated to *rank* 0 on *branch* 1.

Since temporal affinity in accesses is determined to a large extent by placement of cache lines on the underlying architecture that consists of *branches, channels* etc we abstract the notion of distance between cache line pairs based on their relative placement in the memory sub-system. We express this in the form of *Spatial Reference Affinity (SRA)* and *Spatial Location Affinity (SLA)* metrics and use these metrics to analyze the impact of a memory configuration on the end-to-end *delay*.

5.2.2 Predicting Temporal Affinity

We predict the temporal affinity between memory accesses with the aid of the *SRA* and *SLA* metrics. Let us consider cache lines that belong to a single *page*. Adjacent cache line pairs have a higher probability of temporal affinity in accesses compared to those that are further apart in the same *page*. We use the ***SRA Metric*** to capture the temporal affinity arising out of this closeness in space. The *SRA* between a cache line pair (i,j) is measured by the number of cache lines that separate them. Hence *SRA* is zero for adjacent cache line pair and $(s_P / s_{CL} - 1)$ for the first and last cache lines on the *page* where s_P denotes *page* size and s_{CL} denotes cache line size.

Temporal affinity in accesses may have very different impact on *delay* depending on the relative position of the cache line pairs within the memory hierarchy. For example with reference to Figure 3, adjacent cache line pairs (*SRA* = 0) with high temporal affinity may have negligible impact on the *delay* if they are placed on separate memory *branches*. This is because even when they are accessed immediately after one another each access is serviced by a separate *branch* which can be accessed in parallel. We use the ***SLA Metric*** to capture the temporal affinity that arises out of this closeness in physical location of cache line pairs. However, unlike *SRA*, there is no simple way of computing *SLA*. So we manually configured the memory architecture on a server unit to multiple different configurations (e.g – *ranks* on different *branches*, *ranks* on same *branch* different *channels* etc). We then ran SPECjbb2005 and recorded its performance on each of these configurations. From these results we derived the following empirical function to compute *SLA*.

$$SLA_{[B_i,D_i,R_i][B_j,D_j,R_j]} = (w_B * |B_i - B_j| + w_D * |D_i - D_j| + w_R |R_i - R_j| + w_z) \tag{2}$$

where w_B : weight of cache line pair across *branches*, w_D : weight across FB-DIMMs, w_R : weight across *ranks* and w_Z : weight of cache line pair on the same *rank*. These are the empirically derived weights where $w_B > w_D > w_R > w_Z$, $w_B > (w_D + w_R)$ etc. Hence, a cache line pair across separate *branches* B_i and B_j gives a higher *SLA* compared to that across separate *ranks* ($w_B > w_R$).

We now combine these two metrics in order to weigh the impact of one placement strategy over another in terms of their combined impact on the *delay*. We do this with the aid of the ***conflict metric*** $\Delta\psi$ given by equation (3).

$$\Delta\psi(s) = \frac{1}{SLA_{\min}(s) * SRA_{\min}(s)} \tag{3}$$

Let us consider two placement strategies that use different memory sizes. We compute the minimum *SRA* given the minimum *SLA* for each placement strategy. From Equation (3) the strategy that has the smallest *SLA* and smallest *SRA* for that *SLA* has a higher $\Delta\psi$ indicating a higher impact on *delay*. Similarly, for two placement strategies that use the same memory size (number of *ranks*) but different memory configuration (physical location of *ranks*) we compute minimum *SLA* for adjacent cache line pair (minimum *SRA*) for each strategy. Since the minimum *SRA* is fixed for both the strategies, in this case the strategy that gives the smaller *SLA* has a higher $\Delta\psi$ indicating a higher impact on *delay*. For example, each migration strategy of Figure 1 has the same memory size but different memory configuration. Now, for strategy II, *SLA* for minimum *SRA* is w_D but for strategy I it is w_B. Since $w_B > w_D$ $\Delta\psi$ for strategy II is higher than that for I. This explains the drop in performance for SPECjbb2005 for strategy II. Hence we always favor the migration strategy that has the smallest value of $\Delta\psi(s)$.

5.3 Formulating the Optimization Problem for *Performance-per-Watt* Management

We formulate our adaptive interleaving technique as a *performance-per-watt* maximization problem.

$$\text{Maximize } ppw_{t_i} = \frac{1}{d_k * e_k} \text{ such that}$$
$$1. n_k * s_r >= N_{ws} * s_p \tag{4}$$
$$2. d_{\min} \le d_k \le d_{\max}$$
$$3. \sum_{k:1}^{N_s} x_{jk} = 1$$
$$4. \vee x_{jk} = 0 | 1$$

where, ppw_{t_i} is the *performance-per-watt* during interval t_i, d_k is the *delay* and e_k is the energy consumed during target *state* s_k where $e_k = \sum_{k:1}^{N_s} (c_{jk} * \tau_{trans_{jk}} + p_a * n_k * t_{obs}) * x_{jk}$ is the sum of the transition energy consumed ($c_{jk} * \tau_{trans_{jk}}$) and the energy consumed in the target *state* ($p_a * n_k * t_{obs}$), N_s : total number of system *states*, c_{jk} : power consumed in state transition and $\tau_{trans_{jk}}$:time taken for *state* transition, $[d_{\min}, d_{\max}]$: the threshold delay range, x_{jk} : decision variable for transition from *state* s_j to s_k, s_r : size per *rank*. The *state* s_k represents a specific memory configuration given by the number and physical location of the *ranks* in the '*active*' power state. It is defined by two tuple - fixed base power consumption p_k and variable end-to-end *delay* d_k.

The first constraint in Equation 4 states that the target *state* should have enough memory to hold all the N_{ws} *pages*. The second constraint states that in the target *state*, *delay* should stay within the threshold range. The third constraint states that the optimization problem leads to only one decision. The decision variable corresponding to that is 1 and the rest are 0. The fourth constraint states that the decision variable is a 0-1 integer.

Analysis of Transition Overhead: The transition overhead $c * \tau_{trans}$ is the energy spent during *state* transition. We factor this overhead into the objective function to identify *state* transitions that would give the smallest overhead among all possible transitions. We also account for the impact of the transition time τ_{trans} on *delay*. Owing to constraint 2 Equation 4, this prevents *state* transitions when τ_{trans} is too high. Hence this reduces the frequency of *state* transitions and thereby maintains the algorithm sensitivity to workload changes, within acceptable bounds.

The transition time is expressed as $\tau_{trans} = \tau_m + \tau_p$ where τ_p denotes the *rank* power state transition time (see Figure 2) and τ_m denotes the data migration time. Since *ranks* can transition in parallel τ_p is essentially the *rank* reactivation time. τ_m is a sum of the *rank* read time, data transfer time on the link(s), *rank* write time and the time taken to update a hardware indirection data structure that routes accesses to migrated data blocks. Note that τ_m is directly proportional to the amount of data being migrated. We call this the migration data M. In our case, M is the predicted application *working set* (N_{ws}). Hence, maintaining the *working set* in memory not only reduces memory over-provisioning but also reduces migration overhead.

Migration Energy: The energy consumed during migration $c*\tau_{trans}$ can be expressed as $c*\tau_{trans} = n_k * p_t * \tau_p + p_m * \tau_m$ where p_t is the transition power consumed by a *rank* and p_m is the power consumed in the memory sub-system during data migration. p_m can be expressed as a sum of the buffer power, base FB-DIMM power, DRAM refresh power, read power, link power and write power. We assume a close-page policy which is energy-efficient for interleaved memory. Hence we do not account for the energy spent in accessing open pages during migration.

6 Experimental Results

Our test-bed consists of a server with Intel Xeon processors and 5000 series chipset. It has memory architecture similar to that shown in Figure 3. It consists of two *branches,* two *channels* per *branch,* two FB-DIMMs per *channel* and two *ranks* per FB-DIMM. The server can support a total of 8 FB-DIMMs or 16 *ranks* in total.

We studied the *performance-per-watt* for the SPECjbb2005 on our server unit. SPECjbb2005 emulates a 3-tier client/server system with emphasis on the mid-tier business logic engine. It gives the performance of the system in a *throughput* measure called BOPS (business operations per second).

Current server technology does not support dynamic memory interleaving. To get around this problem we emulated dynamic interleaving by manually reconfiguring the memory sub-system as required. Every reconfiguration required a system restart.

6.1 Analysis of *Performance-per-Watt* Improvement for SPECjbb2005

Our algorithm monitored the MRC for SPECjbb2005 as described in Section 5.2. It also monitored the average end-to-end *delay* by using chipset performance counters per *rank*. It used these parameters to trigger a search for an optimal *state* as discussed in section 5.3. We then manually reconfigured the memory sub-system to this optimal configuration and restarted the system. We repeated this process until the application execution was complete. At the end of each phase that required a memory reconfiguration, we recorded the BOPS and the power consumed by the system.

Note that sometimes the algorithm returned a memory configuration that could not be configured in hardware without changing the configuration registers in the MC. For example, since the *channels* were configured to work in lock-step we always needed to populate FB-DIMMs as a pair, one on each *channel*. Hence we could only work with even-numbered FB-DIMMs. In such cases the algorithm returned a second sub-optimal solution that gave a smaller *performance-per-watt* compared to the optimal solution and we reconfigured the memory sub-system accordingly. Figure 4 shows the temporal variation of optimal and sub-optimal *states* given by our algorithm.

In order to compare the *performance-per-watt* improvement given by our algorithm, at the end of each epoch that required a reconfiguration, we reconfigured the memory not only to that desired by our algorithm but also to all other possible memory configurations allowed by the hardware. We ran SPECjbb2005 on each of these configurations and recorded the BOPS as well as the power consumed. Figure 5 shows the temporal variation of *performance-per-watt* (BOPS/Joules) for each such configuration. We observed that our algorithm always determined the memory configuration (configuration IV in Figure 5) that gave the maximum *performance-per-watt* among all possible configurations with the maximum improvement in *performance-per-watt* recorded at 88.48%.

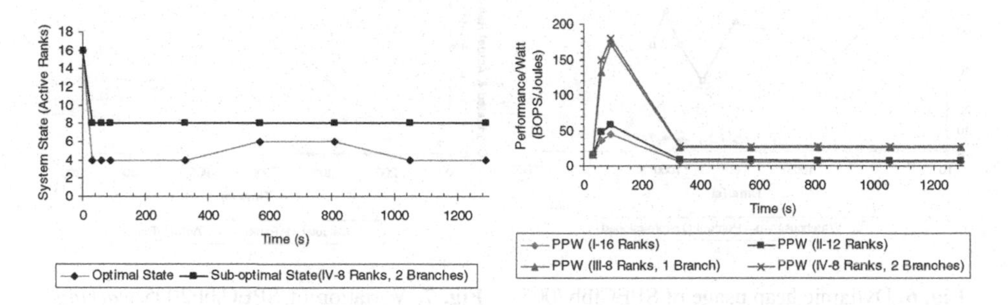

Fig. 4. Optimal and sub-optimal *states*

Fig. 5. Performance-per-watt comparison

On the same server we ran SPECjbb2005 and measured the idle durations between memory accesses to each *rank* by using chipset counters. With a threshold-based power management algorithm, where we transition the *rank* to a low-power *suspend* state when it is "sufficiently" idle to break-even, we got an energy saving of 4.47% (189.6 J). This compares to about 48.8% (26.7 kJ) energy saving with our technique.

6.2 Algorithm Adaptivity to Workload

SPECjbb2005 launches an additional warehouse at the end of each observation epoch that executes randomly selected business operations from an in-memory database of operations. Instead of computing the MRC, our algorithm used the benchmark's heap usage at the end of each warehouse to predict the memory requirements of SPECjbb2005. This is because it was not possible to measure the number of hits per *page* accurately from the OS to compute the MRC. However as can be seen from Figure 5, this approximate approach still gave the memory configurations with the maximum improvement in *performance-per-watt* among all possible configurations. We also instrumented the linux kernel to index the memory *pages* starting at the head of the LRU active list until it was equal in size to the heap used. This is the *working set* for SPECjbb2005 and comprises the migration data M that is to be dynamically interleaved on the memory configuration given by our algorithm. Figure 8 plots these *pages* as a percentage of the total *pages* in the LRU *active* list. As expected, this graph varies inversely as the percentage of over-provisioned memory as plotted in Figure 6.

Figure 7 plots the memory size (in *ranks*) that is predicted to be required by SPECjbb2005 as discussed in Section 5.2. The 'actual *ranks*' in Figure 7 is the ceiling value of the 'calculated *ranks*'. By comparing Figures 6 and 7 we see that the memory size varies inversely with over-provisioned heap as expected. Also notice that the optimal and sub-optimal *ranks* computed by our algorithm (Figure 4]) are always higher than the 'calculated *ranks*' of Figure 7. This is because consolidating the *working set* on these 'calculated *ranks*' maintains the application memory requirements but significantly increases the *delay*. Hence it violates the *delay* constraint which makes these *states* infeasible.

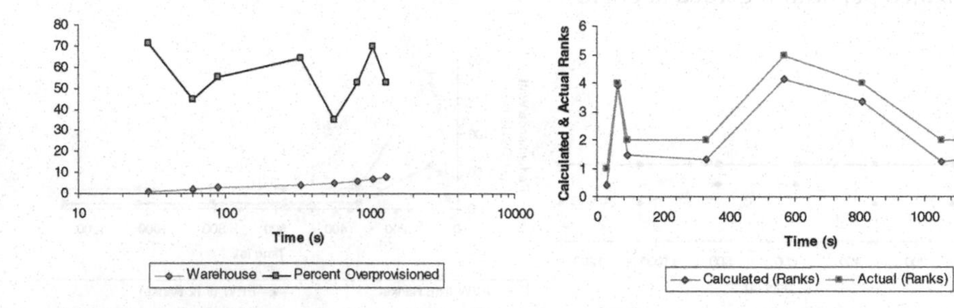

Fig. 6. Dynamic heap usage of SPECjbb2005

Fig. 7. Variation of SPECjbb2005 *working set) pages* (memory *ranks*)

Also note from Figure 6 that around 600 sec into the application execution, the over-provisioning reduces close to 30%. As can be seen from Figure 4, it is around this time that the algorithm increases the memory size from 4 to 6 *ranks* in anticipation of a heavy workload arrival phase. However when the over-provisioning increases around 800 sec the algorithm maintains the same memory size (6 *ranks*). At about 1000 sec, when the over-provisioning further increases, the algorithm reduces the memory size from 6 *ranks* back to 4 *ranks*. The algorithm has a tendency to latch on to previous memory configurations. It initiates reconfigurations only when significant over-provisioning is detected. It works conservatively because it accounts for the overhead involved in *state* transitions. This is discussed in the following section.

6.3 Analysis of Migration Overhead

Figure 8 plots the migration overhead (in milli seconds) associated with the migration data M for SPECjbb2005. This overhead has been computed for solution IV in Figure 5 that gives the maximum *performance-per-watt*. Notice that the overhead is very small at the end of the first SPECjbb2005 warehouse because the migration data is small enough and the *state* transition decision being evaluated was to go from 16 to 8 *ranks*. Consequently, as we see from Figure 4, the algorithm allowed this *state* transition. However, at the other warehouses the transition overhead increased considerably as M increased and the transition decision to be evaluated

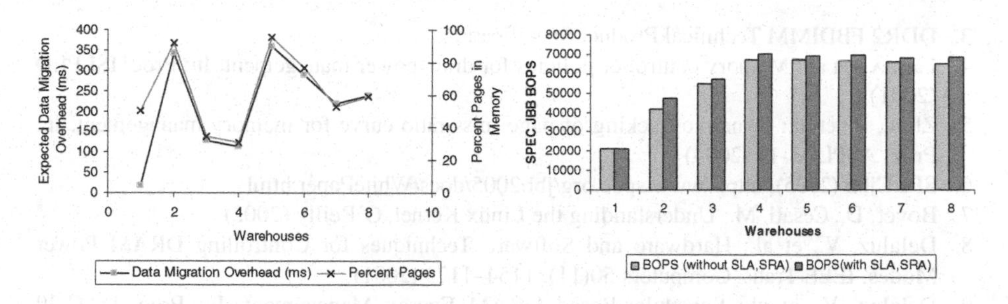

Fig. 8. Migration overhead

Fig. 9. Comparison of migration strategy

was to migrate from 8 to 4 *ranks*. The algorithm did not allow this *state* transition and instead it paid the overhead one time and maintained the memory configuration at a steady state with 8 *ranks* distributed across two *branches*.

6.4 Impact of Migration Strategies on SPECjbb2005 Performance

Figure 5 shows the *performance-per-watt* obtained for two solutions (III-8 *ranks*, 1 *branch* & IV-8 *ranks*, 2 *branches*). Note that these two solutions have the same number of *ranks* but different physical location in the memory hierarchy. However solution IV provides a higher *performance-per-watt* compared to solution III. Figure 9 plots the BOPS measured at the end of each warehouse for both these solutions. Solution III gives a performance drop of 5.72% for SPECjbb2005 when compared to solution IV. Our algorithm is able to effectively identify this with the aid of the temporal affinity prediction technique discussed in Section 5.2.2 and chooses solution IV over solution III thus giving the maximum *performance-per-watt*.

7 Conclusion

In this paper, we presented a technique to optimize the *performance-per-watt* of a fully-interleaved memory sub-system. Our approach yielded an energy saving of about 48.8 % (26.7 kJ) compared to traditional techniques measured at 4.5%. It gave a transition overhead of about 18.6 ms leading to energy saving of 1.44kJ per ms of transition overhead time and a maximum *performance-per-watt* improvement of 88.48%.

We are currently validating our results on different memory traces and studying the algorithm scalability, adaptivity and sensitivity to threshold values. We are applying data mining and rule learning techniques to implement an efficient real-time version of our algorithm that significantly reduces the runtime complexity of the algorithm. We are also extending our technique to servers running multiple applications.

References

1. Lebeck, R., et al.: Power aware page allocation. In: Proc. ASPLOS-9 (2000)
2. Rambus, RDRAM (1999), http://www.rambus.com

3. DDR2 FBDIMM Technical Product Specifications
4. Fan, X., et al.: Memory controller policies for dram power management. In: Proc. ISLPED (2001)
5. Zhou, P., et al.: Dynamic tracking of page miss ratio curve for memory management. In: Proc. ASPLOS-11 (2004)
6. SPECjbb (2005), http://www.spec.org/jbb2005/docs/WhitePaper.html
7. Bovet, D., Cesati, M.: Understanding the Linux Kernel, O'Reilly (2002)
8. Delaluz, V., et al.: Hardware and Software Techniques for Controlling DRAM Power Modes. IEEE Trans. Computers 50(11), 1154–1173 (2001)
9. Delaluz, V., et al.: Scheduler-Based DRAM Energy Management. In: Proc. DAC-39 (2002)
10. Huang, H., et al.: Design and Implementation of Power-Aware Virtual Memory. In: Proc. USENIX Technical Conference, pp. 57–70 (2003)
11. Delaluz, et al.: Automatic Data Migration for Reducing Energy Consumption in Multi-Bank Memory Systems. In: Proc. 39th Design Automation Conf., pp. 213–218. ACM Press, New York (2002)
12. Wang, D., et al.: DRAMsim: A memory-system simulator. SIGARCH Computer Architecture News 33(4), 100–107 (2005)
13. Li, et al.: Performance-directed energy management for main memory and disks. In: Proc. ASPLOS-11, Boston, MA, USA (2004)
14. Diniz, et al.: Limiting the power consumption of main memory. In: Proc. ISCA-34, San Diego, California, USA (2007)

Distributed Algorithms for Lifetime of Wireless Sensor Networks Based on Dependencies Among Cover Sets

Sushil K. Prasad and Akshaye Dhawan

Computer Science Department
Georgia State University
Atlanta, GA 30303, U.S.A
{sprasad, akshaye}@cs.gsu.edu

Abstract. We present a new set of distributed algorithms for scheduling sensors to enhance the total lifetime of a wireless sensor network. These algorithms are based on constructing minimal cover sets each consisting of one or more sensors which can collectively cover the local targets. Some of the covers are heuristically better than others for a sensor trying to decide its own sense-sleep status. This leads to various ways to assign priorities to the covers. The algorithms work by having each sensor transition through these possible prioritized cover sets, settling for the best cover it can negotiate with its neighbors. A local lifetime dependency graph consisting of the cover sets as nodes with any two nodes connected if the corresponding covers intersect captures the interdependencies among the covers. We present several variations of the basic algorithmic framework. The priority function of a cover is derived from its degree or connectedness in the dependency graph - usually lower the better. Lifetime improvement is 10% to 20% over the existing algorithms, while maintaining comparable communication overheads. We also show how previous algorithms can be formulated within our framework.

1 Introduction

Wireless sensor networks (WSNs) are emerging as a key enabling technology for applications domains such as military, homeland security, and environment [7]. For example, in typical security surveillance or environmental monitoring scenarios, tiny low-cost radio-enabled sensors can be deployed in large numbers over a difficult or hostile terrain. Then, they configure themselves into a network to collectively sense and route data to gateway nodes. Interested readers are referred to [1] [7] for detailed background, applications and challenges of WSNs.

The sensors, by design, have limited battery. Therefore, they must conserve power while communicating, sensing, and computing. One key challenge is to utilize the network effectively to maximize the duration of time while all the targets (or alternatively certain area [9]) can be continuously monitored. This duration is called the lifetime of the network, which is the concern of this paper. Only a subset of sensors usually need to be in "sense" or "on" mode at any given time to cover all the targets (henceforth called "covers" – See Section 2 for a formal definition), while others can go into power conserving "sleep" or "off" mode. Therefore, an ideal sense-sleep schedule would choose appropriate covers at various intervals to maximize the lifetime. Thus, at its heart, this is an NP-complete problem [6] [8].

S. Aluru et al. (Eds.): HiPC 2007, LNCS 4873, pp. 381–392, 2007.

Both centralized and distributed heuristics have been proposed for finding the longest lifetime schedule. The premise of centralized algorithms, such a those based on linear programming [2] [5], is that, given the locations of sensors and targets, the lifetime scheduling algorithm can be executed offline at the gateway node or elsewhere and then the sense-sleep schedule can be communicated to the sensors. Thus, the premise of global network information is both an advantage yielding better lifetime and a liability due to the associated communication overheads and loss of distributed robustness. The existing distributed algorithms typically work iteratively in rounds of reshuffles of predetermined duration. At the beginning of each reshuffle round, each sensor negotiates with its neighboring sensors to decide on its sense-sleep status such that all its neighboring targets are covered by the sensor itself or one of its neighbors. The limitations of existing algorithms are their simple greedy approaches, lacking enough insight into the problem structure. Section 5 compares and contrasts some of the distributed algorithms against ours, even showing how they can be formulated using our overall framework.

Our contributions: Although globally there is exponential number of possible covers making the problem intractable, the number of local covers, those minimal subsets of neighboring sensors covering nearby targets, is usually small. This opens up the problem to individual sensors distributively constructing the local covers and employing them as possible local configurations to systematically transition through them to arrive at a good neighborhood sense-sleep decision for each reshuffle round. The concept of local covers captures the collective tension among the neighbors to a certain extent, as opposed to the simplistic sensor vs. its neighbors generally considered. It is assumed that the sensing range is one-half of the communication range. A sensor can construct its local covers by considering one-hop neighbors it can communicate to while trying to cover only those targets within its sensing range or also its neighbors' targets. For a better decision, it can also consider all neighbors up to two hops and their targets at a slightly increased communication cost.

The lifetime of a cover is bounded by its sensor with the weakest battery, the "bottleneck sensor." A sensor can certainly prioritize these covers based on simple heuristics using only the properties of individual covers. For example, a sensor can prefer those covers not involving it or those with longer lifetimes, as the existing algorithms do (albeit indirectly, as they lack the concept of transitioning through the set of possible local covers). What is more interesting, however, is how these covers influence each other. For example, if two covers share one or more sensors, their weakest common sensor is an upper bound on the lifetime of both covers collectively. This is because using ("burning") either cover reduces the battery of the common sensors. To model such interactions, we define the local "lifetime dependency (LD) graph" wherein the minimal covers form the nodes and two such nodes are connected by an edge if there are common sensors between the corresponding two covers (an edge therefore represents the non-empty intersection of the two covers connected). Therefore, an isolated cover is most preferred while a cover which is more densely connected to others can be less desirable candidate as burning it may reduce the life of several other covers. This new algorithmic framework based on the twin concepts of local minimal covers and the lifetime dependency graphs is the primary contribution of this paper. Section 2 describes this framework.

Although the LD graphs expose the dependencies among covers, how the covers be chosen to locally maximize the schedule length can depend on how the weights on edges and degrees of nodes are defined. For example, the edge weight can be its weakest common battery and degree of a node can be the conventional sum of all its adjoining edge weights. This results into a basic heuristic algorithm. However, some variations are also compelling. For example, in the case when the weakest common battery in an edge is larger than the sum of the weakest batteries in the two covers it connects, the edge no longer upper bounds the two covers (the sum does) and, therefore, such an edge weight may be discounted. If two different covers share the same bottleneck sensor, their collective lifetime is bounded by that sensor. If such two covers are connected to a third cover, the latter may choose to ignore one of the two edges when calculating its degree. Thus various priority functions can be derived. Section 3 describes some of the resulting distributed algorithms.

We simulate our algorithms over a range of sensor networks and compare the lifetime of their schedules with the current state-of-art algorithms. Our preliminary results in Section 4 show an improvement of 10-20% in network lifetimes over others, while maintaining the same communication complexity. In Section 5, we take up the key previous distributed algorithms to systematically compare and contrast against ours, even demonstrating how the existing algorithms can be formulated within our framework. Finally, Section 6 contains our conclusions and future work.

2 The Basic Cover Based Algorithmic Framework

Symbols and definitions: Let us begin with a few basic conventions and definitions for this paper. We will use s_1, s_2, etc., for sensors, t_1, t_2, etc., for targets, and C, C', etc., to denote covers. Let us assume we have *n* sensors and *m* targets, both stationary. Consider the sensor network in Figure 1 with $n = 8$, $s = \{s_1, s_2, \ldots, s_8\}$ and $m = 3$ targets, t_1, t_2, and t_3.

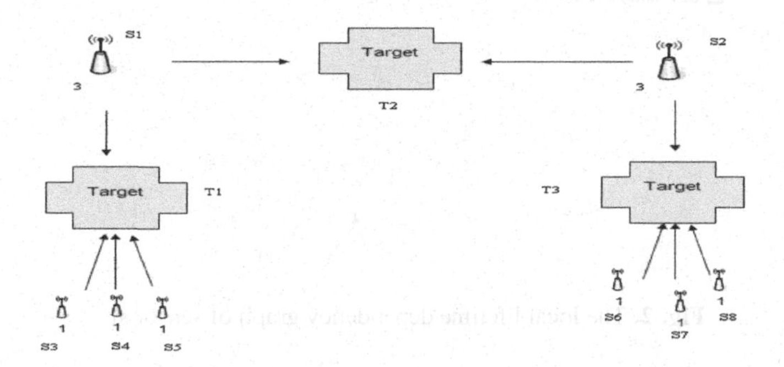

Fig. 1. A Sensor Network

We will employ the following definitions, illustrated using this network.

- $b(s)$: strength of a battery of sensor *s*; for example, $b(s_1) = 3$ while $b(s_3) = 1$.
- $T(s)$: set of targets that sensor *s* can sense; e.g., $T(s_1) = \{t_1, t_2\}$;

- $N(s, k)$: set of neighbors of sensor s at no more than k hops (that s can communicate with using $\leq k$ hops) - this contains s itself; thus, $N(s_1,1) = \{s_1,s_2,s_3,s_4,s_5\}$.
- Cover: C is a cover for targets in set T if (i) for each target $t \in$ T there is at least one sensor in C which can sense t and (ii) C is minimal. For example, the possible (minimal) covers for the two targets of s_1 are $\{s_1\}$, $\{s_2, s_3\}$, $\{s_2, s_4\}$ and $\{s_2, s_5\}$. There are other non-minimal covers as well such as $\{s_1, s_2\}$ which need to be avoided. Likewise, the possible covers for the only target of sensor s_3 are $\{s_1\}$, $\{s_3\}$, $\{s_4\}$ and $\{s_5\}$.
- $lt(C) = \min_{s \in C} b(s)$, the maximum lifetime of a cover. The bottleneck sensor of the cover $\{s_2, s_3\}$ is s_3 with the weakest battery of 1. Therefore, $lt(\{s_2,s_3\}) = 1$.

An optimal lifetime schedule of length 6 for this network is (($\{s_1, s_6\}$, 1), ($\{s_1, s_7\}$, 1), ($\{s_1, s_8\}$, 1), ($\{s_2, s_3\}$, 1), ($\{s_2, s_4\}$, 1), ($\{s_2, s_5\}$, 1)) where each tuple has a cover for the entire network followed by its duration.

Lifetime dependency (LD) graph: Let the local lifetime dependency graph be $G = (V, E)$ where nodes in V denote the local covers and edges in E exist between those pairs of nodes whose corresponding covers share one or more common sensors. For simplicity of reference, we will not distinguish between a cover C and the node representing it, and an edge e between two intersecting covers C and C' and the intersection set $C \cap C'$. Each sensor constructs its local LD graph considering its one- or two-hop neighbors and the corresponding targets. Figure 2 shows the local lifetime dependency graph of sensor s_1 considering its one-hop neighbors $N(s_1,1)$ and its targets $T(s_1)$.

In the LD graph, we will use the following two definitions:

- $w(e) = \min_{s \in e} b(s)$, the weight of an edge e (if e does not exist, i.e., if e is empty, then w(e) is zero).
- $d(C) = \sum_{e \in E \text{ and incident to } C} w(e)$, the degree of a cover C.

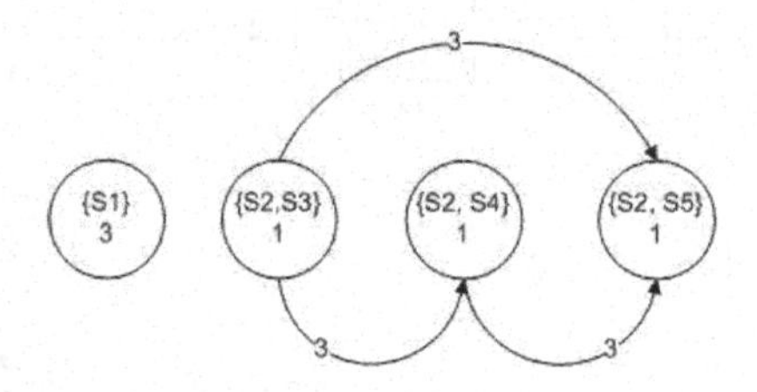

Fig. 2. The local lifetime dependency graph of sensor s_1

In Figure 2, the two local covers $\{s_2, s_3\}$ and $\{s_2, s_4\}$ for the targets of sensor s_1 have s_2 in common, therefore the edge between the two covers is $\{s_2\}$ and $w(\{s_2\}) = 3$. Therefore, s_2's battery of 3 is an upper bound on the lifetime of the two covers collectively. It just so happens that the individual lifetimes of these covers are 1 each due to their bottleneck sensors and, therefore, a tighter upper bound on their total life is

2. In general, given two covers C and C', a tight upper bound on the life of two covers is min $(lt(C) + lt(C'), w(C \cap C'))$.

The basic algorithm

For the purpose of this explanation, without loss of generality, let us assume that the covers are constructed over one-hop neighbors. The algorithm consists of two phases. During the initial setup phase, each sensor calculates and prioritizes the covers. Then, for each reshuffle round of predetermined duration, each sensor decides its on/off status at the beginning, and then those chosen remain on for the rest of the duration.

Initial setup: Each sensor s communicates with each of its neighbor $s' \in N(s,1)$ exchanging mutual locations, battery levels $b(s)$ and $b(s')$, and the targets covered $T(s)$ and $T(s')$. Then it finds all the local covers using the sensors in $N(s,1)$ for the target set being considered. The latter can be solely $T(s)$ or could also include $T(s')$ for all $s' \in N(s,1)$ (see Variant 3 in Section 3). It then constructs the local LD graph $G = (V, E)$ over those covers, and calculates the degree $d(C)$ of each cover $C \in V$ in the graph G.

The *"priority function"* of a cover is based on its degree (lower the better). Ties among covers with same degree are broken first by preferring (i) those with longer lifetimes, then (ii) those which have fewer remaining sensors to be turned on, and finally (iii) by choosing the cover containing the smaller sensor id. A cover which has a sensor turned off becomes infeasible and falls out of contention. Also, a cover whose lifetime falls below the duration of a round is taken out of contention, unless it is the only cover remaining.

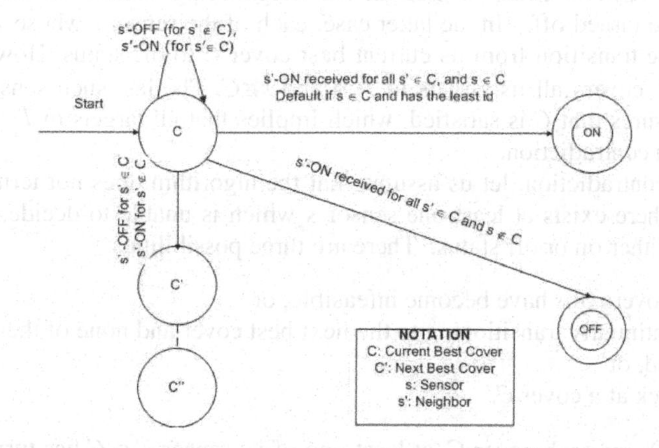

Fig. 3. The state transitions to decide on/off status of a sensor s for reshuffle rounds

Reshuffle rounds: The automaton in Figure 3 captures the algorithm for this phase. A sensor s starts with its highest priority cover C as its most desirable configuration for its neighborhood. If successful, the end result would be switching on all the sensors in C, while others can sleep. Else, it transitions to the next best priority cover C', C'', etc., until a cover gets satisfied.

The transitions are as follows.

- Continue with the best cover *C*: Sensor *s* continues with its current best cover *C* if its neighbor $s' \notin C$ goes off (thus not impacting the chances of ultimately satisfying *C*) or if neighbor $s' \in C$ becomes on (thus improving chances for *C*).
- To on/sense status: If all the neighboring sensors in cover *C* except *s* become on, s switches itself on satisfying the cover *C* for its neighborhood, and sends its on-status to its neighbors.
- To off/sleep status: If all the neighboring sensors in cover *C* become on thus satisfying *C*, and *s* itself is not in cover C, *s* switches itself off, and sends its off-status to its neighbors.
- Transition to the next best cover *C'*: Sensor s transitions to the next best priority cover *C'*, if (i) *C* becomes infeasible because a neighboring sensor $s' \in C$ has turned off, or (ii) priority of *C* is now lower because a sensor $s' \notin C$ has turned on causing another cover *C'*, with same degree and lifetime as *C*, with fewer sensors remaining to be turned on.

The transitions from *C'* are analogous to that from *C*, with the possibility of even going back to *C*.

Correctness: We sketch a proof here that this algorithm ensures that, in each reshuffle round, all the targets are covered and the algorithm itself terminates enabling each sensor to decide and reach on/off status.

For contradiction, let us assume that in a given round a target *t* remains uncovered. This implies that either this target has no neighboring sensor within sensing range and thus network itself is dead, or else all the neighboring sensors which could have covered t have turned off. In the latter case, each of the sensor s whose *T*(s) contains *t* has made the transition from its current best cover *C* to off status. However, *s* only does that if *C* covers all its targets in *T*(s) and $s \notin C$. The last such sensor s to have turned off ensures that *C* is satisfied, which implies that all targets in *T*(s) including *t* are covered, a contradiction.

Next, for contradiction, let us assume that the algorithm does not terminate. This implies that there exists at least one sensor s which is unable to decide, i.e., make a transition to either on or off status. There are three possibilities:

(i) all the covers of *s* have become infeasible, or
(ii) *s* is continually transitioning to the next best cover and none of them are getting satisfied, or
(iii) *s* is stuck at a cover *C*.

For case (i), for each cover *C*, at least one of its sensor $s' \in C$ has turned off. But the set of targets considered by sensor s is no larger than $T' = \cup_{s' \in N(s,1)} T(s')$. Since s itself can cover *T*(s), there exist a target $t \in T' - T(s)$, from *T*(s'), that none of the cover sets at s are able to cover. This implies that *s'* is off, else {s,s'} would have formed part of a cover at *s* covering *t* (given that *s* constructs all possible covers). This leads to the contradiction, as before turning off, *s'* ensures that $t \in T(s')$ is covered.

For case (ii), each transition implies that a neighbor sensor has decided its on/off status, thereby making some of the covers at *s* infeasible and increasingly satisfying portions of some other covers, thus reducing the choices from the finite number of its covers. Eventually, when the last neighbor decides, *s* will be able to decide as well becoming on if any target in *T*(*s*) is still uncovered, else going off.

For case (iii), the possibility that all sensors are stuck at their best initial covers is conventionally broken by a sensor $s \in C$ with least id in its current best cover C proactively becoming on, even though *C* may not be completely satisfied. This is similar to the start-up problem faced by others distributed algorithms such as DEEPS with similar deadlock breaking solutions. At a later stage, if *s* is stuck at C, it means that either all its neighbors have decided or one or more neighbors are all stuck. In the former case, there exists a cover *C* at *s* which will be satisfied with *s* becoming on (case i). The latter case is again resolved by the start-up deadlock breaking rule by either *s* or *s'* proactively becoming on.

Message and time complexities: Let us assume that each sensor *s* constructs the covers over its one-hop neighbors to cover its targets in T(s) only. Let $S=\{s_1,s_2,\ldots,s_n\}$ $\Delta = \max_{s \in S} |N(s,1)|$, the maximum number of neighbors a sensor can communicate with. The communication complexity of the initial setup phase is $O(\Delta)$, assuming that there are constant number of neighboring targets that each sensor can sense. Also, for each reshuffle round, a sensor receives $O(\Delta)$ status messages and sends out one. Assuming Δ is a constant practically implies that message complexity is also a constant. Let maximum number of targets a sensor considers is $\tau = \max_{s \in S} |T(s)|$, a constant. The maximum number of covers constructed by sensor s during its setup phase is $O(\Delta^{\tau})$, as each sensor in *N*(s,1) can potentially cover all its targets considered. Hence the time complexity of setup phase is $O((\Delta^{\tau})^2)$ to construct the LD graph over all covers and calculate the priorities. For example, if $\tau = 3$, the time complexity of the setup phase would be $O(\Delta^6)$. The reshuffle rounds transition through potentially all the covers, hence their time complexity is $O(\Delta^{\tau})$.

3 Variants of the Basic Algorithm

We briefly discussed some of the properties of the LD graph earlier. For example, an edge *e* connecting two covers *C* and *C'* yields an upper bound on the cumulative lifetime of both the covers. However, if *w*(e), which equals *b*(s) for weakest sensor $s \in e$, is larger than the sum of the lifetimes of *C* and *C'*, then the edge e no longer constrains the usage of *C* and *C'*. Therefore, even though *C* and *C'* are connected, they do not influence each other's lifetimes. This leads to our first variant algorithm.

Variant 1: Redefine the edge weight *e* as follows:

$$\text{If } \min_{s \in e} b(s) < lt(C) + lt(C'), \text{ then } w(e) = \min_{s \in e} b(s),$$
$$\text{else } w(e) = 0.$$

Thus, when calculating the degree of a cover, this edge would not be counted when not constraining, thus elevating the cover's priority.

Next, the basic framework is exploiting the degree of a cover to heuristically estimate how much it impacts other covers, and the overall intent is to minimize its impact. Therefore, we sum the edge weights emanating from a cover for its degree. However, if a cover *C* is connected to two covers *C'* and *C"* such that both *C'* and *C"* have the same bottleneck sensor s, s is depleted by burning either C' or C''. That is, in a sense, only one of C' and C'' can really be burned completely, and then the other is rendered unusable because s is completely depleted. Therefore, for all practical purposes, C' and C'' can be collectively seen as one cover. As such, the two edges connecting C to C' and C'' can be thought of as one as well. This yields our second variant algorithm.

Variant 2: Redefine the degree of a cover C in the LD graph as follows. Let a cover C be connected to a set of covers $V' = \{C_1, C_2, \ldots, C_q\}$ in graph G. If there are two covers C_i and C_j in V' sharing a bottleneck sensor s, then if $w(C,C_i) < w(C,C_j)$ then $V' = V' - C_j$ else $V' = V' - C_i$. With this reduced set of neighboring covers V', the degree of cover C is

$$d(C) = \Sigma_{C' \in V'} \; w(C,C')$$

In the basic algorithm, each sensors constructs cover sets using its one-hop neighbors to cover its direct targets $T(s)$. However, with the same message overheads and slightly increased time complexity, a sensor can also consider its neighbors' targets. This will enable it to explore the constraint space of its neighbors as well.

Variant 3: In this variant, each sensor s constructs LD graph over one-hop neighbors $N(s,1)$ and targets in $\cup_{s' \in N(s,1)} T(s')$.

Variant 4: In the basic two-hop algorithm, each sensor s constructs LD graph over two-hop neighbors $N(s, 2)$ and targets in $\cup_{s' \in N(s,1)} T(s')$. In this variant, each sensor s constructs LD graph over two-hop neighbors $N(s, 2)$ and targets in $\cup_{s' \in N(s,2)} T(s')$.

4 Simulation Results

In this section we first evaluate the performance of the one-hop and two-hop versions of the basic algorithm as compared to 1-hop algorithm LBP [2] and 2-hop algorithm DEEPS [4], respectively (see Section V for these algorithms). We also consider the performance of the different variations of the basic algorithm as outlined in Section III.

For the simulation environment, a static wireless network of sensors and targets scattered randomly in 100m x 100m area is considered. We assume that the communication range of each sensor is two times the sensing range. Different variations of the number of targets, number of sensors and energy model are considered for the simulation. The linear energy model is one where the power required to sense a target at distance d is a function of d. In the quadratic energy model the power required to sense the target at distance d is a function of d^2.

One of the key things to note is that the LD graph requires all possible covers for the local targets being considered. However, since the algorithm operates on either 1-hop or 2-hop neighbors, the number of such covers is bounded, if not small. For the purpose of implementation we create a coverage matrix wherein each row represents a sensor and each column a target, and an entry i, j is set to 1 if sensor i covers target j and 0 otherwise. Note that the number of rows of this matrix is given by the size of the 1- or 2-hop neighborhood depending on the version being considered. Iterating through every column of this matrix and adding every covering sensor to all existing covers allows us to construct all combinations of covers for the targets being considered. The covers obtained in this fashion form the nodes of the LD graph at that sensor. Associated with each cover is a lifetime given by the minimum energy sensor of that cover. This forms the node weight of the LD graph. To allow easy construction of edges for the LD graph we implement the covers as sets and their intersection represents the edges.

In order to compare the algorithm against LBP and DEEPS, we use the same experimental setup as employed in [4]. We conduct the simulation with 25 targets randomly deployed, and vary the number of sensors between 40 and 120 with an increment of 20 and each sensor has a maximum sensing range (diameter) of 60m. The energy consumption model is linear. The results are shown in *Figure 4*. As can be seen from the figure, both the basic 1-hop and 2-hop algorithms outperform LBP. The 1-hop algorithm is almost as good as the 2-hop DEEPS algorithm.

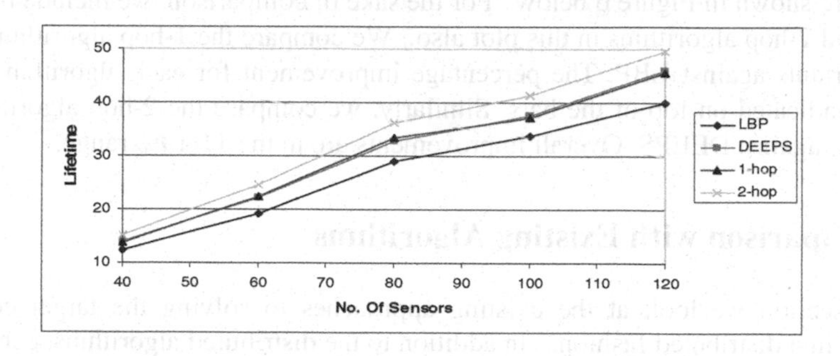

Fig. 4. Network Lifetime for 25 targets with linear energy model

To study variations in targets and energy models, we simulate a network of 60 sensors with 60 m sensing range for both 25 and 50 targets and linear and quadratic energy models. The results are presented in Figure 5. We see a trend consistent with the previous plot with LBP being outperformed by the 1-hop and 2-hop algorithms and DEEPS and the 1-hop version showing similar lifetimes.

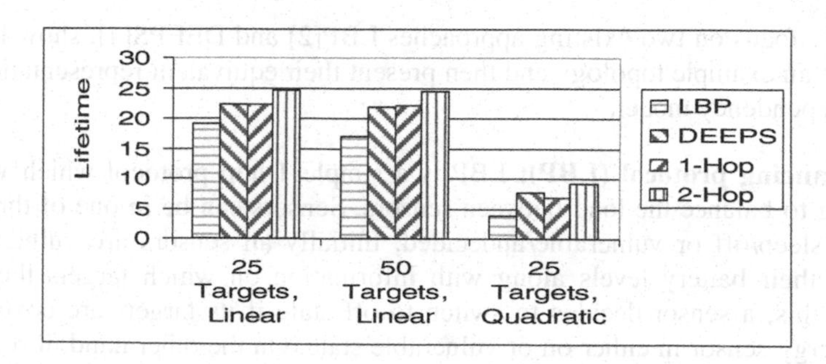

Fig. 5. Lifetime for 60 sensors with varying energy model and targets

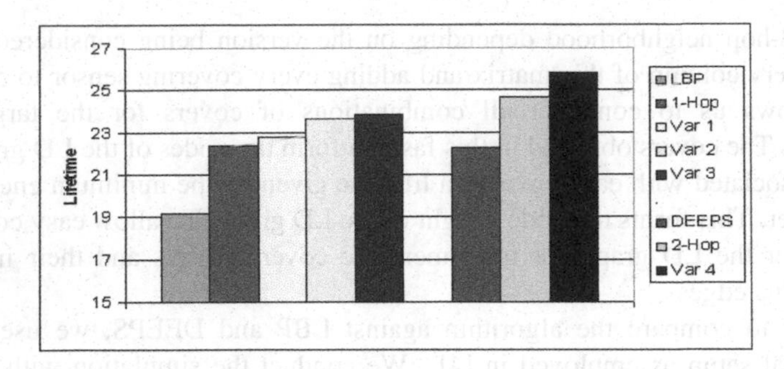

Fig. 6. Performance of Variants of the Basic Algorithm: Network Lifetime for 60 sensors, 25 targets, linear energy model

Finally, we simulate the different variants of the basic algorithm as outlined in Section III for a network of 60 sensors, 25 targets and a linear energy model. The results are shown in Figure 6 below. For the sake of comparison, we include the basic 1-hop and 2-hop algorithms in this plot also. We compare the 1-hop algorithm and its three variants against LBP. The percentage improvement for each algorithm against LBP is indicated on top of the bars. Similarly, we compare the 2-hop algorithm and Variant 4 against DEEPS. Overall improvements are in the 11-19% range.

5 Comparison with Existing Algorithms

In this section we look at the existing approaches to solving the target coverage problem in a distributed fashion. In addition to the distributed algorithms, centralized approaches to the problem have also been considered in [2] [3]. An initial study of the coverage problem with disjoint cover sets was carried out in [10]. [2, 3] have a linear programming based solution that assigns a schedule that covers all targets while attempting to maximize the network lifetime.

A related problem has been studied in [12] in which the authors present a power aware distributed algorithm to construct a connected dominating set. The work is an extension of [11] which presents simple localized rules to allow the distributed construction of a CDS. The CDS provides the network with connectivity in addition to coverage.

We now focus on two existing approaches LBP[2] and DEEPS[4], show how they operate on an example topology and then present their equivalent representation in the lifetime dependency model.

Load balancing protocol (LBP): LBP is a simple 1-hop protocol which works by attempting to balance the load between sensors. Sensors can be in one of three states sense/on, sleep/off or vulnerable/undecided. Initially all sensors are vulnerable and broadcast their battery levels along with information on which targets they cover. Based on this, a sensor decides to switch to off state if its targets are covered by a higher energy sensor in either on or vulnerable state. On the other hand, it remains on if it is the sole sensor covering a target.

Thus, LBP is simplistic and attempts to share the load evenly between sensors instead of balancing the energy for sensors covering a specific target. Hence, for the example shown in Figure 1, LBP picks the sensor s_1 to be active since it is the largest sensor covering the bottom-left target T_2. Similarly, it picks s_2 for the bottom-right target T_3. This results in a total lifetime of 3 units when compared to the optimal of 6 units for the given example. Its schedule is ({ s_1, s_2},3).

Formulation using LD graph framework: LBP can be simulated as a special case in the lifetime dependency graph model as follows. Given a sensor s_i, its local covers for its targets $T(s_i)$ are the singleton sets $\{s\}$, for all $s \in N(s_i,1)$. These singleton sets are then assigned priorities in order to choose which one to use next. There are two defaults: the priority is highest if s_i is the only one covering a target (so s_i must switch on). On the other hand, the priority is lowest if all of s_i's targets are covered, so that s_i can switch off. Otherwise, the priority is assigned based on the battery level preferring to burn those sensors with higher battery, with preference for those covers not containing s_i. Thus, a key limitation of LBP is the lack of collective negotiation captured by non-singleton cover sets in our algorithms.

DEEPS protocol: The maximum duration that a target can be covered, its 'life,' is the sum of the batteries of all its nearby sensors that can cover it. The main intuition behind DEEPS is to try to minimize the energy consumption rate around those targets with smaller lives. A sensor thus has several targets with varying lives. A target is defined as a '*sink*' if it is the shortest-life target for at least one sensor covering that target. Otherwise, it is a '*hill*.' To guard against leaving a target uncovered during a shuffle, each target is assigned an in-charge sensor. For each sink, its in-charge sensor is the one with the largest battery for which this is the shortest-life target. For a hill target, its in-charge is that neighboring sensor whose shortest-life target has the longest life. An in-charge sensor does not switch off unless its targets are covered by someone. Apart from this, the rules are identical as those in LBP protocol. DEEPS relies on two-hop information to make these decisions.

For the example shown in Figure 1, DEEPS achieves a lifetime of 5, assuming a shuffle round duration of 1 since initially both the sensors $s_1$and s_2 are switched on. Its schedule would be {{(s_1, s_2),1}, {(s_1, s_6),1}, {(s_1, s_7),1}, {(s_2, s_3), 1}, {(s_2, s_4),1}} for a total of 5 units.

Formulation using LD graph framework: DEEPS can also be represented with our lifetime dependency graph model. The representation is just like LBP with singleton set covers from $N(s_i,1)$. The priority function of LBP is now modified suitably to account for the concept of in-charge sensors. Specifically, the order of priority preference is if a sensor alone can cover a target, a sensor is in-charge of a target, and then higher battery level. The default least priority is for a sensor if the target it is in-charge of is now covered. Again, singleton cover sets of DEEPS, as in LBP is its key limitation.

6 Conclusion

This paper takes a fundamental look into the problem structure of finding longest lifetime schedule for covering targets by an ad-hoc wireless sensor network. The existing approaches typically employ greedy approaches based on sensor vs. its neighbors in trying to decide which sensors should be sensing and which ones can be in energy-conserving sleep mode. We consider the covers consisting of subsets of a sensor and its neighbors, which can cover all the nearby targets as a primary construct to decide the local on/off configuration of each neighborhood. The dependencies among the covers are modeled using a graph structure over the covers. The priority of a cover is a function of how minimally connected a cover is in this graph. This yields a basic framework, leading to several variants, with superior performance as compared to the existing distributed algorithms. The framework nature is further reinforced by demonstrating how other algorithms can be formulated.

This work has opened up a new way to explore this problem. Even though the number of covers that each sensor transitions through is a function of the number of local neighbors and targets, both expected to be small, a more computationally efficient technique would be desirable. Several other variants of the basic framework are being explored.

References

[1] Akyildiz, I.F., Su, W., Sankarasubramaniam, Y., Cayirci, E.: A Survey on Sensor Networks. In: IEEE Communications Magazine, pp. 102–114 (2002)

[2] Berman, P., Calinescu, G., Shah, C., Zelikovsky, A.: Power Efficient Monitoring Management in Sensor Networks. In: IEEE Wireless Communication and Networking Conference (WCNC 2004), Atlanta, pp. 2329–2334 (March 2004)

[3] Berman, P., Calinescu, G., Shah, C., Zelikovsky, A.: Efficient Energy Management in Sensor Networks. In: Xiao, Y., Pan, Y. (eds.) Ad Hoc and Sensor Networks, Wireless Networks and Mobile Computing, vol. 2, Nova Science Publishers (2005)

[4] Brinza, D., Zelikovsky, A.: DEEPS: Deterministic Energy-Efficient Protocol for Sensor networks. In: ACIS International Workshop on Self-Assembling Wireless Networks (SAWN 2006) Proc. of SNPD, pp. 261–266 (2006)

[5] Cardei, M., Thai, M.T., Li, Y., Wu, W.: Energy-efficient target coverage in wireless sensor networks. In: Proc. of IEEE Infocom (2005)

[6] Cardei, M., Du, D.-Z.: Improving Wireless Sensor Network Lifetime through Power Aware Organization. ACM Wireless Networks 11(3) (May 2005)

[7] Chong, C.-Y., Kumar, S.P.: Sensor Networks: Evolution, Opportunities and Challenges. Proceeding of the IEEE 91(8) (August 2003)

[8] Garey, M.R, Johnson, D.S.: Computers and Intractability: A Guide to the Theory of NP-Completeness. W.H. Freeman, New York (1979)

[9] Kumar, S., Lai, T.H., Balogh, J.: On k-coverage in a mostly sleeping sensor network. In: Proceedings of the 10th annual international conference on Mobile computing and networking, Philadelphia, PA, USA (2004)

[10] Slijepcevic, S., Potkonjak, M.: Power efficient organization of wireless sensor networks. In: Proc. IEEE International Conference on Communications (ICC), pp. 472–476 (2001)

[11] Li, W.: On calculating connected dominating set for efficient routing in ad hoc wireless networks. In: Proceedings of the 3rd international workshop on Discrete algorithms and methods for mobile computing and communications, pp. 7–14 (1999)

[12] Wu, J., Dai, F., Gao, M., Stojmenovic, I.: On Calculating Power-Aware Connected Dominating Sets for Efficient Routing in Ad Hoc Wireless Networks. Journal of Communications and Networks 4(1) (March 2002)

DPS-MAC: An Asynchronous MAC Protocol for Wireless Sensor Networks

Heping Wang[1], Xiaobo Zhang[1], Farid Naït-Abdesselam[2], and Ashfaq Khokhar[1]

[1] University of Illinois at Chicago, USA
{hwang10,xzhang20,ashfaq}@uic.edu
[2] University of Lille – INRIA, France
nait@lifl.fr

Abstract. Asynchronous power efficient communication protocols are crucial to the success of wireless sensor networks (WSNs) as a distributed computing paradigm. This paper presents an improved asynchronous duty-cycled MAC protocol for WSN. It adopts a novel dual preamble sampling (DPS) approach by combining low power listening (LPL) with short strobed preambles to significantly reduce idle listening in existing protocols. In our ns-2 based experiments, the performance of the proposed solution is compared with B-MAC and X-MAC, two most recent and popular asynchronous MAC protocols for WSNs. Depending on the traffic load and preamble length, the proposed DPS-MAC improves energy consumption significantly compared to X-MAC without degrading other network performances such as delivery ratio and latency. For example for the traffic rate of 0.1 packets/s and preamble length of 0.1s, the average improvement in energy consumption compared to X-MAC is about 154%.

Keywords: Energy efficiency, MAC protocol, Wireless sensor network, Duty cycle, Low-power listening, Short strobed preamble, Dual preamble sampling.

1 Introduction

Energy consumption is critical to the lifetime of wireless sensor network (WSN) applications because of the energy constraint in sensor nodes. The low traffic load in WSN applications makes it possible to explore low power designs for the communication protocols. In WSN, radio is the major source of energy consumption, which is used either for regular network functions like data transmission or for implicit operations such as idle listening and overhearing [4]. These implicit radio operations are generally employed by the medium access control (MAC) protocol to guarantee fair channel access for each node, and hence MAC layer is undoubtedly one of the fundamental layers in which low power protocol design is adopted to avoid unnecessary energy waste. Existing protocols are broadly classified into two categories: contention-based [4,5,6,7,8,9] and contention-free protocols [13,14]. Contention-free protocols, while being highly energy efficient, require tight synchronization among nodes and thus are less popular and scalable fo dynamic traffic or mobile WSN applications. This paper deals with the design of energy efficient contention-based asynchronous protocols for WSNs.

In designing such asynchronous MAC protocols the prevailing method to reduce energy consumption is duty cycling, in which nodes wake up for a short period in each

S. Aluru et al. (Eds.): HiPC 2007, LNCS 4873, pp. 393–404, 2007.

cycle to listen the channel for any potential communication. If the channel is found busy, nodes prepare to receive data otherwise they switch themselves off to sleep. This mechanism reduces idle-listening significantly, which has been identified as the main energy waste in WSNs. However, for accurate functioning of the communication among nodes, duty-cycled MAC protocols must provide proper strategies for sender and receiver nodes to rendezvous. These strategies include: scheduling, lower power listening (LPL), scheduled-LPL and short strobed preamble. For convenience, we refer to the corresponding MAC protocols by the rendezvous techniques employed in them. For relative merits of these protocols and a discussion on their shortcomings we refer to section 4.

The newer generation of radio transceivers, such as Chipcon CC2420 [1], introduces an additional radio mode named as IDLE except the popular modes such as transmit (Tx), receive(Rx) and power down (SLEEP) modes. This new radio mode has significantly low power consumption compared to Tx and Rx modes (shown in Table 1), and more importantly radio mode transition from IDLE to Tx or Rx is extremely fast. These transceivers have given rise to the opportunity of further reducing idle listening while still being highly asynchronous.

In this paper, we propose a power efficient asynchronous MAC protocol based on a novel dual preamble sampling (DPS) technique exploiting every opportunity to switch the radio of nodes to idle or sleep mode. DPS combines the strengths of standard low power listening (LPL) and short strobed preamble to realize the MAC protocol, referred to as DPS-MAC. Compared with the most efficient protocols available to date, such as X-MAC [9] and B-MAC [8], ns2 based simulation results show that our DPS-MAC protocol improves energy efficiency of the network significantly, particularly for low traffic applications in WSN. Design of such energy efficient asynchronous communication protocols is crucial to the success of sensor networks as distributed computing paradigms, because any synchronous application of WSN is likely to have limited success due to its synchronization requirement.

2 DPS-MAC Protocol Design

Most of the scheduling and standard LPL based duty-cycled MAC protocols have control overhead issues that emerge either from the schedule maintenance or due to the long preamble. Short strobed preamble is an asynchronous approach adopted in some of the recent MAC protocol proposals to overcome the disadvantages of a long preamble length. To further reduce the idle-listening, we propose a dual preamble sampling (DPS) technique that combines the best features of standard LPL and short strobed preamble to achieve low power operation in MAC layer without any synchronization requirement among nodes.

2.1 Dual Preamble Sampling Overview

In standard LPL (Figure 1a), nodes check the channel status within a short active period P_1 at the beginning of each cycle. If the channel is found busy, it implies that a communication activity is in progress. Therefore all the non-sender nodes stay awake for an extra duration EP to determine the identity of the receiver. X-MAC [9] addresses this overhearing issue by employing a stream of short preamble messages including the target's address which allows for receiver's quick response(named as ACK) to cut down the long preamble. This requires non-sender nodes to be periodically active for a period P with duration long enough for them to capture one of the preamble messages sent by the sender and determine whether to respond the sender. The duration of period P logically ranges from Tpm to T_{pm} to $2T_{pm} + T_{ACK}$, where T_{pm} and T_{ACK} represent the lengths of a short preamble message and an ACK message respectively (Figure 1b). In a low traffic application, this approach tends to waste much more energy due to the idle listening when the duration of period P is closer to $2T_{pm} + T_{ACK}$. On the other hand, a smaller duration for P will decrease the probability of non-sender nodes to catch a preamble message within one check interval.

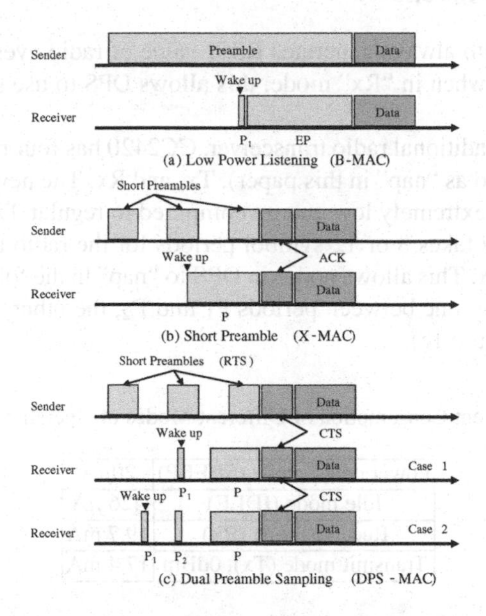

Fig. 1. Sender and Receiver Rendezvous Schemes in BMAC, X-MAC and Proposed DPSC-MAC Protocols

In this paper a dual preamble sampling approach based on low power listening and short strobed preamble is proposed to reduce the idle listening. Our protocol differs in terms of the behavior of the non-sender nodes and exploits the energy efficient radio features of the cutting edge transceivers. A sender in DPS-MAC sends out a series of short preamble messages with the receiver information included in the preamble message(named as RTS), and wait for the response (named as CTS) from the receiver between two adjacent RTS messages. Non-sender nodes periodically check the channel state by polling the channel in one to two separate periods P_1 and P_2 (Figure 1c) at the beginning of each cycle. If the channel is found free in both periods, nodes return to sleep immediately; otherwise nodes "nap" (IDLE mode in CC2420) for a period and proceed to receive an RTS message to determine who is the intended receiver, which is similar to X-MAC (Figure 1b). The period P_2 is needed only when the channel is found free in P_1. This is necessary because the receiver might happen to wake up at the two ends of an RTS period or in a CTS period of the sender and hence miss the last RTS message sent by the sender when performing the channel polling during P_1. Therefore, the receiver should perform another channel polling to be sure of the actual channel status (busy or free, refer to Figure 1c). If period P_2 is applied during a DPS operation, non-sender nodes should also "nap" between P_1 and P_2 for energy efficiency. By properly choosing the durations of P_1 and P_2 as well as the interval $T_{P_1P_2}$ between them, nodes can determine whether a sender is trying to talk to one of them based on the sampled radio RSSI (received signal strength indicator) value within period P_1 or P_2. Chipcon CC2420 [1] is the state of the art radio transceiver for sensor network applications. The following new features provided by CC2420 makes DPS function well in the following aspects:

- The CC2420 radio always generates RSSI value of radio averaged over 8 symbol periods (128μs) when in "Rx" mode; this allows DPS to use short polling periods P_1 and P_2.
- Different from traditional radio transceiver, CC2420 has four radio modes: SLEEP, IDLE (also talked as "nap" in this paper), Tx, and Rx. The newly introduced IDLE mode consumes extremely low energy compared to regular Tx and Rx modes (see Table 1); it only takes 8 or 12 symbol periods for the radio to transit from mode IDLE to Tx or Rx. This allows nodes in DPS to "nap" in the following two intervals for energy saving: one between periods P_1 and P_2, the other one between periods P_1/P_2 and P (Figure 1c).

Table 1. Current Consumption of Different Modes of Operations in CC2420

Mode	Current
Power down mode (SLEEP)	20μA
Idle mode (IDLE)	426 μA
Receive mode (Rx)	19.7 mA
Transmit mode (Tx), 0dBm	17.4 mA

Next we discuss how to determine optimal values for the durations of P_1, P_2 and interval $T_{P_1P_2}$ between P_1 and P_2 to make DPS-MAC work under given sizes of the RTS and CTS messages. The primary consideration to choose these values is to minimize the energy cost used for a DPS operation and allow the sender to rendezvous with the receiver quickly. For the convenience of the discussion, we define the following notations:

- T_{RTS}: the time duration of a RTS message
- T_{CTS}: the time duration of a CTS message
- T_0: minimum period during which the sampled RSSI (received signal strength indication) value of channel indicates a valid channel state (busy or free). T_0 depends on the specific radio transceiver, in CC2420 it is equal to 128μs.
- T_{P_1}: minimum duration of P_1, we simply set T_{P_1} to T_0.
- T_{P_2}: minimum duration of P_2 which must be no less than T_0.

When channel polling period P_1 occurs after time t_1 and before t_3 (Figure 2), channel polling period P_2 must be applied to check the channel activity. This is because in this case the RSSI value sampled during P_1 will indicate a free channel. The following two cases need to be considered to determine the optimal values for $T_{P_1 P_2}$:

- If P_1 occurs just after t_1, the remaning duration of RTS message is less than T_0. Therefore the channel cannot be sampled properly. In this case the minimum value for $T_{P_1 P_2}$ is T_{CTS} so that the channel state can be sampled at the beginning of the next RTS message;
- If P_1 occurs to the left of time t_3, the maximum value for $T_{P_1 P_2}$ is T_{RTS}-$2T_0$ so that the channel state can still be sampled within the current RTS message.

Combining these two cases, we choose T_{P_2} and $T_{P_1 P_2}$ as follows:

$$T_{P_1 P_2} = min(T_{CTS}, T_{RTS} - 2T_0) \tag{1}$$

and

$$T_{P_2} = max(T_{CTS}, T_{RTS} - 2T_0) - min(T_{CTS}, T_{RTS} - 2T_0) + T_0 \tag{2}$$

In the RHS of equation 2, the first two items imply that the gap between the right end of duration $T_{P_1 P_2}$ and time t_2 should be covered by P_2, if needed. In order to minimize the duration of T_{P_2} of the period P_2, in our implementation, we set $T_{RTS} = T_{CTS} + 2T_0$, so T_{P_2} and $T_{P_1 P_2}$ are simply stated as:

$$T_{P_1 P_2} = T_{CTS} \tag{3}$$

and

$$T_{P_2} = T_0 \tag{4}$$

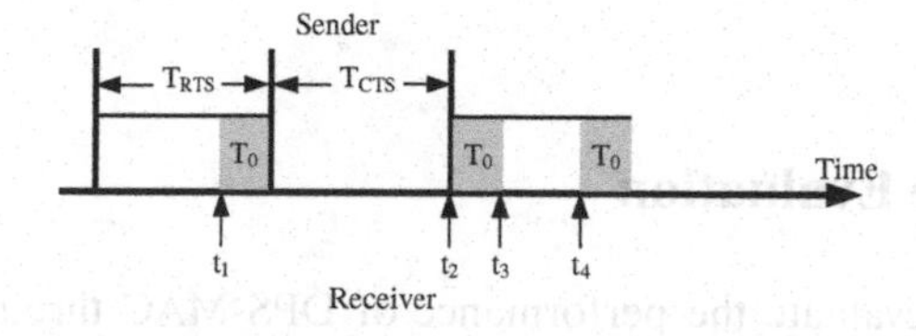

Fig. 2. Possible Occurences of P_1 and P_2 Peroids Relative to RTS and CTS Messages

2.2 DPS-MAC Implementation Details

DPS-MAC is an asynchronous low power MAC protocol based on DPS approach. We summarize it as follow:

1. Nodes in DPS-MAC periodically wake up to perform DPS operation if they have no data to send, and sleep when there is no traffic in network;

2. If a node has data to send, it initiates a preamble session containing a sequence of alternate RTS/CTS message periods. The sender tries to wake up the receiver by sending out RTS messages. The receiver replies the sender with a CTS message as an acknowledgement. Each RTS or CTS message contains both the sender and receiver's address. Before sending the first RTS message, the sender needs to perform carrier sensing to reduce the collision between multiple senders.
3. When any non-sender node samples a busy channel during period P_1 (Case 1 in Figure 1c), or P_2 (Case 2 in Figure 1c), it first "nap" for a duration T_{CTS}, and then prepares to receive an RTS message from the sender; if no RTS is received within duration P with length $T_{RTS}+T_{CTS}$, the node goes to sleep.
4. Polling period P_2 is needed only if the channel is found free in P_1. This might happen when the period P_1 is not fully covered within the duration of an RTS message. Under this situation, the node needs to "nap" till the beginning of P_2 and poll the channel again.
5. If an RTS message is received in step 3, a node checks the embedded target address in the received RTS message, and the intended receiver responds the sender with a CTS message for unicast data.
6. Upon receiving the CTS message, the sender stops sending the remaining RTS messages and transmits the data packet to the intended receiver.

2.3 Broadcast Support in DPS-MAC

Broadcast is an important communication primitive in network protocols. X-MAC does not elaborate this issue in its design. To support one-hop broadcast operation, the sender in DPS-MAC is required to send out two consecutive RTS messages right before the data packet to guarantee the acceptance of the broadcast packet by all its one-hop neighbors. If a broadcast address is found in step 3 described in section B, the node sleeps till the instance of data packet transmission. The sleeping time is calculated based on other embedded information in RTS messages. This information includes the total number of RTS messages for each preamble session and index of each RTS message in the current preamble session.

3 Performance Evaluation

In this section, we evaluate the performance of DPS-MAC through simulations. We implemented DPS-MAC in ns-2 [2] with CMU wireless extensions. The current implementation of ns-2 does not support preamble sampling. We modified the physical layer implementation of ns-2 to provide preamble sampling. For comparison purpose, we have also implemented B-MAC and X-MAC in ns-2. However in our implementations, B-MAC excludes the configuration interfaces, and X-MAC does not contain the adaptive algorithm. These simplifications for both MAC protocols do not lose comparison fairness because we are mainly concerned with how our rendezvous solution achieves better energy performance. Note that DPS-MAC can also use the configuration interfaces proposed in B-MAC or the adaptive algorithm proposed for X-MAC to optimize the performance.

In this paper, we mainly focus on three metrics to evaluate and compare the performance of these three protocols: average energy consumption per node, latency and delivery ratio. All the statistics of the performance metrics are under unicast traffic scenarios.

The simulation parameters have been chosen based on the characteristics of Chipcon radio chip CC2420. The current consumption for different radio modes of CC2420 is listed in Table 1. Other parameters are summarized in Table 2. The sizes of the data and control packets include the control bytes needed for frame control in MAC layer and physical layer. In CC2420, these control bytes can be at least 11 bytes long. Note that the preamble length in the simulation is associated with the duration of check interval for non-sender nodes.

Table 2. Other Simulation Parameters

Channel bandwidth	250 kbps
Simulation duration	1000 s
Data packet size	50 bytes
CTS, ACK, Short Preamble message	16 bytes

3.1 Average Energy Consumption

As in X-MAC, we build a 10-node star network to evaluate the energy efficiency of DPS-MAC. All the nodes are within one hop transmission range of each other. One of the nodes is assumed to be the sink. We change the traffic load by varying the number of senders from 1 to 9. Each sender sends unicast CBR traffic to the sink. The traffic interval is 10s. Figures 3 and 4 show that the average energy consumption per node under two different preamble lengths. We observe that DPS-MAC achieves the best energy performance among these three MAC protocols. This is because the combination of the LPL and short strobed preamble approaches in DPS-MAC reduces the energy cost in the sender compared to the long preamble in B-MAC. It further reduces the idle listening in non-sender nodes in X-MAC during the active period of each cycle. When there is no traffic in the network, nodes in X-MAC still listen to the channel for the preamble message. In a light traffic network, this type of idle listening in X-MAC leads significant energy consumption. In DPS-MAC, nodes only need to poll the channel for two brief periods when there is no traffic in the network. Comparing Figures 3 and 4, we find that the preamble length has significant impact on the energy performance. Figure 5 shows the average energy consumption per node with the increase of the preamble length, assuming 5 transmitters sending unicast CBR traffic at rate 0.1 packets/sec.

3.2 Latency

To study the latency performance, we build a 10-hop chain network. A source and a sink nodes are located at the two ends of the chain, and the source sends CBR traffic to the sink. We measure the average per-hop latency and end-to-end delay. Figure 6 shows the average per-hop latency as the preamble length increases. Figure 7 shows the end-to-end delay as the number of hops increases. From Figures 6 and 7, we can see

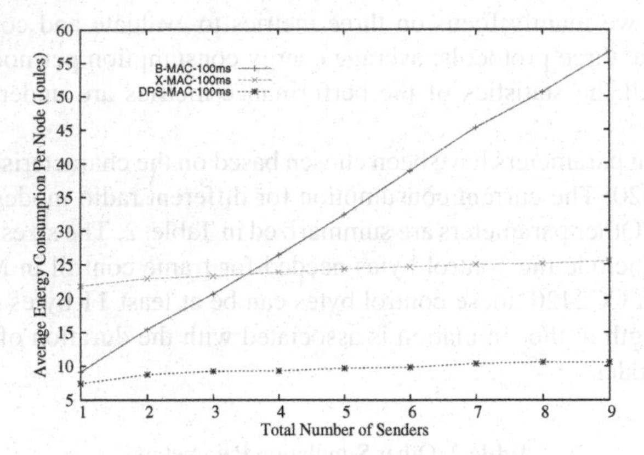

Fig. 3. Average Energy Consumption (Preamble Length = 100ms)

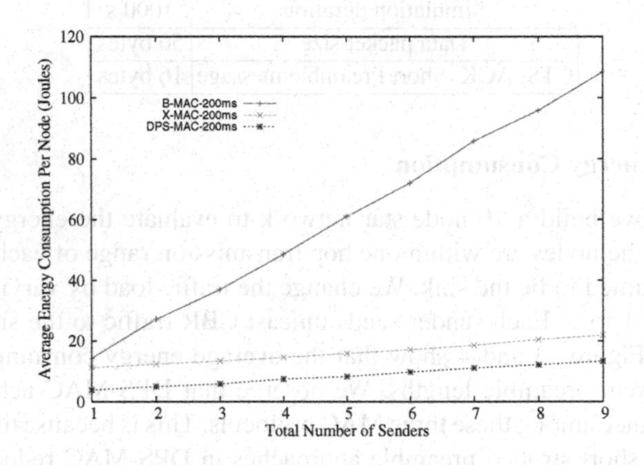

Fig. 4. Average Energy Consumption (Preamble Length = 200ms)

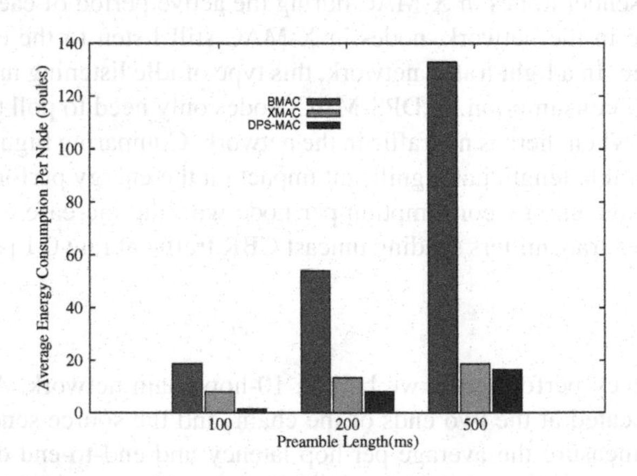

Fig. 5. Average Energy Consumption (5 Senders, 1 Receiver, Traffic Rate = 0.1)

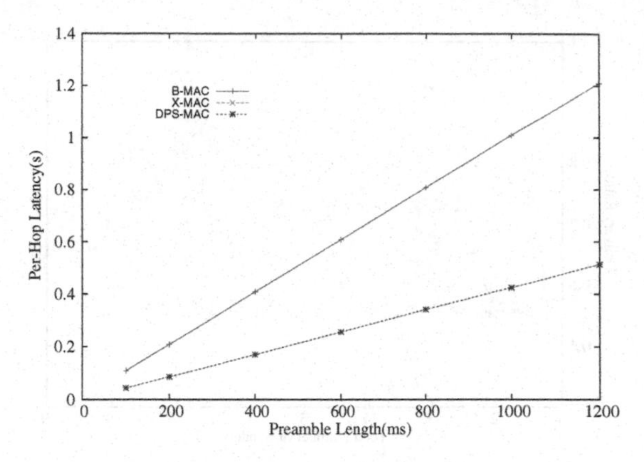

Fig. 6. Average Per-Hop Latency

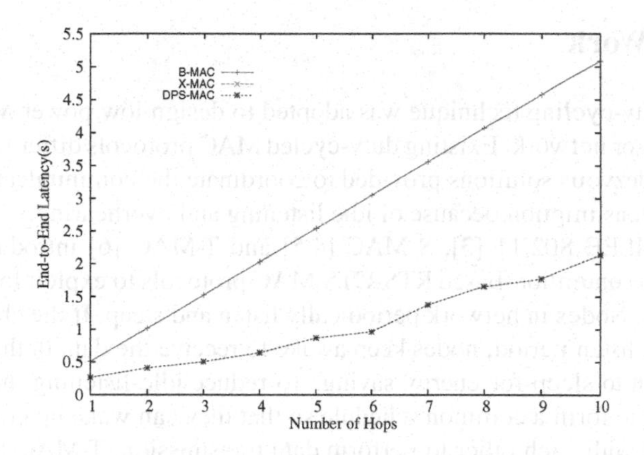

Fig. 7. Average End to End Latency

that DPS-MAC and XMAC achieve similar latency performance. In our experiments, we notice that DPS-MAC induces a little bit increase of latency (less than 1 millisecond at each hop) due to the extra one or two preamble polling periods.

3.3 Delivery Ratio

Delivery ratio corresponds to the number of packets received by a sink node divided by the number of packets sent by all the senders. We use the star network described above to evaluate the delivery ratio. The preamble length is set to 200ms. We also change the traffic load in the network by varying the number of transmitter from 1 to 9. Each transmitter sends CBR traffic with 1s intervals to a unique receiver. From Figure 8, once again we observe very similar performances of DPS-MAC and X-MAC. In contrast, B-MAC loses more packets with the increase in the traffic load. This is because longer preamble in B-MAC leads to packet drop due to collision.

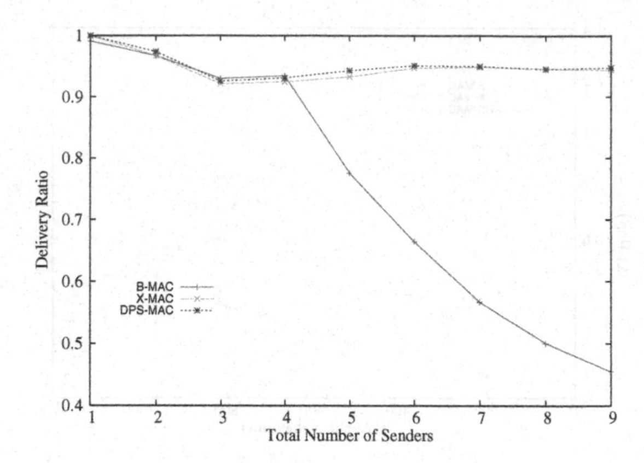

Fig. 8. Packet Delivery Ratio

4 Related Work

In literature, duty-cycling technique was adopted to design low power MAC protocols for wireless sensor network. Existing duty-cycled MAC protocols differ from each other in terms of rendezvous solutions provided to coordinate the communicating nodes and reduce energy consumption because of idle listening and overhearing.

Inspired by IEEE 802.11 [3], S-MAC [4,5] and T-MAC [6] introduce scheduling mechanism into contention-based RTS-CTS MAC protocols to exploit low power operation for WSNs. Nodes in network periodically listen and sleep. If the channel is found busy during the listen period, nodes keep awake to receive the data in the sleep period, otherwise return to sleep for energy saving. To reduce idle-listening, nodes negotiate with each other to form a common schedule so that they can wake up concurrently and then handshake with each other to perform data transmission. T-MAC further reduces idle-listening in S-MAC by dynamically adjusting the duration of the listen period in each listen/sleep cycle.

B-MAC [8] and WiseMAC [7] coordinate data transmission between nodes through preamble sampling. Nodes periodically wake up to check the channel activity briefly without receiving any actual packet. According to the received signal strength indication (RSSI) of the channel, a node determines whether to stay awake to receive the data or go back to sleep. In B-MAC, each node can choose an independent schedule, based on which a node periodically wakes up and sleeps. A sender puts a long preamble before each outgoing packet to wake up the intended receiver. During each data transmission, this asynchronous rendezvous solution incurs much communication cost at the sender and its neighbors due to the long preamble. An optimization in WiseMAC [7] reduced this overhead by allowing senders to record each receiver's next channel checking time so that the sender sends the future packet can be sent right before the receiver's channel checking time with a short preamble.

SCP-MAC [10] replaces the long listen period of S-MAC with a short preamble and proposes a new scheduled channel polling scheme that combines scheduling and LPL together to allow ultra-low duty cycle operations for low traffic applications. The sender eliminates long preambles in LPL by synchronizing the channel polling time of all neighbors. To do this, nodes need to periodically exchange synchronization beacons or piggyback synchronization information on data transmission if each node has periodical data traffic. Synchronization overhead and multiple schedules may also exist in SCP-MAC that might affect its performance.

Similar to STEM [11], X-MAC [9]proposed a short strobed preamble technique to reduce energy consumption associated with overhearing and idle-listening. Before sending a data packet, a sender in X-MAC transmits a series of short preamble messages instead of a long preamble as did in LPL to wake up a receiver; each preamble message contains the receiver's address. Nodes other than the sender periodically wake up for a period to listen to the channel for an incoming preamble message. Upon receiving a preamble message, the intended receiver responds the sender a short ACK message to terminate the preamble sequence; and non-receiver nodes sleep immediately. By successfully doing this, the long preamble in LPL is cut down and the sender transmits data right after the reception of the ACK message from the intended receiver. So X-MAC further reduces energy consumption and latency in LPL. To optimize the energy consumption and latency, X-MAC uses an adaptive algorithm to dynamically tune nodes' check interval based on the estimated traffic load in the network. The difference between STEM and X-MAC is that STEM uses an extra radio to alert the receiver while X-MAC uses only one radio for both preamble message and data transmission. However, ilde-listening in non-sender nodes will be the potential overhead.

5 Conclusions

This paper describes DPS-MAC, a new asynchronous low power MAC protocol for wireless sensor networks. In DPS-MAC, a novel rendezvous strategy was used to efficiently handle node coordination issue, which is very important for nodes to achieve low power operation in low traffic applications of WSNs. By combining LPL and short strobed preamble approaches together, DPS can significantly reduce the idle-listening issue in X-MAC while keeping the advantages of X-MAC. We have implemented DPS-MAC in ns2. Simulations show that DPS-MAC has better energy performance than X-MAC. DPS-MAC also has similar delivery ratio and close latency performance compared to X-MAC.

Our future work will focus on investigating a simpler optimization algorithm to adaptively adjust the preamble length according to the traffic load in the network. We also plan to build DPS-MAC on a real testbed based on Chipcon transceiver CC2420 to evaluate its practical performance.

References

1. Chipcon Inc. CC2420 data sheet, `http://www.chipcon.com/`
2. The network simulator, `http://www.isi.edu/nsman/ns`
3. LAN MAN Standards Committee of the IEEE Computer Society. Wireless LAN medium access control (MAC) and physical layer (PHY) specification, IEEE, New York, USA, IEEE Std 802.11-1999 edition (1999)
4. Ye, W., Heidemann, J., Estrin, D.: An energy-efficient mac protocol for wireless sensor networks. In: Proceedings of the IEEE Infocom, pp. 1567–1576 (June 2002)

5. Ye, W., Heidemann, J., Estrin, D.: Medium access control with coordinated, adaptive sleeping for wireless sensor networks. ACM Transactions on Networking 12(3), 493–506 (2004)
6. Dam, T.V., Langendoen, K.: An Adaptive Energy-Efficient MAC Protocol for Wireless Sensor Networks. In: First ACM Conference on Embedded Networked Sensor Systems (SenSys'03), Los Angeles, CA, USA (November 2003)
7. El-Hoiydi, A., Decotignie, J.-D., Enz, C., Le Roux, E.: WiseMAC: An ultra low power MAC protocol for the wisenet wireless sensor networks (poster abstract). In: Proceedings of the First ACM SenSys Conference, Los Angeles, CA, USA (November 2003)
8. Polastre, J., Hill, J., Culler, D.: Versatile low power media access for wireless sensor networks. In: Proceedings of the 2nd ACM SenSys Conference, Baltimore, MD, USA, pp. 95–107 (November 2004)
9. Buettner, M., Yee, G.V., Anderson, E., Han, R.: X-MAC: A Short Preamble MAC Protocol for Duty-Cycled Wireless Sensor Networks. In: Proceedings of the fourth ACM SenSts Conference, Boulder, CO, USA (November 2006)
10. Ye, W., Silva, F., Heidemann, J.: Ultra-Low Duty Cycle MAC with Scheduled Channel Polling. In: Proceedings of the fourth ACM SenSys Conference, Boulder, CO, USA (November 2006)
11. Schurgers, C., Tsiatsis, V., Ganeriwal, S., Srivastava, M.: Optimizing sensor networks in the energy-latency-density design space. IEEE Transactions on Mobile Computing 1(1), 70–80 (2002)
12. Halkes, G., Dam, T.V., Langendoen, K.: Comparing energy-saving mac protocols for wireless sensor networks. ACM Mobile Networks and Applications 10(5), 783–791 (2005)
13. Rajendran, V., Obraczka, K., Garcia-Luna-Aceves, J.J.: Energy-Efficient, Collision-Free Medium Access Control for Wireless Sensor Networks. In: First ACM Conference on Embedded Networked Sensor Systems (Sensys 2003) (November 2003)
14. Chen, Z., Khokhar, A.: TDMA based Energy Efficient MAC Protocols for Wireless Sensor Networks. In: IEEE SECON 2004 (October 2004)

Compiler-Assisted Instruction Decoder Energy Optimization for Clustered VLIW Architectures

Rahul Nagpal and Y.N. Srikant

Department of Computer Science and Automation,
Indian Institute of Science, Bangalore, India
{rahul,srikant}@csa.iisc.ernet.in

Abstract. Traditionally, an instruction decoder is designed as a monolithic structure that inhibit the leakage energy optimization. In this paper, we consider a split instruction decoder that enable the leakage energy optimization. We also propose a compiler scheduling algorithm that exploits instruction slack to increase the simultaneous active and idle duration in instruction decoder. The proposed compiler-assisted scheme obtains a further 14.5% reduction of energy consumption of instruction decoder over a hardware-only scheme for a VLIW architecture. The benefits are 17.3% and 18.7% in the context of a 2-clustered and a 4-clustered VLIW architecture respectively.

1 Introduction

The ongoing improvements in the semiconductor technology bring along various challenges [5]. One such challenge is the rising level of the leakage energy consumption in the logic. With the 65nm and smaller technologies currently in fabrication, the leakage energy is on par with the dynamic energy consumption. In future technologies the leakage energy will further dominate the overall energy consumption [5].

A significant fraction of the total leakage energy consumption in VLIW architectures is attributed to functional units and instruction decoder. Frequent access to the instruction decoder raise the temperature level and make the leakage energy consumption even worse. A study in the context of Texas instruments' VelociTI architecture [18] attributes more than 50% of energy consumption in instruction fetch and decoding activity [7]. Though, the exact percentage may depends upon the architecture and circuit details, earlier studies clearly indicate that 20% to 25% of the static energy consumption in a VLIW architecture is attributed to instruction decoder. Thus, optimizing leakage energy in instruction decoder is becoming more important by each process generation. Even if we assume that the energy savings in instruction decoder does not translate to huge overall energy savings, it is still desirable to save energy in instruction decoder as it is one of the hot-spot in processor.

VLIW architectures are often designed targeting embedded domains where the real-time performance is of utmost importance. Thus, the design is often

S. Aluru et al. (Eds.): HiPC 2007, LNCS 4873, pp. 405–417, 2007.

optimized for the peak performance and as a result, functional units and the instruction decoder are underutilized due to the inherent variations in the ILP of the programs. Clustered VLIW architectures improve over the VLIW architectures by solving the scalability problem (in order to obtain a better clock rate) by distributing functional units among different clusters [10]. However, contentions for the limited number of slow inter-cluster communication channels introduce many short idle cycles and makes the utilization of instruction decoder worse. The underutilization of functional units and instruction decoder can be exploited to reduce leakage energy consumption.

The traditional monolithic design of instruction decoder inhibit the leakage energy management in instruction decoder. As a result, earlier work only focus on leakage energy management for functional units mostly at a coarser granularity of loop level or block level [13]. However, the rising level of leakage energy in current and future process technologies requires aggressive leakage energy management even for short idle periods. One such purely hardware based scheme for reducing leakage energy in functional units in the context of a superscalar architecture is due to Albonesi et al. [9]. Their scheme utilizes the unique characteristics of dual-threshold domino logic with sleep mode that can transition between active mode and sleep mode without any performance penalty [15]. However, such a fast transition incurs moderate amount of energy penalty. Their scheme called 'MaxSleep' puts any integer ALU into low leakage mode after one cycle of idleness. Their results confirm the benefits of such an aggressive scheme. However, being a purely hardware based scheme, the benefits are severely (on average, by 30%) affected by frequent transitions from active mode to sleep mode and vice-versa because of many short idle periods.

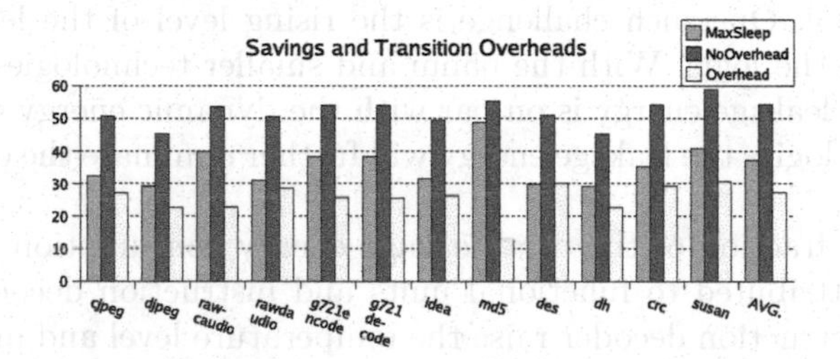

Fig. 1. % Savings for 'MaxSleep' and 'NoOverhead' Policies

In this paper, we consider a split instruction decoder design that enables the use of a hardware based scheme such as [9] for leakage energy savings in instruction decoder. Figure 1 presents the energy savings obtained by 'MaxSleep', energy savings obtained by a 'NoOverhead' scheme which is a hypothetical scheme (same as 'MaxSleep') but does not incur any transition energy overheads and % energy overhead of 'MaxSleep' due to transitions as compared to that of 'NoOverhead' scheme for a split instruction decoder design (that provides facility to decode up to six instructions in parallel) for a 2-cluster configuration. These results clearly indicate that the 'NoOverhead' scheme is able to achieve an average savings of 56.86% in total energy, whereas the average savings for 'MaxSleep' is only 38.92%. 'MaxSleep' has an average energy overhead of 29.37% (due to transitions) as compared to the 'NoOverhead' scheme. Thus, reducing the number of transitions will increase the idleness duration for decoders and improves the energy benefits of a hardware based scheme. Motivated by this, we have developed a scheduling algorithm in the context of VLIW and clustered VLIW architectures. Whereas the purely hardware based scheme suffers from the problem of a limited program view, a compiler can analyze whole program regions and is capable of adjusting the operations decoded every cycle while maintaining the desired performance. The proposed scheme exploits the scheduling slacks of the instructions to maximize the simultaneous idle time and usage of decoders, thereby reducing the number of transitions drastically. This reduction in the number of transitions leads to significant improvements in energy savings over those obtained by a purely hardware based scheme. Moreover, since the proposed scheme keeps a limited number of decoder active and use them as much as possible, it generates a more balanced schedule which helps to reduce the peak power and the step power [19] in instruction decoder.

The rest of the paper is organized as follows. Section 2 describes the split decoder design useful for leakage energy management in instruction decoder. Section 3 describes our new instruction scheduling algorithm and presents an example to show the benefits of the proposed scheme. Section 4 provides detailed experimental results and analysis. Section 5 describes the related work and section 6 concludes this paper with future directions for this work.

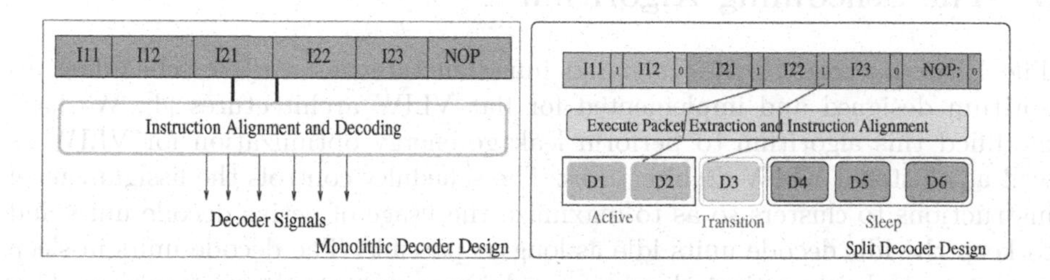

Fig. 2. (a)Traditional Monolithic Decoder Design (b) Split Decoder Design

2 Split Decoder Design

Decoding activity involves dividing a fetch packet into execute packets and then decoding individual micro-instructions in each execute-packet to issue signals. A parallel-bit is dedicated in a VLIW micro-instruction that specifies whether the next micro-instruction is in the same execute-packet (i.e., executes in the same cycle) or starts a new execute-packet. A traditional monolithic design of instruction decoder as shown in Figure 2 (a) inhibits any fine grained control for hibernating parts of the decoder circuit that are idle. A decoder circuit can be easily pipelined and split as shown in Figure 2 (b). This provides the benefits of ease of design and verification of circuit and performance benefits of pipelining

[14] and also enables leakage energy savings at the granularity of individual decoders. The performance benefits of pipelining the decode stage have already been identified and such a design is in use in many high performance commercial DSPs including the Texas Instruments' VelociTI [18]. However, we capitalize on the energy management capability of such a design as follows.

Due to variations in the ILP of the programs, the full issue width of the processor is rarely utilized continuously and hence several decoder will be idle most of the time. The split decoder design can leverage the capabilities of dual-threshold domino logic for fast transition from active mode to sleep mode and vice versa in less than a cycle (as used in [9] for functional units) to save tremendous amount of leakage energy in mostly idle decoder circuit. However, in order to avoid the explicit penalty of activating a sleeping decoder, it is required to issue the activating signal one cycle in advance. Fortunately, the parallel-bit that specifies the parallel instructions in the current execute-packet can be used to drive the activation signal. To avoid introducing any new hardware, in our machine model, we always keep first decoder active, and use the parallel-bits in execute packet to drive the active signal for the required number of decoders. It is important to note that these signals are activated during the first stage of decoding when the execute packet is being extracted and aligned from the fetch packet. Thus, by the time the micro-instructions reach stage 2 for actual decoding, the required number of decoders are in active state to perform the decoding. It is important to note that a fast transition comes at the cost of a moderate energy penalty.

3 The Scheduling Algorithm

The Elcor backend of the Trimaran infrastructure has a cycle scheduling algorithm designed and implemented for flat VLIW architectures [4]. We have modified this algorithm to perform leakage energy optimization for VLIW as well as clustered VLIW architectures. The scheduler controls the assignment of instructions to clusters so as to maximize the usage of active decode units and to keep the idle decode units idle as long as possible. The decode units in sleep mode are explicitly activated only if not doing so impacts the performance. This ensures that decoder energy consumption because of spurious transitions from sleep mode to active mode and vice-versa is reduced. Our integrated scheduling algorithm for leakage energy optimization consists of the three main steps described as follows. Section 3.4 presents an example that illustrates the functioning of the algorithm in detail.

3.1 Prioritizing the Ready Instructions

Instructions in the ReadyList are prioritized using a priority function that uses the instruction slack and the number of consumers of the instruction. Scheduling slack of an instruction is defined as the difference between the earliest start time and the latest finish time of the instruction. Instructions with less slack should be scheduled early and are given higher priority over instructions with more slack to avoid unnecessary stretching of the schedule. Instructions with the same slack values are further ordered in the decreasing order of the number of consumers. An instruction with a large number of successors is more constrained in the sense that its spatial and temporal placement affects scheduling of more number of instructions and hence should be given higher priority. Giving preference to an instruction with many dependent instructions also enables better future scheduling decisions by uncovering a larger portion of the graph.

Traditionally, slack is determined statically during dependence graph analysis before the scheduling begins, assuming a machine with infinite resources of each type. We quantify the slack of instructions while scheduling a region for the specific target machine by taking resource constraints into account. We first schedule the instruction using a simple cycle-by-cycle scheduler. The schedule time of the instructions is stored during this phase. In the second phase, this schedule time (Late cycle) is used to determine the slack of the instruction. In our implementation, slack is dynamically updated for all the operations in the ready list after every cycle. The earliest schedule time of an instruction is set to the current cycle, before scheduling for the current cycle begins (Early cycle). The slack is then determined as a difference of the Early cycle and the Late cycle. The dynamic update of slack after each cycle ensures that any consumed slack is taken into account while scheduling instructions in the future cycles.

3.2 Cluster Assignment

Once an instruction has been selected for scheduling, we make a cluster assignment decision. The primary constraints are :

- The chosen cluster should have at least one free resource of the type needed to perform this operation
- Given the bandwidth of the channels among clusters and their usage, it should be possible to satisfy the communication needs of the operands of this instruction on the cluster by scheduling these communications in the earlier cycles (so that operands are available at the right time).

Note that if we are scheduling for a plain VLIW architecture with no clustering, we assume that there is only one cluster (numbered 0) that is holding all the resources and the same algorithm is used. Selection of a cluster from the set of the feasible clusters is done as follows. A cluster with an active decoder to schedule the operation is given preference. If no such cluster is available or more than one such cluster is available, the one which reduces the communication cost gets preference. The communication cost is computed by determining the number and type of communications needed by a binding in the earlier cycles as well as the communication that will happen in the future. Future communications are determined by considering the successors of this instruction which have one of their parents bound on a cluster different from the cluster under consideration.

This is due to the fact that if the instruction is bound to the cluster under consideration, it will surely lead to communication(s) in the future while scheduling the successors of the instructions. Although, we have experimented with many other heuristics for cluster assignment, the above mentioned heuristic seems to generate the best schedule in almost all cases [17].

3.3 Instruction Binding

A instruction binding scheme decides to bind or defer the chosen instruction to the selected cluster. The algorithm maintains a decoder map that explicitly keeps track of the status of each decode unit. A decode unit is marked to be in sleep mode after one cycle of idleness and is marked as activated on next use. If a decode unit is active in the target cluster, the instruction is bound to that cluster. Otherwise, the available slack of the instruction is considered. If the slack is below a threshold (we use the threshold value of 0 in our experiment), the instruction is bound anyway and the extra decoder unit required by the instruction is automatically woken up during execution. In case the instruction possesses enough slack, its scheduling is deferred to a future cycle and it is put back in the ReadyList. Note that the next time this instruction is picked up for scheduling, its earliest scheduling time and hence the slack get updated. This guarantees that the slack of an instruction reduces monotonically and eventually goes below the threshold ensuring that it is scheduled. Hence the algorithm is guaranteed to terminate.

3.4 An Example

We now present an example to illustrate how the proposed scheduling algorithm gets energy benefits without hurting performance. Figure 3 shows a data dependency graph and Figure 4 shows some schedules. Schedules 1 and 2 are for a plain VLIW architecture having two adders, two multipliers, and 4 decoders. We assume that the latency of an add operation is one cycle and the latency of a multiply operation is two cycles. Schedule 1 is generated by a traditional performance-oriented scheduler which schedules the instructions as early as possible and uses the slack value of instructions to break any contentions for resources and the total schedule length is 8 cycles. Total number of transition for Scheduler 1 are 3 as decoder D4, D3 and D2 each incur one transition in cycle 3, 4 and 5 respectively after idleness of 1 cycle.

Our energy efficient scheduler realizes the criticality of MPY operations and available slack for ADD operations and schedules the same data dependence graph as shown in schedule 2. Since deferring the execution of any MPY operation leads to stretching of schedules, they are scheduled in the same way as in the performance-oriented schedule 1. However, scheduling of ADD operations is delayed as well as serialized, exploiting the available slack of add operations. Notably, the scheduler determines the slack value available in scheduling an operation by first doing a performance-oriented scheduling pass on data-dependence graph and uses the estimate of schedule length from this pass to calculate the

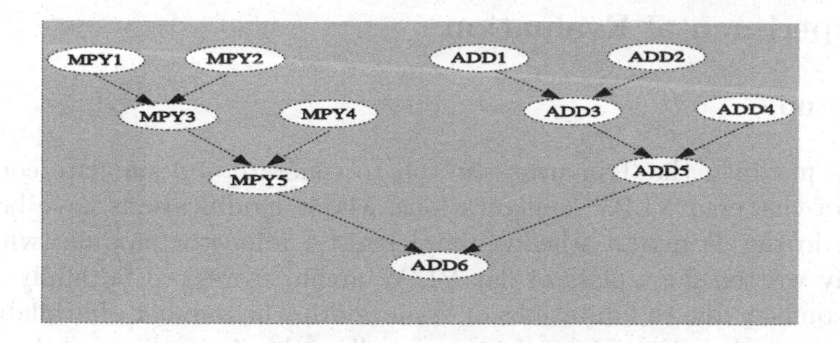

Fig. 3. An Example Data Dependency Graph

	Schedule 1
1	MPY1/D1, MPY2/D2, ADD1/D3, ADD2/D4
2	MPY4/D1, ADD4/D2, ADD3/D3
3	MPY3/D1, ADD5/D2
4	
5	MPY5/D1
6	
7	ADD6/D1
8	

	Schedule 2
1	MPY1/D1, MPY2/D2
2	MPY4/D1, ADD1/D2
3	MPY3/D1,ADD2/D2
4	ADD3/D1,ADD4/D2
5	MPY5/D1,ADD5/D2
6	
7	ADD6/D1
8	

	Schedule 3	
	Cluster 1	Cluster 2
1	MPY1/D1,ADD1/D2	MPY2/D3,ADD2/D4
2	MPY4/D1,ADD4/D2	
3	ADD3/D1	
4	MPY3/D1,ADD5/D2	
5		
6	MPY5/D1	
7		
8	ADD6/D1	
9		

	Schedule 4	
	Cluster 1	Cluster 2
1	MPY1/D1	MPY2/D2
2	MPY4/D1	ADD1/D2
3	ADD2/D1	ADD4/D2
4	MPY3/D1	
5	ADD3/D1	
6	MPY5/D1	
7	ADD5/D1	
8	ADD6/D1	
9		

Fig. 4. (a) Schedule 1 (b) Schedule 2 (c) Schedule 3 (d) Schedule 4

exact slack value available in scheduling an instruction which is used to generate the schedule for energy efficiency. Schedule 2 uses only two decoders and incurs only one transition for D2 in cycle 7. Schedule 2 is also more balanced as compared to schedule 1 in terms of resource usage. The resource usage vector of the first schedule is (4,3,2,0,1,0,1,0) and that of second is (2,2,2,2,2,0,1,0). Thus cycle to cycle variation in resource usage is clearly reduced in schedule 2 as compared to schedule 1, which in turn helps in reducing step power and peak power dissipation [19]. Thus, it is clear that the proposed scheme is capable of reducing leakage energy consumption, transition energy overheads, as well as peak power and step power dissipation without affecting the performance.

Consider schedules 3 and 4 generated for a 2-clustered VLIW architecture (equivalent to above mentioned VLIW architecture) having 1 adder and 1 multiplier in each cluster and a bidirectional bus between the two clusters with 1 cycle transfer latency. Schedule 3 is generated by a performance-oriented scheduler. The extra delay of inter-cluster communication stretches the schedule from 8 cycles to 9 cycles as compared to the schedule 1. Again the total number of decoder transitions are 3.

Scheduling the same set of operations using our energy-efficient scheduler generates schedule 4. The major point to note is that the scheduler leverages the available slack due to inter-cluster communication to achieve the same 9-cycle schedule with only two decoders and only one transition for D2 in cycle 5. Finally schedule 4 is much more balanced : The resource usage vector of first schedule is (4,2,1,2,0,1,0,1,0) and that of the schedule 4 is (2,2,2,1,1,1,1,1,0).

4 Experimental Evaluation

4.1 Setup

We have modified the Trimaran suite [4] to generate and simulate code for a variety of clustered VLIW configurations. Major modifications have been carried out in the Trimaran scheduler and register allocator module (which was originally written for a class of flat VLIW architectures) to faithfully account for the conflicts due to limitations on issue width, the number of available functional units and registers in a cluster as well as the limitations on the number of available cross-paths between clusters. The scheduler has been modified to implement the scheduling algorithm described in the last section. We have used twelve benchmarks out of which nine are from mediabench [1] *(viz. cjpeg, djpeg, rawcaudio, rawdaudio, g721encode, g721decode, md5, des, and idea)*, two from netbench [3] *(viz. crc, and dh)*, and one *(susan)* is from MiBench [2].

We present results for an unclustered, a two-cluster machine and a four-cluster VLIW machine each having an issue width of 6 instructions par cycle (implying 6 decoders). The unclustered VLIW configuration has 4 ALUs, 2 load-store units, 1 branch unit, and 64 registers. The 2-clustered configuration has 2 ALUs, 1-load store units, 1 branch unit and 32 registers in each cluster, whereas the 4-clustered configuration has 1 ALU, 1-load store unit, 1 branch unit and 16 registers in each cluster. The issue width and number of functional units selected for the VLIW configurations are such that the performance achieved using this configuration is within 95% of the peak performance achieved by using many more functional units. This moderate number of functional resources and decode units guarantees that the benefits reported have not been obtained by trivially putting the numerous idle decode units into low leakage mode.

4.2 Energy Model

We have used the same general analytical energy model proposed in [9] for combinational circuits to directly compare the energy benefits of our compiler-assisted scheme over the pure hardware based scheme proposed in [9]. However, unlike [9] that target leakage energy in functional units, we target leakage energy savings in instruction decoder. We briefly describe this model here. The reader is referred to [9] for details. The total energy in a decode unit is determined as follows:

$$\mathbf{E'_{total}} = \mathbf{DynamicEnergy} + \mathbf{LeakageEnergy} + \mathbf{TransitionEnergy} + \mathbf{SleepEnergy}$$

$$\mathbf{E'_{total}} = \mathbf{n_A}(\alpha\mathbf{E_A} + (\mathbf{1} - \mathbf{D})\mathbf{E_{s_1}}) + (\mathbf{n_A}\mathbf{D} + \mathbf{n_{UI}}) * (\alpha\mathbf{E_{s_0}} + (\mathbf{1} - \alpha)\mathbf{E_{s_1}}) + \mathbf{M_z}((\mathbf{1} - \alpha)\mathbf{E_A} + \mathbf{E_{Sleep}}) + \mathbf{n_Z}\mathbf{E_{s0}}$$

Here $\mathbf{n_A}$ is the number of active cycles, $\mathbf{n_{UI}}$ is the number of uncontrolled idle cycles, $\mathbf{n_Z}$ is the number of sleep cycles and $\mathbf{M_z}$ is the number of transitions. We have determined these values differently for each configuration by using the trimaran simulator. $\mathbf{E_{s_0}}$ and $\mathbf{E_{s_1}}$ are low leakage and high leakage energy and are related by the following equations.

$$\mathbf{E_{s_0} = s * E_{S_1}, 0.0001 \leq s \leq 0.01} \text{ and } \mathbf{E_{s_1} = p * E_A, 0 \leq p}$$

Where $\mathbf{p}$ is the ratio of the maximum leakage energy expended to the maximum energy for evaluation per unit of time (1 cycle). After simplifying and normalizing the equations with respect to active energy, the following model for total energy consumption is obtained :

$$\mathbf{E_{total} = n_A(\alpha + (1 - D)p) + (n_A D + n_{UI}) * (\alpha sp + (1 - \alpha)p)}$$
$$\mathbf{+ M_z((1 - \alpha) + E_{Sleep}/E_A) + n_Z sp}$$

The technology parameters that we have used (s=0.01 and $E_{Sleep}/E_A = 0.01$) are also the same as in [9]. Considering the current 65nm fabrication technology where leakage energy is on par with dynamic energy, we set p to 0.5. α is the activity factor and $\mathbf{D}$ is the duty cycle of the clock. We use a typical value of 0.5 for both of these parameters in our simulation.

4.3 Results

We have performed a detailed experimental evaluation of the proposed scheme in terms of the reduction in the number of transitions and the associated energy savings. We present results for the 'AlwaysActive' scheme that doesn't not apply any leakage energy management, the hardware-only scheme from [9] called 'MaxSleep' used in the context of decode units that puts a decode unit into low leakage mode after one cycle of idleness, and our scheduling scheme called 'Optimized' that assists the hardware based scheme by reducing undesirable transitions. The results are presented in comparison with a hypothetical scheme called 'NoOverhead' that is the same as 'MaxSleep' but does not incur any of the energy overheads of transitions. This scheme represents a theoretical ideal against which a leakage energy management scheme can be compared for its effectiveness.

Figure 5 (a) shows the percentage reduction in the number of transitions due to our algorithm as compared to the hardware-only scheme. We observe that the number of transitions reduce by 53%, 58.88%, and 62.74% for VLIW, 2-Clustered

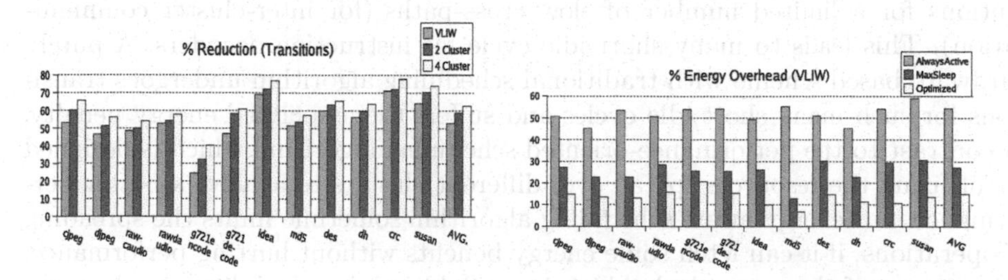

Fig. 5. (a) % Reduction in Transitions with scheduling w.r.t. Hardware only Scheme (b) % Increase in energy w.r.t Hypothetical No-overhead Scheme (VLIW)

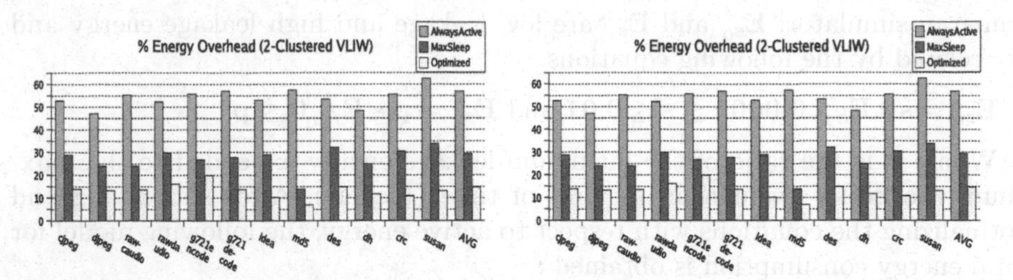

Fig. 6. % Increase in energy w.r.t No-overhead Scheme (a) 2 Cluster (b) 4 Cluster

VLIW, and 4-Clustered VLIW respectively. The reduction in the number of transitions depends on the total available slack in scheduling instructions as well as the distribution of idle cycles in the benchmark. Benchmarks like des, dh, crc, and susan have many short idle cycles and our algorithm is able to exploit the available slack in these applications to avoid many transitions. In the case of g721encode and g721decode, the available slack is relatively less and consequently the reduction is also less.

Figure 5 (b) shows the energy overhead of 'AlwaysActive', 'MaxSleep' and 'Optimized' schemes as compared to the 'NoOverhead' scheme. 'AlwaysActive', 'MaxSleep' and 'Optimized' schemes show average energy overheads of 54.33%, 27.29%, and 14.99% respectively as compared to the 'NoOverhead' scheme. The proposed 'Optimized' scheme reduces the total energy overhead by 14.46% over the 'MaxSleep' scheme which is significant taking into account that it is a purely software based scheme and does not incur any hardware overhead.

The benefits of our scheme are even more pronounced in the context of clustered architectures. In the context of 2-clustered architecture 'AlwaysActive','MaxSleep' and 'Optimized' have average energy overheads of 56.86%, 29.37% and 14.6% respectively as compared to the 'NoOverhead' scheme (Refer Figure 6 (a)). The energy benefits of 'Optimized' over Maxsleep is 17.3% in context of 2-clustered architecture. For a 4-clustered configuration, 'AlwaysActive','MaxSleep', and 'Optimized' incur 57.51%, 29.88%, and 13.7% overhead as compared to 'NoOverhead' scheme (Refer Figure 6 (b)). The 'Optimized' scheme improves over the 'MaxSleep' scheme on the average by 18.74% in the context of 4-clustered architectures. The reasons for more savings in the context of clustered architectures are as follows. Clustering brings along extra contentions for a limited number of slow cross-paths (for inter-cluster communication). This leads to many short idle cycle for instruction decoders. A purely hardware based scheme with traditional scheduling algorithm undergoes transitions for such many short idle cycles and suffers the associated energy penalty. In contrast to the performance-oriented scheduling algorithm which is designed for utilizing the resources spread over different clusters to achieve a better performance, our energy-aware scheduling algorithm sometime limits the spreading of operations, if it can fetch some energy benefits without hurting performance. Thus, some of the extra slack which is available while scheduling for clustered architectures due to contention for inter-cluster communication is utilized to gain energy benefits in our algorithms.

5 Related Work

Earlier proposals for scheduling on clustered VLIW architectures can be classified into two main categories, viz., phase-decoupled approaches and phase-coupled approaches. A phase-decoupled approach to scheduling works on a data flow graph (DFG) and performs partitioning of instructions into clusters to reduce inter-cluster communication while approximately balancing the load among clusters. The annotated DFG is then scheduled using a traditional list scheduler while adhering to earlier spatial decisions [8] [6]. An integrated approach to scheduling combats the phase-ordering problem by combining spatial and temporal scheduling decisions in a single phase [12] [17] [16].

Study of leakage energy management at the architectural level has mostly focused on storage structure such as cache [11]. Some of the earlier work has targeted energy efficiency in functional units. [9] proposes an architectural policy for aggressively controlling leakage energy in integer ALUs. However the overhead of transitions from active mode into low-leakage mode and vice-versa are significant. Zhang et al. [20] have proposed a rescheduling scheme to reduce dynamic and leakage energy in the functional units of a VLIW processor by exploiting the remnant slack of a performance-oriented schedule. Kim et al. [13] have proposed a leakage energy management scheme for VLIW processors.

To the best of our knowledge, the only work for energy optimization in the context of instruction decoder is due to Kuo et al. [14]. Kuo et al. [14] consider instruction decoding as in superscalar architectures and propose to split (horizontally partition) instruction decoder circuitry into two or more sub-decoders based on execution frequencies of different instructions. They also propose to do pipelining (vertical partitioning) of the instruction decoder to achieve energy and area benefits. The experimental results of Kuo et al., based on physical synthesis clearly demonstrates that the horizontal and vertical partitioning of the instruction decoder is in general useful in reducing the design complexity, power consumption, area overhead and delay because of simplification of circuitry. In contrast to the work of Kuo et al. [14], partitioning of instruction decoder in our work is geared more toward VLIW and clustered VLIW architectures that demands decoding of large number of instructions in parallel. Thus, compared to functionally asymmetric partitioning of Kuo et al, we consider partitioning of instruction decoder circuitry into functionally identical individual sub-decoders each of which can be controlled independently. The pipelining of decoder as considered by us is more natural in VLIW context where a fetch packet needs to be broken into execute packets and the current execute packet needs to be aligned before actual decoding can begin.

6 Conclusions and Future Directions

In this work, we consider a split instruction decoder design that enable energy optimization in instruction decoder. We evaluate the purely hardware based scheme to gain energy benefit for short idle cycle delimited by frequent transition. We also propose a new energy-aware instruction scheduling algorithm for VLIW and clustered VLIW architectures that provides significant energy benefits over purely hardware based scheme by reducing the number of transitions using scheduling slack of instructions. In future, we would like to integrate the proposed scheme for leakage energy management with the slack based approach to dynamic and leakage energy management in functional units.

References

1. MediaBench, `http://cares.icsl.ucla.edu/MediaBench/`
2. MiBench, `http://www.eecs.umich.edu/mibench/`
3. NetBench, `http://cares.icsl.ucla.edu/NetBench/`
4. Trimaran System, `http://www.trimaran.org/`
5. Borkar, S.: Design Challenges of Technology Scaling. IEEE Micro 19(4) (1999)
6. Chu, M., Fan, K., Mahlke, S.: Region-based Hierarchical Operation Partitioning for Multicluster Processors. In: SIGPLAN Notices, pp. 300–311 (2003)
7. Cooper, K.D., Waterman, T.: Understanding energy consumption on the c62x. In: Proc. of the Work. on Compilers and Operating Systems for Low Power (2002)
8. Desoli, G.: Instruction Assignment for Clustered VLIW DSP Compilers: A New Approach. Technical Report, Hewlett-Packard (1998)
9. Dropsho, S., Kursun, V., Albonesi, D.H., Dwarkadas, S., Friedman, E.G.: Managing Static Leakage Energy in Microprocessor Functional Units. In: Proc. of the Intl. Symp. on Microarchitecture, Los Alamitos, CA, USA, pp. 321–332 (2002)
10. Faraboschi, P., Brown, G., Fisher, J.A., Desoli, G.: Clustered Instruction-level Parallel Processors. Technical report, Hewlett-Packard (1998)
11. Flautner, K., Kim, N.S., Martin, S., Blaauw, D., Mudge, T.: Drowsy Caches: Simple Techniques for Reducing Leakage Power. In: Proc. of the Intl. Symp. on Computer Architecture, Washington, DC, USA, pp. 148–157 (2002)
12. Kailas, K., Agrawala, A., Ebcioglu, K.: CARS: A New Code Generation Framework for Clustered ILP Processors. In: Proc. of Intl. Symp. on High-Performance Computer Architecture, p. 133 (2001)
13. Kim, H.S., Vijaykrishnan, N., Kandemir, M., Irwin, M.J.: Adapting Instruction Level Parallelism for Optimizing Leakage in VLIW Architectures. In: Proc. of Conf. on Language, Compiler, and Tool for Embedded Systems, pp. 275–283 (2003)
14. Kuo, W.-A., Hwang, T., Wu, A.C.-H.: Decomposition of Instruction Decoders for Low-power Designs. ACM Trans. Des. Autom. Electron. Syst. 11(4) (2006)
15. Kursun, V., Friedman, E.G.: Low swing Dual Threshold Voltage Domino Logic. In: Proc. of the ACM Great Lakes Symp. on VLSI, New York, USA (2002)
16. Nagpal, R., Srikant, Y.N.: A Graph Matching Based Integrated Scheduling Framework for Clustered VLIW Processors. In: Proc. of ICPP Workshop on Compile and Runtime Techniques Parallel Computing, pp. 530–537 (2004)
17. Nagpal, R., Srikant, Y.N.: Integrated Temporal and Spatial Scheduling for Extended Operand Clustered VLIW Processors. In: Proc. of Conf. on computing frontiers, pp. 457–470 (2004)

18. Seshan, N.: High VelociTI Processing. In: IEEE Signal Proc. Magazine (March 1998)
19. Yun, H., Kim, J.: Power-aware Modulo Scheduling for High-Performance VLIW Processors. In: Proc. of 2001 Intl. Symp. on Low Power Electronics and Design, pp. 40–45 (2001)
20. Zhang, W., Vijaykrishnan, N., Kandemir, M., Irwin, M.J., Duarte, D., Tsai, Y.-F.: Exploiting VLIW Schedule Slacks for Dynamic and Leakage Energy Reduction. In: Proc. of Intl. Symp. on Microarchitecture, pp. 102–113 (2001)

P2P Document Tree Management in a Real-Time Collaborative Editing System

Jon A. Preston and Sushil K. Prasad

Department of Computer Science
Georgia State University
Atlanta, GA
jon.preston@acm.org, sprasad@gsu.edu

Abstract. This paper presents our work in combining peer-to-peer dynamic tree management with hierarchical Operational Transformation (OT) over document trees to achieve low computational and communication costs. We discuss our approach in storing the document tree in a peer-to-peer, distributed manner and maintaining convergence, causality preservation, and intention preservation (CCI) via a peer-to-peer caching system. Because changes are sent to other users within the system only as needed (and cached when possible), our approach minimizes communication costs among multiple readers and writers. Our algorithms balance the traffic and computational load among peers. They ensure that users always have the most current/correct copy of the section(s) of the document which they are viewing. Our approach outperforms existing OT techniques that broadcast messages and compute OT for each operation at all peers. This paper presents our algorithms and simulation results demonstrating the efficiencies and load balancing among peers within the system.

Keywords: Concurrent P2P tree, Load Balancing, Lazy updates, Communication efficiency.

1 Introduction

The field of Computer Supported Collaborative Work (CSCW) and the subfield of Real-time Collaborative Editing Systems (RTCES) seek to achieve the goal of providing synchronous (real-time) access to a shared document. RTCES and consistency maintenance within distributed, synchronous RTCES are active research areas within the field of CSCW [18]. Most notable are algorithms [3][8] that work to achieve a high level of concurrent access through optimistic concurrency control and Operational Transformation (OT) to merge changes on other users' copies while ensuring consistency, causality-preservation, and intention-preservation – the widely accepted CCI model [17][18]. There exists a significant opportunity to reduce the computation and communication costs associated with OT [13].

We revisit the idea that locking offers RTCES research opportunities. Locking is one technique used to ensure consistency and data integrity in distributed systems, but locking has the disadvantage of reducing concurrent access. Consequently, the RTCES research community has adopted OT or similar approaches to maintain real-

S. Aluru et al. (Eds.): HiPC 2007, LNCS 4873, pp. 418–431, 2007.

time, high responsiveness while striving to maintain the CCI model. Unfortunately, adopting an existing OT approach is costly in that all changes/operations must be broadcast to all users, and each of these users must "replay" these incoming operations locally. This is costly with respect to communication and computation (since OT algorithms employ history buffers of previous operations which must be maintained and potentially modified).

Motivated by the idea that locking and OT serve complimentary roles [19], we incorporate both techniques such that users are able to make changes locally within locked sections of a document (and cache these changed) while adopting OT when many users make changes to a shared section of the document (multicasting these changes to only those users sharing the section). Whether a section is locked or shared is a user-defined policy that can change depending upon the scenario and on a per-section basis. By dynamically managing lock granularity, we maximize concurrent access among writers and avoid performing OT globally over the entire document. OT can be applied locally to a subsection of the entire document to reduce communication and computation costs.

Our approach employs a relaxed consistency model in which not all users within the CES have the most current copy of the entire document – rather, all users have the most current copy of the sections of the document they are viewing (i.e., the visible/focused portion of the document is always current on a user-by-user basis). By relaxing the consistency constraint heretofore enforced in OT-based systems, we are able to reduce communication and computation costs.

This research extends our previous work on centralized document trees [10]-[13] by distributing the document management among all peers within the CES and allowing the cached changes (history buffers) to be applied at an arbitrary level in the hierarchical document tree. These P2P algorithms for managing document sections and distributing existing OT algorithms are complimentary, superior to the current best practices of existing OT algorithms over linear document representations, and significantly reduce the computational and communication costs.

The remainder of this paper is structured as follows. We begin with a discussion of our hierarchical, tree-based view of the document, and then provide an overview of the data structures used to manage the document among the peers in Section 2. Next, we discuss our peer-to-peer algorithms to manage ownership of the sections of the document in Section 3. Section 4 presents a discussion of load balancing and fault tolerance, and Section 5 presents our simulation results validating our approach. We present related work in Section 6 and conclusions in Section 7.

2 Maintaining Exclusive Access

To avoid blocking users from editing, we use a tree-based view of the document and allow for locking at various levels within the tree such that users manage/lock local portions of the document. When a user desires a section for writing, we dynamically adjust lock granularity via demotion of the lock down in the document tree until the conflict among other users is resolved. When a user leaves a section of the document and makes it available to other users, conflict among users is potentially reduced; as a result, our approach automatically promotes the lock to a higher level within the document tree – maximizing the amount of the document owned for the remaining user [10].

2.1 Representing a Document Via a Tree

By viewing the shared document as a hierarchical structure we are able to better achieve context-specific consistency preservation, and we can reduce communication and computational costs. Based upon the semantic structure of the document, the document may be broken up into sections, subsections, etc. The document tree consists of internal nodes that represent structure, and all document content resides at leaf nodes.

Each portion of a document may consist of content and sub elements (sub sections). Because any section of a document may contain any number of text elements (paragraphs, sentences, etc.), and may contain any number of subsections, we employ algorithms for inserting and removing locks from the collaborative space to work within an n-ary tree data structure that is representative of a shared document [10].

It is important to note that structure of the tree is defined within the document itself and does not depend upon any voting schemes or user involvement. For example, the structure of this paper is defined by the sections, subsections, paragraphs, sentences, etc.; the structure of a CAD file is defined by layers, objects, etc. Our approach does allow for users to establish at what granularity the leaf nodes should be defined, but this could also be defined automatically (defaults depending upon the type of document in use). Once established, the tree is utilized to manage ownership of subsections within the document. Rather than locking the entire document, lock granularity is adjustable, ranging from the entire document (ownership marked at the root of the tree) to an atomic level (ownership marked at a leaf node in the tree). The size of a subsection is not specified within our algorithms, thus it is scalable to accommodate the semantic structure of the document being edited.

2.2 Maintaining Node Coloring, Grey Count, and Black Siblings

To enable efficient management of which user owns/manages each section within the document, we utilize a tree coloring scheme. Each node in the document tree maintains a color (white, black, or grey) to denote whether it is available, currently being written to by another user, or if two or more users are editing sub-trees, respectively. Ownership (black coloring) of a vertex v by user U implies that U owns/manages v and the sub-tree rooted at v. We allow multiple users to concurrently write to the section denoted by a node and employ an OT consistency maintenance algorithm among the writers; thus a black node may have multiple owners/managers. Each node n in the tree maintains a numeric value that denotes how many users are managing nodes in the sub-trees of n. This count is defined as the *grey-count* of the node n. This grey count value is useful in efficiently determining if the node can be colored white or grey when a user leaves or moves and promotion is possible (as explained later).

A grey node v maintains references to the node's children (sub-trees); additionally, if there exists at least one black child node of v, then v also maintains a reference to the first black child node. The black child nodes of v ($b_1, b_2, \ldots b_k$, where k = number of black child nodes of v) are linked together using a doubly-linked list. These sibling references and the reference to the first black child of a grey node are used for fast (O(1)) location of which node may be promoted when all but one user has left the sub-tree (as explained in Section 3.2).

When we say that user U owns a node v, we claim that this user contains the most current cached copy of the document's content represented by the node v. When we say that a set of users $\{U_1, U_2, \ldots, U_n\}$ own a node v, we claim that these users all maintain the most current cached copy of the document's content represented by node v and are employing a localized OT algorithm to maintain consistency on the content of v. Thus the total correct, up-to-date copy of the document is distributed among the peers and can be constructed as needed by requesting the cached sections from the peers.

2.3 Lock Granularity, Localized OT, and Caching Local Changes

It is advantageous to maintain a lock on the largest sub-tree that is permissible; by maximizing the sub-tree that any user owns, we minimize the communication costs of the system by utilizing caching. For example, if a user U_i owns a section of the tree, then all changes to that section can be stored locally in the user's cache and do not have to be transmitted to other users until they desire to enter (read or write) that section.

Each node may employ a sharing policy to either allow or disallow multiple writers. If exclusivity is adopted at a node n (such that we allow only one writer), a lock on a sub-tree rooted at node n is permissible for user U_i so long as no other user has a lock on any node within the tree rooted at node n. If another user U_j enters the system and requests a section of the document, then the section of the tree owned by user U_i is reduced to accommodate the insertion of user U_j (assuming n is not an atomic/leaf node and demotion is possible).

Alternatively, if a multiple-writer policy is adopted at a node n, then multiple users may write to n and maintain consistency via OT. In this case, if n is owned by a single user U_i and another user U_j desires to enter n, then both U_i and U_j own n and OT is employed to maintain consistency. Any changes made by U_i to n before U_j's arrival are transmitted to U_j upon arrival.

3 P2P Algorithms for Distributed Document Management

We agree with Edwards [2] that conflicts are a "naturally-arising side effect of the collaborative process" and "will occur simply because of the semantics of multi-user applications." Further we agree with Handley and Crowcroft [4] that "temporary inconsistencies are necessary to achieve good performance" within collaborative editing systems. Thus, at various points in time, the copies of the document are not consistent, but the distributed, managed copy of the document in its entirety is correct and preserves user intention. We record ownership and change history sufficient to recreate the entire document as needed (i.e., when a user wishes to view any specific section). These changes will be communicated and replayed among local copies as the users move about and view new sections, and changes can also be sent among the users (moving changes up the tree – minimizing communication costs) at specified intervals if desired [14][15]. Selective multicast is employed to improve communication cost [7].

To minimize communication costs, locating which peer is currently managing a section of the document (currently managing a node in the document tree) may be done quickly via a peer location approach such as Chord [16]. Further, whenever a history buffer (Δx) for a node x is being transmitted, reduction of Δx to $\Delta x'$ can also decrease the communication cost.

Given the structure of the document tree and the distributed nature of the tree management, when a user u manages a section denoted by node n, then all changes made to the content of the sections rooted at n are cached locally on u. Thus any structural change to the document tree (such as combining two sections, splitting a section into two sections, or deleting a section) can also be cached locally on u. If a set of users are sharing n and employing OT to maintain consistency of n, then any structural changes within the sub-tree rooted at n will be maintained via the OT algorithm on n among the users sharing n.

As a result, the algorithms specific to handling the placement and management of users within the document tree are the focus of this research. The two algorithms we have developed are USERENTER and USEREXIT. We present each below with a detailed analysis of correctness and efficiency. As in our previous, centralized algorithms [10]-[13], we avoid deadlock among peers by employing handshake locks on parent/child nodes and by always moving downward through the tree.

3.1 User Entering a Section

When a user desires to write to a section s of the shared document, the user must enter the section of the document tree that represents the desired section s. This operation is performed by the USERENTER algorithm that works from top-to-bottom by examining nodes in the path from the root to the desired node. The correct path is determined by first querying the peer who manages the root, and then descending further down by following peers' references to other peers. The coloring of the nodes along the path denote the availability of the sections (see Section 2.2). If a white node is reached, then the node becomes managed by the user (and marked black denoting the ownership). If a grey node is reached, the algorithm proceeds further down the path. If a black node is reached, then the user's entry to the section depends upon the sharing policy adopted at the black node. If the sharing policy is set to exclusive write, then the entry fails (i.e., the user cannot enter the desired section of the document); if the sharing policy is to adopt OT (allowing multiple writers at this node), then the requesting user is added to the set of concurrent writers for the section; finally, we can adopt a *demote* policy. Demotion works by moving the management of the existing user down the tree hierarchy until the conflict among the users is resolved.

The most complicated case of the USERENTER algorithm is when demotion occurs. Figure 1 demonstrates the demotion of U_1 from the section v down to the subsections denoted by $\{w_1, \ldots, w_n\}$ and the injection of U_2 at the section denoted by x. Any history buffer at U_1 to x (denoted by Δx) must be passed to U_2. At this point, U_1

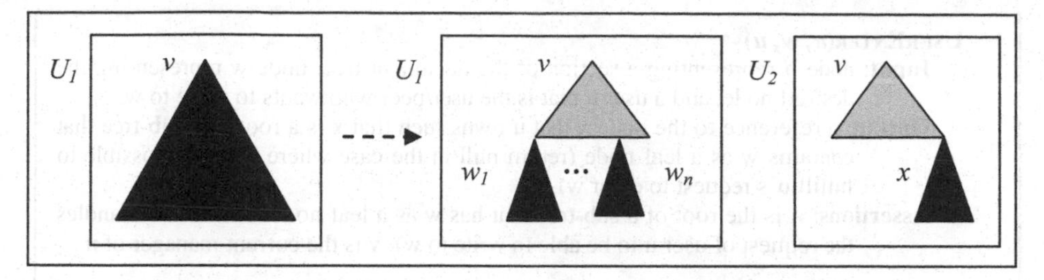

Fig. 1. Adding U_2 by reducing and distributing U_1's management of v

contains the most current copy of the sections $\{w_1, \ldots, w_n\}$, and U_2 contains the most current copy of section x.

Correctness: As there are only three colorings for any node (white, black, and grey), the USERENTER algorithm handles all cases. In the case where n is colored white (1), there are no other users in n, so u obtains n. In the case where n is colored black (2), we have three cases on resolving the entry of u to n: (2.1) OT is adopted at n, so u is added to the user set of n; (2.2) n is not shareable in which case the entry of u to n fails and we must decrement the grey count of the nodes along the path from the root to n to keep the grey count correct (since the node n was not granted to u); and (2.3) when demotion occurs and n is colored grey. In the more complicated case of 2.3, the current manager of n (v) is demoted along the path such that the original request of v (that led to v's management of n) is still fulfilled. This either resolves the conflict (2.3.4) or we repeat the invocation of the algorithm one level down in the tree closer to w (2.3.5). In the case where n is colored grey (3), we increase the grey count (optimistically assuming u's request for w will be successful) and repeat the invocation of the algorithm one level down in the tree closer to w.

Performance Analysis: Since the USERENTER algorithm traverses from the root down to a leaf (or stops earlier if a white or black node is reached), this algorithm must traverse $O(h)$ nodes, where h equals the height of the document tree. The work involved at each node is $O(1)$ since the work in processing an individual node involves updating references/pointers, coloring, and grey count (integer) values. It is possible upon a failure (step 2.2) that the USEREXIT function will be invoked, but this USEREXIT (as discussed below) runs in $O(h)$, thus it is not asymptotically greater than the existing $O(h)$ work for the USERENTER algorithm. Thus the computation cost for the USERENTER algorithm is $O(h)$. As the algorithm traverses down the tree, peers that manage each of the nodes along the path must handle the request; thus as many as $O(h)$ peers must be involved in resolving the request – and this incurs $O(h)$ messages. Communication also occurs when the lock is granted and the history buffer is communicated to the requesting user for a cost of $O(b)$ where b is the size of the single history buffer communicated. In the case where we are adding u to the manager set of n (2.1), this requires $O(n)$ messages where n is the number of users in the OT set of n (since all users in the set must be notified of the user entering the set). Thus the total communication cost for USERENTER is $O(n + b + h)$.

USERENTER(*n*, *w*, *u*)

Input: node n representing a section of the document tree, node w representing the desired node, and a user u that is the user/peer who wants to write to w.

Output: a reference to the node x that u owns such that x is a root of a sub-tree that contains w as a leaf node (return null in the case where it is not possible to fulfill u's request to enter w)

Assertions: n is the root of a sub-tree that has w as a leaf node (such that n handles the request of user u to be able to write to w), v is the current manager of n

```
1.      if n.color = WHITE
            SetManager(n, u, w) // n.color BLACK and n.originalRequest = w
            LinkToSiblings(n)
            Return n
2.      else if n.color = BLACK
2.1.        if n.policy = OT
                Communicate(Δn, u)
                AddToManagers(n, u, w)
                Return n
2.2.        else if n.policy = EXCLUSIVE or n.IsLeaf // demotion not possible
                USEREXIT(root, w, u) // correct artificially-inflated grey counts
                Return null        // failed entry due to lack of sharing on n
2.3.        else // demotion must occur
2.3.1.          n.color = GREY
2.3.2.          n.greyCount = 2
2.3.3.          Demote v down to NextInPath(n, n.OriginalRequest)
2.3.4.          if NextInPath(n, n.OriginalRequest) != NextInPath(n, w)
                    Communicate(Δn, u)
                    Return NextInPath(n,w)
2.3.5.          else return USERENTER(NextInPath(n, w), w, u)
3.      else // color is GREY
            n.greyCount++
            return USERENTER(NextInPath(n,w), w, u)
```

3.2 User Leaving a Section

When a user desires to exit to a section *s* of the shared document, the node in the document tree that represents the *s* must remove the user from its list of managers/writers. This operation is performed by the USEREXIT algorithm that works from top-to-bottom by examining nodes in the path from the root to the node to release. As in the USERENTER algorithm, the correct path is determined by first querying the peer who manages the root, and then descending further down by following peers' references to other peers. Again, the coloring of the nodes along the path indicate how to handle the request to leave. If a black node is reached, then we remove the user from the set of users managing the node and inform all remaining users within the section that the user has left. If a grey node is reached, there are three cases: the grey count is decremented and the algorithm proceeds down the tree to the node next in the path to the desired node; the grey count is decremented to 0 which means the exit is complete as all users have left the section represented by this sub-tree; the third case is when promotion is possible because the grey count is going from

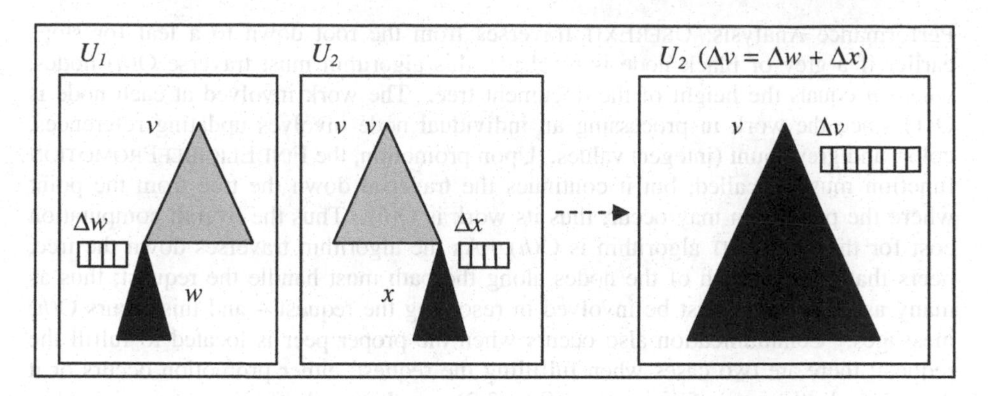

Fig. 2. Removing U_1 by promoting U_2 & communicating Δw to U_2

2 to 1 (denoting there will only be one user remaining after the exit). Promotion works by finding the remaining user (the user not leaving) and moving their management up to the current node being examined. This remaining user can assume ownership of a larger portion of the document (and maximize the caching).

The most complicated case of the USEREXIT algorithm is when promotion occurs. Figure 2 demonstrates the promotion of U_2 to v when U_1 leaves w. As a result, Δw (the history buffer of w) is communicated from U_1 to U_2. At U_2, x is current since U_2 already managed it, and w is now current because Δw has been "replayed" at U_2. Thus U_2 contains a proper and complete, up-to-date version of v since v is defined by w and x (i.e., v is current because $v = w + x$ and $\Delta v = \Delta w + \Delta x$). When Δw is communicated to U_2, U_2 may elect to incorporate Δw into its copy of w, or U_1's changes to w may be rejected. This acceptance or rejection of changes by other users could be done automatically by the system based upon embedded rules or done explicitly by users as prompted by the system.

Correctness: The USEREXIT algorithm handles only two cases for a node's coloring: black and grey. The algorithm does not handle the case where the color is white because a white coloring denotes the node n is not managed by any user, thus no user can exit n. In the case where n is colored black (1), the user u is removed from the user set managing n and removed from the sibling list. Promotion is not possible or it would have occurred earlier in the path from the root to n (as discussed below). In the case where n is colored grey (2), the grey count is decremented to denote u is leaving the sub-tree rooted at n (2.1). We then have three cases on resolving the exit of u from n: (2.2) is the case when a promotion occurs since only one user remains in the sub-tree n, so we promote the only remaining user (other than u) to n and communicate the history buffer of the node u is managing to the remaining user; (2.3) n no longer has any users remaining in it, but this coloring to white was temporarily delayed due to a previous, concurrent invocation of USERENTER that failed; and (2.4) when we repeat the invocation of the algorithm one level down in the tree closer to w

Performance Analysis: USEREXIT traverses from the root down to a leaf (or stops earlier if a grey or black node is reached), this algorithm must traverse $O(h)$ nodes, where h equals the height of the document tree. The work involved at each node is $O(1)$ since the work in processing an individual node involves updating references, color, and grey count (integer) values. Upon promotion, the FINDELIGIBLEPROMOTION function must be called, but it continues the traversal down the tree from the point where the promotion may occur, thus its work is $O(h)$. Thus the overall computation cost for the USEREXIT algorithm is $O(h)$. As the algorithm traverses down the tree, peers that manage each of the nodes along the path must handle the request; thus as many as $O(h)$ peers must be involved in resolving the request – and this incurs $O(h)$ messages. Communication also occurs when the proper peer is located to fulfill the request; there are two cases when fulfilling the request: either promotion occurs or it does not. In the case of promotion (Step 2.2), one history buffer is communicated to the user that is promoted, thus the communication cost is $O(b)$ where b is the size of the history buffer. In the case when promotion does not occur (Step 1), the requesting user must be removed from managing the node. This requires $O(n)$ messages where n is the number of users in the OT set of n (since all users in the set must be notified of the user leaving the set). Thus the total communication cost in USEREXIT is $O(n + b + h)$.

USEREXIT(*n*, *w*, *u*)

Input: node n representing a section of the document tree, node w representing the node no longer desired, and a user u that wants to leave w.

Output: none

Assertions: n is the root of a sub-tree that has w as a leaf node (such that n is fulfilling the request of user u to be able to leave w)

```
1.      if n.color = BLACK
            RemoveFromManagers(n, u) // possibly n.color WHITE (if no more
            managers)
            UnlinkFromSiblings(n)
2.      else if n.color = GREY
2.1.        n.greyCount--
2.2.        if n.greyCount = 1
                a = FindEligiblePromotion(n, w)
                SetManager(n, a.manager, a.originalRequest) // promote a
                b = NextInPath(n,w)
                Communicate(Δb, a.manager)
2.3.        else if n.greyCount = 0 // previous failed UserEnter and two users have
                exited n
                RemoveFromManagers(n, u) // n.color is now WHITE
2.4.        else UserExit(NextInPath(n, w), w, u)
```

4 Load Balancing and Fault Tolerance in the P2P Implementation

One motivation in developing our P2P dynamic locking algorithms was to distribute the work of lock management among the peers. Initially, it would seem that this work and communication is distributed uniformly among the peers, but the problem remains that all messages must be processed from the root down as the grey counts must be modified from the root to the desired node to ensure proper promotion and demotion. Thus if a single peer is responsible for managing each node in the tree, some peer must maintain the root and will then become the bottleneck and have an increase in workload when processing the USERENTER and USEREXIT requests.

We achieve workload balancing by adopting a rotating management of the nodes in the tree. When an USERENTER operation is performed, the user begins management of the nodes along the path in the document tree visited in fulfilling the USERENTER operation. In this manner, we adopt a "most recently requested" policy in that all nodes n_i will be managed by the user who's USERENTER request was fulfilled by passing through n_i (i.e., n_1, n_2, ... n_k is in the path from the root to n_k, where n_k is the desired node or the node at which the lock request is fulfilled). We note that when a USEREXIT operation is performed, this implies that the user is leaving a section and thus it is not advantageous to have the user begin management of nodes.

If such a "most recently requested" policy for lock management is adopted, then a single peer p must serve at most $O(n)$ consecutive lock management operations, where n is the number of peers in the collaboration. This is true because if a USERENTER request is handled, then the node acquires a new manager other than p. Only USEREXIT requests can be fulfilled and keep the same manager p, and there can be at most n consecutive USEREXIT request since any more would necessitate a lock request (i.e., a peer can't release a lock it doesn't have). The workload for a peer is proportional to the number of lock requests for the peer – thus more active peers in the RTCES will be responsible for handling more of the work in maintaining the document tree. And if the peers perform an equivalent number of lock requests, the workload of managing the nodes within the document tree is balanced since the amortized time a peer manages a node should be approximately equal to the amortized time the other peers manage the node.

The time a peer manages a node is proportional to the depth of the node in the document tree (since there are fewer paths that travel through a node at a greater depth than a node at a more shallow depth). Thus the root management should change more often than a near-leaf node. This is good because the workload of more shallow nodes in the tree (closer to the root) is more than the workload of deeper nodes. As a result, the workload in managing the distributed, P2P version of the document tree is balanced among the peers.

Another important property of our P2P approach is increasing the reliability and fault tolerance so that if one peer is dropped from the RTCES, the others can continue without any problems. We may increase the reliability and fault tolerance of the document tree by replicating the top portion of the tree among all peers (or a subset of peers). For reliability, we can adopt an n-way replication of sections of the document and consistency maintained via an OT policy. While this increases the communication cost (since all peers must perform OT to maintain consistency regarding the lock states among the replicated portion of document tree), this approach does overcome the single point of failure of a single manager for a document section. If this n-way replication is adopted, then the sections of the document would implement a queue to store the managers of the section of the document – thus the peer who had managed the replica the longest would rotate out when a new peer's request arrived down the document tree (a LIFO approach to achieve load balancing).

5 Simulation and Results

To validate our P2P distributed document management approach, we implemented the model of the node and the USERENTER and USEREXIT algorithms. We modeled three different document trees containing 14, 28, and 56 leaves, respectively. We simulated concurrent users that were either in a reading or writing state; additionally, the users could move to a new section of the document (moving their cursor position), and this new section to which to move was randomly selected. A total of 96 simulation configurations were performed, varying among the three documents and increasing the number of users from 1 to 32.

The results of the 96 simulation runs are shown in Figure 3. Each column denotes a set of peers varying from 1 peer (in simulation runs 1-3) to 32 peers (in simulation runs 94-96). The workload is measured by how many USERENTER and USEREXIT requests were handled on a per-peer basis, thus each point plotted denotes how much work a single peer handled. Note that the y-axis is logarithmic to enable the variance among the peers within the columns to be visible.

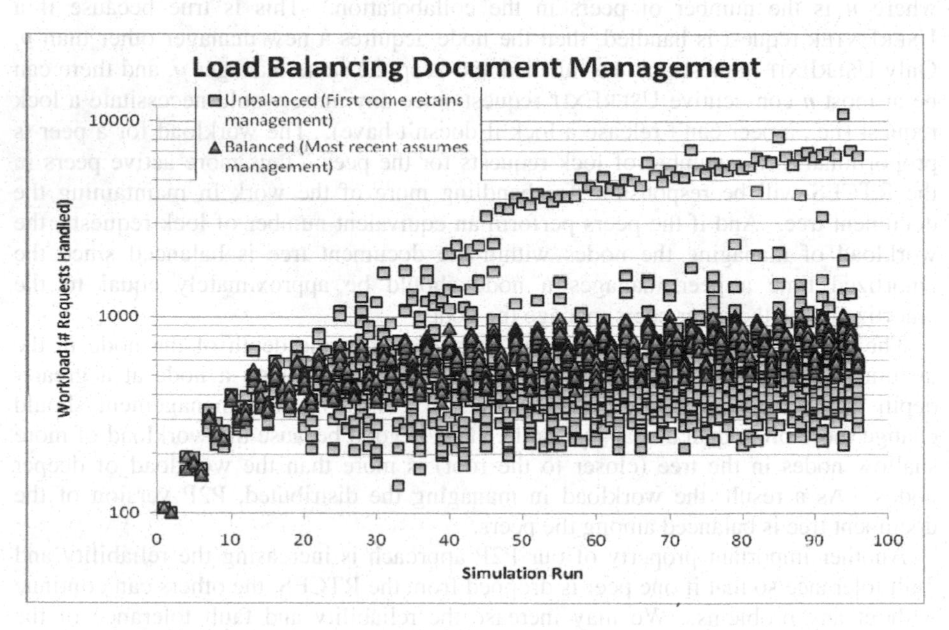

Fig. 3. Balancing the Workload of Document Management among Peers

If we adopt a first-come policy of node management, then as predicted, one (or a small few) peers are unfairly burdened with the bulk of the document management. Notice the high trend line showing the most burdened peer for each simulation run. When the "most-recent," balanced approach is adopted (see Section 4), the work is more fairly distributed among all peers. This is corroborated in that while the total work remains the same, the variance among the peers for any simulation run decreases when a balanced approach is adopted (note the increased clustering). We observe that the total workload decreases when the document size increases. This is

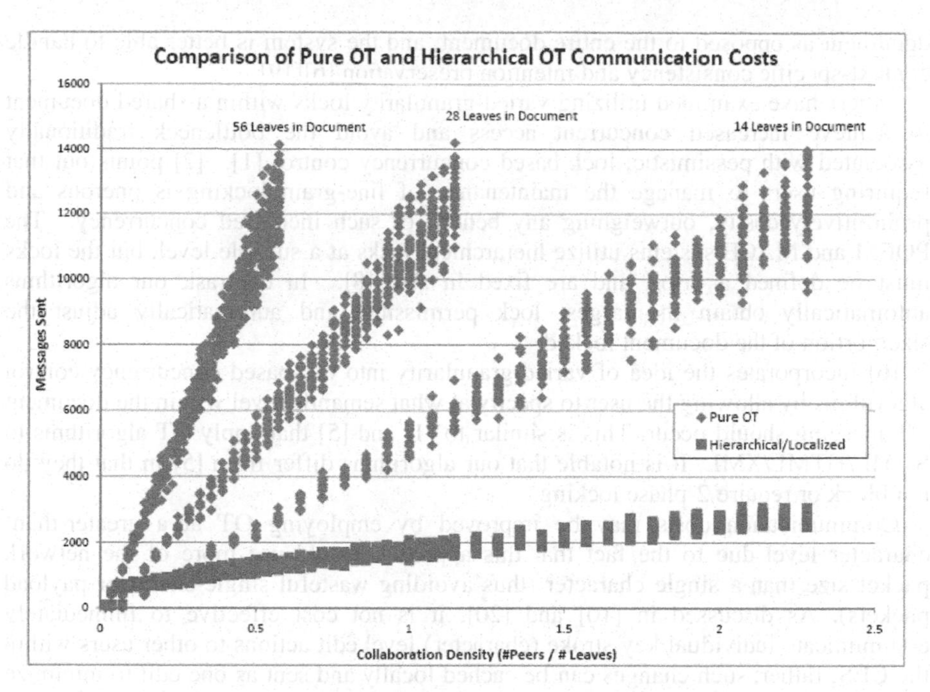

Fig. 4. Pure OT vs. Hierarchical OT Communication Costs

intuitive in that if we increase the document size while retaining the same number of peers, then the opportunity for caching increases under our distributed document management model.

Figure 4 shows how our hierarchical distributed document management approach can reduce the communication costs when compared to a pure OT approach. The ability to cache changes locally and localize OT to a subset of users sharing the same space within the document dramatically decreases the communication costs of the RTCES. We note that as the collaboration density (the average number of peers per section of the document) increases, the communication also increases; this is as expected since more messages will be sent to maintain consistency when more than one peer shares a section of the document.

It is important to note that the size of the document does not affect the workload in managing the collaboration – only the collaboration density affects the workload; thus our algorithms scale to large documents well.

6 Related Work

Traditionally, research within CES has viewed documents to be a linear sequence of data; consequently, OT and other techniques to ensure the CCI model [18] are designed to work on linear content. More recently, others have proposed leveraging the semantic structure of the document and viewing it as a hierarchy [6][12]. Operations to ensure CCI are more efficient when applied to sections of a hierarchical document as opposed to the entire document, and the system is better able to handle context-specific consistency and intention preservation [6][19].

Others have examined utilizing varied-granularity locks within a shared document to achieve increased concurrent access and avoid the bottleneck traditionally associated with pessimistic, lock-based concurrency control [1]. [2] points out that requiring users to manage the maintenance of fine-grain locking is onerous and prohibitively costly, outweighing any benefit of such increased concurrency. The POEM and MACE systems utilize hierarchical locks at a sub-file level, but the locks must be defined a priori and are fixed in size [8]. In contrast, our algorithms automatically obtain the largest lock permissible and automatically adjust the size/portion of the document locked.

[6] incorporates the idea of varied-granularity into OT-based concurrency control algorithms by allowing the user to specify at what semantic level within the document OT merging should occur. This is similar to [1] and [5] that apply OT algorithms to SGML/HTML/XML. It is notable that our algorithms differ from [5] in that they do not block or require 2-phase locking.

Communication costs may be improved by employing OT at a greater-than-character level due to the fact that this approach would use more of the network packet size than a single character (thus avoiding wasteful single-character payload packets). As discussed in [10] and [20], it is not cost effective to immediately communicate individual key-stroke (character) level edit actions to other users within the CES; rather, such changes can be cached locally and sent as one edit to minimize the communication overhead of the network packets.

7 Conclusion

We have presented peer-to-peer algorithms that dynamically manage ownership of a distributed document among peers efficiently, minimizing computation and communication costs. Additionally, we have presented a distributed version of OT algorithms that reduce computation and communication costs as compared to existing multicast-based OT algorithms. Both of these techniques are complimentary and improve the field of CSCW and RTCES.

Our empirical simulation results demonstrate that the work of managing the document and the document tree can be distributed fairly among the peers within the RTCES, and the "most-recent" rotating management policy removes the bottleneck of any single peer in managing the topmost nodes of the tree. Additionally, we have shown that we can achieve fault tolerance by replicating the top portion of the tree among a set of peers and maintain consistency via existing OT algorithms.

References

[1] Davis, A.H., Sun, C., Lu, J.: Generalizing Operational Transformation to the Standard General Markup Language. In: Procs. CSCW 2002, New Orleans, November 16-20, pp. 58–67

[2] Edwards, W.K.: Flexible Conflict Detection and Management In Collaborative Applications. In: Procs. 10th ACM Symp. on User Interface Softw. and Tech (UIST 1997), Banff, Canada, pp. 14–17 (October 14-17, 1997)

[3] Gu, N., Yang, J., Zhang, Q.: Consistency Maintenance Based on the Mark and Retrace Technique in Groupware Systems. In: GROUP 2005, Sanibel Island, FL, pp. 264–273. ACM Press, New York (November 6-9, 2005)

[4] Handley, M., Crowcroft, J.: Network Text Editor (NTE): A scalable text editor for the MBone. In: Procs. ACM SIGCOMM 1997, Cannes, France, pp. 197–208 (August 1997)

[5] Helmer, S., Kanne, C-C., Moerkotte, G.: Evaluating Lock-based Protocols for Cooperation on XML Documents. In: SIGMOD Record (2004)

[6] Ignat, C-L., Norrie, M.C.: Flexible Merging of Hierarchical Documents. In: Procs of the Seventh Intl Workshop on Collaborative Editing, GROUP 2005, Sanibel Island, Florida (November 2005)

[7] Li, D., Zhou, L., Muntz, R.R.: A New Paradigm of User Intention Preservation in Realtime Coollaborative Editing Systems. In: Procs. of the Seventh Intl. Conf. on Parallel and Distributed Systems, Iwate, Japan, pp. 401–408 (2000)

[8] Li, R., Li, D.: A Landmark-Based Transformation Approach to Concurrency Control in Group Editors. In: GROUP 2005, Sanibel Island, FL, pp. 284–293. ACM Press, New York (November 6-9, 2005)

[9] Magnusson, B.: Fine-Grained Version Control in COOP/Orm, European Conference on Computer Supported Cooperative Work 1995. In: Workshop on Version Control in CSCW Applications, Stockholm (September 1995)

[10] Preston, J.A., Prasad, S.K.: A Deadlock-Free Multi-Granular, Hierarchical Locking Scheme for Real-time Collaborative Editing. In: 7th Intl. Workshop on Collaborative Editing Systems, Sanibel Island, FL (2005)

[11] Preston, J.A., Prasad, S.K.: Achieving CCI Efficiently by Combining OT and Dynamic Locking with Lazy Consistency in a Peer-to-Peer CES. In: 8th Intl. Workshop on Collaborative Editing Systems, Banff, Canada (2006)

[12] Preston, J.A., Prasad, S.K.: An Efficient Synchronous Collaborative Editing System Employing Dynamic Locking of Varying Granularity in Generalized Document Trees. In: Procs. 2nd Intl. Conf. on Collaborative Computing: Networking, Appln. and Worksharing, Atlanta (November 2006)

[13] Preston, J.A, Xiaolin, H., Prasad, S.K.: Simulation-based Architectural Design and Implementation of a Real-time Collaborative Editing System. In: Procs. 2007 DEVS Integrative Modeling and Simulation Symposium, Norfolk, VA (2007)

[14] Qin, X.: Delayed Consistency Model for Distributed Interactive Systems with Real-time Continuous Media. Journal of Software, China 13(6), 1029–1039 (2002)

[15] Qin, X., Sun, C.: Recovery Support for Internet-based Real-Time Collaborative Editing Systems. In: Proc. Intl. Conf. on Computer Networks and Mobile Computing (October 2001)

[16] Rao, V.N, Kumar, V.: Concurrent Access of Priority Queues. IEEE Trans. on Comput. 37(12), 1657–1665 (1988)

[17] Sun, C., Jia, X., Zhang, Y., Yang, Y.: A Generic Operational Transformation Scheme for Consistency Maintenance in Real-time Cooperative Editing Systems. In: Procs. of Intl. ACM SIGGROUP Conf. on Supporting Group Work, Phoenix, pp. 425–434 (November 1997)

[18] Sun, C., Jai, X., Zhang, Y., Yang, Y., Chen, D.: Achieving convergence, causality-preservation, and intention-preservation in real-time cooperative editing systems. ACM Trans. on Computer-human Interaction 5(1), 63–108 (1998)

[19] Sun, C., Sosič, R.: Optional locking integrated with operational transformation in distributed real-time group editors. In: Proc. of The 18th ACM Symp. on Principles of Distributed Computing, pp.43-52, Atlanta (May 4-6, 1999)

[20] Yang, Y., Sun, C., Zhang, Y., Jia, X.: Real-Time Cooperative Editing on the Internet. In: IEEE Internet Computing, pp. 18–25 (May/June 2000)

Structuring Unstructured Peer-to-Peer Networks

Stefan Schmid and Roger Wattenhofer

Computer Engineering and Networks Laboratory
ETH Zurich
8092 Zurich, Switzerland

Abstract. Flooding is a fundamental building block of unstructured peer-to-peer (P2P) systems. In this paper, we investigate techniques to improve the performance of flooding. In particular, we present *Clustella*, a novel semi-structured P2P architecture with bounded peer degree. Clustella decomposes the network into different clusters, allowing peers to quickly find those neighbors which contribute much to their routing efficiency. By its link selection strategy, Clustella achieves a good performance in static and dynamic environments.

1 Introduction

While *distributed hash tables* (DHT) are well-studied in the research community, most peer-to-peer (P2P) systems in today's Internet are still *unstructured*. Unstructured architectures are attractive because of their simplicity and their high robustness.

In unstructured systems, there is neither a centralized directory nor any control over the network topology or resource placement. When a new peer joins the P2P network, it forms connections with other peers freely, e.g., it selects *arbitrary* peers as neighbors. In order to publish its resources, a peer usually just stores them locally or places them on randomly chosen peers. Generally, unstructured overlays have loose guarantees for resource discovery, and it is possible that a file is not found although it exists in the network.

It is often believed that—due to the absence of topological constraints—such unstructured systems have a better performance and require less maintenance overhead in highly dynamic environments where peers join and leave frequently and concurrently. Usually, these systems also support richer queries than just search by identifier, for example keyword searches with regular expressions, range queries, etc.

The main Achilles heel of unstructured P2P systems are the underdeveloped routing mechanisms. Basically, there are two fundamental routing operations: *flooding* and *random walks*. In flooding, a search packet with a limited *time-to-live* (TTL)—maximally 10 hops in Gnutella for example—is repeatedly forwarded to all neighbors, while in random walks, a packet is only forwarded to one randomly chosen neighboring peer. While a random walk is usually the less

S. Aluru et al. (Eds.): HiPC 2007, LNCS 4873, pp. 432–442, 2007.

costly alternative in terms of the number of messages sent per query, the flooding approach is more robust and has better response times. This paper focuses on the flooding mechanism. However, we believe that our techniques are also useful in systems based on random walks.

The major concern about flooding is the total number of messages caused per query. More severely, in practice many of these messages are of no use and unnecessarily increase the load on the system. The main reason are redundant retransmissions: If a peer's neighbors are likely to be neighbors as well, the peer receives the same packet multiple times.

In this paper, we introduce a measure to evaluate the efficiency of flooding on a given topology. We believe that this criterion captures the essence of flooding well, and also engenders many interesting theoretical questions. We then identify means to structure the topology in order to improve the quality of floods with respect to this criterion. In particular, we present *Clustella*, a novel semi-structured P2P network.

Clustella is a fully decentralized (local) system with undirected connections only and a limited peer degree. Beacons are used to decompose the network into clusters. Peers can orient themselves using the beacon information, and quickly find other peers which—if a link to them is established—significantly increase the number of peers covered by floods. If the found peer already has full degree, local link rotations are applied and—also in this case—the connection request can be satisfied quickly with a good neighbor. Small cycles in the topology—and hence redundant messages—are avoided. Moreover, Clustella is self-stabilizing and maintains its routing performance also in dynamic environments.

The rest of this paper is organized as follows. After reviewing related work in Section 2, the model in general and the flood coverage criterion in particular are introduced in Section 3. The Clustella architecture is described in Section 4. Simulation results are presented in Section 5. After briefly discussing possible extensions in Section 6, the paper is concluded in Section 7.

2 Related Work

Gnutella [17] is probably the most prominent unstructured P2P network. However, albeit its success, there have been concerns about its scalability from the beginning [19,20]. Indeed, when Napster was unplugged in 2001, Gnutella broke down soon afterwards due to the inrush of former Napster users. A lot of interesting solutions have been proposed since then.

In spite of the large literature about structured *distributed hash tables* [18,21,23,26] (some authors have even proposed to implement unstructured data placement schemes on top of structured systems [2,3]), many researchers have aimed at improving the performance of unstructured systems, acknowledging their simplicity and their predominance in today's Internet. One active thread of research concerns content movement or *replication* [5,6,16,27]. In particular, Cohen and Shenker [5] have shown that search is most efficient if the number of replicas is proportional to the square root of the object's popularity.

Another fruitful field—also for structured networks—is *interest-based locality* [11, 12, 22]. The idea is that instead of (or additionally to) random neighbor connections, peers should establish connections to peers with similar interests. It has been shown that thereby queries can be satisfied with a much smaller flooding radius.

Many recent solutions for unstructured systems use alternatives for the flooding operations, for example random walks (e.g. [4]). But there have also been proposals to improve flooding itself. In [15, 25], the flooding is executed in several successive rounds with increasing TTL, until enough responses are received. While this solution can effectively reduce the message complexity for finding popular files, the response times are worse.

The work which is the closest related to ours is by Jiang et al. [9]. Their scheme aims at minimizing the number of redundant messages by constructing a tree-like sub-overlay on which the packets are propagated. However, in contrast to our work, many connections have to be maintained which are of no use for flooding. This also implies that—compared to the total number of connections in the system—the amount of peers covered by a flooding is low. More severely, the tree-like sub-overlay may result in disconnected components, especially in dynamic environments, reducing the efficiency further. In contrast, in our system, *all* links can be used for flooding, since they have actively been selected in consideration of their quality. Hence, while redundant messages are also rare, the flood coverage is much larger.

Note that our paper is related to literature on *virtual coordinate systems* [7,8]. These systems typically assign coordinates to the different peers in order to estimate distances (in terms of *latency*, rather than number of *hops*) between a node and its (potential) neighbors. However, in this paper, we do not make use of these techniques.

Finally, the idea of structuring unstructured systems is also used by researchers in order to avoid a mismatch of the overlay with the underlying, real network [13].

3 Model

We model the P2P network as an *undirected* graph $G = (V, E)$, where V is the set of peers and E the set of connections between the peers. That is, for $u, v \in V$, $\{u, v\} \in E$ denotes that peers u and v know the IP addresses of each other. The *r-neighborhood* $\Gamma^r(v)$ of a peer $v \in V$ is defined as the set of peers which are at most r hops away from peer $v \in V$, excluding v itself. For v's direct neighbors, we use the short form $\Gamma(v)$ instead of $\Gamma^1(v)$. Moreover, let R be the TTL or flooding radius of the system (e.g., $R \leq 10$ in Gnutella). As will be discussed in Section 4, and unlike some other unstructured systems, in Clustella, every peer has at least δ but no more than Δ neighbors, i.e., for all $v \in V$, $\delta \leq |\Gamma(v)| \leq \Delta$.

In this paper, a pure flooding algorithm is considered where each peer forwards a packet to all its neighbors as long as the packet has a non-zero TTL. In order to maximize the probability of finding a data item or file, a flooding operation should reach—for a given radius or TTL—as many peers as possible. Therefore, $|\Gamma^R(v)|$—the size of the R-neighborhood of a peer v—is a natural criterion to quantify the efficiency of a flooding operation. In the following, we will refer to $|\Gamma^R(v)|$ as v's *flood coverage*. The *flood coverage of a network network* $G = (V, E)$ is defined as the minimal flood coverage of all peers in the network (cf. Definition 3.1).

Definition 3.1. *The* flood coverage $\Xi(G)$ *of a topology* G *for a given flooding radius* R *is defined as*

$$\Xi(G) = \min_{v \in V} |\Gamma^R(v)|.$$

Of course, not every network topology is equally suited for flooding. If a peer has neighbors which are also neighboring, many redundant messages are sent which do not increase the propagation scope. In an optimal topology G, the number of peers reached grows exponentially per hop, and if all peers have degree Δ, it holds that $\Xi(G) = \min\left\{\sum_{i=1}^{i=R} \Delta(\Delta-1)^{i-1}, |V|\right\}$.

Observe that our definition of a network's flood coverage is somehow related to the important criterion of *graph expansion* [24]. However, there are two crucial differences: First, we are not concerned with the expansion of all *subsets* of peers, but of single peers only (subsets of size one). And second, for these peers the entire R-neighborhood is considered (instead of just their immediate neighbors).

An ideal topology achieving maximal flood coverage must have a large *girth*, i.e., large minimal cycles. While finding such graphs is an interesting research area on its own (cf. [10] for an explicit construction), we do not follow these theoretical considerations further but go on and describe our semi-structured P2P system Clustella which strives—in a decentralized manner—for creating topologies with large flood coverage.

4 Clustella

In this section, we introduce the basic techniques used in Clustella for creating topologies with large flood coverage. As described in Section 3, a peer should connect to peers of different areas of the network, such that the shortest path between two neighbors $\pi', \pi'' \in \Gamma(\pi)$ of a peer π—*except* for the one via π itself—is long.

However, in unstructured P2P systems, if a peer π learns about another peer π', π has a priori no information about π''s location in the network, and thus also not about the hop distance between π and π'. Therefore, π can not decide whether it is useful to connect to π', or whether its flood coverage using the existing neighbors is better.

4.1 Clustering Topology and Beacons

A natural way to get rough topological location information is to decompose the network into *clusters* or zones. The idea is that if the neighbors are chosen from different clusters, then they are likely to be distant from each other. In order to be applicable in realistic and dynamic environments, our system must fulfill three properties: (1) The clustering mechanism should not require the peers to perform global operations or gather large amounts of additional topological information. (2) The clustering should be established quickly and in a decentralized manner. (3) The solution should be fair in the sense that all peers incur more or less the same amount of work.

Such a clustering can be achieved by having some peers assuming the role of *beacons*. Concretely, in Clustella, if a peer has no beacon in its R_d-neighborhood, it considers itself a beacon. Furthermore, a protocol ensures that each peer knows the beacons in its R_b-neighborhood. Both R_d and R_b are system parameters which are explained later in more detail. For now, assume that $R_d < R_b$ and $R_b \approx R$. In order to find out whether it is worth establishing a connection to each other, two peers π and π' exchange the information about their beacons. If their neighborhoods look very different, the distance between the peers must be large and thus the connection $\{\pi, \pi'\}$ of good quality.

In Clustella, a beacon peer appends its identifier (IP address) to the packets passing through it, together with a TTL initialized to R_b. The other peers then forward this packet as usual, decrementing the TTL in every step. Since the packet may have travelled several hops before a beacon appends its identifier, and since the beacon parameter R_b can be larger than the packet flooding radius, the packet's TTL can expire before the beacon information is propagated R_b hops far. Therefore, if a peer receives a packet with TTL=0, it buffers the beacon identifiers with non-zero hop count, and piggybacks them on future packets. The propagation of the beacon information becomes independent of the packets' TTLs, and beacons are indeed known in their R_b-neighborhoods. Moreover, note that due to the piggybacking, Clustella itself does not require the transmission of any messages at all.[1] This solution also fulfills the fairness property (Property (3)), as the work is equally divided between both beacon and non-beacon peers. The basic ideas of the clustering topology are illustrated in Figure 1.

4.2 Neighbor Selection

In this section, we first present Clustella's neighbor selection strategy. Afterwards, the issue of replacing existing connections by better ones is addressed.

New Neighbors. Consider a peer π which—for example during the joining process—is given a set of candidate peers from which it has to choose the best neighbor. For each such candidate π', π knows—due to a preceding information exchange—the identifer of π''s closest beacon, and which beacons it has in common with π'. If π itself does not see the closest beacon of π', the hop distance between π and π' must be larger than $R_b - R_d$. In addition to this deterministic guarantee, the amount of beacons which are in *both* π's and π''s R_b-neighborhood

[1] Of course, however, the *size* of the messages becomes larger as they include beacon information.

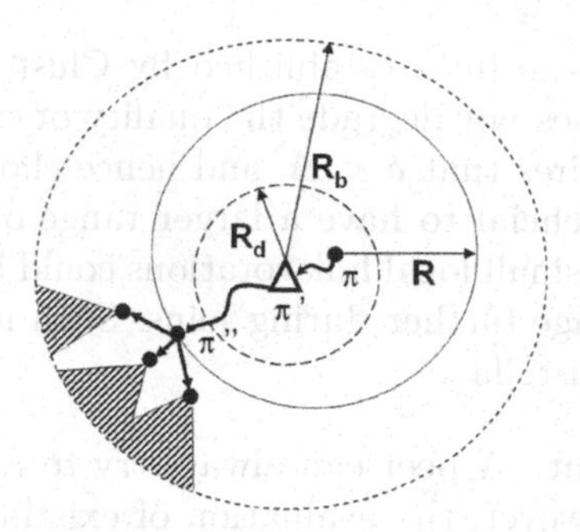

Fig. 1. When the query flooded by peer π arrives at the beacon π', π' appends its identifier. The packet is forwarded until its TTL becomes zero at peer π''. Peer π'' extracts the beacon's identifier and piggybacks it to other packets. Hence, the information about π' is propagated in the entire R_b-neighborhood of π'

correlates roughly with the length of the shortest path between the two peers. In Clustella, the following algorithm is applied to select the best candidate: From all candidates of which π does not see the closest beacon (if possible), π chooses the one with which it has the least beacons in common.[2]

In our system, connections are undirected, and there is an upper bound Δ on the amount of links a peer can have. This poses an interesting problem: What happens if a peer wants to connect to another peer which already has Δ neighbors? To find alternative candidates can be time-consuming, or even impossible if all existing peers have full degree. In the following, we describe Clustella's joining procedure which—in spite of this problem—quickly establishes new connections of good quality (i.e., between formerly distant peers).

First recall that a peer is allowed to have between δ and Δ neighbors. When a peer joins the Clustella network (or if some of its neighbors have crashed), it only looks for new neighbors as long as its degree is smaller than δ. However, a peer π always accepts a connection request from another peer π' if it has less than Δ neighbors. If π already has Δ neighbors when it receives the connection request from π', it checks whether there exists a neighbor $\pi'' \in \Gamma(\pi)$ with degree smaller than Δ. If this is the case, the connection $\{\pi', \pi''\}$ is built. Since π'' is adjacent to π, the quality of the link $\{\pi', \pi''\}$ is similar to the one of $\{\pi', \pi\}$. If on the other hand the degrees of all neighbors are also Δ, π drops a connection with an arbitrary existing neighbor π'' and accepts π''s joining request. While this would already be a good solution, Clustella exploits the situation further and additionally establishes a connection between π' and π''. Note that since π' had a degree of at most $\delta - 1 < \Delta - 2$ before issuing the join request, this operation is legal. More importantly, the link $\{\pi', \pi''\}$ must be of good quality (similar to the one of $\{\pi, \pi''\}$ before it was broken). Finally, this trick also speeds up the joining process. We have the following result.

Theorem 4.1 *In Clustella, by choosing $\delta < \Delta$, a connection request at a peer π can always be satisfied by π itself or by a neighboring peer $\pi' \in \Gamma(\pi)$.*

[2] Note that the distances to the beacons are not considered.

Also observe that since the links established by Clustella are between distant peers only, a new link does not degrade the quality of existing links.

Our system only requires that $\delta < \Delta$, and hence choosing $\delta = \Delta - 1$ is fine. However, it may be beneficial to have a larger range of allowed degrees. With this increased flexibility, small local link rotations could be performed in order to increase the flood coverage further during joins. Such mechanisms are planned for future versions of Clustella.

Neighbor Replacement. A peer can always try to replace its current neighbors by better ones. However, the evaluation of existing neighbors of a peer π poses a problem: Since they are already adjacent to π, they must as well have almost the same set of beacons in their neighborhood.

One simple solution would be to ignore this issue and just change the neighbors once in a while. As a slight improvement, the neighbors could be assigned a score depending on the beacons they had *before* the link has been established; then, depending on these scores, sporadically the worst neighbor is replaced.

In Clustella, a more sophisticated approach is used: Each beacon information record also contains its flooding path, i.e., a peer adds its identifier (at most R_b entries) before forwarding the packet. Thereby, the peers are able to compute via which neighbors they know about a given beacon. In order to evaluate an existing neighbor π', a peer π asks for π''s beacons without the ones known through the link $\{\pi, \pi'\}$. If π' and π still have one or more beacons in common, then the connection has to be replaced.

4.3 Dynamic Failures

Most P2P systems currently in use are very transient and have a high peer turnover rate. Therefore, it is important that Clustella can efficiently handle peers which leave or crash. The creation of beacons—and the propagation of the corresponding information—is a fully decentralized process and therefore our system adapts quickly to changing environments on its own.

However, it is beneficial to employ additional mechanisms to cope with churn. If a neighbor of a peer crashes, a good substitute has to be found—a costly operation if done from scratch. Therefore, in Clustella, a peer stores for each neighbor π some of π's neighbors. When π crashes, a connection to one of π's former neighbors can be established immediately. The quality of this alternative connection is similar to the one of the old connection.

5 Simulation

We have analyzed the flood coverage of the topologies created by Clustella by simulation for up to 1 million peers. Our tests mainly focused on the parameter space $\Delta \in \{4, ..., 7\}$ (each with $\delta = \Delta - 1$) and $R \in \{5, ..., 10\}$.

Although choosing R_d smaller than R_b gives a deterministic guarantee for the minimal girth, this feature can not be exploited fully since it is very expensive:

The amount of beacon information per peer grows quickly as longer shortest paths are enforced this way. However, such a hard guarantee seems not to be necessary, as already $R_d = R_b - 2$ yields very good results: Since $R_d < R_b$, a peer has enough beacons in its neighborhood for orientation, but also not too many even in case of a large beacon radius R_b. Generally, R_b should roughly equal—or be slightly larger than—R, i.e., $R_b \approx R$.

In our simulations, the following neighbor discovery algorithm has been used: In order to find an additional neighbor, a peer π in the Clustella network sends an exploration packet of only a small constant TTL (e.g. 10 hops). This packet contains information about the beacons in the neighborhood of π, plus a section where already visited peers on the path can be stored. A peer π' which receives this packet forwards it to its neighbor π'' which—without the connection $\{\pi', \pi''\}$—shares the least beacons with π. Neighbors which are already contained in the path are avoided, and if several neighbors are equally well-suited, a random one is chosen.

We have compared Clustella to two other strategies: a *Gnutella-like strategy* and a *random walk strategy.* In the Gnutella-like strategy, in order to find new peers it can connect to, a peer asks its neighbors for their neighbors. This procedure is repeated recursively, until the peer reaches its desired degree. In the random walk strategy, a peer sends a discovery packet with the same TTL as Clustella. However, unlike in Clustella, this packet is always forwarded to a random neighbor, and does not benefit from the beacon information for orientation.

In all our simulations, the performance of the Gnutella-like strategy was of course poor: The flood coverage was up to one hundred times worse than the coverage of the other two strategies. The resulting topologies were highly clustered, and the neighbors' neighbors were often adjacent. While the random walk strategy had a much better flood coverage than the Gnutella-like strategy, it was outperformed by Clustella where peers quickly reached much more distant neighbors.

Finally, since existing P2P systems often use longer random walks in order to find new neighbors, we have also studied a different scenario. Thereby, both Clustella and the random walk strategy chose neighbor candidates uniformly and at random *from the entire network.* Such a *uniform sampling* can be achieved by sufficiently long random walks. Clustella outperformed this random graph, albeit only by up to slightly more than 10%. (Of course, for very large graphs where only a small fraction of peers can be reached by a flooding, the difference of the coverage of the two graphs diminishes. Similarly, for very small graphs where all peers can be reached, the flood coverage is the same.) However, we believe that this uniform sampling scenario is not realistic in practice, as the *mixing times* [14]—and thus the length—of the random walks are large and also difficult—or even impossible in dynamic environments!—to compute. (Note that this computation requires a good estimation of the total number of peers in the system.) Furthermore, the probability that an exploration packet is lost is high for long walks, and it is necessary to send several redundant packets in parallel.

Therefore, Clustella does not apply this neighbor discovery strategy but only sends packets with small TTLs, as described above.

In conclusion, our first *in vitro* evaluation results are promising both with respect to the search efficiency and with respect to the messages' sizes. Of course, we are aware that many issues such as the dynamics of the system remain to be analyzed in detail in future work. Moreover, there is a wide variety of other alternative approaches to which have not compared Clustella. Although we plan to perform such comparisons, our focus here is rather on the introduction of novel ideas to improve flooding than on proposing a complete and ready-to-use system; in fact, Clustella can be enhanced by adding many existing heuristics, for example by introducing some form of replication, or by the extensions discussed in the next section.

6 Extensions

The basic system as described in Section 4 can be extended in several ways. In this section, we briefly discuss two possible enhancements.

The first enhancement concerns the clustering. So far, there is only one level of beacons. It might be beneficial to organize the beacons in a *hierarchy* of several levels with increasing radius of responsibility. If a peer looks for a distant peer to connect to, it can choose the candidate with which it has only high-level beacons in common. A small beacon hierarchy (three or four levels) might already do the job: Since floods have a small constant radius, the optimal flood coverage can also be achieved with peers which are not very far away.

There are several challenges. In particular, the hierarchy must be easily maintainable when peers (and thus also beacon peers) join and leave. Moreover, it should be no disadvantage to be a high-level beacon (fairness property), and information about all beacons must be propagated efficiently.

The second enhancement concerns the size of the transmitted messages. Besides general compression mechanisms, the use of *Bloom filters* [1] is appealing: Sending only the Bloom array instead of the entire beacon identifiers can reduce the burden on the system's resources (memory and bandwidth) while still yielding acceptable probabilistic guarantees.

7 Conclusion

This paper has embarked on identifying techniques to make flooding on unstructured P2P topologies much more efficient. Unlike other systems which only combat the symptoms, we strive for avoiding bad connections from the beginning. Moreover, in contrast to many structured P2P systems, Clustella can be started from arbitrary network topologies, i.e., from any connected graph, and develops towards better network structures in a *self-stabilizing manner*. Our first evaluations are promising. Moreover, many heuristics such as smart data replication are orthogonal to our approach and could be integrated to further improve search performance.

An interesting feature of our techniques is that they can *co-exist* in networks with normal clients (e.g., Gnutella clients); it is not necessary for the existing clients to know about our new clients. What is more, the entire network will benefit from our neighbor selection of—possibly a small number of—new clients, as many existing clients will experience a larger fan-out as well.

We plan to investigate efficient flooding topologies further, addressing for instance Clustella's dynamics: Does the system require measures in order to stabilize quickly, and if yes, which mechanisms are best? The ultimate goal is to have a running Clustella client which collaborates seamlessly with other unstructured P2P clients. Finally, we believe that our clustering approach may be interesting in other areas of distributed computing as well.

Acknowledgments

The authors would like to thank the participants of the DCG Mediterranean Seminar in Toscany, Italy for fruitful discussions. This research has been supported in part by the Swiss National Science Foundation (SNF).

References

1. Bloom, B.H.: Space/Time Trade-offs in Hash Coding with Allowable Errors. Commun. ACM 13(7), 422–426 (1970)
2. Castro, M., Costa, M., Rowstron, A.: Should We Build Gnutella on a Structured Overlay? In: Proc. 2nd Workshop on Hot Topics in Networks (HotNets) (2003)
3. Castro, M., Costa, M., Rowstron, A.: Peer-to-Peer Overlays: Structured, Unstructured, or Both? Technical Report MSR-TR-2004-73, Microsoft Research, Cambridge, UK (2004)
4. Chawathe, Y., Ratnasamy, S., Breslau, L., Lanham, N., Shenker, S.: Making Gnutella-like P2P Systems Scalable. In: Proc. Conf. on Applications, Technologies, Architectures, and Protocols for Computer Communications (SIGCOMM) (2003)
5. Cohen, E., Shenker, S.: Replication Strategies in Unstructured Peer-to-Peer Networks. In: Proc. ACM SIGCOMM Conference (2002)
6. Cooper, B.F.: Quickly Routing Searches Without Having to Move Content. In: Castro, M., van Renesse, R. (eds.) IPTPS 2005. LNCS, vol. 3640, pp. 163–172. Springer, Heidelberg (2005)
7. Dabek, F., Cox, R., Kaashoek, F., Morris, R.: Vivaldi: A Decentralized Network Coordinate System. In: Proc. Conf. on Applications, Technologies, Architectures, and Protocols for Computer Communications (SIGCOMM), pp. 15–26 (2004)
8. Francis, P., Jamin, S., Jin, C., Jin, Y., Raz, D., Shavitt, Y., Zhang, L.: IDMaps: A Global Internet Host Distance Estimation Service. IEEE/ACM Trans. Netw. 9(5), 525–540 (2001)
9. Jiang, S., Guo, L., Zhang, X.: LightFlood: An Efficient Flooding Scheme for File Search in Unstructured Peer-to-Peer Systems. In: Proc. Itl. Conf. on Parallel Processing (ICPP) (2003)
10. Lazebnik, F., Ustimenko, V.A.: Explicit Construction of Graphs with Arbitrary Large Girth and of Large Size. Discrete Applied Math. 60, 275–284 (1997)

11. Le Blond, S., Guillaume, J.-L., Latapy, M.: Clustering in P2P Exchanges and Consequences on Performances. In: Castro, M., van Renesse, R. (eds.) IPTPS 2005. LNCS, vol. 3640, Springer, Heidelberg (2005)
12. Le Fessant, F., Handurukande, S., Kermarrec, A.-M., Massoulié., L.: Clustering in Peer-to-Peer File Sharing Workloads. In: Voelker, G.M., Shenker, S. (eds.) IPTPS 2004. LNCS, vol. 3279, Springer, Heidelberg (2005)
13. Liu, X., Xiao, L., Liu, Y., Ni, L.M., Zhang, X.: Location Awareness in Unstructured Peer-to-Peer Systems. IEEE Trans. Parallel Distrib. Syst. 16(2), 163–174 (2005)
14. Lovász, L.: Random Walks on Graphs: A Survey. Combinatorics 2 (1993)
15. Lv, Q., Cao, P., Cohen, E., Li, K., Shenker, S.: Search and Replication in Unstructured Peer-to-Peer Networks. In: Proc. 16th ACM Itl. Conf. on Supercomputing (ICS) (2002)
16. Morselli, R., Bhattacharjee, B., Srinivasan, A., Marsh, M.A.: Efficient Lookup on Unstructured Topologies. In: Proc. 24th Annual Symposium on Principles of Distributed Computing (PODC), pp. 77–86 (2005)
17. Open Source Community. Gnutella (2001), `http://gnutella.wego.com/`
18. Ratnasamy, S., Francis, P., Handley, M., Karp, R., Shenker, S.: A Scalable Content Addressable Network. In: Proc. of ACM SIGCOMM 2001 (2001)
19. Ripeanu, M., Foster, I.: Mapping Gnutella Networks. IEEE Internet Computing , 50–57 (2002)
20. Ritter, J.: Why Gnutella Can't Scale. No, Really (2001)
21. Rowstron, A., Druschel, P.: Pastry: Scalable, Decentralized Object Location and Routing for Large-Scale Peer-to-Peer Systems. In: Proc. 18th IFIP/ACM Int. Conference on Distributed Systems Platforms (Middleware), pp. 329–350 (2001)
22. Sripanidkulchai, K., Maggs, B., Zhang, H.: Efficient Content Location Using Interest-Based Locality in Peer-to-Peer Systems. In: Proc. 22nd IEEE Conf. on Computer Communications (INFOCOM) (2003)
23. Stoica, I., Morris, R., Karger, D., Kaashoek, M.F., Balakrishnan, H.: Chord: A Scalable Peer-to-peer Lookup Service for Internet Applications. In: Proc., A.S. (ed.) Proc. ACM SIGCOMM Conference (2001)
24. Walters, I.: The Ever Expanding Expander Coefficients. Bull. Inst. Combin. Appl. 97 (1996)
25. Yang, B., Garcia-Molina, H.: Improving Search in Peer-to-Peer Systems. In: Proc. 22nd Itl. Conf. on Distributed Computing Systems (ICDCS) (2002)
26. Zhao, B.Y., Huang, L., Stribling, J., Joseph, A.D., Kubiatowicz, J.D.: Tapestry: A Resilient Global-scale Overlay for Service Deployment. IEEE Journal on Selected Areas in Communications 22(1) (2004)
27. Zhong, M., Shen, K.: Popularity-Biased Random Walks for Peer-to-Peer Search under the Square-Root Principle. In: Proc. 5th Intl. Workshop on Peer-to-Peer Systems (IPTPS) (2006)

Multi-objective Peer-to-Peer Neighbor-Selection Strategy Using Genetic Algorithm

Ajith Abraham[1,3], Benxian Yue[2], Chenjing Xian[3], Hongbo Liu[2,3], and Millie Pant[4]

[1] Centre for Quantifiable Quality of Service in Communication Systems, Norwegian University of Science and Technology, N-7491 Trondheim, Norway
ajith.abraham@ieee.org

[2] Department of Computer, Dalian University of Technology, 116023 Dalian, China
yuebenxian@vip.sina.com, lhb@dlut.edu.cn

[3] School of Computer Science, Dalian Maritime University, 116026 Dalian, China
Xcj2003@newmail.dlmu.edu.cn

[4] Department of Paper Technology, Indian Institute of Technology - Roorkee, Saharanpur 247 001, India
millifpt@iitr.ernet.in

Abstract. Peer-to-peer (P2P) topology has significant influence on the performance, search efficiency and functionality, and scalability of the application. In this paper, we present a Genetic Agorithm (GA) approach to the problem of multi-objective Neighbor Selection (NS) in P2P Networks. The encoding representation is from the upper half of the peer-connection matrix through the undirected graph, which reduces the search space dimension. Experiment results indicate that GA usually could obtain better results than Particle Swarm Optimization (PSO).

1 Introduction

Peer-to-peer computing has attracted great interest and attention of the computing industry and gained popularity among computer users and their networked virtual communities [1]. It is no longer just used for sharing music files over the Internet. Many P2P systems have already been built for some new purposes and are being used. An increasing number of P2P systems are used in corporate networks or for public welfare (e.g. providing processing power to fight cancer) [2]. P2P comprises peers and the connections between these peers. These connections may be directed, may have different weights and are comparable to a graph with nodes and vertices connecting these nodes. Defining how these nodes are connected affects many properties of an architecture that is based on a P2P topology, which significantly influences the performance, search efficiency and functionality, and scalability of a system. A common difficulty in the current P2P systems is caused by the dynamic membership of peer hosts. This results in a constant reorganization of the topology [3], [4], [5], [6].

The simplest neighbor selection strategy would be to select a node at random from the candidate nodes. Kurmanowytsch et al. [7] developed the P2P

S. Aluru et al. (Eds.): HiPC 2007, LNCS 4873, pp. 443–451, 2007.

middleware systems to provide an abstraction between the P2P topology and the applications that are built on top of it. These middleware systems offer higher-level services such as distributed P2P search and support for direct communication among peers. The systems often provide a pre-defined topology that is suitable for a certain task (e.g., for exchanging files).

Koulouris et al. [8] presented a framework and an implementation technique for a flexible management of peer-to-peer overlays. The framework provides means for self-organization to yield an enhanced flexibility in instantiating control architectures in dynamic environments, which is regarded as being essential for P2P services to access, routing, topology forming, and application layer resource management. In these P2P applications, a central tracker decides about which peer becomes a neighbor to which other peers.

Koo et al. [9] investigated the neighbor-selection process in the P2P networks, and proposed an efficient single objective neighbor-selection strategy based on Genetic Algorithm (GA). Sun et al. [10] proposed a PSO algorithm for neighbor selection in P2P networks. In this paper, we explore the multi-objective neighbor-selection problem based on GA for P2P Networks.

This paper is organized as follows. We formulate the problem in Section 2. The proposed approach based on genetic algorithm is presented in Section 3. In Section 4, experiment results and discussions are provided in detail, followed by some conclusions in Section 5.

2 Neighbor-Selection Problem in P2P Networks

Kooa et al. modeled the neighborhood selection problem using an undirected graph and attempted to determine the connections between the peers [9], [11]. Given a fixed number of N peers, we use a graph $G = (V, E)$ to denote an overlay network, where the set of vertices $V = \{v_1, \cdots, v_N\}$ represents the N peers and the set of edges $E = \{e_{ij} \in \{0, 1\}, i, j = 1, \cdots, N\}$ represents their connectivities : $e_{ij} = 1$ if peers i and j are connected, and $e_{ij} = 0$ otherwise. For an undirected graph, it is required that $e_{ij} = e_{ji}$ for all $i \neq j$, and $e_{ij} = 0$ when $i = j$. Let C be the entire collection of content fragments, and we denote $\{c_i \subseteq C, i = 1, \cdots, N\}$ to be the collection of the content fragments each peer i has. We further assume that each peer i will be connected to a maximum of d_i neighbors, where $d_i < N$. The disjointness of contents from peer i to peer j is denoted by $c_i \setminus c_j$, which can be calculated as:

$$c_i \setminus c_j = c_i - (c_i \cap c_j). \tag{1}$$

where $\setminus$ denotes the exclusion operator, and $\cap$ intersection operation on sets. This disjointness can be interpreted as the collection of content fragments that peer i has but peer j does not. In other words, it denotes the fragments that peer i can upload to peer j. Moreover, the disjointness operation is not commutative, i.e., $c_i \setminus c_j \neq c_j \setminus c_i$. We also denote $|c_i \setminus c_j|$ to be the cardinality of $c_i \setminus c_j$, which is the number of content fragments peer i can contribute to peer j. In order to maximize the disjointness of content, we want to maximize the number of content fragments each peer can contribute to its neighbors by determining the connections e_{ij}'s. Define ϵ_{ij}'s to be sets such that $\epsilon_{ij} = C$ if $e_{ij} = 1$, and $\epsilon_{ij} = \emptyset$ (null set) otherwise. Therefore the neighbor selection can be formulated as the following optimization problem:

$$\max_E \sum_{j=1}^{N} \left| \bigcup_{i=1}^{N} (c_i \setminus c_j) \cap \epsilon_{ij} \right| \tag{2}$$

It is desirable to select peers with the most mutually disjoint collection of content fragments as neighbors. However, downloading the file fragments between each peer pair would consume away the bandwidth and connect cost, etc. τ_{ij} describes the cost coefficient between peer i and j. The performance of the whole system would be emphasized. The neighbor selection strategy is expected not only to assure maximum content availability but also to minimize the downloading cost to improve the overall throughput of the system. So the objectives are summarized as follows:

$$f_1(x) = \max_E \sum_{j=1}^{N} \left| \bigcup_{i=1}^{N} (c_i \setminus c_j) \cap \epsilon_{ij} \right| \tag{3}$$

$$f_2(x) = \min_E \sum_{j=1}^{N} \sum_{i=1}^{N} \tau_{ij} |(c_i \setminus c_j)| |\epsilon_{ij}| \tag{4}$$

Subject to

$$\sum_{j=1}^{N} e_{ij} \leq d_i \text{ for all } i \tag{5}$$

3 Genetic Algorithm for Multi-objective Neighbor Selection

Multi-objective genetic algoritm has been a very popular multiobjective technique, and it normally exhibits a very good overall performance. Many multiobjective optimization techniques using evolutionary algorithms have been proposed in recent years [12], [13], [14]. Given a P2P state $S = (N, C, M, F)$, in which N is the number of peers, C is the entire collection of content fragments, M is the maximum number of the peers which each peer can connect steadily in the session, F is to goal the number of swap fragments, i.e. to maximize equation (3) and minimize equation (4) with the constraint in equation (5). To apply the genetic algorithm successfully for the NS problem, one of the key issues is the mapping of the problem solution into the search space, which directly affects its feasibility and performance. The neighbor topology in P2P networks is an undirected graph, i.e. $e_{ij} = e_{ji}$ for all $i \neq j$. We set up a search space of D dimension as $N * (N-1)/2$. Accordingly, each individual is represented as a binary bit string of length D. Each dimension maps one undirected connection.

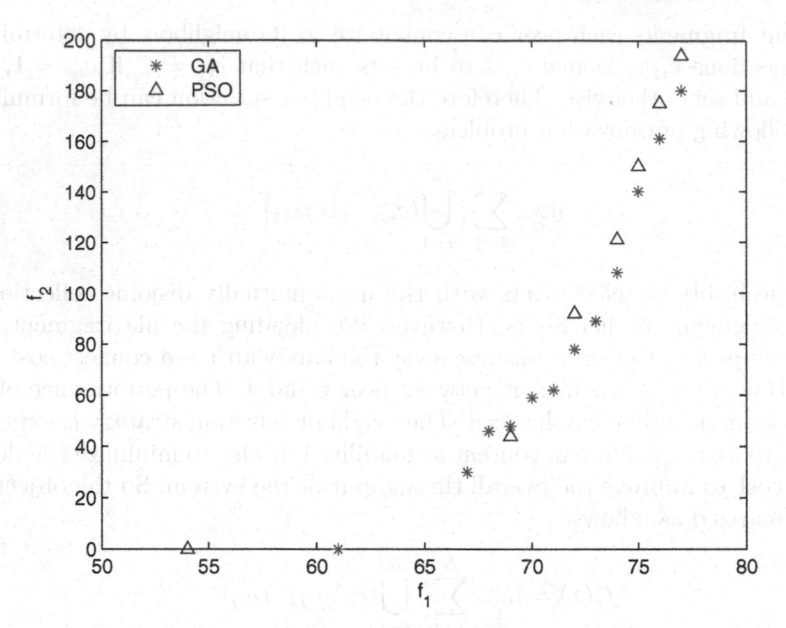

Fig. 1. Performance for the NS (6, 60, 3)

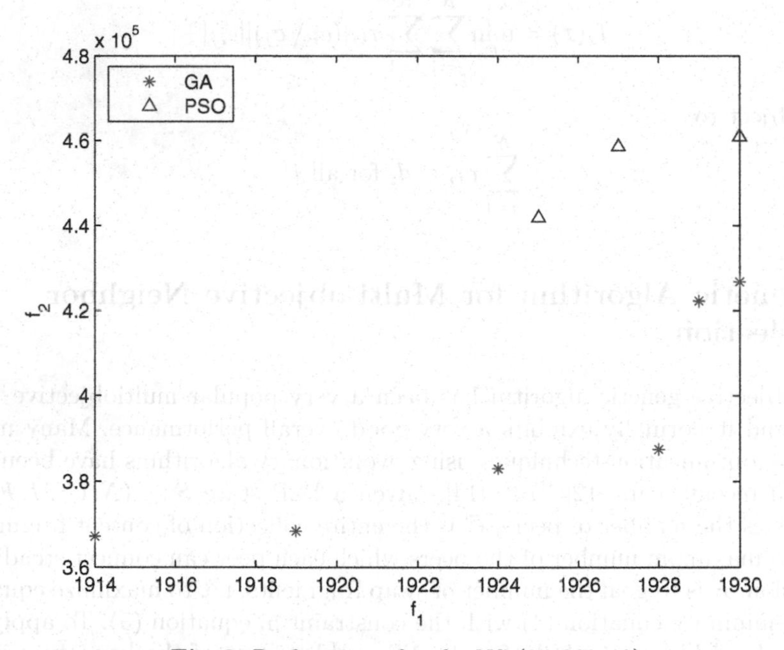

Fig. 2. Performance for the NS (25, 300, 12)

The domain for each dimension is limited to 0 or 1. The binary string has a priority levels according to the order of peers. The sequence of the peers will be not changed during the iteration. It indicates the potential connection state. The pseudo-code for our P2P neighbor selection method is illustrated in Algorithm 1.

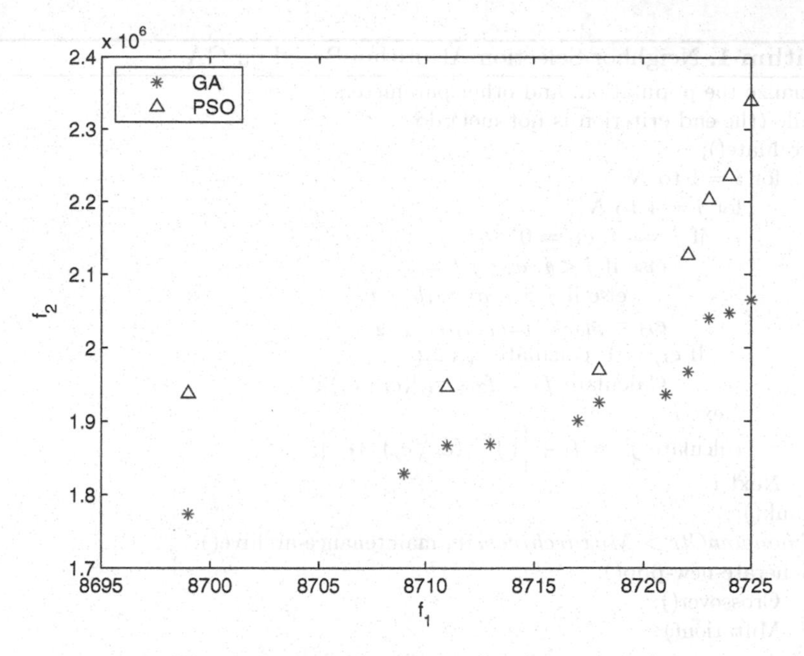

Fig. 3. Performance for the NS (25, 1400, 12)

4 Algorithm Performance Demonstration

To illustrate the effectiveness and performance of our algorithm, we demonstrate an execution trace of the algorithm for the NS problem. A file of size 7 MB is divided into 14 fragments (512 KB each) to distribute, 6 peers download from the P2P networks, and the connecting maximum number of each peer is 3, which is represented as $(6, 14, 3)$ problem. In some session, the state of distributed file fragments is as follows:

$$\begin{bmatrix} 1 & 0 & 0 & 4 & 0 & 6 & 7 & 8 & 0 & 10 & 0 & 12 & 0 & 14 \\ 0 & 0 & 0 & 4 & 5 & 0 & 7 & 0 & 9 & 0 & 11 & 0 & 13 & 0 \\ 0 & 2 & 0 & 0 & 0 & 6 & 0 & 0 & 0 & 0 & 11 & 12 & 0 & 14 \\ 0 & 2 & 3 & 4 & 0 & 6 & 0 & 0 & 0 & 0 & 11 & 0 & 0 & 0 \\ 0 & 2 & 0 & 0 & 0 & 0 & 7 & 8 & 0 & 10 & 0 & 12 & 0 & 14 \\ 1 & 2 & 0 & 0 & 5 & 0 & 0 & 0 & 9 & 10 & 11 & 0 & 13 & 14 \end{bmatrix}$$

The cost matrix is as follows:

$$\begin{bmatrix} 0 & 5 & 2 & 4 & 1 & 0 \\ 5 & 0 & 3 & 0 & 2 & 2 \\ 2 & 3 & 0 & 0 & 0 & 0 \\ 4 & 0 & 0 & 0 & 5 & 2 \\ 1 & 2 & 0 & 5 & 0 & 10 \\ 0 & 2 & 0 & 2 & 10 & 0 \end{bmatrix}$$

Algorithm 1. Neighbor Selection Algorithm Based on GA

01. Initialize the population, and other parameters.
02. While (the end criterion is not met) do
03. Evaluate();
04. for $i = 1$ to N
05. for $j = 1$ to N
06. if $j == i$, $e_{ij} = 0$;
07. else if $j < i$, $a = j; b = i$;
08. else if $j > i$, $a = i; b = j$;
09. $e_{ij} = p_{[a*N+b-(a+1)*(a+2)/2]}$;
10. If $e_{ij} = 1$, calculate $c_i \setminus c_j$;
11. Calculate $f_2 = f_2 + \tau_{ij}|(c_i \setminus c_j)|$;
12. Next j
13. calculate $f_1 = f_1 + \left| \bigcup_{i=1}^{N} (c_i \setminus c_j) \cap \epsilon_{ij} \right|$;
14. Next i
15. Rank();
16. If $nondomCtr > Maxarchivesize$, maintenance-archive();
17. Generate-new-pop();
18. Crossover();
19. Mutation();
20. $t++$;
21. If rank == 1 output the fitness;
22. End While.

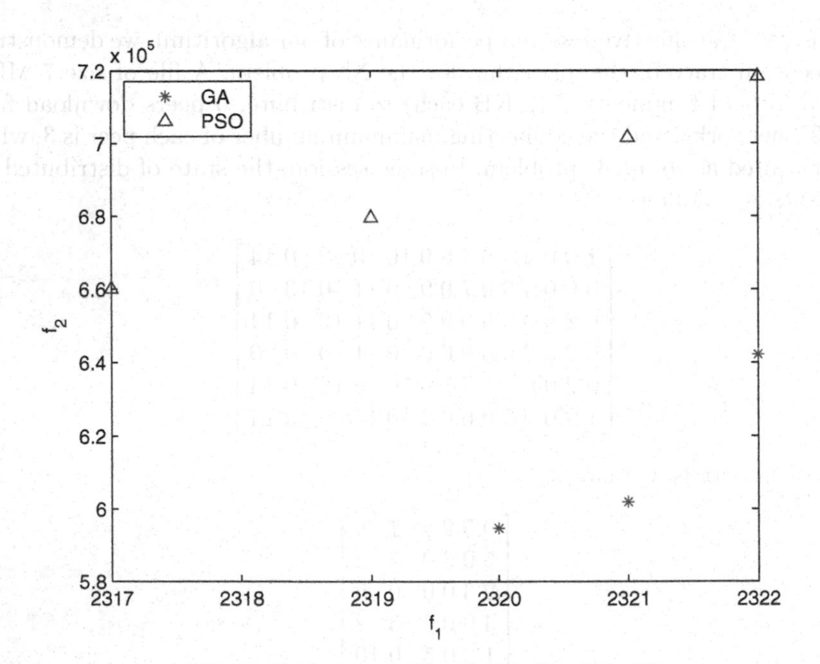

Fig. 4. Performance for the NS $(30, 300, 15)$

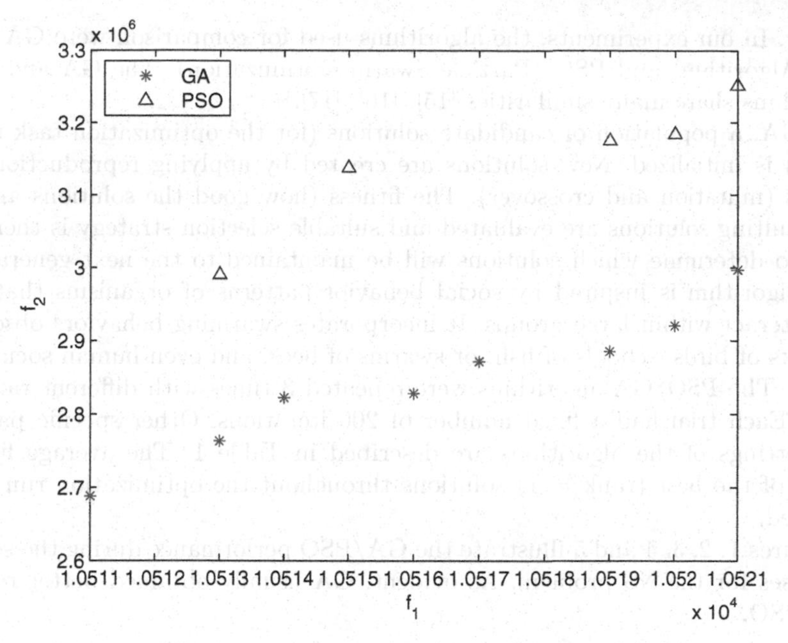

Fig. 5. Performance for the NS (30, 1400, 15)

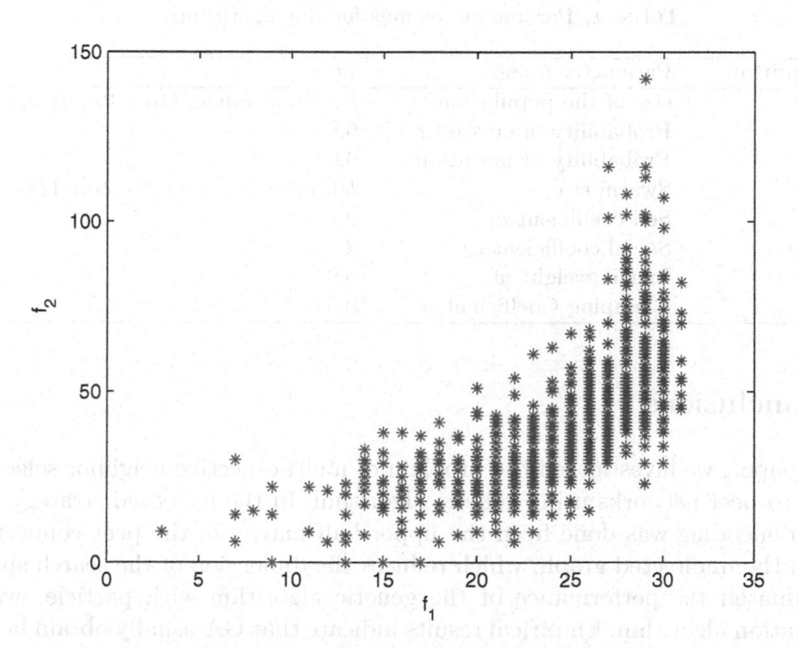

Fig. 6. Performance for the NS (6, 60, 3)

The performance output is illustrated in Figure 6 by the proposed algorithm. We also tested other five representative instances (problem (6,60,3), problem (25,300,12), problem (25,1400,12), problem (30,300,15), problem (30,1400,15)) further. In our experiments, the algorithms used for comparison were GA (Genetic Algorithm) and PSO (Particle Swarm Optimization). The GA and PSO algorithms share many similarities [15], [16], [17].

In GA, a population of candidate solutions (for the optimization task to be solved) is initialized. New solutions are created by applying reproduction operators (mutation and crossover). The fitness (how good the solutions are) of the resulting solutions are evaluated and suitable selection strategy is then applied to determine which solutions will be maintained to the next generation. PSO algorithm is inspired by social behavior patterns of organisms that live and interact within large groups. It incorporates swarming behaviors observed in flocks of birds, schools of fish, or swarms of bees, and even human social behavior. The PSO/GA algorithms were repeated 3 times with different random seeds. Each trial had a fixed number of 200 iterations. Other specific parameter settings of the algorithms are described in Table 1. The average fitness values of the best (rank = 1) solutions throughout the optimization run were recorded.

Figures 1, 2, 3, 4 and 5 illustrate the GA/PSO performance during the search processes for the NS problem. As evident, GA usually obtained better results than PSO.

Table 1. Parameter settings for the algorithms.

Algorithm	Parameter name	value
	size of the population	$left\ even\ number(10 + 2sqrt(D))$
GA	Probability of crossover	0.8
	Probability of mutation	0.08
	Swarm size	$left\ even\ number(10 + 2sqrt(D))$
	Self coefficient c_1	2
PSO	Social coefficient c_2	2
	Inertia weight w	0.9
	Clamping Coefficient ρ	0.5

5 Conclusions

In this paper, we investigated the problem of multi-objective neighbor selection in peer-to-peer networks using genetic algorithm. In the proposed strategy, the solution encoding was done from the upper half matrix of the peer connection through the undirected graph, which reduces the dimension of the search space. We evaluated the performance of the genetic algorithm with particle swarm optimization algorithm. Empirical results indicate that GA usually obtain better results than PSO. The proposed algorithm could be an ideal approach for solving the multi-objective NS problem.

Our future work is targeted to test more complicated instances in an online environment of P2P networks and involve more intelligent/heuristics approaches.

References

1. Kwok, S.: P2P searching trends: 2002-2004. Information Processing and Management 42, 237–247 (2006)
2. Idris, T., Altmann, J.: A Market-managed topology formation algorithm for peer-to-peer file sharing networks. In: Stiller, B., Reichl, P., Tuffin, B. (eds.) ICQT 2006. LNCS, vol. 4033, pp. 61–77. Springer, Heidelberg (2006)
3. Surana, S., Godfrey, B., Lakshminarayanan, K., Karp, R., Stoica, I.: Load balancing in dynamic structured peer-to-peer systems. Performance Evaluation 63, 217–240 (2006)
4. Duan, H., Lu, X., Tang, H., Zhou, X., Zhao, Z.: Proximity Neighbor Selection in Structured P2P Network. In: Proceedings of the Sixth IEEE International Conference on Computer and Information Technology (CIT 2006), vol. 52 (2006)
5. Schollmeier, R.: A Definition of Peer-to-Peer Networking for the Classification of Peer-to-Peer Architectures and Applications. In: Proceedings of the First International August Conference on Peer-to-Peer Computing, pp. 101–102 (2001)
6. Ghosal, D., Poon, B.K., Kong, K.: P2P Contracts: A Framework for Resource and Service Exchange. Future Generation Computer Systems 21, 333–347 (2005)
7. Kurmanowytsch, R., Kirda, E., Kerer, C., Dustdar, S.: OMNIX: A topology-independent P2P middleware. In: Eder, J., Missikoff, M. (eds.) CAiSE 2003. LNCS, vol. 2681, Springer, Heidelberg (2003)
8. Koulouris, T., Henjes, R., Tutschku, K., de Meer, H.: Implementation of adaptive control for P2P overlays. In: Wakamiya, N., Solarski, M., Sterbenz, J.P.G. (eds.) IWAN 2003. LNCS, vol. 2982, pp. 292–306. Springer, Heidelberg (2004)
9. Koo, S.G.M., Kannan, K., Lee, C.S.G.: On neighbor-selection strategy in hybrid peer-to-peer networks. Future Generation Computer Systems 22, 732–741 (2006)
10. Sun, S., Abraham, A., Zhang, G., Liu, H.: A Particle Swarm Optimization Algorithm for Neighbor Selection in Peer-to-Peer Networks. In: 6th International Conference on Computer Information Systems and Industrial Management Applications, pp. 166–172. IEEE Computer Society Press, Los Alamitos (2007)
11. Kessler, J., Rasheed, K., Arpinar, I.B.: Using genetic algorithms to reorganize superpeer structure in peer to peer networks. Applied Intelligence 26, 35–52 (2007)
12. Abraham, A.: Evolutionary Computation. In: Sydenham, P., Thorn, R. (eds.) Handbook for Measurement Systems Design, pp. 920–931. John Wiley and Sons Ltd, London (2005)
13. Abraham, A., Jain, L.: Evolutionary Multiobjective Optimization. In: Abraham, A., Jain, L.C., Goldberg, R. (eds.) Evolutionary Multiobjective Optimization: Theoretical Advances and Applications, ch. 1, pp. 1–9. Springer, London (2005)
14. Srinivas, N., Deb, K.: Multiobjective optimization using nondominated sorting genetic algorithms. Evolutionary Computation 2(3), 221–248 (1994)
15. Clerc, M.: Particle Swarm Optimization. ISTE Publishing Company, London (2006)
16. Abraham, A., Guo, H., Liu, H.: Swarm Intelligence: Foundations, Perspectives and Applications. In: Nedjah, N., Mourelle, L. (eds.) Studies in Computational Intelligence, pp. 3–25. Springer, Heidelberg (2006)
17. Raquel, C., Naval, P.: An Effective Use of Crowding Distance in Multiobjective Particle Swarm Optimization. In: Proceedings of Genetic and Evolutionary Computation Conference (GECCO 2005), pp. 25–29. ACM, New York (2005)

Effect of Dynamicity on Peer to Peer Networks

Bivas Mitra, Sujoy Ghose, and Niloy Ganguly

Department of Computer Science & Engineering
Indian Institute of Technology, Kharagpur, India
{bivasm,sujoy,niloy}@cse.iitkgp.ernet.in

Abstract. In this paper, we propose *an analytical framework* based on percolation theory to assess the robustness of peer to peer networks in face of user churns and/or attacks targeted towards important nodes. It is observed in practice that in spite of churn of peers, superpeer networks show exceptional robustness and do not disintegrate into disconnected components. With the help of the analytical framework developed, we formally measure its stability against user churn and validate the general observation. The effect of intentional attacks upon the superpeer networks is also investigated. Our analysis shows that fraction of superpeers in the network and their connectivity have profound impact upon the stability of the network. The results obtained from the theoretical analysis are validated through simulation. The simulation results and theoretical predictions match with high degree of precision.

Keywords: Superpeer networks, peer dynamics, complex network theory.

1 Introduction

Currently there has been much interest in peer-to-peer (p2p) network based data sharing and content distribution applications. These applications are used by millions of users and they represent a large fraction of the traffic in the Internet [1,2]. Peers in p2p system are connected among themselves by some logical links forming an overlay above the physical network. Superpeer topologies have emerged as the most influencing structure among the overlay networks. Most of the commercial systems like KaZaA have also adopted superpeers in their design [3]. In this system, superpeer nodes with higher bandwidth and connectivity connect to each other forming the upper level in the network hierarchy. Each superpeer works as a server on behalf of the set of client peers who form the lower level of network hierarchy [4,5].

Peers in the superpeer system join and leave the network randomly without any central coordination. This churn of nodes might partition the network into smaller fragments and breakdown communication among peers. But in practice, superpeer overlay networks exhibit stable behavior against churn. Consequently the possible breakdown of the network is a rare event [6]. However the stability of the overlay network can get severely affected through intended attacks targeted towards the important peers [7]. A comprehensive study of stability of the

S. Aluru et al. (Eds.): HiPC 2007, LNCS 4873, pp. 452–463, 2007.

superpeer networks against all these dynamics that take place in the network, is the primary focus of this paper.

A survey of the literature reveals that most of the commercial peer to peer networks can be modeled as complex graphs [4,8,9]. Some analysis of dynamics of complex graphs have been done mainly by the physicists. These approaches can be utilized to understand the various properties of peer to peer networks. Effect of random failures and intentional attacks in various kinds of graphs are discussed in [10,11,12]. In [13], Newman *et al.* introduced the concept of generating function formalism. Using it, Callaway [14] found the exact analytic solutions for percolation on random graphs with arbitrary degree distribution. In this paper, we utilize many of the aforesaid results of percolation theory and propose a generalized equation to measure stability of any given p2p overlay structures in face of churn of peers as well as attacks mounted on them.

The rest of the paper is organized as follows. Section 2 proposes an analytical framework to find the amount of disturbances required to disrupt the giant component [15] of the network. Section 3 models the superpeer topologies as mixed poisson graph and also models the churns and attacks mounted on the network. In section 4 we mathematically analyze the effect of churn in the superpeer networks and validate the results with the help of simulation. In section 5 the effect of targeted attack upon superpeer networks is discussed. Finally section 6 concludes the paper.

2 Developing Analytical Framework Using Generating Function Formalism

In this section, we use generating function to derive the general formula for measuring the stability of overlay structures undergoing any kind of disturbances in the network. We explain the basic concept behind development of the framework without going into mathematical details. Let p_k be the probability that a randomly chosen vertex in the graph has degree k. q_k be the probability that a vertex of degree k be present in the network after the removal of a fraction of nodes. In our formalism f_k $(=1 - q_k)$ and p_k specifies the churn/attack model and network topology respectively whose stability is subjected to examination. The formalism helps us to locate the transition point where the giant component [15] breaks down into smaller components. $p_k.q_k$ specifies the probability of a node having degree k to be present in the network after the process of removal of some portion of nodes is completed. Hence

$$F_0(x) = \sum_{k=0}^{\infty} p_k.q_k x^k$$

becomes the generating function for this distribution. Distribution of the outgoing edges of the first neighbor of a randomly chosen node can be generated by

$$F_1(x) = \frac{\sum_k kp_k q_k x^{k-1}}{\sum_k kp_k} = F_0'(x)/z$$

where z is the average degree [14].

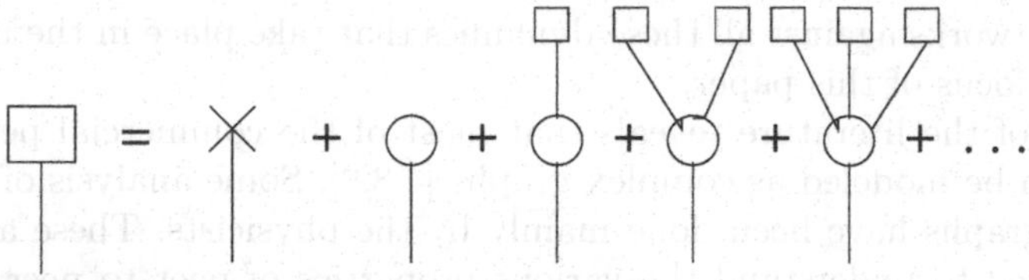

Fig. 1. Schematic representation of the sum rule for the connected component of vertices reached by following a randomly chosen edge [13]

Let $H_1(x)$ be the generating function for the distribution of the component sizes that are reached by choosing a random edge and following it to one of its ends. Except when we are precisely at the phase transition where giant component appears, typical component size is finite. Moreover as chance of a component containing a closed loop of edges goes down exponentially with size of the graph, it becomes negligible for large graph [13]. Therefore the component may be conceptualized as a treelike structure that contain zero node if the node at the other end of the randomly selected edge is removed, which happens with probability $1 - F_1(1)$. The edge may otherwise lead to a node with k other edges leading out of it other than the edge we came in along, distributed according to $F_1(x)$ (Fig. 1). That means $H_1(x)$ satisfies a self-consistency condition of the form [14]

$$H_1(x) = 1 - F_1(1) + xF_1(H_1(x)). \tag{1}$$

The distribution for the component size to which a randomly selected node belongs to is similarly generated by (Fig. 1) $H_0(x)$ where

$$H_0(x) = 1 - F_0(1) + xF_0(H_1(x)). \tag{2}$$

Therefore the average size of the components becomes

$$H_0'(1) = \langle s \rangle = F_0(1) + \frac{F_0'(1)F_1(1)}{1 - F_1'(1)}$$

which diverges when $1 - F_1'(1) = 0$, that is the size of the component becomes infinite. Therefore

$$F_1'(1) = 1 \Rightarrow \sum_{k=0}^{\infty} kp_k(kq_k - q_k - 1) = 0 \tag{3}$$

Significance of the Eq. (3) lies in the fact that it states the critical condition for the stability of giant component with respect to any type of graphs (characterized by p_k) undergoing any type of failure and attack (characterized by q_k). Using this formalism, we investigate the stability of superpeer networks in face of various dynamics of the nodes.

3 Environmental Definition

In this section, we formally model the superpeer networks and churn/attack to utilize the analytical framework. Also we define the stability metric and explain the simulations undertaken to verify the theoretical results.

3.1 Topology of the Superpeer Overlay Networks

The different types of overlay networks can be modeled using the uniform framework of probability distribution p_k, where p_k is the probability that a randomly chosen node has degree k. So the degree distribution p_k signifies the topology of the overlay network. In this paper, we model the superpeer overlay networks as mixed poisson network. In mixed poisson network, interconnection between superpeers are selected to approximate a E-R graph [16] which follows Poisson distribution. Similarly the degree distribution of peers follow Poisson distribution. The average degree of the superpeers are much higher than peers. Mathematically, if r be the fraction of peers in the network and rest are superpeers then degree distribution of the network

$$p_k = rp_{k_{pr}} + (1 - r)p_{k_{spr}}$$

where degree distribution of peers $p_{k_{pr}} = \frac{\langle k_p \rangle^{k_{pr}} e^{-\langle k_p \rangle}}{k_{pr}!}$ and superpeers $p_{k_{spr}} = \frac{\langle k_{sp} \rangle^{k_{spr}} e^{-\langle k_{sp} \rangle}}{k_{spr}!}$ follow Poisson distribution with average degree $\langle k_p \rangle$ and $\langle k_{sp} \rangle$ respectively and $\langle k_p \rangle << \langle k_{sp} \rangle$. The average degree of the mixed poisson network becomes

$$\langle k \rangle = r\langle k_p \rangle + (1 - r)\langle k_{sp} \rangle$$

3.2 Different Kinds of Churn and Attack Models

As defined in the previous section, let q_k be the probability that a vertex of degree k be present in the network after the removal of a fraction of nodes. In our framework q_k is used to specify the churn and attack models.

- In churn, the probability of removal of any randomly chosen node is degree independent and equal (constant) for all other nodes in the graph. Therefore the presence of any randomly chosen node having degree k after this kind of failure is $q_k = q$ (independent of k).
- In targeted attack, the nodes having high degrees are progressively removed. Formally $q_k = 1$ when $k < k_m$ but $0 \leq q_k < 1$ otherwise. This removes a fraction of nodes from the network with degree $\geq k_m$. Formally
  $q_k = 0$ when $k > k_m$
  $0 \leq q_k < 1$ when $k = k_m$
  $q_k = 1$ when $k < k_m$
  This removes all the nodes from the network with degree greater than k_m and a fraction of nodes having degree k_m.

3.3 Stability Metric

The stability of overlay networks are primarily measured in terms of certain fraction of nodes (f_c) called percolation threshold [14,15], removal of which disintegrates the network into large number of small, disconnected components.

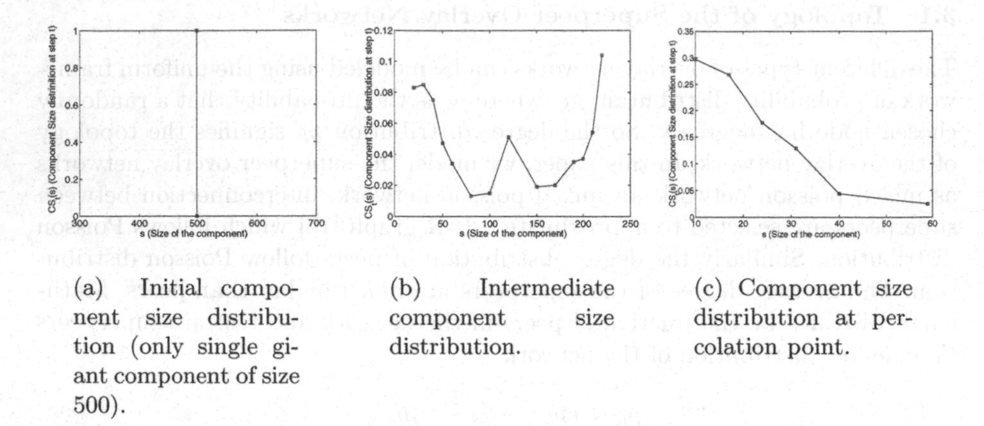

(a) Initial component size distribution (only single giant component of size 500).

(b) Intermediate component size distribution.

(c) Component size distribution at percolation point.

Fig. 2. The above plots represent the change in the component size distribution during percolation process and indicate the percolation point

Below that threshold, there exists a connected component which spans the entire network. This connected component is also termed as the giant component. The value of percolation threshold f_c theoretically signifies the stability of the network, higher value indicates greater stability against churn and attack.

We take cue from condensation theory used by physicists to develop the metric to measure the percolation threshold experimentally [17]. During experiment, we remove a fraction of nodes f_t from the network in step t and check whether we reach the percolation point. After each step t, we find out the status of the network in terms of component size distribution $CS_t(s) = sn_s/\sum_s sn_s$ where s and n_s respectively are the size of the component formed and the number of components of size s. The component size distribution initially exhibits unimodal characteristics confirming a single connected component (Fig. 2(a)) or bimodal character (Fig. 2(b)) confirming a large component alongwith a set of small components. Eventually at a particular step $t = t_n$, $CS_t(s)$ becomes monotonically decreasing function indicating t_n as percolation point (Fig. 2(c)). Therefore t_n is considered as the time step where percolation occurs and the total fraction of nodes removed at that step f_{t_n} specifies the percolation threshold.

3.4 Simulation Environment

The superpeer overlay structure is represented by a simple undirected graph stored as an adjacency list. In order to generate the topology, every node is assigned a degree according to the mixed poisson degree distribution. Thereafter the edges are generated using the "matching method" [18]. Some of the edges are then rewired using "switching method" to generate sufficient randomness in the graph [19]. In our experiment, we simulate the overlay network by generating graphs with 5000 nodes.

Churn or attack on a peer effectively means deletion of the node and its corresponding edges. We implement this phenomena by removing a fraction of nodes in each step depending on the disrupting event in the network. In the case of churn, nodes are randomly selected using a time-seeded pseudo-random number generator and its edges are removed from the adjacency list. For targeted attack, high degree nodes in the network are removed sequentially in each step until the percolation point is reached. We perform each experiment for 500 times and take the average of the percolation threshold.

4 Stability of Superpeer Networks Against Churn

The superpeer networks mostly suffer from the churn of peers which can be modeled by the random failure of nodes in complex graph. In this section, we use our equation to show that stability of the superpeer networks is quite unaffected due to churn of peers. We validate the theoretical results with the help of simulation. At first, we present the result for generalized random graph and then customize it for superpeer networks.

Generalized random graph
In this section, we discuss the effect of random failure in a generalized random graph. If $q = q_r$ is the critical fraction of nodes whose presence in the graph is essential for the stability of the giant component after this kind of failure then according to Eq. (3)

$$\sum_{k=0}^{\infty} kp_k(kq_r - q_r - 1) = 0$$

$$\Rightarrow q_r = \frac{1}{\frac{\sum_{k=0}^{\infty} k^2 p_k}{\sum_{k=0}^{\infty} kp_k} - 1} \Rightarrow q_r = \frac{1}{\frac{\langle k^2 \rangle}{\langle k \rangle} - 1}$$

where $\langle k^2 \rangle = \sum_{k=0}^{\infty} k^2 p_k$ and $\langle k \rangle = \sum_{k=0}^{\infty} kp_k$ are the second and the first moment of the degree distribution respectively. Now if f_r is the critical fraction of nodes whose random removal disintegrates the giant component then $f_r = 1 - q_r$. Therefore percolation threshold

$$f_r = 1 - \frac{1}{\frac{\langle k^2 \rangle}{\langle k \rangle} - 1} \tag{4}$$

This is the well known condition [10] (derived differently) for the disappearance of the giant component due to random failure. Note that, we have reproduced it to show that it can also be derived from the proposed general formula (Eq. (3)).

Superpeer networks
In mixed poisson network, let r be the fraction of peers in the network and rest be superpeers. Superpeer nodes are connected to each other to form an E-R network [16] with average degree $\langle k_{sp} \rangle$. Similarly peers connected with superpeers forms another E-R graph with an average degree $\langle k_p \rangle$ where $\langle k_p \rangle << \langle k_{sp} \rangle$. Now we examine the stability of this kind of superpeer network undergoing churn. In mixed poisson network, first and second moment of the degree distribution

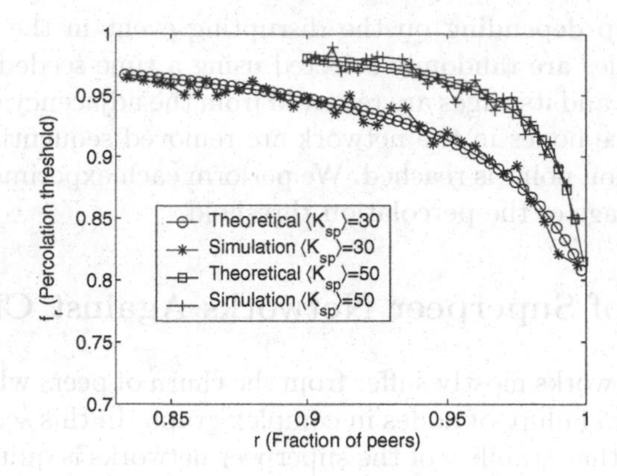

Fig. 3. The above plot represents a comparative study of theoretical and simulation results of stability for two mixed poisson networks undergoing churn

becomes $\langle k \rangle = r\langle k_p \rangle + (1-r)\langle k_{sp} \rangle$ and $\langle k^2 \rangle = r\langle k_p^2 \rangle + (1-r)\langle k_{sp}^2 \rangle$ respectively. If k is a random variable following Poisson distribution then it can be shown that $\langle k^2 \rangle \approx \langle k \rangle^2 + \langle k \rangle$. Hence according to Eq. (4), percolation threshold becomes

$$f_r = 1 - \frac{r\langle k_p \rangle + (1-r)\langle k_{sp} \rangle}{r\langle k_p \rangle^2 + (1-r)\langle k_{sp} \rangle^2}$$

Substituting for $\langle k_p \rangle$, we get

$$f_r = 1 - \frac{\langle k \rangle r}{\langle k \rangle^2 - 2\langle k \rangle(1-r)\langle k_{sp} \rangle + (1-r)^2\langle k_{sp} \rangle^2 + r(1-r)\langle k_{sp} \rangle^2} \tag{5}$$

Feasible fraction of peers: Since the mean peer degree $\langle k_p \rangle$ needs to be > 0 to be connected in the network therefore

$$\frac{\langle k \rangle - (1-r_r)\langle k_{sp} \rangle}{r_r} > 0$$

$$\Rightarrow r_r > 1 - \frac{\langle k \rangle}{\langle k_{sp} \rangle}$$

That means we can form a connected superpeer network with prescribed peer and superpeer degrees only if the fraction of peers in the network is greater than the feasible peer fraction (r_r). For $\langle k_{sp} \rangle = 30, 50$ this feasible fraction r_r becomes 0.833, 0.90 respectively. Below that fraction, there does not exist any network, therefore our theoretical analysis as well as simulations are performed with peer fraction r above the feasible fraction r_r.

Using Eq. (5), we study the variation of percolation threshold (f_r) due to the change in the fraction of peers (r). We validate the analytically derived result with the help of simulation. We perform the simulation on two mixed poisson networks with average superpeer degree $\langle k_{sp} \rangle = 30$ and 50, keeping the average degree $\langle k \rangle = 5$. Comparative study reveals that networks having higher superpeer degree exhibit more robustness than with lower superpeer degree for any peer-superpeer ratio. *It can be observed from Fig. 3 that simulation results match closely with theoretical predictions which shows the success of our theoretical framework.*

Observations:
1. It is important to observe that for the entire range of peer fractions, the percolation threshold f_r is greater than 0.7 which implies that superpeer networks are quite robust against churn. During churn, removal of a significant number of low degree peers alongwith a few high degree superpeers have less impact upon the stability of the networks.
2. Another significant observation is, lower fraction of superpeers in the network (specifically when it is below 5%) results in a sharp fall of f_r, that is the vulnerability of the network increases drastically. When the fraction of superpeers in the network is high, most of the peers are only connected to superpeers (and not within themselves), hence stability of the network depends entirely upon superpeers. As fraction of superpeer reduces below 5%, some peers are not connected to the superpeers at all, but only connected to fellow peers. This produces an avalanche effect during churn which results in a drastic reduction of stability of the network in this region.

5 Stability of Superpeer Networks Against Targeted Attack

Stability of the superpeer networks is challenged by various kinds of attacks on prominent peers or superpeers. The attack model has been formally defined in section 3. In this section, we analyze the effect of this kind of targeted attack upon superpeer networks where r be the fraction of peers and rest are superpeers. In the case of targeted attack two cases may arise

Case 1. Removal of a fraction of superpeers is sufficient to disintegrate the network. This happens when the percentage of superpeers is relatively higher than peers.

Case 2. Removal of all the superpeers is not sufficient to disintegrate the network. Therefore we need to remove some of the peer nodes along with the superpeers.

We analyze these two cases separately with the help of our analytical framework. From Eq. (3) the critical condition for the stability of the giant component can be rewritten as

$$\sum_{k=0}^{\infty} k(k-1)p_k q_k = \langle k \rangle$$

The equation can be further expanded as below to differentiate between peers and superpeers

$$\sum_{k=0}^{k_{max}-1} k(k-1)p_k q_k + \sum_{k=k_{max}}^{\infty} k(k-1)p_k q_k = \langle k \rangle \tag{6}$$

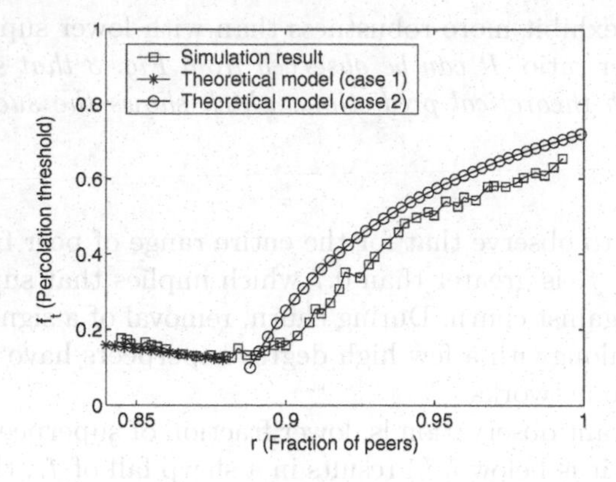

Fig. 4. The above plot represents the behavior of the mixed poisson network in face of targeted attack found experimentally and compares it with the proposed theoretical model. Case 1 and case 2 of the theoretical model represent Eq. (7) and (8) respectively.

where all the nodes having degree less than k_{max} are peers and rest are superpeers.

Case 1: In this case, removal of a fraction of superpeers is sufficient to disintegrate the network. If f_{sp} be the critical fraction of superpeer nodes, removal of which disintegrates the giant component then $q_k = 1$ for $k < k_{max}$ and $q_k = 1 - f_{sp}$ for $k \geq k_{max}$. Hence according to Eq. (6),

$$\sum_{k=0}^{k_{max}-1} k(k-1)p_k + \sum_{k=k_{max}}^{\infty} k(k-1)p_k(1-f_{sp}) = \langle k \rangle$$

$$\Rightarrow f_{sp} = 1 - \frac{\langle k \rangle - \sum_{k=0}^{k_{max}-1} k(k-1)p_k}{\sum_{k=k_{max}}^{\infty} k(k-1)p_k}$$

As the fraction of superpeer nodes in the network is $(1-r)$, then percolation threshold for case 1 becomes $f_t = (1-r) \times f_{sp}$

$$\Rightarrow f_t = (1-r)\left(1 - \frac{\langle k \rangle - \sum_{k=0}^{k_{max}-1} k(k-1)p_k}{\sum_{k=k_{max}}^{\infty} k(k-1)p_k}\right)$$

$$= (1-r)\left(1 - \frac{\langle k \rangle - r\sum_{k=0}^{\langle k_p \rangle+\delta} k(k-1)\frac{\langle k_p \rangle^k e^{-\langle k_p \rangle}}{k!}}{(1-r)\sum_{k=\langle k_p \rangle+\delta+1}^{\infty} k(k-1)\frac{\langle k_{sp} \rangle^k e^{-\langle k_{sp} \rangle}}{k!}}\right) \quad (7)$$

where mean peer degree $\langle k_p \rangle = \frac{\langle k \rangle - (1-r)\langle k_{sp} \rangle}{r}$ and we choose suitable value of δ depending on the standard deviation of the Poisson distribution.

Case 2: Here we have to remove f_p fraction of peer nodes alongwith all the superpeers to breakdown the network. Therefore $q_k = 1 - f_p$ for $k < k_{max}$ and $q_k = 0$ for $k \geq k_{max}$. Hence according to Eq. (6),

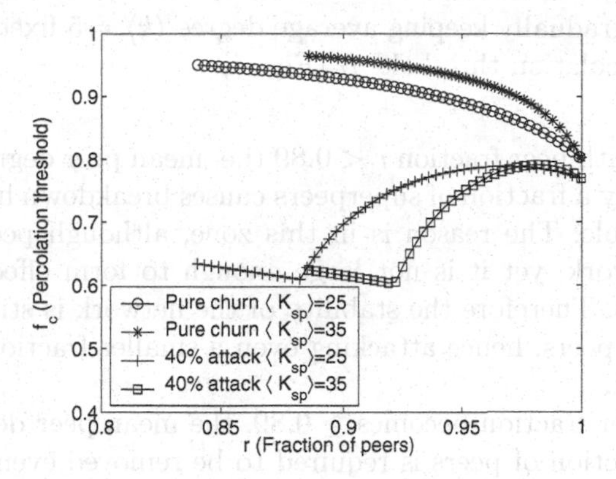

Fig. 5. We plot percolation threshold f_c for various peer fraction r. Two different mixed poisson networks have been considered with average superpeer degree $\langle k_{sp} \rangle = 25, 35$ with fixed average degree $\langle k \rangle = 5$. We compare theoretically the stability of these two networks against pure churn and combination of churn (60%) and attack (40%).

$$\sum_{k=0}^{k_{max}-1} k(k-1)p_k(1-f_p) = \langle k \rangle$$

$$\Rightarrow f_p = 1 - \frac{\langle k \rangle}{\sum_{k=0}^{k_{max}-1} k(k-1)p_k}$$

Therefore the total fraction of nodes required to be removed to disintegrate the network for case 2 becomes $f_t = rf_p + (1-r)$.

$$\begin{aligned} \Rightarrow f_t &= r\left(1 - \frac{\langle k \rangle}{\sum_{k=0}^{k_{max}-1} k(k-1)p_k}\right) + (1-r) \\ &= r\left(1 - \frac{\langle k \rangle}{r\sum_{k=0}^{\langle k_p \rangle+\delta} k(k-1)\frac{\langle k_p \rangle^k e^{-\langle k_p \rangle}}{k!}}\right) + (1-r) \end{aligned} \quad (8)$$

where mean peer degree $\langle k_p \rangle = \frac{\langle k \rangle - (1-r)\langle k_{sp} \rangle}{r}$.

Transition point: The transition from case 1 to case 2 can be easily marked by observing the value of percolation threshold f_t. While calculating using Eq. (7) (case 1), if the percolation threshold f_t exceeds the fraction of superpeers in the network $(1-r)$, it indicates that removal of all the superpeers is not sufficient to disrupt the network. Hence subsequently we enter into case 2 and start using Eq. (8) to find percolation threshold.

We validate our theoretical model of attack on mixed poisson network with the help of simulation. In simulation, we consider a mixed poisson network with average degree $\langle k \rangle = 5$ and mean superpeer degree $\langle k_{sp} \rangle = 30$. We increase the fraction of peers gradually keeping average degree $\langle k \rangle = 5$ fixed and observe the change in the percolation threshold f_t (Fig. 4).

Observations:
In the networks with peer fraction $r < 0.89$ (i.e. mean peer degree $0 < \langle k_p \rangle \leq 2$), the removal of only a fraction of superpeers causes breakdown hence makes these networks vulnerable. The reason is in this zone, although peers have a larger share in the network, yet it is not large enough to form effective connections within themselves. Therefore the stability of the network is still entirely dependent on the superpeers, hence attacking even a smaller fraction of them breaks down the network.

However as peer fraction becomes ≥ 0.89, the mean peer degree increases to 3 and 4 and a fraction of peers is required to be removed even after removal of all the superpeers to dissolve the network. Here the high degree peers strongly connect among themselves which results in the increase of stability of the network.

6 Conclusion

In this paper we have developed a common analytical framework to evaluate the robustness of superpeer networks against various disturbances in the network. We have modeled superpeer networks by mixed poisson degree distribution. We have also modeled the churn of peers as random failure of nodes. It has been observed from both theoretical and simulation results that superpeer networks remain robust for user churn. Next we have analyzed the behavior of superpeer networks in face of targeted attack. Unlike churn, in this case increase of peers improves the stability of the network and the rate of improvement is almost linear to the fraction of peers present in the network.

Our analysis has shown that presence of superpeers impart conflicting advantages for churn and attack. Hence proper mix of fraction of superpeers with peers is necessary to improve the robustness of the network in face of combination of churn and attack. It appears from Fig. 5 that when percentage of attack is 40%, the network having lower superpeer degrees ($\langle k_{sp} \rangle = 25$) performs better than network having higher superpeer degree ($\langle k_{sp} \rangle = 35$). So to obtain optimized performance, it is upto the design engineers to choose the correct superpeer to peer ratio depending on the working environment. The theoretical framework developed in this paper will help them to easily and accurately calculate the ratio.

References

1. Lv, Q., Cao, P., Cohen, E., Li, K., Shenker, S.: Search and Replication in Unstructured Peer-to-Peer Networks. In: ACM International Conference on Supercomputing, New York, USA (2002)
2. Ganguly, N., Deutsch, A.: Developing Efficient Search Algorithms for P2P Networks Using Proliferation and Mutation. In: Proceedings of the International Conference on Artificial Immune Systems, Catania, Italy (September 2004)

3. KaZaA website, `http://www.kazaa.com`
4. Pyun, Y.J., Reeves, D.S.: Constructing a Balanced, log(N)-Diameter Super-peer Topology. In: Proceedings of the 4th International Conference on Peer-to-Peer Computing, Zurich, Switzerland (August 2004)
5. Yang, B., Garcua-Molina, H.: Designing a Super-Peer Networks. In: Proceedings of the International Conference on Data Engineering (ICDE), Los Alamitos, CA (March 2003)
6. Stutzbach, D., Rejaie, R., Sen, S.: Characterizing Unstructured Overlay Topologies in Modern P2P File-Sharing Systems. In: Proceedings of ACM SIGCOMM/USENIX Internet Measurement Conference, Berkeley, CA (October 2005)
7. Saroiu, S., Gummadi, P.K., Gribble, S.D.: Measuring and Analyzing the Characteristics of Napster and Gnutella Hosts. Multimedia Systems Journal 8(5) (2002)
8. Montresor, A.: A Robust Protocol for Building Superpeer Overlay Topologies. In: Proceedings of the 4th International Conference on Peer-to-Peer Computing, Zurich, Switzerland (August 2004)
9. Schollmeier, R.: Signaling And Networkin. In Unstructured Peer-To-Peer Networks, Ph.D. Thesis, Technical University, Munchen, pp. 75–86 (2005)
10. Cohen, R., Erez, K., Avraham, D., Havlin, S.: Resilience of the Internet to Random Breakdown. Physical Review Letters 85(21) (2000)
11. Cohen, R., Erez, K., Avraham, D., Havlin, S.: Resilience of the Internet under Intentional Attack. Physical Review Letters 86(16) (2001)
12. Albert, R., Jhong, H., Barabsi, A.L.: Error and Attack Tolerance of Complex Networks. Nature 406 (2000)
13. Newman, M.E.J., Strogatz, S.H., Watts, D.J.: Random Graphs with Arbitrary Degree Distributions and Their Application. Physical Review E (2001)
14. Callaway, D.S., Newman, M.E.J., Strogatz, S.H., Watts, D.J.: Network Robustness and Fragility: Percolation on Random graphs. Physical Review Letters 85(21) (2000)
15. Molloy, M., Reed, B.: The Size of the Giant Component of a Random Graph with a Given Degree Sequence. Combinatorics, Probability and Computing 7, 295–298 (1998)
16. Erdos, P., Renyi, A.: On the Evolution of Random Graphs. Publ. Math. Inst. Hangar Acad. Sci. 5, 17–61 (1960)
17. Majumdar, S.N., Evans, M.R., Zia, R.K.P.: Nature of the Condensate in Mass Transport Models. Physical Review Letters 94, 180601 (2005)
18. Milo, R., Shen-Orr, S., Itzkovitz, S., Kashtan, N., Chklovskii, D., Alon, U.: Network Motifs: Simple Building Blocks Of Complex Networks. Science 298, 824–827 (2002)
19. Milo, R., Kashtan, N., Itzkovitz, S., Newman, M.E.J., Alon, U.: On the Uniform Generation of Random Graphs with Prescribed Degree Sequences. eprint arXiv:cond-mat/0312028 (2003)

Hierarchical Multicast Routing Scheme for Mobile Ad Hoc Network

A. Shajin Nargunam[1] and M.P. Sebastian[2]

[1] A. Shajin Nargunam is with Department of Computer Applications, N.I College of Engineering, Thuckalay, Tamil Nadu, India
ashajins@yahoo.com
[2] M.P Sebastian is with Department of Computer Engineering, National Institute of Technology Calicut, Kerala, India
sebasmp@nitc.ac.in

Abstract. Ad hoc network is a dynamic multi-hop wireless network that is established by a group of mobile nodes on a shared wireless channel. The plain flooding algorithm provokes a high number of unnecessary packet rebroadcasts, causing contention, packet collisions and ultimately wasting precious limited bandwidth. Another barrier is that routing in ad hoc networks does not scale up as easily as in fixed network. Hierarchical techniques have long been known to afford scalability in networks. By introducing hierarchical routing scheme to ad hoc networks, we can effectively address this problem. Clustering provides a method to build and maintain hierarchical routing scheme in ad hoc networks. By summarizing topology detail via a hierarchical map of the network topology, network nodes are able to conserve memory and link resources. This paper proposes a fully distributed cluster based routing algorithm for mobile ad hoc networks. Non-overlapping clusters are formed using the dynamic cluster creation algorithm. Packets are routed according to the routing information available with each gateway node. The mobility issues are also handled locally in this routing architecture. The proposed cluster maintenance algorithm dynamically adapts to the changes and hence the efficiency is not degraded by node mobility. In addition, our analysis shows that building clustered hierarchies is affordable and that clustering algorithms can also be used to enhance network quality of service.

Keywords: Ad hoc Networks, Clustering, Mobility, Multicast Routing and Quality of Service.

1 Introduction

In any complex distributed system of nodes, clustering of nodes into groups results in simplification of management and administration of the nodes and also gives up better performance since details about the remote nodes of the distributed system can be handled in a collective manner. Thus, obligation of a hierarchical organization is beneficial for the management of a complex system, and results in scalability of operations. The wired Internet, for example, cannot be managed without hierarchical

S. Aluru et al. (Eds.): HiPC 2007, LNCS 4873, pp. 464–475, 2007.

addressing and management. Essential services like routing are highly scalable owing to this hierarchical organization.

In a dynamic mobile environment maintaining the accurate route information of all nodes in the network is impossible. Because the mobile nodes may leave the network boundary region or new mobile nodes may enter into the network boundary region over and over again. This requires *O(ne)* control messages to be sent, where *n* is the number of nodes and *e* is the number of links in the network. Recurrent route changes flood the network with routing information packet which reduces the overall performance of the system. The increase in number of nodes in the network increases the route table size in each node. In order to avoid the routing packet storm the network is divided into non-overlapping groups called clusters. Each cluster members maintains the route information of its cluster members. This greatly reduces control packet traffic and route table size.

2 Related Works

Generally, clustering can be used to reduce the amount of information used to represent the state of the network. By grouping together multiple nodes into a single cluster, one can reduce the state representation of those nodes to that of one cluster. Multicast communication is an important operation for many applications of ad hoc networks. Similar to multicast protocols for wired networks, one of the major goals in designing multicast protocols is to reduce unnecessary packet delivery to other nodes outside the group by having only a subset of nodes participating in multicast data forwarding. In [2] a new multicast routing protocol for ad hoc networks that is based on an adaptive dynamic backbone algorithm is proposed.

A new distributed clustering algorithm based on sending beacons is given in [13]. In this algorithm, mobile nodes compete with each other to become cluster-head based on the number of neighbors. [14] presents the general architecture and the functional entities for location information management and exploitation of an advanced, open, flexible service provision platform for reconfigurable networks. A market-based mechanism for stimulating cooperation in wireless ad hoc networks, where each node can freely decide on the amount of traffic it relays and how much it charges other nodes for relaying their traffic is given in [12]. Wireless sensor networks represent a new frontier in the development of technology to be used in a variety of applications of our daily life in the future. As a new research area, there are several open problems that need to be investigated. One of them is management of those networks. The task of building and deploying management systems in environments where there will be tens of thousands of network elements with particular features and organization is very complex. [8] presents and discusses a management architecture for wireless sensor networks.

Two techniques for modeling data for compression in wireless ad hoc networks are given in [11]. The first technique enables efficient and scalable shortest path routing. [9] proposes a weight based distributed clustering algorithm (WCA) which can dynamically adapt itself with the ever changing topology of ad hoc networks. This approach restricts the number of nodes to be catered by a cluster head so that it does not degrade the MAC functioning.

In mobile ad hoc networks nodes move even during the broadcast process. So it is difficult to maintain up-to-date and consistent local views. A formal framework is used to model inaccurate local views in MANETs, where full coverage is guaranteed if three sufficient conditions, connectivity, link availability, and consistency, are met. [6] proposes a mechanism called aggregated local view to ensure consistent local views.

In most existing localized topology control protocols, it is assumed that the network is connected at all times under a normal transmission range. Each node selects a few logical neighbors from its 1-hop neighbors within the normal transmission range. The selection of logical neighbors is usually based on 1-hop information (i.e., location information of all 1-hop neighbors), although some protocols use only partial 1-hop information such as the direction or location information of nodes within a search region that is smaller than the normal transmission range.

The local positioning system should work based on the coordination of the nodes inside the wireless network, without any assistance from other infrastructure. For a given node distribution the Euclidean distance between two nodes is estimated according to the length of the shortest path obtained by sending a control packet. [5] proposes a self-configurable positioning technique for multi-hop wireless networks.

A fundamentally different approach in the differentiated services framework is the *relative differentiated services*. In this approach, the network traffic is grouped into classes of service which are ordered, such that Class i is better than Class $i+1$. In [3] a proportional delay differentiation model is proposed. Static IP address assignment for MANET nodes is difficult as it needs to be done manually with prior knowledge about the MANET's current network configuration. Dynamic configuration protocols like Dynamic Host Configuration Protocol (DHCP) [1] require the presence of centralized servers. MANETs may not have such dedicated servers. Hence, centralized protocols cannot be used to configure nodes in MANETs. In [10] distributed protocol for dynamic IP address assignment is given. This guarantees unique IP address assignment under a variety of network conditions including message losses, network partitioning and merging. [7] presents a new transport protocol called ATP (ad hoc transport protocol) that is tailored toward the characteristics of ad hoc networks. ATP, by design, is an antithesis of TCP and consists of: rate-based transmissions, quick-start during connection initiation and route switching, network supported congestion detection and control, no retransmission timeouts, decoupled congestion control and reliability, and coarse-grained receiver feedback.

3 Clustering Scheme and Our Assumptions

The *ad hoc* network is represented by means of an undirected graph $G=(V, E)$, where V is the set of nodes in the graph and E is the set of edges in the graph. The number represents the degree of each node. Cluster-ID is declared by the node which initiates the cluster creation process and is denoted by C_i. Each cluster consists of two or more cluster members. Due to the topology changes caused by node insertion, removal, and motion, additional control messages are generated for cluster update. The question is how the control overhead is quantified and reduced as much as possible, that is the major objective in our clustering scheme.

It is assumed that each node is equipped with a single network interface card (NIC) having a transmission radius of R_{TX}. If the distance separating a pair of nodes is less than R_{TX}, then a bidirectional link connects them and they are considered to be neighbors. Otherwise, the nodes are not connected. Each NIC employs carrier sense multiple access with collision avoidance (CSMA/CA).

4 The Clustering Architecture

The key concept in our clustering scheme is to localize control messages to a small set of cluster nodes and minimize the frequency of network-wide flooding. This leads to reduction in control packet transmissions and average delay due to less contention in a network. The second objective of the proposed clustering algorithm is to find an interconnected set of clusters covering the entire node population. The network is divided into small partitions (*clusters*) with independent control. A good clustering scheme must preserve its structure when a few nodes move away and the topology changes slowly. Otherwise, high processing and communications overheads will have to be paid to reconstruct clusters. Within a cluster, it is easy to schedule packet transmissions and to allocate the bandwidth to real time traffic. We have developed a fully distributed hybrid clustering scheme which dynamically group the nodes into non-overlapping clusters. Each mobile node executes cluster creation module to become a cluster member and initialize the cluster member table. Before executing the clustering algorithms, each nodes first exchange hello messages within its neighbors to collect the neighborhood information. The communications between two clusters is accomplished by gateway nodes.

```
1. Construct set S dynamically from the hello messages received from neighbors.
2. If no cluster topology message is received within the initial waiting time
       cluster_id = New_Cluster_ID
       forward to all nodes in S  send_cluster_info(N_ID,C_ID,Loc)
3. while (S != empty)
   on receiving cluster info(N_ID,C_ID,Loc)
         if (cluster_id == UNKNOWN)
              cluster_id = C_ID;
              id=N_ID;
              md = diff(cur_loc – Loc);
         else if (md > diff(cur_loc – Loc)
               and cluster_id <> C_ID)
            cluster_id = C_ID;
              md = diff(cur_loc – Loc);
              call update_gateway_node( id)
         else if (md < diff(cur_loc – Loc)
               and cluster_id <> C_ID)
              call update_gateway_node( id)
         S = S – {N_ID};
         forward to all nodes in S send_cluster_info(N_ID,C_ID,Loc)
```

Fig. 4.1. Dynamic Clustering Algorithm

We can find from this algorithm that each node only broadcasts one cluster init message before the algorithm stops, and the time complexity is $O(|C|)$ where C is the set of nodes grouped together. The clustering algorithm converges quickly. In the worst case, the convergence is linear in the total number of nodes.

4.1 Analysis of Cluster Initialization and Maintenance Parameters

During cluster initialization each node transmits a hello message and listens to the media for reply. All nodes which are within the transmission range of the transmitting node receive the hello message. The overhead involved in the cluster initialization depends on the average number of neighbors. Let A_n be the average number of neighbors. Two rounds of communication are performed in this phase. The second round message travels up to h hop neighbors and the value of h depends on the degree of each node. Let N be the total number of nodes and N_c be the number of clusters. Average number of nodes per cluster is denoted as C_a and $C_a = N / N_c$. Therefore the number of messages exchanged during the cluster creation process per cluster is $2A_nh$ and the average cluster initialization overhead CI_{OH} is:

$$CI_{OH} = 2A_nh\,C_a \tag{4.1}$$

$$= 2A_nh\,(N / N_c)$$

$$= \frac{2A_nhN}{N_c} \tag{4.2}$$

Above equation shows that an increase in number of nodes increases the cluster initialization overhead but the increase in number of clusters decreases the overhead. The other two terms are average number of neighbors and the number of hops separating two extreme nodes of a cluster. To keep the product of this term as constant in our scheme we decrease h when the average number of nodes increases. It shows that in the proposed scheme cluster initialization overhead remains almost constant as the number of nodes in the system increases.

Another major factor that affects the performance of the system is node mobility. Increase in node mobility increases the link change and it initiates the cluster member table updation process. Let A_e be average number of route error messages generated due to link failure and A_v be the average volume (in terms of number of bits) of route update message. During each link change the old cluster members should remove the corresponding entry from the cluster member table and the overhead involved in this is A_vC_a. Similarly when the mobile node becomes the member of a new cluster then the route update message is forwarded to all cluster members and the overhead is A_vC_a. Therefore the cluster maintenance overhead CM_{OH} is

$$CM_{OH} = A_e + 2\,A_vC_a$$

$$= A_e + 2\,A_v(N/N_c)$$

$$= A_e + {2A_vN}/{N_c} \tag{4.3}$$

Above equation shows that even though the cluster maintenance overhead increases as number of nodes increases it can be controlled by increasing the number of clusters.

In the proposed clustering technique, each node requires one or more neighborhood information to execute the cluster creation and maintenance algorithm. The scalability of a routing protocol is assessed in terms of a number of increasing node count (N) and increasing average node density (nodes per unit area). In order to isolate the performance of cluster based routing with respect to increasing node count and node density, we have divided the nodes into non-overlapping clusters and the cluster radius is adjusted according to node density. The average number of node per cluster is almost maintained as constant with respect to increasing node count. The average hop count on the shortest path between an arbitrary pair of nodes in a two-dimensional network consisting of N nodes is proportional to number of clusters.

Here, $N[v]$ is the *neighborhood* of node v, defined as

$$N[v] = \bigcup_{v^l \in V, v^l \neq v} \{v^l \mid dist(v, v^l) < tx_{range}\} \tag{4.4}$$

To determine the inter cluster call forwarding probability that node x member of cluster c_1 forwards a call to node y member of cluster c_2. We rely on our assumption that nodes are random and uniformly distributed in the network. Each node has a transmission radius r and non-overlapping clusters are formed with a cluster radius of r_c.

N: Node x member of cluster c_1 must be the neighbor of node y member of cluster c_2.

$$P(N) = \frac{\pi r^2}{\pi r_c^2} = \frac{r^2}{r_c^2} \tag{4.5}$$

Let g be the average number of gateway nodes per cluster and P(F) is the probability that a gateway node forwards the packet. N_c is the average number of node per cluster.

$$P(F) = \frac{g}{N_c} \tag{4.6}$$

P(E) be the probability that node x forwards the packet before node y's timer expires.

$$P(E) = \frac{1}{2}$$

$$P(N \cap F \cap E) = \frac{r^2}{r_c^2} \frac{1}{2} \frac{g}{N_c} = \frac{gr^2}{2r_c^2 N_c} \tag{4.7}$$

The inter-cluster call forwarding probability for different number of gateway nodes can be calculated using the above equation.

Each node proactively emits h_m hello message and t_m topology broadcast messages during cluster initialization.

Let l_t be the link life time (i.e. the average life time of an edge connecting two vertices). Link life time is inversely proportional to node mobility. The proactive intra-cluster routing overhead is directly proportional to link life time and average number of emissions e_n.

$$R_{OH} \propto l_t e_n \tag{4.8}$$

$$R_{OH} = N_c l_t e_n \tag{4.9}$$

5 Multicast Communication

We have proposed a new on-demand multicast routing protocol for ad hoc networks. The new routing scheme, CBMRP, is based on clustering technique and designed to minimize control overhead. CBMRP also attempts to improve the routing efficiency by giving preference to gateway nodes in establishing a route. After the network has been clustered, multicast routing algorithm constructs the multicast paths which connects other clusters that containing receivers. . By setting up the cluster forwarding through gateway nodes, the multicast path can be easily constructed and the data packets can be forwarded to the receivers efficiently.

A new source initially sends a Join-Request packet. All other members who wish to join the multicast group must send a reply. Multicast group member belongs to the same cluster maintains the multicast membership information of all its cluster members. In our scheme multicast tree construction and maintenance process is fully distributed. Optimum tree linking all intra-cluster multicast group members are constructed using the cluster member table data.

Each gateway node maintains the fitness value of each link connecting other cluster members. The fitness value is proportional to the resource availability of each link. When a gateway node receives a multicast request to be forwarded, it waits for a small amount of time which is inversely proportional to the fitness parameter. This scheme provides fairness and it allows forwarding the request along the link having maximum free recourses (bandwidth).

In the receiving side, if it is the first time to receive this packet, the received gateway node forwards request packet and the multicast-ID of the request is added to the gateway node table. The clusters that contain multicast receivers then send a reply to the corresponding gateway nodes and gateway nodes maintains the multicast membership information.

As soon as the Route Reply packet reaches the cluster of the source *s*, it constructs the multicast tree connecting all multicast members. The multicast receivers can send packet to other multicast members along the reverse path. The new multicast sources should send a Route Request packet to append the previously constructed path. In our scheme Join-Request messages are forwarded only to gateway nodes. But in most of the conventional schemes request massages are flooded using broadcast schemes.

1. Route discovery: Forwarding the route request message through the cluster gateway nodes:
   - Node *s* starts by sending the message M to destination *d* through its cluster gateway using a short-way transmission.
   - Suppose G_s receives the message (d, M) from *s* at time t = 1. Node G_s forwards a route request message (RREQ) at time = 2, to all of its adjacent clusters through the gateway nodes.

- o Each cluster gateway nodes receiving an RREQ for the first time checks whether x is d itself. Otherwise, if k > 1, then it forwards an RREQ with TTL equal to k - 1 and its own id, i.e., node z forwards the RREQ*(d, i, k – 1, x; s)*, to its adjacent clusters through the gateway nodes.
- o Node x keeps the just received RREQ for broadcast round i and discards the stored RREQ from round i-1 , if any to the same destination.
- o Each cluster gateway that still has a stored RREQ (3*i) time steps after the receipt of the RREQ promptly discards the RREQ.

2. Route discovery: Acknowledging receipt of RREQ and selecting (s, d) path:
   - o Node d, upon receiving a RREQ(d, i, k, G_d, s) where G_d is the cluster gateway of node d, sends a path acknowledgment message via a short-gateway transmission.
   - o Node G_d, upon receiving a path acknowledgment notice from node d, sends a long-gateway transmission path acknowledgment message (ACK).
   - o Each cluster gateway nodes G_x, upon receiving an ACK message checks if the request passes through it. If so, then x marks itself as ACTIVE(s, d) and sends an ACK message via long-gateway transmission.

Lemma 1
The Route discovery complexity of the CBMRP algorithm is O (d_{min}). Where d_{min} is the minimum hop distance between source s and destination d.

Proof
We first prove the complexity of the route discovery part of the CBMRP algorithm. Suppose the request reaches node d during the ith broadcast round originated at node G_s. Hence, the distance from G_s to G_d must be at most *(3* i)*. The broadcast rounds out of node G_s will end as soon as an ACK is received by that node. The RREQ message that first reached node d must have been sent before the ith round was completed, since the ith broadcast commences at time step 3(i+1) and takes at most 3 time steps to complete. The ACK sent out of node G_d must have been sent at time t+1. Any ACK sent by the algorithm goes from a node reachable from G_s in h hops to a node reachable from G_s in h-1 hops. Thus, since G_d is reachable from G_s after t<3*i time steps of the ith round, it will take at most 3*i time steps for the ACK originating at G_d to reach G_s. Putting all these costs together, the route discovery takes at most 3(i+1)+3i+2 time steps. (the constant additive term comes from the fact that there may be two additional communication steps between G_d and d). Since we know that d was not reached in the (i-1) round, the distance between G_s and G_d must be at least 3*i hops, implying that the shortest distance between G_s and G_d must also be at 3*i hops. Since any (G_s , G_d) gateway linking path from s,G_s … G_d ,d is a candidate path for being the path between G_s and G_d with the smallest possible number of short hops (which, as we have seen, must be longer than or equal to 3*i), the shortest distance between s and d has to be at least *(3*i)-2=O(i)*. Hence, the route discovery of the CBMRP algorithm takes time which is linearly proportional to d_{min}.

Message transmission: The message transmission phase only involves the nodes in the selected path from G_s to G_d and each node in this path takes one time step to forward each packet M to the next node in the path. We have seen that the selected path from G_s to G_d (and, hence, the extension of this path that goes from s to d) has $O(d_{min})$ hops. Hence, the message and time complexity of the message transmission phase are in order of $O(d_{min} | M)$.

Route Maintenance: Route maintenance is the mechanism by which a node identifies that a link along an active path has broken such that it can no longer forward the packets to destination node *d* through that route. When route maintenance indicates a link is broken, the intermediate node finds an alternate path using local repairing scheme. The local repairing scheme finds the path locally using the cluster member table. The proposed CBMRP discover multicast routes only in the presence of data packets to be delivered to a multicast destination.

6 Simulation Model

The simulator for evaluating the proposed cluster based protocol is implemented using the Network Simulator ns-2. NS-2 is a scalable simulation environment for wireless network systems. The simulation models the network of 50 mobile hosts migrating within a 500 meter x 500 meter space with a transmission radius of five meters. Every node in the network moves in a random fashion. The cluster creation and multicast routing modules were implemented in C++ and it was linked with ns2 via TclClass object. Source nodes and destination nodes were chosen randomly. Source initiates the multicast tree creation process and intermediate gateway nodes multicast table was updated according to the multicast membership information. Multicast source generates 512-byte data packets with constant bit rate.

We have evaluated the performance of CBMRP via simulation. The simulation result shows that CBMRP effectively delivers around 95% of data packets and is robust against frequent topology changes. Moreover, data packet transmissions and control message exchanges are reduced by 25% compared to existing ad hoc multicast routing protocols. We also observed that CBMRP reduces the average packet delay by 15%. A number of movement scenario files were generated and used as inputs to the simulations. Each movement scenario file determines movements of 50 mobile nodes, and the mobile nodes are uniformly distributed. Each intermediate node along the path to the destinations is responsible for forwarding the data packet to the next intermediate node. The acknowledgement is not transmitted back to the source. A packet is dropped when no acknowledgment is received from the neighbor after retransmitting it *'x'* times.

The simulation was run for different scenarios and the results were compared with other multicast ad hoc routing protocols. For comparisons we have selected ODMRP and MAODV.

Parameters of interest are control overhead, end-to-end delay and packet delivery ratio. Figure 6.1 shows the packet delivery ratio of MAODV, ODMRP and CMRP for different mobility scenarios. As the node mobility increases the delivery ratio decreases for all three protocols. Even though the packet delivery ratio of proposed CBMRP is less than ODMRP, for higher mobility scenarios CBMRP packet delivery ratio is better than ODMRP.

Figure 6.2 shows the control overhead incurred by MAODV, ODMRP and CBMRP. In all protocols control messages were generated and forwarded by intermediate nodes during multicast tree creation. If the nodes were less mobile then link failure was also less and number of control messages generated was less. We can observe form the graph increasing mobility makes CBMRP more efficient than MAODV and ODMRP.

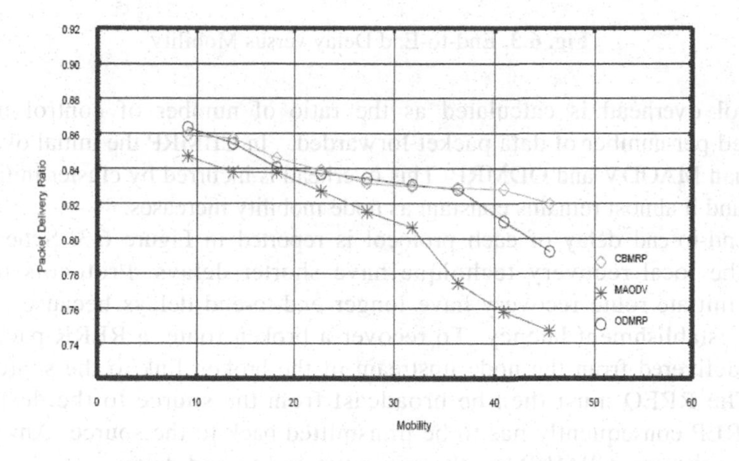

Fig. 6.1. Packet Delivery Ratio versus Mobility

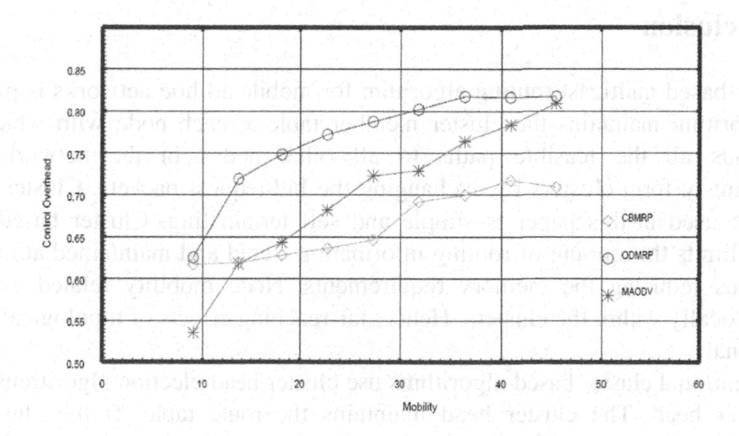

Fig. 6.2. Control Overhead versus Mobility

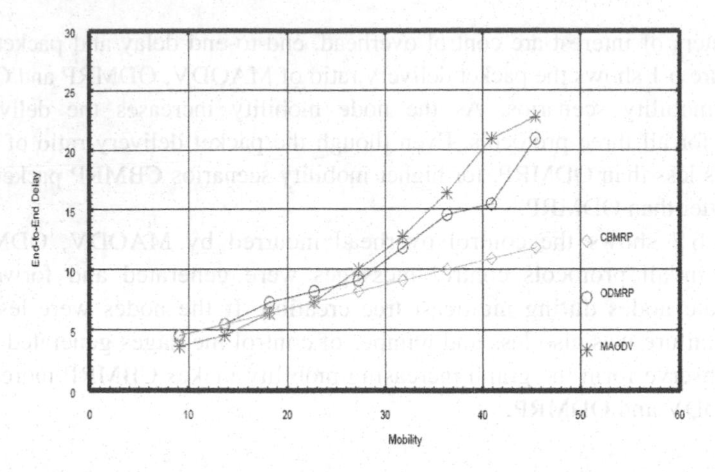

Fig. 6.3. End-to-End Delay versus Mobility

Control overhead is calculated as the ratio of number of control messages forwarded per number of data packet forwarded. . In CBMRP the initial overhead is higher than MAODV and ODMRP. This overhead is incurred by cluster initialization packets and it almost remains constant as node mobility increases.

The end-to-end delay of each protocol is reported in Figure 6.3. Schemes that utilize the local recovery technique have shorter delays. Protocols in which sources initiate route recovery have longer end-to-end delays because of longer route re-establishment latency. To recover a broken route, a RERR packet must first be delivered from the node upstream of the broken link to the source of the route. The RREQ must then be broadcast from the source to the destination, and a RREP consequently has to be transmitted back to the source. Among local recovery schemes, CBMRP has the minimum end-to-end delay.

7 Conclusion

A cluster-based multicast routing algorithm for mobile ad hoc networks is presented. This algorithm maintains the cluster member table at each node with which every node finds all the feasible paths to all other nodes in the network. Nodes autonomously form clusters by exchanging the bid-request packets. Cluster creation technique used in this paper is simple and self terminating. Cluster based routing protocol limits the amount of routing information stored and maintained at individual hosts, thus reducing the memory requirements. Node mobility related events are handled locally within the clusters. Hence, far-reaching effects of topological changes are minimal.

Conventional cluster based algorithms use cluster head election algorithms to elect the cluster head. The cluster head maintains the route table. If the cluster head crashes, all routing information will be lost. Network traffic is also centered towards the cluster head. Such networks are susceptible to single point failures. In the proposed fully distributed architecture, all cluster members maintain the cluster member table, and is free from single point failure. Cluster maintenance is lightweight compared to cluster head election process. Traffic is regulated according to the network traffic conditions.

Analysis shows that the performance of our algorithm is not adversely affected by the increase in network size. Future works includes the implementation and testing of the algorithm in real environments.

References

[1] Futernik, A., Haimovich, A.M., Papavassiliou, S.: An Analytical Model for Measuring QoS in Ad Hoc Wireless Networks. In: Globecom 2003 (2003)

[2] Jaikaeo, C., Shen, C.-C.: Adaptive Backbone-Based Multicast for Ad Hoc Networks. In: Proceedings of the IEEE International Conference on Communication, vol. 5, pp. 3149–3155 (2002)

[3] Dovrolis, C., Stiliadis, D., Ramanathan, P.: Proportional Differentiated Services: Delay Differentiation and Packet Scheduling. IEEE/ACM Transactions on Networking 10(1), 12–26 (2002)

[4] Blough, D.M., Leoncini, M., Resta, G., Santi, P.: The k-Neighbors Approach to Interference Bounded and Symmetric Topology Control in Ad Hoc Networks. IEEE Transactions on Mobile Computing 5(9), 1267–1272 (2006)

[5] Wu, H., Wang, C., Tzeng, N.-F.: Novel Self-Configurable Positioning Technique for Multihop Wireless Networks. IEEE/ACM Transactions on Networking 13(3), 609–621 (2005)

[6] Wu, J., Dai, F.: Efficient Broadcasting with Guaranteed Coverage in Mobile Ad Hoc Networks. IEEE Transactions on Mobile Computing 4(3), 259–270 (2005)

[7] Sundaresan, K., Anantharaman, V., Hsieh, H.-Y., Sivakumar, R.: ATP: A Reliable Transport Protocol for Ad Hoc Networks. IEEE Transactions on Mobile Computing 4(6), 588–603 (2005)

[8] Ruiz, L.B., Nogueira, J.M., Loureiro, A.A.F.: MANNA: A Management Architecture for Wireless Sensor Networks. IEEE Communications Magazine, 116–125 (2003)

[9] Chatterjee, M., Das, S.K., Turgut, D.: WCA: A Weighted Clustering Algorithm for Mobile Ad Hoc Networks. Cluster Computing 5(2), 193–204 (2002)

[10] Thoppian, M.R., Prakash, R.: A Distributed Protocol for Dynamic Address Assignment in Mobile Ad Hoc Networks. IEEE Transactions on Mobile Computing 5(1), 4–19 (2006)

[11] Drinić, M., Kirovski, D., Potkonjak, M.: Model-Based Compression in Wireless Ad Hoc Networks. In: Proceedings of the 1st international conference on Embedded networked sensor systems, pp. 231–242 (2003)

[12] Marbach, P., Qiu, Y.: Cooperation in Wireless Ad Hoc Networks: A Market Based Approach. IEEE/ACM IEEE Transactions on Networking (6), 1325–1338 (2005)

[13] Purtoosi, R., Taheri, H., Mohammadi, A., Foroozan, F.: A Light-Weight Contention-Based Clustering Algorithm for Wireless Ad Hoc Networks. In: Proceedings of the Fourth International Conference on Computer and Information Technology (CIT 2004) (2004)

[14] Panagiotakis, S., Alonistioti, A., Merakos, L.: An Advanced Location Information Scheme for Supporting Flexible Service Provisioning in Reconfigurable Mobile Networks. IEEE Communications Magazine, 88–98 (2003)

The Impact of Noise on the Scaling of Collectives: The Nearest Neighbor Model [Extended Abstract]

Nisheeth K. Vishnoi

Computer Science Division
University of California Berkeley, CA 94720
nisheeth.vishnoi@gmail.com

Abstract. This paper presents a theoretical study of the impact of noise on the scaling of a cluster when the processors participate in "local" collectives with their nearest neighbors. The model considered here is an extension of that introduced in [9] for understanding the effect of noise on the scaling of "global" collectives in large clusters. In this paper, the scaling is studied with respect to three fundamental aspects: (1) the distribution of noise: whether it is heavy or light tailed; (2) the temporal independence of noise; (3) the topology of the cluster. When the noise has a "light" tail and is temporally independent, it is shown that the cluster scales well, i.e., the slowdown per phase is just proportional to the (logarithm of the) maximum degree of the communication topology. This implies that for popular topologies such as grids and toruses the slowdown per phase is just a constant factor, which is independent of the number of processors. In the light tailed case, assuming only a weak temporal independence, a general upper bound is derived in terms of an "expansion" parameter of the communication topology. For grid-like graphs this establishes an exponential speedup compared to what was shown for global collective operations in [9].

1 Introduction

Motivation. It has been observed by several researchers, see [1,2,3], that the "throughput" of several high performance computing (HPC) clusters running scientific applications drops as the number of processors in the cluster increases. It has been suggested in [1,2] that one of the main causes of this is the "noise" in the processors of the cluster in the form of overheads such as daemons and interrupts. Given the exorbitant amount of resources invested in building such systems, it becomes extremely important to understand the reasons for this loss in efficiency and, if possible, rectify it. As a first formal step towards achieving this, the impact of noise on the scaling of these clusters was explained via a stochastic model in [9]. They abstracted a typical scientific parallel application in which each node in the cluster repeatedly performs a "phase" which consists of a computation stage followed by a "global" collective stage. A collective involves all processors coming to a common state once they are through with

S. Aluru et al. (Eds.): HiPC 2007, LNCS 4873, pp. 476–487, 2007.

their computation stage, from where on they resume the next phase. Hence, a large amount of noise in one of the processor, which would result in an untimely completion of the work assigned to it, may end up slowing down everyone. The results of [9] showed, in particular, that even in the most favorable case when the noise in each processor is distributed according to the exponential distribution, the per phase throughput drops by a factor of $\Omega(\ln N)$, where N denotes the number of processors. Or, if each processor would do w work per phase when there was no noise, it would end up doing only $O(w/\ln N)$ work per phase. Hence, as the number of processors tends to infinity, the throughput goes to zero, somewhat defeating the purpose why such clusters are in place. The main objective of such a study is to understand the problem of noise and the ways in which it has the potential to degrade system performance. The end goal being to reduce the impact of this necessary evil, and hence, to improve the performance of HPC systems.

This Work. In this paper we study the impact of noise on the performance of a cluster when the processors participate in "local" collectives, i.e., with their nearest neighbors in the communication topology. We suitably extend the stochastic model of [9] who did exactly this, albeit, for global collectives. Before we go on, let us briefly (at the expense of being imprecise) outline the model.[1] Consider a parallel program with N threads running on a system with N processors. The system is dedicated to running the same program repeatedly, each such run being referred to here as a *phase*. Each phase consists of two stages (1) a computation stage: In this stage each processor is supposed to do w amount of work. (2) A communication stage: In this stage each processor communicates with a specified set of processors, referred to as its *neighbors*. This communication pattern is referred to as the communication topology and is captured by a directed graph on N processors: G. But, because of noise, each processor has to devote an overhead time which in phase j for the i-th processor is $\delta_{i,j}$ and is assumed to be distributed according to a random variable δ with mean wf, where f is a fixed number in the interval $[0,1]$. It can be safely assumed that the noise is independent spatially, i.e., in the absence of coordinated OS policies, the noise across different processors is uncorrelated. It sometimes may be difficult to argue, and not true, that the noise is uncorrelated temporally, i.e., with j, although one may expect noise to be uncorrelated in the short range and long range. In this setup, we explore the scaling, or per phase work per processor with respect to three fundamental aspects: (1) the tail of the distribution of noise: $\mathbf{Pr}[\delta \geq t]$; (2) the temporal independence of noise: the correlation patterns of $\{\delta_{i,j}\}$ with j; (3) the communication topology: G.

Our Results. Our results can be informally summarized in the following points. To see the precise technical statement, the reader is referred to the corresponding theorem.

1. For the case when the noise is temporally independent and light tailed, we prove that time per phase is "sharply concentrated" around the quantity

[1] A detailed description of the model appears in Section 2.

$w + wf(\ln \Delta) + \tau$, here Δ is exactly one more that the largest in-degree of the communication topology G. (See Theorem 7.)
2. For the case when the noise is "weakly" temporally dependent, light tailed and the growth rate of the topology is bounded, we prove that the time per phase is upper bounded by the quantity $w + wf(\ln \ln N) + \tau$. (See Theorem 8.)
3. For the case when the noise is heavy tailed and temporally independent, we show that the topology has little effect and the time taken per phase is almost the same as if the processors were involved in a global barrier. (See Section 4.) This has been proved independently in [11].
4. We also prove sharp results when the communication patterns are multi-round, such as a binary tree, or one that arises in the Fast Fourier Transform. (See Theorems 9 and 10.)

Related Work. Our work is a natural follow-up to that of [9]. The work of [10] takes an in-depth look into the theoretical model of the impact of noise on the collectives studied in [9] by validating its accuracy against data collected from production clusters. Recently, we came across the work of Lipman and Stout [11] who have, independently of this work, considered a problem which reduces to a similar stochastic problem considered in our paper. The main result of their paper is a tight bound for the case when the noise is light tailed, temporally independent and the topology is a directed cycle. This can be seen as a special case of our first result.

The problem on how to alleviate the problem of noise has been looked by several research teams [3,4,5,6,7,8]. Our theoretical understanding could potentially be coupled with ideas from these works to alter the noise in the systems, as a function of the communication topology, and help improve the performance of these systems considerably.

Organization. In Section 2 we present a description of the model considered in the paper. In Section 3 we analyze the model presented in Section 2. The results presented in Section 3 are general and rigorous. In Section 4 we describe the results obtained in Section 3.2 for the light and heavy tailed distributions. The canonical example of light tailed distribution we consider is the exponential distribution, while for the heavy tailed distribution, we consider the Pareto distribution. In Section 5 we present the results pertaining to multiple communications per phase. For the lack of space, most of the proofs are deferred to the full version of this paper.

2 The Setup

2.1 Modeling the Application

In this section we present the stochastic model considered in this paper. Since it is an extension of the model presented in [9], we choose to keep the terminology as similar as possible. The basic setup is a parallel program with N threads running on a system with N processors. Typically, the system is dedicated to running the same program repeatedly, each such run being referred to here as a *phase*, albeit on different inputs every time. Each phase is a composition of two kinds of stages:

- Computation stage: In this stage the processor does computation without any message exchange or I/O activity.
- Communication stage: Once the processor has finished its computation stage, it enters the communication stage. In this stage each processor communicates with a specified set of processors, referred to as its *neighbors*. In this stage there is negligible computation except that associated to communication.

For the sake of simplicity, as has been done in the previous papers on this, we assume that for every processor, each phase just consists of a computation stage followed by a communication stage. Later, we will consider the case of FFT where this is not the case, and each phase comprises of a number of alternations of computation stages with communication stages. Thus, every phase is characterized by the amount of work assigned to each processor and the pattern of communication that occurs between them. Formally, in phase $j \geq 1$, let $W_{i,j}$ be the work assigned to processor $1 \leq i \leq N$, which would be completed by it in (deterministic) time $w_{i,j}$ was there no noise. The communication in phase j can be captured by a directed graph (possibly with loops) $G_j([N], E_j)$. Here, E_j consists of the directed edges along which communication happens. We represent an edge as (i_1, i_2), meaning that i_1 communicates a message to i_2. The time this communication takes is $c_{i_1,i_2,j}$. This completes the overview of a phase. Now we proceed to a detailed quantitative description of a phase.

A Phase. Let $t^s_{i,j}$ denote the time when the i-th thread begins phase j, and let $t^f_{i,j}$ denote the time when it ends the computation stage in the j-th phase. Let $W_{i,j}$ denote the amount of work carried out by thread i in the computation stage of the j-th phase. If the system is noiseless, the time required by processor i to finish work $W_{i,j}$ in its j-th phase will be a deterministic quantity, which we denote by $w_{i,j}$. This quantity typically depends on several characteristics of the processor such as its clock frequency, its architectural parameters, and the state of the node (such as cache contents) just before the j-th phase is entered. Therefore, $t^f_{i,j} - t^s_{i,j} = w_{i,j}$. Due to the presence of noise, the time taken by processor i to finish the work $W_{i,j}$ is typically not a constant. There will be a variable component that represents the time consumed to service the daemons and other asynchronous events. This is captured by a random variable $\delta_{i,j}$. More precisely, $t^f_{i,j} - t^s_{i,j} = w_{i,j} + \delta_{i,j}$. Let $f_{i,j} \in [0, 1]$ be the fraction representing the system overhead for the processor, i.e., let $f_{i,j} := \frac{E[\delta_{i,j}]}{w_{i,j}}$. Thus, we may think of the noise as a random variable $\eta_{i,j} := \frac{\delta_{i,j}}{f_{i,j} w_{i,j}}$ with mean one. Thus, we may write the wall-clock time taken by processor i for the j-th phase as $t^f_{i,j} - t^s_{i,j} = w_{i,j}\left(1 + f_{i,j}\eta_{i,j}\right)$.

For $j \geq 1$, phase $j+1$ starts for the i-th processor when it has completed phase j. The first phase starts at time zero for all processors. The $j+1$-th phase ends for the i-th processor when it has completed its computation in the $j+1$-th phase, as well as, all the processors i' such that $(i', i) \in E_{j+1}$ have ended the computation in their $j+1$-th phase. Define $N_j(i)$ to be the set of processors which have an edge directed towards i in G_j. With this notation, one can define the time taken by the i-th processor to complete the $j+1$-th phase, denoted by $T_{i,j+1}$ as

$$\max\left\{T_{i,j} + w_{i,j}(1 + f_{i,j}\eta_{i,j}), \max_{i' \in N_{j+1}(i)} \{T_{i',j} + w_{i',j}(1 + f_{i',j}\eta_{i',j}) + \tau_{i',i}\}\right\} \quad (1)$$

Here, $\tau_{i',i}$ denotes the communication time between processors i and i'. (We assume this is symmetric and independent of j.) This completes a description of a generic phase in the most general setting.

Performance Measure. Given this description of a phase, a natural measure of performance is the amount of time taken by each processor to complete n phases. More formally, given an $\varepsilon > 0$, to be thought of as very small, one would be interested in "eff"[2] which is defined to be the smallest number such that $\mathbf{Pr}\left[\frac{\max_{i=1}^n\{T_{i,n}\}}{n} \leq \text{eff}\right] \geq 1 - \varepsilon$. Thus, with probability at-least $1 - \varepsilon$, each processor finishes n phases in time $n \cdot \text{eff}$.

2.2 Simplifying Assumptions

Now we present some simplifications, which make the model amenable for theoretical analysis and, yet, not render it unrealistic. The justifications for these assumptions are presented in detail in the paper [9], and we will only discuss them here very briefly. Several assumptions have been verified for real systems in [10]. Of course, one can make the model more and more *real* by removing some of these assumptions, but then the theoretical analysis of the model also becomes considerably difficult.

(1) Identical Communication. $G_j = G$ are the same for all $j \geq 1$. Also, as in [9], we assume that each message transmission between a pair of processors takes time τ, which is referred to as the *one-way latency.*

(2) Balanced Load. $W_{i,j} = W$ for all i, j. This means that each thread is supposed to do the same amount of work in its compute stage. For instance, each thread is supposed to be multiplying two matrices of the same size.

(3) Identical Processors. $w_{i,j} = w$ for all i, j. This means that all the processors are identical in their computational power. Hence, in the noiseless case, given that $W_{i,j} = W$, $w_{i,j} = w$ for all processors.

(4) Stationary and Balanced Overheads. $f_{ij} = f$ for all i, j. In a typical systems, the processors are assigned an application for the lifetime of the application and running any other application on the node is avoided. Thus, the only interference is due to the background processes or daemons. The amount of daemon activity is not expected to change over time. Thus, we may assume $f_{i,j} = f_{i,j'}$, for all i, j

[2] This quantity, to be thought of as the efficiency of the system with respect to the application, will depend on the application and ε.

and j'. We further assume that $f_{i,j} = f_{i',j}$, for all i, i' and j. (See [9] for more on this assumption.)

(5) Identical Noise. $\eta_{i,j} \sim \eta$ for all i, j. Recall that we have arranged η and f such that $\mathbf{E}[\eta] = 1$.

(6) Spatial Independence. $\{\eta_{i,j} : 1 \leq i \leq N\}$ are independent for each j. This assumption is crucial to our results. This can be justified as, in a typical system under consideration, there is no coordinated scheduling policy to synchronize processes across different processors.

(7) t-Temporal Independence. For the simplest of our results, we will assume that the the random variables $\{\eta_{i,j}\}$, are independent, i.e., apart from spatial independence, there is temporal independence as well. This may not be necessarily true as some of the daemons could be somewhat periodic, and we do expect weak correlation patterns between these random variables across different phases. In general we may only assume limited independence. To this effect, we say that the process is t temporal independent, if for all $1 \leq i \leq N$ and $j \geq 1$, the set of random variables $\{\eta_{i,j'} : j \leq j' \leq j + t\}$ are independent. Typically, we will assume that $t \ll N$.

3 Analysis

3.1 The Simplified Problem

In this section we present the problem at hand with the simplifications made in the previous section. Applying assumptions (1)-(5) to Equation (1), we obtain that for $j \geq 0$,

$$T_{i,j+1} := \max\left\{T_{i,j} + w(1 + f\eta_{i,j}),\ \max_{i' \in N_{j+1}(i)} \{T_{i',j} + w(1 + f\eta_{i',j}) + \tau\}\right\}, \quad (2)$$

where $T_{i,0} = 0$ for all $1 \leq i \leq N$. The communication graph $G(V, E)$, and the parameters w, f and τ are fixed for the rest of the paper. This graph contains loop edges of the form (i, i) for all $1 \leq i \leq N$. The graph does not contain multiple edges in the same direction between a pair of vertices. Given $\varepsilon > 0$, recall that the goal is to give tight estimate of the quantity "eff" such that $\mathbf{Pr}\left[\frac{\max_{i=1}^{n}\{T_{i,n}\}}{n} \leq \text{eff}\right] \geq 1 - \varepsilon$. Since w, f and G are fixed, this quantity is just a function of n, ε and η.

3.2 General Results

We proceed to give general bounds on this quantity as a function of the random variable η. First we need a few definitions.

Definition 1. *A walk $\mathcal{W}$ of length n in a directed graph $G(V, E)$ consists of a sequence of (possibly repeated) edges $e_1, e_2, \ldots, e_n$ such that, if $e_k = (i_k, j_k)$, then for all $1 \leq k < n$, $j_k = i_{k+1}$. The starting vertex of a walk is i_1 while the ending vertex is j_n. With abuse of notation, a walk $\mathcal{W}$ will be denoted by $i_1, i_2, \ldots, i_n, i_{n+1}$, where $i_{n+1} := j_n$.*

The maximum in-degree of G is denoted by Δ (this includes the self-loop at each vertex). G is said to be Δ regular if all vertices have in-degree Δ. Let $\mathcal{W}_{i,n}$ denote the number of walks in G of length n that end at i, or $\mathcal{W} = i_1, \ldots, i_{n+1}$ such that $i_{n+1} = i$. It follows from the definitions that $|\mathcal{W}_{i,n}| \leq \Delta^n$, and if G is Δ regular, then this is an equality.

Definition 2. *For a vertex i in G, let $B(i,d)$ denote the ball of radius d centered at i. Formally, $B(i,d)$ contains all vertices i' such that there is a directed path of length at-most d from i' to i. For $G(V,E)$, let $B_G(d)$ denote $\max_{i \in V} |B(i,d)|$.*

Definition 3. *Let η be a random variable such that $\mathbf{E}[\eta] = 1$, and let $\sigma > 0$ and n be given. Let $\{\eta_i\}_{i=1}^n$ be n independent copies of η. Then η is said to be $p(\eta,n,\sigma)$-tailed if $\mathbf{Pr}\left[\sum_{i=1}^n \eta_i \geq \sigma n\right] \leq p(\eta,n,\sigma)$.*

Let M_η^d denote the the random variable which is distributed according to the maximum of d independent copies of η.

*Temporal Independence.*Assuming that the random variables $\eta_{i,j}$ are independent for all i and j, we obtain the following results. The first follows from Equation (2) via a direct application of union bound.

Theorem 4 (Upper Bound). *For $\sigma > 0$ and $n \geq 1$, let η be $p(\eta,n,\sigma)$-tailed, and the maximum in-degree of a vertex in G be Δ. Then, with probability at-least $1 - p(\eta,n,\sigma)\Delta^n$, $\frac{\max_{i=1}^N \{T_{i,n}\}}{n} \leq wf\sigma + w + \tau$.*

This theorem says that if for some σ, $p(\eta,n,\sigma)$ goes to zero faster than $1/\Delta^n$, then the per phase efficiency is at-most $wf\sigma + w + \tau$ with high probability. Indeed, for light tailed distributions, such as exponential distribution, this is true for $\sigma = \ln \Delta$. In fact for such distributions, one can show that this is the best we can hope for. We present a lower bound technique which, when applied to the exponential distribution shows that each phase will take time $wf\sigma + w + \tau$ on an average. (See Theorem 7.)

Now we present a general lower bound which is more convenient to state for regular graphs. This can be generalized to the case when the graph is not regular, but we omit it here.

Theorem 5 (Lower Bound). *Let G be a Δ regular graph, $n \geq 1$ and for $1 \leq j \leq N$, let M_j be i.i.d. M_η^Δ random variable. Then, for all $1 \leq i \leq N$, $T_{i,n} \geq wf \sum_{j=1}^n M_j + wn$.*

It is possible to incorporate the dependency of the lower bound on τ via a slightly more involved argument. We omit the easy proofs of these theorems from this version of the paper and focus on what they imply in Section 4.

Limited Temporal Independence. Consider now the case when the noise random variables $\eta_{i,j}$ t-temporally independent. In this case, Theorem 4 can no longer be expected to hold. Here we provide an upper bound by a *stochastic embedding* technique which takes into account the topology of the communication graph. The basic idea is to consider t phases at a time, which we refer to as a meta-phase. If the graph G has the property that a large delay at a node does not end up affecting processors further than distance t from it, then the meta phase ends much faster. Imagine the following stochastic process. Every processor draws t samples from its noise distribution. Because of the spatial independence and t-temporal independence, all these samples are i.i.d according to η. We denote these noise distributions as $\eta_{i,j}$. The meta-phase ends when all processors have finished t phases. Consider a node i, and recall that $B(i,r)$ denotes the set of processors from which i is reachable by a path of length at-most r. Recall also that $B_G(r)$ denotes the size of the largest ball of radius r in G. Let $\zeta_j := \max_{i' \in B(i,r)} \eta_{i',j}$, where $\eta_{i',j}$ are i.i.d. according to η. Let $p(s) = \mathbf{Pr}[\sum_{j=1}^{t} \zeta_j \geq s]$. Hence, $\mathbf{Pr}[T_{i,t} \geq wfs + wt + t\tau] \leq p(s)$. Thus, using a union bound, one obtains the following theorem.

Theorem 6 (Limited Independence Upper Bound). *Let $\sigma \geq 0$ and $t \geq 1, r \geq 0$ be integers. Let η be the distribution of noise which is t-temporally independent. Further, let $Y_1, \ldots, Y_t$ be i.i.d. according to $M_\eta^{B_G(r)}$, and $Y := \sum_{i=1}^{t} Y_i$. Then $\mathbf{Pr}\left[\frac{\max_{i=1}^{N}\{T_{i,t}\}}{t} \leq wf\sigma + w + \tau\right] \geq 1 - N\mathbf{Pr}[Y \geq \sigma t]$.*

This theorem says that even in the case of limited temporal independence, as long as the noise has the property that the sum of a small number of them have a light tail, one can still upper bound the per phase time by something much better than what one would expect in the global collective case. Of course, this requires that the communication graph is not expanding in the sense that the number of neighbors in a radius r grow slowly as a function of r for all the processors. This is indeed true for d-dimensional grids and toruses for fixed d.

4 Results for Representative Distributions

In this section we explain Theorems 4, 5 and 6 for the context of light tailed and heavy tailed distributions for noise. We pick the canonical examples of these two cases: the exponential and the Pareto respectively.

4.1 Distributions

Exponential. An exponential distribution Exp(1) has the following distribution: $\forall x \geq 0$, $\mathbf{Pr}[\text{Exp}(1) \leq \text{x}] = 1 - \exp(-x)$. First we note some important properties of this distribution. Let $X_1, \ldots, X_n$ be i.i.d. according to Exp(1), then

1. $Y := \sum_{i=1}^{n} X_i$ is distributed according to $\Gamma(n,1)$ which has the following p.d.f. $f(x;n,1) := x^{n-1}\frac{\exp(-1)}{(n-1)!}$, for $x > 0$. The moment generating function is $(1-t)^{-n}$ for $t < 1$. It follows from Chernoff Bounds that for all $\Delta > 1$, for all $0 < \delta \leq \delta_0(\Delta)$, and all $n \geq n_0(\delta)$, $\mathbf{Pr}[Y \geq (1+\delta)n \ln \Delta] \leq \exp(-(1+\delta/2)n \ln \Delta)$.
2. Let $Y := \max_{i=1}^{\Delta} X_i$. Then Y is distributed according to the random variable $\sum_{i=1}^{\Delta} \frac{1}{i} X_i$. (Lemma 1 below). It follows that $\mathbf{E}[Y] = \text{Var[Y]} = \text{H}_\Delta := \sum_{\text{i}=1}^{\Delta} \frac{1}{\text{i}}$. Hence, it follows from Chebyshev's Inequality that if $Y_1, \ldots, Y_n$ are

distributed according to Y, are pairwise independent, then for any $0 < \delta < 1$, $\mathbf{Pr}[\sum_{i=1}^n Y_i \geq (1-\delta)nH_\Delta] \geq 1 - \frac{1}{\delta^2 n H_\Delta}$. It follows from the inequality that $\frac{1}{2\Delta+2} \leq |H_\Delta - \ln\Delta - \gamma| \leq \frac{1}{2\Delta}$ (where $\gamma > 0$ is the Euler-Mascheroni constant) that, for all $0 < \delta \leq \delta_0(\Delta)$, and all $n \geq n_0(\delta)$, $\mathbf{Pr}[\sum_{i=1}^n Y_i \geq (1-\delta/2)n\ln\Delta] \geq 1 - \frac{1}{\delta^2 n H_\Delta}$.

Lemma 1. *Let $X_1, \ldots, X_k$ be i.i.d. according to* Exp(1). *Then $\max_{i=1}^k X_i$ has the same distribution as $\sum_{i=1}^k \frac{X_i}{i}$.*

Spatial and Temporal Independence. The fact (1) above implies that for all $\delta > 0$ small enough, when $\eta \sim$ Exp(1), $p(\eta, n, (1+\delta)\ln\Delta) \leq \frac{1}{\Delta^{(1+\delta/2)n}}$. Hence, by Theorem 4, with probability at-least $1 - \Delta^{-n\delta/2}$, $\frac{\max_{i=1}^N\{T_{i,n}\}}{n} \leq wf(1+\delta)\ln\Delta + w + \tau$. While the fact (2) above combined with Theorem 5 implies that for all $\delta > 0$ small enough, with high probability, every processor finishes n phases in time at-least $(1-\delta/2)n\ln\Delta$. These together imply the following theorem.

Theorem 7. *For all $\delta > 0$ small enough, for $\eta \sim$* Exp(1), *and when the communication topology is given by a (regular) digraph G with in-degree Δ, with high probability, the efficiency or the maximum average time per phase for each processor lies between $wf(1-\delta/2)\ln\Delta + w + \tau$ and $wf(1+\delta)\ln\Delta + w + \tau$.*

Thus, for standard communication topologies such as toruses or meshes, this result is optimal.

Limited Temporal Independence. Now we show that assuming $O(\ln N)$-temporal independence, and the fact that for $r = O(\ln N)$, the communication graph is not expanding, i.e., $B_G(O(\ln N)) = O((\ln N)^{O(1)})$, $\ln N$ phases will finish in time $O(\ln N \ln\ln N)$. Thus giving an efficiency of $O(\ln\ln N)$ per phase with high probability. Compare this to the case when each phase takes $\Theta(\ln N)$ time in the case of global collectives [9]. Formally, we have the following theorem.

Theorem 8. *Let $c_1, c_2 > 0$ be large constants. Let the communication graph G have the property that for all $r \leq c_1 \ln N$, $B_G(r) \leq (\ln N)^{c_2}$. If the noise is distributed according to* Exp(1) *and is $\ln N$-temporally independent, then with probability at-least $1 - 1/N^{100}$ each processor finishes $\ln N$ phases in time at-most $100\ln N(wf\ln\ln N + w + \tau)$. Thus, the average time per phase for each processor is at-most $100wf\ln\ln N + w + \tau$ with high probability.*

The proof of this theorem relies on Theorem 6 and the measure concentration inequality for the random variable distributed according to the maximum of exponential distributions and we will include it in the full version of this paper. This is a significant speed-up compared to $\ln N$ per phase and assumes that noise is temporally independent for only about $\ln N$ phases. It would be interesting to see if this is indeed observed for real systems.

Pareto. In this section we consider the case when the noise has a heavy tail. This is unlike the exponential case and the noise looks more like the uniform distribution. A natural and very popular way to model data which has heavy tail is the so-called Pareto distribution. The Pareto random variable X^a_{par} with parameter a has the following distribution: $\forall x \geq 1, \quad \mathbf{Pr}[X^a_{\text{par}} \leq x] = 1 - \frac{1}{x^a}$. The Pareto distribution has mean $\frac{a}{a-1}$. To make this random variable with unit mean, we let η be $\frac{a-1}{a} X^a_{\text{par}}$. The reason why when the noise is distributed according to Pareto the system will invariably slow down is very simple. After $t \leq N$ phases, a processor i starts depending on processors which are connected to it by a directed path of length t. Thus, the number of independent copies of η after t phases on which i-th processor depends is $P(i,t) := \sum_{r=0}^{t} |B_G(i,r)|$. It follows from the distribution of the maximum of Pareto distribution that, with high probability, there is one of these which will be at-least $\Omega(P(i,t)^{1/a})$. Hence, the i-th processor will take time at-least $\Omega(\frac{1}{t} P(i,t)^{1/a})$ on an average per phase. If $P(i,t) = \Omega(t^\beta)$ for some $\beta > a$, then this quantity is at-least $\Omega(t^{\beta/a-1})$. When G is a ring, as noted in [11], $\beta = 2$, and hence, for $1 < a < 2$, the slowdown is $\Omega(N^{2/a-1})$. We do not discuss it further here as the main idea already appears in the paper of Lipman and Stout [11].

5 Multiple Communications Per Phase

In this section we consider two cases of a complex communication pattern per phase. The first is the complete binary tree and the second is Fast Fourier Transform (FFT). Both are very natural. The binary tree will arise in any divide and conquer type of application, e.g. Merge Sort, while FFT is a standard benchmark for HPCC. For the sake of clarity we would consider the case when the communication delays are negligible. This is just to highlight the impact of noise, and all results can be suitably modified to incorporate the communication component.

5.1 Binary Tree

Consider the case when $N = 2^k$, where the processors are labeled $\{0, 1, \ldots, 2^k - 1\}$. Each phase consists of k rounds. In the i-th round $(1 \leq i \leq k)$, processors j and $j + 2^{i-1}$ communicate for $j = 0 \cdot 2^i, 1 \cdot 2^i, 2 \cdot 2^i, \ldots, \left\lfloor \frac{2^k-1}{2^i} \right\rfloor \cdot 2^i$. An example of such a communication pattern is given in Figure 1. We assume that the one way latency $\tau \sim 0$, and hence, only focus on the delay due to synchronization. Let $\eta_{i,j}$ be the random variable denoting the noise incurred by processor i in the j-th round. Assume $\{\eta\}_{i,j}$ are i.i.d. according to Exp(1), let w be the work by each processor done per round, and the overhead factor per round is f. Thus, time taken to complete one phase is the random variable $wk + wf \max_P \sum_{(i;j) \in P} \eta_{i,j}$. Here the maximization is over all paths that go from a *leaf* to the *root* of the binary tree. There is a path corresponding to each leaf, which is just a processor, and there are exactly 2^k of them. Hence, the quantity that we need to estimate is $B_k := \max_P \sum_{(i;j) \in P} \eta_{i,j}$. We prove the following theorem which establishes a remarkable threshold phenomena in the completion time of each phase.

Theorem 9. *Let* $c_L = 2.678\ldots$ *be a solution to the equation* $\ln 2 + \ln x - x + 1 = 0$. *Then* $B_k/k \to c_L$ *almost surely as* $k \to \infty$. *(Here* B_k *is as defined above when*

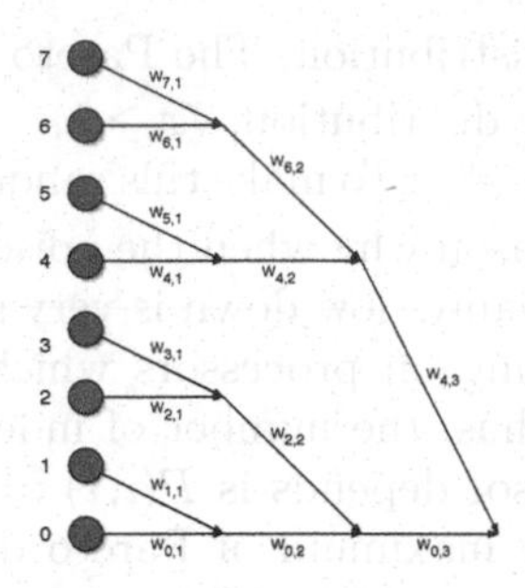

Fig. 1. The communication pattern for a binary tree and a single node in FFT

$\eta_{i,j}$ are i.i.d. according to Exp(1).*) Thus, the time to complete each phase is almost surely* $(1 + fc_L)w \log_2(N + 1)$.

The upper bound proof involves a tight approximation to the distribution obtained by summing k independent copies of the exponential random variable. The lower bound proof is technically interesting as it uses a result from the theory of branching processes on the behavior of a supercritical Galton-Watson process. It is worth noting that the lower bound argument given earlier for the general case (Theorem 5) does not give the optimal constant, and one needs to appeal to theory of branching processes to obtain the optimal constant. Also, a close look at the proof of Theorem 9 yields that a similar threshold result can be obtained when $\eta_{i,j}$ are i.i.d. according to any distribution for which there is a large deviation inequality.

5.2 Fast Fourier Transform

In this section we consider the communication pattern for an application computing the FFT. Here, each phase consists of k rounds. In the i-th round, $1 \leq i \leq k$, processors j and $2^{i-1} + j$ communicate with each other, where $0 \leq j < 2^k$. The communication pattern for one processor is a binary tree, as depicted in Figure 1. A phase consists of a binary tree for each processor, except that these binary trees share edges. For instance, the binary tree for processes j and $2^{k-1} + j$ are the same for all j. Let H_k denote the time taken for a phase to complete when each processor does w work per round, f is the noise overhead, and the noise in each round is distributed according to Exp(1). Then we have the following theorem on H_k.

Theorem 10. *Let $c_L = 2.678\ldots$ be a solution to the equation $\ln 2 + \ln x - x + 1 = 0$, and $c_U = 3.692\ldots$ be a solution to $2\ln 2 + \ln x - x + 1 = 0$. Then the following hold: (1) $\limsup_{k\to\infty} H_k/k \leq c_U$ almost surely. (2) $\liminf_{k\to\infty} H_k/k \geq c_L$ almost surely. Thus, the time to complete each phase is almost surely bounded between $(1 + fc_L)w\log_2(N + 1)$ and $(1 + fc_U)w\log_2(N + 1)$.*

This theorem establishes that in the case of FFT, inspite of the dependencies among the binary trees of the processors, each phase finishes in time $O(\log N)$.

Remark 1. We conjecture that H_k also has threshold behavior as B_k in Theorem 9. The current techniques do not seem powerful enough to resolve this question.

Acknowledgment. I would like to thank Rahul Garg for suggesting to look at this model involving local collectives and Daniel Stefankovic for collaborating on this paper at an early stage.

References

1. Gioiosa, R., Petrini, F., Davis, K., Lebaillif-Delamare, F.: Analysis of System Overhead on Parallel Computers. In: ISSPIT 2004. The 4th IEEE International Symposium on Signal Processing and Information Technology, Rome, Italy (December 2004)
2. Jones, T.R., Brenner, L.B., Fier, J.M.: Impacts of Operating Systems on the Scalibility of Parallel Applications. Tech. Rep. UCRL-MI-202629, Lawrence Livermore National Laboratory, (March 2003)
3. Petrini, F., Kerbyson, D.J., Pakin, S.: The Case of the Missing Supercomputer Performance: Achieving Optimal Performance on the 8,192 Processors of ASCI Q. In: ACM/IEEE Conference on Supercomputing (SC 2003), Phoenix, Arizona, USA (November 2003)
4. Frachtenberg, E., Petrini, F., Fernandez, J., Pakin, S., Coll, S.: STORM: Lightning-Fast Resource Management. In: ACM/IEEE Conference on Supercomputing (SC 2002), Baltimore, Maryland, USA (November 2002)
5. Hori, A., Tezuka, H., Ishikawa, Y.: Highly Efficient Gang Scheduling Implementation. In: SC 1998. ACM/IEEE Conference on Supercomputing, Orlando, FL, USA (November 1998)
6. Jones, T., Dawson, S., Neely, R., Tuel, W., Brenner, L., Fier, J., Blackmore, R., Caffrey, P., Maskell, B., Tomlinson, P., Roberts, M.: Improving the Scalability of Parallel Jobs by adding Parallel Awareness to the Operating System. In: ACM/IEEE Conference on Supercomputing (SC 2003), Phoenix, Arizona, USA (November 2003)
7. Frachtenberg, E., Feitelson, D., Petrini, F., Fernández, J.: Flexible Coscheduling: Mitigating Load Imbalance and Improving Utilization of Heterogeneous Resources. In: IPDPS 2003. International Parallel and Distributed Processing Symposium 2003, Nice, France (April 2003)
8. Agarwal, S., Choi, G.S., Das, C.R., Yoo, A.B., Nagar, S.: Co-ordinated Coscheduling in Time-Sharing Clusters through a Generic Framework. In: CLUSTER 2003. IEEE International Conference on Cluster Computing, Hong Kong (December 2003)
9. Agarwal, S., Garg, R., Vishnoi, N.: The Impact of Noise on the Scaling of Collectives: A Theoretical Approach. In: Bader, D.A., Parashar, M., Sridhar, V., Prasanna, V.K. (eds.) HiPC 2005. LNCS, vol. 3769, pp. 280–289. Springer, Heidelberg (2005)
10. Garg, R., De, P.: Impact of Noise on Scaling of Collectives: An Empirical Evaluation. In: Robert, Y., Parashar, M., Badrinath, R., Prasanna, V.K. (eds.) HiPC 2006. LNCS, vol. 4297, pp. 460–471. Springer, Heidelberg (2006)
11. Lipman, J., Stout, Q.F.: A performance analysis of local synchronization. In: SPAA. Symp. Parallelism in Algorithms and Architectures (2006)

Optimization of Collective Communication in Intra-cell MPI

M.K. Velamati[1], A. Kumar[1], N. Jayam[1], G. Senthilkumar[1], P.K. Baruah[1], R. Sharma[1], S. Kapoor[2], and A. Srinivasan[3]

[1] Dept. of Mathematics and Computer Science, Sri Sathya Sai University
[2] IBM, Austin
[3] Dept. of Computer Science, Florida State University
asriniva@cs.fsu.edu

Abstract. The Cell is a heterogeneous multi-core processor, which has eight co-processors, called SPEs. The SPEs can access a common shared main memory through DMA, and each SPE can directly operate on a small distinct local store. An MPI implementation can use each SPE as if it were a node for an MPI process. In this paper, we discuss the efficient implementation of collective communication operations for intra-Cell MPI, both for cores on a single chip, and for a Cell blade. While we have implemented all the collective operations, we describe in detail the following: barrier, broadcast, and reduce. The main contributions of this work are (i) describing our implementation, which achieves low latencies and high bandwidths using the unique features of the Cell, and (ii) comparing different algorithms, and evaluating the influence of the architectural features of the Cell processor on their effectiveness.

Keywords: Cell Processor, MPI, heterogeneous multicore processor.

1 Introduction

The Cell is a heterogeneous multi-core processor from Sony, Toshiba and IBM. There has been much interest in using it in High Performance Computing, due to the high flop rates it provides. However, applications need significant changes to fully exploit the novel architecture. A few different models of the use of MPI on the Cell have been proposed to deal with the programming difficulty, as explained later. In all these, it is necessary to implement collective communication operations efficiently within each Cell processor or blade.

In this paper, we describe the efficient implementation of a variety of algorithms for a few important collective communication operations, and evaluate their performance. The outline of the rest of the paper is as follows. In §2, we describe the architectural features of the Cell that are relevant to the MPI implementation, and MPI based programming models for the Cell. We explain common features of our implementations in §3.1. We then describe the implementations and evaluate the performance of MPI_Barrier, MPI_Broadcast, and MPI_Reduce in §3.2, §3.3, and §3.4 respectively.

S. Aluru et al. (Eds.): HiPC 2007, LNCS 4873, pp. 488–499, 2007.

We summarize our conclusions in §4. *Further details on this work are available in a technical report* [4].

2 Cell Architecture and MPI Based Programming Models

Architecture. Figure 1 shows an overview of the Cell processor. It consists of a cache coherent PowerPC core (PPE), which controls eight SIMD cores called Synergistic Processing Elements (SPEs). All cores run at 3.2 GHz and execute instructions in-order. The Cell has a 512 MB to 2 GB external main memory, and an XDR memory controller provides access to it at a rate of 25.6 GB/s. The PPE, SPE, DRAM and I/O controllers are all connected via four data rings, collectively known as the EIB. Up to 128 outstanding DMA requests between main storage and SPEs can be in process concurrently on the EIB. The EIB's maximum bandwidth is 204.8 GB/s. The Cell Broadband Engine Interface (BEI) manages data transfers between the EIB and I/O devices. One of its channels can be used to connect to another Cell processor at 25.6 GB/s, creating a Cell blade with a logical global shared memory.

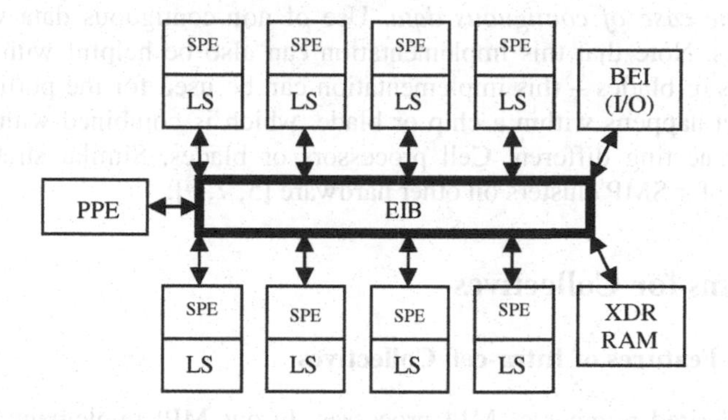

Fig. 1. Overview of the Cell processor

Each SPE has its own 256 KB local store from which it fetches code and reads and writes data, with access latency of 6 cycles. All loads and stores issued from the SPE can only access the SPE's local store. Any main memory data needed by the SPE must be moved into the local store explicitly through a DMA. An SPE can have up to sixteen pending requests in its DMA queue. The maximum DMA size is 16 KB.

The DMAs may execute out-of-order. Partial ordering can be ensured by fenced or barriered DMAs. The former executes only after all previous DMAs with the same tag on the same SPE have completed. The latter has the same guarantee, but also ensures that all subsequent DMAs issued on the same SPE with the same tag execute after it has completed. A DMA list can be used to scatter data to or gather data from multiple locations. It occupies only one slot in the DMA queue.

We observe the following regarding the performance of DMAs [4]: (i) SPE-SPE DMAs are much faster than SPE-main memory DMAs on the same chip, (ii) sending multiple small DMAs is slower than sending fewer long ones from the same SPE,

(iii) latency between SPEs on different chips are significantly higher than those on the same chip, (iv) maximum bandwidth between SPE and main memory is around 7 GB/s, while between SPE and SPE it is around 25 GB/s, (v) latency is higher in the presence of congestion, when multiple SPEs are transferring data, and (vi) the variance of the latency is higher with congestion.

MPI Based Cell Programming Models. The heterogeneous architecture and the small local stores of the SPEs make programming the Cell difficult. Some of the programming models to deal with this challenge are based on MPI. In the MPI microtask model [6], the application is divided into several smaller tasks with small code and data size. A scheduler schedules the tasks on the SPEs. In another model [3], an existing application is ported to the Cell by treating each SPE as if it were a node for an MPI process, using the main memory to store the application data, and the local store as software controlled cache. Large code size can be dealt with by bringing in the code as needed through code overlaying. *This is the model for which we target our MPI implementation, assuming that application data is in main memory, and that the MPI calls are provided the effective addresses of these locations.* If the application data is in local store, then more efficient implementations can be developed. *We also discuss only the case of contiguous data.* Use of non-contiguous data will lead to higher latencies. Note that this implementation can also be helpful with clusters of Cell processors or blades – this implementation can be used for the portion of communication that happens within a chip or blade, which is combined with MPI communication connecting different Cell processors or blades. Similar strategies have been developed for SMP clusters on other hardware [5, 7, 9].

3 Algorithms for Collectives

3.1 Common Features of Intra-cell Collectives

Let P be the desired number of MPI processes. In our MPI implementation, a PPE process spawns P SPE threads, which perform the actual computation. We sometimes refer to an SPE thread as a process, and an SPE as a processor, for the sake of consistency with usual usage in MPI. Each SPE runs one thread at most, and so P SPEs are involved in the computation. Each SPE maintains a metadata array of P elements in its local store. (This memory usage can be decreased for some algorithms.) Each entry is 16 Bytes; smaller space would suffice, but this size is forced by DMA alignment requirements. With the maximum of 16 SPEs on a blade, this requires 256 B, which is small. The barrier call has a separate metadata array to avoid interference with other calls. The implementation also allocates two buffers of 16 KB each on the local store to use as software controlled cache. Timing results [4] indicate that buffers of 4 KB each would yield comparable performance. The implementation tries to minimize data transfers involving the main memory, because of the larger latencies involved in such transfers, compared with that to local store on-chip. The bandwidth to main memory is also the bottleneck to most algorithms, and thus access to it should be minimized. The two buffers above are used instead; the use of multiple buffers helps reduce latency by enabling double buffering – when one buffer has data being transferred out of it, another buffer is used to transfer data into the SPE.

SPE *i* typically transfers data to SPE *j* by DMAing data to metadata array location *i* on SPE *j*. SPE *j* polls this entry, and then DMAs data from a local store buffer on SPE *i* to one on SPE *j*. It then typically acknowledges receipt to SPE *i* by DMAing to metadata entry *j* on SPE *i*. Serial numbers are often used in the metadata entries to deal correctly with multiple transfers. Writing to a metadata entry is atomic, because DMAs of size up to 128 B are atomic. However, the DMAs may be executed out of order, and the data sent may also differ from the data at the time the DMA was queued, if that location was updated in the meantime. We don't discuss the implementation details to ensure correctness in the presence of these issues, in order to present a clearer high level view of the algorithms. In order to simplify some of the implementation, we made the collective calls synchronize at the end of each call, using a barrier. We later show that the barrier implementation on the Cell is very efficient.

The experimental platform was a Cell IBM QS20 revision 5.1 blade at Georgia Tech, running Linux. The *xlc* compiler for the Cell, with optimization flag *–O5*, was used. The timings were performed using the decrementer register on the Cell. This has a resolution of around 70 nano-seconds. The variances of the timing results for collective calls, other than the barrier, were fairly small. The variance for the barrier, however, was somewhat higher.

3.2 Barrier

This call blocks the calling process until all the other members of the group have also called it. It can return at any process only after all the group members have entered the call.

Algorithms. We have implemented three classes of algorithms, with a few variants in one of them.

Gather/Broadcast. In this class of algorithms, one special process, which we call the root, waits to be informed that all the processes have entered the barrier. It then broadcasts this information to all processes. On receiving the information broadcast, a process can exit the barrier. We have implemented the following three algorithms based on this idea. Along with an algorithm's name, we also give an abbreviation which will be used to refer to the algorithm later.

(OTA) One-To-All. Here, an SPE informs the root about its arrival by setting a flag on a metadata entry in the root. The root waits for all its entries to have their flag set, unsets these flags, and then sets a flag on a metadata entry of each SPE. These SPEs poll for this flag to be set, then unset it and exit. Note that polling is quite fast because the metadata entry is in the local store for each SPE performing the polling; the bottlenecks are (i) DMA latency and (ii) processes arriving late. The broadcast phase of this algorithm, where the root sets flags, has two variants. In the first one, the root uses a DMA *put* to transfer data to each SPE. An SPE can have sixteen entries in its own DMA queue, and so the root can post the DMA commands without blocking. In the second variant, the root issues a single *putl DMA List* command.

(SIG) Use a Signal Register. The signal registers on each SPE support one-to-many semantics, whereby data DMAed by an SPE is *OR*ed with the current value. The broadcast phase of this algorithm is as in OTA, but the gather phase differs; each SPE sets a different bit of a signal register in the root, and the root waits for all signals to be received.

(TREE) Tree. This gathers and broadcasts data using the usual tree based algorithm [10]. In the broadcast phase of a binomial tree algorithm, the root starts by setting a metadata flag on another SPE. In each subsequent phase, each process that has its flag set in turn sets the flag of one other SPE. Thus, after i phases, 2^i processes have their flags set. Therefore $\lceil log_2 P \rceil$ phases are executed for P SPEs. In a tree of *degree k* [1, 10], in each phase SPEs, which have their flag set, set the flags of k - 1 other distinct SPEs, leading to $\lceil log_k P \rceil$ phases. The gather step is similar to the broadcast phase, but has the directions reversed.

Pairwise-Exchange (PE). This is a commonly used algorithm for barriers [10]. If P is a power of 2, then we can conceptually think of the SPEs as organized as a hypercube. In each phase, an SPE exchanges messages with its neighbor along a specific dimension. The barrier is complete in $log_2 P$ phases. If P is not a power of two, then a slight modification to this algorithm [10] takes $2 + \lfloor log_2 P \rfloor$ steps.

Dissemination (DIS). This is another commonly used algorithm for barriers [10]. In the i th phase here, SPE j sets a flag on SPE $j+2^i$ *(mod P)* and waits for its flag to be set by SPE $P+j-2^i$ *(mod P)*. This algorithm takes $\lceil log_2 P \rceil$ steps, even if P is not a power of two.

Performance Evaluation. We next evaluate the performance of the above algorithms. We found that the use of DMA lists does not improve the performance of the barrier [4] – in fact, the performance is worse when P is greater than four. We also found that the use of the signal register in the gather phase does not improve performance compared with the use of plain DMAs, which are used in OTA.

Figure 2 (left) evaluates the influence of tree degree in the TREE algorithm. We optimized the implementation when the tree degree is a power of 2, replacing modulo operations with bit-wise operations. This difference is not sufficient to explain the large difference in times seen for degree 2 and 4, compared with other degrees. We believe that the compiler is able to optimize a for loop involved in the computation better with power of two degrees. However, increasing the degree to eight lowers the performance. This can be explained as follows. As the tree degree increases, the number of phases decreases. However, the number of DMA issued by the root increases. Even though it can queue up to sixteen DMAs, and the message sizes are small enough that the bandwidth is not a limiting factor, each DMA in the queue has to wait for its turn. Consequently, having multiple DMAs in the queue can lower the performance. This trend is also shown by DMA timing results not presented here.

The PE and DIS algorithms perform substantially better than the gather/broadcast type of algorithms, with PE being clearly better than DIS when P is greater than eight. Before explaining this, we first discuss a factor that sometimes influences the performance of DIS. In contrast to PE, where pairs of processes exchange information, in

DIS, each process sends and receives messages to different processes. On some networks, exchange between processes is faster than sending and receiving between different processes, which can cause DIS to be slower than PE. This is not the case here. DMA tests show that exchanging data is no faster than communication between different SPEs. The reason for the difference in performance is that when the number of processes is greater than eight, some of the processes are on a different chip. The DMA latency between these is higher. In PE, all the inter-chip DMAs occur in the same phase. In DIS, this occurs in each phase. Thus each phase gets slower, whereas in PE, only one of the phases is slowed down due to this fact. This slower phase also explains the sudden jump in latency from eight to ten processes.

Further details on alternate algorithms and related work are given in [4].

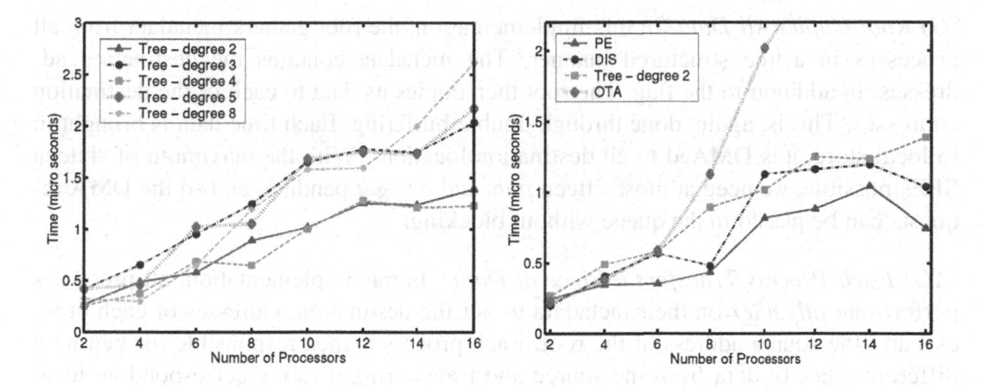

Fig. 2. Barrier latencies. *Left:* Comparison of TREE with different degrees. *Right:* Comparison of four barrier algorithms.

3.3 Broadcast

Algorithms. We discuss below five algorithms for broadcast.

(TREEMM) Send/Receive. This algorithm is the usual tree based Send/Receive algorithm [2, 8], with modifications given below. The tree structure is as in the broadcast phase of TREE for the barrier. However, instead of just setting a flag, a process that sends data also passes the main memory location of its application data. A receiving process copies this data to its own main memory location. This cannot be performed directly, because DMA is possible only between a local store address and an effective address. So, an SPE first copies a chunk of data from the source location to its local store, and then copies this back from the local store to the destination location in main memory. While this seems wasteful, a similar process occurs in regular cache-based processors, where copying a line can involve two or three cache misses. We ameliorate the DMA latency by double buffering. Performance tests on memory to memory copy shows [4] that double buffering yields a significant improvement in performance over single buffering. TREEMM's communication structure is similar to an implementation built on top of MPI_Send and MPI_Recv. However, it avoids the extra overheads of MPI calls by directly implementing the DMA calls in this routine.

Furthermore, it avoids extra copy overheads, and uses double buffering to reduce the memory access latency (the latter is, in some sense, like prefetching to cache).

(OTA) Each SPU Copies its Data. In this implementation, the root broadcasts its metadata, as in the broadcast phase of barrier TREE. It sends the main memory location of the source data, in addition to setting the flag. Once an SPU receives this information, it copies data from the root's locations to its own location in main memory, using double buffering. On some systems, simultaneous access to the same memory location can degrade performance by making this a hotspot. We include a *shift S* to avoid this. That is, SPE i first copies with an offset of $i \times S$, and then copies the initial portion. If $i \times S$ is greater than the data size, then this index wraps around.

(G) Root Copies All Data. In this implementation, the root gathers metadata from all processes in a tree structured manner. The metadata contains the destination addresses, in addition to the flag. The root then copies its data to each of the destination addresses. This is, again, done through double buffering. Each time data is brought in to local store, it is DMAed to all destination locations. With the maximum of sixteen SPEs possible, we need at most fifteen *put*s and one *get* pending, and so the DMA requests can be placed in the queue without blocking.

(AG) Each Process Transfers a Piece of Data. In this implementation, all processes perform an *allgather* on their metadata to get the destination addresses of each process, and the source address of the root. Each process is then responsible for getting a different piece of data from the source and transferring it to the corresponding location in each destination. This is done in a double buffered manner, as with broadcast G. We also specify a *minimum size* for the piece of data any process can handle, because it may be preferable for a few processes to send large DMAs, than for many processes to send small DMAs, when the total data size is not very large. Increasing the minimum size decreases parallelism in the data transfer, with the potential benefit of fewer DMAs, for small messages.

(TREE) Local Store Based Tree. In this implementation, the root gets a piece of data from main memory to its local store, and broadcasts this piece in a tree structured manner to the local stores of all processes. Each piece can be assigned an index, and the broadcast is done by having an SPE with data sending its children (in the tree) metadata containing the index of the latest piece that is available. A child issues a DMA to actually get this data. After receiving the data, the child acknowledges to the parent that the data has been received. Once all children have acknowledged receiving a particular piece, the parent is free to reuse that local store buffer to get another piece of data. A child also DMAs received data to its main memory location, and sends metadata to its children in the tree. In this implementation too, we use double buffering, so that a process can receive a piece into one buffer, while another piece is waiting to be transferred to its children. In this implementation, we denote *pipelined* communication between the local stores by a tree of *degree 1*.

Performance Evaluation. We first determined the optimal parameter for each algorithm, such as the tree degree, shift size, or minimum piece size. We have also evaluated the effect of other implementation choices, such as use of fenced DMAs and DMA lists, but do not discuss these.

We found that on four processors, including some shift to avoid hotspots improves performance of OTA [4], though it is not very sensitive to the actual shift used. On larger numbers of processors, all shifts (including no shift) perform equally well. The likely reason for this is that with more processors, the time at which the DMA requests are executed varies more, and so we don't have a large number of requests arriving at the same time. We also found that on sixteen processors, using a minimum piece size of 2K or 4K in AG yields better performance than larger minimum sizes. (Lower sizes – 1KB and 128 B – perform worse for intermediate data sizes.) At small data sizes, there is only one SPE performing one DMA for all these sizes, and so performances are identical. For large data, all pieces are larger than the minimum, and so this size makes no difference. At intermediate sizes, the smaller number of DMAs does not compensate for the decrease in parallelism for minimum sizes larger than 4 KB. The same trend is observed with smaller numbers of processes.

We next compared the effect of different tree degrees on the performance of the tree-based algorithms [4]. In TREEMM, a tree degree of 3 yields either the best performance, or close to the best, for all process counts. A tree of degree two yields the worst performance, or close to that. However, the relative performances do not differ as much as they do in TREE. Furthermore, the differences show primarily for small messages. Note that for small messages, a higher tree degree lowers the height of the tree, but increases the number of metadata messages certain nodes send (unlike in a Send/Receive implementation, a parent sends only metadata to its children, and not the actual data). It appears that the larger number of messages affects the time more than the benefits gained in decrease of tree heights, beyond tree degree 3. A similar trend is demonstrated in TREE too, though the differences there are greater. For large messages, performances of the different algorithms are similar, though the pipelined implementation is slightly better for very large data sizes. The latter observation is not surprising, because the time taken for the pipeline to fill is then negligible related to the total time, and the number of DMAs issued by any SPE subsequently is lowest for pipelining.

Figure 3 compares the performance of the different algorithms. The trend for the different algorithms on eight processes (not shown here) is similar to that on sixteen processes. We can see that AG has the best, or close to the best, performance for large messages. TREE degree 3 is best for small messages with more than four processes. Up to four processes, broadcast G is best for small messages. Pipelining is also good at large message lengths, and a few other algorithms perform well under specific parameters. As a *good choice of algorithms*, we use broadcast AG for data of size 8192 B or more, broadcast TR degree 3 for small data on more than four processes, and broadcast G from small data on four or fewer processes. The maximum bandwidth that can be served by the main memory controller is 25.6 GB/s. We can see that with this choice of algorithms, we reach close to the peak total bandwidth (for *P-1* writes and one read) for all process counts of four or more, with data size *16 KB* or more. The bandwidth per process can be obtained by dividing the total bandwidth by the number of processes.

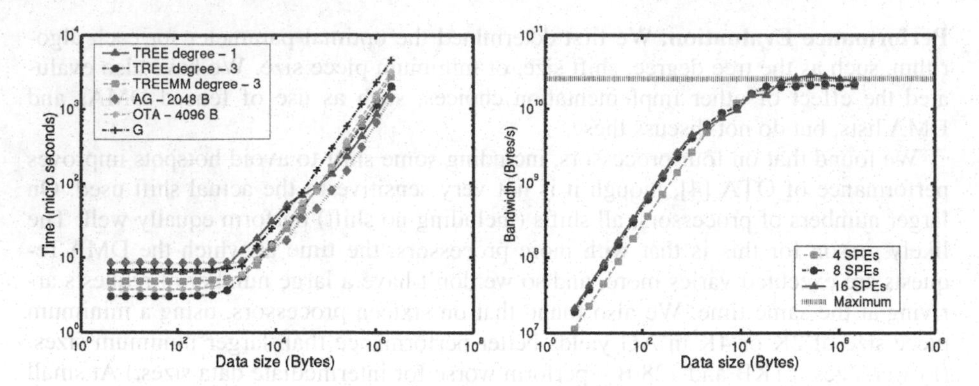

Fig. 3. Broadcast performance. *Left:* Timing on sixteen processes. *Right:* Main memory bandwidth with a "good" choice of algorithms.

3.4 Reduce

In this call, data from all the processes are combined using an associative operator, such as MPI_SUM for addition, and the result placed in the root. Two of the algorithms also assume that the operation is commutative, which is true for all the built-in operators.

Algorithms. The communication structure of this operation is similar to that of the broadcast, but with the directions reversed. In addition, each time a processor gets data, it also applies the operator to a current value and the new data. Since the communication structure is similar to the broadcast, we considered only the types of algorithms that worked well for the broadcast, namely, TREE and AG. In both these algorithms, the computation can also be parallelized efficiently, unlike with OTA.

(TREE) Local Store Based Tree. In this implementation, the communication direction of the broadcast TREE is reversed. A process gets a piece of data from its memory location to local store, gets data from a child's local store to its own local store when that data is available, and combines the two using the specified operator. It repeats this process for each child, except that it does not need to get its own data from main memory for subsequent children. Once it has dealt with all the children, it informs the parent about the availability of the data by DMAing a metadata entry, as in the broadcast. It repeats this for each piece of data in main memory. Double buffering is used to reduce the latency overhead by bringing data from main memory or the next child into a new buffer. Unlike with the broadcast, we need four buffers, two for each operand of a reduce operation, and two more because of double buffering. Due to space constraints on the local store, we used the same buffers as in the broadcast, but conceptually treated them as having half the size (four buffers of 8KB each instead of two buffer of 16K each with broadcast).

(AG) Each Process Handles a Piece of Data. In this implementation, each process is responsible for reducing a different piece of the data, and then writing this to the

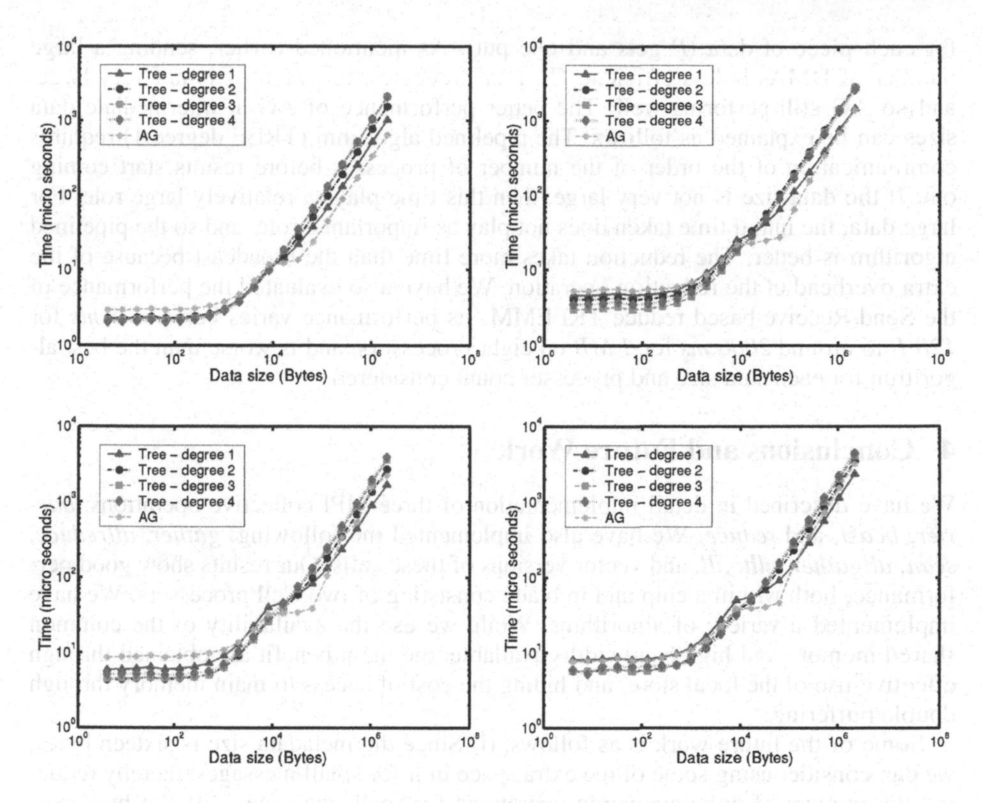

Fig. 4. Reduce timings for MPI_SUM on MPI_INT (four Bytes per int). *Top left:* Four processes. *Top right:* Eight processes. *Bottom left:* Twelve processes. *Bottom right:* Sixteen processes.

destination location of the root. An initial all gather is used to get addresses of all SPEs, as in broadcast.

(TREEMM) Send/Receive. We implemented the usual tree based reduction algorithm on top of our implementation of MPI_Send and MPI_Recv [3] for the Cell processor, just for comparison purposes. The MPI_Send and MPI_Recv operations themselves make effective use of the features of the Cell. However, the extra copying to memory makes the performance worse.

Performance Evaluation. We give performance results in figure 4, for different numbers of processes. We can see that, for small data sizes, TREE degree 3 is either the best, or close to it, on greater than four processes. This is consistent with the behavior expected from the broadcast timings. On four processes, a tree of degree four, which has height 1, performs best. But, degrees 2, 3, and 4 are very close to each other in most cases. Reduce AG is worse for small messages, because of the overhead of all-gathering the metadata initially. TREE degree 1 is best for very large data, except on four processes. Reduce AG is the best at intermediate data sizes, except on four processes, where it is the best even for large messages. This can be explained as follows. Reduce AG parallelizes the computations perfectly, but issues *P+1* DMAs for each piece of data (*P* gets and one put). As mentioned earlier, sending a large number of DMAs is less efficient. On four processes, this number is not very large, and so AG still performs well. The better performance of AG at intermediate data sizes can be explained as follows. The pipelined algorithm (TREE degree 1) requires communication of the order of the number of processes before results start coming out. If the data size is not very large, then this time plays a relatively large role. For large data, the initial time taken does not play as important a role, and so the pipelined algorithm is better. The reduction takes more time than the broadcast because of the extra overhead of the reduction operation. We have also evaluated the performance of the Send/Receive based reduce TREEMM. Its performance varies between *5 ms* for *128 B* to around *2000 ms* for *1 MB* on eight processors, and is worse than the best algorithm for each data size and processor count considered.

4 Conclusions and Future Work

We have described in detail implementation of three MPI collective operations: *barrier*, *bcast*, and *reduce*. We have also implemented the following: *gather*, *allreduce*, *scan*, *allgather*, *alltoall*, and vector versions of these calls. Our results show good performance, both within a chip and in blade consisting of two Cell processors. We have implemented a variety of algorithms. While we use the availability of the common shared memory and high bandwidths available, the main benefit are obtained through effective use of the local store, and hiding the cost of access to main memory through double buffering.

Some of the future work is as follows. (i) Since the metadata size is sixteen bytes, we can consider using some of the extra space in it for small messages, thereby reducing the number of communication operations for small messages. (ii) We have performed barrier synchronization at the end of each collective call, in order to prevent successive calls from interfering with each other. This can be avoided by using counters, as some implementations have done, and may be beneficial when applications reach the collective call at much different times. (iii) It will be useful to consider the integration of this intra-Cell implementation with implementations that connect Cell blades using networks, such as Infiniband. (iv) If the application data is in local store, then our implementation can be made faster. This can be useful, for example, in the MPI microtask model for programming the Cell.

Acknowledgements. We thank the Sony-Toshiba-IBM Cell Center of Competence at Georgia Tech for providing use of their IBM QS20 Cell blades, and IBM for providing access to their Cell blades under the VLP program. Most of all, we express our gratitude to Sri Sathya Sai Baba for bringing us all together to perform this work, and for inspiring and helping us toward our goal.

References

1. Gupta, R., Balaji, P., Panda, D.K., Nieplocha, J.: Efficient Collective Operations Using Remote Memory Operations on VIA-Based Clusters. In: Proceedings of IPDPS (2003)
2. Kini, S.P., Liu, J., Wu, J., Wyckoff, P., Panda, D.K.: Fast and Scalable Barrier Using RDMA and Multicast Mechanisms for Infiniband-Based Clusters. In: Dongarra, J.J., Laforenza, D., Orlando, S. (eds.) PVM/MPI. LNCS, vol. 2840, pp. 369–378. Springer, Heidelberg (2003)

3. Krishna, M., Kumar, A., Jayam, N., Senthilkumar, G., Baruah, P.K., Kapoor, S., Sharma, R., Srinivasan, A.: A Buffered Mode MPI Implementation for the Cell BE Processor. In: ICCS 2007. LNCS, vol. 4487, pp. 603–610. Springer, Heidelberg (2007)
4. Krishna, M., Kumar, A., Jayam, N., Senthilkumar, G., Baruah, P.K., Kapoor, S., Sharma, R., Srinivasan, A.: Optimization of Collective Communication in Intra-Cell MPI, Technical Report TR-070724, Dept. of Computer Science, Florida State University (2007), http://www.cs.fsu.edu/research/reports/TR-070724.pdf
5. Mamidala, A.R., Chai, L., Jin, H-W., Panda, D.K.: Efficient SMP-Aware MPI-Level Broadcast over Infiniband's Hardware Multicast. In: Communication Architecture for Clusters Workshop, in Proceedings of IPDPS (2006)
6. Ohara, M., Inoue, H., Sohda, Y., Komatsu, H., Nakatani, T.: MPI Microtask for Programming the Cell Broadband EngineTM Processor. IBM Systems Journal 45, 85–102 (2006)
7. Sistare, S., vande Vaart, R., Loh, E.: Optimization of MPI Collectives on Clusters of Large-Scale SMP's. In: Proceedings of SC1999 (1999)
8. Thakur, R., Rabenseifner, R., Gropp, W.: Optimization of Collective Communication Operations in MPICH. International Journal of High Performance Computing Applications 19, 49–66 (2005)
9. Tipparaju, V., Nieplocha, J., Panda, D.K.: Fast Collective Operations Using Shared and Remote Memory Access Protocols on Clusters. In: Proceedings of IPDPS (2003)
10. Yu, W., Buntinas, D., Graham, R.L., Panda, D.K.: Efficient and Scalable Barrier over Quadrics and Myrinet with a New NIC-Based Collective Message Passing Protocol. In: Workshop on Communication Architecture for Clusters, in Proceedings of IPDPS (2004)

Routing-Contained Virtualization Based on Up*/Down* Forwarding

Åshild Grønstad Solheim[1,2], Olav Lysne[1,2], Thomas Sødring[1], Tor Skeie[1,2], and Jakob Aleksander Libak[1]

[1] Networks and Distributed Systems Group, Simula Research Laboratory, Lysaker, Norway
[2] Department of Informatics, University of Oslo, Oslo, Norway

Abstract. Virtualization of computing resources is becoming increasingly important both for high-end servers and multi-core CPUs. In a virtualized system, the set of resources that constitute a virtual compute entity should be spatially separated from each other. Dividing the cores on a chip, or the CPUs in a high end server into disjoint sets for each task is a trivial problem. Ensuring that they use disjoint parts of the interconnection network is, however, complex, and in existing methods the requirement of routing-containment of each virtual partition severely degrades the utilization of the system. In this paper, we present an allocation strategy that is based on Up*/Down* routing. Through simulations, we demonstrate increases (in some cases above 30%) in system utilization relative to state-of-the-art in a Dimension Order routed mesh - a topology that is assumed to be widely deployed in Networks on Chip.

1 Introduction

The allocation of subsets of compute resources to incoming tasks[1] was studied quite intensely in the nineties. At the time it was mainly an academic exercise, since the work developed assumed a mode of operation that was hardly found anywhere. This has now changed profoundly for two reasons. Chips with multiple compute cores have emerged and become mainstream, and chips with as many as 256 cores are expected in the not too distant future. This means that operating systems will soon have to allocate parallel tasks to partitions of the chip in an efficient way. The other development that has revitalized this problem area is what we call *Utility Computing Data Centers (UCDCs)*. These are facilities that have large amounts of computing resources at their disposal, and that partition these resources into multiple virtual servers based on the demand of customers. Recently, vendors have introduced industrial solutions to Utility Computing [1,2,3], and several Utility Computing services are now being offered [4,5]. Architectural challenges for an interconnection network [6] in a

[1] A task may also be referred to as a job.

S. Aluru et al. (Eds.): HiPC 2007, LNCS 4873, pp. 500–513, 2007.

UCDC environment are discussed in [7], where flexible partitioning is emphasized as one of the main issues.

A significant number of processor allocation algorithms have been proposed for traditional high performance computing multiprocessors. Although hybrid methods such as [8] have been introduced, the majority of algorithms fall into one of two categories: they may be contiguous [9,10,11,12,13,14,15,16,17,18,19,20] or non-contiguous [21,22,23,24]. A contiguous algorithm designates a set of adjacent processors to a task, whereas a non-contiguous algorithm may designate a set of processors that are not adjacent. The choice between a contiguous and a non-contiguous algorithm is a trade-off between the advantages and disadvantages associated with each of the two categories.

External fragmentation is an inherent issue for contiguous algorithms that can be completely avoided by non-contiguous algorithms. It occurs when a sufficient number of processors is available, but the allocation attempt nevertheless fails due to some restriction (many strategies require that a region of available processors constitutes a sub-mesh). Internal fragmentation occurs if more processors than requested must be allocated to a task (e.g. if the allocated area must be a quadratic sub-mesh for which the side lengths are powers of 2 [25]).

For a non-contiguous algorithm, contention for the link capacity may be unavoidable, but modern interconnect technologies can use virtual channels to separate traffic. Some contiguous resource allocation algorithms have an attractive quality that we refer to as *routing-containment*: the set of resources assigned to a task is selected in accordance with the underlying routing function, such that no links are shared between messages that belong to different tasks. Routing-containment in resource allocation is important for a series of reasons. Most importantly, each task should be guaranteed a fraction of the interconnect capacity regardless of the properties of concurrent tasks. Thus, if one task introduces severe congestion within the interconnection network, other tasks should not be affected. In previous works the notion of routing-containment is often only hinted at. Even so, many strategies, like those that allocate sub-meshes in meshes, will be routing-contained whenever the predominant Dimension Order routing algorithm (DOR) is used. Although not always explicitly stated, this is perhaps the sole reason for restricting oneself to allocating sub-meshes in meshes.

We believe that the advent of topology agnostic routing algorithms that perform almost as well as topology specific routing algorithms (see e.g. [26,27,28]) may form the basis for resource allocation strategies that are more flexible than the established contiguous strategies are. This paper introduces a novel routing-contained resource allocation algorithm, UDFlex, that is based on the topology agnostic Up*/Down* routing algorithm [29]. UDFlex may be used on any topology, and the only restriction on a resource allocation is that the set of resources must constitute a sub-graph of the overall Up*/Down* graph. This provides increased flexibility and potentially reduced external fragmentation when compared to the algorithms that restrict the allocated areas to specific shapes.

The traditional processor allocation algorithms assume a space sharing environment where a task runs on the allocated set of processors without interruption until completion. Other common assumptions are that all resources are equivalent and that compute nodes are the only type of resource eligible for allocation. Although such a homogeneous system may be a simplification of a real system, these assumptions are adopted here for the sake of comparison with the established algorithms. Some of the processor allocation algorithms found in the literature are summarized in Section 2. UDFlex, our novel allocation algorithm, is described in detail in Section 3. In Sections 4 and 5 the performance of the proposed method is evaluated before we conclude in Section 6.

2 Related Work

Several contiguous processor allocation algorithms have been proposed for mesh and k-ary n-cube topologies. As opposed to UDFlex most of them restrict the allocated regions to sub-meshes or sub-cubes. Most papers ([30] is an exception) that propose allocation algorithms for k-ary n-cubes do not discuss the issue that if DOR is used and sub-meshes are allocated in a topology that has wraparound links the shortest path between two nodes may include intermediate nodes that are not part of the allocated region. In such cases, messages that are routed outside the allocated region risk interference with messages that belong to other tasks, and the allocation algorithm is thus not routing-contained.

First Fit and Best Fit [15] were proposed for two-dimensional meshes to solve problems related with the 2D Buddy [25] and Frame Sliding [31] strategies. The applicability of 2D Buddy is restricted to square mesh systems and allocations are restricted to square sub-meshes, where the side lengths of the squares are powers of 2. As a result of the sliding of frames in fixed strides through the mesh, Frame Sliding will not always recognize a free sub-mesh even if one is available. First Fit and Best Fit keep track of free and busy processors, and for each scheduled task they calculate which processors are the bottom left node in a free sub-mesh of a requested size $a \times b$. First Fit then allocates the first free sub-mesh found, whereas Best Fit attempts to reduce external fragmentation by selecting the smallest free region for allocation, and thereby leaving larger free contiguous areas for future and possibly larger tasks.

For an incoming resource request $a \times b$ Adaptive Scan [10] may rotate the original request by 90 degrees and also search for a free sub-mesh $b \times a$. Adaptive Scan does not scan through every node in the mesh, and achieves a complete sub-mesh recognition capability as it does not fix the strides.

Flexfold [11] has complete sub-mesh recognition capability, and searches first for a sub-mesh of size $a \times b$ or $b \times a$, and may in addition (after consideration of a possible communication overhead) fold the originally requested sub-mesh $a \times b$ and search for a sub-mesh of size $\frac{a}{2} \times 2b$, $2a \times \frac{b}{2}$, $2b \times \frac{a}{2}$, or $\frac{b}{2} \times 2a$ (if a or b are odd numbers some of these alternatives are of no interest).

In [20] a strategy is proposed for two-dimensional mesh systems that places an allocation in an available sub-mesh that has its left boundary towards another allocated sub-mesh or towards the edge of the mesh. This principle is also used by the Leapfrog method [14], that introduces a more efficient data structure for faster recognition of free sub-meshes in a mesh topology. In [14] analytical models concerning the execution cost of the allocation process and the probabilities of finding free sub-meshes under different load levels are presented. As opposed to UDFlex, the focus of both [20] and [14] is more on allocation efficiency than on the fragmentation issue, and the results presented in [20] and [14] do not indicate a significant reduction of fragmentation when compared to e.g. Adaptive Scan.

In [16] multi-dimensional sub-meshes are allocated for three-dimensional tori. The strategy has complete sub-mesh recognition capability, does not restrict the orientation of sub-meshes, and may allocate sub-meshes across wraparound links. Routing issues are not discussed in [16]. We observe, however, that the use of DOR cannot ensure routing-containment since the shortest path between two nodes may traverse nodes that are not part of the allocated sub-mesh.

The k-ary Partner strategy [30] uses a tree structure to represent sub-cubes, and the tree is searched with an aim to localize free m-dimensional sub-cubes (slices) of k-ary n-cubes. The strategy is routing-contained, but does not have complete sub-cube recognition capability, and may be affected by internal fragmentation due to the requirement that the allocated sub-cubes have radix k.

The scan search scheme [19] allocates three-dimensional sub-meshes in tori where the sub-meshes may be allocated across wraparound links, has complete sub-mesh recognition capability, allows flexible orientation of the sub-meshes, but is not routing-contained. A particular data structure is used to reduce the three-dimensional information on a torus to two-dimensional information to improve the average allocation time when compared to [16].

The Extended Tree-Collapsing strategy [17] addresses the internal fragmentation issue, and can allocate either cubic or non-cubic sub-cubes in a k-ary n-cube. The scheme is affected by significant external fragmentation, and cannot ensure routing-containment (e.g. for an allocated r-ary m-cube with $r > \frac{k}{2}$).

The main idea of the Isomorphic allocation strategy [12] is to partition a k-ary n-cube recursively into 2^n $\frac{k}{2^i}$-ary n-cubes in the ith step of partitioning. The strategy is, however, not restricted to k-ary n-cube systems nor to the allocation of n-cubes (several n-cubes can be merged to accommodate a resource request of a different topology). The strategy cannot always ensure routing-containment if DOR is used. Consider e.g. an 8-ary 3-cube where the resource requests have the form $x \times y \times z$ where $1 \leq x, y, z \leq 8$, then if either of $x, y, z > \frac{8}{2}$, messages may be routed outside the allocated region due to the wraparound links.

The allocation of sub-tori in high-dimensional tori is the focus of [18] which requires that the sizes of all but one dimension of the torus are powers of 2, and that the tasks request a number of processors that is a power of 2. In contrast, UDFlex, a topology agnostic allocation algorithm, can handle high-dimensional systems without such requirements on the numbers of processors in either systems or resource requests.

For faulty meshes and tori the strategies described above will not perform satisfactory. Therefore particular methods have been proposed such as [13] that identifies and allocates virtual sub-meshes in faulty meshes and [9] that uses a distributed procedure to derive healthy sub-meshes from a faulty mesh or torus.

For UDFlex no particular considerations are needed for faulty meshes and tori due to the topology agnosticism of the method.

Although most strategies that allocate sub-meshes or sub-cubes in meshes or tori do not cause as severe internal fragmentation as [25] does, these strategies are nevertheless restricted to allocate a number of processors that can be expressed as a product of numbers $a \times b \times \ldots$. Consider e.g. a two-dimensional mesh or torus of size $w \times h$ where internal fragmentation may occur if the required number of processors cannot be expressed as a product $a \times b$ where $a \leq w$ and $b \leq h$ and as a result a higher number of processors than required must be allocated. UDFlex, on the other hand, can allocate any number of processors and thus eliminates this source of internal fragmentation.

3 UDFlex

UDFlex is a topology agnostic resource allocation algorithm that can allocate any number of resources. It is based on the Up*/Down* routing algorithm that assigns up and down directions to all the links in the network to form a directed graph rooted in one of the nodes, and that avoids deadlock by prohibiting the turn from a down to an up direction link. Assume that an Up*/Down* graph has been constructed for an interconnection network that connects a number of resources. Then, the main idea behind UDFlex is to allocate resources that form a separate Up*/Down* sub-graph to an incoming task. This ensures routing-containment. UDFlex can recognize a free Up*/Down* sub-graph given that one is available, and the allocation of Up*/Down* sub-graphs allows the allocated regions to form irregular shapes. This increased flexibility can alleviate the fragmentation problem inherent in contiguous resource allocation algorithms.

3.1 Description

We assume a traditional system model where tasks arrive in a queue, and each task has requirements on such aspects as the number of resources, running time etc. A scheduler decides the sequence in which the queued tasks are selected, and an allocator attempts to locate and reserve a set of free resources that meets the requirements of the selected task.

Assume that a task selected by the scheduler requests a number of resources $|R|$, that the allocator uses the UDFlex allocation algorithm, and that the routing algorithm has calculated an Up*/Down* graph G on the topology. Given that a sufficient number of resources are available to support the request, UDFlex first identifies the roots of all free Up*/Down* sub-graphs in the network. Subsequently, the set of roots are searched to identify the smallest sub-graph, g, that has a sufficient number of free resources ($|g| \geq |R|$) to accommodate the request. In case of a tie the sub-graph with the deepest root relative to the root of G is selected as we aim to keep the area close to the root of G free and defragmented to increase the probability of successful accommodation of subsequent tasks with possible high resource requirements.

If none of the free sub-graphs is of sufficient size, the task is returned to the queue, awaiting a future allocation attempt following the termination of one of the already running tasks. This is an example of external fragmentation (the algorithm first checked that a sufficient number of resources were free).

Assume that a valid[2], free, and sufficiently large sub-graph, g, of G has been selected. If $|g| = |R|$, the entire sub-graph g is allocated to the task. If $|g| > |R|$, then in order to avoid internal fragmentation, the redundant number of nodes, r_+, of g must not be included in the allocation. In order to identify a set of r_+ redundant nodes, and with the objectives of keeping free resources defragmented and maintaining valid Up*/Down* sub-graphs for future allocation attempts, we identify the smallest valid Up*/Down* sub-graph g' of g where $|g'| \geq r_+$. In the case of a tie, since we aim to keep the region close to the root of G free and defragmented, the sub-graph with the shallowest root relative to the root of G is selected. If $|g'| = r_+$, we simply allocate $g - g'$. If $|g'| > r_+$, for simplicity, although fragmentation of a region of free resources may result, we apply a breadth-first search from the root of g' to identify a valid Up*/Down* sub-graph g'' where $|g''| = |g'| - r_+$, and allocate $g - g' + g''$.

3.2 Complexity

UDFlex addresses the fragmentation issue inherent in contiguous strategies, and does not address the complexity and running time cost of the allocation algorithm. Although this cost is an important metric for any algorithm, there may be a trade-off between allocation cost and the potentially higher resource utilization of an advanced algorithm. We argue that for a UCDC, where allocated tasks may run for several seconds, minutes or hours, the degree of fragmentation may matter more than the algorithm complexity. The current version of UDFlex includes some optimizations with respect to node-selection, of which the most important are the placement of an allocation in the smallest possible free sub-graph, and the preference of the deepest sub-graph in case of a tie. A "first-fit" approach, where the first possible set of nodes encountered are selected, may decrease the running time at the expense of increased fragmentation.

Assume that the Up*/Down* graph can be expressed as $G = (V, E)$ where V is the set of vertices and E is the set of edges, and that $|V|$ is the number of vertices and $|E|$ is the number of edges. For resource allocation the most complex algorithm step is the search for the smallest free sub-graph that is large enough to hold the resource request. The complexity of this step is formally $O(|V| \times (|E| + |V|))$, but for two-dimensional tori and meshes where $|E| = 2 \times |V|$ and $|E| < 2 \times |V|$, respectively, the complexity becomes $O(|V|^2)$. For practical purposes we observe that in cases where few tasks are running, the number of free sub-graphs is probably significantly less than $|V|$ whereas the size of each sub-graph may be relatively large. If, on the other hand, many tasks are running there may be a higher number of free sub-graphs, but the size of each sub-graph may be smaller. The complexity of the deallocation phase is $O(|V|)$ and consists

[2] A valid sub-graph is a correct Up*/Down* graph that has e.g. only one root node.

of a traversal through the nodes currently allocated to the finishing task to change their state from busy to free.

4 Experiment Setup

We developed a simulator model in the J-Sim [32] environment to compare the performance of UDFlex with that of several traditional processor allocation strategies. Our model consists of the following main components: a queue of infinite size where tasks with certain resource requirements arrive according to an exponential distribution; a FCFS scheduler that upon task arrival or termination of a running task selects the first task in the queue as the next candidate for allocation; and an allocation module that runs a specific allocation algorithm and attempts to localize free resources to meet the demands of the task. When successfully allocated, a task runs without interruption on the allocated resources for an exponentially distributed time referred to as the *service time*, ST. Time is measured in cycles - an abstract time unit.

In this study we have considered both meshes and tori of size 16×16, 32×16, and 32×32. To evaluate the performance of the various allocation algorithms we use the metrics system utilization and queuing time. Assume that a mesh or torus has width w and height h, that the processor in position (i, j) has been busy for the aggregated time $busy_{i,j}$, and that data has been collected for a period of time T. Then according to [15] the *system utilization* is $\frac{\sum_{1 \leq i \leq w, 1 \leq j \leq h} busy_{i,j}}{w \times h \times T}$, and the *system fragmentation* is $1 - system\ utilization$. The *queuing time* is the average time that a task is held in the queue, from the time of the arrival of a task until its requested resources have been allocated.

In the evaluation experiments, routing-contained allocation strategies are used as our main points of reference. We do, however, include a Random allocation strategy in our plots. This is an allocation strategy that fails only when the next task to be served requires more resources than those that are currently vacant (thus fragmentation is not an issue). The reason for including this strategy is to visualize the upper performance benchmark with respect to our metrics (as long as communication overhead is not considered). For practical purposes, communication overhead may significantly reduce the attractiveness of Random.

In these experiments UDFlex uses an Up*/Down* graph that is based on a tree identified by a breadth-first search (as proposed in [29]) and that has the root in the upper left corner of the topology. For meshes we compare UDFlex with the contiguous allocation algorithms First Fit, Best Fit, Adaptive Scan, and Flexfold. For tori we compare UDFlex with a contiguous and recognition-complete scheme that allocates sub-meshes (possibly across wraparound-links). First, allocation of the originally requested sub-mesh $a \times b$ is attempted, and if no such sub-mesh is available the 90 degrees rotation of the original request $(b \times a)$ is attempted. We believe that for the metrics considered, this scheme is a reasonable representative of algorithms that allocate sub-meshes in two-dimensional tori. As with most of the allocation strategies for k-ary n-cubes that were presented in Section 2 this scheme is not routing-contained when DOR is used. Nevertheless, for the evaluation of UDFlex with respect to e.g. system utilization we believe that a comparison with this strategy may be more interesting than a comparison with e.g. the k-ary Partner strategy [30] which is routing-contained, but does not have complete sub-cube recognition capability and may also be affected by internal fragmentation.

For a fair comparison between algorithms that allocate sub-meshes and those that do not, the resource demand of tasks should be equal for either group: each task requests $a \times b$ resource entities (which is considered a sub-mesh or a product of numbers depending on the group of algorithm used). a and b are drawn from separate uniform distributions with maximum sizes a_{max} and b_{max}, respectively. We conducted experiments to observe the effect of tasks with high, medium, and low resource demands: for high resource demand a_{max} is set to w and b_{max} is set to h; for medium resource demand a_{max} is set to $\frac{w}{2}$ and b_{max} is set to $\frac{h}{2}$; and for low resource demand a_{max} is set to $\frac{w}{4}$ and b_{max} is set to $\frac{h}{4}$.

The input *load* for a mesh or torus of width w and height h is $\frac{|R|_{mean} \times ST_{mean}}{w \times h \times IT_{mean}}$ (in accordance with [19]), where $|R|_{mean}$ is the mean number of resources requested by the tasks, ST_{mean} is the mean service time of tasks, and IT_{mean} is the mean inter-arrival time of tasks. In the majority of our experiments ST_{mean} is fixed at 1 000 cycles and the desired load levels result from variation of IT_{mean}.

The experiments were run on a Condor [33] cluster, and each of the experiments was stopped when 20 000 tasks had been allocated and completed. To ensure representative results, both the initial and final 10% of the observations were discarded. The presented values are the mean values that result from 16 repetitions of each experiment (each initialized by a different seed). For each observed mean value a 95% confidence interval is plotted.

Although the fragmentation issue is the primary subject of this study, the effect of possible communication overhead is also considered to verify that UDFlex is advantageous even for communication intensive applications. The efficiency of Up*/Down* routing may be affected by congestion that may form close to the root node and also by the fact that a legal path may not be the shortest path between two nodes. For UDFlex the distance between nodes may be altered compared to sub-mesh or sub-cube allocation (and the use of DOR) due to the allocation of irregularly shaped regions and the routing restrictions of the Up*/Down* algorithm. These issues may result in an increased service time of tasks for UDFlex compared to the routing-contained methods that allocate sub-meshes. However, for an interconnection network that uses cut-through switching [34] a small increase in the number of hops between nodes will have limited impact on message latency.

For tasks with high resource demands, we conducted a set of experiments for meshes to study the level of increase in service time that can be tolerated before the advantage of UDFlex, with respect to fragmentation, is outweighed. In addition to communication overhead the increase in service time may also represent a possible allocation overhead for each task due to the complexity of UDFlex. We have compared the queuing time and throughput (system utilization is barely affected by increased task service time) of UDFlex with the queuing time and throughput of three routing-contained methods that allocate sub-meshes without manipulating the shape of requested sub-meshes - Adaptive Scan, Best Fit and First Fit. This study does not include Flexfold that may alter the original distance between nodes by folding the requested sub-mesh. A load level of 0.9 (with ST_{mean} of 1 000 cycles and IT_{mean} of 314 cycles) was selected as basis for the experiments. For tasks that request a number of resources $|R| > 1$ the service time, ST, that is drawn from an exponential distribution, is modified for UDFlex according to the formula $ST' = ST + \frac{ST \times x \times |R|}{100 \times w \times h}$, where x is percentage increase in task service time, w is the width, and h is the height of the mesh.

5 Results

The system utilization and queuing time versus input load for the 32 × 32 mesh and torus topologies are shown in Figures 1 and 2, respectively. Figures 1(a) and 1(b) show that for the 32 × 32 mesh the system utilization is significantly higher for UDFlex compared to the other contiguous allocation algorithms. As expected Flexfold has the best performance of the algorithms that allocate sub-meshes. For tasks with high resource demand (Figure 1(a)) the utilization for Flexfold for the highest load levels is 0.53, for UDFlex it is 0.68 (28.3% higher than Flexfold), whereas for Random, the theoretical upper bound performance indicator, it is 0.73 (not more than 7.4% higher than UDFlex). For tasks with low resource demand (Figure 1(b)) the utilization for Flexfold for the highest load levels is 0.73, for UDFlex it is 0.86 (17.8% higher than Flexfold), whereas for Random it is 0.98 (14.0% higher than UDFlex). For the 32 × 32 torus the system utilization for tasks with high and low resource demand are shown in Figures 2(a) and 2(b), respectively. As for the mesh topologies UDFlex significantly improves the system utilization compared to the allocation of sub-meshes (with 21.4% and 19.4% for high and low task resource demand, respectively).

The higher system utilization of UDFlex compared to the strategies that allocate sub-meshes is even more pronounced for tasks with medium and high resource demand than for tasks with low resource demand (and this is even more apparent in meshes than in tori). Generally, the benefit of UDFlex over the other contiguous algorithms is higher for larger topologies. We note a particular increase in performance (more than 30% higher system utilization) when we compare the results of UDFlex with the results of Adaptive Scan and Flexfold for tasks with high resource demand for the non-quadratic (32 × 16) mesh or torus. Adaptive Scan and Flexfold perform relatively worse in this case since a resource request for a sub-mesh $a \times b$ cannot be rotated by 90 degrees if $a > 16$, and in addition several of the folded alternatives may not fit the topology.

For both meshes and tori we observe that the knee-point where the maximum system utilization is reached occurs under higher input load for UDFlex than for the other contiguous algorithms. Figures 1(c), 1(d), 2(c), and 2(d) show that as the load increases the queuing time for UDFlex stays significantly lower than

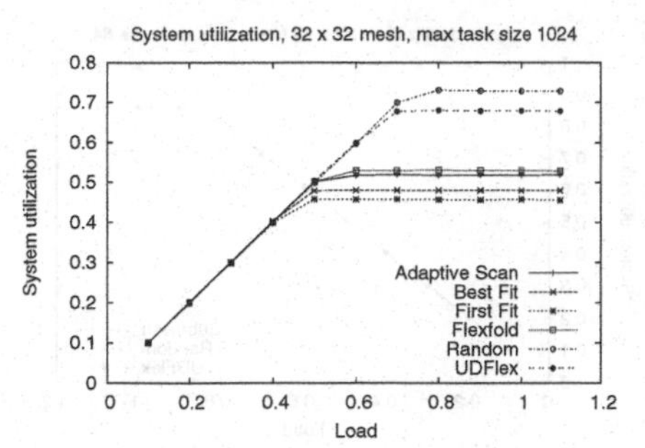

(a) System utilization for tasks with high resource demand.

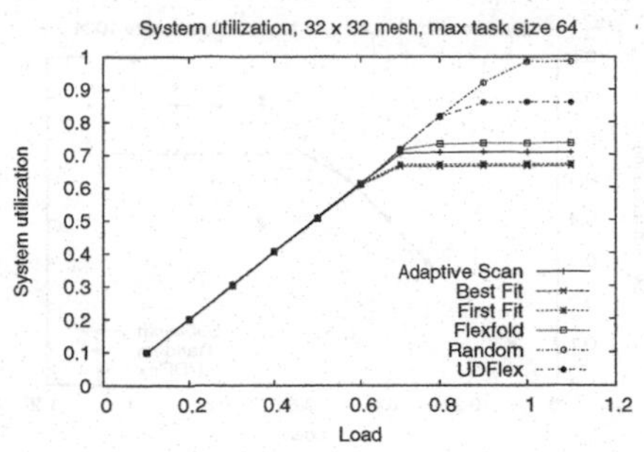

(b) System utilization for tasks with low resource demand.

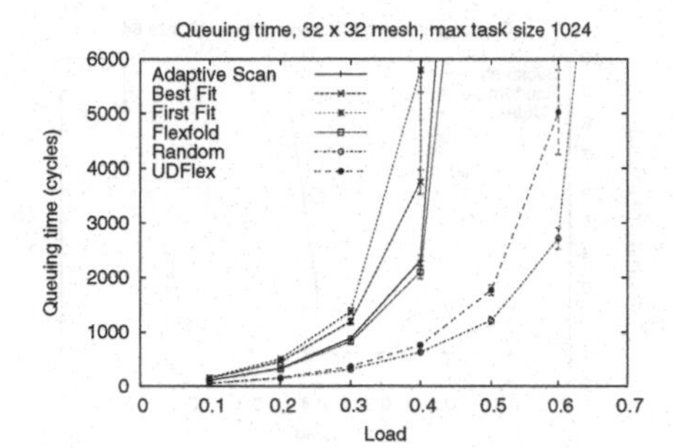

(c) Queuing time for tasks with high resource demand.

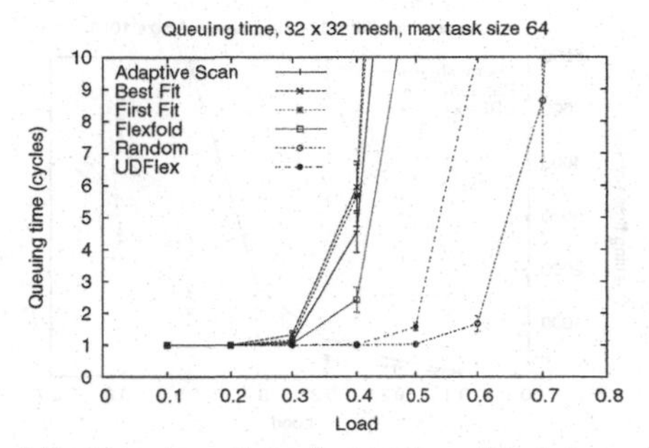

(d) Queuing time for tasks with low resource demand.

Fig. 1. System utilization and queuing time for 32 × 32 mesh

for the algorithms that allocate sub-meshes, and confirm that due to reduced fragmentation UDFlex sustains a significantly higher load before the system saturates. With the exception of the already discussed issue of Adaptive Scan and Flexfold for non-square topologies, the results for the 32×16 and 16×16 mesh and torus are similar to those for the 32×32 mesh and torus, respectively. The system utilization achieved by UDFlex is similar for the mesh and torus experiments even though the Up*/Down* graphs are different for the two topologies.

In most of our experiments ST_{mean} was fixed and the different load levels result from the variation of IT_{mean}. For tasks with high resource demand we also conducted a set of experiments where IT_{mean} was fixed at 500 cycles and ST_{mean} was varied to achieve the desired load levels. For all algorithms and both in meshes and tori the variation of load has the same effect on system utilization regardless of which of the two parameters that was fixed. The effect on queuing time was not the same for the two sets of experiments, however. We observed e.g. that for low load levels the queuing time was lower for the experiments with fixed IT_{mean} than for the experiments with fixed ST_{mean}.

In addition to system utilization and queuing time we also evaluated *throughput* - the number of tasks terminated over a certain period of time - that is $\frac{16\ 000\ tasks}{T}$. These results support the conclusions that were drawn from the considerations of system utilization and queuing time.

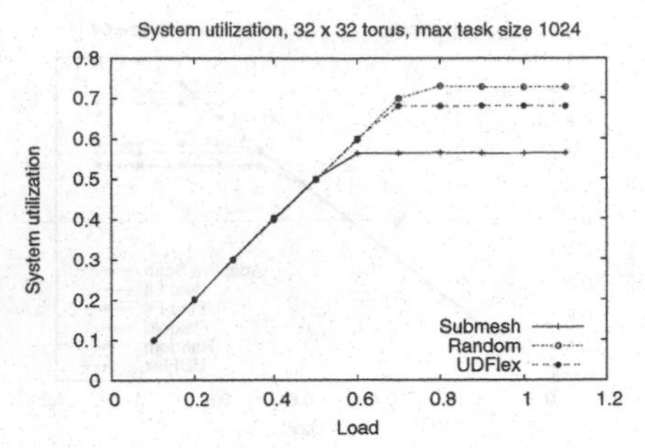

(a) System utilization for tasks with high resource demand.

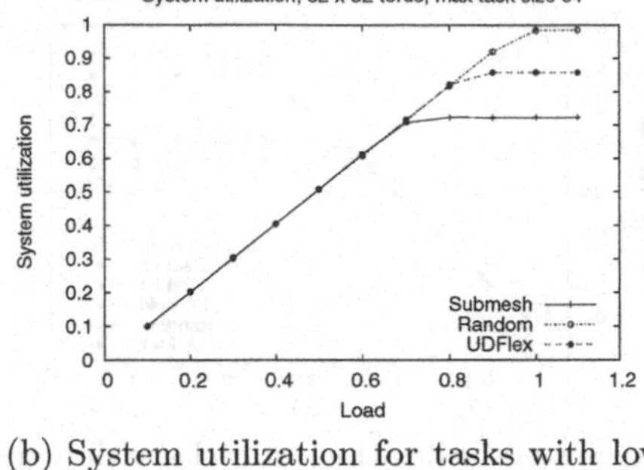

(b) System utilization for tasks with low resource demand.

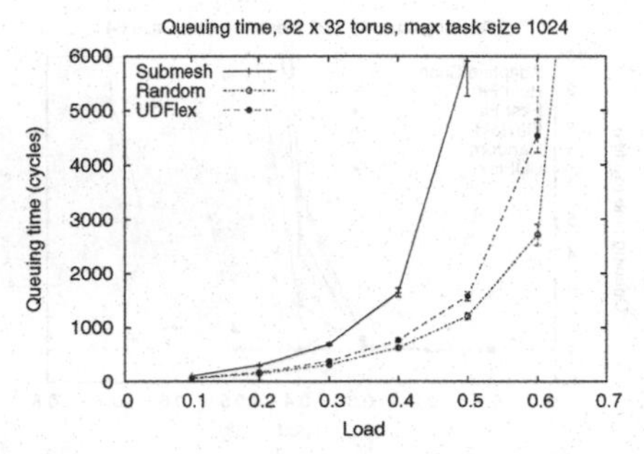

(c) Queuing time for tasks with high resource demand.

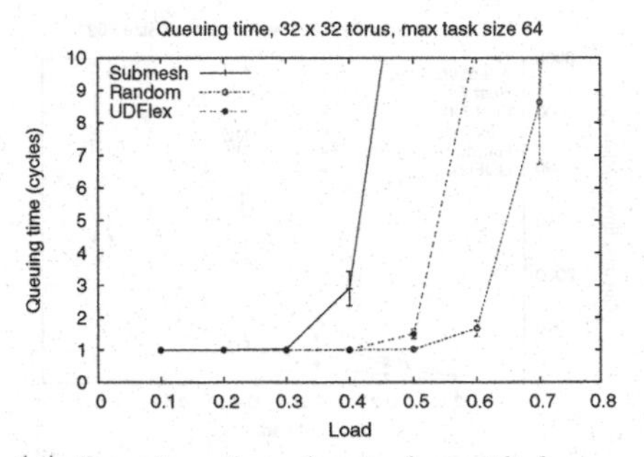

(d) Queuing time for tasks with low resource demand.

Fig. 2. System utilization and queuing time for 32×32 torus

Section 4 provided motivation for the set of experiments where a prolonged duration of tasks for UDFlex is used to represent possible communication and allocation overhead of UDFlex when compared to methods that allocate sub-meshes. For the meshes of size 16×16 and 32×16 Figure 3 shows the effect on queuing time as the service time of tasks is increased for UDFlex (as previously pointed out this increase in task service time is relative to task size). For the 16×16 mesh the task service time of UDFlex may be increased by around 60% before the queuing time of UDFlex crosses that of Adaptive Scan (Figure 3(a)). For the 32×32 mesh the crossing occurs around 70%. As previously discussed the performance of Adaptive Scan deteriorates for a 32×16 mesh for tasks with high resource demand since a resource request $a \times b$ cannot be rotated by 90 degrees if $a > 16$. Figure 3(b) shows that for a 32×16 mesh UDFlex tolerates an increase in task service time of around 90% before the queuing time of UDFlex crosses the queuing time of Best Fit. Compared to First Fit that has the highest queuing time of the methods that allocate sub-meshes, UDFlex tolerates an increase in task service time of at least 100%. The consideration of throughput gives the same conclusions as for queuing time.

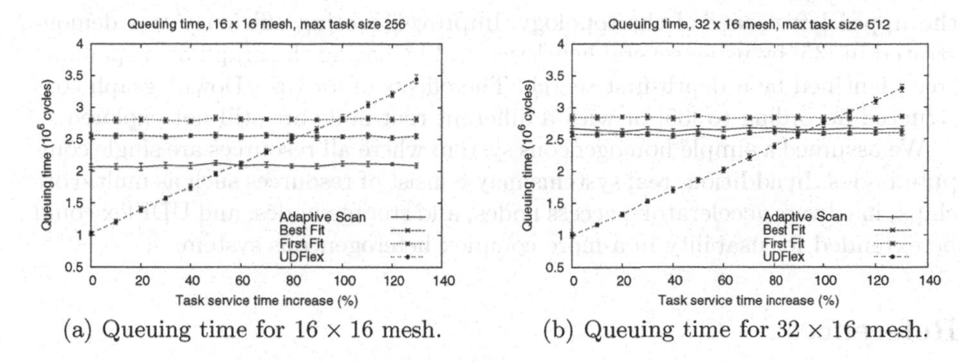

(a) Queuing time for 16 × 16 mesh. (b) Queuing time for 32 × 16 mesh.

Fig. 3. Effect on queuing time of the increased running time of tasks for UDFlex used to represent possible communication and allocation overhead of UDFlex when compared to methods that allocate sub-meshes

6 Conclusion

In systems such as Networks on Chip and Utility Computing Data Centers virtualization of computing resources is an important challenge, and for a resource allocation algorithm two major issues are the minimization of fragmentation and the prevention of any disturbance between messages that belong to different tasks (*routing-containment*). Most of the traditional contiguous strategies for meshes and tori allocate strict sub-meshes or sub-cubes and cannot prevent a high level of fragmentation from occurring, in addition some of the methods cannot ensure routing-containment.

We propose a novel contiguous resource allocation strategy, UDFlex, that can be used on any topology - a particularly attractive property in the face of faults in regular topologies. The strategy recognizes any available Up*/Down* sub-graph, is not affected by internal fragmentation, and does ensure routing-containment without the need for reconfiguration. UDFlex assumes that the interconnection network uses Up*/Down* routing, and the allocation of Up*/Down* subgraphs to tasks enables flexibility with respect to the shapes of the allocated regions. The simulation results show that this flexibility significantly reduces fragmentation and increases the utilization of system resources (in some cases by above 30%) compared to traditional contiguous allocation strategies. In addition, the possible drawbacks of UDFlex with respect to algorithm complexity and communication overhead are shown not to outweigh the advantage of UDFlex until these issues increase the aggregated task allocation and service time by at least 60%. Thus, UDFlex should also be attractive for communication intensive applications.

Several issues concerning the UDFlex resource allocation strategy may be further investigated. With respect to fragmentation, the current version (that includes some optimizations) could be compared both with a simpler approach where the first available Up*/Down* sub-graph is selected for allocation, and with alternative advanced approaches. In our experiments the Up*/Down* graph is based on a breadth-first search and the root of the graph is the node in the upper left corner of the topology. Improved routing efficiency was demonstrated in [35] by using several heuristics and by basing the graph on a spanning tree identified by a depth-first search. The effects of an Up*/Down* graph constructed according to [35] or with a different root node are still not explored.

We assumed a simple homogeneous system where all resources are single compute nodes. In addition, real systems may consist of resources such as multi-core chips, hardware accelerators, access nodes, and storage nodes, and UDFlex could be extended for usability in a more complex heterogeneous system.

References

1. Bulhões, P.T., et al.: N1 grid engine 6 features and capabilities. Sun Microsystems White Paper (2004)
2. HP. Adaptive enterprise: Business and IT synchronized to capitalize on change. HP White Paper (2005)
3. IBM. Unleash the power of e-business on demand. IBM White Paper (2003)
4. Sun Microsystems. Sun grid compute utility - reference guide. Part No. 819-5131-10 (2006)
5. Amazon Elastic Compute Cloud, `amazon.com/gp/browse.html?node=,201590011`
6. Duato, J., Yalamanchili, S., Ni, L.: Interconnection Networks: An Engineering Approach. Morgan Kaufmann, San Francisco (2003)
7. Lysne, O., et al.: The interconnection network - architectural challenges for Utility Computing Data Centres. Computer (submitted, December 2006)
8. Subramani, V., et al.: Selective buddy allocation for scheduling parallel jobs on clusters. In: 4th IEEE Int'l. Conf. Cluster Comp., pp. 107–116 (2002)
9. Chen, H.-L., Hu, S.-H.: Submesh determination in faulty tori and meshes. IEEE Trans. Par. and Dist. Syst. 12(3), 272–282 (2001)
10. Ding, J., Bhuyan, L.N.: An adaptive submesh allocation strategy for two dimensional mesh connected systems. In: Int'l. Conf. Par. Proc., p. 193 (1993)
11. Gupta, V., Jayendran, A.: A flexible processor allocation strategy for mesh connected parallel systems. In: Int'l. Conf. Par. Proc., p. 166 (1996)
12. Kang, M., et al.: Isomorphic strategy for processor allocation in k-ary n-cube systems. IEEE Trans. Comp. 52(5), 645–657 (2003)
13. Kim, G., Yoon, H.: On submesh allocation for mesh multicomputers: A best-fit allocation and a virtual submesh allocation for faulty meshes. IEEE Trans. Par. and Dist. Syst. 9(2), 175–185 (1998)
14. Wu, F., Hsu, C.-C., Chou, L.-P.: Processor allocation in the mesh multiprocessors using the Leapfrog method. IEEE Trans. Par. and Dist. Syst. 14(3), 276–289 (2003)
15. Zhu, Y.: Efficient processor allocation strategies for mesh-connected parallel computers. Jrnl. Par. and Dist. Comp. 16(4), 328–337 (1992)
16. Qiao, W., Ni, L.M.: Efficient processor allocation for 3D tori. In: 9th Int'l. Par. Proc. Symp., pp. 466–471 (1995)
17. Chuang, P.-J., Wu, C.-M.: An efficient recognition-complete processor allocation strategy for k-ary n-cube multiprocessors. IEEE Trans. Par. and Dist. Syst. 11(5), 485–490 (2000)
18. Mao, W., Chen, J., Watson III, W.: Efficient subtorus processor allocation in a multi-dimensional torus. In: 8th Int'l. Conf. High-Perf. Comp. Asia-Pacific Reg., p. 53 (2005)

19. Choo, H., Yoo, S.-M., Youn, H.Y.: Processor scheduling and allocation for 3D torus multicomputer systems. IEEE Trans. Par. and Dist. Syst. 11(5), 475–484 (2000)
20. Chiu, G.-M., Chen, S.-K.: An efficient submesh allocation scheme for two-dimensional meshes with little overhead. IEEE Trans. Par. and Dist. Syst. 10(5), 471 (1999)
21. Bunde, D.P., Leung, V.J., Mache, J.: Communication patterns and allocation strategies. In: 18th Int'l. Par. and Dist. Proc. Symp., p. 248 (2004)
22. Leung, V., et al.: Processor allocation on Cplant: Achieving general processor locality using one-dimensional allocation strategies. In: 4th IEEE Int'l. Conf. Cluster Comp., pp. 296–304 (2002)
23. Lo, V., et al.: Non-contiguous processor allocation algorithms for mesh-connected multicomputers. IEEE Trans. Par. and Dist. Syst. 8(7), 712–726 (1997)
24. Mache, J., Lo, V., Windisch, K.: Minimizing message passing contention in fragmentation free processor allocation. In: 10th Int'l. Conf. Par. and Dist. Comp. Syst., pp. 120–124 (1997)
25. Li, K., Cheng, K.: A two-dimensional buddy system for dynamic resource allocation in a partitionable mesh connected system. Jrnl. Par. and Dist. Comp. 12, 79–83 (1991)
26. Lysne, O., et al.: Layered routing in irregular networks. IEEE Trans. Par. and Dist. Syst. 17(1), 51–65 (2006)
27. Sancho, J.C., et al.: Effective methodology for deadlock-free minimal routing in InfiniBand networks. In: Int'l. Conf. Par. Proc., pp. 409–418 (2002)
28. Skeie, T., et al.: LASH-TOR: A generic transition-oriented routing algorithm. In: 11th Int'l. Conf. Par. and Dist. Syst. (2004)
29. Schroeder, M.D., et al.: Autonet: a high-speed, self-configuring local area network using point-to-point links. SRC Res. Rep. 59, Digital Equipment Corp. (1990)
30. Windisch, K., Lo, V., Bose, B.: Contiguous and non-contiguous processor allocation algorithms for *k*-ary *n*-cubes. In: Int'l. Conf. Par. Proc., p. 164 (1995)
31. Chuang, P.-J., Tzeng, N.-F.: An efficient submesh allocation strategy for mesh computer systems. In: 11th Int'l. Conf. Dist. Comp. Syst., p. 256 (1991)
32. J-Sim, `http://www.j-sim.org`
33. Condor, `http://www.cs.wisc.edu/condor`
34. Kermani, P., Kleinrock, L.: Virtual cut-through: A new computer communication switching technique. Computer Networks 3, 267–286 (1979)
35. Sancho, J.C., Robles, A., Duato, J.: An effective methodology to improve the performance of the Up*/Down* routing algorithm. IEEE Trans. Par. and Dist. Syst. 15(8), 740–754 (2004)

A Routing Methodology for Dynamic Fault Tolerance in Meshes and Tori

Nils Agne Nordbotten[1] and Tor Skeie[1,2]

[1] Simula Research Laboratory, P.O. Box 134, 1325 Lysaker, Norway
{nilsno,tskeie}@simula.no
[2] Department of Informatics, University of Oslo, Norway

Abstract. This paper proposes a fully distributed fault-tolerant routing methodology for tori and meshes. A dynamic fault-model is supported, enabling the network to remain fully operational at all times. Contrary to most previous proposals that support a dynamic fault-model, the methodology is able to tolerate concave fault regions, thereby avoiding disabling healthy nodes in most practical scenarios. The methodology provides high network performance through the use of adaptive routing and provides graceful performance degradation in the presence of faults.

1 Introduction

Interconnection networks are used for a variety of purposes, from connecting the various components of a single device (e.g., connecting the internal units of a chip) to connecting the nodes of massively parallel computers covering hundreds of square meters. Due to their large application area, interconnection networks are found in systems with high requirements for reliability and continued operation. Faults in the interconnection network may potentially leave the remainder of the system disconnected, thus, providing a reliable interconnection network is essential for the overall reliability of the system.

In this paper we consider reliability in interconnection networks with mesh and torus topologies. These two topologies are among the most commonly used in interconnection networks. For instance, all top five spots on the current top 500 list of supercomputers [1] are held by machines using these topologies. Enduring a fault-free network is very difficult in such large systems. Because of the high number of components, there is an increased probability that some components may fail. Thus, for massively parallel computers, fault tolerance is a critical design issue [2][3] that will become increasingly important as systems continue to scale.

Torus and mesh topologies are also found in more commercial architectures, like the Alpha 21364 [4] (2D torus), that are targeted at application domains such as database servers, web servers, and telecommunications. For such commercial applications there are often strict requirements for uninterrupted service, and failure to meet these requirements may have severe economic consequences. Two dimensional mesh/torus topologies are also a popular choice for networks on-chip [5]. In a recent tera-scale prototype from Intel, 80 cores are connected in a

S. Aluru et al. (Eds.): HiPC 2007, LNCS 4873, pp. 514–527, 2007.

2D mesh on a single processor chip [6]. It is considered a requirement for such future interconnects that they are robust in the face of failures [7].

Faults in an interconnection network can be dealt with statically or dynamically. When a static fault-model is used, all the faults need to be known when the system is started. Thus, when a fault occurs, the system has to be restarted. In order to be effective, such a method may need to be combined with checkpointing. If restarting the system when faults occur is not desirable, either because of the overhead imposed by relying on a checkpointing mechanism (i.e., maintaining checkpoints and rolling back to the last checkpoint at system failure/restart) or because continued operation is required, a dynamic fault model should be used. When a dynamic fault model is used, the system remains operational while measures are taken to circumvent the faulty component(s).

By using fault-tolerant routing, fault-tolerance can be provided (without requiring spare components) by utilizing the inherent redundancy of topologies such as mesh and torus. One approach to provide fault-tolerant routing has been to develop adaptive routing algorithms, where the adaptivity can be used to circumvent faulty components. Notice that adaptive routing algorithms are not necessarily fault-tolerant though. A strictly minimal adaptive routing algorithm is not able to handle a single fault, as the source-destination pairs connected by a single minimal path are disconnected by any fault within this path.

Linder and Harden [8] proposed a method providing sufficient adaptivity to tolerate at least one fault. However, the number of virtual channels required by their method increases exponentially with the number of dimensions. Chien and Kim [9] observed that the high number of virtual channels required by Linder and Harden is due to the freedom to traverse dimensions in an arbitrary order. They therefore proposed planar adaptive routing [9], where adaptivity is limited to adaptive routing in two dimensions at a time. This method requires at most three virtual channels for meshes of any dimension, but does not properly handle faults on the edges of the network.

Glass and Ni [10] used the partial adaptivity provided by the turn model [11] to develop a fault-tolerant routing algorithm for meshes. Their method does not require any virtual channels, but tolerates only $n-1$ faults in an n-dimensional mesh and uses non-minimal paths in the fault-free case. The turn model is also utilized by Cunningham and Avresky [12] who provide fault-tolerant routing in two dimensional meshes using two virtual channels. Their method incurs a significant performance loss by a single fault, however, as adaptive routing must be disabled. It also requires healthy nodes to be disabled.

Boppana and Chalasani [13] use local information to create rectangular fault regions in two-dimensional meshes with dimension order routing. The non-faulty nodes and links on the border of a fault region create an *f-ring* or *f-chain* used for rerouting packets around the fault(s). By combining this method with planar adaptive routing it can also be applied to higher dimensional meshes. An improved version by Sui and Wang [14] is able to tolerate overlapping fault regions in meshes using three virtual channels. Using rectangular fault regions has the disadvantage of disabling an unnecessary high number of healthy nodes.

Kim and Han [15] partly address this by extending the method to support overlapped nonconvex fault-regions in meshes, using four virtual channels. Recently, Gu et al. [16] proposed extensions to also support concave fault-regions. This latter method can be applied in combination with previous proposals for handling nonconvex faults in meshes and tori. However, it requires ejecting and reinserting packets when entering/leaving a concave section, thereby increasing latency and occupying memory at the nodes. Park et al. [17] handle simple concave, non-overlapping, fault-regions in meshes without ejecting/reinserting packets, requiring three or four virtual channels depending on the provided fault tolerance. However, this method does not handle faults on the edges of the mesh.

Chalasani and Boppana [18] also proposed a variation of their method for torus, requiring a total of six virtual channels. Shih later improved on this by proposing a method tolerating block faults in tori using three virtual channels [3] and another proposal tolerating nonconvex fault-regions when using four virtual channels [19]. Unless combined with the method of Gu et al., where packets are absorbed and re-injected when entering a concave region, these methods may require healthy nodes to be disabled though. Gómez et al. [2] proposed a fault-tolerant routing methodology based on routing packets via intermediate nodes. This methodology supports fully adaptive routing and is able to tolerate any combination of five faults in three dimensional tori without disabling healthy nodes. Three virtual channels are required in tori when used in combination with the bubble flow control [20]. However, only a static fault-model is supported.

While generic routing algorithms for irregular networks can be used to provide fault tolerance in regular topologies, such strategies generally provide poor network performance compared to using topology specific routing protocols. This disadvantage can be mitigated by using a topology specific routing protocol in the fault-free case, and then switch to a generic routing algorithm once the network becomes faulty. According to this strategy, Puente et al. [21] propose a method providing good network performance in the fault-free case while at the same time providing strong fault-tolerance. The performance in the presence of faults, however, is degraded by non-minimal escape paths, especially in larger networks. Also, the global reconfiguration requires that packet injection is temporarily stopped.

In this paper we propose a routing methodology that is able to handle overlapping concave fault-regions. The methodology requires no virtual channels in meshes, three virtual channels in two dimensional tori, and four virtual channels in three dimensional tori. For all the topologies, fully adaptive routing can be supported by adding at least one additional virtual channel. Because fully adaptive routing significantly improves the network performance, we will focus on using the methodology with fully adaptive routing. The proposed method is inspired by a method proposed for meshes by Skeie [22]. The main differences between the proposal in this paper and the one in the previous paper is that the methodology proposed in this paper is fully distributed, does not require separate control lines, can be applied to both mesh and torus topologies, supports a dynamic fault-model, and provides a higher fault tolerance. Compared

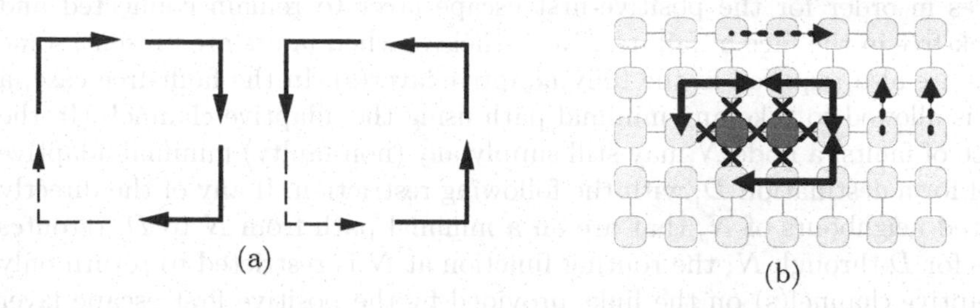

Fig. 1. (a) The solid arrows show the legal turns when positive-first routing is used. No cycles can be created using the legal turns only. (b) Faults can be circumvented using the adaptivity provided by positive-first routing.

to other previous proposals, the method proposed in this paper is able to combine support for a dynamic fault-model and fully adaptive routing, while at the same time not requiring global reconfiguration or stopping packet injection at any time, in a fully distributed manner using a limited number of virtual channels. Furthermore, the method being proposed tolerates faults on the edges of the network and is able to handle concave fault-regions without absorbing and re-injecting packets.

We will now present the fault tolerant routing methodology. Thereafter, in the third section, the network performance of the methodology is evaluated.

2 The Fault-Tolerant Routing Methodology

For simplicity we first assume a two dimensional mesh network, and then later expand this to tori and higher dimensional networks. Because a node fault can be modeled as the failure of all the links of a node, only link faults are considered in this paper. It is assumed that all link faults are bidirectional.

The methodology is based on positive-first routing in order to provide deadlock-freedom. Positive-first is a variation of the turn-model [11], which ensures deadlock freedom in meshes by prohibiting some turns. More specifically, as shown in Figure 1a, the south to east and the west to north transitions are forbidden. In addition, we also require that all paths are minimal in the fault-free case.

In order to improve the network performance, the methodology also supports fully adaptive routing. This is achieved by using positive-first routing as an escape layer for one or more fully adaptive layers, where each layer uses separate virtual channels. Thus, at each hop, a packet may take any minimal path using a fully adaptive channel. If there is no fully adaptive channel free, a positive-first escape channel is used. If wormhole routing is applied, a packet is not allowed to use an adaptive channel after first having used an escape channel. Such a routing function is deadlock-free in accordance with [23].

Because the deadlock freedom of the routing function is provided by the positive-first escape layer, the focus of this section will be on the required measures in order for the positive-first escape layer to remain connected and deadlock-free in the face of failures. Nevertheless, when faults are present, some changes are also required in the fully adaptive layer(s). In the fault-free case, a packet is allowed to take any minimal path using the adaptive channels. In the presence of faults, a node N may still supply any (non-faulty) minimal adaptive channel for a destination D, with the following restriction: If any of the directly connected neighbours of N, that are on a minimal path from N to D, reroutes packets for D through N, the routing function at N is restricted to return only the adaptive channel(s) on the links provided by the positive first escape layer for destination D. This way we avoid loops because of packets switching between the escape and adaptive layers. With this in mind, we will from now on concentrate on the rerouting performed in the positive-first layer.

So, let us turn our attention to how faults are circumvented using the proposed fault-tolerant routing methodology, starting with an example scenario. Figure 1b shows a fault scenario where packets are rerouted around the faults by using only the turns allowed by positive-first routing. As can be seen in the figure, the nodes enclosing the faults form a chain of nodes on which packets can be rerouted around the faults. We will refer to such a chain of nodes as an f-chain. As illustrated, packets on the south side of the faults are routed around the faults counterclockwise, while packets on the west side are routed around the faults in clockwise direction. Packets on the north side are routed clockwise if destined for a destination to the east or south of the faults, while they are routed counterclockwise if destined for a node west (including southwest) of the faults. Finally, packets on the east side of the faults are routed counterclockwise if destined for a destination west or north of the faults, while they are routed south if destined for a packet south (including southwest) of the faults. Special care must be taken on the north and east sides of the faults, for destinations that are rerouted east/north, in order to avoid the illegal turn. Specifically, as illustrated by the dotted arrows in Figure 1b, nodes straight north of the fault(s) must reroute packets eastward if they are to be rerouted around the faults clockwise, so that the illegal turn is not introduced. Similarly, nodes straight east of the fault(s) must reroute packets, which are to be rerouted around the faults counterclockwise, northward. We will refer to such rerouting that is performed by nodes not on the f-chain as secondary reroutes, because these reroutes are required as a result of rerouting performed on the f-chain. Anyway, when rerouting is performed this way, all routing is according to positive-first and is thereby deadlock free.

2.1 Distribution of Status Information

If our method were to be used with a static fault-model, the routing function could simply be calculated based on the network status at system start-up and uploaded to the nodes by a central manager. However, assuming that our method is to be used with a dynamic fault-model in a fully distributed manner, things are more complicated. Under these assumptions, status-information must be distributed through control messages and rerouting decisions must then be taken locally at each node based on this information. This can either be done by distributing the location of the faults and having each node compute its next hops based on this information or by distributing route changes. We will assume that fault information is distributed. If local or non-local faults cause a node to no longer being able to provide one of its directly connected neighbours with a route to a given destination, that neighbour is informed of this change through an update message specifying the location of the faults. Specifically, if node A starts rerouting packets for some destination(s) through node B, node A sends node B an update message with the updated status information. Furthermore, if the change causes node A to reroute some destinations north, the node to the east, that is now to perform a secondary reroute in order to avoid the illegal turn, is notified as well. Likewise, if A reroutes some destinations east, the node to the north must be notified. If changes in the fault status result in A again being able to provide its neighbours with routes to these destinations, the same neighbours should receive this updated status information as well. Notice that if the link connecting two nodes becomes faulty, there is no need to exchange fault information between these two nodes and any status information previously received through the failed link should be discarded.

The distribution of fault information is illustrated in the scenario in Figure 2a. Let us consider the faulty vertical link. We will refer to the node connected to the south end of this link as N_S, and to the node connected to the north end of the link as N_N. Upon detecting that its north link has failed, N_S reroutes packets for destinations relying on this link eastward. Because N_S is now rerouting these destinations eastward, the node east of N_S is notified of the fault through an update message. Because the west-to-north transition is illegal, the node east of N_S does not have any positive-first paths using the failed link. Still, it must restrict the adaptive layer from forwarding such packets westward. That is, for the destinations rerouted east at N_S, the node east of N_S supplies only the adaptive channels of the link provided by the positive-first layer, i.e., the north link. Because these destinations were also forwarded on the north link in the fault-free case, there is no need for further update messages.

The node north of the fault, N_N, also detects that its south link has become faulty and reroutes the destinations relying on this link eastward. N_N must therefore notify the node to the east of the fault through an update message. The node east of N_N had both positive-first and adaptive paths through the failed link, and reroutes these paths south. Because all these destinations were also routed south in the fault-free case, no update message is required to be sent by the node east of N_N. However, because N_N has rerouted some destinations east, it is no longer able to provide its neighbour to the north with paths to these destinations as this would introduce the illegal turn. Thus, N_N must also send an update message to the node to the north, informing about the fault. The node to the north of N_N handles this in a similar manner as if its south link had become faulty, that is, by rerouting the destinations relying on the faulty link eastward and informing its east and north neighbours of the fault. This way information about the fault propagates to all the nodes performing

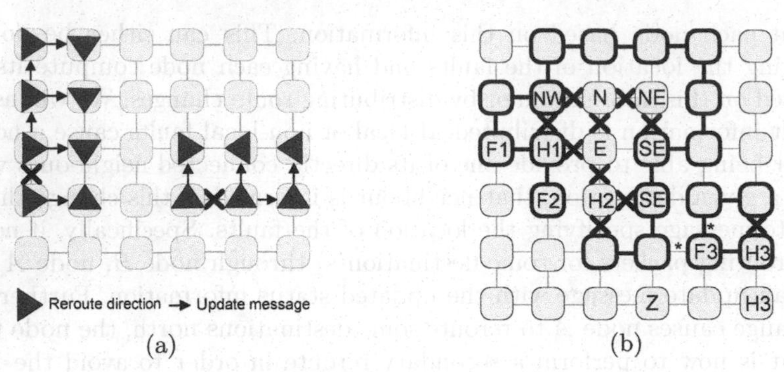

Fig. 2. (a) Rerouting around separate vertical and horizontal link faults. (b) A complex fault-scenario with concavities and hidden areas.

secondary reroutes, resulting in the new routing function using only the turns allowed by positive-first routing. As shown in Figure 2a, the faulty horizontal link is handled in a similar manner. Thus, connectivity is reestablished despite of the faulty links.

2.2 The Dynamic Transition from the Old to the New Routing Function

As shown in the previous sections, the secondary reroutes ensures that the illegal turn is not required by the new routing function. However, because the transition from the old to the new routing function is done dynamically, deadlock is still a concern during the transition phase. The old routing function has become disconnected due to the fault, and thus some packets being routed in the escape layer may have ended up in a situation where they have no legal escape path according to the new routing function. This is for instance the case, in Figure 2a, for packets arriving on the escape channel of N_N's north input link that are to be rerouted on the east link. Forwarding such packets in the escape layer could potentially create a deadlock because it would introduce the illegal turn. For this reason we try to forward such packets using the fully adaptive layer(s). If there is no free buffer space in the fully adaptive layer(s), the packet is dropped. Notice that a packet may only be dropped during transition from the old to the new routing function though, and only at a node that has altered its routing function for the destination of that packet in such a way that an illegal turn would be introduced. Considering that the packets being buffered at a failing node or being transmitted on a failing link are generally lost, it would be unrealistic to try to guarantee that there is no packet loss at all in the face of failures.

2.3 Concave and Nonconvex Fault-Regions and Faults on the Edges of the Network

So far we have considered single link faults and simple collections of faults (i.e., block faults), and shown that these can be circumvented using only the turns allowed by positive-first routing. There are some cases that require additional attention though. While faults on the west and south edges of the mesh are covered by the rules presented in the previous sections, faults on the north and east edges are not. Consider the vertical link failure in the eastmost column of the network in Figure 2b. The nodes south of this fault in the same column, marked $H3$ in the figure, are unreachable from all nodes north of the faulty link, and vice versa, unless the illegal turn is used. Let us define such an area that is unreachable, without using the illegal turn, as a hidden area. A hidden area is enclosed by faulty links, or by one or more faulty links in combination with the edge(s) of the mesh, and can only be entered using a positive (i.e., north/east) channel and only be left using a negative (i.e., west/south) channel. Thus, a hidden area has an opening only on the south and/or west side. If being entered from the west side, the nodes in a hidden area are unreachable from nodes north of the northmost entry to the hidden area without using the illegal turn. Similarly, if being entered from the south side, the nodes in the hidden area are unreachable from nodes east of the eastmost entry to the area. Figure 2b shows some examples of hidden areas. The node labeled $H2$ constitutes a hidden area with opening to the west. This hidden area is nested within another hidden area, consisting of nodes $H1$ and $F2$ in addition to $H2$, with opening to the south and west. The two nodes labeled $H3$, constitutes a hidden area with opening to the west, enclosed by the faulty link and the edges of the mesh.

Let us denote a node that may introduce the illegal turn without risk of deadlock as a free-node. The free-node itself is positioned outside the hidden area, and hidden areas with entry from one side has one free-node while a hidden area with entry from two sides has two free-nodes. When there are two free-nodes for the same hidden area, only one of them may introduce the illegal turn and by convention we chose to introduce the illegal turn on the west side in such cases. When hidden areas are nested within each other, a hidden area may also contain a free-node for entering another hidden area. In Figure 2b, $F1$, $F2$, and $F3$ are free-nodes that may introduce the illegal turn. Because a cyclic dependency can not be created by introducing the illegal turn at these nodes, there is no risk of deadlock.

We define an entry node of a hidden area as a node within the hidden area that is directly connected (by a non-faulty link) to a node outside the hidden area. Based on this, let us more formally define a node F, connected to node F_N through its north link and to node F_E through its east link, as a free-node if one of the following conditions apply:

- F_N is the eastmost entry node of a hidden area with entry from the south, and there is no dependency in the positive-first layer from the north output link of F to the north input link of F.
- F_E is the northmost entry node of a hidden area with entry from the west, and there is no dependency in the positive-first layer from the east output link of F to the east input link of F.

Lemma 1. *The illegal turn can be introduced at one free-node for each hidden area without risk of deadlock.*

Proof. For the illegal turn at F to create a deadlock there needs to be a cyclic dependency, in the positive-first layer, that includes the illegal turn at F. Thus, this cycle must go through both F_N and F_E. Let us consider the case where F_E is the northmost entry node of a hidden area with entry from the west side. Then F_E is inside the hidden area and F_N is outside the hidden area. Also, F_N is further north than F_E and the hidden area can not be left/entered north of F_E. Thus, in order to complete a cycle going through both F_N and F_E, two illegal transitions are required. This can be achieved in two ways, either the cycle uses the illegal turn at F twice (once in each direction), or the cycle must go through an illegal turn at another free-node. If the cycle uses the illegal turn at F twice, there must be a dependency from the east output link of F to the east input link of F. However, then F is by definition not a free-node. Another free-node is therefore required to complete the cycle, meaning another hidden area is required. However, because a hidden area can be entered only through a positive channel and be left only on a negative channel, and there is no cycle using the same free-node twice, there is no negative channel entering a hidden area that depends on a positive channel leaving the hidden area, even when using the illegal turn. Thus, another hidden area can not provide the illegal transition required to complete the cycle. The case where F_N is the eastmost entry node of a hidden area can be proved in a similar manner.

By using the illegal turn according to Lemma 1, faults on both the north and east edges are tolerated. This is illustrated in Figure 2b, where the hidden area $H3$ remains connected through the illegal turn at free-node $F3$. Furthermore, Lemma 1 can be used to handle concavities with entry from the south and/or west sides. This is illustrated in the figure by free-node $F1$ connecting the hidden area with opening to the west and south, and free-node $F2$ connecting the hidden area/node $H2$. When the illegal turn is introduced, this enables the nodes connected to the north and east links of the free-node to provide routes to new destinations. Therefore, fault status information is sent to the next node on the f-chain which again may propagate the information further around the faulty region. If the faulty links causing the illegal turn to be introduced are repaired, the routing function should not be updated to use the repaired links until the illegal turn has been removed.

As can be seen from the figure, a concavity with opening on the north or east side (like N and E in Figure 2b) does not require the use of the illegal turn in order to be connected. The same holds for regions with opening on the north and west sides (like NW in the figure) and for regions with opening on the south and east sides (like SE in the figure).

One special case may occur when there is a concavity on the west side of a collection of faults on the north edge, or on the south side of a collection of faults on the east edge. For instance, if the north and west links of $F3$ (marked with stars in Figure 2b) were faulty, F_3 and H_3 would create such a concavity on the south side. The illegal turn would now be required at node Z southwest of F_3. However, the illegal turn can not be introduced according to Lemma 1 in this case, because there is a dependency from the east link of Z going through F_3 and

H_3 and back to the east link of Z. Thus, in order to satisfy the requirements of Lemma 1 so that the illegal turn can be introduced at Z, we must reroute packets that would normally use the east/west links of the nodes within the concavity (i.e., nodes F_3 and H_3) south. In fact, the link between these two nodes is used only for direct communication between these two nodes. Traffic to/from other nodes goes through the south links, introducing an additional illegal turn south of F_3. Because of the alteration of the routing function at F_3, this turn can be introduced according to Lemma 1. Similar scenarios can also be created away from the edges of the network, but a high number of closely located faults are then required to create these scenarios and the practical use of handling them may therefore be limited. Anyway, such scenarios can be handled in a similar manner but are considered an implementation issue.

2.4 Extension to Tori

Let us now consider what changes are necessary in order to be able to apply the proposed methodology to torus topologies. First of all, positive-first routing alone ensures deadlock freedom only in meshes. Thus, in order to ensure deadlock-free minimal routing in tori, additional virtual channels are required. By always changing virtual layer when crossing a wraparound link, the additional dependencies introduced by the wraparound links are broken and the network remains deadlock free. When minimal routing is used in a two dimensional tori, a packet may use at most two wraparound links. Thus, in order to be able to change virtual layer each time a wraparound link is used, two additional virtual layers are required, for a total of three virtual channels. In three dimensional tori, a total of four virtual channels are required. As before, in order to improve the network performance, one or more fully adaptive layers may be used in addition.

Rerouting packets over wraparound links could potentially result in packets crossing wraparound links more than twice. For instance, if packets were rerouted east to circumvent a faulty link on the east edge, they would cross a wraparound link once when being rerouted east and then again when going back west after having circumvented the faulty link. Now, if such a packet also was to use a north/south wraparound link, there would not be enough virtual channels in order to change virtual layer each time a wraparound link is used. To avoid this problem, packets are generally rerouted the same way in tori as we have previously described for meshes. Specifically, a packet is not rerouted across a wraparound link, unless it would also use a wraparound link in that direction in the fault-free case. This restriction not only ensures that we are always able to change virtual layer when crossing a wraparound link, but also ensures that a packet never encounters the same f-chain/fault(s) more than once.

This way of avoiding rerouting packets over wraparound links also has the implication that a node S, that previously routed packets for a given destination D over a wraparound link, may have to avoid using this wraparound link. This would be the case if some other node, T, that is not allowed to use the wraparound link used by S for destination D, is rerouting packets for D through

S. In this case, packets for D must be rerouted around the faulty region so that a route using only wraparound links allowed by T (and S) can be provided.

Free-nodes and illegal turns are introduced in the same way as for meshes. In fact, the illegal turns introduced in a mesh are introduced at the same places in a torus topology. In addition, all the nodes on the north and east edges of the torus can safely introduce the illegal turn, as these dependencies are broken because of the change in virtual layer when using a wraparound link.

Let us now consider the secondary reroutes that are performed in order to avoid the illegal turn. Because packets are routed minimally, it is not necessary to update the entire row/column to the east/north in a torus. Given that k is the number of nodes in the dimension, only the $k/2$ nodes east/north of the faulty link have minimal paths through this link, thus, only these nodes need to perform secondary reroutes. Furthermore, because packets change virtual layer when crossing a wraparound link, thereby breaking the dependencies, it is not necessary to perform secondary reroutes across wraparound borders. Thus, only the nodes within distance $k/2$ to the east/north, and that are not across a wraparound border, are required to perform secondary reroutes.

If multiple faults have partitioned the mesh, a mesh network would become physically disconnected. A torus network could still be connected through its wraparound link(s) however. Because such cases are relatively rare and requires special handling, our implementation of the methodology does not handle such cases. This is an implementation choice however. The ability to introduce the illegal turn on the north and east edges, and according to Lemma 1, provides sufficient flexibility in order to handle these cases. However, special care must be taken on how the escape layers and wraparound links are used. E.g., if a partition is connected to the remainder of the network through only one edge of the network, the cyclic dependencies introduced by the wraparound links are broken by the faults and it is therefore not required to change virtual layer when using these wraparound links. Instead this saved layer change should be used for packets that now have to cross the same edge of the network twice (once in each direction) in order to enter/leave the partition. Also, the nodes along the edge of the network, from which the partition can be entered, must reroute packets destined for the partition so that they do not cross the wraparound border but instead are routed towards the entry of the partition.

2.5 Three-Dimensional Networks

We will now briefly describe how the proposed methodology can be extended to three-dimensional networks. When another dimension is added, we denote the positive direction in the new dimension up and the new negative direction down. In addition to the transitions already forbidden for two-dimensional networks, the west-to-up, south-to-up, down-to-north, and down-to-east transitions are forbidden according to positive-first routing in order to preserve deadlock freedom. Notice that all the additional forbidden transitions involves the new dimension, thus, all previously used turns are still valid.

A hidden area is still defined as an area that can be entered only through a positive channel (up, east, north) and be left only on a negative channel (down, west, south). The definition of a free-node F can be generalized to that F should be connected to the positivemost entry node of the hidden area. E.g., if the hidden area has opening on the south side, the free-node should be connected to the east-up-most entry node. As before, there should be no dependency between the outgoing positive-first channel, and the incoming positive-first channel, on the link connecting the free-node with the entry node.

Based on this, rerouting can be performed in a similar way as described for two-dimensional networks. In particular, in a three dimensional network, each link is part of two two-dimensional planes. Thus, when a link fails, the previously described rerouting must be performed in both planes.

3 Evaluation

A flit-level event-driven simulator has been used for evaluating the performance of the proposed methodology. For all the simulations, a 16×16 torus topology is used. Virtual cut through routing is applied. Each physical link is divided into five virtual channels, where each virtual channel has enough buffer space to store two packets. Each packet consists of 32 flits. Three virtual channels (i.e., the escape channels) are used for routing packets according to positive-first routing, while two virtual channels are used for fully adaptive routing. Furthermore, there is a virtual channel used for control messages that is given priority above the data channels. A processing delay of 40 cycles is added after receiving an update message or detecting a faulty link. Each simulation has been performed 30 times, thus, each value in the plots represents the average of 30 simulations. In each of these 30 simulations the positions of the faults have been selected randomly, with the restriction that they do not physically disconnect the network or partition the mesh. In Figure 3a, each simulation has first been run for a stabilization period, where a regression analysis is performed to determine if the network has stabilized, thereafter the simulations have been run for 30 000 cycles.

Figure 3a shows the accepted throughput, depending on the number of faults in the network, for two different traffic patterns. The top plot shows the throughput degradation, in the presence of faults, under uniform traffic. With this traffic pattern, the destination of each packet is selected randomly with equal probability for all destinations. The lower plot in the figure shows the throughput under permutation traffic. With this traffic pattern, each source sends all its packets to a single randomly selected destination so that each destination receives packets from exactly one source. As can be seen, the methodology provides graceful degradation under both traffic patterns. Specifically, with uniform traffic, throughput is on average degraded 14.5% in the presence of seven faults. For the permutation traffic pattern, the average performance degradation in the presence of seven faults is 11.2%. The smaller degradation with the permutation traffic pattern may be explained by the fact that the network traffic is already unbalanced in the fault-free case. With the uniform traffic pattern, the traffic first becomes unbalanced when network is faulty. Nevertheless, the use of

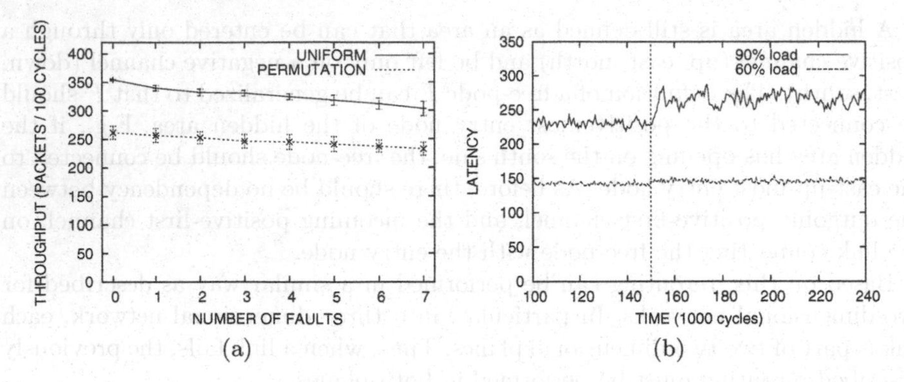

Fig. 3. (a) Throughput in a 16×16 torus network with uniform and permutation traffic. The error bars show the 95% confidence intervals. (b) Latency in a 16×16 torus network with uniform traffic. Three random link faults are introduced at the time signified by the vertical line.

adaptive routing helps mitigate congestion around the faults, maintaining network performance in the presence of faults.

In Figure 3b, three random link faults are injected at the time signified by the vertical line. 30 random scenarios have been simulated under uniform traffic with both 60% and 90% of the maximum accepted load in the fault-free case. As can be seen, with 90% load there is a distinct increase in latency when the faults are introduced and the network approaches saturation. At 60% load the increase in latency is very small. As the network remains below saturation, there is no significant change in throughput in either of the cases. Notice that at no time is network traffic stopped. Packets are forwarded as normal. In the case that a rerouting decision is not yet reached, and there is no non-faulty route, the packet is dropped. Also, if a packet that has been forwarded on an escape channel according to the routing function for the fault-free case can not be legally forwarded according to the new routing function it is forwarded in the adaptive layer if possible, otherwise the packet is dropped. Still, the total packet loss is modest. Specifically, in each of the scenarios with three random link faults under 90% load, there was on average a total packet loss of 19.5 packets (including the loss of the packets occupying the failing links). With 60% network load, only 8.7 packets were lost on average.

4 Conclusions

We have proposed a fault-tolerant routing methodology that tolerates concave fault-regions and provides graceful performance degradation in the presence of faults. The proposed methodology allows the network to remain fully operational in the face of failures, without stopping network traffic at any time, and is therefore suitable for applications with high requirements for availability. Because the network remains continuously operational, the proposed solution enables failures in the interconnection network to be made transparent to the applications.

References

1. Top 500 Supercomputing Sites (2007), http://www.top500.org/lists/2007/06
2. Gómez, M., Nordbotten, N., et al.: A routing methodology for achieving fault tolerance in direct networks. IEEE Trans. Computers 55(4), 400–415 (2006)
3. Shih, J.D.: Fault-tolerant wormhole routing in torus networks with overlapped block faults. IEE Proc. Computers and Digital Techniques 150(1), 29–37 (2003)
4. Mukherjee, S., Bannon, P., Lang, S., Spink, A., Webb, D.: The Alpha 21364 network architecture. IEEE Micro 22(1), 26–35 (2002)
5. Wang, H., et al.: A technology-aware and energy-oriented topology exploration for on-chip networks. In: Design, Automation and Test in Europe, pp. 1238–1243 (2005)
6. Intel Corporation: Tera-scale research prototype, ftp://download.intel.com/research/platform/terascale/tera-scaleresearchprototypebackgrounder.pdf
7. Held, J.: et al.: From a few cores to many: A tera-scale computing research overview, ftp://download.intel.com/research/platform/terascale/
8. Linder, D., Harden, J.: An adaptive and fault tolerant wormhole routing strategy for k-ary n-cubes. IEEE Trans. Computers 40(1), 2–12 (1991)
9. Chien, A., Kim, J.: Planar adaptive routing: Low-cost adaptive networks for multiprocessors. Journal of the ACM 42(1), 91–123 (1995)
10. Glass, C., Ni, L.: Fault-tolerant wormhole routing in meshes without virtual channels. IEEE Trans. Parallel and Distributed Systems 7(6), 620–636 (1996)
11. Glass, C., Ni, L.: The turn model for adaptive routing. Journal of the ACM 41(5), 874–902 (1994)
12. Cunningham, C., Avresky, D.: Fault-tolerant adaptive routing for two dimensional meshes. In: Proc. Symp. High-Performance Comp. Architecture, pp. 122–131 (1995)
13. Boppana, R., Chalasani, S.: Fault-tolerant wormhole routing algorithms for mesh networks. IEEE Trans. Computers 44(7), 848–864 (1995)
14. Sui, P.H., Wang, S.D.: An improved algorithm for fault-tolerant wormhole routing in meshes. IEEE Trans. Computers 46(9), 1040–1042 (1997)
15. Kim, S.P., Han, T.: Fault-tolerant wormhole routing in mesh with overlapped solid fault regions. Parallel Computing 23, 1937–1962 (1997)
16. Gu, H., et al.: A new routing method to tolerate both convex and concave faulty regions in mesh/torus networks. In: Proc. PDCAT, pp. 714–719 (2005)
17. Park, S., et al.: Fault-tolerant wormhole routing algorithms in meshes in the presence of concave faults. In: Proc. Int. Paral. and Dist. Processing Symp. (2000)
18. Chalasani, S., Boppana, R.: Fault-tolerant wormhole routing in tori. In: Proc. ACM Int. Conf. on Supercomputing, pp. 146–155 (1994)
19. Shih, J.D.: A fault-tolerant wormhole routing scheme for torus networks with nonconvex faults. Information Processing Letters 88(6), 271–278 (2003)
20. Carrion, C., et al.: A flow control mechanism to avoid message deadlock in k-ary n-cube networks. In: Int. Conf. High Performance Computing, pp. 322–329 (1997)
21. Puente, V., et al.: Immunet: A cheap and robust fault-tolerant packet routing mechanism. In: Proc. Int. Symp. Computer Architecture, pp. 198–209 (2004)
22. Skeie, T.: Handling multiple faults in wormhole mesh networks. In: Pritchard, D., Reeve, J.S. (eds.) Euro-Par 1998. LNCS, vol. 1470, pp. 1076–1088. Springer, Heidelberg (1998)
23. Duato, J.: A necessary and sufficient condition for deadlock-free adaptive routing in wormhole networks. IEEE Trans. Parallel and Distributed Systems 6(10) (1995)

Fault-Tolerant Topology Adaptation by Localized Distributed Protocol Switching

Sushanta Karmakar and Arobinda Gupta

Department of Computer Science and Engineering,
Indian Institute of Technology, Kharagpur - 721 302, India
{sushantak,agupta}@cse.iitkgp.ernet.in

Abstract. Adaptation is a desirable requirement in a distributed system. For many problems, there exists more than one protocol such that one protocol performs better in one environment while the other performs better in another. In such cases, adaptive distributed systems can be designed by dynamically switching between the protocols as the environment changes. In this work, we present distributed algorithms to switch from a BFS tree to a DFS tree and from a DFS tree to a BFS tree. For low network load, a BFS tree is a better choice for broadcast since it also minimizes delay, whereas for higher network load, a DFS tree may be more suitable to reduce the load on any one node. The proposed switching algorithms can handle arbitrary crash failures. They ensure that switching eventually completes in spite of failures with the desired tree as the output. Also, all messages are correctly broadcast in the absence of failures even in the presence of switching.

Keywords: protocol switching, crash failure, BFS tree, DFS tree, broadcast.

1 Introduction

The performance of a distributed system depends on its environment. However, the environment may change with time. Therefore it is necessary for a distributed system to be adaptive under changing environments. Adaptation is a desirable requirement in any distributed system since it helps the system to perform gracefully under different scenarios. Adaptation can be achieved in various ways. One way is to change the runtime parameters of the algorithm appropriately. For example, Jacobson [1] showed how congestion can be controlled by suitably varying the flow window size in TCP. The ability to adapt may also be in-built into a system. For example, Anderson et. al. [2] provided an adaptive mutual exclusion algorithm where the arbitration time is proportional to the contention for the shared resource. There are other adaptive algorithms for different problems [3][4]. However these techniques are less general and often application specific. In many distributed systems, it may happen that the same problem has multiple protocols, each of which performs differently under different environments. For example, for routing protocols for ad-hoc networks, Das et. al. [5] showed that for high mobility, AODV has lower delay than DSR whereas for low mobility, DSR has lower delay than AODV. In such cases adaptation can be achieved by dynamically switching between them as the environment changes.

S. Aluru et al. (Eds.): HiPC 2007, LNCS 4873, pp. 528–539, 2007.

In this paper, we illustrate the idea of designing adaptive distributed system using protocol switching by presenting an adaptive protocol that dynamically switches between two topologies. The protocol uses either a BFS tree or a DFS tree for broadcast depending on the load of the system, both trees being rooted at the broadcast source. The system load can be monitored by the root node using convergecast on an existing spanning tree. At low load a BFS tree is used as it reduces the broadcast delay since the distance of any node from the root is always minimum in a BFS tree. However at higher load a DFS tree is used to reduce the load on any one node since the degree of a node in a DFS tree is generally lower than that in a BFS tree. So the broadcast adapts to the network load by dynamically switching between a BFS tree and a DFS tree. The message complexity of the algorithm that switches to a DFS tree is $O(|E|)$, where E is the number of links in the system. The algorithm that switches to a BFS tree needs $O(|V||E|)$ messages. The switching is also crash fault-tolerant. More specifically the switching from a BFS tree to a DFS tree can handle arbitrary crash failures and switching eventually completes with a DFS tree as the output. Similarly the switching from a DFS tree to a BFS tree can handle arbitrary crash failures and switching eventually completes with a BFS tree as the output. If a fault happens when no switching is in progress then the original tree is restored using local repair actions. Also under no failure, the algorithms guarantee that each broadcast message is eventually correctly delivered to all the nodes.

Switching between a BFS tree and a DFS tree can be done by computing both the trees in advance and switching between them using distributed reset [6]. However if any of the precomputed outputs becomes invalid due to failure of nodes or links, then the switching fails to establish the correct output. So whenever a switch is needed, the output must be recomputed to establish the correct output after the switching. In this paper, none of the proposed algorithms assume the existence of a precomputed output at any stage. They can switch from an arbitrary graph to a DFS tree or BFS tree. It is to be noted that the recomputation of the desired tree can be done by pausing the broadcast at the root, executing a standard distributed algorithm for the tree formation, and resuming the broadcast after the desired tree is formed. This has the limitation that the broadcast gets stalled in the whole network during switching. In the algorithms proposed in this paper, the broadcast gets stalled only in a small part of the network, thus reducing the delay in broadcasting a message.

Bar-Noy et. al. [7] proposed a method of dynamically changing between different byzantine agreement protocols. Arora et. al. [6] proposed a method to switch from one state to another in a distributed system without requiring a global freeze. Liu et. al. [8] described a method to build a hybrid protocol which adapts by dynamically mapping the state of a process in one protocol to the state in another. Chen et. al. [9] described a software architecture for constructing adaptive software that can react to changes in the execution environment or user requirements by switching algorithms at runtime. Rutti et. al. [10] described an adaptive group communication middleware that switches between different atomic broadcast protocols on the fly. Mocito and Rodrigues [11] proposed another algorithm that dynamically switches between different total order algorithms with negligible interference to the data flow. Our work is different from the earlier works since none of them provided a distributed algorithm that switches from one protocol to another while maintaining some desirable application layer property provided no failure occurs. Also to the best of our knowledge, this is the first work that discusses the fault-tolerance aspect of distributed protocol switching.

The rest of the paper is organized as follows. Section 2 and Section 3 describe non-fault-tolerant algorithms that switch to a DFS tree and BFS tree respectively. We then modify these protocols to present the fault-tolerant switching protocols in Section 4.

2 Non-fault-Tolerant Switching to DFS Tree

We assume an asynchronous, message-passing distributed system represented by a connected, undirected graph $G = (V, E)$ with node set V representing the processes and edge set E representing the links. The links are assumed to be reliable. For the time being, we assume that nodes do not fail. The algorithm switches from an arbitrary graph G to a DFS tree. Let $Child(v)$ denote the set of children of v and $p(v)$ denote the parent of v. For the root node v_r, always $p(v_r) = v_r$. Let $N(v)$ denote the set of neighbors of v. Let E_u denote the set of edges incident on a node u. Let $G_v = (V', E')$ be some subgraph of G where $V' \subseteq \{v\} \cup N(v)$ and $E' \subseteq \bigcup_{u \in V'} E_u$. If T_v is a DFS spanning tree of G_v rooted at v then we call T_v the *local DFS subtree* rooted at v.

The pseudocode of the algorithm is given in Figure 1. There is one node v_r, the initiator, that starts the switching. Also it is the first node to get a *TOKEN*. Let $TSet(v) \subseteq N(v)$ be the set of neighbors of v that have already received a *TOKEN*. Initially for each v, $TSet(v) = \phi$. On receiving a *TOKEN*, a node v takes the following action. If v has already got a *TOKEN* earlier then it sends the *TOKEN* to some $u \in Child(v)$ such that u has not yet received any *TOKEN* message. If no such u exists then v sends the *TOKEN* to $p(v)$. However, if v gets a *TOKEN* for the first time then it constructs a local DFS subtree, rooted at v itself, of the graph induced by the set of nodes $CSet(v) \cup \{v\}$ where $CSet(v) = N(v) - [TSet(v) \cup \{p(v)\}]$. If $CSet(v) = \phi$ then the local DFS subtree to be built is trivial and hence v sends the *TOKEN* to some u as stated earlier. However, if $CSet(v) \neq \phi$ then v starts the local DFS subtree construction by sending $LDFS(v, CSet(v))$ to some $u \in CSet(v)$. Also if $u \notin Child(v)$ then v adds u to $Child(v)$ since u will later change its parent to v to become part of the DFS subtree.

On receiving $LDFS(v, S)$ from w, a node u takes the following action. Suppose u has not yet received a *TOKEN*. So $u \in S$. If u has already received an *LDFS* message then it just forwards $LDFS(v, S)$ to u' where $u' = u''$ if $\exists u''$ such that $u'' \in N(u) - [TSet(u) \cup \{p(u)\}] \wedge u'' \in S$, or $u' = p(u)$ if no such u'' exists. On the contrary, if u receives the *LDFS* message for the first time then u adds v to $TSet(u)$ (since v has got the *TOKEN*). Also u removes itself from S and creates $S' = S - \{u\}$. If $p(u) = w' \neq w$ then u changes its parent from w' to w and sends a *REMOVE_CHILD* message to its old parent w' which in turn sends a *REMOVE_ACK* message back to u as soon as u is removed from $Child(w')$. In any case, u finally sends $LDFS(v, S')$ to an appropriate node u' as stated earlier. However, if u has already received a *TOKEN* and now it receives $LDFS(v, S)$ from w then $u = v$ since v is the only node in the local DFS subtree that has received a *TOKEN*. In this case, if $S = \phi$ then v has completed the local DFS subtree construction and thus v passes the *TOKEN* to an appropriate node. If $S \neq \phi$ then v continues to construct the remaining part of the local DFS subtree by sending $LDFS(v, S)$ to some $u' \in S$. This process continues and finally a DFS

```
UPON RECEIVING TOKEN
  if tokenVisited = false then
    tokenVisited = true
    CSet = N − [TSet ∪ {p}]
    if CSet = φ then passTOKEN( )
    else send LDFS(v, CSet) to u ∈ CSet
         Child = Child ∪ {u}
  else passTOKEN( )

UPON RECEIVING REMOVE_CHILD FROM w
  Child = Child − {w}
  send REMOVE_ACK to w

UPON RECEIVING REMOVE_ACK
  cpflag = false; passLDFS( )

FUNCTION passTOKEN()
  CTRT = Child − Visited
  if CTRT = φ then
    if p ≠ v then send TOKEN to p
  else send TOKEN to u ∈ CTRT
       Visited = Visited ∪ {u}

UPON RECEIVING LDFS(w, S) FROM w'
  ldfsSource = w; CSet = S
  if tokenVisited = true then
    if CSet = φ then passTOKEN( )
    else send LDFS(ldfsSource, CSet) to u ∈ CSet
         Child = Child ∪ {u}
  else if ldfsVisited = false then
    ldfsVisited = true; CSet = CSet − {v}
    TSet = TSet ∪ {ldfsSource}
    if p ≠ w' then cpflag = true
                   send REMOVE_CHILD to p; p = w'
    if cpflag = false then passLDFS( )

FUNCTION passLDFS()
  if ∃u : u ∈ N − [TSet ∪ {p}] ∧ u ∈ CSet then
    send LDFS(ldfsSource, CSet) to u
    Child = Child ∪ {u}
  else ldfsVisited = false
    send LDFS(ldfsSource, CSet) to p
```

Fig. 1. Algorithm for switching to a DFS tree for node v

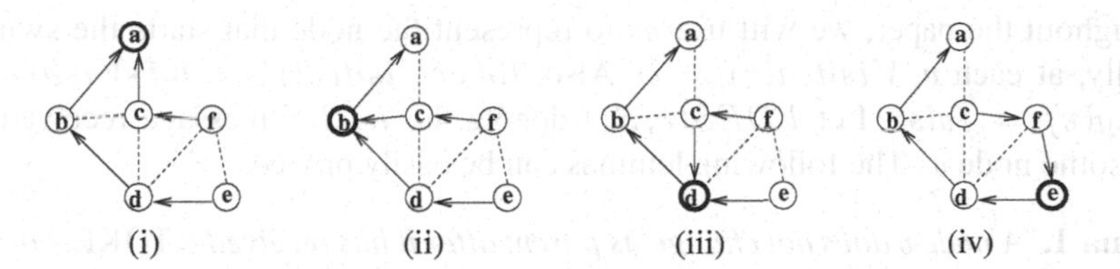

Fig. 2. An illustration of switching from a BFS tree to a DFS tree

tree results when the algorithm terminates. If some spanning tree T of G rooted at v_r exists (at the initial state or from some intermediate state) then the algorithm switches from T to T', a DFS tree of G rooted at v_r, while always maintaining some spanning tree T_s of G rooted at v_r during switching. Initially $T_s = T$ and finally $T_s = T'$. This reduces the delay and the number of messages required for broadcast during the switching. An example illustrating the switching from a BFS tree to a DFS tree is shown in Figure 2. The arrows represent the tree edges and the dashed lines represent the non-tree edges in the graph. Initially a BFS tree of the graph G exists as shown in Figure 2(i). After the switching completes, a DFS tree of the graph G is formed as shown in Figure 2(iv). Figure 2(ii) and Figure 2(iii) show the intermediate states. The node holding the $TOKEN$ at any instant is circled with bold.

2.1 Correct Delivery of Broadcast Messages

Figure 2 shows that the underlying topology changes with time until the DFS tree is formed. So using the topology for broadcast is difficult during the switching. We give one approach to deal with the situation. Let each broadcast message from the root be sequentially timestamped. When a node v decides that a protocol switch is needed (i.e. $TOKEN$ received and $tokenVisited = false$), it starts buffering all broadcast messages having timestamp greater than T, when T is the timestamp of the last broadcast message sent by v. Also v sends a special control message $\mathcal{P}$ to each $u \in U = CSet(v)$. When u gets $\mathcal{P}$ from v, it starts buffering all broadcast messages having timestamp greater than T and sends a control message $\mathcal{Q}$ to v if all broadcast messages having timestamp less than or equal to T have been received. After getting the control message $\mathcal{Q}$ from each $u \in U$, v actually switches the protocol (for example, by local DFS subtree formation). At the end of the switching v sends a control message $\mathcal{R}$ to each $u \in U$ and also stops buffering of any broadcast message. On receiving $\mathcal{R}$, u stops buffering of any broadcast message. From now on, each broadcast message from the root eventually reaches each node $u \in V$ using the new topology that results at the end of the protocol switching at v. Also v forwards the switching message (the $TOKEN$) to an appropriate node $w \in N(v)$. So the broadcast can continue even during switching and it gets stalled only in a small subset $V' = \{v\} \cup CSet(v) \subseteq V$ of nodes at a time. Also it is proved below that each broadcast message from the root is eventually delivered to each $u \in V$.

2.2 Outline of Proof of Correctness

Throughout the paper, we will use v_r to represent the node that starts the switching. Initially, at each v, $Visited(v) = \phi$. Also, $tokenVisited(v) = ldfsVisited(v) = cpflag(v) = false$. Let $LDFS(v, S_u)$ denote the $LDFS$ message received by u, from some node w. The following lemmas can be easily proved.

Lemma 1. *A node v does not change its parent after it has received a* TOKEN *message.*

Lemma 2. *Each node v receives the $TOKEN$ from a neighbor u exactly once.*

Lemma 3. *If a node v sends $LDFS(v, S_u)$ to some $u \in S_u$ then $\forall w \in S_u$, w eventually receives $LDFS(v, S_w)$ and v eventually receives $LDFS(v, \phi)$.*

Lemma 4. *If a node v receives a* TOKEN *message for the first time then each edge incident on v eventually becomes either a* tree *edge or a* back *edge before it sends the* TOKEN *to some other node.*

Proof. Suppose v gets a $TOKEN$ message for the first time. So $tokenVisited(v) = false$. So v computes $CSet(v) = N(v) - [TSet(v) \cup \{p(v)\}]$. Let $p(v) = v'$. So by Lemma 1, v does not change $p(v)$ anymore. So (v', v) remains a *tree* edge. Again $\forall v'' \in TSet(v)$ where $v'' \neq v'$, (v, v'') is a *back* edge since $v'' \notin Child(v)$. Therefore $\forall u \in TSet(v) \cup \{p(v)\}$, (v, u) is either a *tree* edge or a *back* edge. So $\forall u \in CSet(v)$, (u, v) is either a *tree* edge ($u \in Child(v)$) or *non-tree* edge ($u \notin Child(v)$). We show that $\forall u \in CSet(v)$, (u, v) eventually becomes either a *tree* edge or a *back* edge. Let $CSet(v) = S$. If $S \neq \phi$ then by Lemma 3, $\forall u \in S$, u eventually receives $LDFS(v, S_u)$. Node u receives $LDFS(v, S_u)$ either directly from v or from some other node $w \in S$ where $w \neq v$. Let $x \rightsquigarrow y$ mean there exists a path from x to y. The following cases are possible. **(i)** $u \in Child(v)$ and u gets $LDFS(v, S_u)$ from $w = v$. So $p(u) = v = w$. Hence u does not change $p(u)$. So (u, v) remains a *tree* edge. **(ii)** $u \in Child(v)$ and u gets $LDFS(v, S_u)$ from $w \neq v$. So $p(u) = v \neq w$. Thus u changes $p(u)$ from v to w

and thus $u \rightsquigarrow w$ holds. So (u, v) becomes a *non-tree* edge. Since u got $LDFS(v, S_u)$ from $w \neq v$, $w \rightsquigarrow v$ exists. $u \rightsquigarrow w \wedge w \rightsquigarrow v \Rightarrow u \rightsquigarrow v$. (u,v) is a *non-tree* edge and $u \rightsquigarrow v$ implies that (u, v) is a *back* edge. **(iii)** $u \notin Child(v)$ and u gets $LDFS(v, S_u)$ from $w = v$. So $p(u) \neq v = w$. Therefore u changes $p(u)$ from u' (say) to v. Hence (u, v) becomes a *tree* edge. **(iv)** $u \notin Child(v)$ and u gets an $LDFS(v, S_u)$ from $w \neq v$. So either $p(u) = w$ or $p(u) \neq w$. In either case $p(u)$ becomes w and therefore $u \rightsquigarrow w$ holds. Again since u got $LDFS(v, S_u)$ from $w \neq v$, $w \rightsquigarrow v$ exists. $u \rightsquigarrow w \wedge w \rightsquigarrow v \Rightarrow u \rightsquigarrow v$. Therefore (u, v) is a *back* edge. So $\forall u \in CSet(v)$, (u, v) eventually becomes a *tree* edge or a *back* edge. So $\forall u \in N(v)$, (u, v) eventually becomes either a *tree* edge or a *back* edge. □

Theorem 1. *The switching algorithm eventually terminates with a DFS spanning tree.*

Proof. By Lemma 2, v gets the $TOKEN$ from a neighbor u exactly once. Each node v has at most $N(v)$ neighbors. So v can send or receive the $TOKEN$ at most $N(v)$ times. Hence eventually the system will reach a state after which no $TOKEN$ message will be sent by any node and hence no other messages will be sent. Thus the switching algorithm eventually terminates. Also by Lemma 4, each edge incident on v eventually becomes a *tree* edge or a *back* edge before it sends the *TOKEN* to some other node u. So when the switching algorithm terminates each edge $e \in E$ is either a *tree* edge or a *back* edge. Hence the switching algorithm terminates with a DFS spanning tree of G rooted at v_r. □

Theorem 2. *The message complexity of the switching algorithm is $O(|E|)$.*

Proof. Let d_v be the degree of a node v . So v sends d_v number of $TOKEN$ messages. Hence total number of $TOKEN$ messages sent is, $\mathcal{M}_T = \sum_{v \in V} d_v = 2|E|$. If v receives a $TOKEN$ message for the first time then it creates a local DFS subtree. During local DFS subtree formation at v, at most $O(d_v)$ number of messages are sent. So by Lemma 2, the total number of messages sent by all the nodes in the worst case is given by, $\mathcal{M} = \mathcal{M}_T + \sum_{v \in V} O(d_v) = 2|E| + O(|E|) = O(|E|)$. □

Theorem 3. *Each message broadcast during the switching is eventually delivered to all the nodes.*

Proof. If v gets the $TOKEN$ and $tokenVisited(v) = false$ then the set of nodes $\{v\} \cup CSet(v)$ are involved in switching. For a node u, if the broadcast path from v_r to u does not pass through any one of $\{v\} \cup CSet(v)$ then it receives all broadcast messages from v_r despite switching. But if the broadcast path from v_r to u has some $w \in \{v\} \cup CSet(v)$ as the intermediate node then according to the control algorithm proposed in Section 2.1, each messages having timestamp less than or equal to T is received by u where T is the timestamp of the last broadcast message sent by v. But each messages having timestamp greater than T is buffered at v until the switching is over. At the end of the switching at v, each $w \in \{v\} \cup CSet(v)$ gets all the buffered and current broadcast messages and therefore each u eventually gets all broadcast messages having timestamp greater than T. □

3 Non-fault-Tolerant Switching to BFS Tree

Let $G_v = (V', E')$ be a subgraph of G where $V' \subseteq \{v\} \cup N(v)$ and $E' \subseteq \bigcup_{u \in V'} E_u$. E_u is the set of edges incident on a node u. If T_v is a BFS spanning tree of G_v rooted at v then we call T_v the *local BFS subtree* rooted at v. The pseudocode of the algorithm is given in Figure 3. There is one node v_r, the initiator, that starts the switching. Local BFS subtree construction starts at v when it receives a *TOKEN* message. We assume that v_r is the first node to get a *TOKEN*. Let $h(v)$ denote the height of v from v_r. Initially for each $u \in V$, $h(u) = 0$. As v gets a *TOKEN* it takes the following steps. If v has already received a *TOKEN* then it simply forwards the *TOKEN* to some $u \in Child(v)$ where u has not yet received a *TOKEN*. If no such u exists then v sends the *TOKEN* to $p(v)$. Unlike the previous switching algorithm, even if v completes the local BFS subtree construction, it may have to repeat the same. This is because v may reduce its height further if it receives an *LBFS* message from some neighbor w where $w \neq p(v)$. In that case, for each $u \in Child(v)$, the height of u should be reduced. That is why, after sending the *TOKEN* to $p(v)$, v assumes that neither v nor any of its children has received a *TOKEN*. On the other hand, if v gets a *TOKEN* for the first time then it constructs a local BFS subtree, rooted at v, of the graph induced by the set of nodes $CSet(v) \cup \{v\}$ where $CSet(v) = N(v) - \{p(v)\}$. If $CSet(v) = \phi$ then the local BFS subtree to be formed is trivial and thus v sends the *TOKEN* to an appropriate node u as stated earlier. On the other hand, if $CSet(v) \neq \phi$ then v starts the local BFS subtree formation by sending $LBFS(v, h(v))$ to each $u \in CSet(v)$.

On receiving $LBFS(v, h')$, u joins v as a child only if it has received the *LBFS* message for the first time (i.e. $h(u) = 0$) or joining v reduces $h(u)$. If u joins v then u may change its parent (in case $p(u) \neq v$). In that case, u sends a *REMOVE_CHILD* message to its old parent u' which in turn acknowledges the removal of u from $Child(u')$ by a *REMOVE_ACK* message. In any case, u sends an *ACCEPT* message to v if it joins v. Otherwise u sends a *REJECT* message to v. When v receives an *ACCEPT* or *REJECT* message corresponding to each *LBFS* message it sent, the local BFS subtree formation at v is complete and thus v forwards the *TOKEN* to an appropriate node as stated earlier. It can be proved that the algorithm terminates with a BFS tree rooted at v_r, and the worst case message complexity is $O(|V||E|)$. If some spanning tree of the graph initially exists then the algorithm always maintains some spanning tree during switching. This reduces the delay and the number of messages required for broadcast during the switching. The proof of the algorithm is omitted here due to lack of space. The scheme as proposed in Section 2.1 should be used to ensure that every broadcast message is correctly delivered to all the nodes even during switching.

4 Fault-Tolerant Distributed Protocol Switching

In this section we modify the earlier algorithms to propose crash-tolerant switching algorithms. Faults can occur at any time irrespective of whether switching is in progress or not. We first discuss local repair techniques for a BFS or DFS tree under arbitrary crash faults. Using this we next propose switching algorithms that can tolerate arbitrary crash faults. It is assumed that the root node r does not fail, and the graph always remains connected even after the faults.

```
UPON RECEIVING TOKEN
    if tokenVisited = false then
        tokenVisited = true; CSet = N − {p}
        if CSet = φ then passTOKEN( )
        else ∀u ∈ CSet, send LBFS(v, h) to u
    else passTOKEN( )

UPON RECEIVING LBFS(u, h')
    if (h = 0) ∨ (h > h' + 1) then
        h = h' + 1
        if p = u then send ACCEPT to u
        else send REMOVE_CHILD to p; p = u
    else send REJECT to u

UPON RECEIVING REMOVE_CHILD FROM u
    Child = Child − {u}
    send REMOVE_ACK to u

UPON RECEIVING REMOVE_ACK
    send ACCEPT to p

UPON RECEIVING ACCEPT FROM u
    Child = Child ∪ {u}
    msgCount = msgCount + 1
    if msgCount = |CSet| then
        msgCount = 0; passTOKEN( )

UPON RECEIVING REJECT FROM u
    msgCount = msgCount + 1
    if msgCount = |CSet| then
        msgCount = 0; passTOKEN( )

FUNCTION passTOKEN()
    CTRT = Child − Visited
    if CTRT = φ then
        if p ≠ v then
            send TOKEN to p
            tokenVisited = false
            Visited = φ
    else send TOKEN to u ∈ CTRT
        Visited = Visited ∪ {u}
```

Fig. 3. Algorithm for switching to a BFS tree for node v

4.1 Local Repair of Faults in a BFS Tree

Let some node v in the BFS tree T crash. So $\forall u \in N(v) : p(u) = v$, there does not exist a valid path from any node w in T_u, the subtree rooted at u, to the root node r. So each node w in T_u must adjust $p(w)$ properly so that there exists a valid path from w to the root node r. Also this path must be the shortest path for the reconstructed tree to be a BFS tree. The following distributed algorithm gives an outline to repair the fault locally. Let each node x maintain a path $\mathcal{P}_{x \leadsto r} = \{x, u_1, u_2, \ldots, u_m, r\}$ where $p(x) = u_1$, $p(u_1) = u_2$, and so on and finally $p(u_m) = r$. This is created at each node during the construction of the BFS tree. So whenever a node v crashes, each node x in each T_u can know, by a reset wave generated at u, whether the crashed node $v \in \mathcal{P}_{x \leadsto r}$. Each node x that receives the reset wave updates its level L_x only among those neighbors (say y) whose current path $\mathcal{P}_{y \leadsto r}$ do not contain the crashed node v. Also a node x, on updating its L_x and $p(x)$ values, corrects $\mathcal{P}_{x \leadsto r}$ (by appending x to $\mathcal{P}_{p(x) \leadsto r}$ or symbolically $\mathcal{P}_{x \leadsto r} = \mathcal{P}_{p(x) \leadsto r} \odot x$) so that $\mathcal{P}_{x \leadsto r}$ does not contain the faulty node v. Also x asks each neighbor $z : z \neq p(x)$ to update L_z, and $p(z)$, and $\mathcal{P}_{z \leadsto r}$ provided L_x gets changed due to an update. It is assumed that the crash of a node v can be detected by each of its neighbors. Let $Crash(v)$ denote the detection of the crash of v by a neighbor u. On detecting crash of a neighbor v, a node u executes the function *BfsCrashAction(v)* shown in Figure 4. Again execution of *BfsCrashAction(v)* may cause the send of a *ResetLevel(v)* message. When a node u receives a *ResetLevel(v)* message, it executes the function *ResetLevelAction(v)* shown in Figure 5. It can be easily proved that the local repair algorithm eventually terminates with a BFS tree T' of the graph G.

4.2 Local Repair of Faults in a DFS Tree

Let some node v in the DFS tree T crash. So $\forall u \in N(v) : p(u) = v$, the subtree T_u, rooted at u becomes orphan as there does not exist a valid path from any node w

BFSCRASHACTION(V)	DFSCRASHACTION(V)
$N(u) = N(u) - \{v\}$	$N(u) = N(u) - \{v\}$
if $p(u) = v$ **then**	**if** $p(u) = v$ **then**
ResetLevelAction(v)	ChangePathAction(v)

Fig. 4. Actions invoked on detecting the crash of a node v by node u

in T_u to the root r. Also since its a DFS tree, the orphan subtrees are disjoint. Each node w in each subtree T_u rooted at u must reassign $p(w)$ so that there exists a path from w to r. Also this should be done in such a way that eventually a DFS tree of the graph results. The following distributed algorithm gives an outline to repair the fault locally. Let each node x maintain a path $\mathcal{P}_{x \rightsquigarrow r}$ as described earlier. This is created at each node during the construction of the DFS tree. So whenever a node v crashes, each node x in each T_u, subtree rooted at $u : p(u) = v$, can know, by a reset wave generated at node u, whether the crashed node v is in the path from x to the root node r. Each node x that receives the reset wave updates its $p(x)$ such that eventually the path $\mathcal{P}_{x \rightsquigarrow r}$ does not contain the crashed node v. Also in that case, x asks each neighbor y to update $p(y)$ and $\mathcal{P}_{y \rightsquigarrow r}$ so that eventually a DFS tree of the entire graph results. On detecting the crash of v, node u executes the function *DfsCrashAction(v)* shown in Figure 4. Again execution of *DfsCrashAction(v)* may cause the send of a *ChangePath(v)* message. When a node u receives a *ChangePath(v)* message, it executes the function *ChangePathAction(v)* shown in Figure 5.

Each node x orders the edges incident on it by some arbitrary ordering β_x. For an edge e incident on x and whose other end is y, the edge index is $\beta_x(y)$ from the point of view of x. The same edge has a different edge index $\beta_y(x)$ from the point of view of y. However node x can read $\beta_y(x)$. Each simple loopless path in the DFS tree is a sequence of edge indices starting from the root node. Each node x with degree δ_x can have δ_x paths from the root. Each node x has a path variable named $path_x$. We define a lexicographic ordering ($\prec$) among the paths and the minimum of the δ_x paths is the path that x uses in its DFS path variable $path_x$. It can be shown [12] that following the minimum path at each node as defined by the lexicographic order generates a DFS tree of the graph. It can be easily proved that the local repair algorithm eventually terminates with a DFS tree T' of the graph G.

4.3 Fault-Tolerant Switching from BFS to DFS

In this section we discuss fault-tolerant switching from a BFS tree to a DFS tree. The algorithm is presented in the form of guarded statements shown in Figure 6. A guarded statement consists of a guard and an action, and the action is executed only if the corresponding guard is true. As per discussions in Section 2, at any instant, the entire graph can be partitioned into two regions – a partial DFS tree, and a partial BFS tree. Those nodes that have received the $TOKEN$ at least once form a partial DFS tree and $tokenVisited(u) = true$ at each node u in the partial DFS tree. Similarly the nodes that have not yet received the $TOKEN$ form a partial BFS tree and $tokenVisited(u) = false$ at each node u in the partial BFS tree. For BFS to DFS switching, initially the entire graph is a BFS tree and the partial DFS tree is

```
RESETLEVELACTION(V)
𝒩_u = {x : x ∈ N(u) ∧ v ∉ 𝒫_{x⇝r}}
if 𝒩_u = φ then
  ∀w ∈ N(u) : p(w) = u, send ResetLevel(v) to w
else
  if v ∈ 𝒫_{u⇝r} then
    L_u = ∞
  endif
  L'_u = L_y + 1 where y ∈ 𝒩_u ∧ L_y = min_{z∈𝒩_u}{L_z}
  if L'_u < L_u then
    p(u) = y
    L_u = L'_u
    𝒫_{u⇝r} = 𝒫_{y⇝r} ⊙ u
    ∀w ∈ N(u) − {p(u)}, send ResetLevel(v) to w
  endif
endif
```

```
CHANGEPATHACTION(V)
𝒩_u = {x : x ∈ N(u) ∧ v ∉ 𝒫_{x⇝r}}
if 𝒩_u = φ then
  ∀w ∈ N(u) : p(w) = u, send ChangePath(v) to w
else
  path_u = min_≺{path_x ⊙ β_x(u) : x ∈ 𝒩_u}
  oldp(u) = p(u)
  p(u) = f(path_u)
  if oldp(u) ≠ p(u) then
    𝒫_{u⇝r} = 𝒫_{p(u)⇝r} ⊙ u
    ∀w ∈ N(u) − {p(u)}, send ChangePath(v) to w
  endif
endif
```

Fig. 5. Details of ResetLevelAction(v) and ChangePathAction(v) for node u

empty. Similarly when the switching completes, the entire graph is a DFS tree and the partial BFS tree is empty. Each node has a variable called $tokenHolder$ and it is $true$ only at the node which is presently holding the $TOKEN$. At all other nodes $tokenHolder = false$.

There can be arbitrary number of node crashes in the graph. If any node v crashes then $\forall u \in N(v)$, u executes $DfsCrashAction(v)$ if $tokenVisited(u) = true$. Again if u receives a $ChangePath(v)$ message and $tokenVisited(u) = true$ then u executes $ChangePathAction(v)$. Similarly if $tokenVisited(u) = false$ then u executes $BfsCrashAction(v)$ or $ResetLevelAction(v)$ depending on whether it detects the crash of v or gets a $ResetLevel(v)$ message respectively. However, if u receives a $ChangePath(v)$ message and $tokenVisited(u) = false$ then it remembers the receipt of the message. As $tokenVisited(u)$ becomes $true$ due to switching, it initiates the action corresponding to the $ChangePath(v)$ message received earlier. But if $tokenVisited(u) = true$ and u receives a $ResetLevel(v)$ message then no action is taken by u since the BFS to DFS switching will automatically take care of the failure in the partial BFS tree.

The node currently holding the $TOKEN$ may crash. In this case, a $TOKEN$ must be regenerated at an appropriate node such that switching can be reinitiated. To generate a single $TOKEN$ at a proper node, the following scheme is proposed. Let each node u maintain a variable $tdir$ that remembers the neighbor where the $TOKEN$ has been sent by u. Initially $tdir = \phi$ at each node u. Node u can send the $TOKEN$ either to one of its children or to its parent. Since u has sent the $TOKEN$, $tokenVisited(u) = true$ and $tokenHolder(u) = false$. On detecting that a neighbor v has crashed, u checks whether $tdir(u) = v$ and $tdir(u) \neq p(u)$. This indicates that u sent the $TOKEN$ to v and has not got it back from v. So even if the $TOKEN$ was not at v during the crash of v, it cannot return to u since v has crashed. So a new $TOKEN$ must be generated at u. In this case, u simulates the generation of a $TOKEN$ by making $tokenHolder = true$. Also u restarts the switching at u by setting $tokenVisited = false$.

Also, consider a node u which is holding the $TOKEN$ and executing the switching actions. So $CSet(u)$ is the set of nodes other than u which are involved in switching. If a node $v \in CSet(u)$ crashes then the switching must be reinitiated. To do so each node $w \in CSet(u) - \{v\}$ must reset its local variables ($ldfsVisited$, $cpflag$ etc) to

(S_1) $tokenVisited \wedge Crash(v) \rightarrow DfsCrashAction(v)$
(S_2) $tokenVisited \wedge received\ ChangePath(v) \rightarrow ChangePathAction(v)$
(S_3) $\neg tokenVisited \wedge Crash(v) \rightarrow BfsCrashAction(v)$
(S_4) $\neg tokenVisited \wedge received\ ResetLevel(v) \rightarrow ResetLevelAction(v)$
(S_5) $\neg tokenVisited \wedge received\ ChangePath(v) \rightarrow ChangePathFlag = 1; ID = v$
(S_6) $tokenVisited \wedge ChangePathFlag = 1 \rightarrow ChangePathFlag = 0$
$\quad ChangePathAction(ID)$
(S_7) $tokenVisited \wedge \neg tokenHolder \wedge Crash(v) \wedge tdir = v \wedge tdir \neq p(u) \rightarrow$
$\quad tokenHolder = true$
$\quad tokenVisited = false$
$\quad \forall w \in CSet, reset(w)$
(S_8) $tokenVisited \wedge tokenHolder \wedge Crash(v) \wedge v \in CSet \rightarrow$
$\quad tokenVisited = false$
$\quad \forall w \in CSet, reset(w)$

Fig. 6. Fault-tolerant actions for switching from BFS to DFS for node u

their default values. This is done by u by setting $tokenVisited(u)$ to $false$ and then sending a reset message to each $w \in CSet(u) - \{v\}$, which then resets itself (shown as an abstract function $reset(w)$ executed at u). The outline of the proof of correctness of the fault-tolerant actions is omitted in this paper due to lack of space. However the following lemma can be easily proved.

Lemma 5. *Under arbitrary crash failure, the algorithm eventually terminates with a DFS tree of the graph.*

4.4 Fault-Tolerant Switching from DFS to BFS

The fault-tolerant switching from a DFS tree to a BFS tree is similar to the algorithm for fault-tolerant switching from a BFS tree to a DFS tree and is shown in Figure 7. As before, the entire graph can be partitioned into a partial BFS tree, and a partial DFS tree. For a node u, $tokenVisited(u) = true$ if u is in the partial BFS tree and $tokenVisited(u) = false$ if u is in the partial DFS tree. Each node has a variable called $tokenHolder$ and $tokenHolder(u) = true$ if u is presently holding the $TOKEN$. If a node v crashes and $u \in N(v)$ such that $tokenVisited(u) = true$ then u executes the $BfsCrashAction(v)$ or $ResetLevelAction(v)$ depending on whether it detects the crash of v or receives a $ResetLevel(v)$ message respectively. Similarly if $tokenVisited(u) = false$ then u executes $DfsCrashAction(v)$ or $ChangePathAction(v)$ depending on whether it detects the crash of v or gets a $ChangePath(v)$ message respectively. If $tokenVisited(u) = false$ and u receives a $ResetLevel(v)$ message then it waits until $tokenVisited(u)$ becomes $true$ due to switching. Then u initiates the actions corresponding to the $ResetLevel(v)$ message received earlier. But if $tokenVisited(u) = true$ and u receives a $ChangePath(v)$ message then no action is taken by u as the DFS to BFS switching will automatically take care of the failure in the partial DFS tree. In case the node holding the $TOKEN$ crashes, a $TOKEN$ is regenerated at a proper node in a similar manner as discussed in

(S_9) $tokenVisited \wedge Crash(v) \rightarrow BfsCrashAction(v)$
(S_{10}) $tokenVisited \wedge received\ ResetLevel(v) \rightarrow ResetLevelAction(v)$
(S_{11}) $\neg tokenVisited \wedge Crash(v) \rightarrow DfsCrashAction(v)$
(S_{12}) $\neg tokenVisited \wedge received\ ChangePath(v) \rightarrow ChangePathAction(v)$
(S_{13}) $\neg tokenVisited \wedge received\ ResetLevel(v) \rightarrow ResetLevelFlag = 1; ID = v$
(S_{14}) $tokenVisited \wedge ResetLevelFlag = 1 \rightarrow ResetLevelFlag = 0$
$ResetLevelAction(ID)$
(S_{15}) $tokenVisited \wedge \neg tokenHolder \wedge Crash(v) \wedge tdir = v \wedge tdir \neq p(u) \rightarrow$
$tokenHolder = true$
$tokenVisited = false$
$\forall w \in CSet, reset(w)$
(S_{16}) $tokenVisited \wedge tokenHolder \wedge Crash(v) \wedge v \in CSet \rightarrow$
$tokenVisited = false$
$\forall w \in CSet, reset(w)$

Fig. 7. Fault-tolerant actions for switching from DFS to BFS for node u

Section 4.3. Again a node u which is holding the $TOKEN$ and executing the switching actions may detect the crash of one of its neighbors. In this case, node u reinitiates the switching in a similar way as discussed in Section 4.3. It can be proved that in spite of arbitrary crash faults the algorithm eventually terminates with a BFS tree as the output. The proof of correctness is omitted in this paper due to lack of space.

References

1. Jacobson, V.: Congestion avoidance and control. In: ACM SIGCOMM Symp. on Communications Architectures and Protocols (1988)
2. Anderson, J., Kim, Y.J.: Adaptive mutual exclusion with local spinning. In: DISC (2000)
3. Heinzelman, W., Kulik, J., Balakrishnan, H.: Adaptive protocols for information dissemination in wireless sensor networks. In: ACM/IEEE MobiCom (1999)
4. Son, S.H.: An adaptive checkpointing scheme for distributed databases with mixed types of transactions. IEEE Transactions on Knowledge and Data Engineering 1(4) (1989)
5. Das, S.R., Perkins, C.E., Royer, E.M.: Performance comparison of two on-demand routing protocols for ad hoc networks. In: IEEE INFOCOM (2000)
6. Arora, A., Gouda, M.: Distributed reset. IEEE Transactions on Computers 43(9) (1994)
7. Bar-Noy, A., Dolev, D., Dwork, C., Strong, R.: Shifting gears: Changing algorithms on the fly to expedite byzantine agreement. Information and Computation 97, 205–233 (1992)
8. Liu, X., van Renesse, R.: Brief announcement: Fast protocol transition in a distributed environment. In: ACM PODC (2000)
9. Chen, W.K., Hiltunen, M., Schlichting, R.: Constructing adaptive software in distributed systems. In: ICDCS (2001)
10. Rutti, O., Wojciechowski, P., Schiper, A.: Structural and algorithmic issues of dynamic protocol update. In: IEEE IPDPS (2006)
11. Mocito, J., Rodrigues, L.: Run-time switching between total order algorithms. In: Nagel, W.E., Walter, W.V., Lehner, W. (eds.) Euro-Par 2006. LNCS, vol. 4128, pp. 582–591. Springer, Heidelberg (2006)
12. Collin, Z., Dolev, S.: Self-stabilizing depth-first search. Information Processing Letters 49(6), 297–301 (1994)

Accomplishing Approximate FCFS Fairness Without Queues

K. Subramani[1,⋆] and K. Madduri[2,⋆⋆]

[1] LDCSEE,
West Virginia University,
Morgantown, WV
ksmani@csee.wvu.edu
[2] College of Computing,
Georgia Institute of Technology,
Atlanta, GA
kamesh@cc.gatech.edu

Abstract. First Come First Served (FCFS) is a policy that is accepted for implementing fairness in a number of application domains such as scheduling in Operating Systems, scheduling web requests, and so on. We also have orthogonal applications of FCFS policies in proving correctness of search algorithms such as Breadth-First Search, the Bellman-Ford FIFO implementation for finding single-source shortest paths, program verification and static analysis. The data structure used to implementing FCFS policies, the *queue*, suffers from two principal drawbacks, viz., non-trivial verifiability and lack of scalability. In case of large distributed networks, maintaining an explicit queue to enforce FCFS is prohibitively expensive. The question of interest then, is whether queues are *required* to implement FCFS policies; this paper provides empirical evidence answering this question in the negative. The principal contribution of this paper is the design and analysis of a randomized protocol to implement approximate FCFS policies without queues. From the Software Engineering perspective, the techniques that are developed find direct applications in program verification, model checking, in the implementation of distributed queues and in the design of incremental algorithms for Shortest path problems.

1 Introduction

FCFS is a policy used to ensure fairness in a number of application domains such as scheduling [1] and Operating Systems [2]. The motivating factor underlying this form of fairness, especially in the servicing of requests, is to preserve

⋆ The research of the first author was supported by the Air-Force Office of Scientific Research under contract FA9550-06-1-0050.

⋆⋆ The research of the second author was supported in part NSF CAREER CCF-0611589, NSF DBI-0420513, ITR EF/BIO 03-31654, and NASA (NP-2005-07-375-HQ).

S. Aluru et al. (Eds.): HiPC 2007, LNCS 4873, pp. 540–551, 2007.

order, i.e., if request A has a smaller time-stamp than request B, then request A should be serviced before request B. The only known method of implementing this policy is through the use of a "queue" data structure, in which elements are inserted at the rear and removed from the front. In practice, queues are realized through circular arrays or linked lists; the code for maintaining queues, while conceptually simple, mandates the checking of a number of conditions and is therefore non-trivial [3]. Additionally, queues do not scale well; in distributed applications, maintaining a FCFS queue is prohibitively expensive. The question of interest then, is whether FCFS fairness (or at least an approximation of FCFS) can be implemented without queues; this paper is devoted towards answering this question. We provide empirical evidence that conclusively demonstrates that queue structures are not necessary to achieve FCFS fairness; indeed, even implicit queues (See Section 2) need not be used. All that is needed is a source for random bits; our algorithm exploits the existence of such a source to effectively simulate a standard queue. Our work establishes that randomization can serve as an effective substitute to order, insofar as establishing FCFS fairness is concerned. We also point out that our algorithm is the first of its kind to explicitly introduce randomization in reachability problems.

The main advantage of our approach is simplicity; as per the literature, maintaining queues, especially in distributed applications is a non-trivial task. Accordingly, if it is possible to achieve the FCFS effect without queues then that possibility must be explored.

2 Statement of Problem

The problem that we are interested in is as follows: $\mathbf{P_1}$: *Given a sequence of requests, $S = \langle s_1, s_2, \ldots, s_n \rangle$, which are totally ordered by time, can we service the requests in approximately the same order as their arrival, without using a queue to store the requests?*

Observe that if we are told that the requests must be served in *exactly* the same order as their arrival, then a queue is necessary. It is the relaxation of this requirement to "approximately, in the order of arrival" that permits us to use randomization and eliminate queues.

2.1 Breadth-First Search

We now argue that the FCFS problem is simulated by performing the Breadth-First Search on an arbitrary graph; note that we use the algorithm described in [4].

Consider an arbitrary level-based labeling of the vertices in G; with vertices in level i getting a lower label than the vertices in level $(i+1)$. Vertices in the same level are numbered arbitrarily. We can think of the queue $\mathbf{Q}$ as being populated by requests, with the vertices representing the requests; we say that request v_i precedes request v_j, if v_i enters $\mathbf{Q}$ before v_j. When a vertex is deleted from $\mathbf{Q}$, it is said to be serviced. Observe that $\mathbf{Q}$ implements FCFS fairness in that all requests at a particular level are serviced before requests at higher levels.

Function TRAD-BFS($G =< V, E >$, s)
1: {The output of the Algorithm is the Set of all vertices $v \in V$ reachable from s, and $d[v]$, the shortest path from s to v, given all edges to be unit weight}
2: Construct an empty queue Q and initialize it to s.
3: $d[s] = 0$
4: $color[s] = black$
5: **for** ($v \in V, v \neq s$) **do**
6: $d[v] = \infty$
7: $color[v] = white$
8: **end for**
9: **while** $|\mathbf{Q}| \neq \emptyset$ **do**
10: Let $v = head(\mathbf{Q})$
11: Delete $head(\mathbf{Q})$
12: **for** (each vertex u adjacent to v) **do**
13: **if** ($color[u] = white$) **then**
14: Add vertex u to **Q**
15: $color[u] = black$
16: $d[u] = d[v] + 1$
17: **end if**
18: **end for**
19: **end while**

Algorithm 2.1. Traditional Breadth-First Search

As discussed above, there are other methods through which FCFS fairness can be implemented. For an arbitrary protocol $\mathcal{A}$, using storage structure $\mathcal{R}$, we define the *wait time of a vertex* as the number of times it is inserted in $\mathcal{R}$. Likewise, we define the *wait time* of the protocol as the maximum wait time of any vertex.

In the traditional BFS protocol, $\mathcal{R}$ is a queue and each vertex is inserted precisely once in $\mathcal{R}$. Accordingly, the wait time of every vertex is 1 and the wait time of traditional BFS is also 1. The wait time of an arbitrary protocol is a measure of how accurately it implements FCFS, in that larger the wait time, the more it deviates away from FCFS.

For a randomized protocol, the metrics of interested are the *expected wait time* of a vertex and the *expected wait time* of the protocol respectively.

3 Motivation and Related Work

The motivation for our work arises from the following design domains, viz.,

(a) Program Testing - The implementation of a queue mandates the testing of buffer overflow at each insertion and buffer underflow at each deletion, regardless of how the queue is implemented. It has been frequently observed that these bound checks are either ignored or incorrectly implemented (from a logical perspective) leading to program crashes. Our protocol on the other hand, uses a simple "set" data structure with only membership queries permitted. This structure can be implemented by using a counter and is hence trivial to test. It is well-known that reasoning about heap-allocated structures is challenging; although tools exist for detecting errors in software [5], they are known to be imprecise when dealing with heap-allocated structures. It is to be noted that set semantics can be implemented statically, without utilizing the memory heap at run-time.

(b) Program Verification - A typical program is represented as Control-Flow graph (CFG); typical questions involving the reachability of unsafe states, non-termination of loops and so on can be answered through BFS [6]. Reachability analysis is also used in model-checking LTL formulae [7], understanding secure information flow, and timed automata [8]. Likewise, reachability analysis is used to test properties of Timed Automata [9]. Consequently, the technique that we have proposed will find immediate applications in these domains. We would like to point out that our algorithmic paradigm is not part of the existing literature, to the best of our knowledge.

(c) Distributed Queues - Consider a web-server A, which is connected to a number of satellite servers $A_1, A_2, \ldots, A_n$. Requests from the external world to A are routed through the satellite servers. Each request comes with a time-stamp and is stored locally at the satellite server. If we were required to serve requests in strict order of time-stamps, we would have to poll all the servers to determine the request with the smallest time-stamp. Our strategy here demonstrates that it is sufficient to choose a satellite server at random *and* a request at random from the requests in that server. Indeed, one of the fundamental strengths of our technique, is that it can be applied to distributed computing applications.

(d) Incrementality - Incremental algorithms are concerned with maintaining reachability information under edge insertions and deletions. Incremental algorithms for BFS have been studied from both the theoretical [10] and the practical perspectives. The exact complexity of this problem is unknown, although there have been attempts to categorize it [11]. In program analysis and verification, incremental algorithms for reachability analysis are of paramount importance [12]. The Randomized BFS algorithm is incremental in nature and exploits the fact that the BFS tree is itself constituted of BFS subtrees.

(e) Constraint Solving - Constraint solving is an integral component of Program Verification [13]. The approach described in this paper can be integrated into program verification tools such as the ones described in [14]. It is to be noted that our approach is easily and very efficiently parallelizable, which is the necessary in modern day constraint solvers.

4 The Randomized Breadth-First Search Algorithm

Algorithm 4.1 describes the workings of the Randomized Breadth-First Search algorithm.

Function RANDOM-BFS($G =< V, E >$, s, **d**, **Q**)

1: {The output of the Algorithm is the Set of all vertices $v \in V$ reachable from s, and $d[v]$, the shortest path from s to v, given all edges to be unit weight}
2: **if** $|\mathbf{Q}| = \emptyset$ **then**
3:     **return**
4: **else**
5:     Pick a vertex v uniformly and at random from **Q**
6:     **for** (each vertex u adjacent to v) **do**
7:         **if** $(d[u] > d[v] + 1)$ **then**
8:             $d[u] = d[v] + 1$
9:             **if** $(u \notin \mathbf{Q})$ **then**
10:                 Add vertex u to **Q**
11:             **end if**
12:         **end if**
13:     **end for**
14: **end if**
15: RANDOM-BFS($G =< V, E >$, s, **d**, **Q**)

Algorithm 4.1. Randomized Breadth-First Search

The algorithm is initialized as $d[s] = 0$ and $d[v] = \infty, \forall v \neq s$. Further, $\mathbf{Q} = \{s\}$.

We reiterate that the above algorithm is the first of its kind to explicitly introduce randomization in the vertex selection process and therefore represents a fundamentally distinct design paradigm (cf. the Shortest path algorithms discussed in [15].)

4.1 Worst-Case Analysis

Let the graph have m edges and n vertices. Let $t(v_i)$ denote the time spent on processing vertex v_i. We know that $1 \leq d[v_i] \leq (n-1)$ for all $v_i \in V$ and hence v_i can be inserted in **Q** at most n times. Each time v_i is deleted from **Q**, we spend $degree(v_i)$ time in processing its neighbors. Accordingly, $t(v_i) \leq n \cdot degree(v_i)$ and hence the total time spent in processing all the vertices is $T(n) = \sum_{i=1}^{n} t(i) = \sum_{i=1}^{n} n \cdot degree(i) = O(m \cdot n)$. It must be noted that the above analysis is extremely pessimistic and that our experiments show that $T(n)$ is a linear function of n.

We conjecture that $E[T(n)] \leq c_1 \cdot (n+m)$, for some fixed constant c_1; however, a formal proof will require the development of new theoretical techniques.

Let $\delta(v)$ denote the true shortest path distance of vertex v from the source s. As discussed above, Algorithm 4.1 terminates, since each vertex enters **Q** at most n times.

The correctness of Algorithm (4.1) follows from the following lemma.

Lemma 1. *If $d[v_i] > \delta(v_i)$, and $v_i \notin \mathbf{Q}$, then v_i will be inserted into* **Q**.

Proof. We use induction on the true distance $\delta(v_i)$. If $\delta(v_i) = 1$, the lemma is clearly true, since at the first call, the source s is extracted and all its neighbors are inserted into $\mathbf{Q}$. Assume that the lemma is true, whenever $\delta(v_i) \leq k$, $k > 1$. Now observe the sequence of events when $\delta(v_i) = k+1$. By the inductive hypothesis, all vertices v_j, such that $\delta(v_j) \leq k$, will be inserted into $\mathbf{Q}$, till $d[v_j] = \delta(v_j)$. One of these vertices is v_r, the predecessor of v_i in the BFS tree; observe that $\delta(v_r) = k$. If it is the case that $d[v_i] \neq \delta(v_i)$, then $d[v_i] > (k+1)$ and hence $d[v_i] > d[v_r]+1$. When v_r is extracted from $\mathbf{Q}$ for the final time, Lines 7 through 12 of Algorithm 4.1, ensure that v_i is inserted into $\mathbf{Q}$, if it does not already belong there. □

Thus, as long as $d[v] > \delta(v)$ for some vertex v, Algorithm 4.1 will continue to recurse; since no vertex can be inserted into $\mathbf{Q}$, more than n times, it follows that when $|\mathbf{Q}| = \emptyset$, $d[v_i] = \delta(v_i)$, $\forall i = 1, 2, \ldots, n$.

It is important to note that Algorithm 4.1 is a complete procedure in that it can be used to detect the Single Source Shortest Paths in an arbitrarily weighted graph (both positive and negative weights on the arcs). The only modification is to lines 9 and 10, which should be replaced by: *if* $(d[u] > d[v] + c(v, u))$, *then* $d[u] = d[v] + c(v, u)$. The proof of this fact is similar to the above proof and will be presented in an extended version of this paper [16].

Theorem 1. *Algorithm 4.1 can be modified to solve the Single Source Shortest Paths problem.*

5 Experimental Study: Sequential Performance

This section presents the performance results of the sequential Random-BFS algorithm.

Our reference platform for evaluating sequential performance is a 3.2 GHz 64-bit Intel Xeon machine with 6GB memory and 1MB L2 cache. For implementation details, please refer to the extended version of this paper [16].

We test our Random-BFS implementation on a variety of synthetic graph families. These generators and graph instances are part of the DIMACS Shortest Path Implementation Challenge network collection [17]:

- *random graphs*: Random graphs are generated by first constructing a Hamiltonian cycle, and then adding $m - n$ edges to the graph at random. The generator may produce parallel edges as well as self-loops. By varying the parameters m and n, we can generate both sparse as well as dense random graphs.
- *mesh networks*: This synthetic generator produces regular two-dimensional square meshes, where $m = 4n$.
- *scale-free graphs*: we use the R-MAT graph model [18] to generate graphs with power-law degree distributions and small-world characteristics.

The above graph families are frequently used for the experimental evaluation of graph algorithms [19,20]. Random graphs have a Poisson degree distribution, low diameter, and low clustering co-efficients. BFS on sparse random graphs has poor cache locality, and so this is a hard test instance for performance comparison on cache-based architectures. Scale-free networks are sparse graphs characterized by low average distance, high local density, and heavy-tailed power law degree distributions. In contrast, the two-dimensional mesh network is a regular graph with a high diameter. These three families differ in the number of BFS phases, as well as the average number of vertices in each phase.

We report results averaged over ten runs, excluding the best and worst values and any outliers. Our first set of experiments estimate the randomization overhead in Alg. 2.1. We do this by calculating the following metrics defined in Sec. 2.1:

- *Expected wait time of a vertex* (EWT): the average number of times a vertex in the graph is inserted into the queue.
- *Expected wait time of protocol* (WTP): the maximum number of times any vertex in the graph is added to the queue.

Figure 1 plots EWT and WTP for various sparse and dense graph instances.

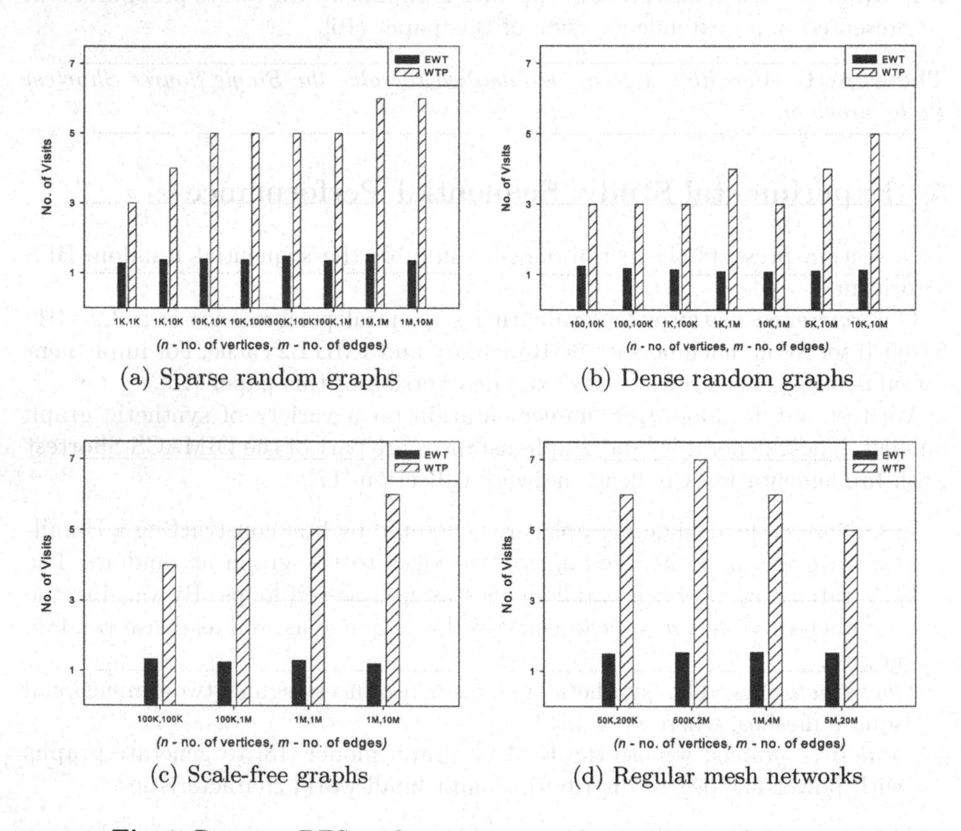

(a) Sparse random graphs

(b) Dense random graphs

(c) Scale-free graphs

(d) Regular mesh networks

Fig. 1. RANDOM-BFS performance counts for various graph instances

For performance comparison, we also implemented the traditional queue-based BFS (Alg. 4.1). Figure 2 plots the execution time of Trad-BFS and Random-BFS for synthetic graphs of different problem sizes.

5.1 Observations

FCFS policy metrics. For random graph instances in Figure 1, the *expected wait time* value (EWT) varies from 1.13 to 1.56. Also note that EWT appears to be independent of the problem size for the graphs we considered. The *expected wait time for the protocol* (WTP) value varies from 3 to 6, with a slight increase for large instances. We observe a similar behavior in case of the mesh and scale-free networks (Figures 1(d) and 1(c) respectively) also. The EWT value for sparse instances (about 1.35 on an average) is higher than the value for dense graphs (averaging 1.15).

Execution Times. Figure 2 gives the running times of Trad-BFS and Random-BFS on the reference sequential platform. While Trad-BFS is faster than

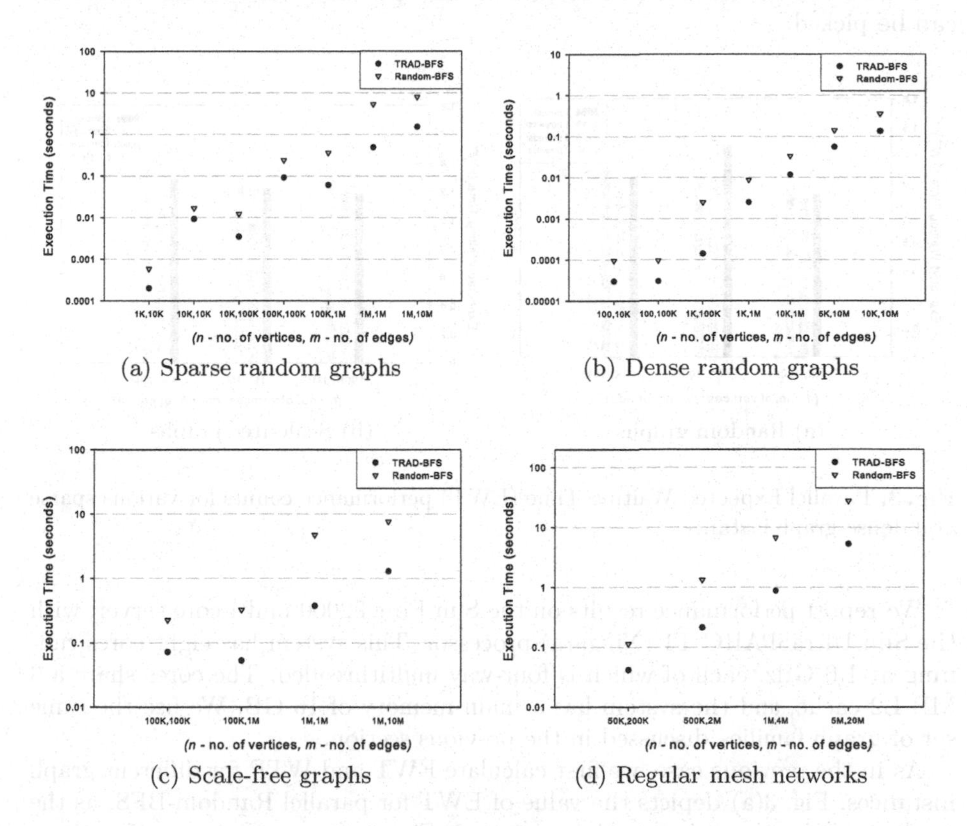

(a) Sparse random graphs

(b) Dense random graphs

(c) Scale-free graphs

(d) Regular mesh networks

Fig. 2. TRAD-BFS and RANDOM-BFS execution time comparison for various graph instances

Random-BFS in all cases, the running times only differ by a constant factor. We observe that Trad-BFS is on an average six times faster than Random-BFS for sparse graphs, and thrice as fast for dense graphs. The running times are also directly correlated to the EWT and WTP values, which are both greater than 1. The execution time trends are similar for all three graph families.

6 Experimental Study: Parallel Performance

We also implement parallel shared-memory versions of Trad-BFS and Random-BFS. In case of Trad-BFS, we employ a level-synchronized approach to parallelization that exploits concurrency at two key steps:

1. All vertices at a given *level* (distance from source vertex) in the graph can be processed simultaneously, instead of just picking the vertex at the head of the queue.
2. Adjacencies of each vertex can be inspected in parallel.

In case of Random-BFS, we further assume that any vertex in the visited set can be picked.

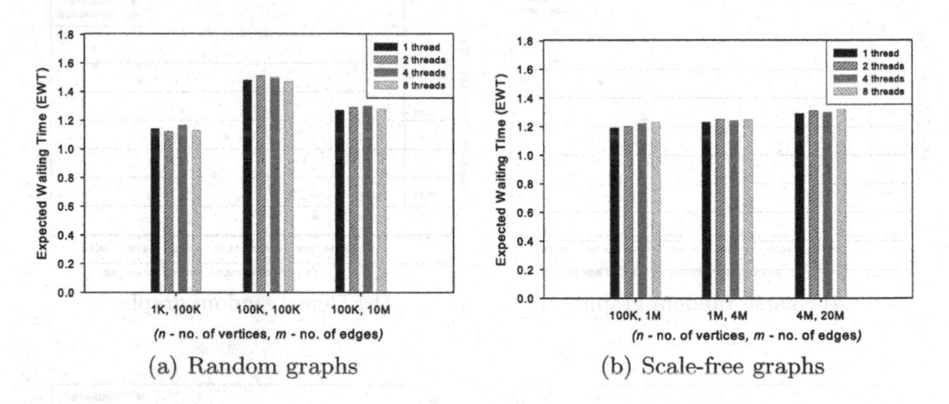

(a) Random graphs (b) Scale-free graphs

Fig. 3. Parallel Expected Waiting Time (EWT) performance counts for various sparse and dense graph instances

We report performance results on the Sun Fire T2000 multi-core server, with the Sun UltraSPARC T1 (Niagara) processor. This system has eight cores running at 1.0 GHz, each of which is four-way multithreaded. The cores share a 3 MB L2 cache, and the system has a main memory of 16 GB. We use the same set of graph families discussed in the previous section.

As in the previous case, we first calculate EWT and WTP for different graph instances. Fig. 3(a) depicts the value of EWT for parallel Random-BFS, as the number of processors is varied from 1 to 8. Figure 4 compares execution times of Trad-BFS and Random-BFS for various problem sizes, as the number of processors is varied from 1 to 8.

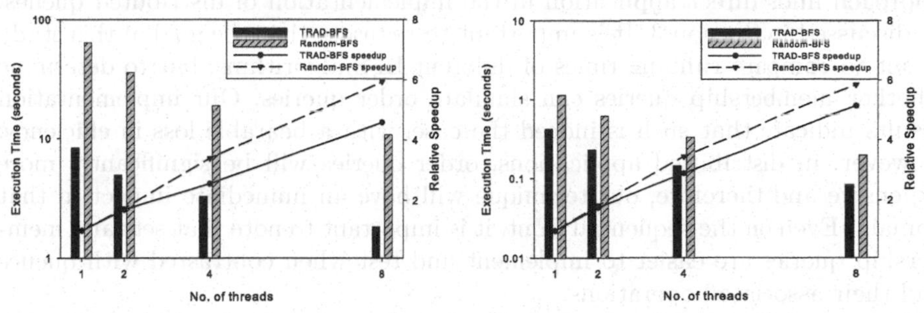

(a) Sparse Scale-free graph (1M vertices, 12M edges)

(b) Dense random graph (100K vertices, 10M edges

Fig. 4. Parallel Execution time and Speedup comparison

6.1 Observations

FCFS policy metrics. Figure 3 gives EWT for graph instances of different sizes from the *random* and *scale-free* families. We again note that EWT is higher for sparse graphs. But the key observation here is that there is no performance drop for parallel Random-BFS. There is very little variation in the EWT value across parallel runs.

Running Time. Figure 4 gives the execution time and speedup acheived for multiprocessor runs. We observe that Trad-BFS is faster than Random-BFS up to 8 threads. The speedup on random graphs is lower than that on dense graphs, in case of both BFS and Random-BFS. The parallel performance on sparse random graphs is similar to the scale-free graph performance reported here. An important observation is that the relative speedup of Random-BFS is greater than Trad-BFS in both the cases. This is expected, as there is more concurrency in parallel Random-BFS than in parallel Trad-BFS. On larger multiprocessor systems, we expect the performance of Random-BFS to match Trad-BFS performance. From these results, we can also expect that Random-BFS would perform favorably on distributed memory systems, as there is no overhead of maintaining the FCFS queue.

7 Conclusion

The primary objective of this paper was to investigate whether FCFS fairness can be accomplished without queues, for the reasons described in previous sections. This is a fundamental problem in Program Verification, inasmuch as verifying the correctness of programs using queue structures is non-trivial. We have succeeded in showing (empirically) that randomization does indeed achieve the "queue" effect in that service requests are met on an almost First Come First Served basis. Our Randomized BFS algorithm is the first of its kind, in that it explicitly introduces randomization in the selection of requests to be serviced. This approach finds direct application in the implementation of distributed queues, as discussed in Section 3. It is important to reiterate that the goal of this study is not to compare running times of different BFS algorithms, but to determine whether membership queries can simulate order queries. Our implementation results indicate that such is indeed the case, with a bearable loss in efficiency. However, in distributed applications, order queries will be significantly more expensive and therefore, our technique will have an immediate impact in that domain. Even on the sequential front, it is important to note that sets and membership queries are easier to implement and test when contrasted with queues and their associated operations.

A number of interesting research problems have arisen out of this work:

(a) As argued in Section 4, the RBFS algorithm is a complete procedure for finding Single Source shortest paths on *arbitrarily weighted* graphs. It would be instructive to study the performance profile of this algorithm for instances of the Single Source Shortest path problem.
(b) We would like to implement this technique in actual program verification tools, such as the ones discussed in [21].
(c) We are currently engaged in developing an expected case analysis of the RBFS algorithm. Our goal is to analytically establish that the expected number of times that an arbitrarily chosen vertex is inserted into the set S is constant.
(d) We are also studying the performance of our algorithm in distributed applications, wherein the simplicity of our approach will lead to performance gains over the more traditional algorithms for implementing FCFS fairness.

References

1. Pinedo, M.: Scheduling: theory, algorithms, and systems. Prentice-Hall, Englewood Cliffs (1995)
2. Levi, S.T., Tripathi, S.K., Carson, S.D., Agrawala, A.K.: The Maruti Hard Real-Time Operating System. ACM Special Interest Group on Operating Systems 23(3), 90–106 (1989)
3. Goodrich, M.T., Tamassia, R.: Algorithm Design: Foundations, Analysis and Internet Examples. John Wiley & Sons, Chichester (2002)
4. Cormen, T.H., Leiserson, C.E., Rivest, R.L., Stein, C.: Introduction to Algorithms. MIT Press, Cambridge (2001)
5. Barnett, M., Leino, K.R.M.: Weakest-precondition of unstructured programs. In: PASTE, pp. 82–87 (2005)
6. Louden, K.C.: Programming Languages: Principles and Practice. Brooks/Cole (2002)
7. Clarke, E.M.: Automatic verification of sequential circuit designs. In: Agnew, D., Claesen, L., Camposano, R. (eds.) Proceedings of the 11th International Conference on Computer Hardware Description Languages and their Applications (CHDL 1993), Amsterdam, The Netherlands, North-Holland. IFIP Transactions A: Computer Science and Technology, vol. 32, pp. 163–166 (1993)
8. Alur, R., Dill, D.L.: A theory of timed automata. Theoretical Computer Science (Fundamental Study) 126(2), 183–235 (1994)

9. Aceto, L., Bouyer, P., Burgueño, A., Larsen, K.G.: The power of reachability testing for timed automata. Theor. Comput. Sci. 300(1-3), 411–475 (2003)
10. Demtrescu, C.: A new approach to dynamic all pairs shortest paths. Journal of the ACM 51(6), 968–992 (2004)
11. Hesse, W.: The dynamic complexity of transitive closure is in dyntc^0. Theor. Comput. Sci. 3(296), 473–485 (2003)
12. Ramalingam, G., Reps, T.W.: On the computational complexity of dynamic graph problems. Theor. Comput. Sci. 158(1&2), 233–277 (1996)
13. Revesz, P.: Safe query languages for constraint databases. ACM Transactions on Database Systems 23(1), 58–99 (1998)
14. Revesz, P.: Introduction to Constraint Databases. Springer, Heidelberg (2002)
15. Ahuja, R.K., Magnanti, T.L., Orlin, J.B.: Network Flows: Theory, Algorithms and Applications. Prentice-Hall, Englewood Cliffs (1993)
16. Subramani, K., Madduri, K.: A randomized, queueless algorithm for breadth-first search. International Journal of Computers and their Applications (accepted, 2007)
17. Demetrescu, C., Goldberg, A., Johnson, D.: 9th DIMACS implementation challenge – Shortest Paths (2005), `http://www.dis.uniroma1.it/~challenge9/`
18. Chakrabarti, D., Zhan, Y., Faloutsos, C.: R-MAT: A recursive model for graph mining. In: Proc. 4th SIAM Intl. Conf. on Data Mining, Florida, USA (2004)
19. Pettie, S., Ramachandran, V., Sridhar, S.: Experimental evaluation of a new shortest path algorithm. In: Mount, D.M., Stein, C. (eds.) ALENEX 2002. LNCS, vol. 2409, pp. 126–142. Springer, Heidelberg (2002)
20. Bader, D.A., Cong, G.: A fast, parallel spanning tree algorithm for symmetric multiprocessors (SMPs). J. Parallel & Distributed Comput. 65(9), 994–1006 (2005)
21. Conway, C.L., Namjoshi, K.S., Dams, D., Edwards, S.A.: Incremental algorithms for inter-procedural analysis of safety properties. In: Computer-Aided Verification, pp. 449–461 (2005)

A Novel Force Matrix Transformation with Optimal Load-Balance for 3-Body Potential Based Parallel Molecular Dynamics Using Atom-Decomposition in a Heterogeneous Cluster Environment

Sumanth J.V, David Swanson, and Hong Jiang

Department of Computer Science and Engineering
University of Nebraska-Lincoln, Lincoln NE 68588, USA
{sumanth, dswanson, jiang}@cse.unl.edu

Abstract. Evaluating the Force Matrix constitutes the most computationally intensive part of a Molecular Dynamics (MD) simulation. In three-body MD simulations, the total energy of the system is determined by the energy of every unique triple in the system and the force matrix is three-dimensional. The execution time of a three-body MD algorithm is thus proportional to the cube of the number of atoms in the system. Fortunately, there exist symmetries in the Force Matrix that can be exploited to improve the running time of the algorithm. While this optimization is straight forward to implement in the case of sequential code, it has proven to be nontrivial for parallel code even in a homogeneous environment.

In this paper, we present a force matrix transformation that is capable of exploiting the symmetries in the force matrix in both a homogeneous and a heterogeneous environment while balancing the load among all the participating processors. The proposed transformation distributes the number of interactions to be computed uniformly among all the slices of the force matrix along any of the axes. The transformed matrix can be scheduled using any well known heterogeneous slice-level scheduling technique. We also derive theoretical bounds for efficiency and load balance for prior work in the literature. We then prove some interesting and useful properties of our transformation and evaluate its advantages and disadvantages. A loop reordering optimization for the symmetric transformation is described. The performance of an MPI implementation of the transformation is studied in terms of the Step Time Variation Ratio (STVR) in a homogeneous and heterogeneous environment.

1 Introduction

Molecular Dynamics (MD) is a powerful technique used to obtain static or dynamic properties of liquids and solids. It can be more formally defined as a computer simulation technique where the time evolution of a set of interacting

S. Aluru et al. (Eds.): HiPC 2007, LNCS 4873, pp. 552–565, 2007.

atoms is followed by integrating their equations of motion [1]. From the motion of the ensemble of atoms, a variety of useful microscopic and macroscopic information can be extracted such as transport coefficients, phase diagrams, and structural properties. The physics of the model is contained in a potential energy functional for the system from which the individual force equations for each atom are derived. There are numerous applications for MD simulations in diverse fields of science and technology such as chemistry, astronomy, biophysics, solid-state physics, material science and fluid dynamics, to mention a few.

MD simulations are not very memory intensive. Their space complexity grows linearly with the number of atoms being simulated. However, their time complexity grows cubically with the number of atoms being simulated (assuming a 3-body potential is being used). Being a very computationally intensive application [8], various solutions to improve execution times have been investigated. The most common methods for improving performance are parallelization[2] and using custom-designed special purpose hardware [11] [10].

In this paper, we describe a force matrix transformation that allows for the parallelization of a 3-body MD simulation and takes advantage of symmetries in the force matrix in a heterogeneous cluster environment. We focus on the 3-body component of the Webber-Stillinger potential. The load balancing properties of our approach are mathematically analyzed. A detailed analysis of previous work in this area by Li et al. [6] is performed and closed form efficiency upper bounds for their approach is determined. However, their cyclic decomposition technique targets a homogeneous cluster environment whereas our technique can also be used in a heterogeneous environment. We then, discuss optimization techniques that can be used to implement our technique. Performance is evaluated using the Step Time Variation Ratio (STVR)[6]. In the homogeneous case, we compare our technique to the cyclic distribution technique.

The rest of this paper is organized as follows: in Section 2, we give an overview of the computational aspects of an MD simulation and describe various parallelization techniques. In Section 3, we describe prior work in this area. Section 4 reviews an existing technique and derives theoretical efficiency bounds for it. In Section 5, our symmetric transformation and its useful properties are described and proved. Section 6 describes a loop optimization technique for the implementation of our transformation. We evaluate the performance of our technique and discuss the pros and cons of our approach in Section 7 and conclude the paper in Section 8.

2 Molecular Dynamics

2.1 Computational Aspects

The computational task in an MD simulation is to perform the time integration of a set of coupled differential equations (Newton's equations) given by

$$m_i \frac{\partial^2 \boldsymbol{r}_i}{\partial t^2} = \sum_j F_2(\boldsymbol{r}_i, \boldsymbol{r}_j) + \sum_j \sum_k F_3(\boldsymbol{r}_i, \boldsymbol{r}_j, \boldsymbol{r}_k) + \dots$$

where, m_i is the mass of atom i, $\boldsymbol{r}_i$ is its position vector, F_2 is a two-body force function and F_3 is a three-body force function. In this paper, we perform the time-integration of Equation 2.1 using the velocity-verlet algorithm [13]. The force functions F_2 and F_3 are computed as the negative gradients of potential functions in which the energy of atom i is typically written as a function of the positions of itself and other atoms.

For concreteness, for the 3-body form we use the popular Weber-Stillinger potential. [9]. The total potential is expressed as two sums, one for unique pair interactions, and another for unique triplet interactions.

$$U = \sum_{i<j} v_2(r_{ij}) + \sum_{i<j<k} v_3(\boldsymbol{r}_i, \boldsymbol{r}_j, \boldsymbol{r}_k)$$

When using a three-body potential, the three-dimensional force matrix is symmetric since $F_{ijk} = F_{ikj}$, where F_{ijk} is the $(i, j, k)^{th}$ element of the force matrix F representing the force exerted on atom i by atoms j and k.

For a system with N particles, the total number of unique triples to be evaluated is $N(N-1)(N-2)/6$. A triple (i, j, k), where $i \neq j \neq k$ is contained in 6 force elements. Each single particle can be involved in $(N-1)(N-2)/2$ triples, therefore the evaluation of the total 3-body force acting on a particle requires a sum over all the $(N-1)(N-2)/2$ triples. However, due to the symmetry described above, only three independent force elements are actually evaluated for each triple.

2.2 Classification

Many-body simulations can be characterized by the range of forces being modeled. If the forces are long-range, like gravitational or Coulombic forces, then each particle is affected by all others in the simulation. If the forces are short-range like LJ and Webber-Stillinger, then each particle is only influenced at each timestep by a limited number of neighboring particles.

Many-body simulations can also be classified by whether they use direct or indirect methods. Simulations that compute each interaction of Equation 2.1 explicitly use what are known as direct methods. Conversely, if the simulation approximates some interactions, it is known as an approximate method. Approximate methods such as particle-mesh algorithms, hierarchical algorithms and fast multipole methods are typically used with long-range forces. However, approximate methods are much harder to implement than direct methods, particularly in parallel machines and systems [4]. Because of this complexity, approximate methods are typically not faster than direct methods until the number of atoms reaches a certain threshold value which can be quite large. In parallel implementations, the performance of approximate methods can suffer further from the fact that the work load can be difficult to balance among processors when the particle density is spatially and/or temporally non-uniform leading to underutilization of available resources.

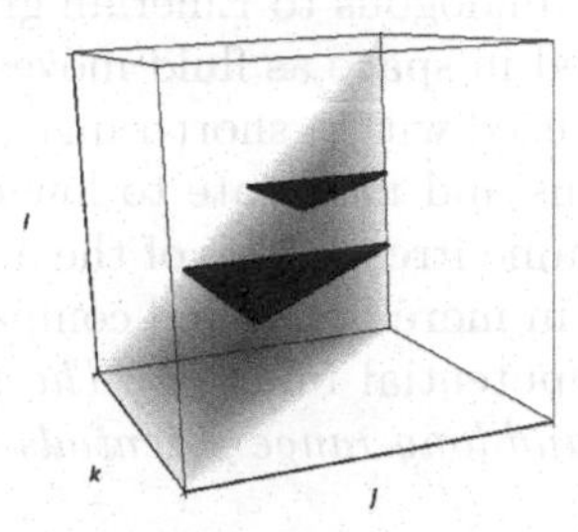

(a) Transformed force matrix when using the cyclic distribution. The dark triangular surfaces represent the number of interactions in that slice.

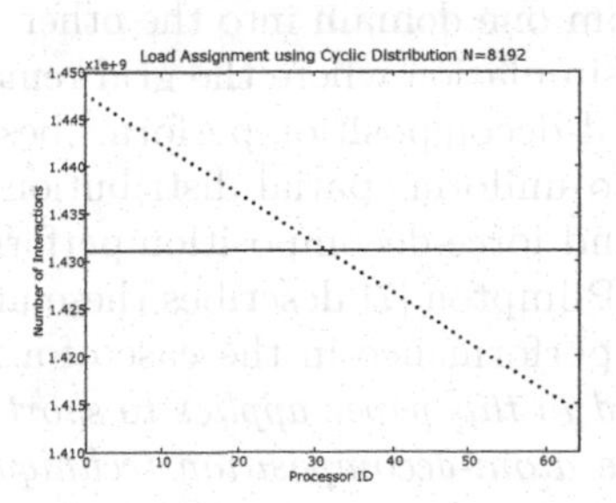

(b) Assignment of interactions among 64 processors when using the cyclic distribution technique. The dark points represent the load assigned to each of the 64 processors. The horizontal line represents the optimal load assignment.

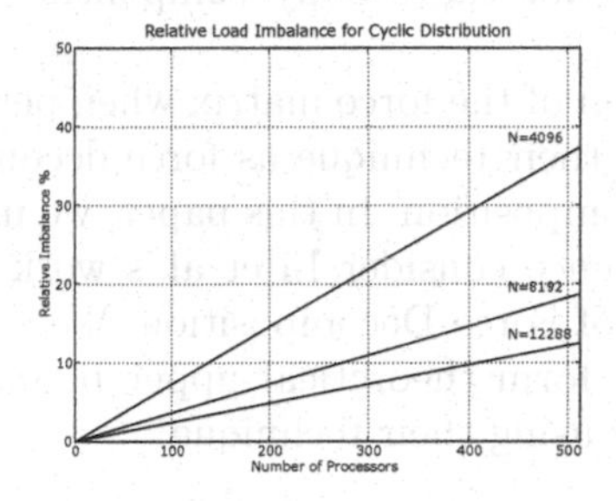

(c) Relative Load Imbalance when using the cyclic distribution.

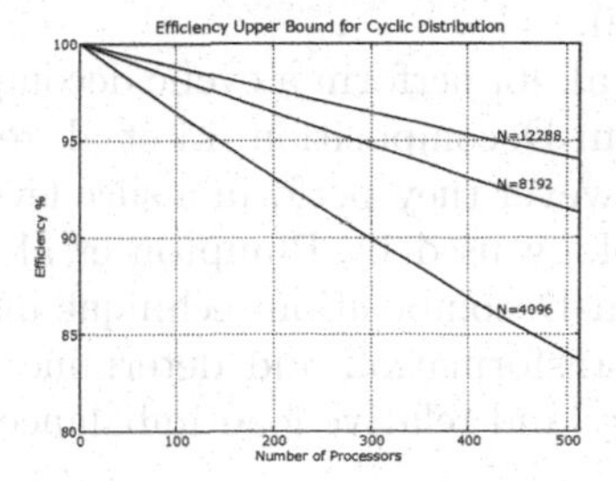

(d) Efficiency Upper Bound due to load imbalance when using the cyclic distribution.

Fig. 1. Cyclic Distribution Properties

2.3 Parallelization

The most common parallelization techniques of the direct methods are atom-, force- and spatial-decomposition. Atom-decomposition involves assigning the force computations of a subgroup of atoms to each processor. Force-decomposition generalizes this approach by assigning a subset of the force loops to each processor. Both of these decompositions are analogous to Lagrangian gridding in a fluids simulation where the grid cells move with the fluid. Further, in the above two techniques, the assignments of atoms to processors remain fixed throughout the entire simulation. Atom-decomposition requires the entire position array to be present at all the processors. Force-decomposition requires only a subset of the position array to be present at each processor leading to better scaling of the communication requirements.

On the other hand, spatial-decomposition works by assigning a portion of the physical simulation domain to each processor. Processors exchange atoms as they move from one domain into the other. This is analogous to Eulerian gridding in a fluids simulation where the grid remains fixed in space as fluid moves through it. Spatial-decomposition performs best when used with a short-range potential, a close-to-uniform spatial distribution of atoms and moderate to low densities. Atom- and force-decomposition perform the same irrespective of the above conditions. Plimpton [7] describes these methods in more detail and compares their relative performance in the case of a 2-body potential function. *The technique developed in this paper applies to short-range and long-range potentials evaluated using the atom-decomposition technique.*

3 Related Work

Plimpton et al [7] [4] have determined a force matrix transformation that exploits the symmetry in a 2-body force matrix. The two-body component of the Webber-Stillinger potential can use this transformation to exploit symmetries in the 2D force matrix. Our technique is to be used for the 3-body component of the potential.

Li et al. [6] perform a cyclic decomposition of the force matrix when performing Atom-Decomposition. Li et al. refer to their technique as force decomposition, however they perform a slice level decomposition. In this paper, we use the terminology used by Plimpton et al. [7], hence consider Li et al.'s work to be an Atom-Decomposition technique instead of Force-Decomposition. We analyze their transformation and determine closed form theoretical upper bounds for efficiency and relative load imbalance when using their technique.

4 Cyclic Distribution

The cyclic distribution technique decomposes the 3D force matrix into slices along one dimension. A triple (i, j, k) is evaluated only if $i < j < k$. This gives rise to a force matrix that looks like Figure 1(a). The interactions to be computed in the force matrix are in the shape of a tetrahedron. It is evident from the figure that lower slices of the matrix evaluate a larger number of tuples than higher slices. In a homogeneous cluster with P processors, the ith slice is computed by the processor with rank $mod(i, P)$. This strategy attempts to balance the load among the available processors in a homogeneous cluster environment.

It is obvious that the cyclic distribution technique assigns processors with lower id's a slightly larger load than processors with higher id's. We quantify this imbalance in section 4.1. This imbalance can be seen in Figure 1(b).

4.1 Analysis of the Cyclic Distribution

We now derive some interesting properties of the cyclic distribution. For the rest of this paper, we assume that P divides N. Proofs and other details of our work are available at [5].

Lemma 1. *For an ensemble of N atoms, if the cyclic distribution technique is used on P homogeneous processors, I_P, the number of interactions evaluated by processor $X, 0 \leq X < P$, is given by $K(N,P) - \frac{XN}{2P}(N+P-3-X)$, where $K(N,P) = \frac{N}{12 \cdot P}(N(2N-9)+P(P+3N-9)+12)$.*

Definition 1. *The Absolute Load Imbalance (ALI) is defined as the difference between the maximum number of interactions assigned to any processor and the minimum number of interactions assigned to any processor.*

Definition 2. *The Relative Load Imbalance (RLI) is defined as the ALI divided by the optimal load assignment. The optimal load assignment for a system with N atoms executing on P processors using a 3-body potential is $\frac{N(N-1)(N-2)}{6P}$.*

Theorem 1. *For an ensemble of N atoms, if the cyclic distribution technique is used on P homogeneous processors, the RLI is $\frac{3(P-1)}{N-1}$.*

Figure 1(c) depicts the effect of varying N and P on the RLI. It can be seen that for a fixed number of atoms, the RLI increases with an increase in number of processors. On the other hand, for a fixed number of processors, an increase in the number of atoms causes a decrease in the RLI.

Definition 3. *Efficiency of a parallel homogeneous system is defined as*

$$\eta = \frac{t_{seq}}{P \cdot t_{par}}$$

where, t_{seq} is the execution time of the entire simulation on a single processor and t_{par} is the execution time of the entire simulation using P processors.

Theorem 2. *For an ensemble of N atoms, if the cyclic distribution technique is used on P homogeneous processors, the upper bound for efficiency (η') due to load imbalance is given by $\frac{N(N-1)(N-2)}{6 \cdot K(N,P)}$, where $K(N,P) = \frac{N}{12 \cdot P}(N(2N-9)+P(P+3N-9)+12)$.*

It must be noted that the efficiency upper bound determined in Theorem 2 is only due to load imbalance. In practice, the bound will be lower due to communication overhead. Figure 1(d) depicts the effect of varying N and P on the above determined efficiency upper bound η'. It can be seen that for a fixed number of processors, increasing the number of atoms improves η'. For a fixed number of atoms, increasing the number of processors decreases the efficiency.

5 Symmetric Distribution

In this section, we propose a novel transformation that assigns an equal number of interactions to each slice of the force matrix. The transformed matrix F' is constructed as follows

$$F'_{ijk} = \begin{cases} F_{ijk}, & (i > j > k) \wedge (i + j + k \equiv 1 \bmod 3) \\ F_{ijk}, & (j > k > i) \wedge (i + j + k \equiv 2 \bmod 3) \\ F_{ijk}, & (k > i > j) \wedge (i + j + k \equiv 0 \bmod 3) \\ 0, & \text{otherwise} \end{cases} \quad (1)$$

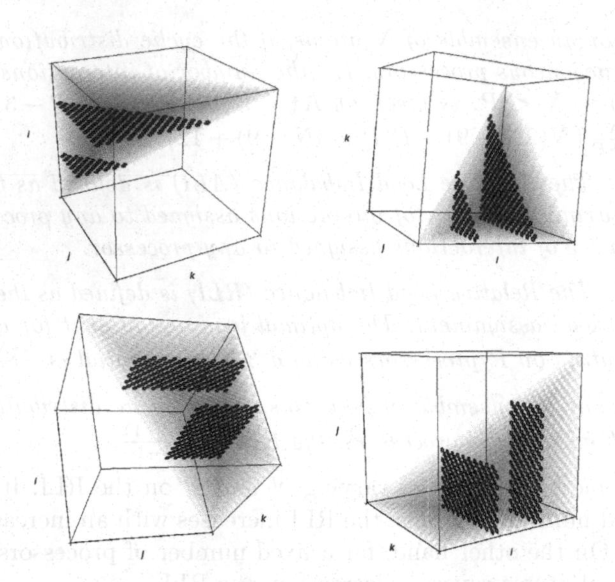

Fig. 2. The first two figures illustrate condition 1 of equation 1. The next two figures illustrate condition 3 of equation 1.

Only the non-zero elements of the transformed force matrix are to be evaluated in the force evaluation routine. The correctness of this transformation is proved in Theorem 3. The modulo component of the transformation acts like a 3D checkerboard pattern with three colors. Assume that a cell (i, j, k) is either cyan(C), red(R) or yellow(Y) depending on whether $i + j + k \equiv 0, 1, 2 (\text{mod } 3)$ respectively. Row 0 of Slice 0 of the board will be of the form CRYCRY..., Row 1 of Slice 0 will be of the form RYCRYC, Row 3 of Slice 0 will be of the form YCRYCR..., Row 4 of Slice 0 will be of the form CRYCRY...., Row 0 of Slice 1 will be of the form RYCRYC.... and so on.

A cube can be packed with 6 suitably sized tetrahedrons. The inequalities in each of the components of the transformation pick 3 of these tetrahedrons as illustrated in Figure 3. The spacing in the cells constituting the tetrahedrons is due to the modulo function. Figure 4 illustrates the interactions that are computed in two slices (parallel to the jk plane) of the transformed force matrix. The plane intersects two of the tetrahedrons such that the intersecting part of the plane with each of the tetrahedrons forms a triangle (case 1 and 2 of Equation 1). The third tetrahedron (case 3 of Equation 1) is cut such that the intersecting part of the plane forms a rectangle. The cutting of the tetrahedron to form a rectangle and triangle is shown in Figure 2. As we move along the i axis, the areas of the triangles and squares change but the sum of their areas remains constant. This ensures that irrespective of where a slice is made, the number of interactions in the slice is constant. This property is true even if the slicing is performed along either of the other two axes. In Figure 4, the black, gray and white elements of the slices correspond to the first, second and third cases in Equation 1. From now on, we refer to the interactions as black, gray and white

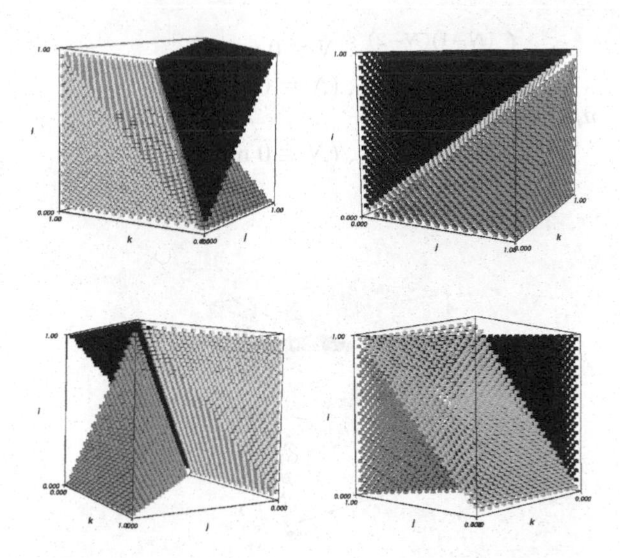

Fig. 3. Various views of the transformed force matrix when using the symmetric distribution

depending on whether they satisfy cases 1, 2 or 3 of Equation 1. The triangles and the rectangles get closest to each other at the focal point $i = j = k$ i.e (i, i) on the jk plane. As the height of the slice increases, the focal point moves along the jk diagonal toward the $(N-1, N-1)$ point on the slice. We refer to the intersection of the diagonal $i = j = k$ of the cube with the slice as the focal point.

5.1 Symmetric Distribution Properties

Proofs and other details of our work are available at [5].

Theorem 3. *Evaluating the non-zero elements of a symmetrically transformed $N \times N \times N$ force matrix results in computing exactly all the required force components of a 3-body potential.*

Lemma 2. *In an $N \times N \times N$ symmetrically transformed force matrix, the number of black, gray and white interactions in slice $m, 0 \leq m < N$ along the i axis are $\left\lfloor \frac{m(m-1)}{6} \right\rfloor$, $\left\lfloor \frac{(N-1-m)(N-2-m)}{6} \right\rfloor$ and $\left\lfloor \frac{m(N-1-m)}{3} \right\rfloor$ respectively.*

Theorem 4. *In an $N \times N \times N$ symmetrically transformed force matrix, the unique interactions to be computed are uniformly distributed among the slices of the force matrix along any axis and n_m, the number of interactions in slice $m, 0 \leq m < N$ along the $i-$axis is*

$$n_m = \begin{cases} \frac{(N-1)(N-2)}{6}, & N \not\equiv 0 \bmod 3 \\ \frac{N(N-3)}{6}, & (N \equiv 0 \bmod 3) \\ & \wedge (m \not\equiv 1 \bmod 3) \\ \frac{N(N-3)}{6} + 1, & (N \equiv 0 \bmod 3) \\ & \wedge (m \equiv 1 \bmod 3) \end{cases} \quad (2)$$

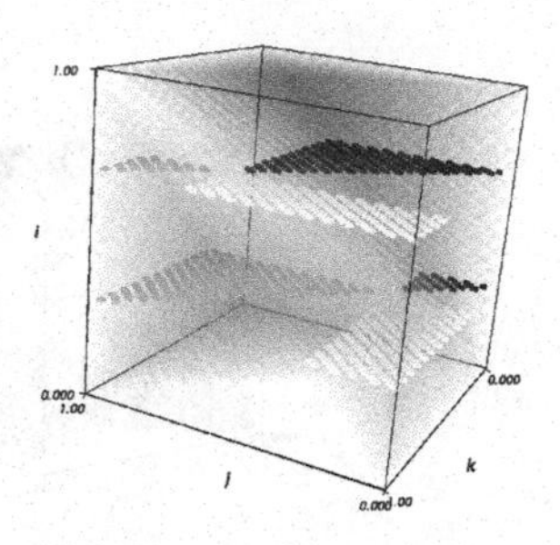

Fig. 4. Two slices of the transformed force matrix when using the symmetric distribution. The two cut planes depict the interactions that are computed in the corresponding slices. Only the visible parts of the plane are computed. The elements of the matrix have been made transcluscent to enable better visibility of the cut slices. The co-ordinates of the force matrix are normalized.

Corollary 1. *In an $N \times N \times N$ symmetrically transformed force matrix, the total number of black (N_b), gray (N_g) and white interactions (N_w) is given by*

$$N_b = N_g = \begin{cases} \frac{N(N-2)^2}{18}, & N \equiv 0 \bmod 3 \\ \frac{(N-1)(N^2-2N-2)}{18}, & N \equiv 1 \bmod 3 \\ \frac{(N+1)(N-2)^2}{18}, & N \equiv 2 \bmod 3 \end{cases}$$

$$N_w = \begin{cases} \frac{N(N^2-3N+6)}{18}, & N \equiv 0 \bmod 3 \\ \frac{(N-1)(N^2-2N+4)}{18}, & N \equiv 1 \bmod 3 \\ \frac{(N-2)(N^2-N+4)}{18}, & N \equiv 2 \bmod 3 \end{cases}$$

5.2 Load Distribution

From Theorem 4, it is evident that when using a symmetrically transformed force matrix, the computational load assigned to each processor is perfectly balanced when using the atom-decomposition algorithm. This implies that when using a symmetrically transformed force matrix, the primary factor affecting the parallel efficiency is communication overhead.

The cyclic distribution assigns slices that are separated by a distance of P slices to each processor, where P is the number of processors. This technique assumes a homogeneous cluster environment. On the other hand, when using the symmetric distribution with the Atom Decomposition algorithm, we can assign contiguous slices of the force matrix to the processors. This allows us to arbitrarily slice up the force matrix when using the symmetric distribution. *This is an important property that allows us to use the symmetric distribution in a heterogeneous environment where the number of slices assigned to each processor is not always N/P.*

6 Implementation

We implemented the atom-decomposition algorithm using both the cyclic distribution and symmetric distribution techniques using the MPICH implementation [3] of MPI.

Due to the symmetry in the force matrix, the *if* statement in the innner-most loop is taken only $\frac{1}{6}^{\text{th}}$ of the time i.e. it is not taken 83.3% of the time if a naive triple nested loop is used. To avoid this, we use loop unrolling and jamming techniques to re-order the loops and eliminate the need for a conditional statement in the inner-most loop. It is not possible for the compiler to perform this optimization since this optimization involves modulo arithemetic to determine the bounds of the loops.

The loop bounds are selected such that they satisfy the inequalities in Equation 1. Consider the implementation of Condition 1 of Equation 1. Determining the loop bounds of the i and j loops are trivial. The initial value of the k loop k_0 can be either $j-1$, $j-2$ or $j-3$ depending on the divisibility constraints. We consider the three cases:

1. $k_0 = j-1$: If k_0 is supposed to be $j-1$, it means that $i+j+(j-1) \equiv 1(\bmod 3)$. Hence, if $i+j+j \equiv 2(\text{mod}3)$, $k_0 = j-1$.
2. $k_0 = j-2$: If k_0 is supposed to be $j-2$, it means that $i+j+(j-2) \equiv 1(\bmod 3)$. Hence, if $i+j+j \equiv 0(\text{mod}3)$, $k_0 = j-2$.
3. $k_0 = j-3$: Similarly, if $i+j+j \equiv 1(\text{mod}3)$, $k_0 = j-2$.

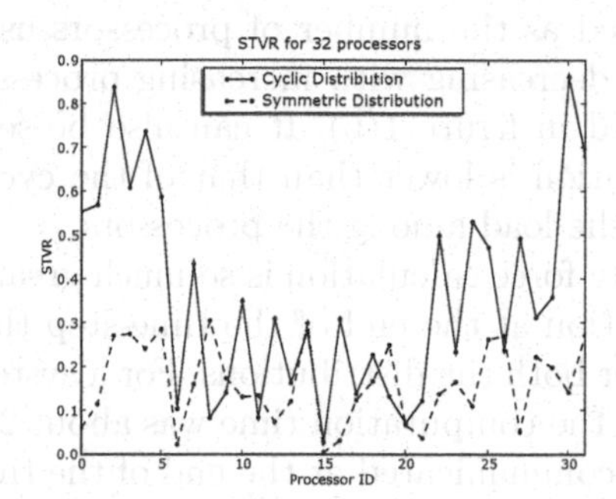

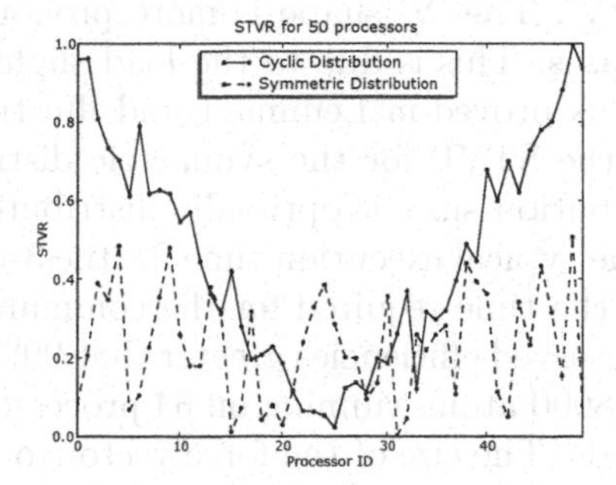

Fig. 5. Comparison of STVR for Cyclic Distribution and Symmetric Distribution in a homogeneous cluster environment. A system with 8000 atoms was used for this experiment.

To combine the above three cases, we need to make the remainders in the derived expressions increase monotonically. This can be done by adding one to both sides of the equivalence equations. Hence, if $i+j+j+1 \equiv 0(\bmod 3)$ then $k_0 = j-1$, if $i+j+j+1 \equiv 1(\bmod 3)$ then $k_0 = j-2$ and if $i+j+j+1 \equiv 2(\bmod 3)$ then $k_0 = j-3$. *Thus,* $k_0 = j-t$, *where* $t = (i+j+j+1) \bmod 3$. Similarly, the loop bounds for the other conditions can be determined.

We have found that the optimized version of the symmetric distribution performs on average 13.8% faster than the unoptimized version. Pseudo-code for the optimized loops is shown in [5].

7 Results and Discussion

To evaluate the load balance properties of the transformations, we use the Step Time Variation Ratio (STVR) [6]. The STVR for processor $i, 0 \le i < P$ is given by

$$\mathrm{STVR}_i = \left| \frac{T_i - \frac{1}{P}\sum_{j=0}^{P-1} T_j}{\frac{1}{P}\sum_{j=0}^{P-1} T_j} \right| \tag{3}$$

where, P processors are used to evaluate an $N \times N \times N$ force matrix and T_i is the average time to perform one time-step of the MD simulation on processor $i, 0 \le i < P$. The average step time T_i corresponds to the execution time of the parallelized force routine on each of the processors. It ignores the sequential execution time since the $O(N^3)$ computational intensity of the force evaluation is much greater than the linear time sequential portions of the code. The T_i values do not consider the communication time. The obtained STVR is thus an index of the imbalance in the force transformation algorithm.

To evaluate the performance of both the distributions in a homogeneous cluster environment, the following tests were performed on an Opteron 275 cluster with a Myrinet backplane. Figure 5 plots the STVR for the cyclic and symmetric distributions. The STVR in the case of the cyclic distribution is in the shape of a 'V'. The 'V' shape is more pronounced as the number of processors used increases. This is due to the load slightly decreasing with increasing processor rank as proved in Lemma 1 and illustrated in figure 1(b). It can also be seen that the STVR for the symmetric distribution is lower than that of the cyclic distribution since it optimally distributes the load among the processors.

The typical execution time for the 3-body force calculation is so much greater than the time required for the communication at the end of the time-step that we observed efficiencies greater that 99% for both the distributions. For a system with 8000 atoms running on 64 processors, the computation time was about 270 seconds. The size of the force vector to be communicated at the end of the time step is only about 187 kilobytes when a system with 8000 atoms is simulated. It takes only about 20 milliseconds to perform the required communication tasks at the end of the time-step. For larger number of atoms, this difference between the

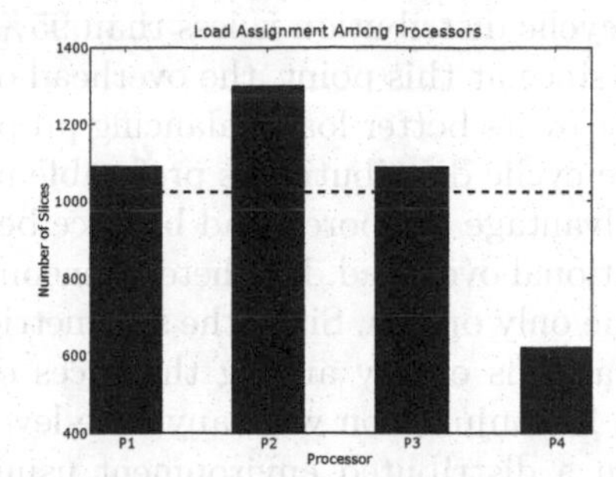

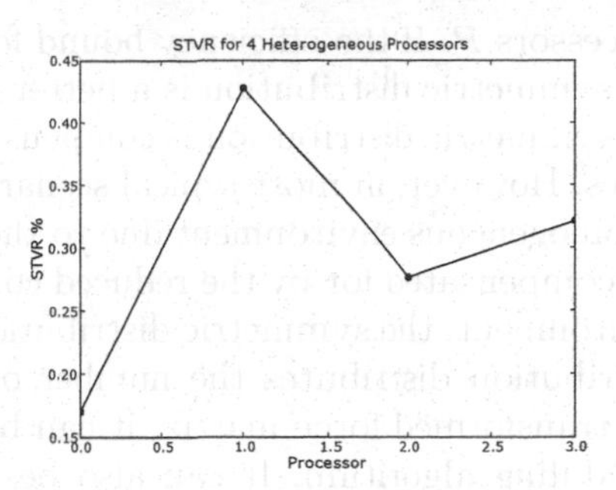

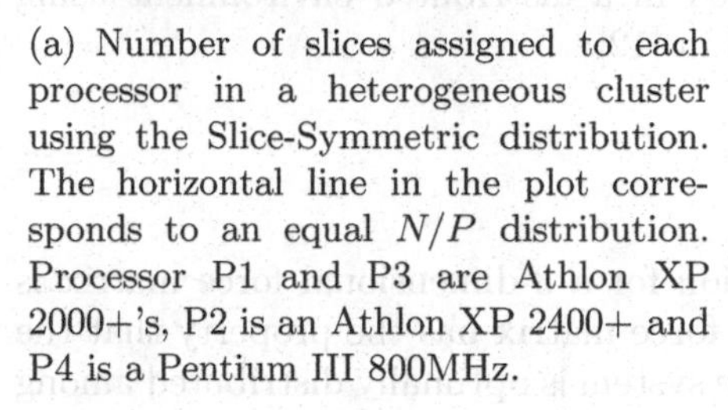

(a) Number of slices assigned to each processor in a heterogeneous cluster using the Slice-Symmetric distribution. The horizontal line in the plot corresponds to an equal N/P distribution. Processor P1 and P3 are Athlon XP 2000+'s, P2 is an Athlon XP 2400+ and P4 is a Pentium III 800MHz.

(b) Corresponding STVR for the heterogeneous cluster when using Slice-Symmetric transformation.

Fig. 6. Performance of the Symmetric Distribution in a heterogeneous environment

computation and communication times reduces further since the communication time scales as $O(N)$ while the computation time scales as $O(N^3)$.

Since the cyclic distribution works only in a homogeneous environment, we evaluate the STVR in a heterogeneous environment only for the symmetric distribution. For this test, we used a 4 node cluster comprising of two Athlon XP 2000+'s, an Athlon XP 2400+ and a Pentium III 800 MHz. We used the initial force calculation step as a benchmark for all the available processors and used the initial step execution time of each processor $t_i, 0 \leq i < P$ to determine the number of slices to be assigned to processor i for the future time-steps as

$$N_i = \frac{N}{t_i \sum_{j=0}^{P-1} \frac{1}{t_j}} \tag{4}$$

where, N is the number of atoms in the simulation. Figure 6(a) illustrates that the number of slices assigned to each of the processors is proportional to the CPU performance. It can be seen from Figure 6(b) that the STVR is below 0.5% even in the case of a heterogeneous environment.

However, it must be noted that our symmetric distribution is more complex in terms of computational intensity than the cyclic distribution. We have found that when using the optimized versions of both the distributions described in Section 6 the execution time of the cyclic distribution is 5.1% lower than that of the symmetric distribution.

The bounds that we derived in sections 4.1 and 5.2 can help in determining which distribution to use for a given number of atoms N and number of processors P. If the efficiency bound for the cyclic distribution is less than 95%, the symmetric distribution is a better choice since at this point, the overhead of the symmetric distribution is compensated for by its better load balancing properties. However, in most typical scenarios, the cyclic distribution is preferable in a homogeneous environment due to the disadvantage of poorer load balance being compensated for by the reduced computational overhead. In a heterogeneous environment, the symmetric distribution is the only option. Since the symmetric distribution distributes the number of interactions evenly among the slices of the transformed force matrix, it can be used in conjunction with any slice-level scheduling algorithm. It can also be used in a distributed environment using scheduling techniques that we developed in [12].

8 Conclusion

In this paper, a symmetric transformation for a 3-dimensional force matrix is described. A symmetrically transformed force matrix has the property that the total number of unique interactions in the system is optimally distributed among the slices of the force matrix along any dimension. This allows any heterogeneous slice level scheduling algorithm to be used on the transformed force matrix. Theoretical upper bounds for efficiency and relative load imbalance when using the cyclic distribution have been established. We have also proved interesting properties about the cyclic and symmetric distributions. An efficient loop reordering optimization for the force calculation routine in a 3-body potential when using the symmetric transformation has been described. Unlike the cyclic transformation, the symmetric transformation can also be used in a heterogeneous environment. We have also evaluated the performance of an MPI implementation of both transformations in a homogeneous environment and the symmetric transformation in a heterogeneous environment. When using the cyclic distribution, the execution time is on average 5.1% lower than that of the symmetric distribution. In the case of a homogeneous system, the bounds described in this paper can be used to determine which transformation technique is optimal for a given number of processors and number of atoms. For a hetereogeneous system, the symmetric distribution is the only option. Nonetheless, only the symmetric distribution provides a viable option for the processing power found in heterogeneous systems such as the computational grid.

We are currently investigating how to incorporate our transformation into the highly optimized and well known MD packages such as GROMACS, NAMD, AMBER etc.

Acknowledgement

This work was completed utilizing the Research Computing Facility with the associated USCMS Tier-2 site at the University of Nebraska-Lincoln.

References

1. Allen, M.P., Tildesley, D.J.: Computer simulation of liquids. Oxford University Press, New York (1987)
2. Clark, T.W., Hanxleden, R.V., McCammon, J.A., Scott, L.R.: Parallelizing molecular dynamics using spatial decomposition. In: Clark, T.W. (ed.) Proceedings of the Scalable High–Performance Computing Conference, pp. 95–102. IEEE Computer Society Press, Los Alamitos (1994)
3. Gropp, W., Lusk, E., Doss, N., Skjellum, A.: A high-performance, portable implementation of the MPI message passing interface standard. Parallel Computing 22(6), 789–828 (1996)
4. Hendrickson, B., Plimpton, S.: Parallel many-body simulations without all-to-all communication. J. Parallel Distrib. Comput. 27(1), 15–25 (1995)
5. Sumanth, J.V,: Novel force matrix transformations with optimal load-balance for 3-body potential based parallel molecular dynamics in a heterogeneous cluster environment. Technical Report TR-UNL-CSE-2007-0010, University of Nebraska-Lincoln (2007)
6. Li, J., Zhou, Z., Sadus, R.J.: A cyclic force decomposition algorithm for parallelising three-body interactions in molecular dynamics simulations. In: Ni, J., Dongarra, J. (eds.) IMSCCS (1), pp. 338–343. IEEE Computer Society, Los Alamitos (2006)
7. Plimpton, S.: Fast parallel algorithms for short-range molecular dynamics. J. Comput. Phys. 117(1), 1–19 (1995)
8. Gupta, S.: Computing aspects of molecular dynamics simulations. J.Comp.Phys.Comm. 70, 243–270 (1992)
9. Stillinger, F.H., Weber, T.A.: Computer simulation of local order in condensed phases of silicon. Physical Review B (Condensed Matter) 31, 5262–5271 (April 1985)
10. Sumanth, J.V., Swanson, D.R., Jiang, H.: Performance and Cost Effectiveness of a Cluster of Workstations and MD-GRAPE 2 for MD Simulations. In: Williams, M. (ed.) ISPDC, pp. 244–249. IEEE Computer Society, Los Alamitos (2003)
11. Sumanth, J.V., Swanson, D.R., Jiang, H.: Scheduling Many-Body Short Range MD Simulations on a Cluster of Workstations and Custom VLSI hardware. In: Bougé, L., Prasanna, V.K. (eds.) HiPC 2004. LNCS, vol. 3296, pp. 166–175. Springer, Heidelberg (2004)
12. Sumanth, J.V., Swanson, D.R., Jiang, H.: Adaptive Load Balancing for Long-Range MD Simulations in A Distributed Environment. In: Feng, W.C. (ed.) ICPP, pp. 135–146. IEEE Computer Society, Los Alamitos (2006)
13. Verlet, L.: Computer experiments on classical fluids i. thermodynamical properties of lennard-jones molecules. Phys. Rev. 159, 98–103 (1967)

Grid'BnB: A Parallel Branch and Bound Framework for Grids

Denis Caromel[1], Alexandre di Costanzo[1],
Laurent Baduel[2], and Satoshi Matsuoka[2]

[1] INRIA - I3S - CNRS - UNSA, France
[2] Tokyo Institute of Technology, Japan

Abstract. This article presents *Grid'BnB*, a parallel branch and bound framework for grids. Branch and bound (B&B) algorithms find optimal solutions of search problems and NP-hard optimization problems.

Grid'BnB is a Java framework that helps programmers to distribute problems over grids by hiding distribution issues. It is built over a master-worker approach and provides a transparent communication system among tasks. This work also introduces a new mechanism to localize computational nodes on the deployed grid. With this mechanism, we can determine if two nodes are on the same cluster. This mechanism is used in *Grid'BnB* to reduce inter-cluster communications. We run experiments on a nationwide grid. With this test bed, we analyze the behavior of a communicant application deployed on a large-scale grid that solves the flow-shop problem.

1 Introduction

Branch and bound (B&B) algorithm is a technique for solving search problems and NP-hard optimization problems. B&B aims to find the optimal solution and to prove that no ones are better. The algorithm splits the original problem into sub-problems of smaller size and then, for each sub-problem, the *objective function* computes the lower/upper bounds.

Because of the large size of handled problems (enumerations size and/or NP-hard class), finding an optimal solution for a problem can be impossible on a single machine. However, it is relatively easy to provide parallel implementations of B&B. Many previous work deal with parallel B&B as reported in [1].

Grids gather large amount of heterogeneous resources across geographically distributed sites to a single virtual organization. Resources are usually organized in clusters, which are managed by different administrative domains (labs, universities, etc.). Thanks to the huge number of resources grids provide, they seem to be well adapted for solving very large problems with B&B. Nevertheless, grids introduce new challenges such as deployment, heterogeneity, fault-tolerance, communication, and scalability.

We present *Grid'BnB*, a parallel B&B framework for grids. *Grid'BnB* aims to hide grid difficulties to users, especially fault-tolerance, communication, and scalability problems. The framework is built over a master-worker approach and

S. Aluru et al. (Eds.): HiPC 2007, LNCS 4873, pp. 566–579, 2007.

provides a transparent communication system among tasks. Local communications between processes optimize the exploration of the problem. *Grid'BnB* is implemented in Java within the *ProActive* [2] Grid middleware. Our second contribution is an extension of the *ProActive* deployment mechanism to localize computational resources on grids. We detect locality at runtime providing the grid topology to applications in order to improve scalability and performance.

2 *Grid'BnB*: Branch and Bound Framework

2.1 Principles

Branch and bound is an algorithmic technique for solving optimization problems. B&B aims to solve problems by finding the optimal solution and by proving that no other ones are better. The original problem is split in sub-problems of smaller sizes. Then, the *objective function* [3] computes the lower/upper bounds for each sub-problem. Thus for an optimization problem the objective function determines how good a solution is. The upper bound is the worst value for the potential optimal solution, the lower bound is the best value. Therefore, if V is the optimal solution for a given problem and $f(x)$ the objective function, then $\textit{lower bound} \leq f(V) \leq \textit{upper bound}$. Problems aim to minimize or maximize the objective function, in this paper we assume that problems minimize.

B&B organizes the problem as a tree, called *search tree*. The root node of this tree is the original problem and the rest of the tree is dynamically constructed by sequencing two operations: *branching* and *bounding*. Branching consists in recursively splitting the original problem in sub-problems. Each node of the tree is a sub-problem and has as ancestor a branched sub-problem. Thereby, the original problem is the parent of all sub-problems: it is named the root node. The second operation, bounding, computes for each tree node the lower/upper bounds. The entire tree maintains a *global upper bound* (GUB): this is the best upper bound of all nodes. Nodes with a lower bound higher than GUB are eliminated from the tree because branching these sub-problems will not lead to the optimal solution; this action is called *pruning*. Conceptually it is relatively easy to provide parallel implementations of B&B. Many previous work use the master-worker paradigm [4,5].

The optimization problem is represented as a dynamic set of tasks. A first task (the root node of the search tree) is passed to the master and branched. The result is a set of sub-tasks to branch and to bound. Even in parallel generating and exploring the entire search tree leads to performance issues. Parallelism allows to branch and to bound a large number of feasible regions at the same time, but the pruning action seriously impacts the execution time. The efficiency of the pruning operation depends on the GUB updates. The more GUB is close to the optimal solution, the more sub-trees are pruned. The GUB's updates are determined by how the tree is generated and explored. Therefore, a framework for grid B&B has to propose several exploration strategies such as *breadth-first search* or *depth-first search* (more details in Section 2.2).

Other issues related to pruning in grids are concurrency and scalability. All workers must share the GUB as a common global data. GUB has multiple parallel accesses in read (get the value) and write (set the value). A solution for sharing GUB is to maintain a local copy on all workers and when a better upper bound than GUB is found the worker broadcasts the new value to others.

In addition, for grid environments, which are composed of numerous heterogeneous machines and which are managed by different administrative domains, the probability of having faulted nodes during an execution is not negligible. Therefore, a B&B for grids has to manage fault-tolerance. A solution may for instance be that the master handles worker failures and the state of the search tree is frequently saved in a file.

2.2 Architecture

Grids lead to scalability issues owing to the large number of resources. Aida and al. [6] show that running a parallel B&B application based on a hierarchical master-worker architecture scales on grids. For that reason we choose to provide *Grid'BnB* with a hierarchical master-worker. Our hierarchical master-worker is composed of four kind of entities: *master*, *sub-master*, *worker*, and *leader*.

The *master* is the unique entry point: it receives the entire problem to solve as a single task (it is the *root task*). At the end, once the optimal solution is found, the master returns the solution to the user. Thus, the master is responsible for branching the root task, managing task allocation to sub-masters and/or workers, and handling failures. *Sub-masters* are intermediary entities whose role is to ensure scalability. They are hierarchically organized and forward tasks from the master to workers and vice versa by returning results to the master (or their sub-master parent). The role of the *workers* is to execute tasks. They are also the link between the tasks and the master. Indeed when a task does branching, sub-tasks are created into the worker that sent them to the master for remote allocation. *Leader* is specific role for workers. Leaders are in charge of forwarding messages between clusters (more details further).

Users who want to solve problems have to implement the task interface provided by the *Grid'BnB* API. Figure 1 shows the task interface and the worker interface implemented by the framework. The task interface contains two fields: `GUB` is a local copy of the global upper bound; and `worker` is a reference on the associated local process, handling the task execution. The objective function that users have to implement is `explore`. The result of this method must be the optimal solution for the feasible region represented by the task. `V` is a Java 1.5 generics: the user defines the real type. The branching operation is implemented by the `split` method. In order to not always send to the master all branched sub-problems, the *Grid'BnB* framework provides, via the `worker` field, the method `availableWorkers`, which allows users to check how many workers are currently available. Depending on the result of this method, users can decide to do branching and to locally continue the exploration of the sub-problem. To help users to structure their codes, we introduced two methods to initialize bounds: `initLowerBound` and `initUpperBound`. These two methods

```
public abstract class Task<V> {
    protected V GUB;
    protected Worker worker;
    public abstract V explore(Object[] params);
    public abstract ArrayList<?extends Task<V>> split();
    public abstract void initLowerBound();
    public abstract void initUpperBound();
    public abstract V gather(V[] values);    }

public interface Worker {
    public int availableWorkers();    }
```

Fig. 1. The task and worker Java interfaces

are called for each task just before the objective function `explore`, and they are not mandatory. The last method to implement is `gather`: the (sub-)master calls this method when all its tasks are solved. The method returns the best results from all tasks, *i.e.* the optimal solution.

The root task is passed to the master that performs the first branching. Then when a task is allocated to a worker that starts to explore it. As soon as a worker is available, a new task can be allocated. The worker starts by heuristic methods to initialize lower/upper bounds for the current feasible region, then it calls the objective function. Within the objective function, the user can decide whenever to branch the current region with the help of the `availableWorkers` method, which returns the current number of free workers.

The master and the search tree strategy handle task allocation; thereby the master works as a queue for task scheduling. The exploration algorithm of the search tree is important regarding performances. Therefore, *Grid'BnB* allows users to choose adapted algorithms to solve their problems. We propose four algorithms: *breadth-first search* explores the tree in larger, *depth-first search* explores all branches one by one, *first-in-first-out* (FIFO) explores the tree following the order tasks have been sent to the master, and *priority* explores in priority branches that updated the GUB the most frequently. If none of those algorithms satisfy the problem, users can implement their owns.

The tasks produce new GUB candidates while they are computed by workers. The GUB must be available to all tasks to prune the maximum of none promising branches of the search tree. The strategy for sharing GUB is to use a local copy of GUB on all workers and to broadcast updated value. Figure 2 shows the process of updating GUB when a worker finds a new better upper bound. To be efficient, a B&B framework has to broadcast the GUB as fast as possible. With a large number of workers, directly broadcasting GUB to every worker cannot scale. For that reason *Grid'BnB* organizes workers in groups.

Groups are sets of workers, which can efficiently broadcast GUB between them. The master is in charge of building groups. Thus, the main criterion to put workers in the same group is their localization on the same cluster. Clusters usually provide a high performance environment for communication. The master

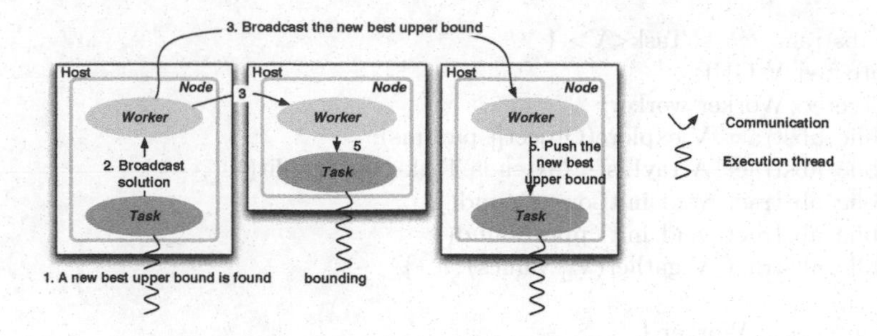

Fig. 2. Update the best global upper bound

elects a worker as *leader* in each group. This leader has a reference to all other group leaders. When a leader receives a communication from outside its group, it broadcasts the communication to its group. Inversely when the leader receives a communication from a member of its group, it broadcasts the communication to the other leaders but only if the new upper bound is better than its own GUB value. Figure 3 shows an example of broadcasting GUB between groups.

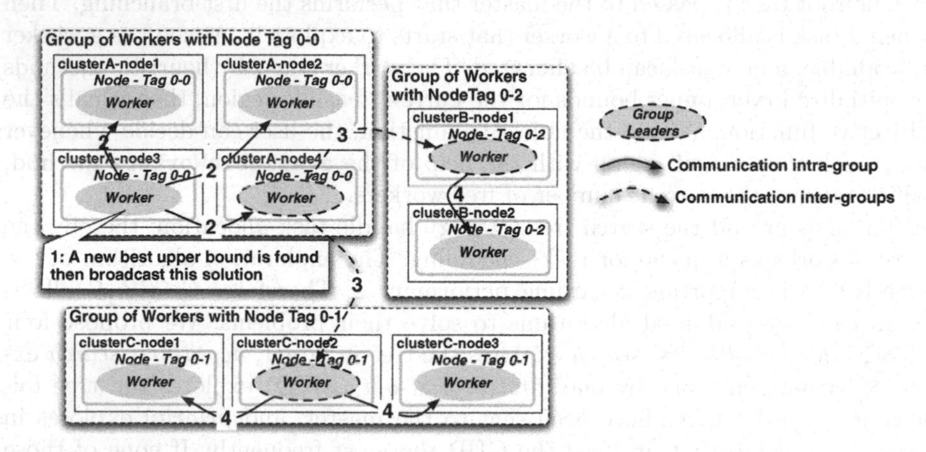

Fig. 3. Broadcasting solution between groups of workers

Within the user code, errors can occurs, such as uncaught exceptions. Workers handle user exceptions. When a worker catches an exception, it forwards it to the master, and then the master stops the whole computation and returns the exception to the user.

The last feature of *Grid'BnB* is the fault-tolerance. Fault-tolerance is a real issue of grid environments; the large number of resources that are distributed on different administration domains implies a high probability of faults, such as hardware failures, networks down time, or maintenance.

Master and sub-masters hierarchically manage infrastructure failures, such as host failures. The monitoring consists of frequently pinging entities. When the ping call fails (communication timeout, network errors, etc.), the remote host is considered as unreachable and down. In that case, the master re-allocates the task to an available worker. If for the same task several results are returned to the master (worker considered down for network problem and come back), only the first one is kept, others are flushed. Masters handle the fault of their sub-masters: if a sub-master does not answer to a ping call, the master chooses a free worker and re-instantiates it as a sub-master. Masters also handle the fault of leaders; the master frequently pings leaders. When a leader is unreachable, the master elects a new leader in the group.

The master must be deployed on a stable machine, because it is at the top of the monitoring hierarchy. As opposed to sub-masters and workers, master host failures cannot be dynamically handled by the framework but require users intervention. The status of the current execution (GUB and all tasks) is frequently saved on disk. Thus for long-running problem, if the master node faults the user can restart the solving at a recent state of the execution.

Grid'BnB provides a high level-programming model for solving problems with parallel B&B. From the users points of view, the framework handles all issues related to distribution/parallelism and fault-tolerance.

2.3 Implementation

Grid'BnB is designed for grids and is implemented with Java, which allows to use a large kinds of resources, operating systems, and machine architectures. More of Java, *Grid'BnB* is implemented within the *ProActive* Grid middleware.

ProActive [2] is a Java library for concurrent, distributed and mobile computing.*ProActive* features transparent remote active objects, asynchronous two-way communications with transparent futures, high-level synchronization mechanisms, and migration of active objects with pending calls. As *ProActive* is built on top of standard Java APIs, neither does it require any modification to the standard Java execution environment, nor does it make use of a special compiler, preprocessor or modified Java Virtual Machine (JVM). A distributed or concurrent application built using *ProActive* is composed of a number of medium-grained entities called *active objects*. Method calls sent to active objects are asynchronous with transparent *future objects* and synchronization is handled by a mechanism known as wait-by-necessity.*ProActive* provides typed *group communication*, an important feature for high-performance and grid computing. The group communication [7] extends the *ProActive* elementary mechanism for asynchronous remote method invocation and automatic futures.

In *Grid'BnB*, master, sub-masters, and workers are active objects. Each active object serves remote calls in FIFO order. Master manages futures on current executing tasks. Then, groups of workers are *ProActive* groups. Leaders are also member of a *ProActive* group. Thereby, hierarchical *ProActive* groups represent workers. A hierarchical group is indeed a group of groups. Finally, to optimize communication between workers to solve more rapidly problems, the management of workers in groups lay to the *ProActive* deployment framework. *ProActive* features a system for the deployment of applications on grids. The next section explains the deployment mechanism and how we improved it to manage organization of workers in groups of communications.

3 Grid Node Localization

The key principle of *ProActive* deployment [8] is to eliminate from the source code the following elements: *machine names*, *creation protocols*, *registry*, and *lookup protocols*. It allows to deploy any application anywhere without modifying the source code. The deployment sites are called *nodes* and correspond to JVMs, which host active objects. The deployment framework uses *Virtual Nodes* (VNs). VNs are the deployment abstractions for the applications; they are defined in the program source and after activation they are mapped to a set of nodes. The deployment framework relies on XML descriptors. They are composed of two parts: mapping and infrastructure. The VN, which is the deployment abstraction for applications, is mapped to nodes in the deployment descriptors, and nodes are mapped to physical resources, *i.e.* to the infrastructure. Nodes are created using remote connection and creation protocols. Deployment descriptors allow combining these protocols in order to seamlessly create remote JVMs.

In Section 2 we proposed to organize workers in groups for optimizing communication. The selection criterion for group acceptance for a worker is its physical localization on a cluster. Therefore, the node localization on the grid is important for an efficient implementation of our *Grid'BnB* framework. The *ProActive* deployment framework provides a high-level abstraction of the underlying physical infrastructure. Once deployed, the application cannot easily access to the topology of the physical infrastructure. For instance, programmers have to compare node addresses for determining if two nodes are deployed on the same cluster. Nevertheless, two nodes may have the same sub-net address on different clusters, with network of NATs. Hence, programmers may use metrics, such as latency, to determine if nodes are "close". Consequently, organizing workers in group by clusters and optimizing communication between clusters is a very difficult and complicated task. For that reason we introduced a new mechanism in the *ProActive* deployment framework to identify nodes, which are deployed on the same cluster or even on the same machine.

The creation of a node is the result of a deployment graph (a directed acyclic graph: DAG) with connection protocols. This deployment graph is specified within the XML deployment descriptor. Our deployment node tagging mechanism aims to tag nodes in regard of the deployment graph on which they are mapped in the deployment descriptor. This tag will allow the application to organize groups in regard to the deployment process that created nodes. With this mechanism, all deployed nodes are tagged with an identifier at deployment time. Nodes that have the same tag value have been deployed by the same deployment process. As a result, they have a high probability to be located in the same the same local network.

Figure 4 shows the process of tagging nodes. The tag is built by a concatenation of identifiers at each level of the deployment graph. At the beginning of the

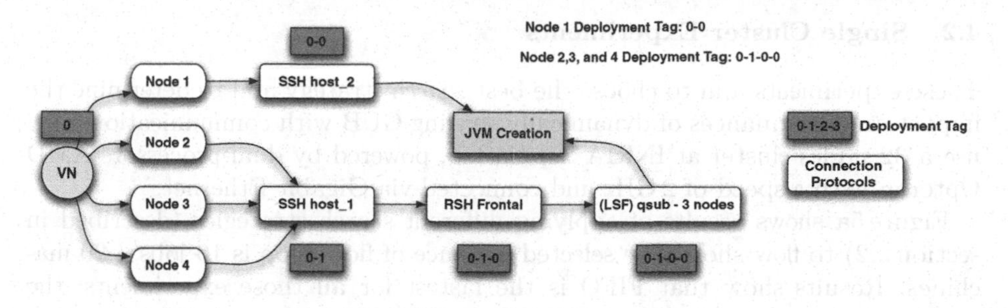

Fig. 4. Deployment tag mechanism

deployment, a new tag is instantiated for each virtual node. For leaf nodes of the DAG, which are JVM creations, no identifier is added. Therefore, all nodes deployed with the same path in the DAG have the same tag.

The tag is an abstraction of the physical infrastructure; it provides more information about how nodes have been deployed. It is now possible to know at the application level that the same deployment graph has deployed two nodes. The deployment tag can be used for instance by applications to optimize communication between nodes or to do data localization. More especially *Grid'BnB* uses the deployment tag to dynamically organize worker communications between clusters. Figure 3 shows the deployment result of a single virtual node on three clusters. The deployment has returned nine nodes: four nodes on *clusterA*, two on *clusterB*, and three on *clusterC*. The node tag mechanism has tagged the nodes *0-0* on *clusterA*, *0-1* on *clusterC*, and *0-2* on *clusterB*. Tags are finally used to organize workers in groups of communication to optimize communication between clusters.

4 Experiments

4.1 The Flow-Shop Problem

Flow-shop is a NP-complete permutation optimization problem. The flow-shop problem consists in finding the optimal schedule of n jobs on m machines. The set of jobs is represented by $J = \{j_1, j_2, \dots j_n\}$, each j_i is a set of operations $j_i = \{o_{i1}, o_{i2}, \dots o_{im}\}$ where o_{im} is the time taken on machine m and the set of machines is represented by $M = \{m_1, m_2, \dots m_m\}$.

The operation o_{ij} must be processed by the machine m_j. The sequence of jobs are the same on every machines, e.g. if j_3 is treated in position 2 on the first machine, j_3 is also executed in position 2 on all machines.

We consider the mono-objective case, which aims to minimize the overall completion time of all jobs, *i.e. makespan*. The makespan is the total execution time of a complete sequence of jobs. Thus, the mono-objective goal is to find the sequence of jobs that takes the shortest time to complete.

4.2 Single Cluster Experiments

These experiments aim to choose the best search strategy and to determine the impact on performances of dynamically sharing GUB with communications. We use a 32 nodes cluster at INRIA Sophia lab, powered by dual-processors AMD Opteron with a speed of 2 GHz and connected via Gigabit Ethernet.

Figure 5a shows results of applying different search strategies (described in section 2.2) to flow-shop. The selected instance of flow-shop is 16 jobs / 20 machines. Results show that FIFO is the fastest for all those experiments; the speedup between 20 CPUs and 60 CPUs is 4.63. This is a super linear speedup owing to increase the total of CPUs allows a larger generation of the search tree in parallel and thereby, improving the GUB faster to prune more branches. Breadth-first search scales with a very good speedup, the speedup between 20 CPUs and 60 CPUs is 5.44, also super linear. The high speedup is normal because more breadth-first search is deployed on nodes the more the tree is explored in parallel. Depth-first search speedup is linear, 3.00, and for priority search the speedup is 1.73. The speedup is particularly high with all these experiments, because with 60 CPUs the chosen flow-shop instance can be widely explored in parallel whatever the search strategy. The built search tree rapidly provides the best solution as upper bound, thus each process can delete many branches.

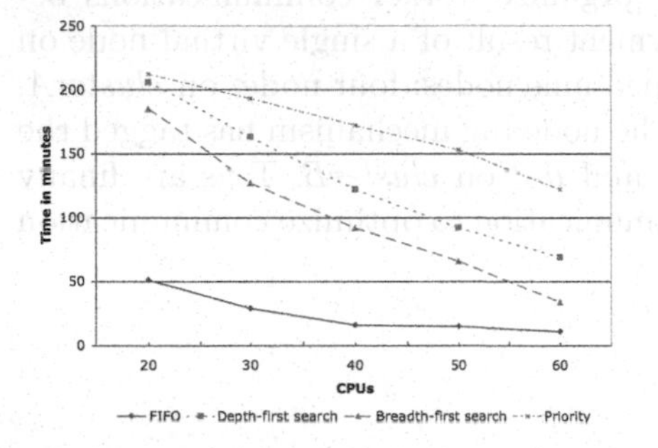

(a) Benchmarking search tree strategies

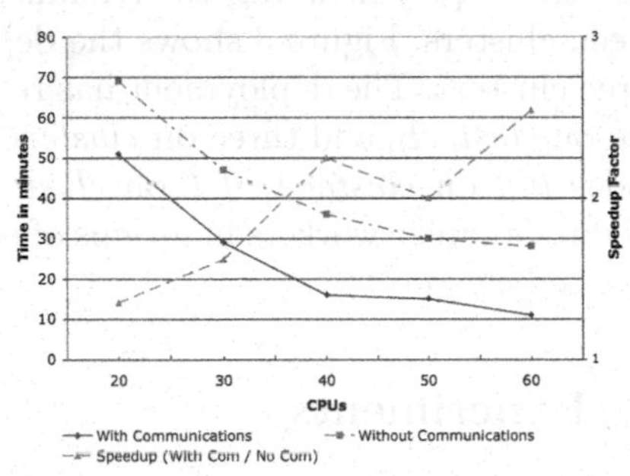

(b) Dynamic GUB sharing vs. no sharing

Fig. 5. Single cluster experiments: flow-shop $n = 16$, $m = 20$

With the same instance of flow-shop and with the FIFO strategy, we now benchmark the impact of dynamically sharing GUB with communications. We benchmark flow-shop with communications between workers for sharing GUB and without dynamically sharing GUB between workers (no communication). In the case of no communication, the master keeps the GUB up-to-date with all results from computed tasks; and when a task is allocated to a worker by the master, it sets the current GUB value to the task. Figure 5b shows the results. Using communications to share GUB improves performance. But the speedup, $\frac{T\ No\ Communication}{T\ Communications}$, is lower for 50 CPUs than 40 CPUs, this decrease comes from the fact that since 40 CPUs this flow-shop instance has enough CPUs to explore the whole tree in parallel, *i.e.* it is the optimal deployment.

These experiments on a single cluster show that dynamically sharing GUB with communications between workers improve execution time, and that choosing the right search strategy considerably affects performances.

4.3 Large Scale Experiments

In order to experiment *Grid'BnB* on grids, we used a large-scale nationwide infrastructure for grid research, *Grid'5000* (G5K) [9]. The G5K project aims at building a highly reconfigurable, controllable and monitorable experimental grid platform gathering 9 sites geographically distributed in France currently featuring a total of about 3000 CPUs. G5K is composed of a large number of machines, which have different kinds of CPUs (dual-core architecture, AMD Opteron 64 bits, PowerPC G5 64 bits, Intel Itanium 2 64 bits, Intel Xeon 64 bits), of operating systems (Debian, Fedora Core 3 & 4, MacOs X, *etc.*), of supported JVMs (Sun 1.5 64 bits and 32 bits, and Apple 1.4.2), and of network connection (Gigabit Ethernet and Myrinet).

Grid experiments run with the same implementation of flow-shop, as previous single cluster experiments. The instance of flow-shop is now a larger problem: 17 jobs / 17 machines. The search tree strategy is FIFO and communications are used to dynamically share GUB. Results of experiments with G5K are summarized in Figure 6a and Table 1.

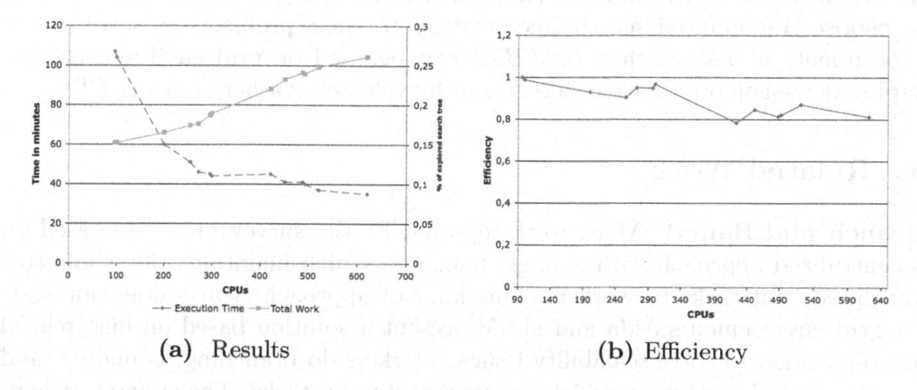

(a) Results (b) Efficiency

Fig. 6. Large scale experiments: flow-shop $n = 17$, $m = 17$

The broken line in Figure 6a shows that the execution time strongly decreases until 272 CPUs, the speedup between 96 CPUs and 272 CPUs is 2.32. From 272 to 621 CPUs the execution time is almost constant, the speedup between 272 and 621 CPUs is 1.31. Then, the global speedup, between 96 and 621 CPUs, is 3.01. Our *Grid'BnB* flow-shop scales well up to 272 (close to linear speedup). However, for more than 272 CPUs, the execution time decreases slowly. Nevertheless, the solid line shows the percentage of branches explored in the search tree, *i.e.* total number of tested permutations, this line increases with the number of CPUs. This line is indeed the total work done by the computation.

Table 1. Large scale experiments results

CPUs	Sites	Execution time	Tasks	% of explored search tree	Gathered time
100	1	104 m	1567	0.152%	167 h
200	1	60 m	2515	0.165%	181 h
300	2	44 m	3729	0.189%	196 h
492	4	40 m	5447	0.239%	251 h
621	5	35 m	6968	0.261%	267 h

Figure 6b shows the efficiency E, this value estimates how CPUs are utilized for the computation. Values of E are between 0 and 1, a single-processor computation and linear speedup have $E = 1$. Here, we consider the execution time (T) efficiency corrected with the work (W: total number of tested permutations) because *Grid'BnB* computes more work with increasing CPUs. Thus, the efficiency for n CPUs: $E_n = \frac{T_n/T_{96} * W_{96}/W_n}{96/n}$. The figure shows that between 96 and 300 CPUs, E is close to 1 (0.9), which is very good. However, for 422 and more, E decreases to 0.8, it is still a good value. This decrease can be explain by the fact that for experiments with less than 422 CPUs are done on 1 or 2 grid sites and for 422 and more 3 up to 5 sites nationally-distributed. In addition, grid sites are heterogeneous in regards of CPUs power and inter-site network connections.

Experiments on single cluster and large scale grid show that it is better to use communications to dynamically share GUB, and that it is important for users to choose the adapted search tree strategy to their problems to solve. Large experiments also show that *Grid'BnB* can be used on grid environments, we deploy flow-shop on a nationwide grid of five clusters gathering a 621 CPUs.

5 Related Work

Branch and Bound. Many work reported by the survey in [1] are based on a centralized approach with a single manager, which maintains the whole tree and hands out tasks to workers. This kind of approach clearly does not scale for grid environments.Aida and al. [5] present a solution based on hierarchical master-worker to solve scalability issues. Workers do branching, bounding, and pruning on sub-problems, which are represented by tasks. The supervisor handles the sharing of the best current upper bound. Supervisor and sub-masters gather results from workers and are in charge to hierarchically update the best upper bound on all workers. We show in section 4.2 that using dynamic communications rather than using the master to share GUB allows to complete the computation faster. In [6] Aida and Osumi propose a study of their hierarchical master-worker framework implemented using GridRPC middleware [10] and Ninf-G [11]. The authors discuss the granularity of tasks, notably when tasks are fine-grain the communication overhead is too high compared to the computation of tasks. Thereby, *Grid'BnB* introduces a method to check how many workers are available. This method helps users to program tasks and to dynamically determine the most appropriate granularity of the tasks.

Iamnitchi and Foster [12] proposes a solution to do B&B over grids that differs from *Grid'BnB* and others because it does not base on master-worker paradigm, but on a decentralized architecture that manages resources through a membership protocol. Each process maintains a pool of problems to solve. When the pool is empty, the process asks for work to other processes. The sharing of the best upper bound is handled by circulating a message among processes. The fault-tolerance issue is addressed by propagating all completed sub-problem to all processes. This approach may result in significant overhead, in terms of both duplicated work and messages.

ParadisEO [13] is an open source framework for flexible parallel and distributed design of hybrid meta-heuristics. Moreover, it supplies different natural hybridization mechanisms mainly for meta-heuristics including evolutionary algorithms and local search methods. All these mechanisms can be used for solving optimization problem. Like *Grid'BnB*, the grid version of ParadisEO is based on the master-worker paradigm. ParadisEO splits the optimization problem in tasks. Then, the task allocation is handled by MW [4], a tool for scheduling master-worker applications over Condor [14], which is a grid resource manager. Unlike *Grid'BnB*, ParadisEO just provides mechanisms for searching algorithms.

Skeletons. The common architecture used for B&B on grids is "master-worker". For parallel programming, the master-worker pattern is called *farm skeleton* [15]. Muskel [16] is a Java skeleton framework for grids that provide farm. Skeleton frameworks usually provide task allocation and fault-tolerance. Thus, skeletons seem well adapted for implementing B&B for grids. Like *Grid'BnB* users just have to focus on the implementation of the problem to solve all other issue related to grid and tasks managing are handled by the framework. However in farm skeletons tasks cannot share data, such as a global upper bound to prune more promising branches of the search tree to find more rapidly the optimal solution. In addition, another skeleton that fits B&B algorithm is the *divide-and-conquer skeleton*. This skeleton allows to dynamically split task, *i.e.* branching, but like farm it is not possible to share the global upper bound between task.

Divide-and-Conquer. Conceptually, B&B technique fits the divide-and-conquer paradigm. The search tree can be divided into sub-trees, and each sub-tree is then assigned to an available computational resource. This is done recursively until the task is small enough to be solved directly.

Satin [17] is a system for divide-and-conquer programming on grid platforms. Satin express divide-and-conquer parallelism entirely in the Java language itself, without requiring any new language constructs. Satin uses so-called marker interfaces to indicate that certain method invocations need to be considered for parallel execution, called spawned. A mechanism is also needed to synchronize with spawned method invocations. Satin can be used directly to implement B&B. Thus, users can mark branching methods to be executed in parallel. Like *Grid'BnB*, Satin is in charge to distribute sub-problems through grids. But unlike our framework, Satin does not provide any mechanisms for sharing global upper bound and more generally no mechanism for communication between parallel executed sub-problems.

6 Conclusion and Perspectives

We described *Grid'BnB* a parallel B&B framework for grids. *Grid'BnB* provides a framework to help users to solve optimization problems hiding grids, parallelism, and distribution related issues. It is based on a hierarchical master-worker architecture enhanced with communications between processes to share the best global upper bound thus exploring less parts of the search tree and decreasing the execution time. Because grids provide a large-scale parallel environment, we propose to organize workers in groups of communications. Groups reflect grid topology. This feature aims to optimize inter-cluster communications and to update more rapidly the global upper bound on all processes. *Grid'BnB* proposes different search tree algorithms to help users to choose the most adapted one for the problem to solve. Finally, the framework allows fault-tolerance for long-running executions. In addition, we introduced a new mechanism, *deployment node tagging*, to localize deployed nodes on grids. The deployment node tagging allows *Grid'BnB* to identify nodes, which are on the same cluster, and to optimize group communications between processes. This mechanism is integrated in the deployment framework of the *ProActive* grid middleware.Experiments show that *Grid'BnB* scales on a real nationwide grid, such as Grid'5000. We were able to deploy a permutation optimization problem, flow-shop, on up to 621 CPUs distributed on five sites.

In future work, we plan to improve our flow-shop implementation with a better objective function, such as the technique proposed by Lageweg [18]. Likewise, we want to run larger scale experiments on a worldwide grid, by mixing clusters located in France and Japan. We believe that *Grid'BnB* can used for more than B&B. Without modification of the framework it may be used to do divide-and-conquer or as farm skeleton. *Grid'BnB* is framework for parallel programming that targets all embarrassingly parallel problems.

References

1. Gendron, B., Crainic, T.: Parallel Branch-And-Bound Algorithms: Survey and Synthesis. Operations Research 42(6), 1042–1066 (1994)
2. Caromel, D., Delbé, C., di Costanzo, A., Leyton, M.: Proactive: an integrated platform for programming and running applications on grids and p2p systems. Computational Methods in Science and Technology 12(1), 69–77 (2006)
3. Atallah, M.: Algorithms and theory of computation handbook. CRC Press, USA (1999)
4. Goux, J., Kulkarni, S., Linderoth, J., Yoder, M.: An Enabling Framework for Master-Worker Applications on the Computational Grid. In: Proc. 9th IEEE Symp. on High Performance Distributed Computing (2000)

5. Aida, K., Natsume, W., Futakata, Y.: Distributed computing with hierarchical master-worker paradigm for parallel branch and bound algorithm. In: CCGrid 2003. 3rd IEEE/ACM International Symposium on Cluster Computing and the Grid, pp. 156–163 (2003)
6. Aida, K., Osumi, T.: A Case Study in Running a Parallel Branch and Bound Application on the Grid. In: SAINT2005. Proc. IEEE/IPSJ The 2005 Symposium on Applications & the Internet, pp. 164–173 (2005)
7. Baduel, L., Baude, F., Caromel, D.: Efficient, Flexible, and Typed Group Communications in Java. In: Joint ACM Java Grande - ISCOPE 2002 Conference, Seattle, pp. 28–36. ACM Press, New York (2002)
8. Baude, F., Caromel, D., Mestre, L., Huet, F., Vayssière, J.: Interactive and descriptor-based deployment of object-oriented grid applications. In: Proceedings of the 11th IEEE International Symposium on High Performance Distributed Computing, pp. 93–102. IEEE Computer Society Press, Los Alamitos (2002)
9. Cappello, F., Caron, E., Dayde, M., Desprez, F., Jeannot, E., Jegou, Y., Lanteri, S., Leduc, J., Melab, N., Mornet, G., Namyst, R., Primet, P., Richard, O.: Grid 2005: a large scale, reconfigurable, controlable and monitorable Grid platform. In: Grid 2005 Workshop, Seattle, USA, IEEE/ACM (to appear)
10. Seymour, K., Nakada, H., Matsuoka, S., Dongarra, J., Lee, C., Casanova, H.: Overview of GridRPC: A Remote Procedure Call API for Grid Computing. In: 3rd International Workshop on Grid Computing (November 2002)
11. Tanaka, Y., Nakada, H., Sekiguchi, S., Suzumura, T., Matsuoka, S.: Ninf-G: A Reference Implementation of RPC-based Programming Middleware for Grid Computing. Journal of Grid Computing 1(1), 41–51 (2003)
12. Iamnitchi, A., Foster, I.: A Problem-Specific Fault-Tolerance Mechanism for Asynchronous, Distributed Systems. In: ICPP. 29th International Conference on Parallel Processing, Toronto, Canada, pp. 21–24 (August 2000)
13. Cahon, S., Talbi, E.G., Melab, N.: Paradiseo: A framework for parallel and distributed metaheuristics. In: IPDPS 2003. Proceedings of the 17th International Symposium on Parallel and Distributed Processing, p. 144.1. IEEE Computer Society, Washington, DC, USA (2003)
14. Litzkow, M., Livny, M., Mutka, M.: Condor - a hunter of idle workstations. In: Proc. of the 8th International Conference of Distributed Computing Systems (1988)
15. Cole, M.: Algorithmic skeletons: structured management of parallel computation. MIT Press, Cambridge, MA, USA (1991)
16. Danelutto, M.: Qos in parallel programming through application managers. In: PDP 2005. Proceedings of the 13th Euromicro Conference on Parallel, Distributed and Network-Based Processing (PDP 2005), pp. 282–289. IEEE Computer Society, Washington, DC, USA (2005)
17. van Nieuwpoort, R.V., Maassen, J., Thilo Kielmann, G.W., Bal, H.E.: Satin: Simple and efficient java-based grid programming. Journal of Parallel and Distribute d Computing Practices (accepted)
18. Lageweg, B., Lenstra, J., Kan, A.: A General Bounding Scheme for the Permutation Flow-Shop Problem. Operations Research 26(1), 53–67 (1978)

The CMS Remote Analysis Builder (CRAB)

D. Spiga, S. Lacaprara, W. Bacchi, M. Cinquilli, G. Codispoti, M. Corvo, A. Dorigo, A. Fanfani, F. Fanzago, F. Farina, M. Merlo, O. Gutsche, L. Servoli, and C. Kavka

University of Perugia and INFN of Perugia
06100 via pascoli 10
Perugia, Italy

Abstract. The CMS experiment will produce several Pbytes of data every year, to be distributed over many computing centers geographically distributed in different countries. Analysis of this data will be also performed in a distributed way, using grid infrastructure. CRAB (CMS Remote Analysis Builder) is a specific tool, designed and developed by the CMS collaboration, that allows a transparent access to distributed data to end physicist. Very limited knowledge of underlying technicalities are required to the user. CRAB interacts with the local user environment, the CMS Data Management services and with the Grid middleware. It is able to use WLCG, gLite and OSG middleware. CRAB has been in production and in routine use by end-users since Spring 2004. It has been extensively used in studies to prepare the Physics Technical Design Report (PTDR) and in the analysis of reconstructed event samples generated during the Computing Software and Analysis Challenge (CSA06). This involved generating thousands of jobs per day at peak rates. In this paper we discuss the current implementation of CRAB, the experience with using it in production and the plans to improve it in the immediate future.

1 Introduction

The CMS experiment (Compact Muon Solenoid) [1] is one of the four physics experiments that will collect data at the Large Hadron Collider (LHC) [2] located at CERN.

The expected rate of event to disk will be about 150 Hz, so few PBytes of data per year will be stored and processed. At the same time, the experiment needs also to use the computational resources for the simulated data generation. The choice of CMS to cover all these needs is a distributed architecture and the use of the grid middleware components.

A hierarchy of computing regional centers, called Tiers, is defined in the CMS Computing Model. The system is geographically distributed and includes a single Tier-0 center at CERN, a CMS Analysis Facility also at CERN, few Tier-1 centers and many Tier-2 centers with different level of resources. In order to manage the data and optimize the use of the distributed resources, a combination of generic Grid tools, provided by the LCG (LHC Computing Grid) [3] and OSG (Open Science Grid) [4] projects, as well as specialized CMS tools are used together.

CMS has also to provide a single interface to physicists, capable to operate with all grid components and different back-ends in a transparent way. In this paper the CMS analysis model and the tools used to perform it are discussed.

S. Aluru et al. (Eds.): HiPC 2007, LNCS 4873, pp. 580–586, 2007.

2 CMS Data Model

The CMS computing model defines that the data collected by the CMS online data acquisition system is sent to the Tier-0 center at CERN where raw data is archived. A prompt reconstruction is performed and a first version of the Analysis Object Data (AOD) is produced.

All the high-level physics objects are stored in the AOD together with the information sufficient to support typical analysis. Raw and first pass reconstructed events are distributed from the Tier-0 to a Tier-1 centers which takes custodial responsibility for those while the AOD are also transferred to all Tier-1. The Tier-1 centers provide services for data archiving, re-processing, calibration, skimming and other data-intensive analysis tasks. All AOD and a fraction of the first pass reconstructed events and RAW data are transferred to Tier-2 centers which provide resources for physics analysis.

The computing model foresees that all CMS users must use the Grid in order to perform its own analysis.

3 User Analysis and CRAB

CRAB has been designed to allow physicists to access efficiently distributed data hiding the complexity of Grid infrastructure. Following the analysis model, the user runs interactively over small data samples in order to develop and test his code, using CMS analysis framework (CMSSW). Once ready, the user uses CRAB from a User Interface,

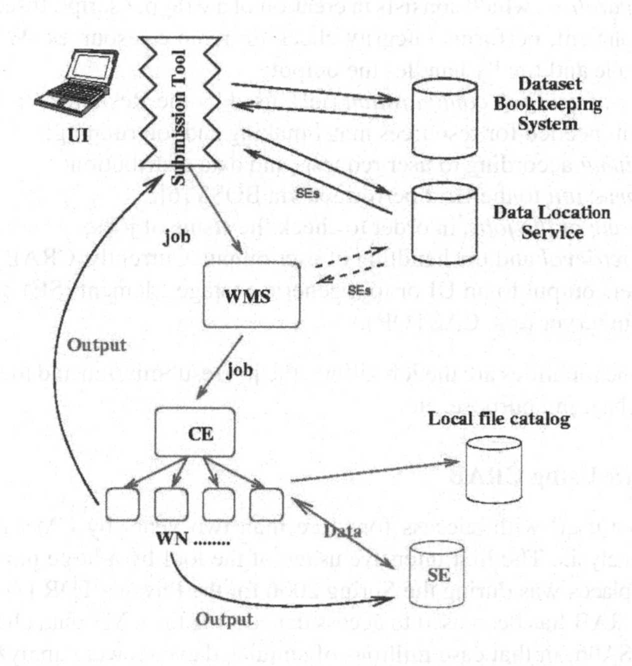

Fig. 1. CRAB work flow and interaction with the Grid and Data Management

where grid client middleware is available, to access to all data available on all remote sites. The work flow covered by CRAB can be factorized on three main steps:

- The Data Discovery step, interacting with the CMS data management infrastructure, to know if required data are found and where are located;
- The interaction with CMSSW on local machine, so that the very same environment can be reproduced on the remote resources;
- The Grid specific step, where all actions, from submission to output retrieval, are performed;

A user interacts with CRAB through simple configuration file which is arranged in several specific sections. Here all job specific parameters are defined, such as: the dataset that the user wants to access, the name of CMSSW specific configuration file, the job splitting parameters, how to manage the produced output, etc.

It is important to note that the very same CMSSW configuration, which the user has used interactively can also be used for remote data access. CRAB will take care to apply any changes needed to run on selected dataset on remote sites. The typical work flow (figure 1) is the following:

- *the input data discovery* to determine the Storage Elements (SE) of sites storing data. They are found interacting with CMS specific services, Data Bookkeeping and Location Services (DBS and DLS);
- *the packaging of user code*, to create a tar archive with user code which contains executable, library and user data files, as found on user local environment on User Interface;
- *the job preparation*, which consists in creation of a wrapper script. It sets up the running environment, performs integrity check on remote resources (WN), launches the executable and finally handles the output;
- *the creation of grid job configuration* (jdl), used by the Resource Broker (RB) the requirements needed for resources matchmaking and job running;
- *the job splitting* according to user requests and data distribution;
- *the job submission* to the Grid performed via BOSS [6];
- *the monitoring of the jobs*, in order to check the status of jobs;
- *the output retrieval* and the handling of user output. Currently, CRAB supports the copy of users output to an UI or to a generic Storage Element (SE) or to any host with a gsiftp server (e.g. CASTOR).

Other useful functionalities are the job killing, the job resubmission and the postmortem analysis, for debugging purpose, etc.

3.1 Experience Using CRAB

CRAB has been used with success for more than two years by CMS physicists, to perform data analysis. The first intensive usage of the tool by a large number of users from different places was during the Spring 2006 for the Physics TDR [7] preparation.

Moreover, CRAB has been used to access data during the CMS data challenges. The last one was CSA06, in that case millions of simulated events were analyzed, reaching peaks of 100'000 submitted jobs per month.

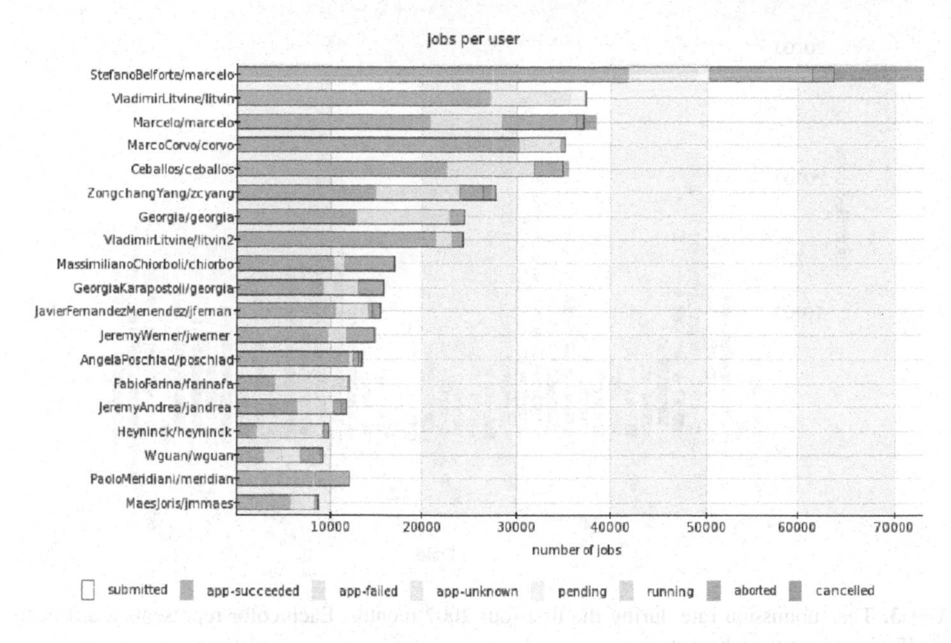

Fig. 2. The top 20 users during the first four 2007 months

During the Magnet Test Cosmic Challenge (MTCC), CRAB was even successfully used to access the real data distributed among several sites, the first real data for CMS. There are about $\sim$ 300 CRAB users distributed around the world, that means a daily rate of submitted jobs which reaches peaks of 10'000 jobs. In figure 2 the top twenty CRAB users in the first part of 2007, sorted by number of submitted jobs, are shown. The different color of the bars represents the different job status. In figure 3 the submission rate is shown, referred to the same period. In this plots each color represents a different site where the jobs run. The total efficiency is currently of order of 80%. The most important causes of the failure rate are related to the input data and to the middleware infrastructure. So the resulting infefficiency is not directly dependent on CRAB which indeed doesn't introduce a relevant fraction of jobs failures.

3.2 CRAB Improvements: Motivations and Implementation

The actual work flow of CRAB is based on a direct submission from the UI, where the user is working, to LCG and OSG via RB. This *standalone* model has the advantage of simplicity, but it lacks some features, which can be provided by a more advanced architecture *client-server* where a server is placed between the user and the Grid to perform a set of actions for him. The main goals of the client-server architecture is to automate as much as possible the whole analysis work flow and to improve the scalability of the system. The aim of the project is also to create a tool which is easy to use for physicists and easy to maintain for administrators.

The client-server implementation is transparent to the end users: the interface, the installation, the configuration procedure and the usage remains exactly the same as for standalone. The general CRAB client-server architecture is shown in figure 4.

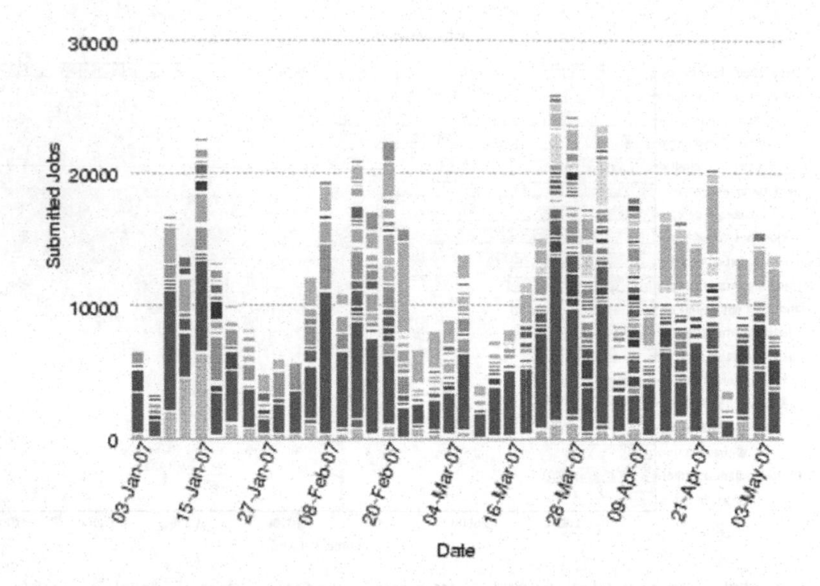

Fig. 3. The submission rate during the first four 2007 months. Each color represents a different CMS site where the jobs run.

The server architecture is based on components implemented as independent agents communicating through an asynchronous and persistent message service, plus a gridftp server which allows all communication from client to server. The user proxy is shipped using a WS-Delegation compliant service for proxy delegation.

The server core is a MySQL [8] DB of which the distinct components publish and subscribe messages to communicate. This architecture is as similar as possible to the CMS production system ProductionAgent [9] sharing components where possible, allowing an easier maintenance of the WorkLoad Management tools.

The role of the client is to interact with DBL/DLS for the data discovery, to prepare the jobs reading the local environment and finally to send the user proxy and the prepared task to the server.

The server manages the project interacting with the Grid from the job submission to the output retrieval on the name of the user. The actual server implementation provides the following components:

DropBoxGuardian: to check the dropBox, which is the container where the client put the user tasks, for new stuff to be managed. It also monitors the delegation service to check new proxy arrivals;

ProxyTarAssociator: Associates the task to the right user proxy and localizes the task configuration files w.r.t. the specific server instance;

CrabWorker: Submits jobs to the Grid in the CRAB style. It fully reuses the submitter components of CRAB (which supports: EDG, glite, glite-bulk, condor_g) It has some task-level resubmission features;

TaskTracking: Keeps general informations about all tasks under execution (e.g. status of the task, percentage of completed jobs in the task, etc);

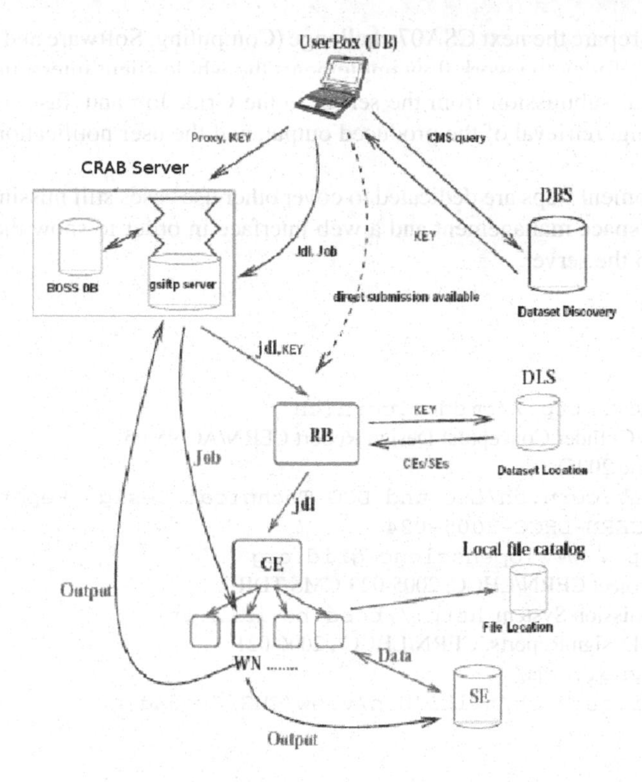

Fig. 4. CRAB server work flow and interaction with the Grid and Data Management

Notification: Notifies via e-mail the users when their tasks are ready/failed;

JobTracking: Tracks the status for every single job querying the grid and caching the infos on the local DataBase;

ErrorHandler: Handles errors of jobs. Depending on the type of error (e.g. job run error) it will initiate the appropriate error handler and update the job state;

JobSubmitter: Resubmits single jobs if needed;

RSSFeeder: Provides multiple RSS channels which can be used to forward important informations to the server administrator.

4 Conclusions

CRAB has been used since summer 2005 by several hundreds of end users distributed all over the world, it has reached more then 100'000 jobs/month with a daily record of 10'000 jobs. CRAB was also extensively used by hundreds of physicists to access data for the Physics TDR preparation, which was published during spring 2006. The CRAB project is still under development in order to satisfy the users and the whole CMS computing requirements for the data analysis.

At the moment the key effort of the development activity is devoted to the client-server implementation. The first version is already released and it is now under commissioning to prepare the next CSA07 challenge (Computing, Software and Analysis CMS challenge). Today the work flow implements the whole client interaction with the server, the actual submission from the server to the Grid, Job and Task tracking, simple error handling, retrieval of the produced output, and the user notification of the task status.

The next development steps are dedicated to cover other use cases still missing, such as job killing, disk space management and a web interface in order to show the status of jobs submitted to the server.

References

1. The CMS experiment, `http://cmsdoc.cern.ch`
2. The Large Hadron Collider Conceptual Design Report CERN/AC/95-05
3. L.C.G Project: (June 2005), `http://lcg.web.cern.ch/LCG and LCG Technical Design Report, CERN-TDR-01 CERN-LHCC-2005-024`
4. OSG Project: `http://www.opensciencegrid.org`
5. The Computing Project CERN/LHCC, 2005-023 CMS TDR 7
6. Batch Object Submission System, `http://boss.bo.infn.it/`
7. Physics Technical Design Reports, CERN/LHCC/, 2006-021
8. `http://www.mysql.com/`
9. `https://twiki.cern.ch/twiki/bin/view/CMS/ProdAgent`

Applying Internet Random Early Detection Strategies to Scheduling in Grid Environments

Manuel Brugnoli[1], Steven Willmott[2], Elisa Heymann[1], Paul Hurley[3], and Miquel A. Senar[1]

[1] Universitat Autònoma de Barcelona, Spain
manuel.brugnoli@aomail.uab.es,
{Elisa.Heymann, MiquelAngel.Senar}@uab.es
[2] Universitat Politècnica de Catalunya, Spain
steve@lsi.upc.edu
[3] IBM Research, Zurich
pah@zurich.ibm.com

Abstract. Resource Allocation in Grid environments to date is generally carried out under the assumption that there is one primary scheduling system scheduling jobs. However, as environments tend towards larger open "utility" Grids it becomes increasingly likely that deployments will involve multiple independent schedulers allocating jobs over the same resources.

In this paper we show that, if using current standard scheduling approaches, such multi-scheduler environments may well be prone to serious oscillation problems in resource allocation similar to those commonly found in IP network traffic. Further we demonstrate how common techniques from IP networks – in particularly approaches based on Random Early Detection (RED) buffer management and its subsequent extensions / variations – may provide an effective way to damp or eliminate such oscillations. The paper describes the analogy between multi-scheduler Grid resource allocation and IP network routing and explores the impact of oscillation and RED methods by simulation.

1 Introduction

Grid computing environments are a powerful model for virtualized resource sharing in wide-area network environments - making it possible for many users from different organisations to transparently access computing resources. However, the resulting resource allocation problem is highly challenging and becomes particularly so when: a) processors are widely distributed – leading to lengthy communication delays between sites and resource controllers having to work with outdated load information, and b) users are associated with multiple centres of authority and are also distributed – making it impractical to centralise resource scheduling for scalability, performance and administrative reasons.

In such environments, resource scheduling involves multiple independent schedulers, each serving a subset of users and able to place jobs on resources without direct communication with other schedulers. Essentially the schedulers are multiple

S. Aluru et al. (Eds.): HiPC 2007, LNCS 4873, pp. 587–598, 2007.

actors working over the same resources in parallel with no central oversight. Furthermore, the information the schedulers have to work with is time delayed and potentially inaccurate. While great advances have been made in resource scheduling in recent years (from resource controllers [1, 2, 3] which now work very effectively in environments where the scheduler is the primary actor to approaches such as application based scheduling) it may be non-trivial to extend these techniques to larger, multi-scheduler environments. In this paper we argue that such problems are strongly analogous to certain types of control problems found in today's packet switched IP (Internet Protocol) networks and structurally very different to those in which a single actor can be assumed. In particular, in this paper:

- We show that multiple scheduler Grid environments do indeed show potentially complex and pathological dynamics sometimes seen in IP networks.
- That one way to tackle such problems may be to borrow techniques from IP networks themselves through the use of particular intelligent buffer management techniques such as Random Early Detection (RED) [4, 5] and (later) [6, 7].

Results show that under certain conditions with multiple schedulers, oscillations clearly exist and heavily impact resource usage. Further, the work presents two RED based strategies (Static RED and dynamic RED) for resource scheduling which could be used to mitigate such effects, in particular when augmented to function with dynamic thresholds.

2 Problem Definition

In order to draw on techniques from network buffer management we define a generic multiple scheduler Grid resource allocation problem as follows:

- A set of n_s schedulers s_i *in* S which do not communicate with each other and each of which emits a series of jobs $(j_{i1} .. j_{in})$ *in* J over time and aims to assign them to processors.
- A set of n_r processors p_i *in* P, each of which has a processing queue with limited capacity **pcap** and is able to (on request) send out information on the status of its current processing queue (length in items or time) ***pinf***.
- Processors handle jobs in a FIFO (First In First Out) manner. Once they have accepted a job it is always eventually processed (processors are assumed not to fail).
- Each scheduler regularly polls all processors to get the latest ***pinf*** information and caches this between updates. [1]
- For simplicity, the time delay t for a message to travel from any s_i to any r_j and back again is the same and is non-zero. (Hence, if an s_i sends a request to an r_j at time T_1, the r_j will receive the message at time T_1+t and s_i the response at T_1+2t.)
- For a scheduler s_i to assign a job to a processor p_j it sends a message to the processor placing the job. The resource then responds either with an *acceptance* of the job or a *rejection*.

[1] More realistically the delays between each resource and scheduler would be determined by the network topology and would not always be the same.

- Once a job has been assigned to and accepted by a processor it cannot be re-assigned. If the job is rejected by the processor the scheduler may try to assign it elsewhere.

In this environment, it is therefore assumed that each scheduler uses the polling mechanism to produce a regularly updated snapshot of the current resource state of the network. The scheduler subsequently makes individual decisions on which processor to place its own incoming jobs. No direct communication is assumed between schedulers since such communication would take at least as long as interactions with processors and generate a strong synchronisation dependency on schedulers.

3 Oscillations and RED in IP Networks

In IP networks, flow control mechanisms such the TCP/IP protocol [8] provide signalling to communicate packet arrivals at the receiver to the sender. In the case of congestion (overloaded nodes at any point on the path between the sender and the receiver) packet losses occur, which induce a "backing-off" (reduction) of the sending rate at the sender. The sender then slowly increases its sending rate once more. Such mechanism (as well as adaptive protocols such as OSPF [9]) make it possible for all endpoints in the network to open and close TCP/IP packet flows to other endpoints without direct communication with each other. In other words, there is no central resource control. Furthermore, congestion is managed using feedback signals generated by the TCP/IP protocol (or similar mechanisms) which indicate to senders concerned when congestion occurs. These mechanisms provide for a baseline in congestion management. However, unfortunately they also exhibit pathological properties under certain (common) network conditions. In particular, the two most serious issues are:

- **TCP/IP Backoff synchronisation [4]:** this phenomenon occurs when one or more nodes in a network become congested and the TCP protocol signals all senders with streams passing through the congested area to "slow down" sending within a short space of time. The result is a quick drop in all traffic through the node (creating a hole in resource use) followed by a gradual speed-up of sending by all sources (potentially creating a further congestion peak, another hole and so on).
- **Route Oscillation:** in networks which use dynamic route allocation and adjust paths through the network based on round-trip times, congestion may cause routes to change to avoid a congested area. However, in general this means changing all traffic between two endpoints – hence, such route shifts may therefore create congestion elsewhere, a consequent shift back to the original route and so on.

These phenomena are well documented in the communication networks literature (see [10] for example) and have been identified in many complex variations. While future Grid networks are unlikely to have as many independent schedulers as communication networks have endpoints, as the number of independent schedulers grows, they may well exhibit similar potential oscillation problems since individual schedulers may naturally be programmed to "back-off" usage of congested processors and/or switch jobs to less loaded machines.

3.1 RED for Oscillation Damping in Networks and Grids

In the face of these oscillation problems, one of the main solutions developed in communications networks to improve performance is a family of mechanisms based on the idea of Random Early Detection (RED) policies. RED policies function by adopting the following underlying principle – *begin sending congestion messages to some endpoints before maximum capacity is reached* [5]. A basic RED approach is defined by two thresholds – MinQ and MaxQ as a %age of the maximum physical buffer. Subsequently, for each incoming packet, if the buffer contains less than MinQ packets, the incoming packet is always accepted, if the buffer contains more than MaxQ, the incoming packet is rejected. For buffer fullness value between MinQ and MaxQ, the system randomly rejects incoming packets with a probability determined by a linear interpolation between 0% (at MinQ) to 100% (at MaxQ).[2]

Hence, the node begins to randomly send a small number of congestion signals to some sources well before the full capacity of the buffer is reached. Subsequently, a wide range of variations on RED were produced in order to optimise a variety of behavioural properties of the network.

In Grid environments, As noted above, while multiple scheduler Grids may not have as many independent schedulers as IP networks have potential senders, there may still be strong analogies between the environments. In particular,

- As soon as multiple independent schedulers are present they will receive feedback from the same environment and hence the potential for oscillation driving feedback mechanism exists.
- Even if schedulers are not directly competitive, they have no strong incentive to cooperate with other schedulers as soon as they are not part of the same organisations, hence rational behaviour would lead each scheduler to be configured to gain maximum individual benefit from the Grid.

In the remainder of the paper we therefore pursue a simple mapping of IP network structures and the RED approach to the generic Grid scheduling problems described in Section 2:

- Processors P are analogous to IP network nodes.
- Jobs (tasks) are analogous to data packets.
- Job rejections are analogous to TCP/IP packet drop messages.
- Processors implement buffer management policies including RED.
- MinQ and MaxQ are set to fractions of the buffer length (see varied parameters in the next section).

[2] Note that, in practical implementations, RED uses an exponentially-weighted moving average on the buffer size rather does not work on the buffer directly to provide greater control, however for simplicity in this paper we consider the buffer directly.

4 Experimental Analysis

In order to test these hypotheses, this section covers a set of simulations (using the SMPL simulation engine [11]) of the performance of the random and shortest queue scheduling policies with and without RED. Of interest are how RED strategies perform on average compared to simple standard scheduling policies, and their effect on the local queues of the available resources.

4.1 Scheduling Strategies

The set of scheduling strategies studied were the following:

- *Random without RED (R):* Jobs ready to be executed are assigned randomly to the machines, regardless of the local queue load. This strategy represents the case of a pure dynamic method that has no information on the processors or the application.
- *Shortest Queue without RED (SQ):* Upon receiving a request for executing a job, each scheduler will use the information provided by a monitor system such as the EGEE Information Supermarket to select the processors with the shortest queue. It should be noted that this monitoring information is accurate as sent by the processors to the schedulers, however there is a time delay before it arrives at each scheduler – the information is therefore gathered asynchronously, hence independent from the scheduling process and is updated every second.
- *Random with RED (R-RED):* This strategy consists of applying RED to the jobs when scheduled using the random strategy. The thresholds of RED are kept constant throughout all simulation runs.
- *Shortest Queue with RED (SQ-RED):* This strategy consists of applying RED to the jobs when scheduled using the SQ strategy. Once the machine with the shortest queue has been chosen to execute a job, it may reject that job depending on its queue occupancy. If the queue length is below MinQ of it capacity, the job will be accepted. Otherwise it may be rejected with a probability proportional to the queue occupancy, i.e. the closer the queue is to MaxQ, the higher the probability of rejecting the job.
- *(All RED Schedulers) backoff mechanism*: in addition to these behaviors, all schedulers working in RED environments implement a simple backoff policy in which, once rejected from a particular resource, they ignore this resource for a number of time steps before trying again (meanwhile they may send jobs, including the rejected job to other servers).

4.2 Simulation Framework

All described scheduling strategies are simulated in scenarios determined by the following parameters:

- *Number of Machines:* This represents the number of available processors on which the jobs are executed. We performed simulations using 20 homogeneous machines.
- *Number of Schedulers:* This represents the number of schedulers that receive requests for executing jobs, and the ones who schedule jobs using either the

random or the shortest queue policies. In our experiments the number of schedulers considered was 5.

- *Number of jobs*: Our experiments considered three types of jobs with execution times of 10, 20 and 30 time units that were uniformly distributed in all simulation. Jobs are generated according to a exponential distribution with rate 5600 (resulting in approximately 5400 jobs).
- *Queue Length*: Each processor has an associated queue. In our experiments the maximum capacity of the queue was 30 jobs.
- *RED Thresholds:* RED will always accept jobs if the queue occupancy is up to 5 jobs. If the queue occupancy is bigger than this, RED will drop jobs with a probability proportional to the queue occupancy.
- *Backoff time*: the backoff time for all schedulers is set to 4x(Queue length minus MinQ).

We assumed that communication among the schedulers and the processors takes a fixed amount of time (10 time units for a job to reach the executing machine queue, and 5 time units for a rejection notification to reach the submitting scheduler). 8 time units are added to allow for the time required to clean the job from the system. The results show total job total execution time (from arrival at the scheduler to completion) and processor queue behaviour for the strategies considered.

4.3 Simulation Results – Oscillations and RED

In each of the figures the x-axis shows execution time in simulation units. The y-axis represents the queue occupancy in number of tasks or jobs for each of the machines. Fig. 1(a) shows the evolution of 4 of the 20 processor queues for the random scheduling policy (R), while Fig. 1(b) shows the queue evolution for Random when RED is applied (R-RED). We show only 4 queues for the sake of clarity (the behaviour of these queues being typical).

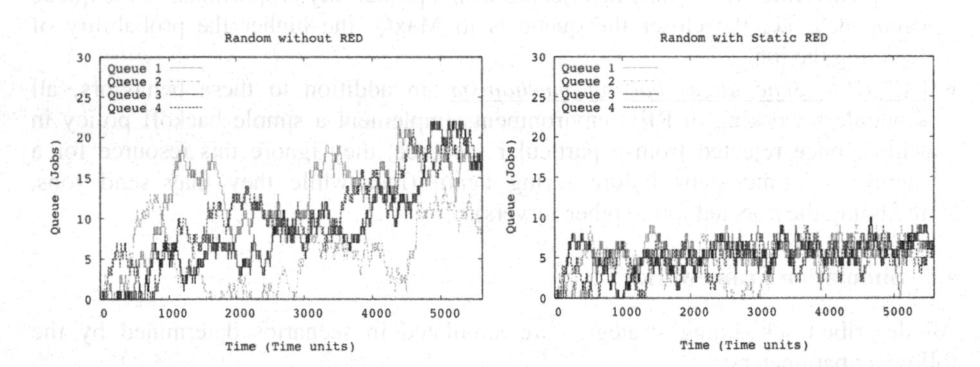

Fig. 1. Comparing R scheduling with and without RED. The introduction of Static RED behavior (b) for the resources in the system hugely reduces the variable nature of resource usage in different systems queues – keeping average queue lengths much lower than without RED.

In the simulation runs shown, the striking feature is the high variability in queue length shown by the simple random strategy without RED as well as the clear mirroring behaviour of some queues of one another (Q3 decreasing and increasing in opposing oscillation to Q1 for example). The application of RED, by contrast, shows a radically different picture, with average queue occupancy more uniform across queues. The results shown are typical of many runs with variations of similar parameters and in terms of results that under R, a total (in the run shown) of 5332 jobs, the average turnaround time was 223.75 time units with a standard deviation on turnaround time of 136.58. For R-RED, results for the run shown (with the same parameters) show for 5348 jobs and average turnaround time of 134.32 time units (almost half of the scenario with no RED) and a standard deviation of 45.74.

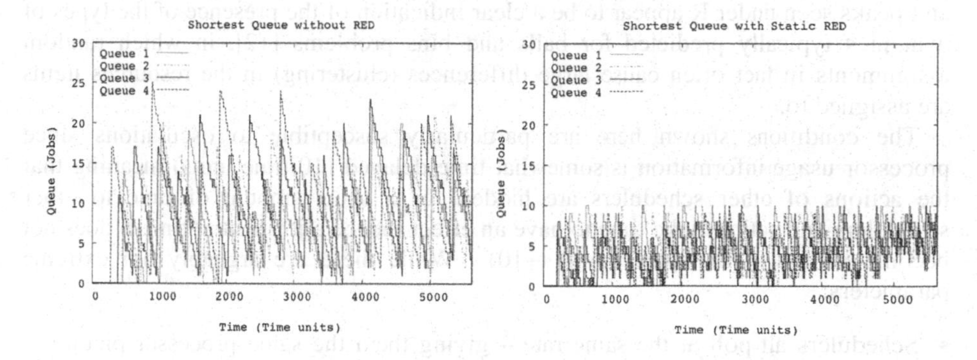

Fig. 2. Comparing SQ scheduling with and without RED. As with random, the introduction of RED shows a radical change in system behavior.

As with random scheduling policies, shortest path allocation also shows a major change in results from non-RED to RED environments. In terms of raw numbers, on its 5384 jobs, SQ showed an average turnaround time of 220.53 time units and a standard deviation of 106.90 – both marginally improved in comparison with the R strategy. With RED however, average turnaround times and standard deviations drop to 133.01 and 52.32 time units respectively, also in each case almost halving the times involved.

While only one plot is shown, many experiments with different random number seeds and varied parameters were executed – all of these showed the same underlying dynamics (examples include significant variations in RED MinQ / MaxQ parameters, increases in job execution time to double their current value with reduction in arrival rate, as well as variations in the number of schedulers [10-100 and 15-300 respectively]).

In particular, for both sets of results shown here (and for all results more generally), key significant features include:

- The average job turnaround time drops dramatically – to nearly half of the values seen without RED. On closer analysis, the longest waiting jobs under both R-RED and SQ-RED are processed much more quickly than under R and SQ, bringing a very large reduction in the standard deviation in job execution time.

- RED's effect is to smooth off the peaks of oscillations as they build and avoid jobs regularly forming unnecessarily long queues at resources which due to delayed load information appear to be attractive to all schedulers.
- This positive result is seen despite the fact that results include extra costs incurred by some jobs as they are resubmitted.

The results (the examples shown and more generally) also show that SQ performs only marginally better than R without RED and performance is essentially identical when using RED. While this appears to be counter-intuitive (since SQ would appear to be the "smarter" scheduling strategy), this effect reflects a cancelling out of the positive effects of queue selection by the negative impact of oscillations brought on by unintended synchronisation among job submissions by SQ. The synchronisations and peaks seen under R appear to be a clear indication of the presence of the types of dynamics typically predicted for balls and bins problems [12], in which random assignments in fact often cause large differences (clustering) in the resources items are assigned to.

The conditions shown here are particularly susceptible to oscillations since processor usage information is somewhat time delayed (10 time units) meaning that the actions of other schedulers are hidden for a short amount of time to other schedulers and actions take time to have an effect (hence a job sent at time t, does not become visible in the queue until t+10). While these are arguably not extreme parameters:

- Schedulers all poll at the same rate – giving them the same processor picture at each decision step.
- Schedulers all use the same assignment policy (shortest queue).

To some extent these factors therefore accentuate the possibility of the emergence of oscillations. However, it is also clear that even if some conditions are softened oscillations would still be expected:

- Modelling different polling rates for information updates for different schedulers mean less synchronized world models. However, these would still ultimately be (time-delayed) models of the same reality – hence ultimately such scenarios also show synchronisation.
- While schedulers may use different heuristics for scheduling, shortest queue is the rational individual choice – making it unlikely that an individual scheduler would choose another policy.
- The impact of RED on the random case is illustrative, since even here the extra early rejection factor manages to cut out the build up of long queues.

4.4 Dynamic RED

In order to address some of the limitations seen in static RED, further experiments were carried out in which the thresholds which govern RED behaviour were adapted dynamically to system load. In particular, Dynamic RED thresholds are calculated in the following way:

$$MaxQ = (Queue\ Length - Current\ Queue) \cdot \alpha$$
$$MinQ = MaxQ \cdot \alpha \tag{1}$$

where, as before, *MaxQ* is the maximum threshold, *MinQ* is the minimum threshold, α is a value between 0 and 1, *Queue Length* the queue maximum capacity and *Current Queue* the current queue occupancy. Depending on system-load, the administrator can set the α value in order to adjust/tune system performance. However, in all experiments reported we set α to 0.2. The effect of the changes to RED is to create a mechanism which tightens and loosens control on a resource based on load. On the basis of this mechanism two more scheduling cases are studied:

- *Random with Dynamic RED (R-DRED):* which consists of applying Dynamic RED to the jobs when scheduled using the random strategy.
- *Shortest Queue with Dynamic RED (SQ-DRED):* applying Dynamic RED to the jobs when scheduled using the SQ strategy.

Results for experiments using the same configurations as used in the previous section are shown in Fig. 3.

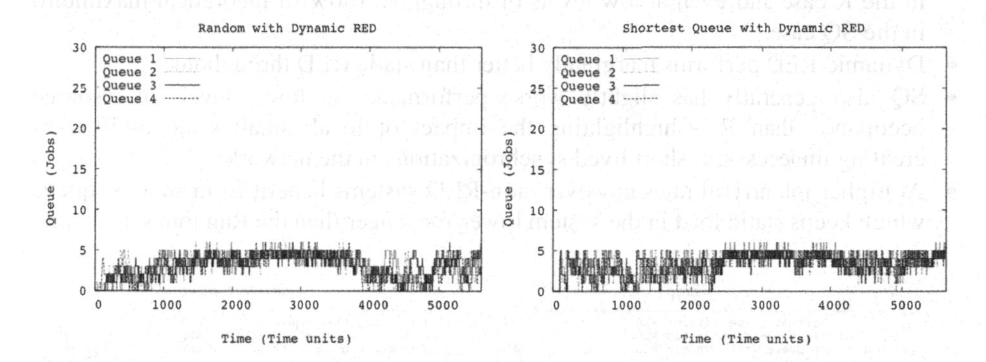

Fig. 3. Showing scheduling cases random with dynamic RED (left) and shortest queue with Dynamic RED (right). Results show a further improvement on the static RED case – further reducing round trip times and standard deviations.

Using Dynamic RED, results improve further (though not dramatically since the major step shown even with static RED is the cut through the majority of delays caused by oscillations):

- For R-DRED, the run shown of 5338 jobs has an average turnaround time of 105.71 time units and a standard deviation of 37.93 – a significant advance on R-RED.
- For SQ-DRED, the run shown of 5486 jobs has an average 112.99 time units and a standard deviation of 34.05, representing a small but significant improvement for SQ (which can be seen consistently over other runs with similar parameters).

5 Comparative Analysis of Results

In order to illustrate the difference in operation between the techniques in question, Fig. 4 compares the average number of jobs per resource currently waiting in the system under gradually increasing job arrival rates. The graph for each scheduler type and resource policy (no RED, static RED and dynamic RED) hence shows the point at which a particular system becomes saturated by latent jobs in the system. Since jobs could be queued at a resource (the maximum buffer length at each is 30) and at the scheduler (if rejected by a resource), the graph sums both figures to create an average.

Load is generated according to the same exponential distribution used in the previous experiments but the number of jobs per second is gradually increased from 0.5 jobs per time unit (50% of system theoretical maximum throughput) through 1 job per time unit (100% of theoretical maximum throughput) at 10000 time units to 1.1 jobs per time unit (110% of capacity) after 12000 time units. The results of this comparison show:

- A clear difference between RED and non-RED strategies at high levels of throughput (from 5000 time units / job arrival rate at 75% of theoretical maximum) in the R case and even at low levels of throughput (50% of theoretical maximum) in the SQ case.
- Dynamic RED performs marginally better than static RED throughout.
- SQ also generally has slightly worse performance at lower levels of resource occupancy than R – highlighting the impact of local, small scale oscillations creating unnecessary, short lived synchronizations in the network.
- At higher job arrival rates however, non-RED systems benefit from shortest queue which keeps static load in the system lower for longer than the Random scheduler.

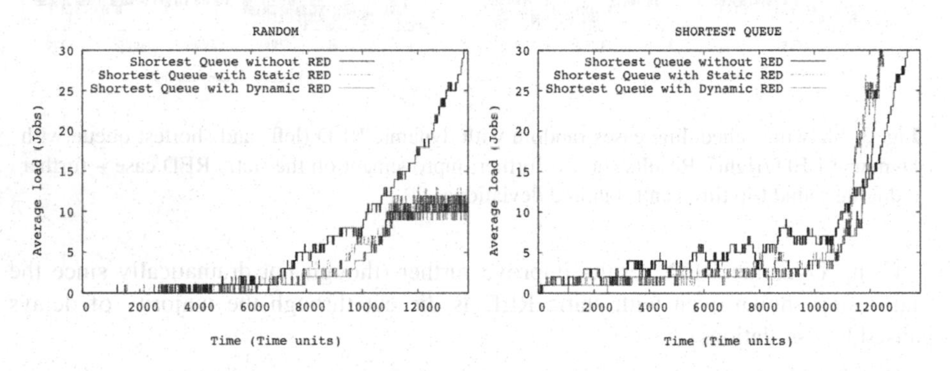

Fig. 4. Comparative behavior under increasing system load for Random (left) and Shortest Queue scheduling. With both R and SQ approaches, RED approaches keep residual jobs in the system significantly lower for longer – keeping more jobs flowing through the system.

In terms of concrete figures for these simulation runs, random without RED showed an average turnaround time of 101.5092 time units, with standard deviation of 80.8737. With static RED, this drooped to an average turnaround of 82.3545 time units and a standard deviation of 45.6551. With dynamic RED, figures drop to

72.1976 time units and 36.5258 respectively. Shortest Queue results follow the same pattern, with non-RED average turnaround time at 153.4731 time units (deviation 88.1687), static RED dropping this to 109.7881 time units and 53.6473 and dynamic RED to average turnaround time of 90.2004 time units and standard deviation of 48.0186.

6 Conclusions

Whilst production Grid environments with many multiple independent scheduling processes may still be way time off, they are likely to grow significantly in importance. In this paper, we argue that such environments may suffer from similar oscillation problems to IP network traffic management scenarios and that, further, approaches based on Random Early Detection policies may be a strong candidate for mitigating some of these pathologies. While results are preliminary, they arguably show promise:

- The oscillations seen are potentially an underlying phenomenon which will have a significant impact on the design and use of multi-scheduler systems in large Grids.
- RED shows significant promise for tackling such oscillations and further, if validated, has a number of attractive properties for deployment in Grids including no need for centralised management, deployment at individual processor sites only and a wide range of flexibility on the precise drop strategy.
- IP network scheduling and routing techniques more generally may hence be a *significant source of inspiration* for scheduling challenges in next generation Grids.

Future work includes testing additional simulation parameters in order to further validate results and exploring new directions that the link between Grid resource scheduling and TCP/IP network management opens up. A longer version of this paper is available in technical report form [13].

References

1. IBM, Load Leveler; User's Guide, IBM (1993)
2. Thain, D., Tannenbaum, T., Livny, M.: Distributed Computing in Practice: The Condor Experience. In: Concurrency and Computation, Practice and Experience (2004)
3. Berman, F., et al.: The GrADS Project: Software Support for High-Level Grid Application Development. International Journal of High Performance Computing Applications 15(4), 327–344 (2001)
4. Floyd, S., Jacobson, V.: Random Early Detection gateways for Congestion Avoidance. ACM Transactions on Networking 1(4), 397–413 (1993)
5. Braden, B., et al.: Recommendations on Queue Management and Congestion Avoidance in the Internet. IETF RFC 2309 (1998), http://www.ietf.org/rfc/rfc2309.txt
6. Kelly, F., Maulloo, A., Tan, D.: Rate control for communication networks: Shadow prices, proportional fairness and stability. Journal of the Operational Research Society 49, 237–252 (1998)

7. Low, S., Lapsley, D.: Optimization flow control–I: basic algorithm and convergence. IEEE/ACM Transactions on Networking 7, 861–874 (1999)
8. Postel, J., (ed.): Transmission Control Protocol (TCP), IETF RFC 793 (1981), http://www.ibiblio.org/pub/docs/rfc/rfc793.txt
9. Partridge, C.: Isochronous Applications Do Not Require Jitter-Controlled Networks. IETF RFC 1257 (1991), http://tools.ietf.org/html/rfc1257
10. Hollot, C.V., Misra, V., Towsley, D., Gong, W.: A Control Theoretic Analysis of RED. In: Proceedings of INFOCOM, pp. 1510–1519 (2001)
11. MacDougall, M.: Simulating Computer Systems, Techniques and Tools. MIT Press, Cambridge (1992)
12. Raab, M., Steger, A.: Balls into Bins - A Simple and Tight Analysis. In: Rolim, J.D.P., Serna, M.J., Luby, M. (eds.) RANDOM 1998. LNCS, vol. 1518, pp. 159–170. Springer, Heidelberg (1998)
13. Willmott, S., Brugnoli, M., Heymann, E., Hurley, P., Senar, M.A.: Applying Internet Random Early Detection Strategies to Scheduling in Grid Environments, Technical Report LSI-07-34-R. Universitat Politècnica de Catalunya, Dep. Llenguatges i Sistemes Informàtics. Spain (2007)

A Consistent Checkpointing-Recovery Protocol for Minimal Number of Nodes in Mobile Computing System

Chandreyee Chowdhury and Sarmistha Neogy

Department of Computer Science and Engineering,
Jadavpur University, India
sarmisthaneogy@yahoo.com

Abstract. The vast computing potential of mobile computing systems is often hampered by their susceptibility to transient and independent failures. To add reliability and high availability to such systems, checkpoint based rollback recovery is one of the widely used ones for scientific computing, database, telecommunication and mission critical applications. This paper presents a coordinated nonblocking checkpointing and recovery technique for such systems that handles the constraints posed by the underlying wireless network, efficiently. Here an initiator (an MSS) sends checkpoint requests to all other MSSs and the MSSs send this request only to those MHs, which have communicated in the last checkpointing interval (relieving the wireless network from synchronization overhead). Also all acknowledged messages are logged at the home station of the receiver MH so that only the faulty MHs need to recover in case of failure and no other process is affected by this fault and subsequent recovery.

Keywords: Mobile computing system, Checkpointing, Recovery, consistency, message logging.

1 Introduction

A mobile computing system consists of both Mobile Hosts (MH) and static Mobile Support Stations nodes (MSS). A set of dynamic and wireless communication links can be established between an MH and an MSS, and a set of high-speed communication link is assumed between the MSSs. An MSS may communicate with a number of MHs but an MH at a time communicates with only one MSS. An MH communicates with the rest of the system via the MSS it is connected to. The links in the static network may support FIFO message communication [14]. Moreover, as long as an MH is connected to an MSS, the channel between them ensures FIFO communication in both directions. Message transmission through these links takes an unpredictable but finite amount of time. Reliable message delivery is assumed during normal operation that is there is no message loss or modification.

Distributed computation in mobile computing environment is performed by a set of processes executing concurrently on MHs and MSSs in the network. The processes communicate asynchronously with each other. A process experiences a sequence of state transitions, called *event,* during its execution. The event having no interaction

S. Aluru et al. (Eds.): HiPC 2007, LNCS 4873, pp. 599–611, 2007.

with another process is called an *internal event*; the message sending and receipt are *external events. Computation* is a sequence of state transitions within a process and the messages generated by the running application are called *computation* message.

2 Related Works

Acharya et al. [1] were the first to present an asynchronous uncoordinated snapshot collection algorithm for distributed applications on mobile computing systems where an MH takes a local checkpoint depending on communication pattern. Prakash and Singhal's algorithm [13] tried to combine the orthogonal views of nonintrusiveness and efficiency. This synchronous algorithm uses transitive dependency to determine the minimal set of nodes for checkpointing. Sometimes this algorithm gives inconsistent results. Cao and Singhal [7] removed inconsistency and proved that the two views of nonintrusiveness and efficiency can never be combined. But their algorithm may sometimes flood the network with messages and return messages because of the online dependency tracking. In [8] Cao and Singhal proposed a nonblocking algorithm using mutable checkpoints to avoid storage overhead. Here weights are used as in [13] to indicate completion of the checkpointing task though there is discrepancy regarding initialization of the weight.

Gass and Gupta [12] in their blocking algorithm take communication induced, local and forced checkpoints to be stored in volatile storage but only one of them (globally consistent one) is made permanent. But longer algorithm invocation period may exhaust local memory with multiple inconsistent checkpoints. Recovery is slower as failure information may not reach all fault free processes within finite time.

For fast recovery the checkpoints and logs of an MH are moved as the MH performs handoff between cells in [5] increasing failure free communication overhead. This is improved in [3] by keeping recovery information in the home of that MH. But if the MH is far from its home, transfer cost increases. Park et. al in [15] proposed an algorithm for efficient recovery based on pessimistic message logging that keeps a distance vector or a handoff frequency counter to decide when to shift the checkpoint between MSSs such that the path to be traced during recovery remains manageable. But the MH may roam around a particular area causing to shift the checkpoint unnecessarily. Badrinath et al in [2] proposed an algorithm based on two-tier principle for cellular networks to reduce computation work on the MH and the overall message transfer.

It is improved by Byun et al. in [9]. Here coordinated checkpointing is used between the MSSs which record the state of the MHs. The messages directed towards an MH are also logged in the MSS to which the MH is connected. The communication cost is higher when the number of MHs is small. The algorithm does not clearly state the location management policy used to detect the foreign agent of an MH by a home agent.

In [16] fixed stations take checkpoints synchronously but an MH takes (inconsistent) checkpoint independent of the other MHs. Due to the inherent nature of communication, each message is broadcasted and hence the MSSs can easily get the message and log it so that during recovery a state beyond the latest checkpoint can be reached. But how the failure information is communicated to the checkpoint coordinator is not clear from the protocol. To take advantages of both optimistic and pessimistic logging schemes, family based logging scheme has been introduced in [17]. Here the sender (parent) partially logs the message before sending. The message becomes relevant when the receiver (child) depending on the received message, sends a message to another process (grand child) which logs the piggybacked receive sequence number of the message (the child received). Now if the child fails, during recovery, the parent and the grandchild cooperate to replay the received messages, making the performance of this protocol depend on the communication pattern. In [18] causal logging is used along with checkpointing to solve this problem. Here, each MH takes checkpoint independently and the entire state information is sent to the local mobility agent for storage deleting the previous checkpoint.

3 System Model and Assumptions

The MSSs are assumed to be fault-tolerant because it is quiet feasible to apply hardware fault tolerance techniques like hotswap at the MSSs. An MH can communicate with the rest of the system via the MSS it is connected to, which may be referred to as the *home* station (HS) of that MH. If an MH moves to the cell of another base station, wireless channel to the old MSS is disconnected and a wireless channel in the new MSS is allocated. The state of the MH at the time of disconnection is available from the old MSS.

There is no shared memory or common clock among the nodes and communication and synchronization between the nodes is via message-passing only. Mobile IP is used as the underlying protocol for message transmission. Hence an MH communicates all messages via its HS. During disconnection interval only local events take place at MH. However, it is assumed that checkpointing requests as well as computation messages from other MHs may be queued at the old MSS during this disconnection interval. All these issues are needed to be addressed during checkpointing [4]. The time interval between two consecutive checkpoints is *checkpointing interval.* Fail-stop model of communication is assumed.

4 The Checkpointing-Recovery Scheme

The scheme proposes that a checkpoint initiator (each MSS takes turn in acting as initiator) sends checkpointing requests from time to time to all MSSs only. After receiving checkpointing request, an MSS finds out whether the MHs to which it is the HS needs to take checkpoint or not. Each MSS also maintains an account of the communication activities in the current checkpointing interval of the concerned MHs. This helps in determining whether an MH would take a checkpoint or not in the present initiation. Hence an MSS forwards the checkpointing request only to those MHs (to which it is HS) if it finds that those particular MHs were active during the current checkpointing interval. Hence only a few selective MHs are able to take checkpoints after the checkpointing request reaches them. Moreover, all MSSs also take checkpoints at every initiation. This decision by MSSs saves considerable amount of computation at each MH and hence conserves energy. The checkpointing overhead in terms of request messages is also minimized thereby saving network bandwidth.

Messages received are logged in the stable storage of the corresponding HS such that during recovery only the failed process needs to restart its computation from its last saved checkpoint while other processes can execute computation without any interruption. Only unacknowledged messages are saved in the HS of the sender MH. These two logs are sufficient for maintaining consistency during recovery. The recovering once-faulty process informs other processes (only the MSSs via its HS) that it is recovering. This is required in case there is any message that had been sent to it but remained unacknowledged at the HS of the sender MH and hence logged in its buffer. Upon getting the recovery message the HS of the sender MH will resend the message from its log. The sender MH continues its execution and it also does not have to interfere since it never receives any such recovery message. Thus overhead during recovery is also kept at a minimum.

4.1 Example Scenario

Let us describe the situation with an example of five MHs. (figure 1). At the end of each checkpointing interval all MSSs receive checkpointing request from initiator MSS. Let us assume that MSS_1 is acting as the HS for MHs P_0, P_3 and P_4 and MSS_2 for the MHs P_1 and P_2. When MSS_1 receives checkpointing request it first finds that MHs P_0, P_3 and P_4 need to take their (k+1)th, (j+1)th and (x+1)th checkpoints respectively since P_0 has received m_2 and sent m_1, P_3 has received m_3 and m_4 and P_4 has sent m_4 in their respective last checkpointing intervals. Hence MSS_1 sends checkpointing requests to P_0, P_3 and P_4, which upon receiving such request duly take their checkpoints. Then MSS_1 takes its own checkpoint. MSS_2 after receiving checkpointing request finds that only P_1 needs to take its (h+1) th checkpoint since it has sent m_2 and m_3 in its last checkpointing interval. But P_2 does not need to take any checkpoint since it has not communicated since its last checkpoint CP_2^i.

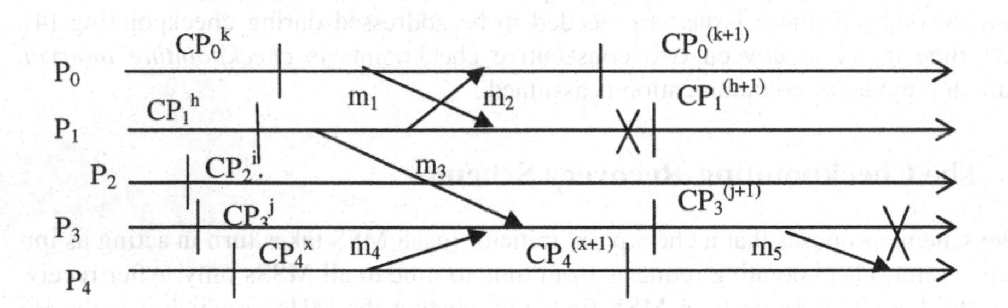

Fig. 1. A mobile computing system consisting of 5 MHs

As in [14], the MHs and MSSs are not blocked during the checkpointing activity. Each MSS logs unacknowledged messages (of the sender MHs for whom it is the HS) in its checkpoint besides its own state. Also, positively acknowledged received messages of these MHs are saved for recovery. Thus unacknowledged messages are logged at the sender end and acknowledged messages are saved at the receiver end. During recovery the received messages may be replayed in order relieving sender MHs from any interruption. Moreover, unacknowledged messages may be replayed from the log to prevent the occurrence of orphan messages. Now if P_3 fails and the HS of P_3 detects that failure it sends the message logs to P_3 and recovers from its last checkpoint $CP_3^{(j+1)}$ and no other processes need to recover unlike [14].

For example, in figure 1 above, when CP_0^{k+1} is created, CP_0^k is deleted and information related to m_1 and m_2 are transferred to the old log. Logs related to the current interval (k+1, k+2) are then maintained. So, even if P_1 fails at the "X" (as shown in the timeline of P_1) it will resume from CP_1^h and its HS (MSS_2) will correctly replay m_1 and m_2 when necessary. P_0 will discard m_2 and P_3 will discard m_3 treating these as duplicate messages by looking into the old logs. In fact, m_2 and m_3 can be discarded at the respective HS.

Thus it is evident that only the faulty process needs to recover to maintain consistency since all information regarding the checkpointing interval in which the process failed is with the HS of that faulty process. Hence reconstruction of the scenario from the faulty process' latest checkpoint will be based on locally available information.

5 Checkpointing Algorithms

5.1 Required Data Structures

Mess_Record[]: This vector of dimension k is maintained by an MSS for k MHs to whom it is HS. It is initialized to zero at the beginning of a checkpointing interval. If MH_j sends/receives computation messages then Mess_Record[j] is set to 1.
Ack_Record[][]: This matrix has k as one of the dimensions and number of messages as the other dimension. It is initialized to zero at the beginning of a checkpointing interval. If MH_i sends m-th computation message to MH_j, then Ack_Record[i][m] is set to 1 till the corresponding acknowledgement arrives.

snt-mess-buff: A buffer that stores unacknowledged messages of the current checkpointing interval with sender id, destination id, sequence number and content.
receive-mess-buff: Positively acknowledged received messages of the current checkpoint interval are stored with sender id, destination id, sequence number and content.
old_snt-mess-buff: A buffer that stores unacknowledged messages of the last checkpointing interval with sender id, destination id, sequence number and content.
old_receive-mess-buff: Positively acknowledged received messages of the last checkpoint interval are stored with sender id, destination id, sequence number and content.
csn_k: Contains the checkpoint sequence number of MH_k that its HS expects.

5.2 Algorithmic Details

Algorithm 1. *//Algorithm for taking checkpoint in Mss*
1. Receive checkpoint request message from initiator
//finds out whether each of the p MHs that are connected to it has communicated
2. For all MH_k $(0 < k < p)$
 2.1 Checks entries of $Mess\_Record_k[]$
 2.2 If any $Mess\_Record_k[]$ is not zero
  2.2.1Send checkpoint request to MH_k, update checkpoint sequence number, csn_k
  2.2.2 While Ack_Record[k][] is not exhausted //checks if ack has come

2.2.2.1 If there is 1 in Ack_Record[k][] //writes unacknowledged message in log
2.2.2.1.1 MSS keeps a copy of the message in snt-mess-buff of MH_k
2.2.3 Transfers contents of receive-mess-buff and snt_mess_buff of MH_k respectively to old_ receive-mess-buff and old_snt_mess_buff
3. MSS takes a checkpoint

Algorithm 2. *// Algorithm for Taking Checkpoint in Mh P_i*
1. Receive checkpoint request message from the HS
2. P_i takes a checkpoint

Algorithm 3. *// Algorithm for receiving computational messages in Mh*
1. Receive message from sender (via HS)
2. Send acknowledgement

Algorithm 4. *// Algorithm for receiving computational messages in Mss*
1. Receive message (on behalf of MH connected to this MSS) from sender
2. If it is not duplicate (by checking the message sequence no with the sequence numbers of messages in the receive_mess_buff)
2.1 Forward it to appropriate MH
2.2 Save it in receive-mess-buff of the corresponding MH
3. Else if it is duplicate
3.1 Discard

Algorithm 5. *// Algorithm for recovery of Mh_k after failure at Mh_k*
1. Load latest checkpoint and send its sequence number to the HS
2. Receive the logged computation messages from MSS and replay them in order.

Algorithm 6. *// Algorithm for recovery of Mh_k after failure at Mss*
1. If the received csn matches with latest csn_k then,
1.1 Transfer current content
2. Else
2.1 Transfer old content

6 Brief Proof of Correctness

Theorem 1. *The checkpointing protocol ensures a consistent recoverable state in the mobile environment.*

Proof: We prove this theorem by contradiction. Let the checkpointing protocol be inconsistent. Then during recovery either of the following cases should occur:

Case 1: There is a missing message m that is recorded 'sent' in sender's checkpoint CP_s^k but not recorded 'received' in the respective receiver's checkpoint CP_r^h or
Case 2: There is an orphan message m as m is recorded 'received' in receiver's checkpoint CP_r^k but not recorded 'sent' in the respective sender's checkpoint CP_s^h.

A brief outline of proof is given below. A message is drawn as an arrow from sender to the receiver of the message. The acknowledgement is shown as a dotted line with an arrow from receiver of the message to the sender as shown below (figure 2).

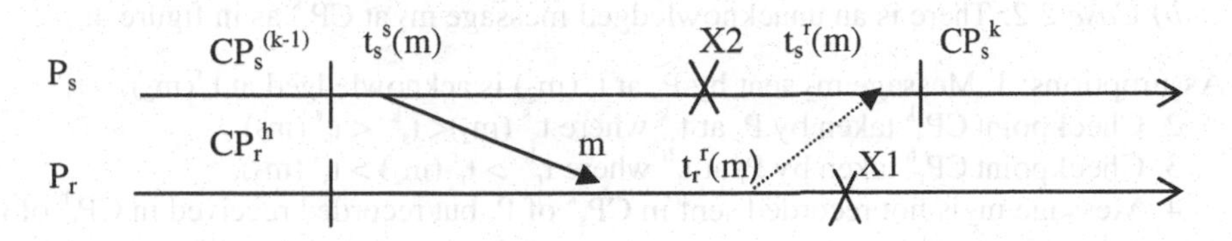

Fig. 2. Message *m* recorded 'sent' (Case 1)

Case 1: Let us consider message m in figure 2. Since m is recorded sent, therefore acknowledgement of m has reached P_s before CP_s^k.

Assumptions: 1. Message m sent by P_s at $t_s^s(m)$ is acknowledged at $t_s^r(m)$,
2. Checkpoint CP_s^k taken by P_s at t_s^k where $t_s^k > t_s^r(m) > t_s^s(m)$,
3. Checkpoint CP_r^h taken by P_r at t_r^h where $t_r^h < t_r^r(m)$,
4. Message m is recorded sent in CP_s^k of P_s but not recorded received in CP_r^h of P_r.

Since acknowledgement is sent by step 2 of Algorithm 3, when P_s takes new checkpoint CP_s^k, m is recorded sent and is logged in the receive_mess_buff of P_r.If failure occurs at X_1, P_r resumes from CP_r^h with m being replayed. If failure occurs at X_2, P_s resumes from $CP_s^{(k-1)}$ with m being sent again but discarded at P_r by steps 3 and 3.1 of Algorithm 4. So even if a message is recorded sent but not received in the corresponding pair of checkpoints, the system is in consistent state.

Case 2: Let us consider message m_1 in figure 3. Since m_1 is not recorded sent, either of the following has occurred: *a) Case 2.1:* m_1 is sent after CP_s^h.
Assumptions: 1. Message m_1 sent by P_s at $t_s^s(m_1)$ is acknowledged at $t_s^r(m_1)$,
2. Checkpoint CP_s^h taken by P_s at t_s^h where $t_s^h < t_s^s(m1) < t_s^r(m1)$,
3. Checkpoint CP_r^k taken by P_r at t_r^k where $t_s^s(m1) < t_r^r(m1) < t_r^k$,
4. Message m_1 is not recorded sent in CP_s^h of P_s but recorded received in CP_r^k of P_r.

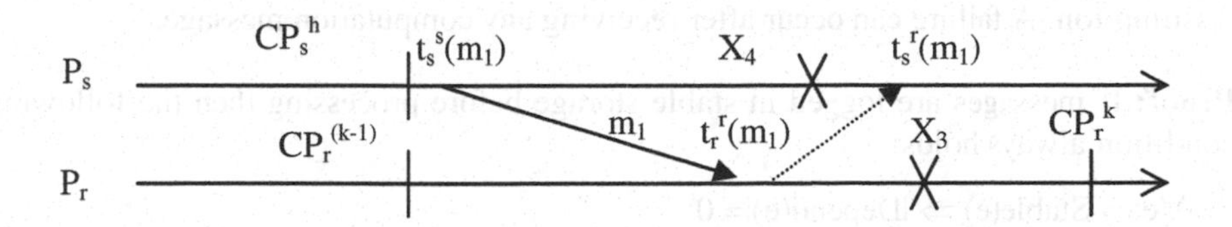

Fig. 3. Message m1 not recorded 'sent' (Case 2.1)

m_1 is recorded in the receive_mess_buff of P_r. If failure occurs at X_3, P_r resumes from $CP_r^{(k-1)}$ with m_1 being replayed from its receive_mess_buff.

If failure occurs at X_4, P_s resumes from CP_s^h with m_1 being sent once again but discarded at P_r by steps 3 and 3.1 of Algorithm 4.

Thus even though there is a message that is not recorded sent but recorded received in the corresponding pair of checkpoints, the system is in consistent state.

b) Case 2.2: There is an unacknowledged message m_2 at CP_s^k as in figure 4.

Assumptions: 1. Message m_2 sent by P_s at $t_s^s(m_2)$ is acknowledged at $t_s^r(m_2)$,
2. Checkpoint CP_s^k taken by P_s at t_s^k where $t_s^s(m_2) < t_s^k < t_s^r(m_2)$,
3. Checkpoint CP_r^h taken by P_r at t_r^h where $t_r^h > t_r^r(m_2) > t_s^s(m_2)$,
4. Message m_2 is not recorded sent in CP_s^k of P_s but recorded received in CP_r^h of P_r.

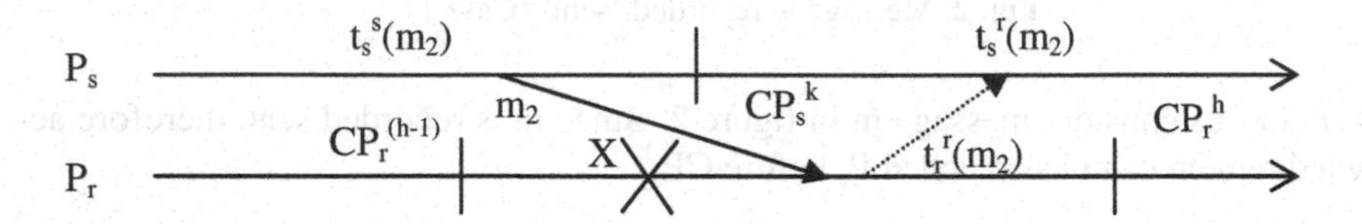

Fig. 4. Message m2 not recorded 'sent' (Case 2.2)

Since acknowledgement of m_2 is due at t_s^k, by steps 2.2.2.1 - 2.2.2.1.1 of Algorithm 1 the MSS keeps m_2 in *snt-mess-buff.* If failure occurs at X in figure 4, during recovery P_r resumes from $CP_r^{(h-1)}$. P_r will issue a message to all the MSSs via the HS that it is recovering. The MSSs thereby will check their own *old_snt_mess_buff* and *snt_mess_buff* and send if there is any unacknowledged message that was sent to this recovering process. There will be no problem with the messages received by P_r since these will be replayed from its own *receive_mess_buff.* Since P_s will not recover it may not directly send m_2 once again and that will be played from the log by its HS. It may be noted here that had P_r failed after $t_r^r(m_2)$, m_2 may have been replayed from its own *receive_mess_buff* and any duplicate m_2 would have been discarded.

Thus even though there is a message that is not recorded sent but recorded received in the corresponding pair of checkpoints, the system is in consistent state.

Lemma 1. *If a message is logged in stable storage before processing it, then during recovery only the process that has failed needs to recover.*

Assumption: A failure can occur after receiving any computation message.

Proof: If messages are logged in stable storage before processing then the following condition always holds:

$\forall e: \neg \text{Stable}(e) \Rightarrow |\text{Depend}(e)| = 0$

This property stipulates that if an event has not been logged on stable storage, then no process can depend on it.

Here, e: receipt of a message by process p.

Depend(e): the set of processes that are affected by event e. This set consists of p, and any process whose state depends on the event e according to Lamport's happened before relation [10].

Log(e): the set of processes that have logged a copy of e's determinant in their volatile memory.

Stable(e): a predicate that is true if e's determinant is logged on stable storage.

Let us consider figure 5. During failure-free operation the logs of processes P_0, P_1 and P_2 contain the determinants needed to replay messages $\{m_0, m_4, m_7\}$, $\{m_1, m_3, m_6\}$ and $\{m_2, m_5\}$ respectively. Let us suppose processes P_1 and P_2 fail as shown, restart from checkpoints B and C, and roll forward using their determinant logs to deliver again the same sequence of messages as in the pre-failure execution. This guarantees that P_1 and P_2 will repeat exactly their pre-failure execution and re-send the same messages. Hence, once recovery is complete, both processes will be consistent with the state of P_0 that includes the receipt of message m_7 from P_1.

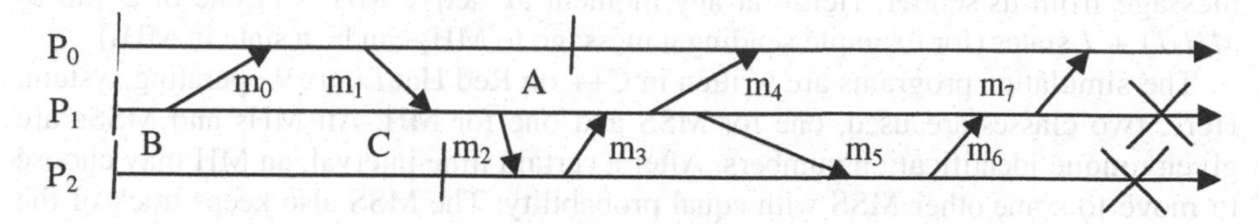

Fig. 5. Message logging scheme used in the protocol

Hence the effects of a failure are confined only to the failed processes. Functioning processes continue to operate and never become orphans.

Theorem 2. *Upon failure only the failed process needs to restart from its most recent checkpoint in order to guarantee consistency.*

The protocol logs every message received, at the receiver's HS by steps of 1-3 of Algorithm 4. Thus this theorem directly follows from lemma 1.

7 Salient Features of the Proposed Scheme

- The problem of concurrent initiation does not arise as the MSSs take turn to act as the initiator. Hence initiator never becomes a bottleneck. The coordinated checkpointing overhead is also minimized since the present scheme is not like the two-phase commit protocol. Hence bandwidth, power both is conserved.
- Only the faulty process needs to recover leaving the others unaffected.
- This scheme conserves energy and bandwidth since not all MHs need to take checkpoint. Also this decision is taken by the HS of the MHs thereby relieving the MHs from executing an algorithm saving the battery power.
- Memory constraint of mobile nodes is considered here and the entire message logs required by the protocol are kept at the stable storage of the HS. This does not incur any extra overhead since the underlying network protocol (mobile IP) ensures that all communication is usually done via the HS.

8 Comparison of Checkpointing Techniques

Simulation experiments were carried out to study the behavior of checkpointing protocols described in [9], [12] and the technique proposed in this paper. We have chosen a very general application platform where a process can undergo any of the following three states: send, receive and compute. These states are uniformly distributed. In compute mode, a process can execute local computations. The computation time follows exponential distribution and is given by: *–log (1-(random number between 1 to no of MH)/no of MH)/6*. In the send mode a process sends computation messages to any other process in the system and in receive mode it receives messages (if any) from other processes. States are generated in such a way that for every send operation there is a corresponding blocking receive. That is the receiver waits until it has got a message from its sender. Hence at any moment an active MH is in one of *2*(no of MH-1) + 1* states (for example sending a message to MH_2 can be a state in MH_3).

The simulation programs are written in C++ on Red Hat Linux 9 operating system. Here, two classes are used, one for MSS and one for MH. All MHs and MSSs are given unique identification numbers. After a certain time interval, an MH may choose to move to some other MSS with equal probability. The MSS also keeps track of the MHs, which are currently directly connected to this MSS. The simulation is run for N processes (running on MHs) where N is an integer.

8.1 Basis of Comparison

The algorithm presented here is compared with the two-tier approach [9] and the protocol by Gass and Gupta [12]. These two schemes are the representatives of two classes of checkpointing schemes - the first one ([9]) is based on the coordinated checkpointing approach and the next one ([12]) is based on communication induced checkpointing approach.

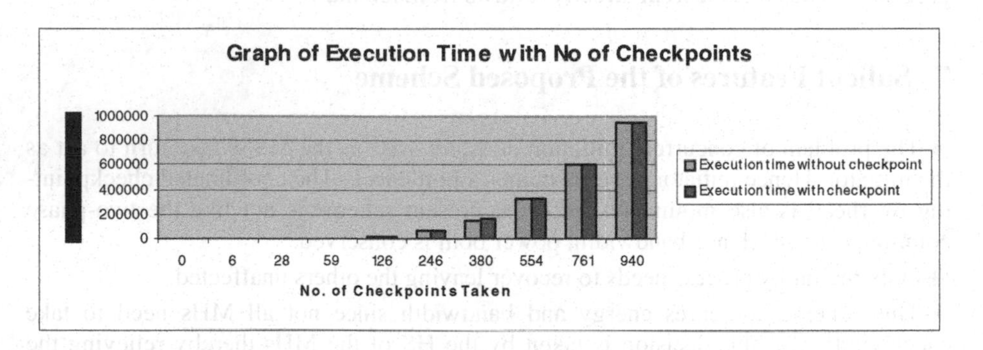

Fig. 6. Graph showing checkpoint overhead

The checkpoint overhead is calculated to be the difference between the execution times with checkpoints and without checkpoints. Hence figure 6 show that our algorithm has very low checkpoint overhead. Now, the checkpoint overhead of our algorithm is compared with the approaches presented in [9] and [12]. No of states of execution is taken in the x-axis. Each of these states can be local computation, sending a message to some other MH or receiving a message from another MH. Results of the simulation are shown in Table 1. Figure 7 shows the plot of the results.

Table 1. Results of simulation of techniques discussed in [9], [12] and in this paper

	Two Tier[9]		GasnGupta [12]		OurAlgorithm	
No of states	No. of msgs	Checkpoint overhead	No. of msgs	Checkpoint overhead	No. of msgs	Checkpoint overhead
10	8	15	7	40	9	0
20	17	95	14	80	19	30
50	40	325	35	265	42	140
100	85	655	82	635	92	295
200	177	1385	173	1375	186	629.99
400	369	2569.99	357	2759.99	390	1230.02
600	561	3874.91	545	4174.97	596	1899.97
900	832	5770.03	813	6225.16	875	2770
1200	1112	7710.06	1083	8255.16	1198	3805
1500	1388	9640.07	1349	10310.24	1497	4700

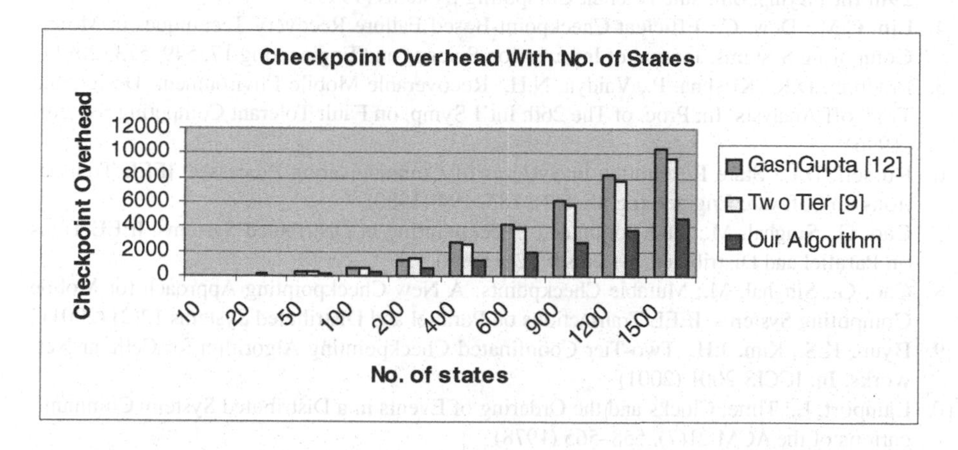

Fig. 7. Graph showing checkpoint overhead against no of states

9 Conclusion

In the proposed nonblocking protocol all synchronization activities are handled by the MSSs, which are connected by high speed cable links freeing the wireless network from such overheads. As all the MHs may not take checkpoints always, battery power is saved. It must be mentioned here that during disconnection interval of an MH, messages meant for it (computation, checkpoint-request message) are queued with the HS MSS till the new MSS is known. The location of MSSs and routing of messages in the network are assumed to follow standard algorithms in practice.

Since the simplicity of pessimistic logging makes it attractive for practical applications this concept is used here. The failure free performance overhead is negligible as the received message logs are kept at the MSSs which are assumed to have large stable storage and are connected by high speed cable links. The unacknowledged message logs are maintained at respective MSSs and hence the MHs have to save only the process states thus taking care of the limited storage of the MHs.

A detailed analysis of the possible case studies (section 6) shows that the proposed scheme always maintains a consistent global system state. The simulation results show that the performance is much better than the two other schemes. The present work also handles the challenges of mobile computing systems in an adept manner.

References

1. Acharya, A., Badrinath, B.R.: Checkpointing Distributed Applications on Mobile Computers. In: Proc. Third Int'l Conf. Parallel and Distributed Information Systems (September 1994)
2. Badrinath, B.R., Acharya, A., Imielinski, T.: Designing Distributed Algorithms for Mobile Computing Networks. Computer Communications 19(4) (1996)
3. Yao, B., Ssu, K., Fuchs, W.K.: Message Logging in Mobile Computing. In: Proc. of the 29th Int'l Symp. on Fault Tolerant Computing Systems (1999)
4. Lin, C.M., Dow, C.: Efficient Checkpoint-Based Failure Recovery Techniques in Mobile Computing Systems. Journal of Information Science and Engineering 17, 549–573 (2001)
5. Pradhan, D.K., Krishna, P., Vaidya, N.H.: Recoverable Mobile Environment: Design and Tradeoff Analysis. In: Proc. of The 26th Int'l Symp. on Fault Tolerant Computing Systems (1996)
6. Russell, D.L.: State Restoration in Systems of Communicating Processes. IEEE Transactions on Software Engineering SE-6(2), 183–194 (1980)
7. Cao, G., Singhal, M.: On Coordinated Checkpointing in Distributed Systems. IEEE Trans on Parallel and Distributed Systems 9(12) (1998)
8. Cao, G., Singhal, M.: Mutable Checkpoints: A New Checkpointing Approach for Mobile Computing Systems. IEEE Transactions on Parallel and Distributed Systems 12(2) (2001)
9. Byun, K.S., Kim, J.H.: Two-Tier Coordinated Checkpointing Algorithm for Cellular Networks. In: ICCIS 2001 (2001)
10. Lamport, L.: Time, Clocks and the Ordering of Events in a Distributed System. Communications of the ACM 21(7), 558–565 (1978)
11. Krishna, P., Vaidya, N.H., Pradhan, D.K.: Recovery in Distributed Mobile Environments. In: IEEE Workshop on Advances in Parallel and Distributed Systems (1993)
12. Gass, R.C., Gupta, B.: An Efficient Checkpointing Scheme for Mobile Computing Systems. In: Proceedings of ISCA 13th international conference on Computer applications in industry and engineering, Honolulu, USA, pp. 323–328 (November 2000)
13. Prakash, R., Singhal, M.: Low-Cost Checkpointing and Failure Recovery in Mobile Computing Systems. IEEE Transactions on Parallel and Distributed Systems 7(10) (1996)
14. Neogy, S.: A Checkpointing Protocol for a Minimum set of Processes in Mobile Computing Systems. In: IASTED PDCS 2004. Proceedings of the International Conference on Parallel and Distributed Computing Systems, MIT, USA (2004)
15. Park, T., Woo, N., Yeom, H.Y.: An Efficient Recovery Scheme for Mobile Computing Environment. In: Proc. of ICPADS 2001, pp. 53–60 (2001)

16. Morita, Y., Higaki, H.: Checkpoint-Recovery for Mobile Computing Systems. In: Proc. of Distributed Computing Systems Workshop, pp. 479–484 (2001)
17. Alvisi, L., Hoppe, B., Marzullo, K.: Nonblocking and Orphan-free Message Logging Protocols. In: Proc.of The Twenty-Third International Symposium on Fault-Tolerant Computing FTCS-23. Digest of Papers, pp. 145–154 (1993)
18. Ahn, J., Min, S., Hwang, C.: A Causal Message Logging Protocol for Mobile Nodes in Mobile Computing System. Future Generation Computer Systems 20(4), 663–686 (2004)

MASD: Mobile Agent Based Service Discovery in Ad Hoc Networks

Neeraj Nehra[1], R.B. Patel[2], and V.K. Bhat[3]

[1] Shri Mata Vaishno Devi University, Katra(J&K),India
[2] M M University, Mullana(Ambala),Haryana,India
[3] Shri Mata Vaishno Devi University, Katra(J&K),India
nehra04@yahoo.co.in, patel_r_b@yahoo.com,
vijaykumarbhat2000@yahoo.com

Abstract. Service discovery is an integral part of Mobile Ad Hoc Networks (MANETs). While several service discovery protocols such as Service Location Protocol [1] and Universal Plug and Play (UPnP)[2] have been developed, most of them are designed for infrastructure based networks and thus not suitable to be used in MANETs. This paper proposes Mobile Agent Based Service Discovery (MASD) for Ad Hoc network. It is a policy driven agent based mechanism that facilitates cross platform service discovery in ad-hoc environments. The various agents are chosen in MASD which executes the predefined policies. Our approach achieves a high degree of flexibility in adapting itself to changes in ad-hoc environments and is aware of common problems associated with structured compound formation in mobile Ad Hoc environments. Mechanism consists of grouping mobile nodes into clusters while a gateway in each cluster is responsible for routing. We evaluated the performance of the scheme by running simulations on Glomosim simulator. Also the motion of agents is taken care of by PMADE (Platform for Mobile Agent Distribution and Execution).The results obtained shows that MASD is quite effective for successful service discovery in MANETs.

Keywords: Service Discovery, Mobile Agent, Cluster, Gateway.

1 Introduction

Mobile Ad Hoc networks (MANETs) is a networks formed by a group of wireless nodes with limited power and transmission range. These networks do not need any existing infrastructure but can form the network on fly. As these networks begin to grow in size, an efficient mechanism is needed to locate the services distributed with them. A service may be a computation, storage, a communication channel to another user, a software, or hardware device needed by another user [3].

There are currently a number of existing protocols for service discovery [2, 4, 5, 6]. These protocols except [2] are centralized, registration oriented with the assumption that a centralized database of services can be maintained and accessed by every node. However these existing strategies do not work well for MANETs, because nodes can

S. Aluru et al. (Eds.): HiPC 2007, LNCS 4873, pp. 612–624, 2007.

join or leave the networks quickly. Also because of the dynamic nature of network, every time a service leaves or join the network it has to inform the centralized server about its presence and this raises scalability issues. Also agent platform used in e commerce [7] have been designed to facilitate flexible service/agent discovery with an agent community. In this type of platforms, an agent that belongs to a certain platform registers itself and its services to some service/agent management/ registration component like a Directory Facilitator (DF) and an Agent Management System (AMS).

Current service discovery protocols designed for MANETs resolve issues of scalability by using directories. Directories manage service advertisements and respond to service queries. Protocols using directories are more scalable since they can easily accommodate an increase in user demand by establishing more directories.

MANET environments present issues that challenge the use of directories. For instance, in most of the cases directories are statically assigned and made known to users beforehand. In other words, service discovery protocols that use directories were designed on the assumption of a static network topology and stable network functions such as multicasting and broadcasting. However, a MANET is a wireless and infrastructure less network that is characterized by dynamic topology changes and multi hop routing. Such infrastructure less networks contradicts the idea of using pre-established directory services. In order to utilize directories in service discovery protocols in MANETs, Kozat et al. [8] pointed out the similarity between directory formation and maintenance, and the idea of clustering used in some MANET network routing protocols. In other words, service discovery protocols can implement directories by using a network layer protocol that already supports clustering and assigns routing responsibility to a node in each cluster called gateway node. Although clustering based routing protocols usually incur high overhead for maintaining the clusters, [8] argues that the efficiency of the whole system nevertheless improves by using directories in the service discovery protocol.

The motivation of this work is that most current service discovery protocols concentrate only on discovering available services. More specifically, they do not solve the problem of how to select a service when several services are available, and they do not provide mechanisms for specifying and performing interaction. Service selection is an important aspect of service discovery especially in a heterogeneous environment containing duplicate services. Therefore, it is important to provide a service selection mechanism. We propose a solution that uses MAs to collect and evaluate information form neighboring nodes and forward this information to the other nodes with the help of other agents without much delay. Additionally, most of the currently developed service discovery protocols support only one method of interaction, whereas the solution in this paper allows service providers to specify interaction methods to access the desired services. Furthermore, this solution provides mechanisms that can utilize to interact with these services with XML file.

Rest of the paper is organized as follows, Section 2 provides an overview of PMADE (Platform for Mobile Agent Distribution and Execution), Section 3 describes MASD architecture and components, Section 4 describes the step wise working of MASD, Section 5 provides the implementation and results obtained, Section 6 discuss the related work, and finally Section 7 concludes the article.

2 Overview of PMADE

Figure 1 shows the basic block diagram of PMADE (Platform for Mobile Agent Distribution and Execution). Each node of the network has an Agent Host (AH), which is responsible for accepting and executing incoming autonomous Java agents and an Agent Submitter (AS) [9], which submits the MA on behalf of the user to the AH. A user, who wants to perform a task, submits the MA designed to perform that task, to the AS on the user system. The AS then tries to establish a connection with the specified AH, where the user already holds an account. If the connection is established, the AS submits the MA to it and then goes offline. The AH examines the nature of the received agent and executes it. On completion of execution, the agent submits its results to the AH, which in turn stores the results until the remote AS retrieves them for the user.

The AH is the key component of PMADE. It consists of the manager modules and the Host Driver. The Host Driver lies at the base of the PMADE architecture and the manager modules reside above it. Details of the managers and their functions are provided in [9]. PMADE provides weak mobility to its agents and allows one-hop, two-hop and multi-hop agents [9]. PMADE has focused on Flexibility, Persistence, Security, Collaboration, and Reliability [10].

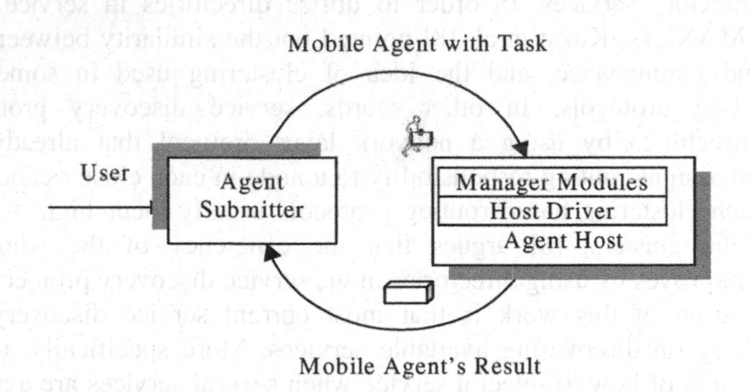

Fig. 1. Block Architecture of PMADE

3 MASD Architecture and Components

The architecture of MASD along with its components is shown in Figure 2. We represent an ad hoc network as a undirected graph $G = (V, E)$, where V is the set of N nodes and E is the set of bidirectional links. We assume that network can be embedded in a two dimensional area A. For two nodes i and j, $dist(i, j)$ is a function that returns the distance between two nodes and $link(i, j)$ is a function that returns true if $dist(i, j) \leq r$, where r is the transmission range of each node and false otherwise. $N(i) = \{k \epsilon V | link(i, k)\}$ is the set of one hop neighbors of node i. For a source node s to successfully broadcast a packet, all nodes in $N(i) - \{s\}$ must not transmit concurrently to prevent collision at i.

3.1 Policy Manager

The Policy Manager is responsible for administration and enforcement of policies chosen for MASD. On initialization, policies are registered with the Policy Manager.

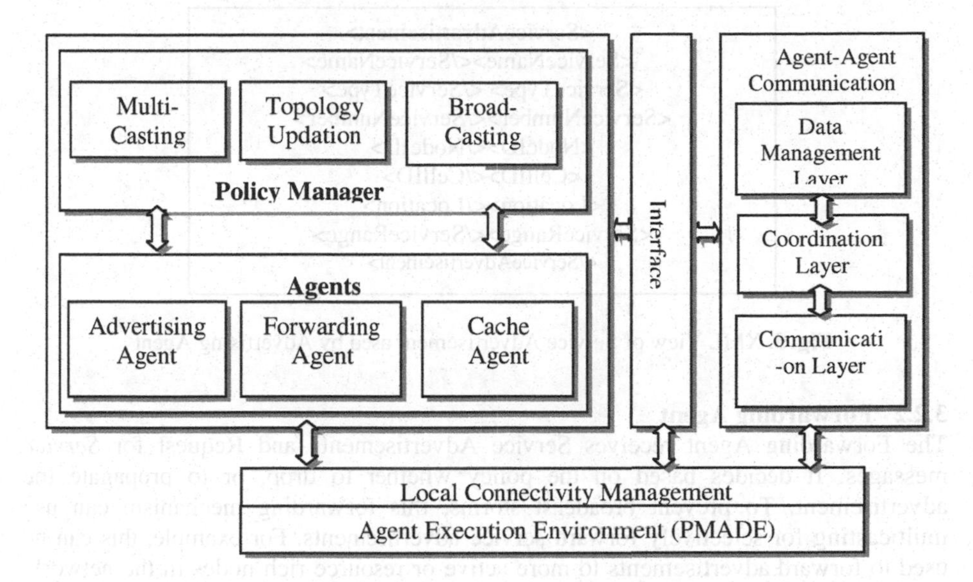

Fig. 2. System Architecture of MASD

The policy can also be modified during runtime and since it is referenced by agents, the changes in policy dynamically propagate to these agents. Agents running on the platform would coordinate all their activities through the Policy Manager. For sake of simplicity we have chosen three different types of policies for MASD namely multicasting, topology updation, broadcasting

3.2 Agents

The following agents are chosen as a part of architecture described above.

3.2.1 Advertising Agent

The Advertising Agent actively broadcasts service descriptions already registered. The Policy Manager controls the rate of advertisements. Various policies can be employed to adjust the rate of advertisement. For example, if the network is fairly static, the advertisement rate can be slowed down. Also policy could be event driven, events here being the arrival or departure of nodes in cluster. Advertisements can also be assigned different priorities.

Figure 3 shows an example of a service advertisement. Service description is represented by Service Key Words that categorize the service that is offered. This description is used to match service queries to the advertisements. Service Range is the number of cells radiuses that allows a service provider to control how far its advertisement reaches. Directories use the Cell ID and Service Range to determine if they are still within the service range. Directories drop the advertisement packet if it is outside the indicated service range.

<ServiceAdvertisement>
<ServiceName></ServiceName>
<ServiceType> </ServiceType>
<ServiceNumber></ServiceNumber>
<NodeID></NodeID>
<CellID></CellID>
<Location></Location>
<ServiceRange></ServiceRange>
</ServiceAdvertisement>

Fig. 3. XML View of Service Advertisement used by Advertising Agent

3.2.2 Forwarding Agent

The Forwarding Agent receives Service Advertisements and Request for *Service* messages. It decides based on the policy whether to drop, or to propagate the advertisement. To prevent broadcast storms, this forwarding mechanism can use multicasting for selectively forward service advertisements. For example, this can be used to forward advertisements to more active or resource rich nodes in the network. To avoid a problem of duplicate messages flooding the network, the Forwarding Agent uses sequence number based mechanism

3.2.3 Cache Agent

The Cache Agent is responsible for handling remote advertisements, storing remote advertisements of services, handling requests to match services present in the cache. The Forwarding Agent, on receiving an advertisement might also decide to forward it to other nodes or broadcast the advertisement to all other nodes. Each advertisement contains a lifetime. When a new advertisement is received by a Cache Agent, the Cache Agent decides to either accept it or reject it. An advertisement is accepted only when there is sufficient space in the cache to hold this advertisement or when an old advertisement can be removed from the cache based on policy chosen.

4 Working of MASD

Once the components namely agents and policies are defined in the architecture, our goal is to make service registry, service advertisement, caching, discovery, service query and reply using above defined MAs. Our methodology consists of dividing the networks into cluster and gateway nodes. In clustering based approach, mobile nodes are grouped together and a node in each group is selected as the gateway to handle routing tasks. These gateways establish connection with each other to form a virtual backbone. All packets in a clustering based approach are routed by the gateways through the virtual backbone. There are three steps that are required to form clusters and to establish the virtual backbone. In the first step, nodes are organized in groups, usually based on distance. Then a node in each group is assigned as the gateway by some election process. In the last step, the virtual backbone is formed by establishing communication between these gateways. The gateways are responsible for forwarding incoming packets to other gateways, if the packets are destined to other networks.

The function of such a gateway is similar to directories used in service discovery protocols [7]. That means a node functioning as a gateway can also handle the task of a directory. Having gateways also function as directories imposes only a small amount of overhead. The proposed solution divides the whole networks into clusters and gateways, while performing routing through the gateway as shown in Figure 4.

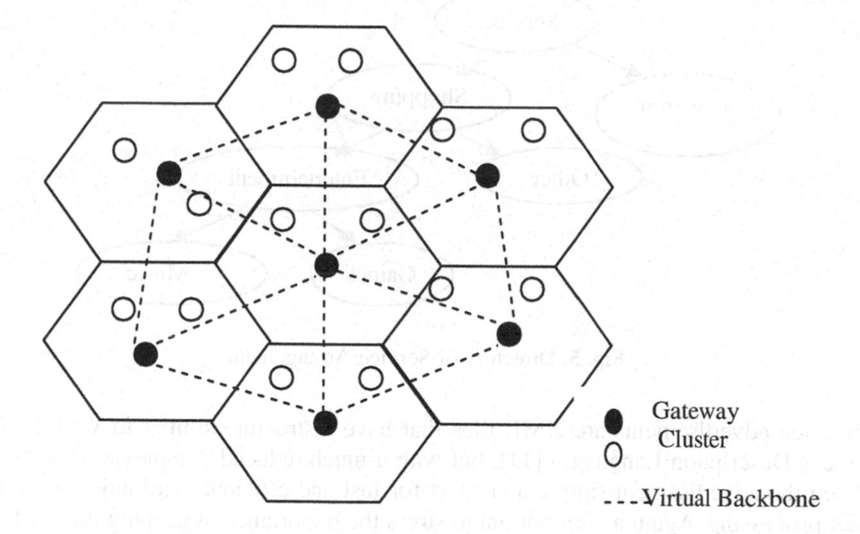

Fig. 4. Formation of Cluster and Gateway in MASD

4.1 Service Registry

Service registry is a structure that enables nodes to store their local services. It also allows them to maintain information about services that they might have discovered or received via advertisements. The MASD maintains a tree based structure as registry. It allows the tree paths to be directly used in service query and advertisement messages for service scoping or classification.

The tree has a number of levels that represent service classification. As we move down the tree from root to the leaves, services become more specific. The service tree is useful to nodes when acting as client as well as server. This tree is used by a server to register local services that it wishes to offer, to advertise the registered services at any level, i.e., all, generic or specific, and to respond to client discovery requests. MA uses the tree to discover all, generic, or specific services, and manage these services. Use of inclusive or exclusive filtering options at different levels of the tree makes it easy for users to manage services of interest Each service is associated with a lease time, that is, the time for which the service is expected to remain available. This time is specified as time to live (TTL), which is part of service registration or advertisement information. Services should be refreshed before their TTL expires. Otherwise, they are removed from the registry. This scheme makes the discovery system robust against unexpected failures.

4.2 Service Advertisement

Service advertisement is a XML file and is advertised by advertising agent whose main responsibility is to describe a service's capabilities. When a node inside a domain decides to provide a service, it forwards the service advertisement to advertising agent.

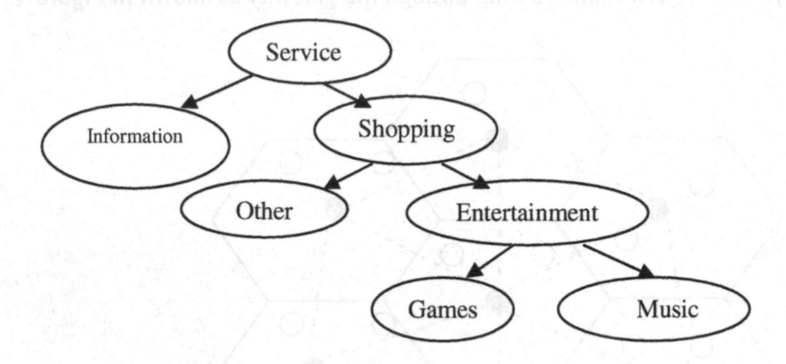

Fig. 5. Directory of Service Arrangement

Service advertisements are XML files that have a structure similar to WSDL (Web Services Description Language) [11], but with a much reduced complexity. The goal is to keep the advertisement simple and short for fast and efficient distribution as well as easier processing. Again it is important to stress the importance of keeping the burden of directories low by using MA since this kind of node is also handling the task of gateways.

4.3 Caching

Each mobile node that is connected to other heterogeneous nodes in the network maintains a cache to store service descriptions of remote services. The cache is controlled by Cache Agent. Each Advertisement Agent in network, after a finite time interval broadcasts a list of its local services to other peer nodes around it. To do so, it uses the Forwarding Agent. The Forwarding Agent receives all messages from the network, then based on the local policy it decides whether to process the message or not. The Forwarding Agent passes the advertisement to the Cache Agent. A Cache Agent in a remote node on receiving the advertisement decides based on its policy whether to cache it or not.

4.4 Service Discovery

Each node upon receiving a request message can chose to drop the message or process it. The local policy is used to make this decision. If the node chooses to process the request message, it checks its database to see if it is hosting the desired service. If the service is local then the service description is sent to the requesting node. If the node is not hosting the needed service, it could check the local cache. If the cache contains information about the service then the MA replies with hint. Otherwise, the request is sent to the Forwarding Agent. The Forwarding Agent decides to forward the request to other nodes, depending on the local policy.

4.5 Service Query and Reply

The service query message is packaged using XML file as shown in Figure 6. The query message contains QueryID, ServiceDescriptionKeyword, QueryType, MinimumAdvertisement, ReplySize, QueryTimeout. QueryID identifies this Service Query message and is used to detect duplicate messages. QueryType tells the directory how to handle this query message. QueryTimeout tells a directory the time before which this type of query has to be replied.

```
<ServiceQuery>
<QueryID>Identification of Query</QueryID>
<ServiceDescriptionKeywords>Description of Service
</ ServiceDescriptionKeywords >
<QueryType>Type of Query< /QueryType>
<QueryTimeout>Time To Live Query</QueryTimeout>
</ServiceQuery>
```

Fig. 6. XML View of Service Query and Reply

Upon receiving a Service Query message from within its own domain, the directory first checks the Query Type to see how it should handle the query.

The directory replies immediately with a Service Reply containing any advertisement the directory finds in its service directory tree. If there are no sufficient advertisements then the Service Query message is stored in the pending service list, and a timer is set according to the Query Timeout. The directory handling this message then broadcasts a Service Query message to surrounding neighbor directories requesting advertisements of the same category. This Service Query message is further forwarded to other directories for a system wide search using MAs.

4.6 Inter Agent communication

Agents in MASD communicate with different layers in architecture which are defined as follows:

Communication and Coordination Layers: Agents in the system communicate with each other or with users using mobile group approach for coordination of MAs. The request an agent receives from the communication layer should be explained and submitted to the coordination layer, which decides how the agent should act on the request according to its own knowledge. Agents communicate by exchanging messages through reliable communications channels, i.e., transmitted messages are received uncorrupted and in the sequential sent order, as long as the message sender does not crash until the message is received [12]. Let *L* denote the set of all possible locations. Let *P* be the set of all possible agents. A mobile group is denoted by the set of agents $g = \{p_1, p_2, p_3, ..., p_n\}$, $g \subset P$. On a mobile group, five operations are defined:

- J*oin (g)*: issued by an agent, when it wants to join group *g*.
- *Leave (g)*: issued by an agent, when it wants to leave group *g*.
- M*ove (g, l)*: issued when an agent wants to move from its current location-to location *l*.
- *Send (g, m)*: issued by an agent when it wants to multicast a message *m* to the members of group *g*.
- R*eceive (g, m)*: issued by an agent to receive a message *m* multicast from the group *g*.

4.7 Local Connectivity Manager and Data Management Layer

This manager manages the local connectivity of nodes at the management layer. Nodes learn about their neighbors in one of two ways. Whenever a node receives a broadcast from a neighbor, it updates its local connectivity information in its Neighborhood table to ensure that it includes this neighbor. In the event that a node has not sent data packets to any of its active neighbors within a predefined timeout, it broadcasts a hello message to its neighbors, containing its identity and activity. This message is prevented from being rebroadcast outside the neighborhood of the node. Neighbors that receive this packet update their local connectivity information in their Neighborhood tables. In this way local connectivity management is maintained by this local connectivity manager along with data management layer defined in the hierarchy.

5 Implementation and Results Obtained

We implemented MASD using Glomosim [13], which is a component-based simulation environment. In our simulation, wireless IEEE 802.11 is used for the MAC layer of each node. The range of the wireless transmission is set to 250 meters and the bandwidth is 10 Mbits/second. The radius of the hexagon cells is set to 80 meters, so that any node inside of a cell can reach any node in its immediate neighbor cell. Each node is assumed to establish bidirectional communication with all nodes within its range. The size of the network is 1000 meters by 1000 meters, and there are 50 nodes in the network. We used random waypoint mobility pattern that was represent by *P (S)* where *P* is a period of time (in seconds) for which a node paused once it arrives to a new location and *S* is node's speed (in m/s). The direction of the node's movement and the destination is generated randomly. Service requests were generated at regular time intervals.

Figure 7 shows the rate of successful service discovery over the maximum node speed for different numbers of combinations of service providers and users. A successful service discovery involves discovering available services and finally selecting a service before timeout. Results show that the success ratio is higher in the case of a node moving at a slow speed. Furthermore, the ratio drops faster when the speed approaches 8 meters/second. This is because nodes are moving to different cells at a higher rate.

The average delay of service discovery is shown in Figure 8. The Figure shows the major difference in delay between having less and more service providers than the minimum number required before performing service selection. In this simulation, when a directory receives a service request it passes that request to the corresponding agents and sets a timeout for gathering service advertisements. If the receiving directory could discover enough service advertisements, it replies to the requester and resets the timeout. Note that with more service providers in the whole system, on average MA has to visit more service providers before the selection process can determine the one with the highest rating. In addition, with the increase of node speed, routes are less dependable and the delay time increases considerably. Moreover, the rate of increase in average delay jumps considerably when there are more potential service providers to visit.

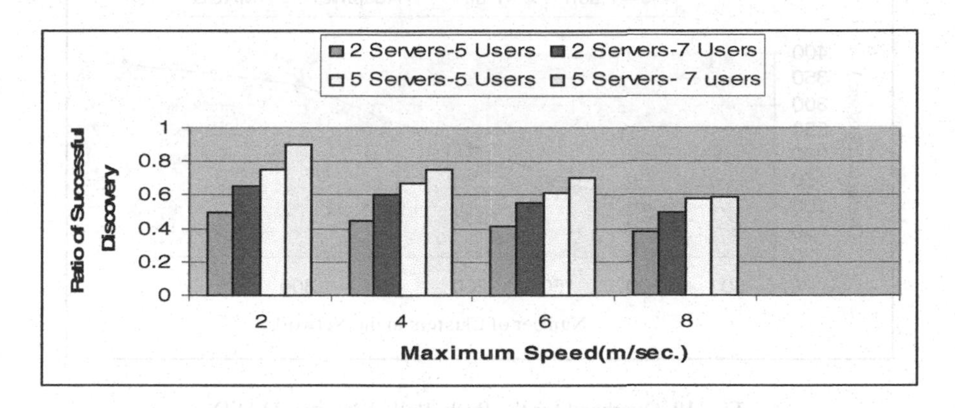

Fig. 7. Successful Service Discovery in MASD

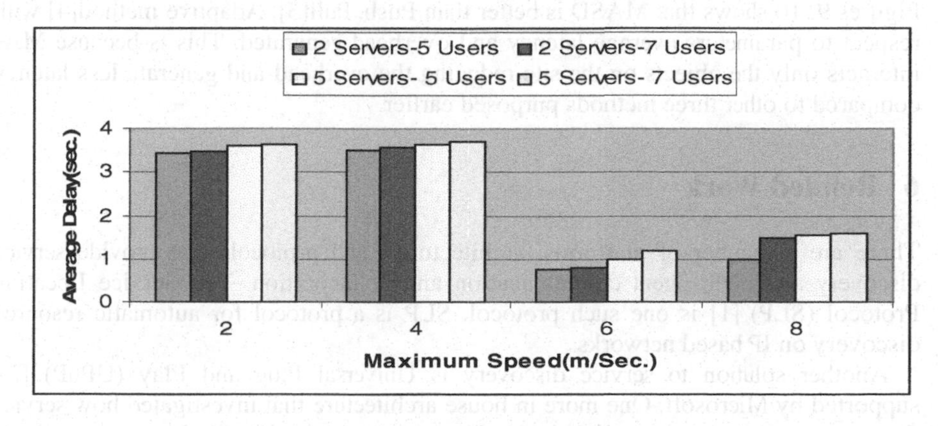

Fig. 8. Delay of Service Discovery in MASD

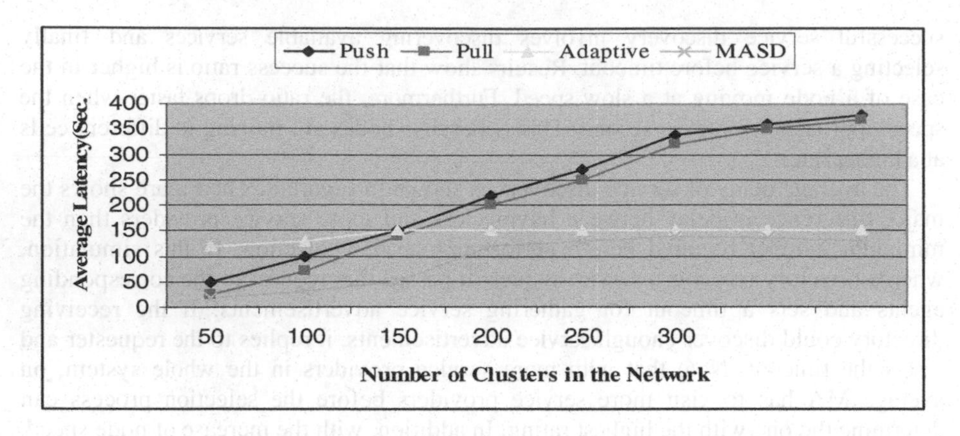

Fig. 9. Average Latency for the Push, Pull, Adaptive, MASD

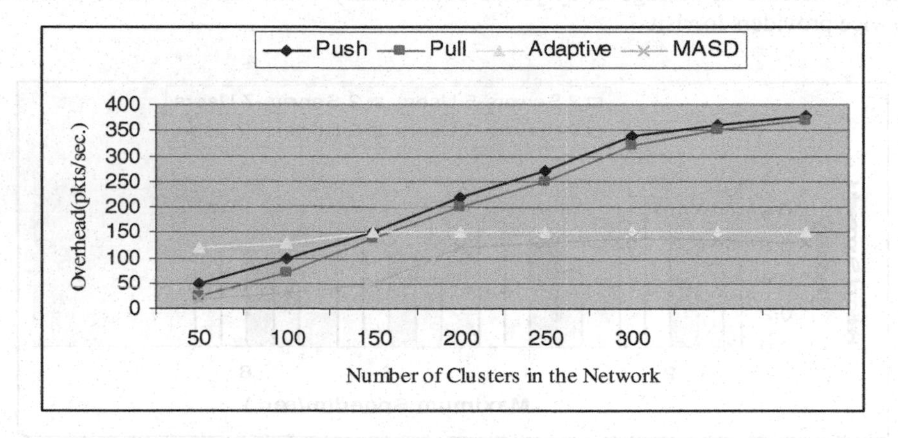

Fig. 10. Overhead for the Push, Pull, Adaptive, MASD

Figures 9, 10 shows that MASD is better than Push, Pull[3], Adaptive method[4] with respect to parameters average latency and overhead generated. This is because MAs interacts only the objects on the site reducing the overhead and generate less latency compared to other three methods purposed earlier.

6 Related Work

There are a number of platforms, architectures, and protocols that provide service discovery and multi agent communication and collaboration. The Service Location Protocol (SLP) [1] is one such protocol. SLP is a protocol for automatic resource discovery on IP based networks.

Another solution to service discovery is Universal Plug and Play (UPnP) [2], supported by Microsoft. One more in house architecture that investigated how service discovery can be taken beyond their simple syntax based service matching approaches is

XReggie [14]. XReggie adds the facilities to describe service functionalities and capabilities using Extended Markup Language (XML). Several concepts have been proposed to resolve the issue of resource discovery. For example, [15] has suggested using the virtual backbone to constantly probe and cache context information to provide a better performance. Belligemine [16] has suggested using a rating system that evaluates context information by placing different weights on different context information. This method has the advantage of allowing users to customize their own ranking system based on their individual requirements. A Service Discovery solution proposed by Cheng [3] uses push and pull methods for advertisement. Extension of Cheng solution is the scalable service discovery solution is given in [4].

There are several FIPA compliant platforms that facilitate service discovery. One such platform is JADE [17]. Another platform is FIPA-OS [18]. The limitations of JADE and FIPA-OS were recognized and LEAP [19] and Micro FIPA OS [20] were developed to extend the functionality and allow mobile devices to participate.

7 Conclusions and Future Work

In this paper, we have presented a peer-to-peer caching based and policy-driven approach for Service discovery in Ad Hoc network. Approach is agent driven and hence, facilitates agent-service discovery. We have proposed MASD which follows a policy-driven approach for advertisements, caching, and request forwarding thus, letting user preferences to be taken into consideration. We use MAs for advertisement, registry, caching, service discovery and query and reply purposes.

We compared *MASD* performance with the performance of push and pull method and adaptive scalable solution. The implementation and the simulation results efficiently demonstrate the flexibility of MASD in Ad Hoc environments. In future, we also plan to carry out comprehensive experiments by changing other policy parameters like advertisement frequency, cache size and study the effect they have on the MASD.

References

[1] Veizadas, J., Guttaman, E., Perkins, C.: SLP: Service location protocol. Internet engineering task force: RFC 2165 (1997)

[2] Consortium Members of the UPnP forum: Universal Plug and Play Device Architecture. Version 0.91 (2000), http://www.upnp.org

[3] Cheng, L.: Service advertisement and discovery in mobile ad hoc networks. In: workshop on ad hoc communication and collaboration in ubiquitous computing environements, pp. 123–128 (2002)

[4] Mohan, U., Almeroth, K., Belding-Royer, E.M.: Scalable Service Discovery. In: Mobile Ad Hoc networks, IFIP networking 2004, Athens, Greece, pp. 234–240 (2004)

[5] Arnold, K., Wollrath, A., O'Sullina, B.: The Jini Specification, Ma, USA. Adddison - Wesley, London, UK (1999)

[6] Members of the Salutation Consortium: Salutation Consortium Homepage (2000), http://www.saluatation.org

[7] FIPA Agent Management Specification, http://www.fipa.org/specs/fipa00023/

[8] Kozat, U.C., Tassiulas, L.: Network layer support for service discovery in mobile ad hoc networks. In: Proc. of IEEE/INFOCOM-2003, San Francisco, pp. 250–257 (April 2003)
[9] Patel, R.B., Garg, K.: PMADE - A Platform for mobile agent Distribution & Execution. In: Proceedings of 5th World Multi Conference on Systemics, Cybernetics and Informatics (SCI2001) and 7th International Conference on Information System Analysis and Synthesis (ISAS 2001), Orlando, Florida, USA, July 22-25, vol. 4, pp. 287–293 (2001)
[10] Patel, R.B., Garg, K.: A new paradigm for mobile agent computing. WSEAS Trans. Computers 3, 57–64 (2004)
[11] W3C Consortium, WSDL (2001), http://www.w3.org/TR/wsdl
[12] Raimundo, J., Macêdo, A., Assis Silva, F.M.: The mobile groups approach for the coordination of mobile agents. J. Parallel Distributed Computing 65, 275–288 (2005)
[13] Global Mobile Information Systems Simulation Library, GLOMOSIM, http://pcl.cs.ucla.edu/projects/glomosim/
[14] Chen, H., Chakraborty, D., Xu, L., Joshi, A., Finin, T.: Service discovery in the future electronic market. In: Working Notes of 17th National Conference on Artificial Intelligence, 11th Innovative Applications of AI Conference, Austin, TX, pp. 178–184 (July 2000)
[15] Johnson, D.B., Maltz, D.A.: The dynamic source routing protocol for mobile ad-hoc networks. In: Imielinski, T., Korth, H. (eds.) Mobile Computing, pp. 153–181. Kluwer, Dordrecht (1996)
[16] Belligemine, F., Rimassa, G.: Jade pa-compliant agent framework. In: Proc. of PAAM 1999, London, pp. 97–108 (1999)
[17] FIPA-OS Website, Emorphia limited (December 27, 2001), http://fipa-os.sourceforge.net
[18] Bergenti, F., Poggi, A.: Leap: A platform for hand held and mobile devices. In: Proc. of ATAL, pp. 156–165 (2001)
[19] University of Helsinki, MicroFIPA-OS small footprint extension to FIPA-OS (2001), available at http://fipa-os.sourceforge.net/
[20] Perkins, C.E., Bhagwat, P.: Highly dynamic Destination-Sequenced Distance-Vector routing (DSDV) for mobile computers. Computer Comm. Rev., 234–244 (October 1994)

Channel Adaptive Real-Time MAC Protocols for a Two-Level Heterogeneous Wireless Network

Kavitha Balasubramanian, G.S. Anil Kumar, and G. Manimaran

Dept. of Electrical and Computer Engineering
Iowa State University, Ames, IA 50011, USA
{kavitha,anil,gmani}@iastate.edu

Abstract. Wireless technology is becoming an attractive mode of communication for real-time applications in typical settings such as in an industrial setup because of the tremendous advantages it is capable of offering. However, the high bit error rate characteristics of wireless channel due to conditions like attenuation, noise, fading and interference seriously impact the timeliness and guarantees that need to be provided for real-time traffic. Existing wireless protocols either do not adapt well to the erroneous channel conditions or do not provide real-time guarantees. The goal of our work is to design and evaluate novel real-time MAC (Medium Access Control) protocols for scheduling messages in a 2-level hierarchical wireless industrial network taking into account the time-varying channel condition. Our objective is to minimize the loss rate of messages using the slot exchange protocol[9] that actively combats the erroneous channel conditions and maximize the channel utilization by enabling parallel transmissions in a collision-free manner. Unfortunately, these two goals have inherent conflicts in shared medium wireless networks. We propose a distributed protocol, called the Adaptive protocol that arbitrates between these two design criteria in order to resolve the inherent conflict between them. Through simulation studies, we show that the proposed Adaptive protocol achieves significant improvement in deadline miss ratio compared to the baseline protocols that exploit complete parallelism and full exchange, for a wide range of channel conditions.

1 Introduction

The term industrial traffic refers to the transfer of messages in applications such industrial automation, process control, communication systems in automobiles etc. Such communication must be performed under stringent hard real-time and reliability constraints since missing a deadline can be disastrous. For guaranteeing this low level of stringent real-time requirements, fieldbuses are able to support time-critical communication between sensors, actuators, programmable logic controllers and operator workstations. These networks are traditionally based on wired technology and a deterministic medium access control. However, the current wired infrastructure is plagued by problems of limited mobility and high deployment and maintenance costs that constrains the viability of any smart real-time system.

S. Aluru et al. (Eds.): HiPC 2007, LNCS 4873, pp. 625–636, 2007.

The wireless evolution offers numerous benefits for industrial applications, where wired solutions have prohibitive problems in terms of cost and feasibility. The growing popularity of wireless communication in different fields including home and office environments has led to its increased dependability, performance improvement and cost reduction. With its widespread standardization, it is very likely in the near future, there will be a proliferation of wireless implementations of factory communication systems. This has motivated the strong research into the use of wireless medium for real-time industrial applications.

In spite of having such clear benefits, wireless technology has its own drawbacks arising due to the unreliable characteristics of the wireless medium which makes it, in its current state, unsuitable for supporting real-time communication for industrial applications. Wireless links are more error prone than their wired counterparts due to noisy channel conditions that vary with time. Occurrences of outages lasting for several seconds, during which no packet transmission are possible over a channel, is not uncommon and there exists a large variability in the distributions of length of error bursts and error free periods[1] making the channel behavior unpredictable. The high error rates over wireless links occur due to different phenomena such as interference, multipath fading, path loss and electromagnetic noise that cause bit errors and packet losses that tend to occur in bursts[2]. Hence, measures to substantially improve transmission reliability overcoming the above mentioned challenges need to be developed so that real-time and reliability requirements demanded by the industrial applications can be guaranteed.

2 Related Work

Providing real-time services over a wireless network requires addressing two different issues. First is to tackle the scheduling issues at the MAC layer to provide real-time guarantees. Second is to combat the time varying and erroneous wireless channel conditions.

In the context of the first issue, several works in literature have explored the applicability of existing standardized MAC protocols such as 802.11 and Bluetooth(BT) or extended the wired real-time protocols such as PROFIBUS for the wireless environment. Works in [3,2,4] propose and analyze the use of master-slave based polling protocols over a wireless network. In [3], the authors investigate the use of IEEE 802.11 for industrial communication by analyzing the possibility of implementing protocols based on master-slave architecture of traditional field buses on a IEEE 802.11 PHY. In [2], the adaptive-intervals MAC protocol has been proposed that uses a polling-based approach combined with group testing feature for improving the delay in low load conditions. However, in [10], the authors show the serious stability issues due to the loss of token frames leading to unsatisfying real-time performance arising from running the existing PROFIBUS MAC and link-layer token-passing master-slave protocols over an 802.11 DSSS PHY. In [11,14], several practical experiments for the performance evaluation of BT have been performed and the authors conclude that channel adaptive error correction protocols will increase the applicability of BT. Hence it follows that even with standardized wireless protocols, mechanisms that address the time-varying channel conditions are necessary to provide real-time guarantees for deadline constrained traffic.

Several channel-adaptive protocols are being proposed for improved reliability over wireless links addressing the second issue. In [5,13], the authors propose modifications to the ARQ protocol and in [1], the author introduce the concept of antenna redundancy that uses multiple antennas for re-transmission. However, the ARQ schemes proposed don't work well at high error rates and Antenna redundancy requires additional hardware in all communicating devices if any-to-any communication needs to be implemented. In [6] and [7], the authors present techniques that make use of the wireless channel conditions while making packet dispatching decisions in a wireless LAN. However, the traffic considered in [6] is best-effort and the protocol in [7] is based on accurate channel estimation that is unfeasible. Most of the existing works on channel adaptive protocols either require additional infrastructure or fail to provide real-time services in the context of the network level scheduling problem.

To summarize, none of the existing works jointly address adapting to the erroneous wireless channel conditions and the real-time scheduling problem which is the main focus of this paper. Specifically, we address the problem of scheduling real-time messages in a 2-level hierarchical shared medium wireless network. *The objective is to minimize the deadline miss ratio of messages.* We achieve this objective by maximizing the channel utilization enabling parallel transmissions and using techniques to overcome the bursty erroneous channel conditions while ensuring that collision-free transmissions are maintained at all times.

3 System Model and Problem Statement

Network Model: We consider an industrial setup where all machines are grouped into cells to form a 2-level hierarchical network based on their functionality and communication range as shown in Figure 1 (a). Smaller machines such as sensors and actuators with short communication range are called the intra-cell nodes and form Level-1 of the network. These intra-cell nodes communicate with other intra-cell nodes within their range using intra-cell messages indicated by a_i. To communicate with other devices out of their range, there exists a controller, also called an inter-cell node, in each cell, which is a more powerful node with a larger communication range than the intra-cell nodes. Message communication between the inter-cell nodes occur by passing inter-cell messages indicated by m_i. Communication between cells only occur through inter-cell messages. The dashed lines indicate periodic messages(m_i and a_i) whose periods are known apriori and the dotted lines indicate aperiodic messages(ap_i) which arrive dynamically at any node. There is a deadline associated with each of these hard real-time messages. Each message is transmitted with the minimum energy required to reach the destination that determines the range of the message. Messages outside of this range can go on in parallel. The source and destination of all messages as well

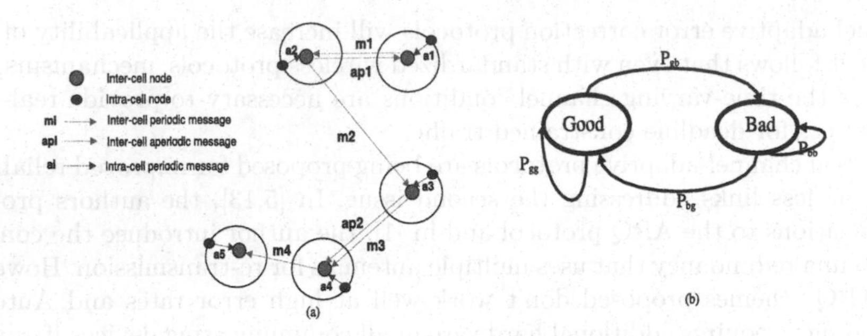

Fig. 1. Network and Channel Model

as the location of the nodes are know beforehand. All messages are assumed to be of the same size.

We consider transmissions to follow a superframe structure comprising of several slots and that repeats itself. We work with slotted time and each message occupies m_s slots in the super-frame equal to its message size. Each slot represents the time in which the source can transmit the smallest atomic unit of a message and receive an implicit acknowledgement(ACK).

Channel Model: Measurements in a wireless industrial environment indicate that the channel causes errors that occur in bursts which can be captured by the Gillbert Elliot model[1] that consists of the good and bad state as indicated in Figure 1 (b). When the channel is in the good state, P_{gg} is the probability that the channel continues to remain in the good state and P_{gb} is the probability with which the channel moves in the bad state. Likewise, when the channel is in the bad state, with a probability P_{bb}, the channel remains in the bad state and P_{bg} is the probability of moving into the good state. We assume every message is transmitted on a channel that that follows its own Gilbert Elliot model and the channel condition changes at the end of every slot. All transmissions that occur in a slot when a channel is in the bad state is considered to have failed that is indicated by the lack of an ACK for that slot.

4 Background Information

4.1 Vertex Coloring

A parallel, conflict-free schedule for a set of messages in a wireless network that maximizes the channel utilization can be obtained using a vertex coloring approach. The parallel schedule is formed by constructing a conflict graph G = (V,E) as shown in Figure 2 (a). In this graph, each vertex represents a transmission(both periodic and aperiodic) and an edge exists between two vertices iff the two transmissions (vertices) cannot be scheduled simultaneously. In our case since every data transmission unit is followed by an ACK, a conflict occurs if either the data transmission from the source or the ACK from the destination of any message conflicts with any other data transmission or ACK. By coloring

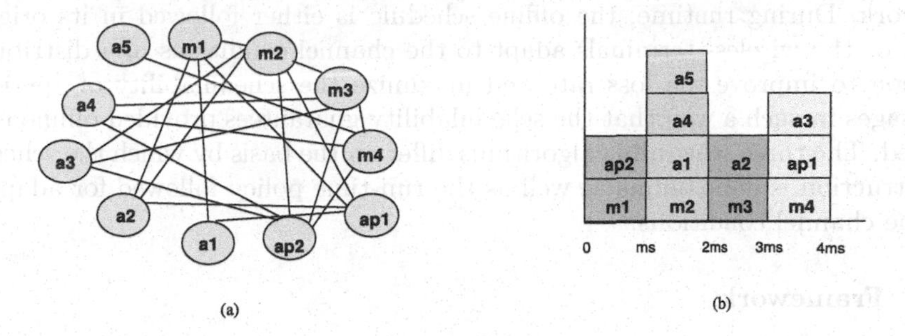

Fig. 2. Conflict graph and parallel schedule with all messages in the system

the vertices of the graph with minimum colors such that no two adjacent nodes connected by an edge have the same color and grouping similar colors together, we can construct a parallel schedule that ensures that all the message complete within the shortest time. Since the problem of finding a minimum coloring for a graph is NP-hard[15], we use a heuristic[8] to construct the parallel collision-free transmission schedule as shown in Figure 2 (b).

4.2 Exchange Protocol and Exchange Setup

The slot exchange protocol[9] aims to improve the success rate of messages meeting their deadline in a wireless network. The main idea is that when a transmission on a channel begins to fail because of erroneous channel conditions, the message exchanges its currently scheduled time slots with another message that uses a different transmission channel. The original message is scheduled during the scheduled time slots of the exchanged message by which time its channel would have moved to the good state. The protocol incurs some control overheads to perform the exchange and these slots are recovered in a distributed conflict-free manner. In this way, the slot exchange protocol dynamically adapts to the channel conditions at run-time for improving the success rate of messages and it preserves the schedulability guarantee given to the messages. In this work, we restrict exchanges to each level i.e. inter-cell messages can exchange with other inter-cell messages and intra-cell messages exchange with other intra-cell messages in the same cell.

In this paper, we extend the exchange scheme to a 2-level heterogeneous hierarchical wireless network as well as maximize the number of messages that can be sent simultaneously in a collision-free manner.

5 Scheduling Algorithms

We propose three scheduling algorithms to schedule periodic and aperiodic messages in a two-level hierarchical wireless network as described in Section 3. Each scheduling algorithm works in two stages. There exists an offline phase where a collision-free schedule is constructed and distributed to all the nodes in the network. During runtime, the offline schedule is either followed in its original form or the wireless terminals adapt to the channel conditions in a distributed fashion to improve the loss rate and maximize the schedulability of aperiodic messages in such a way that the schedulability guarantees provided offline is retained. The three scheduling algorithms differ on the basis by which the schedule construction is done offline as well as the run-time policy followed for adapting to the channel conditions.

5.1 Framework

The basic framework is shown in Figure 3. There exists a control phase, an offline schedule construction phase, schedule transmission phase and a distributed data transmission phase for each of the protocols. In the control phase, all the messages in the system are sent to the central scheduler which performs an admission test that provides the set of messages that needs to be scheduled in every super-frame. The admission test preserves message guarantees by allocating exclusive collision-free time slots for each message. The scheduler then applies an offline schedule construction algorithm to these set of messages based on the protocol being used and produces an offline schedule that is broadcasted to all the nodes in the system. During run-time(online), at each of the nodes, a distributed online-scheduling algorithm works on the offline schedule produced to dynamically adapt to the channel conditions while transmitting the messages. The offline schedule construction algorithm and the online scheduling algorithm varies for each of the three protocols which we explain in detail.

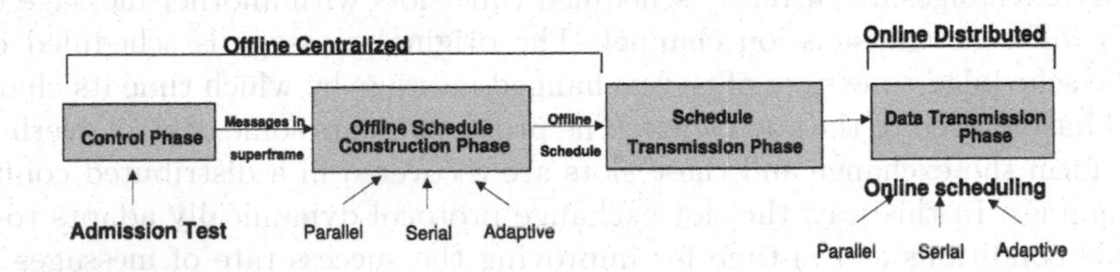

Fig. 3. Framework

5.2 Parallel Protocol

Parallel Offline Schedule Construction Phase: A vertex coloring approach is applied on all the messages in the system to form a parallel schedule. If the superframe is not completely filled with this schedule, a parallel schedule for aperiodic messages is followed for the remaining slots as shown in Figure 4 (a).

Parallel Online Scheduling Phase: During runtime, all the nodes transmit as per the offline schedule. If an aperiodic message is not ready by its scheduled slots, it is transmitted during its scheduled slots in the next superframe.

The above mechanism schedules both periodic and aperiodic messages allowing as many parallel messages as possible to be transmitted simultaneously with

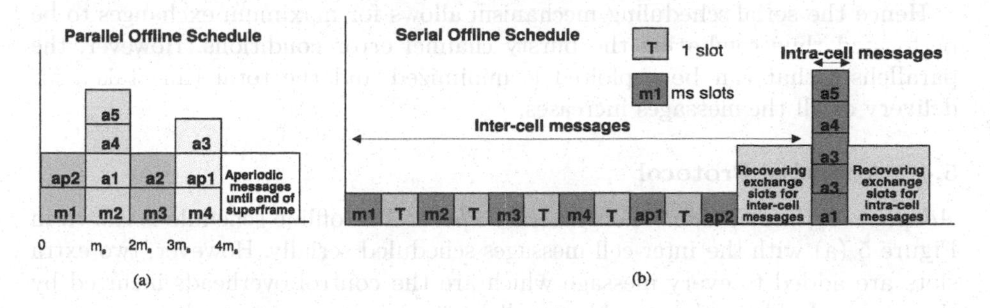

Fig. 4. Parallel and Serial Offline Schedules

the messages completing in the shortest time. However, such a static completely parallel schedule offers no potential for adapting to the channel conditions by performing an exchange since there can be no exchanged message that can co-exists with an existing parallel message transmission without any collissions. Hence, this scheme is not entirely suitable for scheduling messages over a wireless network where the channel conditions vary with time.

5.3 Serial Protocol

Serial Offline Schedule Construction Phase: The offline schedule produces a completely serial schedule for all inter-cell messages that permit all possible combinations exchanges between them as shown in the Figure 4 (b). Intra-cell messages in each cell occur in parallel. The number of exchanges is fixed at 1 for each message and as many slots are allocated at the end of the inter-cell and intra-cell message schedules for each message to transmit data exclusively to make up for the slots that were used as exchange slots for performing an exchange[9]. Since no parallelism between inter-cell message is exploited, the time taken for completing the message transmissions is longer than both the other schemes and hence the superframe length formed by the serial scheduling policy is uses as the basis for the other schemes too.

Serial Online Scheduling Phase: Every node starts transmitting their message in the scheduled time slots as per the offline schedule. However, when any message transmission begins to fail because of erroneous channel conditions, they request an exchange to be performed. A set of slots equal to the number of exchanges, called exchange pool, is reserved at the end of each intra-cell and inter-cell message schedules. Whenever an exchange is performed, an exchange slot is consumed where a data transmission slot is used for initiating an exchange. In order to preserve the schedulability guarantees, each node that performed the exchange exclusively gets a slot for data transmission from the exchange pool[9]. To facilitate this, a transfer slot is needed at the end of each message for keeping track of the number of exchanges that has been performed in the system. A similar approach is followed for intra-cell messages with transmissions and exchanges in each cell occurring in parallel.

Hence the serial scheduling mechanism allows for maximum exchanges to be performed thus combating the bursty channel error conditions. However, the parallelism that can be exploited is minimized and the total time taken for delivery of all the messages increases.

5.4 Adaptive Protocol

Adaptive Offline Schedule Construction Phase: The offline schedule is shown in Figure 5 (a) with the inter-cell messages scheduled serially. However, two extra slots are added to every message which are the control overheads incurred by the protocol. These slots enable parallel transmission of inter-cell messages to occur at run-time if the channel conditions are favorable and exchange is not initiated. At the same time, it gives the capability to perform an exchange during unfavorable channel conditions. In addition to this schedule, the parallel schedule produced by the parallel protocol is also stored for reference which is followed if exchange is not initiated.

Adaptive Online Scheduling Phase: During run-time, all the nodes adapt to the channel conditions in a distributed fashion as shown in Figure 5 (b). The following steps are followed by the nodes in the system during the online phase.

- In the first slot, the source of the scheduled inter-cell message as per the offline schedule transmits an unit of data to the destination.
- In the second slot, the scheduled data transmission continues if the transmission is successful in the first slot. If not, an exchange is initiated.

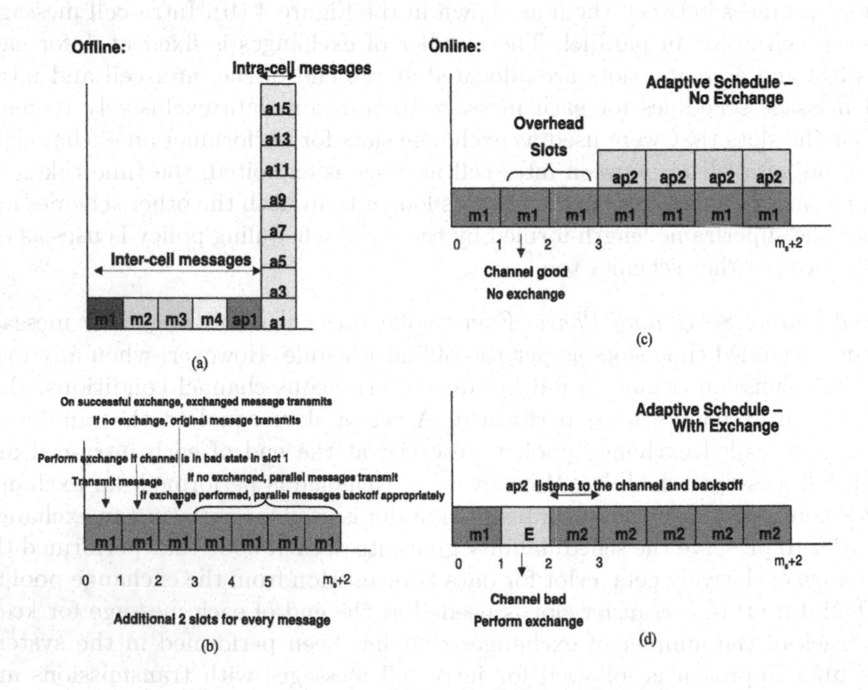

Fig. 5. Adaptive Offline and Online Schedules

- In the third slot, the scheduled data transmission continues through all the scheduled slots until the end of the message transmission as per the offline schedule if no exchange was initiated and the channel was good in the first slot or if an exchange was unsuccessful in the second slot. If an exchange was successful in the second slot, the exchanged message transmits in the third slot.
- Nodes that are scheduled to transmit in parallel with this inter-cell message according to the parallel schedule listen to the channel in the third slot. If they find that no transmission in the vicinity occurs, the start transmitting in parallel with the inter-cell message. If any of the parallel messages observe a different transmission going on within their transmission range, they back-off from transmitting.

Consider the scheduled transmission of m_1 as shown in Figure 5 (c). When the channel is good in the first slot, no exchange is initiated and the parallel message ap_2 according to the schedule in Figure 4 (a) also transmits from the fourth slot. When an exchange is initiated by m_1 with m_4 based on the bad channel state in slot 1, the source of ap_2 hears the transmission of m_4 in the third slot and backs off avoiding collisions. Collision-free transmissions are ensured by the source of the parallel message hearing the exchanged message transmission in the vicinity and backing off or the destination of the parallel message avoiding sending an ACK when it hears the exchanged transmission in the third slot. For intra-cell messages, a decision to exchange with other intra-cell messages in the same cell is done based on the state of the channel in the first slot and transmissions in each cell occur in parallel. If any message has completed its transmission early by transmitting in parallel with other messages, its scheduled slots are used for re-sending any of the failed slots or for sending aperiodic messages between the same source-destination pair. In case of intra-cell messages, aperiodic message to any destination in the cell can be accommodated.

The two control slots used by the protocol are equivalent to the serial protocol that enables one exchange to be performed by each message using a transfer and exchange slot. In the serial protocol, exchange can be initiated at any point in time during the message transmission. Even though the allocation of exchange slot is one for each message, one message can perform several exchanges. However, in the adaptive scheme, only one exchange can be performed by each message and an exchange is performed based on the state of the channel in the first slot. Thus the adaptive protocol removes the drawbacks of the serial protocol that produces a trashing effect while performing multiple exchanges when all the channel conditions are bad that is not fruitful. Also, observing the channel condition in the first slot provides for a good estimation of the channel conditions during the entire message transmission as can be seen from the Simulation studies. In addition, the adaptive protocol used parallel transmissions that increases the number of messages meeting their deadline and provides for better channel utilization.

Hence, the algorithm adopted for the offline schedule construction has an impact on the run-time channel adaptation techniques that can be used. A completely parallel offline schedule yields no exchanges and hence no channel adaptation at run-time and a completely serial schedule permitting exchanges offers very little parallelism during run-time. The adaptive algorithm switches between these two modes dynamically based on the channel conditions, starting off with a pessimistic offline serial schedule and moving towards the parallel schedule at run-time. We show the resultant performance gains by means of our simulation studies.

6 Simulation Studies

We simulate the hierarchical network comprising of 48 nodes randomly distributed in a 100*100 sq.meter region following an uniform distribution. The clustered communication network comprised of 16 cells with 8 periodic inter-cell messages and 2 intra-cell messages per cell. The message size is 10 slots(1500 bits) and the total slots in the superframe is 156. The periodic messages are ready at the beginning of the super-frame and the end of the super-frame is the deadline for all messages. Our objective is to compare the relative performance of the three protocols and observe the success rate of periodic messages for varying channel conditions. Each of the protocol was simulated for 10,000 super-frame runs.

Effect of channel conditions on the success rate of periodic messages: As P_{bb} decreases, the success rate of periodic messages increases due to the

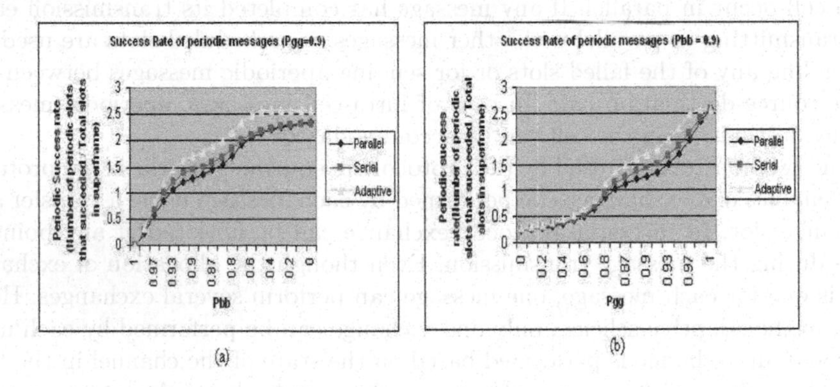

Fig. 6. Success rate of periodic messages with variation in P_{bb} and P_{gg}

decrease in the error burst size as shown in Figure 6 (a). The success rate is greater than 1 for all the 3 schemes because of the parallel transmissions capability exploited by all the 3 schemes (Even in serial, the intra-cell messages are transmitted in parallel). Throughout the range, the adaptive schemes performance is significantly higher since it exploits both parallelism and exchange. The performance gap narrows as P_{bb} approaches 0 since the channel conditions are good and all the transmissions succeed for all the schemes. The success rate

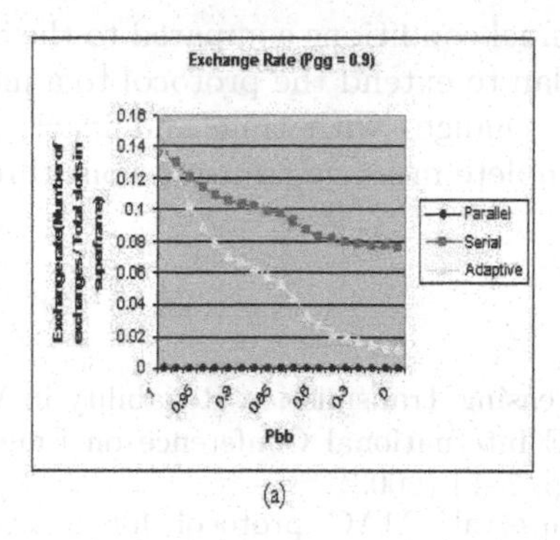

(a)

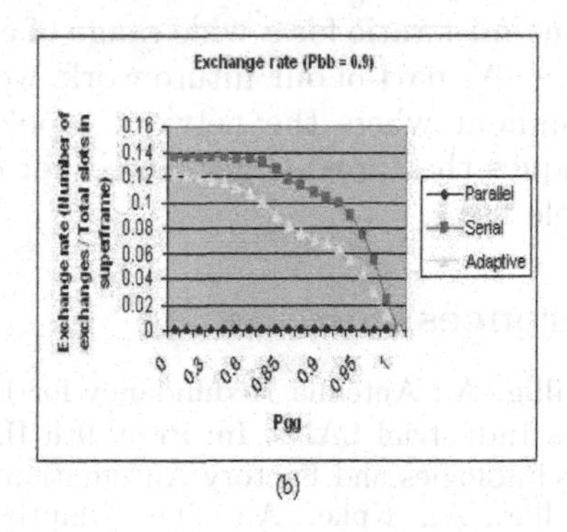

(b)

Fig. 7. Exchange rate of messages with variation in P_{bb} and P_{gg}

of periodic messages with variation in P_{gg} also confirms this trend as shown in Figure 6 (b). The Adaptive scheme gives an average periodic success rate improvement of 24.77% over the Parallel scheme and 16.87% over the serial scheme.

Effect of channel conditions on the exchange rate of messages: As P_{bb} decreases, the channel stays in the bad state for lesser time and since an exchange is initiated only when the channel is in bad state, the exchange rate decreases as shown in Figure 7 (a). No exchanges are performed in the parallel scheme and its exchange rate is 0. When P_{bb} approaches 1, the exchange rate is the same for both the serial and adaptive scheme since in the first few timeslots of every message, an exchange is performed. The adaptive scheme performs exchanges only if the first time slot of a message transmission is noisy while the serial scheme can perform exchange at any point during the message schedule. Hence as P_{bb} decreases, the probability that the first time slot of any message is unsuccessful decreases and hence the exchange rate of the adaptive scheme is lesser than the serial scheme. This is the trend that we want since we do not want too many exchanges happening when the channel occasionally moves to the bad state while for large periods, it remains in the good state (P_{bb} approaching 0). The serial scheme does not cache on this and performs exchanges every time the channel moves to the bad state increasing the overhead. The effect of varying P_{gg} on the exchange rate of messages is shown in Figure 7 (b) which shows a similar trend.

7 Conclusions and Future Work

In this paper, we developed a framework for scheduling real-time messages over a hierarchical industrial network and extended the slot exchange protocol of our previous work to this network setup. We proposed the Adaptive protocol that bridges the tradeoff between maximizing the channel utilization and using the slot exchange protocol to combat the bursty channel error conditions. We performed simulation studies comparing the Adaptive protocol with two other baseline protocols and our results show that it delivers significant reduction in deadline miss ratio for a wide range of channel conditions compared to the other schemes. As part of our future work, we plan to extend the protocol to a mobile environment where the network topology changes with time and incorporate techniques that provide guarantees for complete message transmissions through multiple hops.

References

1. Willig, A.: Antenna Redundancy for Increasing Transmission Reliability in Wireless Industrial LANs. In: Proc. 9th IEEE International Conference on Emerging Technologies and Factory Automation, pp. 7–14 (2003)
2. Willig, A., Kpke, A.: The Adaptive-Intervals MAC protocol for a wireless PROFIBUS. In: Proc. of 2002 IEEE International Symposium on Industrial Electronics, L'Aquila, Italy (July 2002)
3. Miorandi, D., Vitturi, S.: Analysis of Master-Slave Protocols for Real-Time Industrial Communications over IEEE802.11 WLANs. In: Proc. of INDIN 2004 (June 2004)
4. Willig, A.: Polling-based MAC Protocols for Improving Realtime Performance in a Wireless PROFIBUS. In: IEEE Trans. on Industrial Electronics (2003)
5. Willig, A.: Exploiting redundancy concepts to increase transmission reliability in wireless industrial LANS. In: IEEE Trans. on Industrial Electronics (November 2003)
6. Bhagwat, P., Bhattacharya, P., Krishna, A., Tripathi, S.: Enhancing throughput over wireless LANs using Channel State Dependent Packet Scheduling. In: Proc. of the IEEE INFOCOM (March 1996)
7. Willig, A.: A MAC protocol and a scheduling approach as elements of a lower layers architecture in wireless industrial LANs. In: 2nd IEEE Intl. Workshop on Factory Communication Systems, Barcelona (1997)
8. Brelaz, D.: New Methods to Color the Vertices of a Graph. In: Communications of the ACM (April 1979)
9. Kavitha, B., Anil Kumar, G.S., Maniraman, G., Wang, Z.: A Novel Real-Time MAC Protocol Exploiting Spatial and Temporal Channel Diversity in Wireless Industrial Networks. In: Intl. Conference on High Performance Computing (December 2006)
10. Willig, A.: Analysis of the PROFIBUS token passing protocol over wireless links. In: Proc. of the 2002 IEEE Intl. Symposium on Industrial Electronics (June 2002)
11. Bilstrup, U., Wiberg, P.: Bluetooth in industrial environment. In: Proceedings of the IEEE International Workshop on Factory Communication Systems (June 2000)
12. Uhlemann, E., Aulin, T., Rasmussen, T., Wiberg, L.K.: Concatenated hybrid ARQ - a flexible scheme for wireless real-time communication. In: Proc. of the Eighth IEEE Symposium on Real-Time and Embedded Technology and Applications (February 2002)
13. Vedral, A., Wollert, J.F.: Analysis of error and time behavior of the IEEE 802.15.4 phy-layer in an industrial environment. In: IEEE International Workshop on Factory Communication Systems (June 2006)
14. Lee, J.S.: An experiment on performance study of IEEE 802.15.4 wireless networks. In: 10th IEEE Conference on Emerging Technologies and Factory Automation (September 2005)
15. Garey, M.R., Johnson, D.S.: Computers and Intractability - A Guide to the Theory of NP-Completeness. W.H. Freeman and Company, New York (1979)

Modeling Hierarchical Mobile Agent Security Protocol Using CP Nets

Nimesh Desai[1], Kumkum Garg[1], Manoj Misra[1], and Veeravalli Bharadwaj[2]

[1] E&CE Dept, IIT Roorkee, India
[2] Computer Networks and Distributed Systems (CNDS) Lab, Dept of Electrical and Computer Engineering, National University of Singapore, Singapore
nimesh.d.desai@gmail.com, kgargfec@iitr.ernet.in,
manojfec@iitr.ernet.in, elebv@nus.edu.sg

Abstract. Mobile agents are very useful in low bandwidth ad hoc network environments. But various attacks are possible against both agent and agent platforms. In this paper we have modeled hierarchical scheme for protecting mobile agent environment which is applicable to both wired and wireless networks. Colored Petri Net is used as a modeling tool which effectively models platform behavior and agent mobility through the movement of tokens and timed firing of transitions. We have considered centralized security scheme for our model where Trusted-Third-Party plays a central role in the communication.

Keywords: Mobile Agent, Security, Modeling, Colored Petri Net.

1 Introduction

Mobile agent security has been an area of research since mobile agents were introduced. Though it is very useful and effective to send agents which can perform tasks on behalf of the user, attacks to such agents and agent platforms are very common and can cause many serious problems. Securing mobile agent communication is essential for its reliability and usefulness among naive end users. Attacks such as masquerade, eavesdropping, replay, repudiation, denial of service are possible against agent environment [10].

In our earlier work [1], we proposed a security scheme for mobile agent environment. That scheme was based on single domain system, but this situation may not be feasible when number of MAPs is quite large. To solve this problem, we extended our work to provide hierarchical scheme. In this scheme multiple domains can exist scattered across large geographical region. Each domain has a domain controller or Trust Server (TS). The TS is responsible for enforcing security within the domain. Each platform must register with the TS before initiating agent communication. A platform can only be registered to at most one domain at a time. The TS relies on public key crypto-system and message passing for agent security. Our present scheme relies on central Trusted-Third- Party (TTP) for inter-domain security. Also for the purpose of this paper, we have only considered passive mobility of agents. In our work, we modeled this security scheme using Colored Petri Nets [11] using CPN tool [5, 6]. We

S. Aluru et al. (Eds.): HiPC 2007, LNCS 4873, pp. 637–649, 2007.

divided the whole system into different levels, and each level has been modeled in separate page. Substitution transitions are then used to combine these pages together.

Attacks on Mobile agent environment have been prime obstacles in its wide spread deployment. V. Hassler [10] outlines various attacks possible against agent infrastructure. L. Ma et. al [9] gives model of Trust Server based security system for mobile agents using EEOS. Our work extends their work by introducing hierarchical model. In their research they have modeled agents as object net. We omitted this part as our main focus was to obtain hierarchical system. D. Xu et. al [2] and M. Kohle et. al [4] gives another approach to model mobile agents and their mobility. L. M. Kristensen et. al [3] shows how to use CP Nets to model distributed systems.

The rest of this paper is structures as follows. Section 3 describes the hierarchical security scheme. Section 4 gives detailed understanding of our model. Section 5 presents various analysis of the model. Section 6 presents our conclusions and describes future work.

2 Design of a Centralized Security Scheme for Mobile Agents

While designing hierarchical security scheme for mobile agents, two approaches are possible. One is a distributed approach, in which there is no central control. Mobile nodes/platforms are moving in an ad hoc manner and are registered to at most one domain at a time. Each domain has a domain controller (or Trust Server). Trust Server handles the activities within the domain. In distributed approach, different Trust Servers exchange registered hosts information at regular intervals and communicate with each other directly for inter-domain transfers. Another approach is a centralized approach, in which there is a central Trusted-Third-Party (TTP) which handles inter-domain communication. Trust Servers communicate to TTP only. In this paper, we have designed and modeled this centralized scheme for agent security.

2.1 Intra-Domain Communication

Figure 1 depicts the overview of a typical domain. To start communication, a mobile agent platform (MAP) must register itself with the TS. The way MAP registers itself with TS resembles the working of Mobile IP [8]. Each TS continuously broadcasts its presence by sending broadcast messages periodically. The TS keeps a unique random number *R* and includes that number in each such a broadcast message. An MAP can register itself with TS by replying to this message from TS. In its reply message, MAP includes its public key that will be used to encrypt mobile agents later. MAP stores the *ts_id* in its Security Base and this *ts_id* will be used later to exchange agents/messages with TS.

When TS receives *Registration Request* from MAP, it stores its Id and public key in its Security Base. Thus each TS maintains a *registered_host_list* which is dynamically updated. Then it replies MAP to confirm its registration. Only registered MAPs can send agents to TS, agents arriving from other hosts are discarded by TS. Random number *R* is used to limit the registration period of each MAP. After some time, TS updates *R* and also clears the registered-host list. This requires MAPs to re-register themselves with the TS to continue communication. So those MAPs who have left the domain meanwhile will automatically be unregistered.

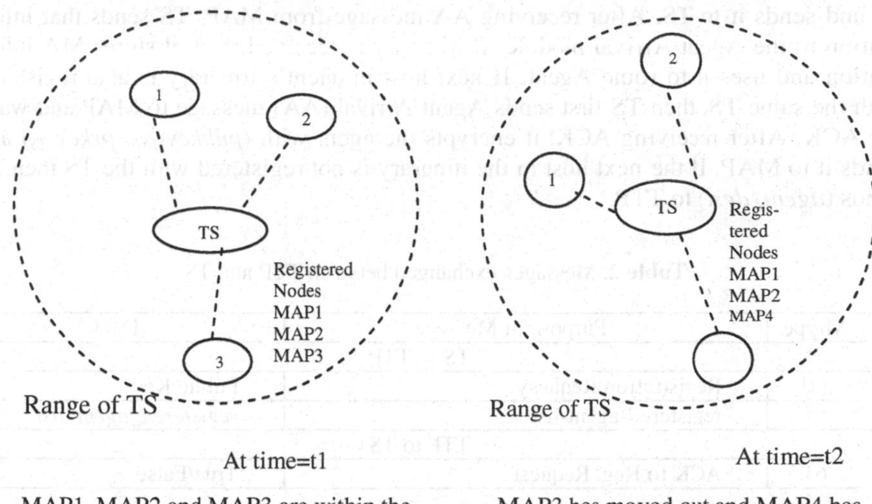

Fig. 1. Domain Overview

When MAP sends an agent to TS, it appends *R* along with it. If TS receives an agent whose *R* value is older than the value currently held by the TS then TS will discard the agent as it is from an unregistered host. Table 1 shows the messages exchanged between an MAP and a TS.

Table 1. Messages exchanged between MAP and TS

mtype	Purpose of Message	Data
	TS to MAP	
0	Announcement	Random Number R
11	Reply to Registration Request from MAP	True/False
21	ACK to AA/AI (2/3) from MAP	Header
4	Agent Arrival(AA)	Header
	MAP to TS	
1	Registration Request Response to 0	Public key
2	Agent Arrival(AA)	Header of Mobile Agent
3	Agent Itinerary(AI)	Storage(Header*ITNR)
41	ACK to AA from TS	Header

2.2 Agent Transfer

Before sending Mobile agent to TS, Security Base of MAP processes the outgoing mobile agent. Security Base sends Agent Arrival (AA) message to TS and waits for its ACK. If MA is created by this MAP then it sends Agent Itinerary (AI) message to TS. After receiving ACK from TS, MAP encrypts the agent with *(prkey* $_{MAP}$*, pubkey*

$_{TS}$) and sends it to TS. After receiving AA message from MAP, TS sends that information to the Agent Arrival module. If itinerary is received then, it stores MA information and uses it to route Agent. If next host in agent's itinerary is also registered with the same TS, then TS first sends Agent Arrival (AA) message to MAP and waits for ACK. After receiving ACK, it encrypts the agent with *(pubkey$_{MAP}$, prkey $_{TS}$)* and sends it to MAP. If the next host in the itinerary is not registered with the TS then TS sends (*agent, dest*) to TTP.

Table 2. Messages exchanged between TTP and TS

Mtype	Purpose of Message	Data
	TS to TTP	
60	Registration Requesy	Public Key
51	registered-agent-list	*registered_agent_list*
	TTP to TS	
61	ACK to Reg. Request	True/False

2.3 Inter-domain Communication

All Trust Severs (TS) are connected by a wired link to the central TTP. Thus the TTP is central to inter-domain communication. Initially, each TS registers itself with the TTP by sending its public key. After receiving public key from the TS, TTP stores that information along with the *ts_id* in its Security Base. Then it sends ACK message back to the TS. All registered TS periodically send their *registered_host_list* to TTP. TTP upon receiving this list updates its knowledge base. TTP always maintains a table of (*ts_id*, *registered_host_list*) which is referred here as *TS_host_list*.

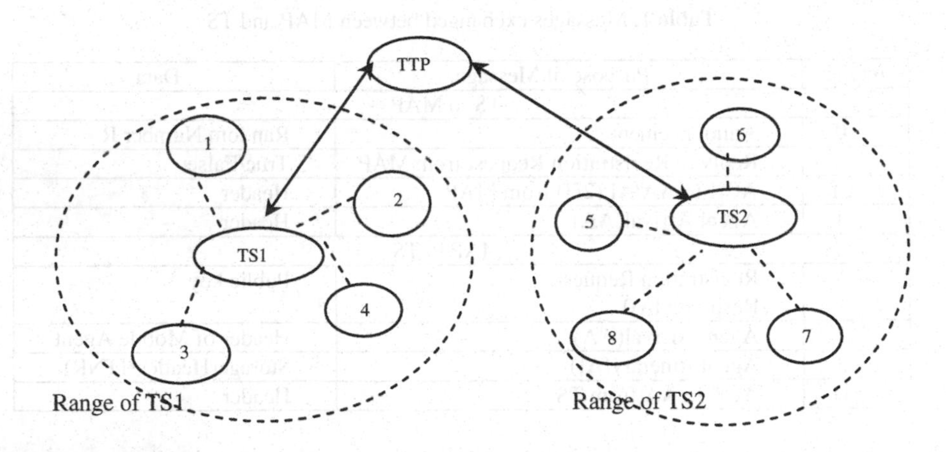

Fig. 2. Inter-domain architecture

As mentioned in the section 2.1, when destination host is not its *registered_host_list,* the TS encrypts the agent with *(prkey $_{TS}$, pubkey $_{TTP}$)* and sends (*agent, dest*) pair to TTP. TTP checks in its *TS_host_list* and finds whether there is any domain to which given destination is registered. If it finds such a domain, then it encrypts the agent with *(prkey $_{TTP}$, pubkey $_{TS}$)* and sends (*agent, dest*) to the TS of that domain and waits for the agent to return. When agent returns after finishing its execution at specified destination host, TTP sends the agent back to its home domain. To distinguish agents received from the TTP, TS maintains *rcvd_list*. In a similar way, TS also maintain a *sent_list* of agents sent to TTP.

2.4 Secure Mobile Agent Transfer

This security work is a natural extension of our previous work [1] and also of the solution suggested by [9]. Combination of TS and TTP help achieve secure mobile agent transfer. To lessen the burden on single TS, domain based system is considered. Each TS can limit the number of host registered in that domain. This way load on single TS can be kept under sustainable limits. Also having multiple domains help achieve scalability. Adding new TS to the system is easy and can be dynamically configured.

Though this work does not ensure that the agent arrives at the destination host securely, but it can detect whether the incoming agent has been tempered with or not. As the agent always passes through TS before reaching MAP, isolating malicious MAP is easy. This avoids other hosts from being directly affected by malicious host. But at the same time, this system does not prevent agents from being lost or modified.

Outgoing agent is always encrypted with private key of sending host and public key of receiving host. If during agent authentication, MAP or TS finds violation of security then it puts the agent into a 'Prison'[9]. To make agent transfer more secure and reliable, messages are exchanged between communication entities before actual transfer. Once an agreement or confirmation is received, then only actual agent transfer can begin. If agent is received without prior confirmation then it is considered malicious and put into 'Prison'. The TS maintains a "malicious MAP list". If a platform is found malicious, the TS will record it into this list. Any further requests by this platform will be discarded by the TS in future then.

3 CPN Modeling

Figure 3 shows the top-level page hierarchy of the CPN model. At the top-level, two domains are connected to TTP. Each node in the figure represents a page in CPN hierarchy in which *Top Level* is the top-most page. To reduce the complexities, only two domains have been considered right now. Each domain houses four MAPs and TS. This system can easily be expanded to include more number of domains, but the complexity and memory requirements will increase significantly with each additional domain.

There are actually separate communication links for agents and messages. Also there are separate links for inter-domain and intra-domain communication. This has been done to simplify the communication and also to facilitate collection of individual statistics and analysis.

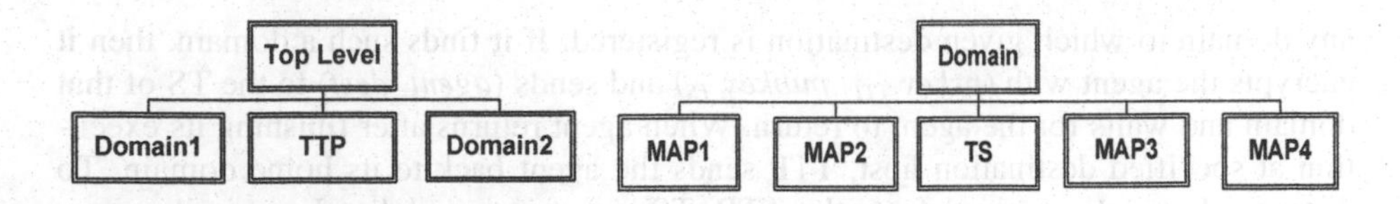

Fig. 3. Top-level Page Hierarchy **Fig. 4.** Hierarchy page for a Domain

3.1 Modeling Agent and Message

The color sets BASEAG, AG and EAG are used to model the mobile agent during different phases of communication. Raw mobile agent is shown by BASEAG, which contains code and data regions. For convenience these regions are considered strings but it reality it can be anything. After creation, each agent is assigned a unique pair of identifier which consists of platform id and agent id. *(platform_id, agent_id, code, data)* makes a unique mobile agent which is modeled by Color set AG. During communication, agent is always packed and encrypted. Color set EAG represents this packed and encrypted mobile agent.

Message communication plays a very important role in this protocol. Different types of messages containing variety of data values are exchanged between TTP, TS and MAPs. MSG colour set represents the structure of the message. It consists of *(dest, source, mtype, data)*. *mtype* signifies the purpose of the message. DATA color set is a union of all the colors that can be used as the data. For different *mtypes* different color sets are used as data items.

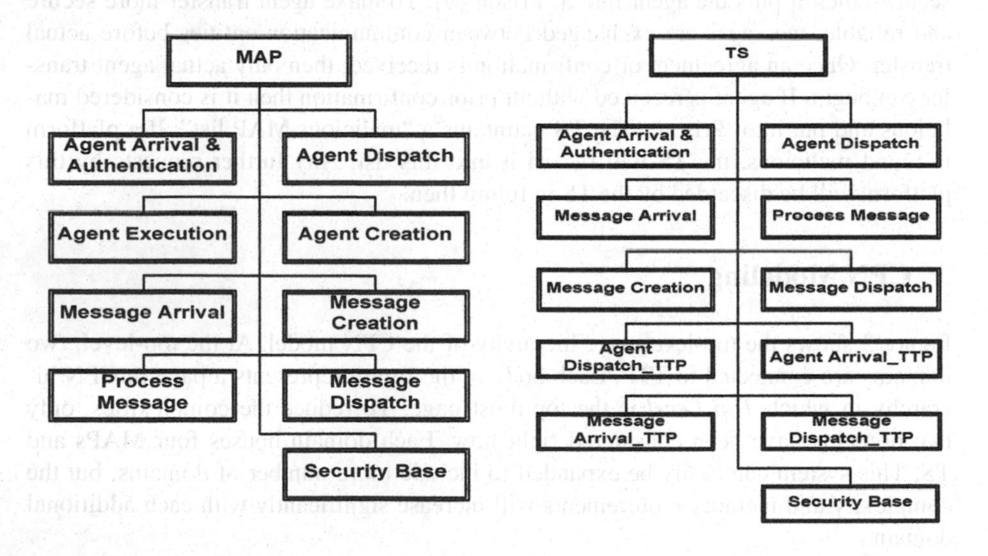

Hierarchy page for Mobile Agent Platform(MAP) Hierarchy page for Trust Server(TS)

Fig. 5. CPN model of MAP and TS

3.2 Modeling Domain

Each domain consists of a TS and some number of MAPs. These MAPs are registered to the TS, which acts as a Domain controller. In the model that we created, each domain consists of a TS and four MAPs. Each MAP has a different start time and after some time they reset their settings. This feature effectively models dynamic nature of the MAPs. Figure 4 shows the hierarchy page of a typical Domain.

3.3 Modeling MAP

Figure 5 depicts the hierarchy page of the CPN model for Mobile agent platform. For each function, a separate page is created in CP Net. A separate page for Security and Knowledge base is also created which handles the data part of the platform. Figure 6 shows working of the MAP.

Agent arrival and Authentication page authenticates incoming agents. If the incoming agent fails authentication, it is sent to 'Prison'. The CPN model for *Agent Arrival and Authentication* executes decryption functions on the incoming agent to check its authenticity. *Agent Dispatch* page encrypts the outgoing agent using encryption function and values provided by *Security Base*. *Agent Creation* handles the creation and loading of mobile agents in the platform. Firing of transition reads agents from file and creates them. Timed firing is used to create different agents at different times. Once agent is created, its itinerary along with header *(platform_id, agent_id)* is sent to the *Security Base*.

Security Base page handles the internal processing and database management for MAP. It has one ID place which contains unique identifier of that MAP. Common fusion place is use to assign each MAP a unique identifier. This fusion place has a list of all Ids as initial marking and ensures single firing for each MAP. When it receives registration request message from TS, it stores the *ts_id* and sends it a registration request

3.4 Modeling TS

Figure 7 shows working of the TS. In case of TS, incoming agent can be either from an MAP or from the TTP. Separate colour sets are used to differentiate agents arriving from these two communication channel. Message handling is very much similar to that of MAP. Pages *Message Arrival*, *Process Message*, *Message Creation* and *Dispatch Message* handle the communication of messages. The major difference is having separate channels for communication with registered MAPs and with the TTP.

Security Base page plays an important role in the design of TS. It provides required information to other pages and also initiates message creation. At start, each TS obtains a unique identifier using the similar fusion set mechanism used in MAP. After that it generates a *random number R* and broadcasts announcement messages periodically. When it receives the registration request from an MAP in response to its announcement, it updates its *registered_host_list* and sends ACK back. After sufficient number of announcements (in this case 20), it updates *R* and clears the *registered_host_list*.

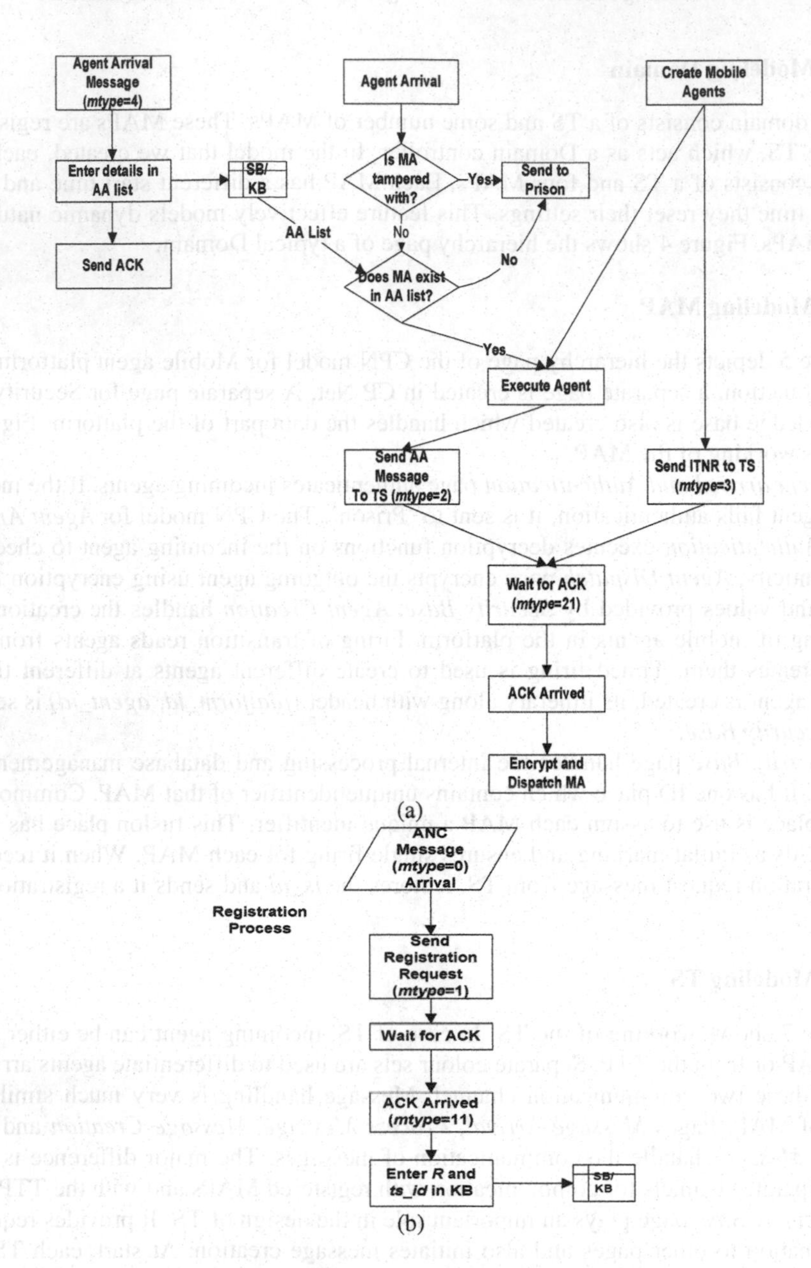

Fig. 6. Flowcharts showing working of MAPs

Once an MAP is registered with the TS, it can send an agent to it. For that TS must receive an agent itinerary message (*mtype*=3) from the home platform of the agent before the agent arrival message (*mtype*=2). While sending the agent out, TS checks the agent itinerary to find the next *dest_id*. If the *dest_id* is available in the *registered_host_list* then TS sends a message *mtype=4* to that host and waits for its ACK

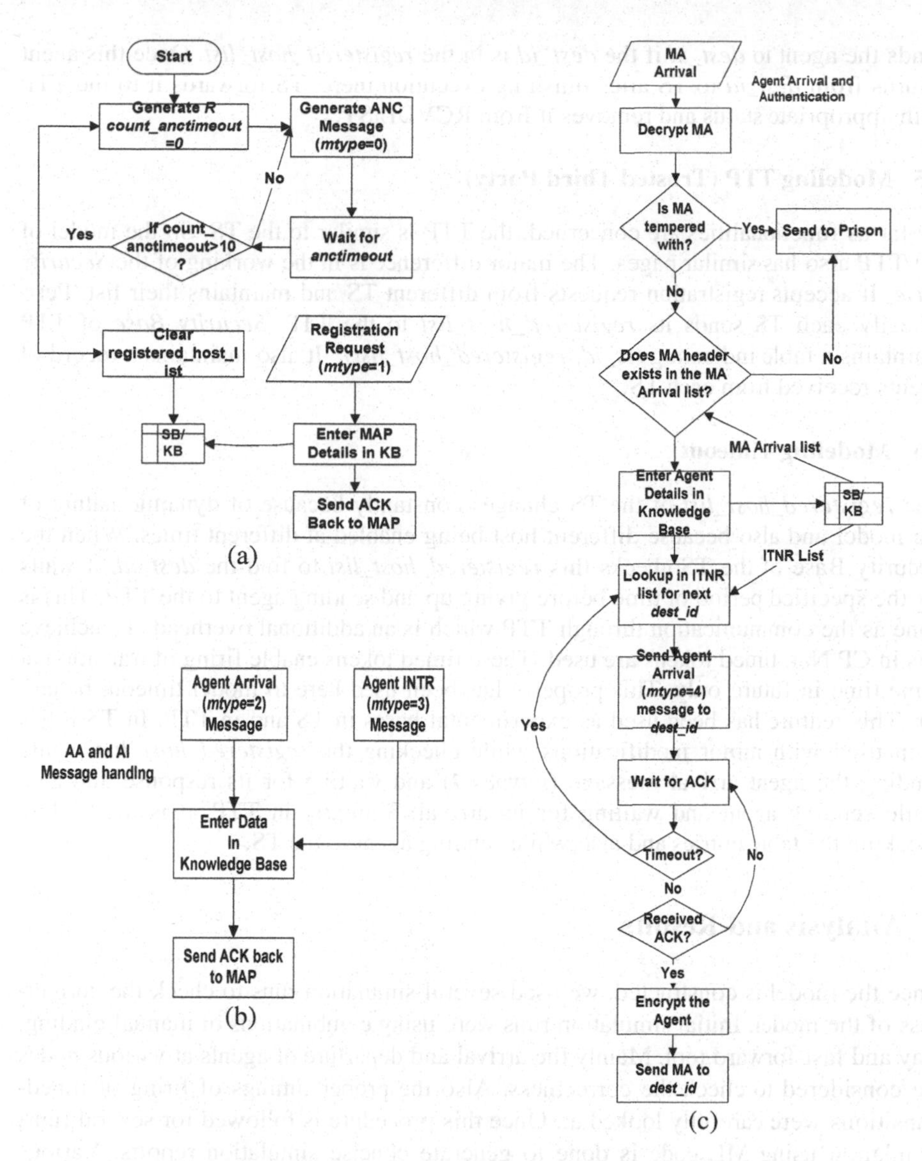

Fig. 7. Flow Charts showing working of Trust-Server(TS)

(*mtype=41*) before sending the agent. If the required *dest_id* is not found in the *registered_host_list* then TS sends the agent to TTP, adds an entry into SENTLIST and waits for the agent to come back.

When an agent arrives at the TS from the TTP, it carries *dest_id* field along with it. If agent header is present in SENTLIST then *dest_id* indicates success (-1) or failure (-2) in finding require destination. In either case, TS checks the itinerary of the agent and starts the same procedure mentioned above. If the *dest_id* is neither -1 nor -2 then this agent belongs to another domain. Thus TS makes an entry in RCVDLIST and sends the agent to *dest_id* if the *dest_id* is in the *registered_host_list*. Once this agent returns from *dest_id* to TS after finishing execution there, TS forwards it to the TTP with appropriate status and removes it from RCVDLIST.

3.5 Modeling TTP (Trusted Third Party)

As far as functionalities are concerned, the TTP is similar to the TS. So the model of the TTP also has similar pages. The major difference is in the working of the *Security Base*. It accepts registration requests from different TS and maintains their list. Periodically each TS sends its *registered_host_list* to the TTP. *Security Base* of TTP maintains a table indicating (*ts_id, registered_host_list).* It also maintains a record of agents received from each TS.

3.6 Modeling Timeout

The *registered_host_list* of the TS changes constantly because of dynamic nature of the model and also because different host being enabled at different times. When the Security Base of the TS checks this *registered_host_list* to find the *dest_id*, it waits for the specified period of time before giving up and sending agent to the TTP. This is done as the communication through TTP which is an additional overhead. To achieve this in CP Net, timed tokens are used. These timed tokens enable firing of transition at some time in future only. This property has been used here to model timeout behavior. This feature has been used as experimental bases in TS and in TTP. In TS it has been used with minor modifications while checking the *registered_host_list*, while sending the agent arrival message (*mtype=4*) and waiting for its response and also while sending agent and waiting for its arrival. Similarly in TTP, it is used while checking the table entries and also while sending agent to the TS.

4 Analysis and Results

Once the model is constructed, we used several simulation runs to check the correctness of the model. Initial simulation runs were using combinations of manual binding, play and fast-forward tool. Mainly the arrival and departure of agents at various nodes are considered to check the correctness. Also the proper timings of firing of timed-transitions were carefully looked at. Once this procedure is followed for several runs, simulation using ML code is done to generate concise simulation reports. Various data collector monitors are used to gather statistics and to check certain properties of the CP Net. Figure 8 shows such a report generated during such a simulation run.

Because of dynamic identifier assignment, each MAP is registered to different TS during different simulation runs. This changes the behavior of agent as the location of destination MAP changes. As figure 8 shows, on average 26 agents participate in inter-domain communication i.e. they move from the TS to the TTP. Also TS receives on average 22-24 agents from intra-domain MAPs and 14 agents from the TTP. These figures are consistent with the agent itinerary and their bindings. Figure 9 shows GNU plot [5] of *registered_host_list* maintained by the TS 1. Similar list can be obtained for TS 2 also. Initially the list is empty. As the MAPs are enabled at different times,

Number of replications: 5

Statistics					
Name	Count	Sum	Avrg	Min	Max
Marking_size_Agent__Arrival_and_Auth_TTPc'PT1_1					
count_iid	5	134	26.80	20	46
Marking_size_Security_Base__TS'PT1_1					
count_iid	5	110	22.00	10	60
Marking_size_Security_Base__TS'PT1_2					
count_iid	5	124	24.80	16	38
Marking_size_Security_Base__TS'PT2_1					
count_iid	5	72	14.40	10	18
Marking_size_Security_Base__TS'PT2_2					
count_iid	5	72	14.40	6	30

Fig. 8. Performance Statistics obtained using Data Collector Monitors

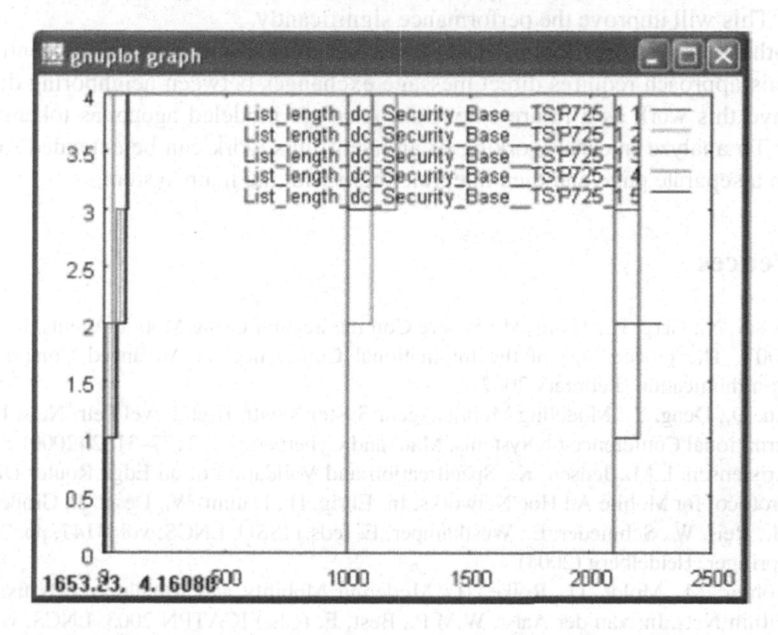

Fig. 9. GNUplot showing registered_host_list length at different times(TS 1)

they register to the TS at different times. This is indicated by the gradual increase in the list length. At time 1000, list is cleared and new announcement starts. This is done to avoid stale registrations.

Table 3 shows the number of agents arrived at and departed from TS 1 and TS 2 during sample simulation runs. The values in the table are averaged over 3 simulation runs. It can be easily observed that almost 60% communication is intra-domain.

Table 3. Agent Movements during Test Run

	Agents Arrived (Intra-Domain)	Agents Arrived (Inter-Domain)	Agents Dispatched (Intra-Domain)	Agents Dispatched (Inter-Domain)
TS 1	26	8	20	10
TS 2	28	10	24	12

5 Conclusion and Future Work

This paper shows how CPN modeling can be used to model mobile agents. The critical part of mobile agent system is its dynamic behavior and agent mobility. The paper shows tokens can be used to model agents. Also Timed-CP Nets can be configured to achieve dynamic behavior by enabling firing of transitions at different point in time.

We have modeled hierarchical security scheme for mobile agents. This scheme relies on central control in the form of TTP. With careful choices of number of domains and number of MAPs registered to each domain, inter-domain activities can be minimized. This will improve the performance significantly.

Another approach for hierarchical system requires elimination of the central control. This approach requires direct message exchanges between neighboring domains. We leave this work as a future extension. Also we modeled agents as tokens in this model. To analyze internal working of an agent, this work can be extended to model them in a separate page and then integrate them with the main system.

References

[1] Desai, N., Garg, K., Misra, M.: Secure Communication Using Mobile Agents. In: ICACC 2007. The proceedings of the International Conference on Advanced Computing and Communication (February 2007)

[2] Xu, D., Deng, Y.: Modeling Mobile Agent Systems with High Level Petri Nets. IEEE International Conference on Systems, Man, and Cybernetics 5, 3177–3182 (2000)

[3] Kristensen, L.M., Jensen, K.: Specification and Validation of an Edge Router Discovery Protocol for Mobile Ad Hoc Networks. In: Ehrig, H., Damm, W., Desel, J., Große-Rhode, M., Reif, W., Schnieder, E., Westkämper, E. (eds.) ISSU. LNCS, vol. 3147, pp. 248–269. Springer, Heidelberg (2004)

[4] Köhler, M., Moldt, D., Rölke, H.: Modeling Mobility and Mobile Agents using Nets within Nets. In: van der Aalst, W.M.P., Best, E. (eds.) ICATPN 2003. LNCS, vol. 2679, pp. 121–139. Springer, Heidelberg (2003)

[5] CPN Tool website: www.daimi.au.dk/CPNtools

[6] Michel, B., Wendy, M., Peter, E.A., Paul, J., Mads, J., Michael, L., Kasper, L., Kjeld, M., Stephanie, M., Anne, R., Katrine, R., Soren, C., Kurt, J.: CPN/ Tools: APost- WIMP Interface for Editing and Simulating Coloured Petri Nets. In: Colom, J.-M., Koutny, M. (eds.) ICATPN 2001. LNCS, vol. 2075, pp. 71–80. Springer, Heidelberg (2001)

[7] Christensen, S., Jorgensen, J.B., Kristensen, L.M.: Design/CPN- A Computer Tool for Coloured Petri Nets. In: Brinksma, E. (ed.) TACAS 1997. LNCS, vol. 1217, pp. 209–223. Springer, Heidelberg (1997)

[8] Mobile IP RFC (2002), www.ietf.org/rfc/rfc, txt
[9] Ma, L., Tsai, J.J.P.: Security modeling and analysis of mobile agent systems. Imperial College Press, London
[10] Hassler, V.: Security fundamentals for e-commerce. Artech House Publishers, ISBN-10: 1580531083
[11] Jensen, K.: Coloured Petri Nets. Basic Concepts, Analysis Methods and Practical Use. In: Monographs in Theoretical Computer Science, vol. 1,2 and 3, Springer, Heidelberg. ISBN: 3-540-60943-1, 3-540-58276-2, 3-540-62867-3.

Single Lock Manager Approach for Achieving Concurrency Control in Mobile Environments

Salman Abdul Moiz[1] and Lakshmi Rajamani[2]

[1] Associate Professor, CSE, Muffakham Jah College of Engineering & Technology, Hyderabad, India

[2] Professor, CSE, University College of Engineering, Osmania University, Hyderabad, India

salmanmca@gmail.com, lakshmiraja@yahoo.com

Abstract. In a mobile computing environment, users can perform on-line transaction processing independent of their physical location. In a mobile environment, multiple mobile hosts may update the data simultaneously which may result in inconsistency of data. To solve such problems many concurrency control techniques have been proposed. The traditional two phase locking protocol has some inherent problems such as deadlocks & long unpredictable blocking. In this paper we propose a concurrency control mechanism with dynamic timer adjustment which helps in reducing the communication overhead and enhances the transaction throughput. The simulation results specify the performance trade off metrics.

Keywords: Mobile Host, Fixed Host, Transaction, Timer, Commit, Rollback.

1 Introduction

Technological advances in wireless communication and satellite services have lead to the emergence of mobile computing environments. The possibility of accessing and processing information on move, everywhere and at any time has been one of the major reasons for huge acceptance and success [7]. Though this mechanism provides the utmost convenience in using various mobile applications, it suffers from various limitations such as unreliable communication, limited battery power, variable bandwidth, reduced storage capacity etc.,

Concurrency control is one of the most important features of transaction management. Research in concurrency control for mobile databases has led to the development of various concurrency control algorithms. Most of these proposals are based on three mechanisms viz., locking, timestamps and optimistic concurrency control. Though these schemes are well suited for traditional database applications, they don't work efficiently in mobile environments.

Due to various constraints in the mobile environment and nature of different online applications, traditional concurrency control mechanism may not work effectively. Transactions initiated by a mobile client that disconnects for longer time period may lead to an unacceptable long locking which may decrease the throughput [2].

S. Aluru et al. (Eds.): HiPC 2007, LNCS 4873, pp. 650–660, 2007.

In this paper we propose a single lock manager approach for achieving concurrency control in Mobile Environments, which is based on timeout mechanism. Timer is dynamically adjusted which helps in maintaining application specific timers. The remaining part of this paper is organized as follows. Section 2 summarizes survey of existing techniques for concurrency control, section 3 gives an overview of the mobile database environment, section 4 specifies the drawbacks of concurrent access in mobile environments, section 5 specifies the proposed concurrency control strategy and the section 6 specifies the performance metrics.

2 Related Work

Concurrent execution of transactions by multiple mobile clients accessing the same data item may lead to inconsistency. Several valuable attempts have been made to efficiently implement concurrency control. However each attempt considers only a subset of the operational requirements.

To achieve concurrency control, two phase locking protocol was used in the traditional environment. However this protocol requires clients to communicate continuously with the server to obtain locks and detect the conflicts. Hence it is not suitable for mobile environments [8]. An optimistic concurrency control technique detects and resolves data conflicts in the phase of transaction validation. In a mobile environment of the transaction validation is done on the server, it may lead to delayed response causing overhead at the server [9]. In [10], A Timeout based Mobile Transaction Commitment Protocol uses timeouts to provide non-blocking protocol with restrained communication. It faces the problem of the time lag between local and global commit. In [6] the proposed Mobile 2PC protocol preserves the 2PC principle and minimizes the impact of unreliable wireless communication. This protocol assumes that all communicating partners are stationary hosts, equipped with sufficient computing resources and power supply with permanently available bandwidth.

An Optimistic Concurrency Control with Dynamic Time stamp Adjustment Protocol requires client side write operations. However because of the delay in execution of a transaction, it may never be executed [3]. In [5, 1], the conventional optimistic concurrency control algorithm in enhanced with an early termination mechanism on conflicting transactions. However because of early termination a transaction need to be initiated again and again. In [4] Mobile speculative locking protocol is introduces to reduce the blocking of transaction if two phase locking is employed. This approach requires extra resources at the mobile host to carry out speculative execution.

This paper proposes a single lock manager approach for achieving concurrency in the wireless environment using dynamic timer management. This technique helps in reducing the upward link as the request for locking of a particular data item is mostly managed by a fixed host. Further the server or a fixed host will not be overloaded as the non-conflicting data items are read by the mobile host and the execution is done by the mobile clients.

3 Mobile Database Environment

The following figure shows the architectural view of the mobile computing environment.

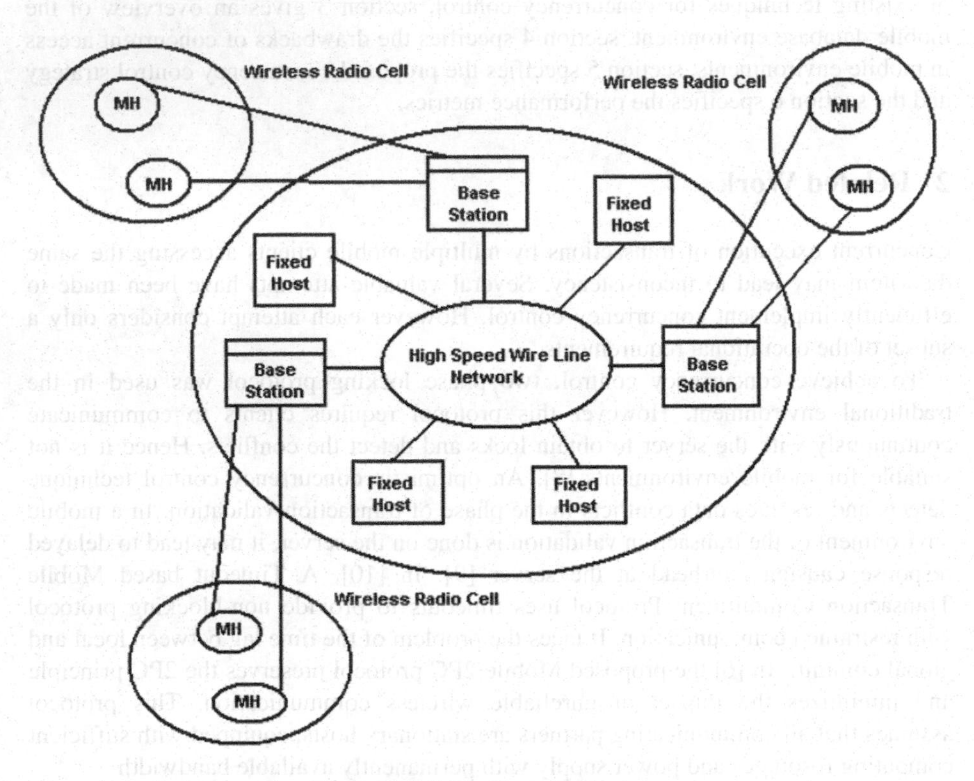

Fig. 1. Mobile Database Environment

It consists of two major components i.e a Mobile Host (MH) and a Fixed Host (FH). Transactions are initiated at a mobile host may be executed at fixed host or mobile host. A Mobile unit connects to a fixed host through a wireless link A Base station connects to a mobile unit and is equipped with a wireless interface. It is also known as a Mobile Support Station.

Mobile Hosts (MH) may not always be connected to the fixed network. They may be disconnected for different reasons. Mobile host may differ with respect to the computing power and storage space; however MH can run a DBMS module. Mobile host may be disconnected to save battery consumption. Hence disconnections are handled as normal situations and not as failures [7].

Mobile Clients may vary from thin to full clients depending on their characteristics. In the thin client, request for a transaction is initiated by the MH and is executed on the fixed host whereas the full clients implement a part of server functionality without being strongly connected to the server

In this paper the transaction is initiated at the mobile host, the required data items are read and transaction is executed by the mobile host. Finally the results are updated on to the fixed host.

4 Concurrent Access Anomalies in Mobile Environments

The traditional database uses locking protocol to overcome the problems of concurrent access. These protocols are not suitable for preserving consistency in presence of concurrent access of mobile hosts. A transaction initiated at a mobile host locks the required data items. If another mobile host requires the same data item it needs to wait till the mobile that requested the data items first unlocks the data item. If the mobile host doesn't commit the transaction, the second mobile host waiting in a queue has to wait for invariant time.

In order to overcome this problem, timeout mechanism was introduced in mobile environments. A mobile host after acquiring the data items has to complete the transaction within the stipulated time. After the expiry of the timer the transaction is rolled back forcefully such that the data items are released and is used by the other mobile host waiting in a queue [8]. This mechanism may not be well suited for the wireless environments because this requires clients to communicate continuously with the server to obtain locks and detect the conflicts. Further the setting of a static timer value may result in many rollbacks because the transaction may never be executed within the stipulated time. Hence the throughput decreases. In the next section dynamic timer adjustment mechanism is proposed which will increase the throughput.

5 Concurrency Control in Mobile Environments

5.1 Single Lock- Manager Approach

In this approach, the mobile environment maintains a single lock manager which resides on the fixed host F_i. All lock and unlock requests are made at the fixed host F_i. When the transaction initiated, the mobile client needs to lock a data item, it sends a lock request to F_i. The lock manager informs the respective mobile client that has initiated the lock request. Otherwise the request is delayed until it can be granted. The transaction initiated by the mobile client reads the data items after locking it. In case of a write all the sites where a replica of data resides must be updated. This scheme reduces multiple messages being exchanged. As the lock and unlock request are made at only one site deadlock handling will be easy.

5.2 Dynamic Adjustment of Time Quantum for Achieving Concurrency Control

The data items needed by mobile clients for execution of a transaction are locked at fixed host and the execution of the transaction is done at the mobile host. Further a job queue is maintained based on first cum first serve mechanism. Once the data items are locked at the stationary/fixed host, the data needed for execution of the transaction is copied at the mobile host. This reduces the congestion at the mobile host.

For example, if A and B are joint account holders, they can't perform the transaction at the same time. If A locks the data item belonging to his account then B has to wait till the data item is unlocked. But if some other user (say C) wants to lock its account for the transaction he can do so without affecting the first transaction.

Table 1. List of user's requests for data items

Users	Items
A	I1
B	I1
C	I2
D	I1
E	I3

If A has locked the data item I1 and is not able to commit the transaction due to low bandwidth, delay in execution of a transaction, disconnections etc, then B and D have to wait. To overcome this problem a timer is set when the user A initiates the transaction.

If A doesn't commit the transaction before the expiry of timer, the transaction at A is rolled back and automatically the mobile client waiting in the FCFS job queue will lock the data item. If the transaction initiated by mobile host A is committed, then the database is updated and immediately the transaction at B is initiated. Row locking mechanism is used to simulate the locking of a particular tuple in the relation.

Since the user C is not requesting for the data item I1, it can execute the transaction irrespective of the state of user A, B or D. Hence C locks the data item I2. Moreover E can also lock I3, irrespective of the execution of the existing transactions. Hence transactions initiated by three users are executed at a time. But as user C is only requesting for the data item I2 and after the expiry of timer it may be rolled back which could be a drawback i.e the transaction at C may not be executed at all as the transaction is rolled back unnecessarily. Hence there is a need of adjustment of dynamic timers.

Multiple mobile hosts can connect to the fixed host and can execute the transactions in parallel when they are referring to different data items. As the transaction execution is distributed at the mobile and the fixed host respectively, there is a less possibility of delay in execution of a transaction.

In the time out mechanism for execution of a transaction, a transaction may be rolled back because of various reasons. More time may be needed to execute the transaction as such the transaction may not at all be executed and is rolled back after the expiry of the timer. Hence a new request for the execution of the same transaction has to be initiated by the mobile client.

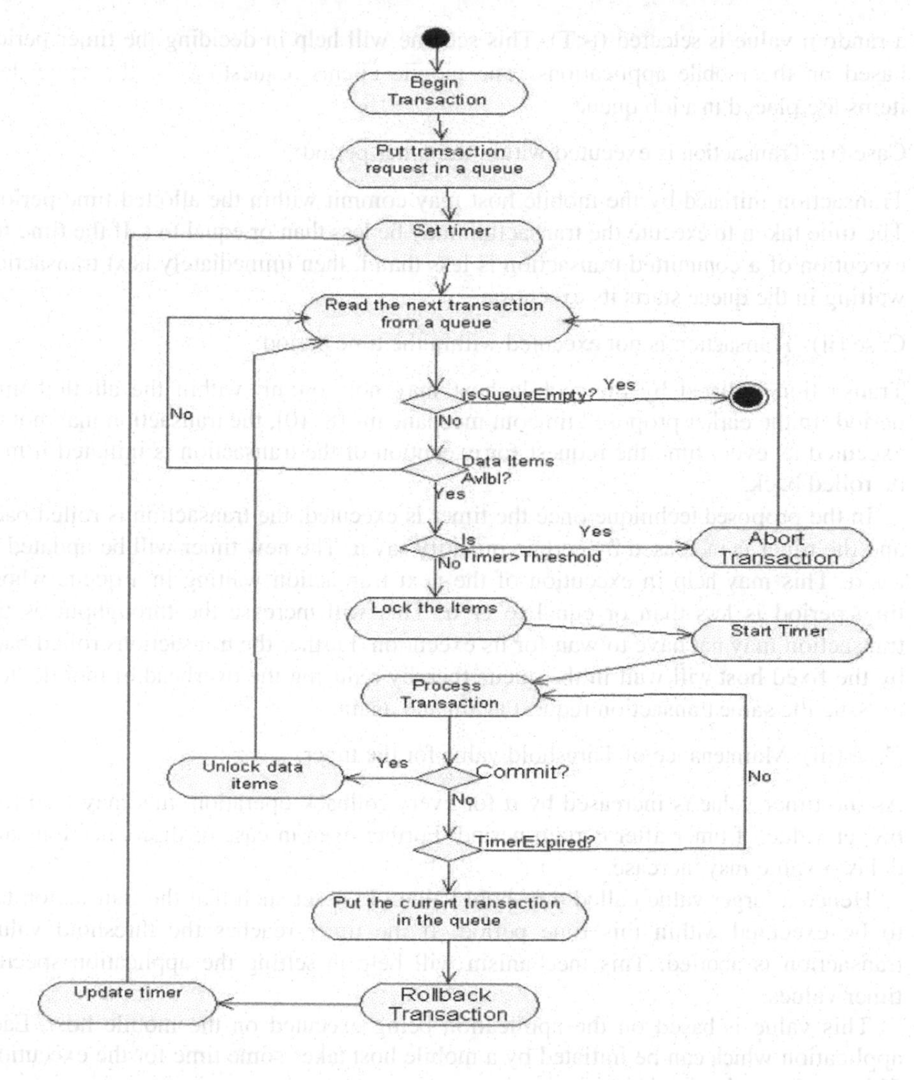

Fig. 2. Activity diagram depicting the dynamic adjustment time quantum for achieving concurrency control

The Fixed host maintains a queue for the execution of the transactions initiated at the mobile host. As such those transactions which were not committed will be maintained in the queue in the round robin approach. This results in reducing the number of requests for the execution of the same transaction initiated by the mobile host.

For any mobile application, a threshold value (T) need to be fixed. The transaction must be executed within this time which is a larger value. Initially the static timer (t), a random value is selected (t<T). This scheme will help in deciding the timer period based on the mobile applications. The mobile clients requesting for the same data items are placed in a job queue.

Case (i): Transaction is executed within the timer period:

Transaction initiated by the mobile host may commit within the allotted time period. The time taken to execute the transaction may be less than or equal to t. If the time for execution of a committed transaction is less than t, then immediately next transaction waiting in the queue starts its execution.

Case (ii): Transaction is not executed within the time period:

Transaction initiated by the mobile host may not commit within the allotted time period. In the earlier proposed timeout mechanisms [8, 10], the transaction may not be executed as every time the request for execution of the transaction is initiated it may be rolled back.

In the proposed technique once the timer is executed, the transaction is rolled back and the timer is increased by certain quantity say α. The new timer will be updated to $t + \alpha$. This may help in execution of the next transaction waiting in a queue whose time period is less than or equal to $t+ \alpha$. This will increase the throughput as the transaction may not have to wait for its execution. Further the transactions rolled back by the fixed host will wait in the queue thereby reducing the overhead of mobile host to issue the same transaction request again and again.

Case (iii): Maintenance of Threshold value for the timer:

As the timer value is increased by α for every rollback operation, this may lead to a bigger value of timer after certain period. Further even in case of disconnections and delay α value may increase.

Hence a larger value called threshold value (T) is set such that the transaction has to be executed within this time period. If the timer reaches the threshold value, transaction is aborted. This mechanism will help in setting the application specific timer values.

This value is based on the application being executed on the mobile host. Each application which can be initiated by a mobile host takes some time for the execution. This can be set by the domain experts.

5.3 Transaction States in Mobile Environments

Based on the proposed concurrency control technique, transaction can be in any of the given states: New, Ready, Pending, Active, Commit or Abort.

A transaction is said to be in a *New* state if the request is initiated by the mobile host but not yet received by the base station. A transaction is said to be in a *Ready* state if the request is received by the fixed host and is placed in a queue. A transaction is said to be in a Active state if it is under execution. If the transaction is rolled back after completion of the timer it is in *Pending* state, then the request is again placed in

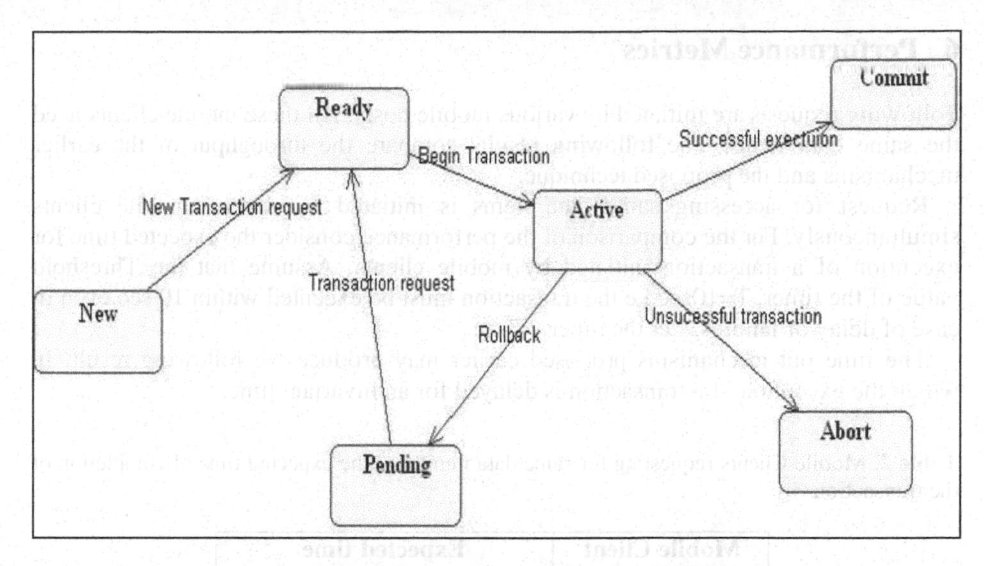

Fig. 3. Transaction State Diagram for Mobile Environments

the queue. An active transaction which is completed successfully is said to be in a *Commit* state. Incomplete transaction after reaching the threshold value will be in an *Aborted* state.

5.4 Failure of the Fixed Host

Since all lock and unlock requests are made at only one stationary host, deadlocks can be easily handled. However if the fixed host fails, then the concurrency control manager may fail. If the fixed host Fi fails then the processing must continue by another site. This is possible if a backup is maintained. We assume that a Primary copy replication is used. Once the primary or master copy is updated all other nodes reflect the changes to the master. If the fixed host fails, another node which maintains the backup information will now act as the coordinator.

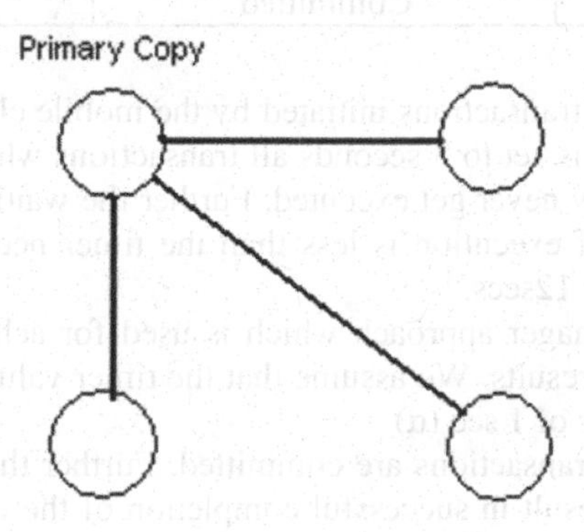

Fig. 4. Primary Copy Replication

6 Performance Metrics

Following requests are initiated by various mobile hosts. All these mobile clients need the same data items. The following results compare the throughput of the earlier mechanisms and the proposed technique.

Request for accessing same data items is initiated by several mobile clients simultaneously. For the comparison of the performance consider the expected time for execution of a transaction initiated by mobile clients. Assume that the Threshold value of the timer, T=10sec i.e the transaction must be executed within 10 sec even in case of delay or failures. Let the timer t=3sec

The time out mechanisms proposed earlier may produce the following result. In which the execution of a transaction is delayed for an invariant time.

Table 2. Mobile Clients requesting for same data items and the expected time of completion of the transaction

Mobile Client	Expected time (in secs)
M1	3
M2	3
M3	4
M4	6
M5	2

Table 3. Performance of the mobile clients using Time out mechanism

Mobile Client	Status	Waiting Time (Seconds)
M1	Commit	3
M2	Commit	6
M3	Aborted (Timer expired)	9
M4	Aborted (Timer expired)	12
M5	Committed	15

In this approach the transactions initiated by the mobile clients M3 and M4 are not executed. As the timer is set to 3 seconds all transactions whose time for execution is more that the timer may never get executed. Further the waiting time for those mobile requests whose time of execution is less than the timer needs to wait. In the above case M5 has to wait for 12secs.

The single lock manager approach which is used for achieving concurrency may produce the following results. We assume that the timer value in case of rollback will be increased by a factor of 1 sec (α)

In this case all the transactions are committed. Further the increase in timer value by a factor of α may result in successful completion of the transaction. Further if the dynamic timer value exceeds the threshold value i.e. the value in 3rd column exceeds 10 seconds (Threshold value) then the transaction is aborted.

Table 4. Performance of the mobile clients using dynamic Time out mechanism

Mobile Client	Status	Timer (Dynamic)	Waiting Time (Seconds)
M1	Commit	3	3
M2	Commit	3	6
M3	Rolled back	4	9
M4	Rolled back	5	13
M5	Committed	5	15
M3	Committed	5	19
M4	Rolled back	6	25
M4	Committed	6	31

7 Conclusion

The Single lock manager approach for achieving concurrency control in mobile environments helps in reducing the messages being exchanged between the mobile clients and fixed hosts. It usually requires only two messages one for handling lock request and another for unlocking. Since the lock and unlock requests are handled at only one place handling deadlocks will be efficient. With the introduction of dynamic timer management, mobile clients may not request for the execution of the same transaction request again and again. This scheme also helps in setting of the application specific timer values.

References

1. Yendluri, A., Hou, W.-C., Wang, C.-F.: Improving Concurrency Control in Mobile Databases. In: Lee, Y., Li, J., Whang, K.-Y., Lee, D. (eds.) DASFAA 2004. LNCS, vol. 2973, pp. 642–655. Springer, Heidelberg (2004)
2. Turker, C., Zini, G.: A Survey of Academic & Commercial Approaches to Transaction Support in Mobile Computing Environments (2003)
3. Choi, H.-J., Jeong, B.-S.: A Timestamp- Based Optimistic Concurrency Control for Handling Mobile Transactions. In: Gavrilova, M., Gervasi, O., Kumar, V., Tan, C.J.K., Taniar, D., Laganà, A., Mun, Y., Choo, H. (eds.) ICCSA 2006. LNCS, vol. 3981, pp. 796–805. Springer, Heidelberg (2006)
4. Krishna Reddy, P., Kitsuregawa, M.: Speculative Lock Management to Increase Concurrency in Mobile Environments. In: Leong, H.V., Li, B., Lee, W.-C., Yin, L. (eds.) MDA 1999. LNCS, vol. 1748, pp. 82–96. Springer, Heidelberg (1999)
5. Lee, M., Helal, S.: HiCoMo: High Commit Mobile Transactions. In: Distributed and Parallel Databases, vol. 11, pp. 73–92. Kluwer Academic Publishers, Dordrecht (2002)
6. Nouali, N., Doucet, A., Drias, H.: A Two-Phase Commit Protocol for Mobile Wireless Environment. In: 16th Australasian Database Conference, vol. 39 (2005)
7. Serrano-Alvarado, P., Roncancio, C., Adiba, M.: A Survey of Mobile Transactions. In: Distributed and Parallel databases, vol. 16, pp. 193–230. Kluwer Academic Publishers, Dordrecht (2004)

8. Moiz, S.A., Rajamani, L.: An Algorithmic approach for achieving Concurrency in Mobile Environment. In: INDIACom, pp.209–211 (2007)
9. Victor, C.S., wa Lam, K., Son, S.H.: Concurrency Control Using Timestamp Ordering in Broadcast Environments. The Computer Journal 45(4), 410–422 (2002)
10. Kumar, V., Prabhu, N., Dunham, M., Seydim, A.Y.: TCOT- A Timeout based Mobile Transaction Commitment Protocol. In: IIS 9979453 (2004)

Author Index

Lecture Notes in Computer Science

Sublibrary 1: Theoretical Computer Science and General Issues

For information about Vols. 1–4507
please contact your bookseller or Springer

Vol. 4873: S. Aluru, M. Parashar, R. Badrinath, V.K. Prasanna (Eds.), High Performance Computing – HiPC 2007. XXIV, 663 pages. 2007.

Vol. 4863: A. Bonato, F.R.K. Chung (Eds.), Algorithms and Models for the Web-Graph. X, 217 pages. 2007.

Vol. 4855: V. Arvind, S. Prasad (Eds.), FSTTCS 2007: Foundations of Software Technology and Theoretical Computer Science. XIV, 558 pages. 2007.

Vol. 4847: M. Xu, Y. Zhan, J. Cao, Y. Liu (Eds.), Advanced Parallel Processing Technologies. XIX, 767 pages. 2007.

Vol. 4846: I. Cervesato (Ed.), Advances in Computer Science – ASIAN 2007. XI, 313 pages. 2007.

Vol. 4838: T. Masuzawa, S. Tixeuil (Eds.), Stabilization, Safety, and Security of Distributed Systems. XIII, 409 pages. 2007.

Vol. 4835: T. Tokuyama (Ed.), Algorithms and Computation. XVII, 929 pages. 2007.

Vol. 4783: J. Holub, J. Žďárek (Eds.), Implementation and Application of Automata. XIII, 324 pages. 2007.

Vol. 4782: R. Perrott, B.M. Chapman, J. Subhlok, R.F. de Mello, L.T. Yang (Eds.), High Performance Computing and Communications. XIX, 823 pages. 2007.

Vol. 4771: T. Bartz-Beielstein, M.J. Blesa Aguilera, C. Blum, B. Naujoks, A. Roli, G. Rudolph, M. Sampels (Eds.), Hybrid Metaheuristics. X, 202 pages. 2007.

Vol. 4770: V.G. Ganzha, E.W. Mayr, E.V. Vorozhtsov (Eds.), Computer Algebra in Scientific Computing. XIII, 460 pages. 2007.

Vol. 4763: J.-F. Raskin, P.S. Thiagarajan (Eds.), Formal Modeling and Analysis of Timed Systems. X, 369 pages. 2007.

Vol. 4746: A. Bondavalli, F. Brasileiro, S. Rajsbaum (Eds.), Dependable Computing. XV, 239 pages. 2007.

Vol. 4743: P. Thulasiraman, X. He, T.L. Xu, M.K. Denko, R.K. Thulasiram, L.T. Yang (Eds.), Frontiers of High Performance Computing and Networking ISPA 2007 Workshops. XXIX, 536 pages. 2007.

Vol. 4742: I. Stojmenovic, R.K. Thulasiram, L.T. Yang, W. Jia, M. Guo, R.F. de Mello (Eds.), Parallel and Distributed Processing and Applications. XX, 995 pages. 2007.

Vol. 4739: R. Moreno Díaz, F. Pichler, A. Quesada Arencibia (Eds.), Computer Aided Systems Theory – EUROCAST 2007. XIX, 1233 pages. 2007.

Vol. 4736: S. Winter, M. Duckham, L. Kulik, B. Kuipers (Eds.), Spatial Information Theory. XV, 455 pages. 2007.

Vol. 4732: K. Schneider, J. Brandt (Eds.), Theorem Proving in Higher Order Logics. IX, 401 pages. 2007.

Vol. 4731: A. Pelc (Ed.), Distributed Computing. XVI, 510 pages. 2007.

Vol. 4726: N. Ziviani, R. Baeza-Yates (Eds.), String Processing and Information Retrieval. XII, 311 pages. 2007.

Vol. 4711: C.B. Jones, Z. Liu, J. Woodcock (Eds.), Theoretical Aspects of Computing – ICTAC 2007. XI, 483 pages. 2007.

Vol. 4710: C.W. George, Z. Liu, J. Woodcock (Eds.), Domain Modeling and the Duration Calculus. XI, 237 pages. 2007.

Vol. 4708: L. Kučera, A. Kučera (Eds.), Mathematical Foundations of Computer Science 2007. XVIII, 764 pages. 2007.

Vol. 4707: O. Gervasi, M.L. Gavrilova (Eds.), Computational Science and Its Applications – ICCSA 2007, Part III. XXIV, 1205 pages. 2007.

Vol. 4706: O. Gervasi, M.L. Gavrilova (Eds.), Computational Science and Its Applications – ICCSA 2007, Part II. XXIII, 1129 pages. 2007.

Vol. 4705: O. Gervasi, M.L. Gavrilova (Eds.), Computational Science and Its Applications – ICCSA 2007, Part I. XLIV, 1169 pages. 2007.

Vol. 4703: L. Caires, V.T. Vasconcelos (Eds.), CONCUR 2007 – Concurrency Theory. XIII, 507 pages. 2007.

Vol. 4700: C.B. Jones, Z. Liu, J. Woodcock (Eds.), Formal Methods and Hybrid Real-Time Systems. XVI, 539 pages. 2007.

Vol. 4699: B. Kågström, E. Elmroth, J. Dongarra, J. Waśniewski (Eds.), Applied Parallel Computing. XXIX, 1192 pages. 2007.

Vol. 4698: L. Arge, M. Hoffmann, E. Welzl (Eds.), Algorithms – ESA 2007. XV, 769 pages. 2007.

Vol. 4697: L. Choi, Y. Paek, S. Cho (Eds.), Advances in Computer Systems Architecture. XIII, 400 pages. 2007.

Vol. 4688: K. Li, M. Fei, G.W. Irwin, S. Ma (Eds.), Bio-Inspired Computational Intelligence and Applications. XIX, 805 pages. 2007.

Vol. 4684: L. Kang, Y. Liu, S. Zeng (Eds.), Evolvable Systems: From Biology to Hardware. XIV, 446 pages. 2007.

Vol. 4683: L. Kang, Y. Liu, S. Zeng (Eds.), Advances in Computation and Intelligence. XVII, 663 pages. 2007.

Vol. 4681: D.-S. Huang, L. Heutte, M. Loog (Eds.), Advanced Intelligent Computing Theories and Applications. XXVI, 1379 pages. 2007.

Vol. 4672: K. Li, C. Jesshope, H. Jin, J.-L. Gaudiot (Eds.), Network and Parallel Computing. XVIII, 558 pages. 2007.

Vol. 4671: V.E. Malyshkin (Ed.), Parallel Computing Technologies. XIV, 635 pages. 2007.

Vol. 4669: J.M. de Sá, L.A. Alexandre, W. Duch, D. Mandic (Eds.), Artificial Neural Networks – ICANN 2007, Part II. XXXI, 990 pages. 2007.

Vol. 4668: J.M. de Sá, L.A. Alexandre, W. Duch, D. Mandic (Eds.), Artificial Neural Networks – ICANN 2007, Part I. XXXI, 978 pages. 2007.

Vol. 4666: M.E. Davies, C.J. James, S.A. Abdallah, M.D. Plumbley (Eds.), Independent Component Analysis and Blind Signal Separation. XIX, 847 pages. 2007.

Vol. 4665: J. Hromkovič, R. Královič, M. Nunkesser, P. Widmayer (Eds.), Stochastic Algorithms: Foundations and Applications. X, 167 pages. 2007.

Vol. 4664: J. Durand-Lose, M. Margenstern (Eds.), Machines, Computations, and Universality. X, 325 pages. 2007.

Vol. 4661: U. Montanari, D. Sannella, R. Bruni (Eds.), Trustworthy Global Computing. X, 339 pages. 2007.

Vol. 4649: V. Diekert, M.V. Volkov, A. Voronkov (Eds.), Computer Science – Theory and Applications. XIII, 420 pages. 2007.

Vol. 4647: R. Martin, M.A. Sabin, J.R. Winkler (Eds.), Mathematics of Surfaces XII. IX, 509 pages. 2007.

Vol. 4646: J. Duparc, T.A. Henzinger (Eds.), Computer Science Logic. XIV, 600 pages. 2007.

Vol. 4644: N. Azémard, L. Svensson (Eds.), Integrated Circuit and System Design. XIV, 583 pages. 2007.

Vol. 4641: A.-M. Kermarrec, L. Bougé, T. Priol (Eds.), Euro-Par 2007 Parallel Processing. XXVII, 974 pages. 2007.

Vol. 4639: E. Csuhaj-Varjú, Z. Ésik (Eds.), Fundamentals of Computation Theory. XIV, 508 pages. 2007.

Vol. 4638: T. Stützle, M. Birattari, H. H. Hoos (Eds.), Engineering Stochastic Local Search Algorithms. X, 223 pages. 2007.

Vol. 4630: H.J. van den Herik, P. Ciancarini, H.H.L.M.(J.) Donkers (Eds.), Computers and Games. XII, 283 pages. 2007.

Vol. 4628: L.N. de Castro, F.J. Von Zuben, H. Knidel (Eds.), Artificial Immune Systems. XII, 438 pages. 2007.

Vol. 4627: M. Charikar, K. Jansen, O. Reingold, J.D.P. Rolim (Eds.), Approximation, Randomization, and Combinatorial Optimization. XII, 626 pages. 2007.

Vol. 4624: T. Mossakowski, U. Montanari, M. Haveraaen (Eds.), Algebra and Coalgebra in Computer Science. XI, 463 pages. 2007.

Vol. 4623: M. Collard (Ed.), Ontologies-Based Databases and Information Systems. X, 153 pages. 2007.

Vol. 4621: D. Wagner, R. Wattenhofer (Eds.), Algorithms for Sensor and Ad Hoc Networks. XIII, 415 pages. 2007.

Vol. 4619: F. Dehne, J.-R. Sack, N. Zeh (Eds.), Algorithms and Data Structures. XVI, 662 pages. 2007.

Vol. 4618: S.G. Akl, C.S. Calude, M.J. Dinneen, G. Rozenberg, H.T. Wareham (Eds.), Unconventional Computation. X, 243 pages. 2007.

Vol. 4616: A.W.M. Dress, Y. Xu, B. Zhu (Eds.), Combinatorial Optimization and Applications. XI, 390 pages. 2007.

Vol. 4614: B. Chen, M. Paterson, G. Zhang (Eds.), Combinatorics, Algorithms, Probabilistic and Experimental Methodologies. XII, 530 pages. 2007.

Vol. 4613: F.P. Preparata, Q. Fang (Eds.), Frontiers in Algorithmics. XI, 348 pages. 2007.

Vol. 4600: H. Comon-Lundh, C. Kirchner, H. Kirchner (Eds.), Rewriting, Computation and Proof. XVI, 273 pages. 2007.

Vol. 4599: S. Vassiliadis, M. Berekovič, T.D. Hämäläinen (Eds.), Embedded Computer Systems: Architectures, Modeling, and Simulation. XVIII, 466 pages. 2007.

Vol. 4598: G. Lin (Ed.), Computing and Combinatorics. XII, 570 pages. 2007.

Vol. 4596: L. Arge, C. Cachin, T. Jurdziński, A. Tarlecki (Eds.), Automata, Languages and Programming. XVII, 953 pages. 2007.

Vol. 4595: D. Bošnački, S. Edelkamp (Eds.), Model Checking Software. X, 285 pages. 2007.

Vol. 4590: W. Damm, H. Hermanns (Eds.), Computer Aided Verification. XV, 562 pages. 2007.

Vol. 4588: T. Harju, J. Karhumäki, A. Lepistö (Eds.), Developments in Language Theory. XI, 423 pages. 2007.

Vol. 4583: S.R. Della Rocca (Ed.), Typed Lambda Calculi and Applications. X, 397 pages. 2007.

Vol. 4580: B. Ma, K. Zhang (Eds.), Combinatorial Pattern Matching. XII, 366 pages. 2007.

Vol. 4576: D. Leivant, R. de Queiroz (Eds.), Logic, Language, Information and Computation. X, 363 pages. 2007.

Vol. 4547: C. Carlet, B. Sunar (Eds.), Arithmetic of Finite Fields. XI, 355 pages. 2007.

Vol. 4546: J. Kleijn, A. Yakovlev (Eds.), Petri Nets and Other Models of Concurrency – ICATPN 2007. XI, 515 pages. 2007.

Vol. 4545: H. Anai, K. Horimoto, T. Kutsia (Eds.), Algebraic Biology. XIII, 379 pages. 2007.

Vol. 4533: F. Baader (Ed.), Term Rewriting and Applications. XII, 419 pages. 2007.

Vol. 4528: J. Mira, J.R. Álvarez (Eds.), Nature Inspired Problem-Solving Methods in Knowledge Engineering, Part II. XXII, 650 pages. 2007.

Vol. 4527: J. Mira, J.R. Álvarez (Eds.), Bio-inspired Modeling of Cognitive Tasks, Part I. XXII, 630 pages. 2007.

Vol. 4525: C. Demetrescu (Ed.), Experimental Algorithms. XIII, 448 pages. 2007.

Vol. 4514: S.N. Artemov, A. Nerode (Eds.), Logical Foundations of Computer Science. XI, 513 pages. 2007.

Vol. 4513: M. Fischetti, D.P. Williamson (Eds.), Integer Programming and Combinatorial Optimization. IX, 500 pages. 2007.

Vol. 4510: P. Van Hentenryck, L.A. Wolsey (Eds.), Integration of AI and OR Techniques in Constraint Programming for Combinatorial Optimization Problems. X, 391 pages. 2007.